Lecture Notes in Mechanical Engineering

Lecture Notes in Mechanical Engineering (LNME) publishes the latest developments in Mechanical Engineering—quickly, informally and with high quality. Original research reported in proceedings and post-proceedings represents the core of LNME. Volumes published in LNME embrace all aspects, subfields and new challenges of mechanical engineering. Topics in the series include:

- Engineering Design
- Machinery and Machine Elements
- Mechanical Structures and Stress Analysis
- Automotive Engineering
- Engine Technology
- Aerospace Technology and Astronautics
- Nanotechnology and Microengineering
- Control, Robotics, Mechatronics
- MEMS
- Theoretical and Applied Mechanics
- Dynamical Systems, Control
- Fluid Mechanics
- Engineering Thermodynamics, Heat and Mass Transfer
- Manufacturing
- Precision Engineering, Instrumentation, Measurement
- Materials Engineering
- Tribology and Surface Technology

To submit a proposal or request further information, please contact the Springer Editor of your location:

China: Dr. Mengchu Huang at mengchu.huang@springer.com
India: Priya Vyas at priya.vyas@springer.com
Rest of Asia, Australia, New Zealand: Swati Meherishi at swati.meherishi@springer.com
All other countries: Dr. Leontina Di Cecco at Leontina.dicecco@springer.com

To submit a proposal for a monograph, please check our Springer Tracts in Mechanical Engineering at http://www.springer.com/series/11693 or contact Leontina.dicecco@springer.com

Indexed by SCOPUS. The books of the series are submitted for indexing to Web of Science.

More information about this series at http://www.springer.com/series/11236

Ubaidillah Sabino · Fitrian Imaduddin ·
Aditya Rio Prabowo
Editors

Proceedings of the 6th International Conference and Exhibition on Sustainable Energy and Advanced Materials

ICE-SEAM 2019, 16–17 October 2019,
Surakarta, Indonesia

Springer

Editors
Ubaidillah Sabino
Mechanical Engineering Program
Faculty of Engineering
Universitas Sebelas Maret
Surakarta, Central Java, Indonesia

Fitrian Imaduddin
Mechanical Engineering Program
Faculty of Engineering
Universitas Sebelas Maret
Surakarta, Central Java, Indonesia

Aditya Rio Prabowo
Mechanical Engineering Program
Faculty of Engineering
Universitas Sebelas Maret
Surakarta, Central Java, Indonesia

ISSN 2195-4356 ISSN 2195-4364 (electronic)
Lecture Notes in Mechanical Engineering
ISBN 978-981-15-4480-4 ISBN 978-981-15-4481-1 (eBook)
https://doi.org/10.1007/978-981-15-4481-1

This Springer imprint is published by the registered company Springer Nature Singapore Pte Ltd.
The registered company address is: 152 Beach Road, #21-01/04 Gateway East, Singapore 189721, Singapore

Contents

Micromechanical Analysis of Elastic Modulus and Tensile Strength on Randomised Discontinuous Alkali and Heat Treated Kenaf Fiber—Unsaturated Polyester Composites

Dody Ariawan, Dharu Feby Smaradhana, and Hammar Ilham Akbar

Abstract In this research, randomised discontinuous kenaf fibre mat/unsaturated polyester composites were fabricated using resin transfer moulding. The effect of kenaf fibre volume fraction on the untreated, heat treated and alkali treated kenaf fibre composites tensile properties were investigated. Experimental results showed that highest tensile properties were attained at volume fraction 30% of alkali treated kenaf fibre composites. The performance of the Tsai-Pagano micromechanical model in predicting the composites' elastic modulus was compared with the Christensen, Manera and Cox-Krenchel model. Meanwhile the predictions of tensile strength were using Hirchs and Bowyer Bader model. Results showed that the consideration of elastic modulus in Cox-Krenchel model and tensile strength in Bowyer Bader model obtained a good prediction of both treated fibre composites.

Keywords Elastic modulus · Kenaf fibre · Micromechanical model · Alkali treated · Heat treated

1 Introduction

Nowadays, there has been many works on Kenaf Fibre (KF) polymer composites, but studies on the modelling of randomized discontinue KF composites are scarce, especially on alkali and heat treated KF. Randomised discontinue fibre composites are typically lower in strength and stiffness than continuous fibre composites, nevertheless, it has several advantages, such as nearly isotropic, low cost, ease of fabricating complex parts [1]. For composites application, there are several efforts to create a high-quality natural fibre composites product, such as selection of certain quality of fibre, effective treatment and appropriate manufacturing process.

D. Ariawan (✉) · D. F. Smaradhana · H. I. Akbar
Mechanical Engineering Department, Engineering Faculty, Universitas Sebelas Maret, Surakarta, Jawa Tengah, Indonesia
e-mail: dodyariawan@staff.uns.ac.id

U. Sabino et al. (eds.), *Proceedings of the 6th International Conference and Exhibition on Sustainable Energy and Advanced Materials*, Lecture Notes in Mechanical Engineering,
https://doi.org/10.1007/978-981-15-4481-1_1

Previous studies [2, 3] stated that various surface modifications methods, such as silane, alkali, peroxide, permanganate and heat treatments, could increase the interfacial bond strength—both critical factors for obtaining better mechanical properties of the composites. The alkali treatment is proven to increase the adhesion between cellulosic fibre and polymeric matrices and significantly enhances in tensile and flexural strength were observed for alkali treated natural fibre composites [3]. Heat treatment is another modification that could affect the mechanical properties, dimensional stability, colour, flexural strength, hardness, biological properties, durability and chemical properties of wood fibres [2].

The elastic properties of randomised discontinue fibre composites can be obtained by experimental methods or derived from micromechanical models. Micromechanical models of composite are derived based on the properties of the individual composite constituent and their arrangement, such as the elastic modulus (E), tensile strength (σ), *Poisson's ratio* (υ) and the volume fractions (V) of both fibre and matrix [4]. In some cases, fibre orientation and fibre aspect ratio are also included [5].

These studies applied the following model to predict the tensile elastic modulus, for example, Tsai-Pagano, Christensen, Puck/Manera, Krenchel, and tensile strength of KF—unsaturated polyester composites including Bowyer Bader and Hirsch's. This research used the two-dimensional mode for all of modelling calculation due to the lengths of the fibre, which are longer than the composite thickness. Furthermore, the fibre length assumed is much longer than critical fibre length ($l_f \gg l_c$), wherein needle punched KF used has 70 mm of fibre length.

2 Experimental

2.1 Materials

KF mats were fabricated using the needle-punching method with an areal density of 1100 g/m^2 and stitching density of 50 cm^2. The average of fiber length was 40.10 ± 17.40 mm with the diameter distribution of 63.79 ± 37.37 μm. The large variation of standard deviation is a common occurrence in natural fibres [6]. Consequently, these produces significant variations in KF composites.

The properties of tensile properties of KF were taken based on previous author studies [7], wherein alkali treatment of KF in 6% NaOH for 3 H and heat treatments of KF at 140 °C for 10 h resulted in the optimum mechanical properties of the randomised discontinue KF needle punched mat. The properties of fibre used for this modelling, i.e. length of fibre (l_f), strength of fibre (σ_f), tensile modulus of fibre (E_f), and diameter of fibre (D_f), are displayed in Table 1, while the unsaturated polyester type orthophthalic (Reversol P-9565) supplied by Synthomer Sdn. Bhd properties is displayed in Table 1.

Table 1 Tensile properties of KF and matrix [7]

Property	Unit	Untreated KF	Heat treated KF	Alkali treated KF	Unsaturated polyester matrix
Tensile strength	MPa	251.43	310.68	384.69	59.07 ± 3.8
Elastic modulus	GPa	17.40	20.42	29.40	2.91 ± 0.18
Diameter	μm	140.1	110.3	108.6	

2.2 *Fabrication*

The composites were fabricated by using the RTM type Innovator Megaject 3250 (8000) with two reciprocating pumps. The fabrication process was set at 1.3 bar for pump pressure, 0.6 bar for mould pressure and 0.55 bar for vacuum pump pressure. The ratio of hardener (MEKP) to unsaturated polyester was 1:100. Post curing was carried out for 24 h at room temperature and followed by 3 h at 80 °C. The composite samples were prepared with KF content at 10, 20 and 30% of KF volume fraction (V_f). Untreated, heat treated and alkali treated nonwoven KF composites are referred to as the UKUPC, HKUPC and AKUPC, respectively

2.3 *Tensile Test*

The tensile properties of the composites were evaluated by using an INSTRON 5969 universal testing machine according to the ASTM D 3039 standard. The test machine used a 50 kN load cell and operated at a rate of 2 mm/min.

3 Results and Discussion

3.1 *Models of Randomised Discontinue Fibre Composite Models*

Tsai-Pagano model uses the modulus of elasticity, the volume fraction of the composite component, as well as the aspect ratio of the fibre as consideration for Eq. (1) [8].

$$\frac{E}{E_m} = \frac{1+\xi\eta V_f}{1-\eta V_f} \text{ and } \eta = \frac{\frac{E_f}{E_m}-1}{\frac{E_f}{E_m}+\xi} \tag{1}$$

where, E corresponds to the direction of modulus (E_{11} or E_{22}) [9]. E_f and E_m represent the moduli of the fibre and the matrix, respectively. For longitudinal modulus, ξ is $2(l/d)$ while ξ is 2 for transverse modulus.

Based on the calculations of E_{11} and E_{22} by using Eq. (1), the in-plane modulus of a random fibre composite as proposed by Tsai and Pagano [1] could be taken to determine the modulus of the random fibre composite that uses the same fibre/resin:

$$E = \frac{3}{8}E_{11} + \frac{5}{8}E_{22} \tag{2}$$

where E_{11} is the elastic modulus of the longitudinal direction loading and E_{22} is the elastic modulus of the transverse direction loading.

Christensen model considers the effect of fibre orientation and fibre-matrix interaction which be simplified as an equation [10]:

$$E_c = \frac{V_f E_f}{3} + \frac{(1 - V_f)E_m}{3} + \frac{19}{27}E_m\left[\frac{E_f(1 + V_f) + E_m(1 - V_f)}{E_f(1 - V_f) + E_m(1 + V_f)}\right] \tag{3}$$

where, V_f is the volume fraction of fibre, E_m is the matrix modulus, E_f is the fibre modulus, and E_c is the composite modulus.

Manera model is suitable for the elastic modulus calculation of composites with high fibre aspect ratio (>300), two-dimensional random distribution of fibres and laminates with an infinite number of layers oriented in all directions [11]. The approximate equations can be expressed as [12]:

$$E_c = V_f\left(\frac{16}{45}E_f + 2E_m\right) + \frac{8}{9}E_m \tag{4}$$

where, V_f is the volume fraction of fibre, E_m is the matrix modulus, E_f is the fibre modulus, and E_c is the composite modulus. Manera [12] reported that this model was intended for volume fraction (V_f) within the range $0.1 \le V_f \le 0.4$ and $\mathrm{E_m}$ within the range 2 GPa $\le E_m \le$ 4 GPa.

The Cox-Krenchel model for elastic modulus of randomized discontinue composites is given by Eq. (5) [13].

$$E = \eta_l \eta_o E_f V_f + E_m V_m \tag{5}$$

where η_o is the distribution factor of fiber orientation and η_l is the distribution factor of fiber length. The distribution factor of fiber orientation was determined based on Krenchel's theory [6].

$$\eta_o = \cos^4(\alpha_o) \tag{6}$$

where η_o take the value of 3/8 for random in-plane fibre orientation laminates [14], while η_l was calculated based on the Cox shear lag analysis, which is given by [13]:

$$\eta_l = 1 - \frac{\tanh\left(\frac{\beta l_f}{2}\right)}{\frac{\beta l_f}{2}} \tag{7}$$

where β is defined as

$$\beta = \frac{2}{d}\sqrt{\frac{E_m}{E_{f1}(1 - v_m)\ln\left(\frac{\pi}{x_i V_f}\right)}} \tag{8}$$

where β point to the stress concentration rate coefficient at fibers' end, l_f is the length of fibre, v_m is the Poisson's ratio of matrix and x_i is 4 [6].

To calculate the tensile properties of randomized discontinuous fibre composites, this research had used the Bowyer–Bader and Hirsch models for untreated, alkali treated and heat treated KF [1]. The first model used is the Hirsch model which combines the parallel and series elements of phases [15] as follows:

$$\sigma_C = x\left(\sigma_m^1 V_m + \sigma_f V_f\right) + (1 - x)\frac{\sigma_m^1 \sigma_f}{\sigma_m^1 V_f + \sigma_f V_m} \tag{9}$$

where based on an agreement between experimental and theoretical values, x is 0.1 for randomly oriented fibre composites. The second model is the Bowyer Bader model, which derives from the tensile strength of the composite (σ_C) as follows:

$$\sigma_C = \sigma_f K_1 K_2 V_f + \sigma_m V_m \tag{10}$$

where σ_f is the fibre ultimate tensile strength while σ_m^1 is the tensile stress in the matrix at fibre failure strain, V_m is the matrix volume fraction and K_1 represents the fibre orientation which takes the value of 3/8 for random in-plane fibre orientation (2-dimensional). In the case of fibre length that is greater than the critical length ($l_f > l_c$) [16], the K_2 is defined as follows:

$$K_2 = \frac{l_f - l_c}{2l_f} \tag{11}$$

while for fibre length that is shorter than the critical length ($l_f < l_c$), K_2 can be derived from the following equation:

$$K_2 = \frac{l_f}{2l_c} \tag{12}$$

This study used several assumptions for prediction model calculation from natural fibre properties to fulfil the modelling requirement, i.e., interfacial shear strength (IFSS), Poisson ratio of matrix (υ) and packing factor (Φ). IFSS value for this study is 9.9 MPa which was taken from previous research for untreated hemp fibre (as natural fibre) to unsaturated polyester IFSS [16]. Unsaturated polyester is an isotropic material which has Poisson ratio by 0.2–0.33 [4]. Based on these values, the 0.3 value was chosen for the modelling. Furthermore, the critical length of KF that was obtained by using the interfacial shear strength (IFSS) of KF toward Unsaturated Polyester is 11.6 mm. The average length of the KF used had been 40.10 mm longer than its critical fibre length (11.63 mm). Therefore, by using Eq. (11), K_2 had been determined as 1.22.

3.2 *Elastic Modulus Prediction Modelling*

Figure 1a–c shows the correlation between the predictions and the experimental results. The elastic modulus of both prediction and experimental results increases

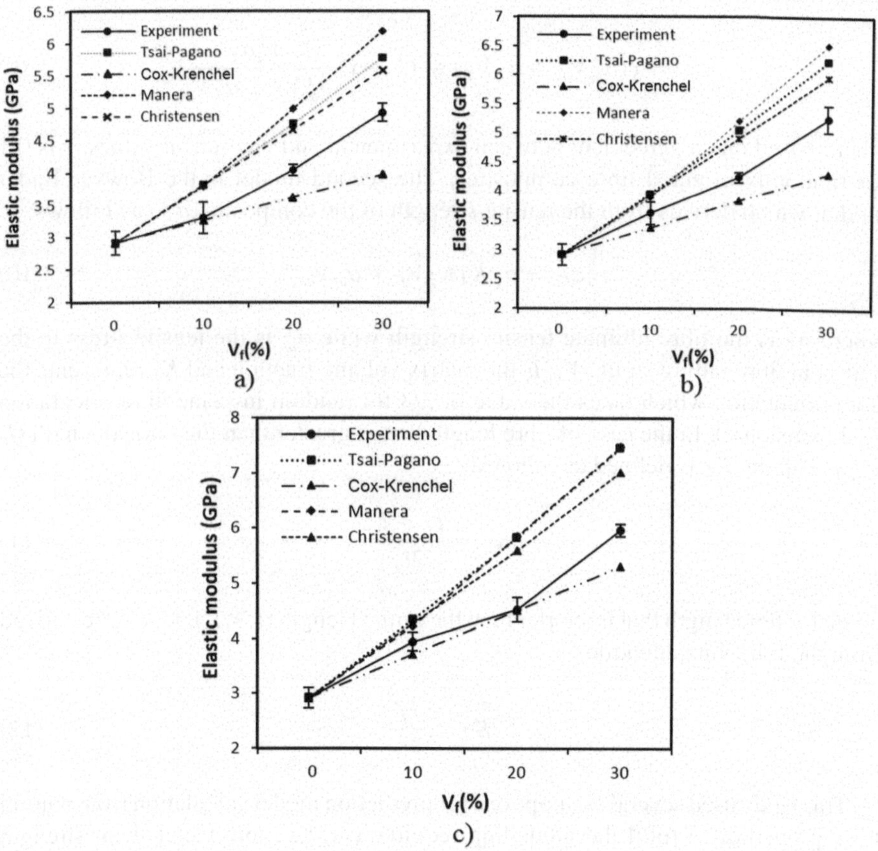

Fig. 1 Elastic modulus prediction for **a** UKUPC, **b** HKUPC, **c** AKUPC in different volume fraction

linearly with the enhancement of fibre volume fractions. The micromechanics modelling of Tsai-Pagano was calculated by using Eq. (2) combined with Eqs. (1) and (2), while the Christensen and Manera modelling used Eqs. (3) and (4). The micromechanics modelling of Cox Krenchel was calculated using Eq. (5) combined with Eqs. (6), (7) and (8) to determine η_l.

The predictions by Christensen, Tsai-Pagano, and Manera are considerably higher than the experimental results and Modified Rule of Mixtures (ROM) has lower value than the experimental results. The results of elastic modulus are Manera > Tsai-Pagano > Christensen > Experimental Result > Cox Krenchel. Both HKUPC and AKUPC have similar trend tendencies. Every prediction for each treatment has different error values (refer Table 2).

The results shown in Table 2 indicate that the Cox-Krenchel equation produced the lowest total error factor. The error levels of Manera, Tsai-Pagano and Christensen model predictions are >10% from experimental results while the error levels for the Cox Krenchel models are lower than 0%. High errors factor of experimental data were shown in previous reports [5, 17]. The error factor in the experimental result can be attributed to several factors, such as uniformity of fibre distribution and strength of fibre, complex geometric of the KF, void, punching direction, interlocking of fibre in mat, interfacial adhesion, etc.

3.3 *Tensile Strength Prediction Modelling*

Figure 2a–c shows that the strength of the composite is less than the strength of UPR, both in the experimental result and prediction model. However the strength of composites increased with the increase of fibre volume fractions. This trend occurs on UKUPC, HKUPC and AKUPC with different values. Andre et al. [18] reported similar findings, wherein KF/unsaturated polyester composites produced by RTM resulted in lower tensile strength and higher elastic modulus than the neat resin at all fibre volume fractions. In micromechanics theory for uni-directional composite [1, 4], a shorter elongation of the composite component would dominantly affect the failure of the composites. In this research, the KF elongations (σ_{fu}) were 1.46–1.83%, which were lower than the average neat resin elongation ($\sigma_m = 4.02\%$) or $\sigma_{fu} > \sigma_m$. This indicated that the matrix would only support the composites up until a limiting value of strength when the elongation of the KF was in the maximum (σ_m^1).

Figure 2 shows that the Hirsch model had higher the KF composites tensile strength at each volume fraction (V_f) values. The significant deviation of Hirsch model results are displayed in Table 3. Hirsch's model has a different trend to the experimental of tensile strength results, thus this model could not precisely predict the tensile strength of composites for untreated, heat treated and alkali treated KF composites. The Bowyer Bader model are capable of capturing the main experimental trends without undue complexity for untreated, heat treated and alkali treated KF composites. Facca et al. [17] stated that the close prediction in this model could probably be reached if factors such as fibre angle and length distribution corrections

Table 2 Error percentage of elastic modulus modelling toward elastic modulus of experimental results

Sample	Error (%)								
	UKUPC			HKUPC			AKUPC		
Volume fraction	10	20	30	10	20	30	10	20	30
Tsai-Pagano	15.5	18.1	17.5	9.20	19.3	18.8	10.4	29.5	25.4
Cox-Krenchel	−1.0	−10.1	−19.0	−6.60	−9.0	−17.5	−5.4	0.1	−10.4
Manera	14.7	23.4	25.5	7.51	22.7	23.9	7.0	29.1	25.3
Christensen	15.5	15.4	13.2	8.6	15.7	13.4	8.4	23.5	17.5

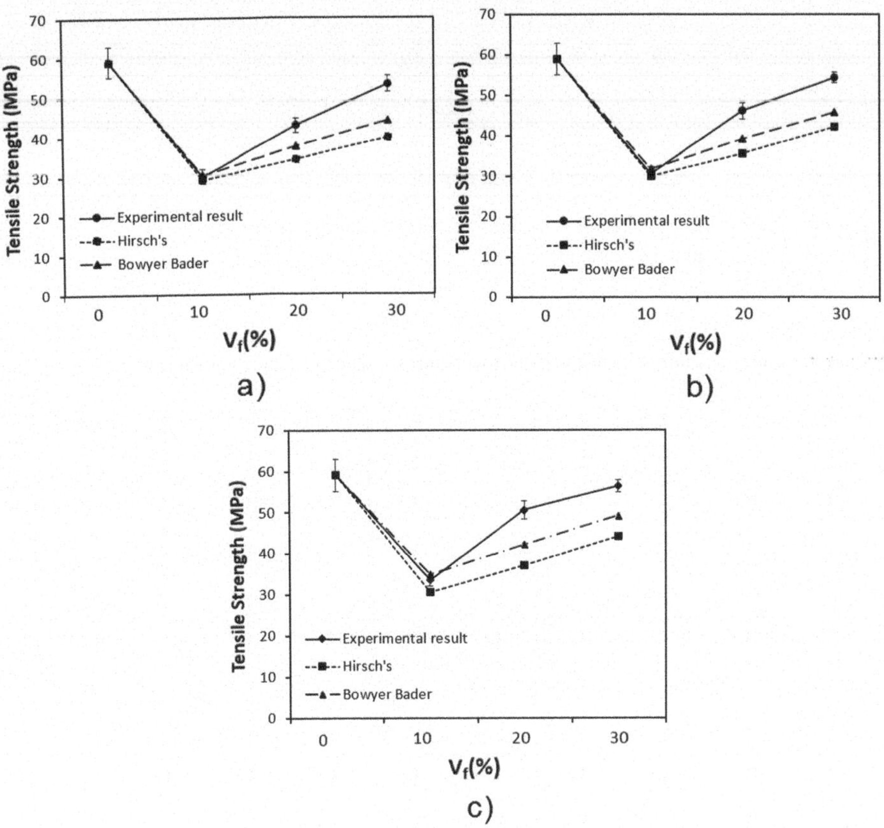

Fig. 2 Tensile strength prediction for **a** UKUPC, **b** HKUPC, **c** AKUPC

were included. Furthermore, the effect of stress transfer through the surface and the ends of fibres were excluded in the calculation.

Table 3 displays the error factor of tensile strength modelling toward tensile strength of experimental results calculated by using Eq. (3.11). The error factor of Bowyer Bader models is less than 20.33% while the error factor of the Hirsch's model are higher than 56%. The experimental tensile strength at 10 and 20% volume fraction showed closer results to Bowyer Bader prediction (refer to Fig. 2a–c and Table 3). Nevertheless, the declining deviation of experimental results occurred at $V_f = 30\%$, especially for HKUPC and AKUPC. These deviations can be induced by many factors, i.e., increase of void, inhomogeneous KF orientation, etc., in all the composites [17].

Table 3 Error percentage of tensile strength modelling toward tensile strength of experimental results

Sample	Error (%)								
	UKUPC			HKUPC			AKUPC		
Volume fraction	10	20	30	10	20	30	10	20	30
Hirsch's	−2.78	−20.2	−25.4	−2.5	−22.6	−22.5	−8.6	−26.7	−21.6
Bowyer Bader	1.13	−12.6	−17.4	3.8	−14.9	−15.85	4.29	−16.4	−12.7

4 Conclusions

The analysis of the effect of KF fibre treatment on composite properties signified that the increasing pattern of tensile properties are AKUPC > HKUPC > UKUPC in all volume fractions. In micromechanical modelling analysis, the experimental results of tensile test have fairly similar trend of results to the Bowyer Bader and Hirsch's modelling results at 10, 20 and 30% of volume fraction (V_f), wherein Bowyer Bader model produce better prediction than Hirsch's model. The experimental results composites have lower elastic modulus than the Tsai-Pagano, Manera and Christensen models at all V_f, excluding Cox Krenchel model. However the Cox Krenchel provide the lowest error.

References

1. Gibson R (2011) Principles of composite material mechanics. CRC Press, Boca Raton
2. Rong MZ, Zhang MQ, Liu Y, Yang GC, Zeng HM (2001) The effect of fiber treatment on the mechanical properties of unidirectional sisal-reinforced epoxy composites. Compos Sci Technol 61:1437–1447
3. Sreekumar PA, Thomas SP, Saiter JM, Joseph K, Unnikrishnan G, Thomas S (2009) Effect of fiber surface modification on the mechanical and water absorption characteristics of sisal/polyester composites fabricated by resin transfer molding. Compos Part A Appl Sci Manuf 40:1777–1784
4. Chawla KK (2012) Composite materials: science and engineering. Springer, New York
5. Epaarachchi J, Ku H, Gohel K (2009) A simplified empirical model for prediction of mechanical properties of random short fiber/vinylester composites. J Compos Mater
6. Andre N, Ariawan D, Ishak ZM (2016) Elastic anisotropy of kenaf fibre and micromechanical modeling of nonwoven kenaf fibre/epoxy composites. J Reinf Plast Compos 35:1424–1433
7. Ariawan D, Salim MS, Mat Taib R, Ahmad Thirmizir MZ, Ishak M, Zainal A (2018) Durability of alkali and heat-treated kenaf fiber/unsaturated polyester composite fabricated by resin transfer molding under natural weathering exposure. Adv Polym Technol 37:1420–1434
8. Islam M, Begum K (2011) Prediction models for the elastic modulus of fiber-reinforced polymer composites: an analysis. J Sci Res 3:225–238
9. Tucker CL III, Liang E (1999) Stiffness predictions for unidirectional short-fiber composites: review and evaluation. Compos Sci Technol 59:655–671
10. Christensen R (1976) Asymptotic modulus results for composites containing randomly oriented fibers. Int J Solids Struct 12:537–544
11. Puck A, Schürmann H (1998) Failure analysis of FRP laminates by means of physically based phenomenological models. Compos Sci Technol 58:1045–1067
12. Manera M (1977) Elastic properties of randomly oriented short fiber-glass composites. J Compos Mater 11:235–247
13. Cox HL (1952) The elasticity and strength of paper and other fibrous materials. Br J Appl Phys 3:72
14. Thomason JL, Vlug MA (1996) Influence of fibre length and concentration on the properties of glass fibre-reinforced polypropylene: 1. Tensile and flexural modulus. Compos Part A Appl Sci Manuf 27:477–484
15. Hirsch TJ (1962) Modulus of elasticity iof concrete affected by elastic moduli of cement paste matrix and aggregate. J Proc 59

16. Sawpan MA, Pickering KL, Fernyhough A (2011) Effect of fibre treatments on interfacial shear strength of hemp fibre reinforced polylactide and unsaturated polyester composites. Compos Part A Appl Sci Manuf 42:1189–1196
17. Facca AG, Kortschot MT, Yan N (2006) Predicting the elastic modulus of natural fibre reinforced thermoplastics. Compos Part A Appl Sci Manuf 37:1660–1671
18. Andre NG, Ariawan D, Ishak ZM (2017) Mechanical properties and micromechanical analysis of nonwoven kenaf fibre/epoxy composites produced by resin transfer moulding. J Compos Mater 51:1875–1885

Tribological Properties of 3D-Printed ABS Under Paraffin Oil Lubrication

Mohd Fadzli Bin Abdollah, Hilmi Amiruddin, and Norjannatul Ainah Norashid

Abstract Recently, 3D printing has been broadly utilised for the design and manufacturing of different parts of systems and mechanisms. Part orientation during 3D printing has its effect on the surface finish and, subsequently, it influences the tribological properties of the tribopairs. However, very little work has been done to comprehend the tribological properties of 3D printing components, which may prompt uncontrolled wear of tribopairs surfaces. Along these lines, the purpose of this preliminary study is to investigate the tribological properties of a commonly used polymer in 3D printing, which is acrylonitrile butadiene styrene (ABS). The sample of ABS was fabricated by using the technique of fused deposition modelling (FDM) 3D printing. The tribology test was performed utilizing a ball-on-disc tribometer under paraffin oil lubrication. The test was kept running at different applied loads with a constant sliding distance, speed and temperature. The outcomes demonstrate that the wear rate of the 3D-printed ABS decreases with the load. Be that as it may, a high coefficient of friction is observed with the applied load.

Keywords 3D printing · ABS · Tribology · Paraffin oil

1 Introduction

Heightened expenses and decreased accessibility of fossil fuels has caused an expansion of the importance of high-performing, light-weight components which not just contribute in saving of energy and reduction of emission of CO_2 yet are competent in

---

M. F. B. Abdollah (✉) · H. Amiruddin · N. A. Norashid
Fakulti Kejuruteraan Mekanikal, Universiti Teknikal Malaysia Melaka, Hang Tuah Jaya, 76100 Durian Tunggal, Melaka, Malaysia
e-mail: mohdfadzli@utem.edu.my

H. Amiruddin
e-mail: hilmi@utem.edu.my

M. F. B. Abdollah · H. Amiruddin
Centre for Advanced Research on Energy, Universiti Teknikal Malaysia Melaka, Hang Tuah Jaya, 76100 Durian Tunggal, Melaka, Malaysia

U. Sabino et al. (eds.), *Proceedings of the 6th International Conference and Exhibition on Sustainable Energy and Advanced Materials*, Lecture Notes in Mechanical Engineering,
https://doi.org/10.1007/978-981-15-4481-1_2

the application-specific demands regarding the physical and mechanical characteristics. With this regard, the utilization of composites and materials based on polymer has come in the limelight of research recently [1–5].

Acrylonitrile butadiene styrene (ABS) is a popular industrial thermoplastic due to its because of its one of kind characteristics, such as chemical resistance, outstanding mechanical response, fine surface finish, and improved processing characteristics. It is widely used in several industries making mechanical elements like gears, mechanical links, lead screws, etc. ABS can be effectively created utilizing 3D printing. One of the advantages of utilizing the 3D printing strategy, for example, fused deposition modelling (FDM) is that it is fast and financially savvy for customised jobs and minimal batch size production. The behaviour and performance of this technique are incredibly affected by a few printing parameters like gap, scaffolding angle, slice thickness and deposition speed. In any case, the present concern for the industrial use of the FDM technique is the quality of the components produced utilizing this technique. There have been a few examinations on the quality of components made through FDM and these qualities include dimensional precision, tensile strength, etc. concerning the process parameters of the FDM technique [6–9]. Nevertheless, the tribological properties of these components have not been examined satisfactorily, which may cause uncontrolled wear of these surfaces. Just a few examinations have been conducted concerning the tribology of FDM components [10, 11].

To bridge this gap, a preliminary study was carried out on the 3D-printed ABS regarding its tribological properties and was examined under paraffin oil lubrication.

2 Methodology

ABS material was utilised to fabricate a sample having a disc of diameter 72 mm and thickness 4 mm using the FDM method (printer model: Cubepro 3D printer). Table 1 displays the material properties of both materials prior to testing. The printing process parameters are depicted in Table 2.

Absorption test was performed by submerging the samples in the paraffin oil. The samples were dried at 100 °C in a vacuum stove for 10–12 h before the process. At that point, they were subjected to complete immersion for 6 h, at which time their weight was measured at 10 min, 20 min, 40 min and 1 h intervals for the entire 6 h. After this, the sample was expelled from the oil, evaporated, and instantly its weight was measured using a microbalance. The absorption is represented as a fraction of weight gained, which can be computed using Eq. (1).

Table 1 Material properties of 3D-printed ABS before testing

Properties	3D-printed ABS
Hardness, H (GPa)	5.742
Surface roughness, R_a (μm)	1.8838
Density, ρ (g/cm^3)	0.9

Table 2 Printing process parameters

Parameter	Value
Layer height	0.18 mm
Print speed	60 mm/s
Fill density	100%
Fill pattern	Line
Temperature (extruder)	220 °C

$$\text{Weight gained} = \frac{\text{Wet weight} - \text{Conditioned weight}}{\text{Conditioned weight}} \times 100 \quad (1)$$

The tribological assessment was completed by utilizing a ball-on-disc tribometer against EN31 chrome steel ball-bearing with diameter 12.7 mm at room temperature under paraffin oil lubrication. All tests were performed at different loads changing from 20 to 55 N, sliding distance of 1000 m and sliding speed of 100 rpm. A computerized magnifying instrument was used to quantify the diameter of the wear scar. The force of friction was defined by utilizing an information recording system connected to a computer. The coefficient of friction (*COF*) was computed from Eq. (2), while the specific wear rate (*k*) was computed based on the spherical geometry as given in Eqs. (3) to (5).

$$\mathrm{COF} = \frac{\mathrm{F}}{\mathrm{W}} \quad (2)$$

$$\mathrm{V_{loss}} = 2\pi\mathrm{R} \times \pi\left(\mathrm{a}^2 + \mathrm{h}^2\right) \quad (3)$$

$$\mathrm{h} = \mathrm{r} - \sqrt{\mathrm{r}^2 - \mathrm{a}^2} \quad (4)$$

$$\mathrm{k} = \frac{\mathrm{V_{loss}}}{\mathrm{W} \times \mathrm{L}} \quad (5)$$

where F is friction force [N], W is applied load [N], V_{loss} is wear volume loss [mm^3], L is sliding distance [mm], R is wear track radius [mm], a is wear scar radius [mm], h is spherical cap height/wear depth [mm] and r is ball bearing radius [mm].

Worn surfaces were inspected with Scanning Electron Microscopy (SEM).

3 Results and Discussion

Figure 1 shows the result of absorption test for 3D-printed ABS. The inherent porosity of the samples that were 3D printed contributed significantly to this study. By structure up the material one layer after another, the FDM process does not make samples having solid interiors, despite the fact that the printer was commanded to deliver a

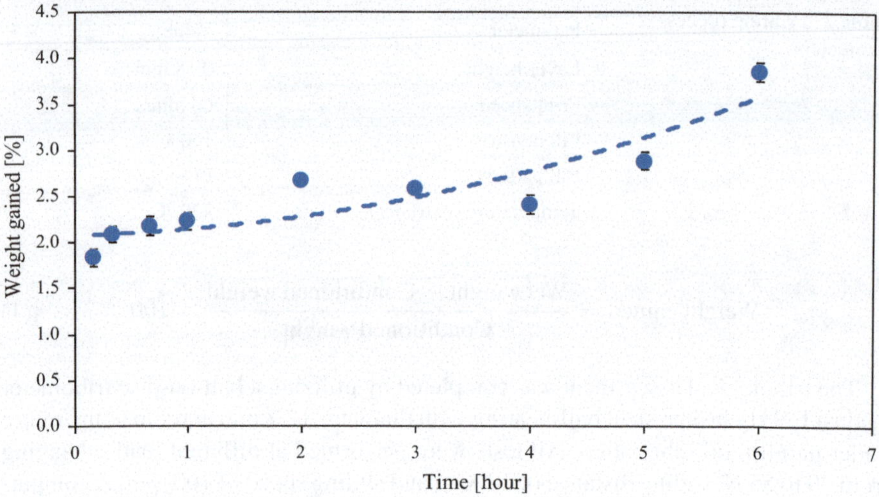

Fig. 1 Absorption test results on increasing weight over 6 h immersion in paraffin oil. The error bar is for the standard deviation

100% infill. It caused intrinsic micro-sized hollow spaces inside the samples which improve the fluid absorption capacity. The absorbent materials contribute essentially to their excellent properties like as extraordinary availability of active locations, high surface area, and improved diffusion and mass transport. It is extremely vital for high performance in a wide range of applications [12].

Figure 2a displays changes in the COF of the 3D-printed ABS for the duration of the test at various applied loads. Figure 2b displays average steady-state *COF* values over the loads that are procured subsequent to sliding 200 m ignoring eventual unsteadiness at running-in and showed a decrease in *COF* (Fig. 2a). It is conceivable on account of the gradual absorption of liquid as shown in Fig. 1.

As can be found in Fig. 2b, *COF* is increased as the applied load increased. The sliding motion occurs in quite small regions at asperity peaks and over a long time. In this manner, an increase in the load causes breaks or ruptures of asperity and causes an increase in the real contact area. This occurrence brings about an increase in the temperature and decrease in the material's strength, which in turn brings about an increase in friction. Nonetheless, *COF* of 3D-printed ABS is still within the polymer's acceptable range of *COF*. It may be related with the surface finish (Table 1), where coarser 3D-printed ABS surface causes more breaks or ruptures of asperity. This phenomenon may increase the real contact area and causes an increase in *COF*.

Figure 3 shows that the wear rate of 3D-printed ABS reduces with increase in the load because of a higher rate of absorption, which makes the surface progressively lubricated.

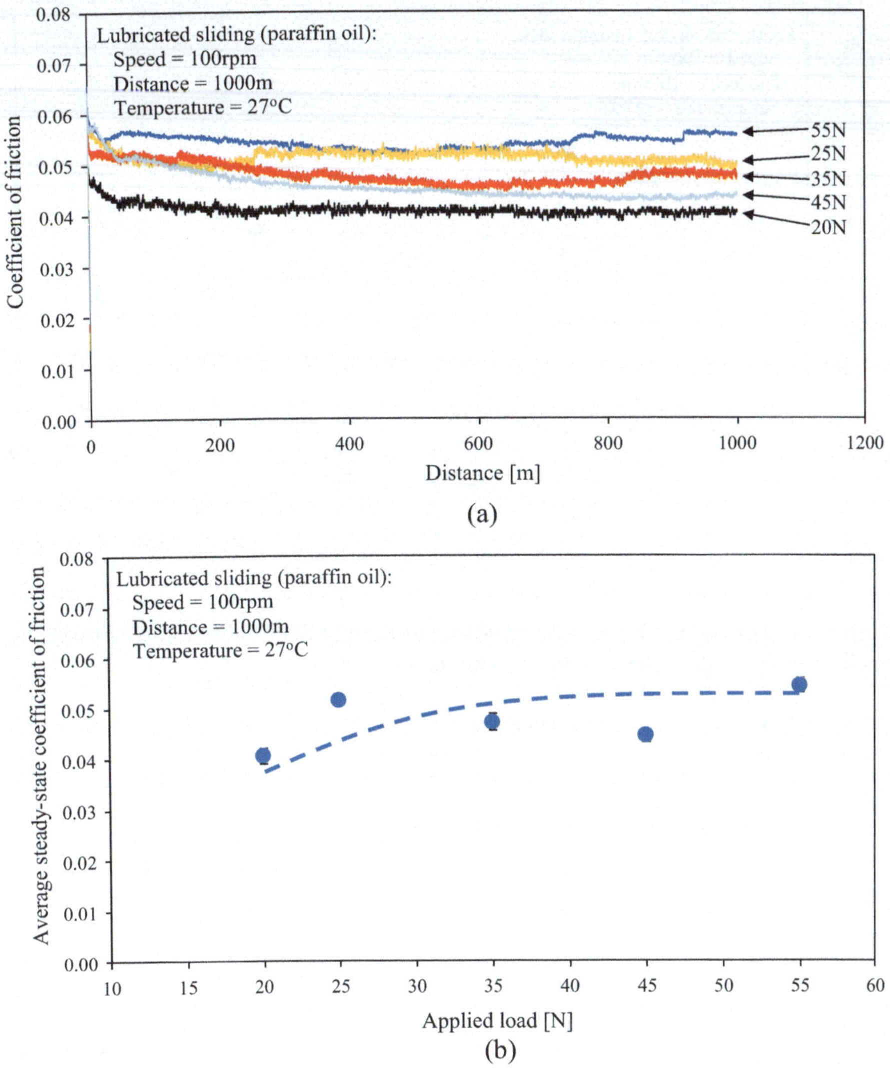

Fig. 2 Coefficient of friction of 3D-printed ABS changes with the applied load. **a** Dynamic coefficient of friction and **b** average steady-state values of coefficient of friction. The error bar is for the standard deviation

4 Conclusions

To whole it up, *COF* increases with an increase in the applied load. Moreover, the wear rate of the ABS that is 3D-printed decreases with a decrease in the applied load. The intrinsic surface finish and the resultant high value of *COF* for the surfaces that are 3D-printed is as yet a challenge that is to be overcome. Enhancements in the

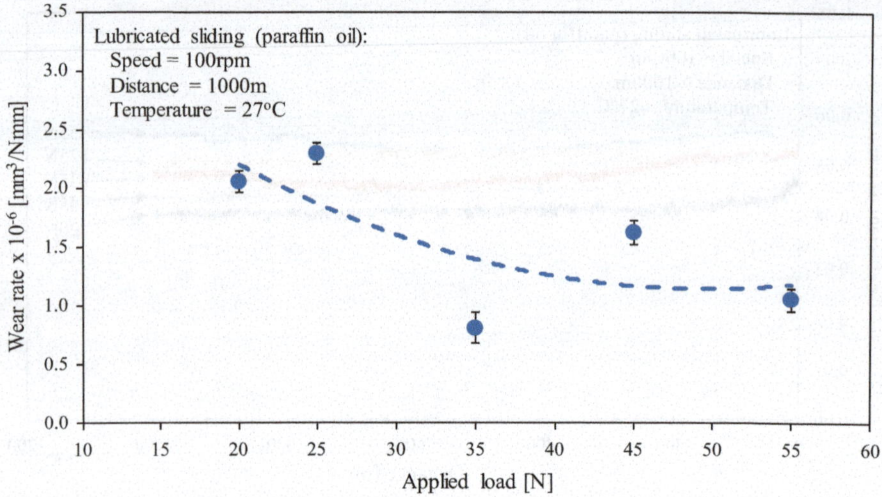

Fig. 3 Wear rate of 3D-printed ABS changes with the applied load. The error bar is for the standard deviation

method of FDM should be possible. The surfaces could likewise be treated physically or chemically so that they can be smoothened.

Acknowledgements The authors gratefully acknowledge the use of the services and facilities of the Fakulti Kejuruteraan Mekanikal, Universiti Teknikal Malaysia Melaka (UTeM). The author also acknowledges Fakulti Kejuruteraan Mekanikal, UTeM for the registration fee support to participate in ICE-SEAM2019. This research has been supported by the grant from Universiti Teknikal Malaysia Melaka (Grant no.: JURNAL/2018/FKM/Q00005).

References

1. Briscoe BJ, Sinha SK (2002) Wear of polymers. Proc Inst Mech Eng, Part J: J Eng Tribol 216(6):401–413
2. Wieleba W (2007) The mechanism of tribological wear of thermoplastic materials. Arch Civ Mech Eng 7(4):185–199
3. Jumahat A, Kasolang S, Bahari MT (2015) Wear properties of nanosilica filled epoxy polymers and FRP composites. J Tribol 6:24–36
4. Mustafa A, Abdollah MFB, Amiruddin H, Shuhimi FF, Muhammad N (2017) Optimization of friction properties of kenaf polymer composite as an alternative friction material. Ind Lubr Tribol 69(2):259–266
5. Mohmad M, Abdollah MFB, Khudhair AQ, Tamaldin N, Amiruddin H, Zin MRBM (2018) Physical-mechanical properties of palm kernel activated carbon reinforced polymeric composite: potential as a self-lubricating material. J Tribol 17:77–92
6. Lee BH, Abdullah J, Khan ZA (2005) Optimization of rapid prototyping parameters for production of flexible ABS object. J Mater Process Technol 169(1):54–61
7. Ahn D, Kweon JH, Kwon S, Song J, Lee S (2009) Representation of surface roughness in fused deposition modeling. J Mater Process Technol 209(15–16):5593–5600

8. Sood AK, Ohdar RK, Mahapatra SS (2010) Parametric appraisal of fused deposition modelling process using the grey Taguchi method. Proc Inst Mech Eng Part B J Eng Manuf 224(1):135–145
9. Singh R, Kumar R, Ahuja IPS (2018) Mechanical, thermal and melt flow of aluminum-reinforced PA6/ABS blend feedstock filament for fused deposition modeling. Rapid Prototyp J 24(9):1455–1468
10. Dawoud M, Taha I, Ebeid SJ (2015) Effect of processing parameters and graphite content on the tribological behaviour of 3D printed acrylonitrile butadiene styrene (Einfluss von Prozessparametern und Graphitgehalt auf das tribologische Verhalten von 3D-Druck Acrylnitril-Butadien-Styrol Bauteilen). Materialwiss Werkstofftech 46(12):1185–1195
11. Beg MD, Khan MS, Khan A (2017) Investigation on tribological behavior of FDM printed ABS polymer. Int J Tech Res Appl 5:75–77
12. Sun MH, Huang SZ, Chen LH, Li Y, Yang XY, Yuan ZY, Su BL (2016) Applications of hierarchically structured porous materials from energy storage and conversion, catalysis, photocatalysis, adsorption, separation, and sensing to biomedicine. Chem Soc Rev 45(12):3479–3563

Comparative Investigation of Matrix and Fiber Orientation Composite Ramie

Komang Astana Widi, Gerald Pohan, Wayan Sujana, Tutut Nani, and Luh Dina Ekasari

Abstract The development on the use of composite materials in the field of engineering is increasing. This is due to its superior properties compared to conventional materials; the ratio between strength and densities is quite high, having high rigidity, the simple manufacturing process, and its resistance to corrosion and fatigue. The purpose of this study was to determine the effect of ramie fibers orientation with Epoxy and polypropylene Resin Matrix on increasing tensile strength, bending strength, impact strength and fatigue strength. However, the utilization of ramie fiber composites is not optimally done because natural fiber composites are very susceptible to failure due to the changes in the orientation of the fiber and the ability of fiber adhesion with a less optimal type of matrix. The object of this research is a composite with ramie fiber using Epoxy Resin and polypropylene as the binding material. The presence of both ramie which is reinforcing fibers with epoxy and polypropylene matrix is greatly influenced by the orientation of the fiber. The tensile strength test results in the epoxy matrix increased up to 140% while the increase in polypropylene matrix could reach 187%. Based on SEM observations, the reduction of composite mechanical properties was due to several factors, namely porosity, adhesiveness between fiber-matrices and between matrices, overlapping fibers and broken fibers.

Keywords Ramie · Fiber orientation · Bending strength · Tensile strength · Impact strength · Fatigue strength

1 Introduction

The price of the component is largely determined by the quality of the material and the manufacturing process. One element of determining quality is based on service life. Composites produced from natural fiber materials offer the potential as

K. A. Widi (✉) · G. Pohan · W. Sujana · T. Nani
National Institute of Technology of Malang (ITN Malang), East Java, Indonesia
e-mail: aswidi@yahoo.com

L. D. Ekasari
Tribhuwana Tunggadewi University, East Java, Indonesia

U. Sabino et al. (eds.), *Proceedings of the 6th International Conference and Exhibition on Sustainable Energy and Advanced Materials*, Lecture Notes in Mechanical Engineering,
https://doi.org/10.1007/978-981-15-4481-1_3

a reinforcing element to increase the service life of components with a better price (efficiency) by using natural fiber materials of ramie (*Boehmeria nivea*) fiber which is very abundant and widely grown in tropical regions such as Indonesia. Besides, this material is easily available as long as the farmers can continue to produce and will automatically improve the standard of living of Indonesian farmers. In line with these considerations, automotive components have become a major target market for the development of ramie natural fiber materials as a substitute and development of automotive components.

Some of the problems in using ramie fiber as a composite reinforcing material on a product are shows that the results of tensile strength testing in the epoxy matrix were 140% while the PP strength increased by 187%.

2 Literature Review

The development of technical materials especially polymer composites, which include high-tech materials, manufacturing processes and the application of materials this decade has shown an increase [1]. Crawford, 1989 has defined and classified fibers based on the dimensions of length in variations in the diameter. Fiber diameter ranges from 5 to 100 μm. It has a continuous long shape and relatively short fiber pieces. Based on the diameter and length fiber is divided into three phases, namely whiskers, fiber and wires. Crawford also stated that if fiber reinforcement blends well with a matrix or resin which has a very strong adhesive bond, it can produce mechanical properties that meet a technical application. The orientation of the type of reinforcement fiber in the matrix of a composite material is divided into four models including (a) continuous fiber, (b) woven fiber, (c) chopped fiber and (d) hybrid [2].

Polymer material is an arrangement of hydrocarbon atomic bonds which have a repeated chain bond configuration. This material is divided into two groups: (a) thermoplastics and (b) thermosets. Thermoplastics are a group of plastic polymers that are able to be repeatedly heated to become soft and hardened again by a cooling process. Thermoplastic types include polymer styrene, acrylics, cellulostics, polyethylene, vinyls, nylons and flourocarbon types. The thermoset is a group of plastic polymers which will be degraded (unable to return to its original molecular arrangement) due to heating process. This group includes amino (melamine and urea), polyester, alkyds, epoxy and phenolics [3].

According to research by Wargadiputra, 2005, the failure of polyester composites with ramie boosters caused by the variation of load in the form of normal tensile stresses and shear stresses, which is due to the bond between ramie fibers and polyester metrics is not strong enough. Whereas Saidah's research in 2005 which analyzed the predictions of tensile loading on fiber orientation showed that angle orientation of 0°, 0°, 90° had a tensile stress greater than the angle orientation of 0°, 90°, 0°. From those research data, Ansys software will be employed in this study to predict stress based on component design and the effect of fiber orientation.

Previous research has been conducted using the hand lay-up method. The disadvantage of this method is the need for the manufacturer skills so that the resulting composite specimens have minimum void levels. Research conducted by Thomson, 1995 found that void were the main factors affecting the quality of glass/polyester fiber composite materials. The use of a suitable rolling tool can determine the quality of the composite panel made. This can be seen on the surface of a composite specimen whether it is distributed evenly or not.

Fillers are materials that are added to the matrix to reduce the amount of matrix volume in a composite material. The cheap price of filler materials will cause the price of the final product to be lower. Materials commonly used as fillers are between 10 nm to macroscopic size [4].

3 Research Method

Ramie fiber originates from Malang, East Java. The epoxy matrix will be made as a constituent material using the hand lay up manufacturing process. While polypropylene as a specimen will be made by injection molding manufacturing process. The processing of these components will use molds specifically designed for composite materials therefore it is necessary to add design tools to install the position of the fibers. Preparation matrices Epoxy and polypropilene. Reinforcement with ramie fibers orientation (random, woven and continues). Process of Making Test Specimens with Process Parameters with composition: Fiber 30% and Epoxy: 70%. Process Variable used in these research are ramie fiber orientation: linear orientation, Random and weaving, matrix material: Epoxy and Polypropylene, process: Hand lay-up. Mechanical universal testing machine used to knew their strength and elongation. Microscope and SEM (Scanning electron microscope) applied for the analysis.

4 Result and Discussion

A tensile test is carried out to determine the tensile strength and elongation of the tested specimens. Tensile test data are presented in Graphs 1, 2.

Test results from the tensile strength and elongation on epoxy and polypropilene electric specimens showed different phenomena where the elongation values in epoxy electric specimens showed a very high value compared to the elongation formed in the polypropylene matrix as shown in Fig. 1, in which the elongation value was highly influenced by the type of matrix. Meanwhile the tensile strength will be more determined by the reinforcing fibers. It is proved by the value of the tensile strength of epoxy and polypropylene electrified specimens having a difference that is not too high when compared to the elongation values of the two matrices. This phenomenon is easier to find in bending testing as the graph model is generally formed on test specimens which is given loading in a perpendicular direction such as bending testing.

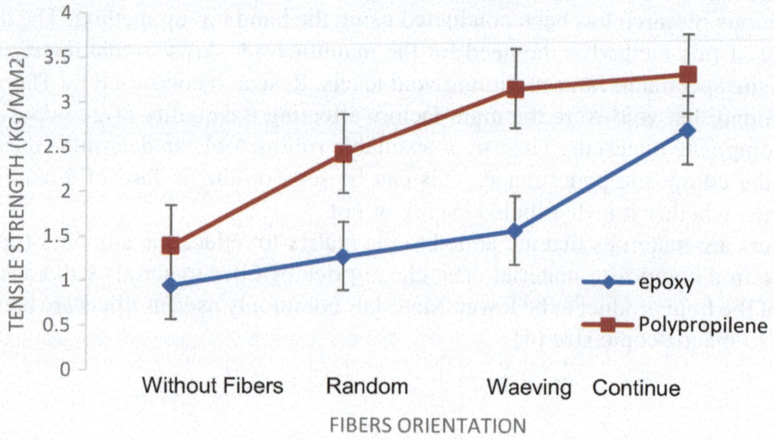

Graph 1 Tensile strength of epoxy and polypropilene matrix products test results with and without ramie fiber

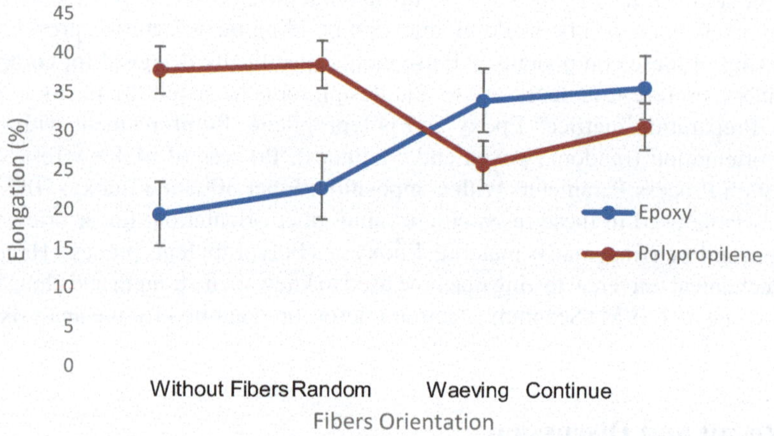

Graph 2 Strain of epoxy and polypropylene matrix product tensile results with and without ramie fiber

As for in tensile testing, this is formed because the tensile test on the test specimen gets the initial loading in perpendicular direction to the specimen but right after that it will experience pulling loading at a parallel direction to the fiber.

The results of the analysis of observations of composite ramie yield results with reinforced epoxy and PP showed different actual strength differences with the results of the analysis predicted by using Ansys software on the specimen tensile test. Experimental results and analysis have differences in fiber distribution; theoretically the mechanics of fiber layers contained in resin can be predicted by equations based on fiber distribution, including square and hexagonal distributions. However, the actual

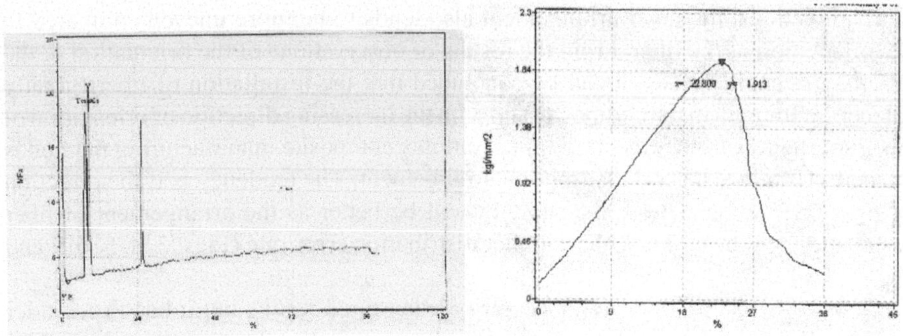

Fig. 1 Graph model of tensile strength towards elongation in epoxy and PP matrix specimens [5]

distribution is very difficult to predict therefore it results in significant deviations. The layered fiber distribution model inside the matrix showed in Fig. 2.

According to the fracture position, the specimens carried out by tensile testing are generally broken in areas that have the smallest volume of component size. By reducing the volume of components, the amount of fiber volume fraction will be greater than other regions and thus the fiber density will be greater so that the strength will be low.

In general the highest strength occurs in the direction of the direct fiber and the lowest strength in the weaving orientation. This can be observed based on the analysis of fiber distribution in the matrix where the phenomenon of failure will be easier to occur in fibers which have a shorter arrangement of spacing between fibers, namely the order of random, woven and then directional orientation. With a shorter distance between fibers, fiber density will increase and the strength will decrease due to decreased fiber strength. Besides, it will result in uneven strength as there are parts of fiber that have shorter spacing between fibers so that the difference in strength increases. It shows an increasing stress concentration. This phenomenon will be easier to occur in random fiber orientation. Whereas for the woven fiber, this phenomenon can also occur because by using two fiber orientations namely the x

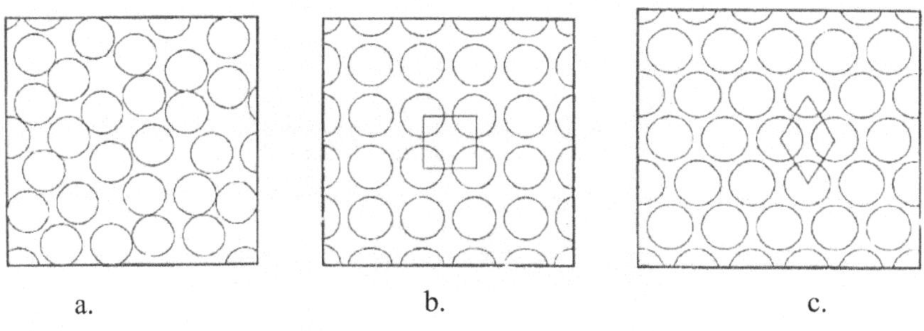

Fig. 2 Fiber distribution in the matrix **a** actual distribution, **b** square distribution, **c** hexagonal distribution [2]

and y directions, the layer arrangement also tends to be more uneven compared to the orientation of the fiber. From the results of observations of the orientation of the resulting strength values, it can be concluded that the installation of fibers greatly affects the strength in the composite and with the increasing direction of orientation of fiber installation arrangements, a more complex composite manufacturing method is needed to produce a good distribution arrangement. The tendency is if the direction of fiber orientation is less, the strength will be better as the arrangement of fiber installation will be easier with a smaller distribution error rate (Figs. 3, 4, 5, 6, 7 and 8).

Composite testing does not necessarily bring good results but it has to consider the cause of the decrease in strength. The decrease in tensile strength is caused by several factors, namely (Figs. 9, 10, 11 and 12).

- In composite strength measurement tests caused by the lack of uniform fiber conditions and uneven mixture of epoxy resin and ramie fiber in the mold.
- Decrease in composite strength is also due to the presence of voids in the composite that cause damage before testing occurs.
- In addition, it is also due to the influence of the position of non-interconnected fibers which makes it easy to crack the composite in the matrix. In the theory of fiber length also affects its strength, short fiber strength is smaller than long fibers.

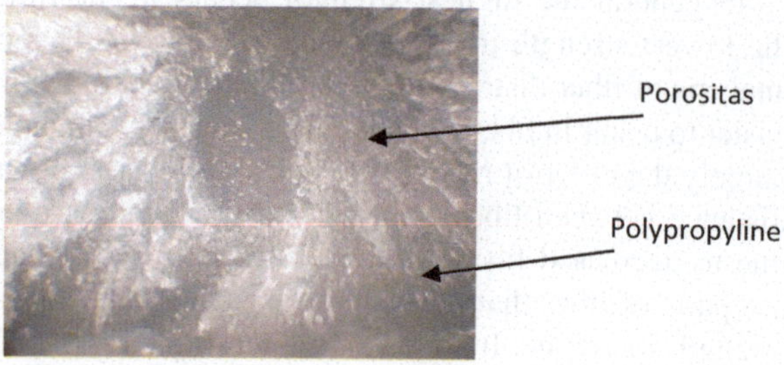

Fig. 3 Macro porosity in the PP matrix is the concentration or initial crack

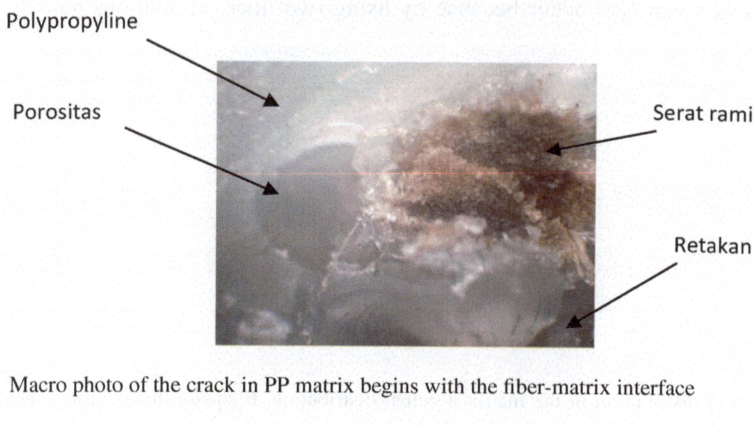

Fig. 4 Macro photo of the crack in PP matrix begins with the fiber-matrix interface

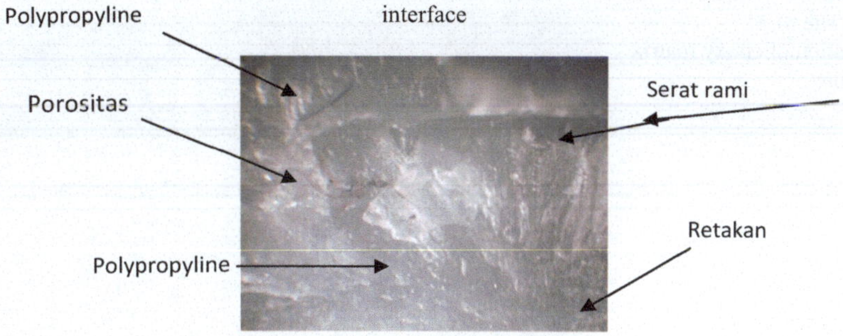

Fig. 5 The presence of crack propagation in the PP matrix

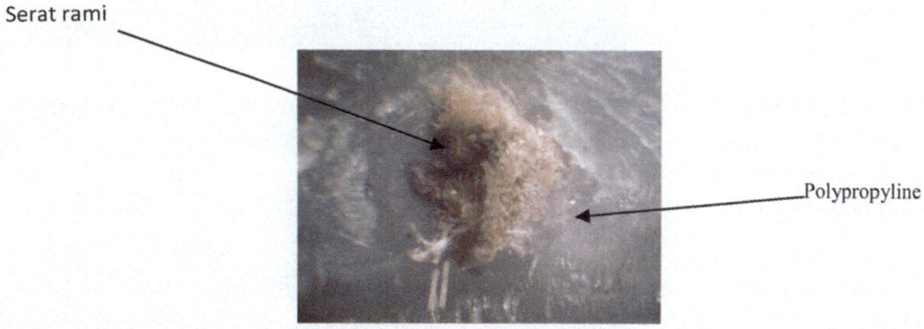

Fig. 6 Ramie fiber shows a burning condition

Fig. 7 PP matrix fracture model in the presence of porosity

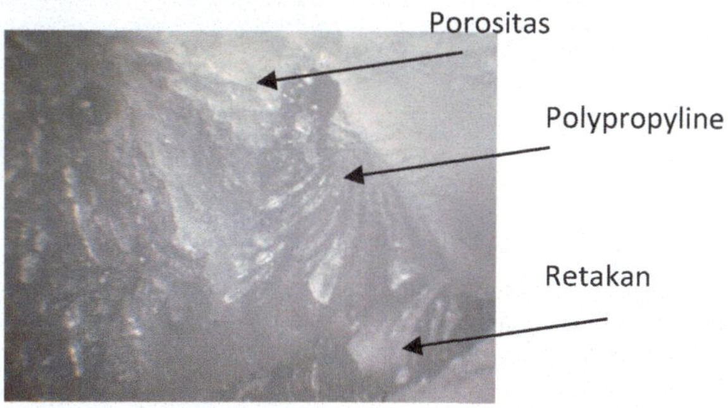

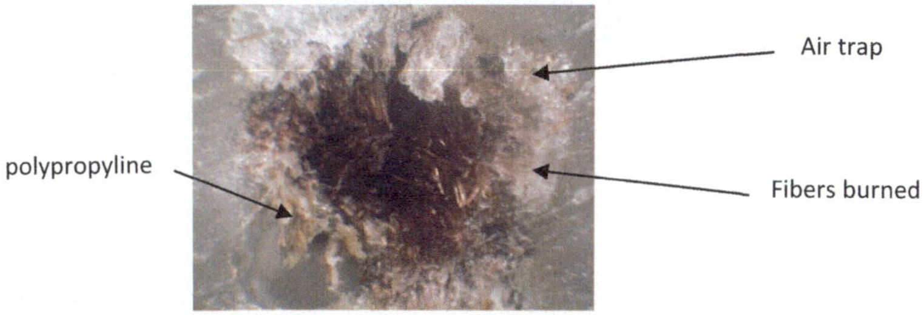

Fig. 8 Ramie fiber that burns tends to be followed by the presence of air trapped inside

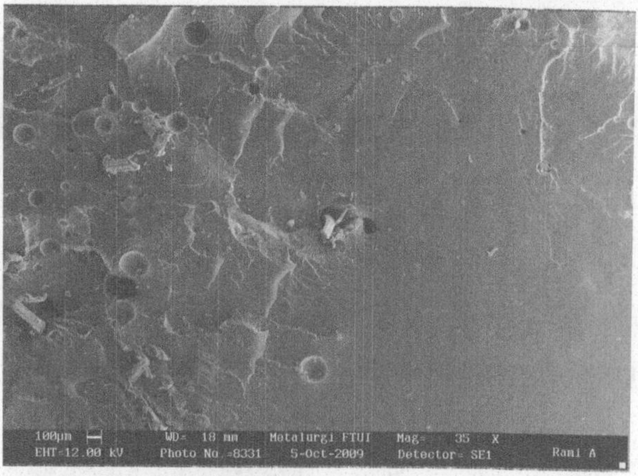

Fig. 9 The SEM examination of epoxy matrix composite

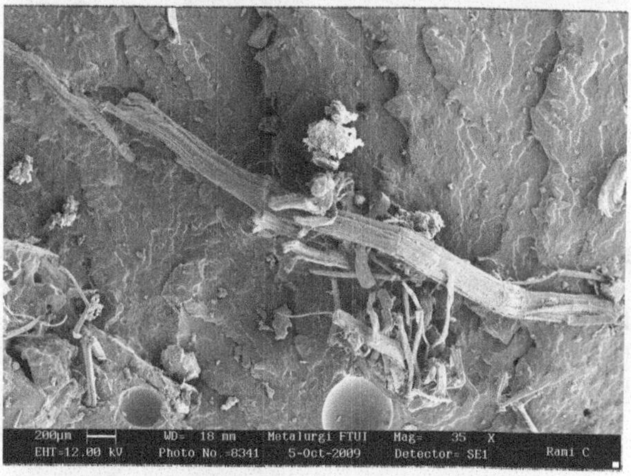

Fig. 10 Results of SEM testing of 35× enlargement in composite specimens with a woven orientation epoxy matrix

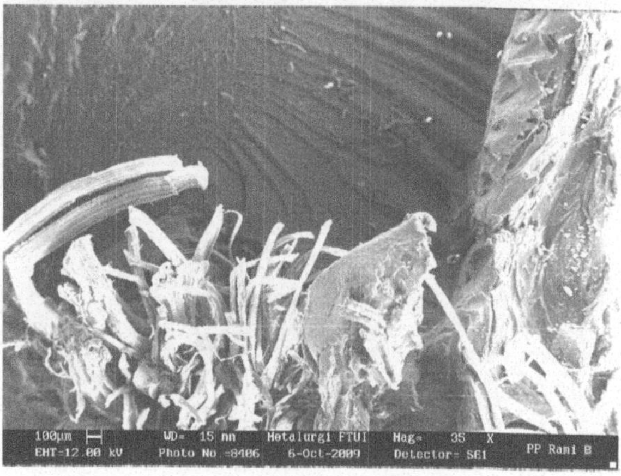

Fig. 11 Results of SEM testing of 35× magnification in composite specimens with a polypropylene matrix

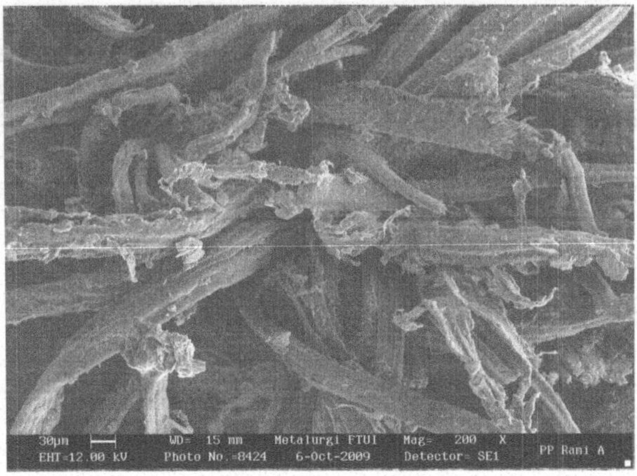

Fig. 12 Results of testing of 200× magnification SEM on composite specimens with the polypropylene matrix

Overall, the specimens with the epoxy matrix in continuous orientation show that the fault characteristics are caused by several factors, including (Fig. 9) in the form of formed porosity which tends to form like a bubble (trapped gas/air) which is cause by poor processing fabrication of specimens. In addition, the presence of this porosity, especially those that form around the fiber, will cause the fault model to become filamentous. The presence of withdrawn fiber is indicated by voids and fibers that are separated from the matrix. This fiber separation is due to the rupture of the fiber.

In the orientation of the woven fiber, Fig. 10 generally illustrated the presence of fibers that are transverse and/or overlapping with perpendicular fibers will produce fiber-matrix bond strength as indicated by the matrix fault groove along the transverse fiber which begins at the position of the overlapping fiber. The overlapping and subsequent crack propagation will be easier to occur along the transverse fiber. It will cause the strength of the woven position composites to have a lower strength than composites with continuous fiber orientation. The more overlapping fibers will cause the composite strength to be lower.

Fibers tend to accumulate because in the injection process the orientation of the fibers is arranged in ramie. The pressure on the manufacturing process of the specimens causes fibers to often gather on the opposite side of the injection hole. This causes the fiber-matrix interface to be very low. It is shown on the surface of the fault that most fibers (collected) will be separated from the matrix. The low bond is indicated by the presence of a gap formed in the fiber-matrix interface area and on the fiber surface there is no matrix attached to the fiber. Meanwhile, fibers which are generally found in the matrix as well as in the epoxy matrix, are not shown in SEM observations with polypropylene. This is due to the position of the collected fiber has the weakest bond therefore it acts as the beginning of the crack and then the crack propagation occurs in the matrix.

The results of tensile testing on composite specimens (fiber reinforced) showed good performance in epoxy and polypropylene matrices in which specimens given ramie reinforcement fibers had better properties (especially in direct orientation)

than test specimens without ramie reinforcement fibers. Natural composites with a PP matrix that are given an injection process tend to produce poorer fiber orientation because the pressure will cause the fibers to collect in the area formed which results in the spread of uneven fibers so that the concentration formed will initiate cracks in the specimen. On the other hand, the injection process by utilizing the temperature will also cause the fiber to burn especially if the fibers are spreading.

5 Conclusion and Recommendation

5.1 Conclusion

From the results of all the mechanical tests, it can be concluded that the presence of ramie reinforcing fibers in both specimens with epoxy and polypropylene electricity is strongly influenced by the orientation of the fibers. The results of tensile strength testing in the epoxy matrix were 140% while the PP strength increased by 187%. Based on observations of microstructure and SEM for all tests in general showed a reduction in the properties of composite tensile strength due to several things including porosity, adhesion between fibers-matrix and between matrices, the presence of overlapping fibers and broken fibers.

5.2 Recommendation

For further research, it is expected to focus more on the manufacturing process of a product as the results of observations which is shows that strength can be optimized by reducing defects formed during the fabrication. The main obstacle in processing components by utilizing materials from natural fiber composites is that the result of these products (quality and quantity) are very varied.

References

1. Jang B (1994) Polymer composites for automotive aplications, advanced polymer composite. ASM International, London
2. Gibson RF (1994) Principles of composite material mechanics. McGraw Hill International, Singapore
3. Callister WD (2000) Materials science and engineering: an introduction, 5th edn. Wiley, New York
4. Abdullah SB (2000) Serat Ijuk Sebagai Pengganti Serat Gelas Dalam Pembuatan Komposit Fiberglass. Laporan Penelitian Fakultas Teknik, Universitas Syiah Kuala, Tahun
5. Crawford RJ (1989) Plastics engineering, 2nd edn. Maxwell Macmillan, Singapore

Potential Application of LiCl/H_2O-CNTs Nanofluids for Liquid Desiccant Cooling System (LDCS): A Preliminary Study Using Numerical Approach

B. Kristiawan, A. T. Wijayanta, and T. Miyazaki

Abstract In the preliminary study, the combination of LiCl/H_2O-CNTs nanofluids and a porous medium has been investigated numerically. By using ANSYS Fluent, it assumes thermal equilibrium between the porous media solids and fluid flows. The interactions between a porous medium and shock waves are not considered. The results show that high temperature entering the inlet section and flowing through porous media decreases significantly when exits from the outlet section until 39 °C. It can be seen that porous media can effectively reduce temperature. In addition to the effect of a porous media, the observed nanofluids also play a key role as a liquid desiccant cooling system (LDCS).

Keywords Nanofluids · Porous media · Lithium chloride · Carbon nanotubes · Desiccant

1 Introduction

Energy usage in the world has extremely increased in recent years. Increased income and quality of life have contributed to the accretion of robust energy necessity. The use of air conditioning (AC) will need a lot of electricity in the world. Fossil fuels are dominant in the primary energy mix approximately 75% of the global energy mix in 2016 as shown in Fig. 1. For energy necessity, Indonesia can be categorized as the country that uses the highest energy in ASEAN region more than 35%.

B. Kristiawan (✉) · A. T. Wijayanta
Department of Mechanical Engineering, Engineering Faculty, Universitas Sebelas Maret, Jl. Ir. Sutami 36A Kentingan, Surakarta 57126, Indonesia
e-mail: budi_k@staff.uns.ac.id

T. Miyazaki
Department of Energy and Environmental Engineering, Interdisciplinary Graduate School of Engineering Sciences, Kyushu University, 6-1 Kasuga-koen, Kasuga-shi, Fukuoka 816-8580, Japan

International Institute for Carbon-Neutral Energy Research, Kyushu University, 744 Motooka, Nishi-ku, Fukuoka 819-0395, Japan

U. Sabino et al. (eds.), *Proceedings of the 6th International Conference and Exhibition on Sustainable Energy and Advanced Materials*, Lecture Notes in Mechanical Engineering,
https://doi.org/10.1007/978-981-15-4481-1_4

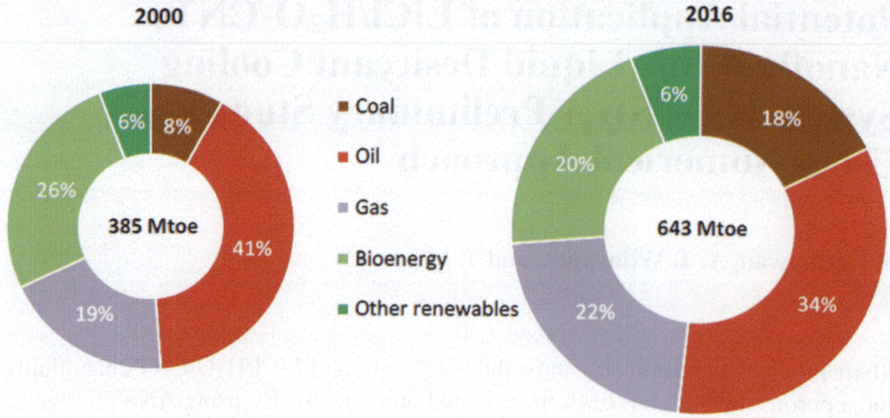

Fig. 1 The global energy mix in ASEAN region [1]

According to the prediction of the International Energy Agency (IEA), electricity consumption for AC will be the main trigger for the surge in world electricity demand in 2050. By 2050 electricity consumption for AC will contribute about 35% to the growth in electricity demand in Southeast Asia, including Indonesia. Global electricity consumption will also be triggered by the industrial sector (26%), lights and equipment (21%), buildings (11%), and other uses (7%). Therefore, the efficiency of using air conditioning is very important. According to the IEA data, about 9% of households in Indonesia use air conditioning. This is still below Thailand where the households in this country have air conditioning is 30%. In the modern country such as United States and Japan, it is almost 90% of households that have air conditioning.

Indonesia stands on the equator with an average temperature of 30 °C and relative humidity above 80%. This climate is stiflingly hot and causes uncomfortable conditions to work especially indoor activities. To get the comfortable condition, it is needed the external devices such as air conditioning and electric fan that require a lot of electricity consumption. Hence, there is needed an intensive effort to get an efficient thermal system to be able to reduce energy usage. Currently, vapor compression refrigeration system is the most widely used for air conditioning applications [2]. Not only air conditioning on indoor usage but electronics cooling has also been developed [3]. This air conditioning device system generally uses a refrigerant as a working fluid. However, refrigerants have Global Warming Potential (GWP) in which it depends on the type of refrigerant. In this numerical study, a liquid desiccant cooling system using advanced heat transfer fluids called nanofluids was selected. The usage of nanofluids in the liquid desiccant cooling system constitutes the novel technique of this work. Hitherto, the previous literatures discussing on this topic are still very limited.

2 Liquid Desiccant Cooling System (LDCS) Using Nanofluids

Desiccant cooling system is defined as the heat-driven cooling cycle that commonly utilizes water as working fluid circulating and contacting directly with air. Liquid desiccant cooling can be used as a sorption matter of cooling cycle besides solid material such as lithium chloride salt, silica gel and zeolite. Between the mentioned material types of cooling technique, using liquid desiccant as a working fluid is highly considered nowadays [4]. Thermophysical properties of liquid desiccant extremely prescribe to evaluate cooling capacity of liquid desiccant application. The vapor pressure of surface constitutes the key thermophysical property inducing heat and mass transfer. In addition to the thermophysical properties, the important requirements of liquid desiccant are non-toxic, odor free, non-flammable and unexpensive. Lithium chloride (LiCl), lithium bromide (LiBr) and calcium chloride ($CaCl_2$) owing to the low vapor pressure are generally used in liquid desiccant cooling system (LDCS). Lithium chloride (LiCl) constitutes the preferred choice due to the lowest vapor pressure and stable liquid [5].

LDCS has been simulated for air conditioning system, namely the Liquid desiccant air-conditioning (LDAC) technique in commercial buildings. The results demonstrated that the performance of LDAC was earnestly influenced by the building's ratio [6]. The LDCS was one of the popular techniques providing cooling, particularly on less thermal sources. This cooling system has attracted by many researchers especially in providing low cooling space at low energy demand [7]. Indonesia is a tropical country that was abundant solar energy and this energy source can change the existing fossil fuel. The humidity of air can also be utilized for cooling system called as evaporative cooling (EC) or air cooler having generally low electric energy compared to vapor compression system (VCS) [8]. The LDCS can be used widely into the remote are under a humid climate with mixing dry space for air dehumidification before evaporation process. The combination between air dehumidification and evaporative cooling enlarges for comfort cooling [9]. Nowadays, dehumidifier design using absorptive material has attracted in the developing of desiccant technology [10]. As a result, the price and size of drying devices continue to decline because it has found a market. However, this cooling technique has also several weaknesses. Due to the psychrometric process, there is a limited temperature that can be obtained. Consequently, practical applications and commercial desiccants on solar energy are relatively slight. Enhancement of the heat transfer efficiency could be conducted by expanding surface area. However, the weaknesses of this method are the larger design of thermal devices. Solid materials commonly have higher the thermal conductivity compared to the liquids, therefore, usage solid particle into liquid has been considered recently to increase heat transfer of the liquid [11]. However, the augmented pressure drop, sedimentation as well as fouling factor are obstacles causing this work less interesting. To overcome these problems, the novel nanomaterial's technology has been utilized by generating in nano sized particles.

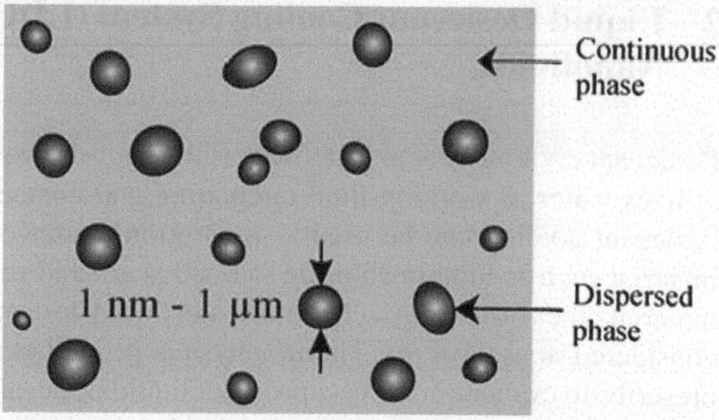

Fig. 2 The scheme of a dispersed colloidal system [19]

Nanofluid is a colloidal system consisting of a carrier medium phase and dispersed phase as shown schematically in Fig. 2. In this case, the carrier medium phase and dispersed phase are base fluids and nanoparticles, respectively. Nanofluid was first introduced by Choi [12] to describe novel properties nanotechnology-based heat transfer fluids although the previous studies have reported an earlier and independent experimental result on thermal conductivity and viscosity of nanoparticles dispersed into the liquid [13]. Nanoparticles can be engineered from metal, metal oxide, and carbon material in the form of spherical, rod-like or tubular shape. The inherently poor thermal conductivity of conventional heat transfer for improving heat transfer performance is the main reason to develop nanofluids. The previous literature reported that the thermal conductivity and thermal performance have been increased owing to the existence of nanoparticles into base fluids [14, 15] and nanofluids can be applied on the many applications [16–18].

Mass transfer accomplishment can extremely be increased by dispersing nanoparticles like carbon nanotubes (CNTs) and adding surfactant concurrently [20]. The effect of nanoparticles on absorption was investigated. The findings indicated that the absorption level can enhance when nanoparticles disperse into bas fluid. The previous study has revealed that the absorption rate increases with increasing concentration of nanoparticles [21]. Some works has been studied regarding the mass transfer using nanoparticles dispersion for bubble [20–24] and absorption refrigeration [25, 26]. This present work focuses on the mass transfer performance of $LiCl/H_2O$ suspension with dispersing carbon nanotubes (CNTs).

3 Governing Equations

The governing equations based on the Euler-Euler multiphase approach involve mass, momentum and energy conservation. For phase q, these equations can be expressed as follows fluent [27].

- *Conservation of mass*

$$\frac{\partial}{\partial t}(\gamma\alpha_q\rho_q) + \nabla\cdot(\gamma\alpha_q\rho_q v_q) = S_{mass,q} \tag{1}$$

Where porosity, volume fraction, density and velocity of phase are represented by γ, α_q, and v_q, respectively. For only liquid, phase index q is 1 and for the vapor phase, the value index is 2. $S_{mass,q}$ describes evaporation and condensation.

- *Conservation of momentum*

$$\frac{\partial}{\partial t}(\gamma\alpha_q\rho_q) + \nabla\cdot(\gamma\alpha_q\rho_q v_q v_q) = S_{M,q} \tag{2}$$

Where the source term of $S_{M,q}$ constitutes interphase momentum transfer, lift force, and virtual mass force. The effect of pressure gradient, gravitation and turbulence are also considered.

- *Conservation of energy*

$$\frac{\partial}{\partial t}(\gamma\alpha_q\rho_q\rho_q) + \nabla\cdot(\gamma\alpha_q\rho_q v_q h_q) = S_{E,q} \tag{3}$$

Where the specific enthalpy of the phase q is noted by h_q. The source term of $S_{E,q}$ describes the interfacial heat exchange and heat source from the porous media containing the effect of turbulence and changing pressure.

4 Numerical Procedure

A preliminary numerical study was performed on the model as shown in Fig. 3. A numerical procedure was listed in Table 1. The inlet/outlet section was 100 mm in

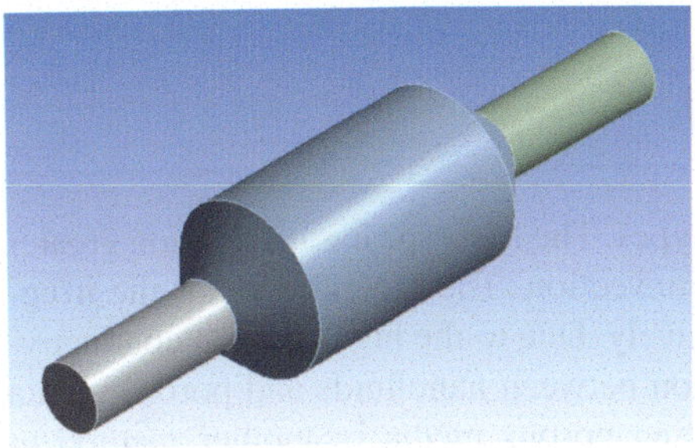

(a) Porous media model

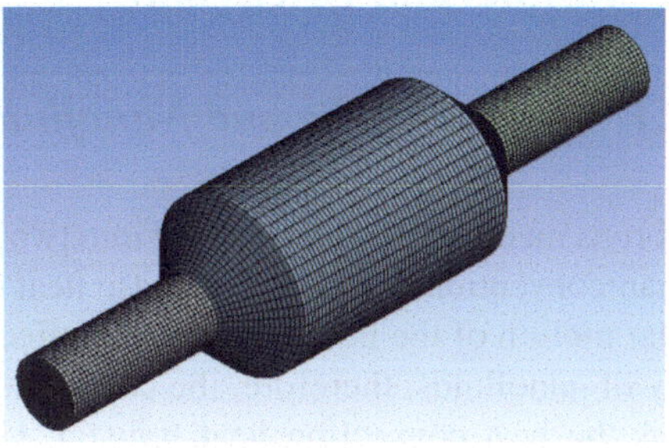

(b) Meshing of model

Fig. 3 A preliminary numerical model

Table 1 Numerical procedure

Nodes	67,032
Elements	63,070
Model	Single-phase model for LiCl (30 wt%)
	Multiphase model for LiCl/CNTs
Boundary condition	Constant heat flux of 1500 W/m^2 at wall inlet section
Viscous model	Laminar. Standard k-epsilon
Near wall treatment	Standard wall function

Table 2 Properties of LiCl at 298 K and 30 wt%

Density (kg/m^3)	1180
Viscosity (Pas)	0.00359
Surface tension (N/m)	0.0893
Specific heat capacity (J/kg K)	2933

Table 3 Conditions of the fluids

Parameters	Desiccant
Temperature inlet (°C)	92
Inlet velocity (m/s)	0.07
Mass concentration (wt%)	30

length, whereas, porous zone was 200 mm in length. Meanwhile, inlet/outlet and the porous zone diameters were 40 mm and 100 mm, respectively.

In this study, LiCl dispersed into H_2O of 30 wt% is used to investigate at porous medium. Properties of LiCl are evaluated at 298 K and mass fraction of 30 wt% and conditions of the fluids as shown in Tables 2 and 3, respectively.

5 Results and Discussion

5.1 Porous Media and Nanofluids

Porous media can be categorized into two types. The first type has dissipation greater than conventional fins that increase heat convection. The second one has the irregular motion of the fluid flow mixing effectively. Due to the high thermal conductivity of nanofluids, therefore, the combination between nanofluids and porous media has the best convection heat transfer. Using porous media is another method to enhance the thermal performance because it provides higher surface improving convection heat transfer rate. Five phenomena such as adsorption, desorption, blocking, transportation, and aggregation will be seen.

Temperature contour map and temperature distribution at radial along the axial direction are showed in Figs. 4 and 5, respectively. Heat flux of 1500 W/m^2 boundary condition applied in the inlet section causes increasing in inlet temperature. High temperature entering the inlet section and flowing through porous media decreases significantly when exits from the outlet section until 39 °C. It can be seen that porous media can effectively reduce temperature. In addition to the effect of a porous media, LiCl/H_2O-CNTs nanofluids also play a key role as liquid desiccant cooling system (LDCS). Local temperature single-phase model for LiCl (30 wt%) of the observed nanofluids are also simulated in Fig. 5. It can be demonstrated that nanofluids temperature changes with radial direction. It indicates that porous media and nanofluids can be combined to reduce higher temperatures into lower temperatures. For single-phase

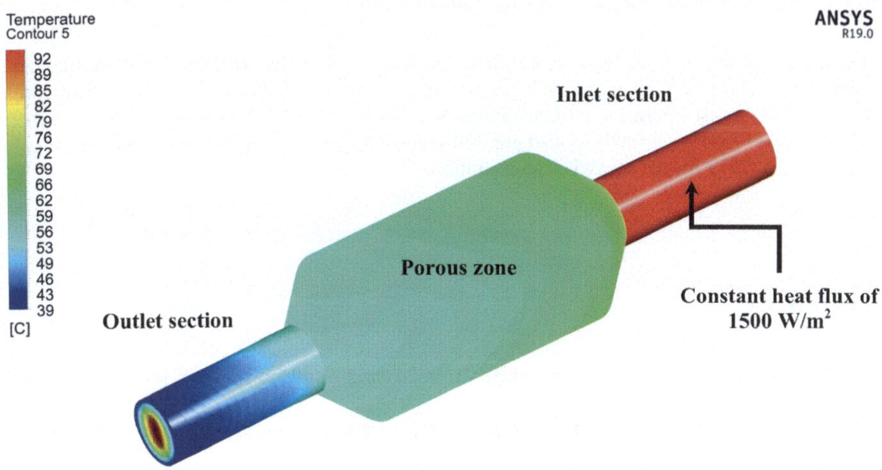

Fig. 4 Temperature contour map of single-phase model for LiCl (30 wt%)

Fig. 5 Local temperature single-phase model for LiCl (30 wt%)

assumption, the porosity of porous media is an isotropic and it can change on space and time. The interactions between porous media and shock waves is neglected.

6 Conclusions

LiCl/H_2O-CNTs nanofluids flowing through a porous medium have been investigated numerically. The results show that high temperature entering the inlet section and flowing through porous media decreases significantly when exits from the outlet section until 39 °C. It can be seen that porous media can effectively reduce temperature. In addition to the effect of a porous media, the observed nanofluids also play a key role as liquid desiccant cooling system (LDCS).

Acknowledgements The authors would like to thank you to the Institute of Research and Community Service of Universitas Sebelas Maret for supporting this work under the research scheme of Kolaborasi Internasional Universitas Sebelas Maret (KI-UNS) under contract number 516/UN27.21/PP/2019. The authors also gratefully thank to JSPS (Japan Society for the Promotion of Science) FY 2019 for the overwhelming support.

References

1. Southeast Asia Energy Outlook 2017 (2017) World energy outlook special report. International Energy Agency, Paris
2. Badr O, O'Callaghan PW, Probert SD (1990) Vapour-compression refrigeration systems. Appl Energy 36(4):303–331
3. Poachaiyapoom A, Leardkun R, Mounkong J, Wongwises S (2019) Miniature vapor compression refrigeration system for electronics cooling. Case Stud Therm Eng 13:100365
4. Sahlot M, Riffat SB (2016) Desiccant cooling systems: a review. Int J Low-Carbon Technol 11:489–505
5. Mei L, Dai YJ (2008) A technical review on use of liquid-desiccant dehumidification for air-conditioning application. Renew Sustain Energy Rev 12(3):662–689
6. Qi R, Lu L, Huang Y (2014) Energy performance of solar-assisted liquid desiccant air-conditioning system for commercial building in main climate zones. Energy Convers Manag 88:749–757
7. Elsarrag E, Igobo O, Alhorr Y, Davies P (2016) Solar pond powered liquid desiccant evaporative cooling. Renew Sustain Energy Rev 58:124–140
8. Gebrehiwot B, Dhiman N, Rajagopalan K, Agonafer D, Kannan N, Hoverson J, Kaler M (2013) CFD modeling of indirect/direct evaporative cooling unit for modular data center applications
9. Bom GB, Foster R, Dijkstra E, Tunmers M (1999) Evaporative air conditioning, applications for environmental friendly cooling. Technical paper no. 421. World Bank, USA
10. Dieng AO, Wang RZ (2001) Literature review on solar adsorption technologies for ice-making and air-conditioning purposes and recent developments in solar technology. Renew Sustain Energy Rev 5(4):313–342
11. Saghafifar M, Gadalla M (2015) Innovative inlet air cooling technology for gas turbine power plants using integrated solid desiccant and Maisotsenko cooler. Energy 87:663–677

12. Choi SUS (1995) Enhancing thermal conductivity of fluids with nanoparticles. In: Siginer DA, Wang HP (eds) Developments and applications of non-Newtonian flows. American Society of Mechanical Engineers (ASME), New York
13. Kristiawan B, Kamal S, Suhanan Y (2016) Thermo-hydraulic characteristics of anatase titania nanofluids flowing through a circular conduit. J Nanosci Nanotechnol 16:6078–6085
14. Sundar LS, Farooky MH, Sarada SN, Singh MK (2013) Experimental thermal conductivity of ethylene glycol and water mixture based low volume concentration of Al_2O_3 and CuO nanofluids. Int Commun Heat Mass Transfer 41:41–46
15. Rizvi I, Jain A, Ghosh S, Mukherjee PS (2013) Mathematical modelling of thermal conductivity for nanofluid considering interfacial nano-layer. Heat Mass Transfer 49(4):595–600
16. Kristiawan B, Wijayanta AT, Enoki K, Miyazaki T, Aziz M (2019) Heat transfer enhancement of TiO_2/water nanofluids flowing inside a square minichannel with a microfin structure: a numerical investigation. Energies 12(16):3041
17. Wijayanta AT, Pranowo M, Kristiawan B, Aziz M (2019) Internal flow in an enhanced tube having square-cut twisted tape insert. Energies 12(2):1–12
18. Kristiawan B, Kamal S, Suhanan Y (2015) Convective heat transfer of titanium (IV) oxide nanofluids under turbulent flow condition. ARPN J Eng Appl Sci 10(10):4385–4395
19. Butt HJ, Kappl M (2010) Surface and interfacial forces, 1st edn. Wiley-VCH Verlag GmbH & Co. KgaA, Federal Republic of Germany
20. Ma X, Su F, Chen J, Bai T, Han Z (2009) Enhancement of bubble absorption process using a CNTs-ammonia binary nanofluid. Int Commun Heat Mass Transfer 36(7):657–660
21. Kim JK, Jung JY, Kang YT (2007) Absorption performance enhancement by nanoparticles and chemical surfactants in binary nanofluids. Int J Refrig 30(1):50–57
22. Kim J-K, Jung JY, Kang YT (2006) The effect of nano-particles on the bubble absorption performance in a binary nanofluid. Int J Refrig 29(1):22–29
23. Lee JW, Jung J-Y, Lee S-G, Kang YT (2011) CO_2 bubble absorption enhancement in methanol-based nanofluids. Int J Refrig 34(8):1727–1733
24. Pang C, Wu W, Sheng W, Zhang H, Kang YT (2012) Mass transfer enhancement by binary nanofluids (NH_3/H_2O + Ag nanoparticles) for bubble absorption process. Int J Refrig 35(8):2240–2247
25. Yang L, Du K, Niu XF, Cheng B, Jiang YF (2011) Experimental study on enhancement of ammonia–water falling film absorption by adding nano-particles. Int J Refrig 34(3):640–647
26. Kim H, Jeong J, Kang YT (2012) Heat and mass transfer enhancement for falling film absorption process by SiO_2 binary nanofluids. Int J Refrig 35(3):645–651
27. ANSYS (2013) ANSYS fluent theory guide. SAS IP, Inc., USA

Strengthening Governance and Research and Community Service Capacity (P2M) UNS Faculty of Engineering Lecturers

Zainal Arifin, Sholihin As'ad, Wahyudi Sutopo, Dody Ariawan, Singgih Dwi Prasetyo, and Catur Harsito

Abstract The strategy to strengthen governance is one of the drivers for achieving the vision, mission and objectives of the Faculty of Engineering. This study aims to find a strategy to strengthen the quality and quantity of Research and Community Service (P2M) at the Faculty of Engineering. The method used in this research is action research with research data of the past five years. Data were analyzed using the SWOT method, then data was presented and conclusions were drawn. Data analysis was conducted to see trends in the number of studies and lecturer services received and funded in the Faculty of Engineering. The achievement indicators are marked by leading activities that have an impact on increasing the publicity index, increasing revenue, generating income and increasing roles in solving development problems in the local, national and global context. The results showed that the proposal approved for funding had decreased. The number of service proposals in the last five years has increased, but the proposals received are relatively constant. Therefore, the optimization of development strategies in strengthening governance can increase the number of research and services received and funded.

Keywords Governance · SWOT · Research · Service

1 Introduction

The research conducted aims to find a strategy to strengthen the quality and quantity of research performance profiles and community service from the Sebelas Maret University (UNS) Faculty of Engineering lecturers so that they are competitive and

Z. Arifin (✉) · D. Ariawan · S. D. Prasetyo · C. Harsito
Department of Mechanical Engineering, Sebelas Maret University, Surakarta, Indonesia
e-mail: zainal_arifin@staff.uns.ac.id

S. As'ad
Department of Civil Engineering, Sebelas Maret University, Surakarta, Indonesia

W. Sutopo
Department of Industrial Engineering, Sebelas Maret University, Surakarta, Indonesia

U. Sabino et al. (eds.), *Proceedings of the 6th International Conference and Exhibition on Sustainable Energy and Advanced Materials*, Lecture Notes in Mechanical Engineering,
https://doi.org/10.1007/978-981-15-4481-1_5

applicable for community development. The strategic position of Research and Community Service (hereinafter referred to as P2M) in achieving the Vision of the Faculty of Engineering UNS: "Becoming an excellent institution of higher education at the international level in the development of engineering based on national cultural noble values." Represented in the Faculty Mission Engineering, namely: (i) to organize education and teaching in engineering by always encouraging lecturers' self-development and student independence in gaining knowledge, skills, and attitudes. (ii) To carry out research that produces new discoveries in the field of engineering. (iii) To carry out community service activities in the field of engineering that are oriented towards community empowerment. While the aims of the Faculty of Engineering are: (i) The creation of an educational and learning environment in engineering that produce reliable graduates who are devoted to God Almighty, virtuous, intelligent, skilled, independent, physically healthy, spiritually, and socially minded. (ii) The creation of a vehicle for the development of the engineering field by conducting research activities that are empowered and effective through dissemination so that a sustainable technological transformation occurs. (iii) The creation of a vehicle for the application of engineering by carrying out community service activities for a better life [1].

Although the commitment of the Faculty of Engineering to strengthen P2M's competitiveness is clear in its vision, mission and goals, there are still problems to be faced. P2M's problems cannot be seen separately from the general condition of the Faculty of Engineering. Based on previous research contained in the Government Institution Performance Accountability Report (LAKIP) and the Institute of Technical Faculty Forms, the main problems faced by the Faculty of Engineering related to the development of P2M are the few proposals funded and the small number of educators compiling proposals for research and community service. A related matter is the lack of competitiveness of publications in indexed or international reputation [2, 3]. By conducting a review, through collecting and analyzing data about P2M. This research focuses on providing solutions to existing problems.

2 Methods

The method used in this research is action research, which means research as well as intervention and analysis of the problems found. Data collection techniques using document studies and SWOT analysis methods (strengths, weaknesses, opportunities, and threats) from document sources that record P2M grant acquisition data and performance data or track records of research and service outcomes in the last 4 years in the form of proceedings or journals [4]. Includes identification from the Iris0311 website/director corner/Report on PNBP and Non PNBP data for 2015, 2016, 2017, 2018 and 2019 according to the number of proposals or articles [5, 6]. In general, the research conducted is described in the following research flowchart.

3 Results and Discussion

From the available data, the number of proposals for the Non-Tax State Revenue (PNBP) research proposal from the Faculty of Engineering lecturers and increased again in 2019 was 111 proposals, which had decreased in 2018, namely 84 proposals. In general, for the last 5 years it has been relatively constant, as shown in Fig. 1. The same phenomenon occurs for the PNBP fund community service scheme, as in Fig. 2, which has increased again in 2019, which is 40 proposals after in 2018 which had 31 proposals. In contrast to research, community service has increased over the past 5 years [5, 6].

On the other hand, submission of proposals funded in general has decreased over the past 5 years, from 64 proposals (2017) to 38 proposals (2018 and 2019). In contrast to general service devotion, which is relatively constant, from 9 proposals (2018) to 14 proposals (2019), as shown in Figs. 1 and 2 [5, 6].

Fig. 1 Research flowchart

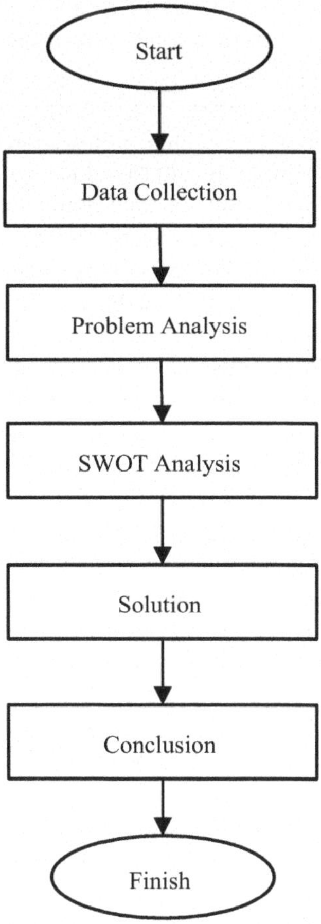

Fig. 2 Number of PNBP research proposals per year

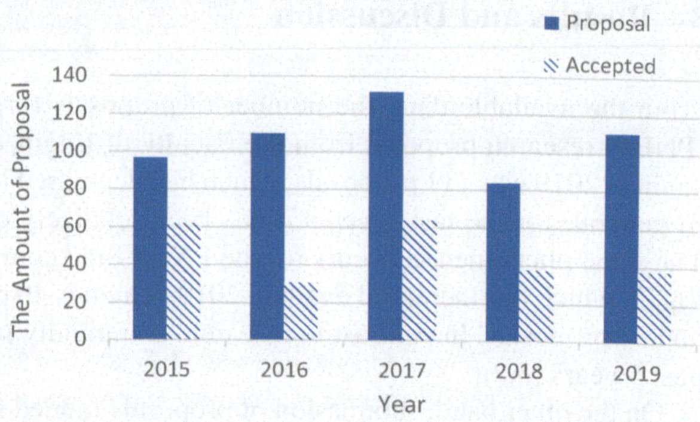

It can be seen that there are fewer devotees of community service compared to research enthusiasts. This condition does not reflect the objective of P2M in the downstream aspect, including increasing revenue in generating income and increasing role in solving development problems in the local, national and global context. Therefore, a solution has been found using SWOT analysis as shown in Tables 1 and 2 [4].

It can be seen that the low number of proposals funded is due to the lack of attention of the Faculties. Faculties can make a good proposal writing workshop for lecturers to improve the quality of proposals submitted. In increasing the number and quality of community service proposals, the faculty must help to develop a network of cooperation between institutions or partners so that lecturers can utilize the collaboration to provide community service.

Based on data, the performance of publications on the results of research and community service in national and international proceedings in general has decreased. In the most national proceedings in 2015 there were 64 articles, in international proceedings indexed by Scopus were 75 articles in 2016, and proceedings international were not indexed in Scopus as many as 64 articles in 2015 [5, 6]. In general, proceedings in publications decreased, as in Fig. 3.

The publication of research and service results through the publication of national and international journals in general also decreased. Publication through national journals has the most value compared to international journal publications. The national journal publications are 149 articles (2015), 183 articles (2016), 142 articles (2017) and 37 articles (2018). International publications are 18 articles (2015), 29 articles (2016), 11 articles (2017), and 6 articles (2018). Publications to international journals indexed by Scopus are 21 articles (2015), 13 articles (2016), and 7 articles (2017) [5, 6]. The most significant decrease occurred in publications in national journals, as shown in Figs. 4 and 5.

The decrease in the index of the number of publications in proceedings and journals due to the absence of data on the Iris0311 website indicates that there are still very few lecturers updating the results of publications made. In addition, the small number of articles published indicates the incompatibility between research and community service to the results to be published in proceedings or journals. This can be

Table 1 SWOT analysis to identify strategy of research performance improvement in society (P2M) in 2015–2019

Component	Strength	Weakness
Internal and external	1. The amount of proposals submitted is always there 2. There is already a research proposal funded 3. The lecturers have publications in accredited national journals and indexed reputable international journals, and/or indexed proceedings 4. The experience writing scientific of lecturers in reputed international journals and accredited national journals	1. Not all lecturers can get funding competitively 2. Not all lecturers can publish in accredited national journals or reputable international journals 3. Research intensity of some lecturers in not optimal 4. Not all lecturers have experience in accredited national journals and reputed international journals The comparison of the number of submissions and proposals received is still small
Opportunity	SO (strength-opportunity)	WO (weakness-opportunity)
1. Journals publication are the key lecturer performance indicator 2. Network from other institutions to get a journals paper	1. Increase the frequency of clinic forums writing journals and journals management 2. Allocating funds for lecturers who not yet succeeded in competition fund	1. Optimizing lecturers from overseas graduates to build network that support the fulfillment of reputable international journal requirements 2. Increasing interaction with LPPM activities for research and service forums 3. Participate in specialized networks with other institutions that have the same aim towards publishing accredited or reputable journals Optimizing available funds Lack of proper and correct proposal writing training
Threat	ST (strength-threat)	WT (weakness-threat)
1. The requirements of accredited journals and reputable international journals are is difficult 2. Research outputs must be published in accredited journals and reputable journals	1. Increase special incentives for journal management teams to have specific skills and focus on journal management towards accreditation or reputation	1. Opportunities to improve the ability of lecturers to mastering international languages to support the scientific journals 2. To give incentives for lecturers involved in international forums to build a network of journal contributor writer's partners

Table 2 SWOT analysis to identify the strategies for improving community service performance in 2015–2019

Component	Strength	Weakness
Internal and external	1. The amount of proposals has an upward trendline 2. Every year there are a number of lecturers who win the Community Service Skim competition	1. Relationships with parties outside the campus are still not optimal (Ministries/Institutions, Local Governments, Companies, Non-Government Institutions) 2. The intensity of community service of some lecturers is less than optimal 3. Lack of publication of the results of community service 4. The comparison of the number of submissions and proposals received is still small
Opportunity	SO (strength-opportunity)	WO (weakness-opportunity)
1. Availability of a budget for community service in LPPM and Higher Education Research	1. Faculty facilitates the digitalization of the lecturer community service results and it is published on the Faculty of Engineering website	1. Faculty facilitates the mechanism for the publication of the expertise of Faculty of Engineering Lecturers to a network of potential partners 2. Lack of proper and correct proposal writing training
Threat	ST (strength-threat)	WT (weakness-threat)
1. Administrative requirements for the management of community service activities in rigid and technical communities (especially financial administration), causing lecturers to be less interested	1. Faculty provides assistance for financial administration accordance with regulations to solve the problem of lack of understanding lecturers about administration finances state	1. Faculty facilitates collaboration with partners will be located of an integrated community service partnership from Faculty of Engineering lecturers

seen in the SWOT analysis that has been done, as shown in Table 3 [4]. The small number of publications will affect on the submitted proposal which can be funded. Therefore, the Faculty must play an active role to assist lecturers in the publication of proceedings and journals by conducting workshops and mentoring regularly. In addition, the lecturers should always play an active role in updating the data on the website.

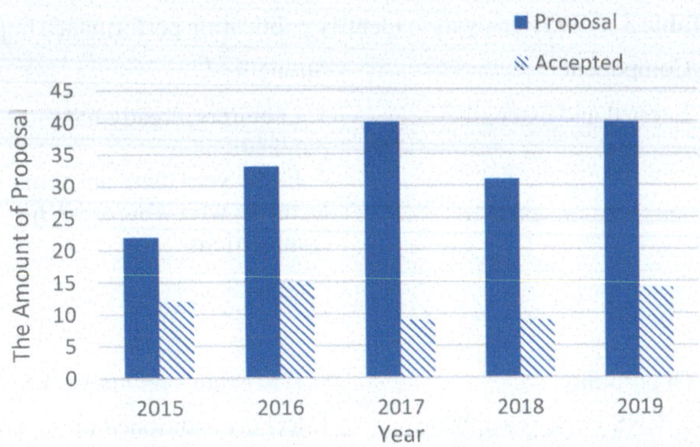

Fig. 3 Number of community service PNBP fund per year

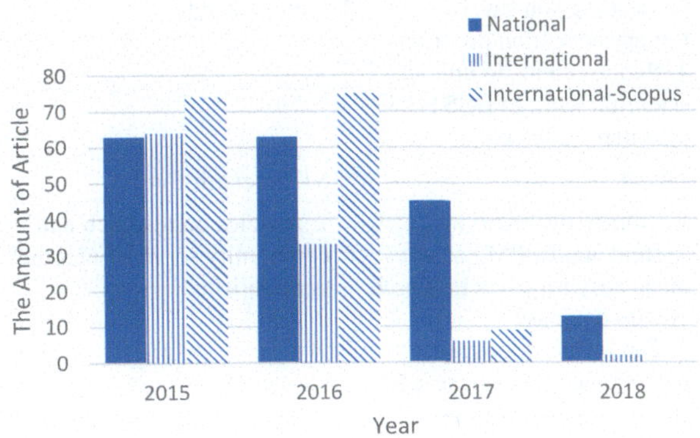

Fig. 4 The amount of proceeding seminar articles per year

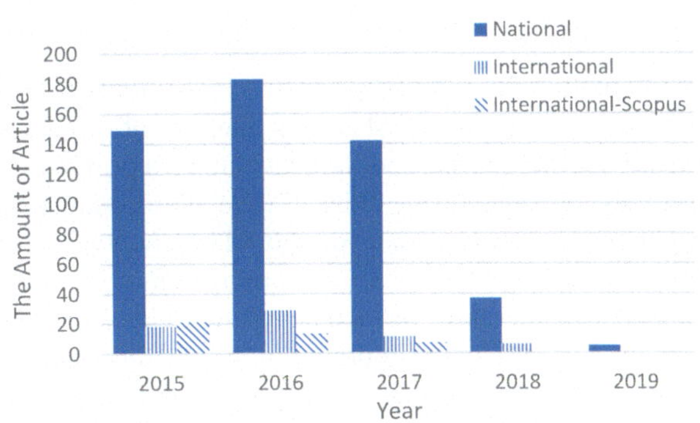

Fig. 5 The amount of articles journal per year

4 Conclusions

The Faculty of Engineering has conducted successfully Research and Community Service (P2M), and published the results of P2M in the proceedings of an internationally reputed conference. However, proposals are funded and scientific publications

Table 3 SWOT analysis to identify publication performance improvement strategies in 2015–2019

Component	Strength	Weakness
Internal and external	1. Lecturers already have publications 2. Every year there are some lecturers who win research competitions	1. Some lecturers are not diligent in uploading their scientific work 2. There are incompetent publication articles 3. The results of research and community service outcomes are not published
Opportunity	SO (strength-opportunity)	WO (weakness-opportunity)
1. UNS Press is available 2. Centralized UNS library 3. Annual international proceedings on several tertiary education programs 4. Requirements to upload scientific work to IRIS1103 information system	1. RG is conditioned to publish the books together as a publication of the results of their research	1. The faculty assigns a digitizing team of works, proceeding articles and lecturer journal to upload on the engineering faculty website 2. Lack of article writing training
Threat	ST (strength-threat)	WT (weakness-threat)
1. Library Information System and LPPM has not made easy for users to search faculty database 2. The lecturer does not publish the results of research and community service outcomes in proceedings or journals	1. Articles on research and community service outcomes not published	1. Add a directory link of digital PHO with the UNS and LPPM library link 2. The lecturer index decreases 3. Proposals submitted are not funded

in internationally reputed journal Scopus indexed are still few and not recorded on the website. The low P2M results that are followed up with publication in proceedings or journal journals are indicated to be a problem. At least proceeding articles or journals can affect on the index of funded proposals which will decrease. In overcoming these problems, assistance should be given in writing journals or articles having international reputation. So, the proportion of lecturers to submit competitive P2M funded will increase. In increasing the quality of proposals submitted, the faculty can conduct assistance in preparing P2M proposals.

In increasing the quality of proposals submitted, the faculty can conduct assistance in preparing P2M proposals. In increasing revenue generating income and increasing roles in solving development problems, Faculties and lecturers need to develop a network of cooperation among institutions or partners. Hence, lecturers can take advantage of this collaboration in conducting research and community service. The establishment of a research peer group will make it easier to provide information related to research collaboration and community service. It can also be a notification for data updates that must be uploaded on the website. Therefore, it will create institutional strengthening that is aligned between lecturers and faculty.

Acknowledgements The authors thank the Rector of Sebelas Maret University for supporting the funding of the research No. 516/UN27.21/PP/2019 for FY 2019.

References

1. Saputro S et al (2019) Panduan Pelaksanaan Penelitian dan Pengabdian kepada Masyarakat, VI ed. Lembaga Penelitian dan Pengabdian kepada Masyarakat Universitas Sebelas Maret
2. Saputro S et al (2018) Buku Borang Institusi Fakultas Teknik Universitas Sebelas Maret
3. Saputro S et al (2018) Laporan Akuntabilitas Kinerja Instansi Pemerintah Tahun 2018. Fakultas Teknik Universitas Sebelas Maret
4. Nisak Z (2013) Analisis SWOT untuk menentukan strategi kompetitif. J Ekbis 9(2):468–476
5. Iris0311/director corner/Report PNBP & Non PNBP. http://www.iris1103.uns.ac.id. Last accessed 2019/09/06
6. Arifin Z (2019) Buku Profil Kinerja Penelitian & Pengabdian Pada Masyarakat Fakultas Teknik Universitas Sebelas Maret 2016–2019

Identifying Geothermal Power Plant Institutional Barrier and External Factors in Indonesia

Tabratas Tharom and Hendro Sasongko Hadi

Abstract Geothermal Power Plant (GPP) installed capacity has not been optimally optimized in Indonesia, although Geothermal is abundant. In fact, the institutional barrier and external factors are amongst factors determining GPP utilization and maintenance. To realize such potential, the Government of Indonesia (GoI) has conducted some initiatives to improve this condition. Particularly, it applies better and proper incentives such as fiscal incentives and conducts Preliminary Survey and Exploration Assignment (PSPE) to properly assesses the potential. The output of PSPE will provide more credible data to ease the risk and cost of the geothermal exploration phase. But there is also an inherent danger in the exploration phase, and the exploration cost tends to rise, makes the GPP less feasible. The trend of declining coal price is also a threat to GPP since it is more favorable to generate electricity from the cheap resource (coal-fired power plant). Thus, in order to lower the risk, technology adoption, and acquisition to reduce the cost and risk associated with exploration is essential. On the other hand, from the GoI's view, Trilemma energy policy applied in Indonesia is the essential/basic consideration to regulate the energy business. For renewable energy, it is emphasized on the affordability aspect, thus in analyzing the institution's policy, the set of the power purchase agreement and related rules remain as the decisive factors. The paper attempts to explain and assess the government initiatives on the issues mentioned above, by using SWOT and Policy Gap analysis to describe the Indonesian geothermal energy condition and analysis. Thus, this paper is constructed to provide an alternative solution to mitigate and ease such institutional barriers and external factors in Indonesia's GPP.

Keywords Acquisition · Adoption · Cost · Exploration · External factors · Geothermal · Indonesia · Institutional barriers · Policy gap analysis · Power plant · SWOT · Technology · Trilemma energy

T. Tharom (✉)
Grenoble Ecole de Management, Grenoble 38000, France
e-mail: tabratas.tharom@uzmagroup.com

Uzma Engineering Sdn. Bhd., 47820 Selangor, Malaysia

H. S. Hadi
Global Guru Indonesia, Jalan Mataram No. 8, Jakarta 12110, Indonesia
e-mail: hendroshadi@gmail.com

U. Sabino et al. (eds.), *Proceedings of the 6th International Conference and Exhibition on Sustainable Energy and Advanced Materials*, Lecture Notes in Mechanical Engineering,
https://doi.org/10.1007/978-981-15-4481-1_6

1 Background

Geothermal is a huge amount of potential energy resources in the world, especially in volcanic or in the path of fire countries like Indonesia, Iceland, New Zealand. But the implementation of geothermal energy-based electricity generation is quite complex. Technical, policy and economic motives play a major role in the geothermal energy business.

In the technical aspects, the costly drilling work is not guaranteed to find such a proven reserve. The number of well to explore, the higher the success rate to find suitable reserves. It is predicted it is about 60% in the development stage and higher than 90% in the operational phase [1]. It is also estimated that the upfront cost to develop well and the field will be about 40% [2]. The high risk and costly explorations are still the main considerations to develop geothermal energy. Technology adoption and acquisition is the successful key to make the exploration phase economically feasible as the proven reserve is accurate and insurable in order to meet the loan requirement. The bank could not provide such a loan, which is inaccurate and non-insurable for high-risk credit like geothermal energy.

In the policy aspect, there are many improvements in Indonesia. Regardless of to survey for Indonesian energy, which expressed pessimism over Indonesia's energy business climate and policy [3], MoF [4] and MoEF [5] had developed such deregulations and incentives to developed geothermal energy investment. These improvements include the incentives, subsidy, and permit, especially for geothermal exploration in the conservation area for an environmental issue. MEMR [19] has been developing PSPE which in turn collects informative data for geothermal engineering aspects to reduce the risk of the exploration phase. Unfortunately, it hasn't been materialized yet to make PPA more insurable and bankable due to limited budget and insufficient data profile to have the geothermal proven reserve to produce electricity.

In the financial view, all the improvements in technology and policy issues could be useless if there are significant changes in coal prices. The declining coal price could encourage the national electric energy operator (PLN) to maintain steam-based electricity generation to lower the operational cost due to its lowest electricity generation cost after hydropower. It could happen since the cost of generating electricity is the heaviest financial burden. The recent decline of coal price could incentivize the operator to use coal-fired power plants in order to reduce the cost more. That is why the PPA for the sustainability of electricity generation plays an important role in price volatility caused by external factors. The GoI has been already supporting PLN by giving subsidies to ease the problem in order to make electricity prices affordable.

The technical, policy and economic considerations could be described as integrated into the institutional barrier and external factors model for Indonesian GPP. Thus by applying SWOT and Policy Gap Analysis methodology, this paper aims to analyze GPP and solution and recommendations to ease institutional barriers and external factors of GPP in Indonesia.

Recent energy development in late 2019 in Indonesia is that electric power is oversupply due to low energy demand as stated in low energy consumption is simply jeopardizing renewable energy absorption.

2 Literature Review

SMI [6] stated that geothermal exploration, especially the drilling work, takes the biggest portion on the geothermal project, it could be up to 40%. Gehringer and Loksha [7] showed how high expenditure and risk profile of drilling cost and any cost related to the exploration phase, which takes a big share on the project. This heavy technical cost is only one of two more considerations, such as policy and economic conditions. The policy regarding the law to rule the geothermal business affects all the stakeholders such as local government, investor, the local community in the aspects of the environment, finance, geothermal operation, and firm/investor strategy. The external factors also play an important role, ADB and WB [8] showed any change in oil/gas and coal prices, exchange rate exposure could disrupt both investors and PLN, from geothermal energy focus due to profitability and cost consideration factors. Here are the selected references regarding institutional and external factors:

1. Geolec [9], found that there are problems within many European countries such as geothermal licensing, fiscal burden, complicated monitoring and reporting, lack of transparency. This policy burden leads to delays and the rise of geothermal electricity projects. Geolec had proposed a proper procedure for licensing even the use of English and standardization because of the diversity of European Union citizens and local systems. Geolec also describes the lack of statistical basis to predict the successful probability drilling, leads towards less proper insurability, which in turn makes the project less bankable.
2. Asia Pacific Energy Research Center (APERC) [10] had advised the guided policy in Asia Pacific countries, including Indonesia, which is viewed on track. It mentioned some pending matters as follows:

    a. Emphasized on harmonizing regulations and authorities in terms of geothermal utilization between government and local government.
    b. Standardization and procedures implementation for geothermal resources development.

3. Arena [11], recommended the approach to make the geothermal better investment by collaboration with the oil and gas matter because any improvement in policy was perceived hard to become the principal and cost-competitive, non-fossil-fuel-based energy source caused mostly by the highly risky exploration stage.
4. The survey of geothermal investors shows that most of the investors view the geothermal energy policy does have an inconsistency to implement such as risk reallocation, which makes the project non-bankable [3]. However, GoI had

already made progress, such as the geothermal field is not considered as a mining field and the escalation of geothermal price in the special case.

5. The US views geothermal energy in terms of Enhanced Geothermal System, not in the conventional hydrothermal reservoir, which has fluid injected to the well to increase permeability and increase the probability of having proven reserve in turn. It could lower the price of energy delivery [12]. Unfortunately, this technology is still in the beginning phase, and it could become a game-changer once it has reached its technology maturity.
6. Low coal prices could ruin the geothermal and other renewable energy resources, especially if there is an abundant potential of coal resources. Although it is known that the ash and other pollutant caused by coal are harmful, it is still considered to be operated in such long term. Coal could stand to operate in a low carbon era by applying CSS (Carbon Captured Storage) in order to reduce environmental damages while promoting efficient coal-fired power plants [13].
7. In order to have better risk management, the GPP project should be insured in the critical stage. The stage is exploration and drilling phase which could be handled by developer, IPP and supported by government's guarantee and small investor/venture capital, the final stage such as power plant construction and operation could be handled by institutional investor and large Independent Power Producer (IPP) [14]. At the beginning and medium stage, insurance takes the biggest portion, and then in the late stage, the bank takes the biggest portion.
8. New technology development, such as EGS (Enhanced Geothermal System), could provide a more successful exploration rate. Water dominated systems such as the Dieng reservoir could be treated by PAA-CO_2 fracturing fluid EGS system to provide more efficient geothermal heat in order to produce more useful heat and electricity [15].
9. The latest geothermal research has been beginning to have EGS commercially viable in 2030, thus supporting the US electricity grid more than three folds. The technology will become more affordable, and the technology to exploit geothermal resource energy will become less risky [16]. It is the chance that Indonesia should be able to participate since Indonesia is the biggest geothermal potential in the world to provide more affordable and large geothermal power plant.

10. Indonesia is the largest geothermal potential in the world (40% of world reserve), and within regulations of Minister of MEMR about the selling prices (more PPA price adjustment), geothermal funds and other incentives, given to the geothermal industry, the GPP will be more attractive [17].

3 Analysis

Despite the fact that the progress of geothermal installed capacity like Saralua, the target of 7000 MW, is still far to reach. By using SWOT Analysis to the references related to geothermal energy, this paper concludes and assess the current condition as follows:

3.1 *SWOT Analysis*

Strength: High Capacity Factor It is widely known that GPP has the highest capacity factor amongst renewable energy. It acts as a baseload provider in the electric grid and startup time is relatively fast compared to the steam power plant. It is the most reliable power energy beside hydropower, and favorable to become dominant clean energy. Despite criticism and feedback for renewable energy policy, Indonesia has been already recognized for improvement in trilemma energy, as Table 1 shows. Indonesia had better environmental Sustainability ranking in 2018.

Weakness: Insufficient Data and Geothermal Wells Profile for Proven Reserve In order to have a reliable electric power supply, the exploration to find such a proven reserve is difficult because of the insufficiency of data and geothermal wells profile—this insufficiency caused by risky and costly exploration related activities. There is always the possibility to fail in order to find a geothermal proven reserve.

Opportunity: Bigger Market GoI has already planned a big share of geothermal energy, as written in RUPTL for 2019–2028 (Electric Supply Business Plan).

Table 1 Trilemma Report 2018 (World Energy Trilemma in 2018 [18])

	2016	2017	2018	Trend	Score
Overall rank and balance score	85	75	71	Flat	BBB
Energy performance					
Energy security	63	59	61	Flat	B
Energy equity	88	82	82	Flat	B
Environmental sustainability	77	68	57	Up	B
Contextual performance	77	83	80	Flat	

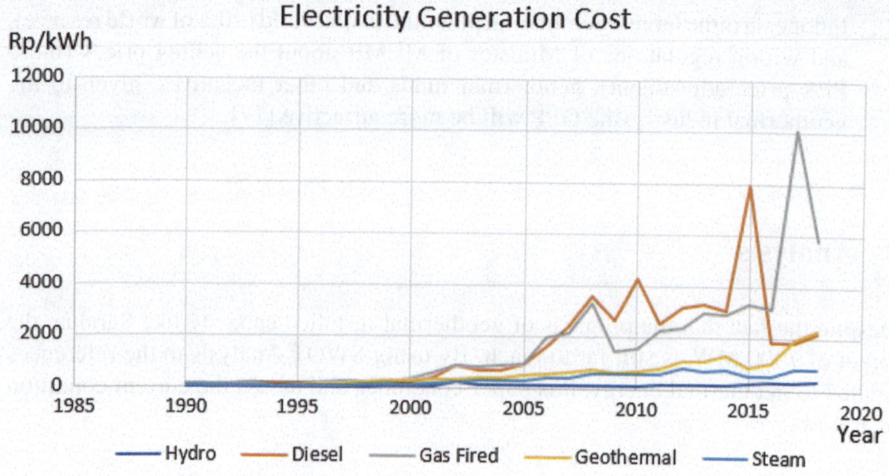

Fig. 1 PLN (Indonesia National Electric Operator) Electricity Generation Cost

The geothermal and hydropower plant will dominate the share of renewable energy resources for electricity generation. This open opportunity should attract investors or developers to expand the business as well as its capacity.

Threat: Rising Geothermal Based Electricity Generation Cost Unfortunately, recent geothermal based electricity generation cost has already. It is known that more complex exploration technology doesn't fit to cost reduction program in the geothermal sector. It will make geothermal energy unfavorable to become baseload since its generating cost even higher than fossil fuels based power plants like diesel plants. Figure 1 shows the statistics of the power plant in Indonesia:

Source PT PLN (Persero)'s Statistics Year 1984–2019

3.2 Policy Gap Analysis

This policy gap will analyze the geothermal power project based on the regulator's view regarding trilemma energy and the firm regarding its business strategy as follows:

1. Strength Aspect

    a. Regulator's view regarding strong aspect Trilemma Energy security: support energy security.
    b. Firm's view related to the strong aspect: Value proposition and brand image for the public.
    c. Policy gap analysis: there is a lack of GPP socialization due to local resistance to geothermal development.

2. Weakness Aspect
    a. Regulator's view regarding weakness aspect Trilemma Energy affordability: costly and will impede the GPP induction to the electric grid. Affordability needs improvement.
    b. Firm's view related to the weakness aspect: the biggest obstacle to handle.
    c. Policy gap analysis: there are no incentives to lower the risk and reduce the cost of exploration.
3. Opportunity Aspect
    a. Regulator's view regarding Opportunity aspect Trilemma Energy security: Successful GPP capacity installation could boost energy security.
    b. Firm's view related to opportunity aspect: more incentives to invest in geothermal energy.
    c. Policy gap analysis: society and the market are not yet prepared to integrate GPP in their daily life.
4. Threat Aspect
    a. Regulator's view regarding threat aspect Trilemma Energy affordability: rising electricity cost and high exploration cost is unacceptable.
    b. Electricity generation is oversupply reducing GPP absorption.
    c. Firm's view related to threat aspect: facing a penalty if the proven reserve could not be found nor explored.
    d. Policy gap analysis: proven sustainable reserve is essential for PLN as the national electric operator to prepare and adjust the grid, especially the transmission and distribution to serve demand, but this proven reserve is hard to find due to high exploration risk and cost.

3.3 *Summarize the SWOT and Policy Gap Analysis*

	Strength	Weakness	Opportunity	Threat
	• High capacity factor • Better renewable energy policy in reliability • It has the potency to become the largest GPP in the world	• Lack of well information and profile • High risk • The affordability remains relatively flat as the result of PPA (power purchase agreement) limitation	• Large energy market • More affordable and reliable due to the maturity of EGS in 2030 • Support grid reliability	• Increasing exploration and maintenance cost lead to increasing electricity generation cost • Community rejection • Declining coal price • Electricity is oversupply
Trilemma energy	• Support energy reliability and environmental sustainability	• Unaffordable for rising exploration cost in terms of rising PPA		• Unaffordable
Firm's view	• Reliable and predicted energy sales	• Risky investment	• Lucrative business	• Business sustainability
Policy gap analysis	• More tax incentives to strengthen energy security	• Lack of research for reliable well drilling information and profile • Rising electricity generation cost	• Pending direct use	• Insurance and bank support to ease exploration risk • Lack of research for exploration cost reduction, technology adoption, and acquisition • Lack of community development • Lack of control of coal-fired/steam power plant • Need incentives for direct use instead of electricity energy

4 Solution and Recommendation

Indeed, the institutional and external factors are getting better, but it is not sufficient to sustain the geothermal energy business. Therefore, this paper aims to give alternative solutions and recommendations based on SWOT and Policy gap analysis above. The solution to cover the strength: GPP should be positioned as the energy provider of heat, electricity and supporting the food industry, tourism industry, residentials, to use more benefit from geothermal energy, providing more welfare aspect to society. The solution to cover the weakness: More incentives for technology adoption and acquisition in order to lower the risk and reduce the cost associated with geothermal energy. The solution to cover the opportunity: Firms need government support for promotion to the society in order to position GPP as a clean energy provider as well as local economic development support related to the industry, residential, and tourism.

4.1 *The Solution to Cover the Threat*

1. If the proven reserve is less than the needs, than the geothermal energy could be used as heat and limited electric power for the local community, which could absorb such reserve. Other business development such as providing heat and electricity to limited industry or residence, should be made feasible.
2. If the proven reserve suits to PPA requirement, so it will need insurance to ensure the bank to provide loan to implement GPP.
3. Green credit should finance the exploration to find the proven reserve whether it is sufficient or insufficient to provide electricity for the PPA requirement.
4. Incentives for more geothermal energy direct use instead of electricity due to electricity oversupply.

4.2 *Recommendations*

1. Incentives for exploration related technology and adoption are intensifying research through inter-department collaboration among MEMR—Menristek/LPDP/BPPT and geothermal developer.
2. Incentives for Operational cost reduction program—MEMR—MoF and geothermal Developer.
3. Incentives for direct use for industrial/residential to strengthen local development.
4. Exploration and operational Insurance and Fund integrated team among MoF, MEMR, PLN, insurance, bank, and geothermal developer.

Strict monitoring of electricity generation for the steam (coal-fired) power plant.

References

1. Sanyal SK, Morrow JW (2011) An investigation of drilling success in geothermal exploration, development, and operation. GRC Trans 35
2. Thorhallsson S, Sveinbjornsson BM (2012) Geothermal cost and drilling effectiveness. UNU-GTP. https://www.researchgate.net/publication/269631557
3. Price Waterhouse Cooper (2018) Alternating currents: Indonesian power industry survey 2018, 2nd edn, July 2018
4. MoF (Ministry of Finance) (2010) Tax subsidy on. PMK No 21/PMK.011/2010
5. MoEF (Ministry of Environment and Forestry) (2018) State of Indonesia's forest. SK.25/MENLHK/SETJEN/PLA.2/2018
6. SMI (2015) SMI insight (state owned company for infrastructure financing)
7. Gehringer M, Loksha V (2012) Geothermal handbook: planning and financing power generation. ESMAP technical report no. 002/12. World Bank http://documents.worldbank.org/curated/en396091468330258187/Geothermal-handbook-planning-and-financing-power-generation
8. Asian Development Bank and The World Bank (ADB and WB) (2015) Unlocking Indonesia's geothermal potential
9. Geolec (2019) Factsheets on geothermal electricity. EGEC (European Geothermal Energy Council)
10. Asia Pacific Energy Research Center (APERC) (2015) Policy success factors for geothermal electricity development in the APEC region assessment of public policies on geothermal electricity in 6 APEC economies. Institute of Energy Economics Japan. http://aperc.ieej.or.jp
11. Australian Renewable Energy Agency (Arena) (2014) Looking forward: barrier, risk, and rewards of the Australian geothermal sector to 2020 and 2030
12. INL (2006) The future of geothermal energy
13. MIT (2007) The future of coal options for carbon-constrained world. Massachusetts Institute of Technology. ISBN 978-0-615-14092-6
14. Munich Re (2015) Exploration risk assurance—Munich Re's experience in Turkey, 22 Sept 2015
15. Riska E et al (2017) Potensi EGS (enhanced geothermal system) di Indonesia Perbandingan Aplikasi EGS di Lapangan Kamojang dan Dieng. Peran Penelitian Ilmu Kebumian Dalam Pembangunan Infrastruktur di Indonesia. In: Universitas Diponegoro Proceeding, Seminar Nasional Kebumian ke-19, 13–14 Sept 2017, Grha Sabha Pramana
16. Augustine C et al (2019) GeoVision analysis supporting task force report: electric sector potential to penetration. National Renewable Energy Laboratory, Golden, CO. NREL/TP-6A20-71833. https://www.nrel.gov/docs/fy19osti/71833.pdf
17. Setiawan H (2014) Geothermal energy development in Indonesia: progress, challenges, and prospects. Ministry of Finance
18. World Energy Council (2018) World energy trilemma 2018
19. MEMR (2018) MEMR Regulation Number 37 Year 2018 about Preliminary Survey Assignment and Exploration

Frictional Characteristic Evaluation of Composite Brake Block Using a Reduced-Scale Brake Dynamometer

Yunus Ari Rokhim, Eko Surojo, Nurul Muhayat, and Wijang Wisnu Raharjo

Abstract Brake blocks are important components that function to decelerate the speed of train. The safety of railway transportation depends on the reliability of this component. One of the important characteristics of the brake block material is the friction coefficient. The brake block material should have the required friction coefficient. The friction coefficient of material is measured by friction test. Previous research showed that the braking conditions including contact pressure and sliding speed affect the friction coefficient of a material. Many friction methods have been developed to study the characteristics of friction material. In contrast to previous studies, in this study, frictional characteristic of composite brake block with resin phenolic matrix was evaluated using a reduced-scale brake dynamometer. By pressing the brake block specimen against the wheel tread friction test was conducted. The size of the brake block specimen and the train wheel were reduced to ¼ original size. Friction measurements were carried out at contact pressure of 0.10, 0.15, 0.20, and 0.25 (MPa) and sliding speed of 3, 6, 9, and 12 (m/s). The results showed that the friction coefficient of the composite brake block decreased with increasing sliding speed, but increased with increasing contact pressure. The friction coefficient of composite brake blocks addressed values that was vary from 0.296 to 0.404. Meanwhile, wheel temperature increased with increasing sliding speed and contact pressure.

Keywords Friction test · Composite brake block · Dynamometer

1 Introduction

Brake block is a braking system component which presses against a wheel tread to decelerate the speed of a train. The safety of rail transportation is affected by the

Y. A. Rokhim · E. Surojo (✉) · N. Muhayat · W. W. Raharjo
Department of Mechanical Engineering, Faculty of Engineering, Universitas Sebelas Maret, Surakarta, Indonesia
e-mail: esurojo@ft.uns.ac.id

E. Surojo
National Center for Sustainable Transportation Technology, Bandung, Indonesia

U. Sabino et al. (eds.), *Proceedings of the 6th International Conference and Exhibition on Sustainable Energy and Advanced Materials*, Lecture Notes in Mechanical Engineering,
https://doi.org/10.1007/978-981-15-4481-1_7

reliability of this component. The brake block is made of composites or cast iron. Composite brake blocks have lower density and longer service life compared to cast iron brake blocks. One important characteristic of brake block material is the friction coefficient. The brake block material should have the required friction coefficient. The friction coefficient can reflect the ability of the component to produce friction forces when the brake block is pressed to the rotating of the wheel tread surface.

The friction coefficient of brake block material is measured by friction test. Previous research showed that the braking conditions including contact pressure and sliding speed affect the friction coefficient of a material. Vakkalagadda et al. [1] conducted a study of the characteristics of the cast iron brake block friction test as well as K and L type composites on locomotives and carriages. The results showed that the friction distance, sliding speed and brake load affect the friction coefficient of the brake block.

Many friction methods have been developed to study the characteristics of friction material. Until now, most of the friction and wear characteristics of composite brake blocks were tested by the pin-on-disk method [2], block-on-disc [3], ball-on-disk [4] or full-scale dynamometers [1]. In contrast to previous studies, in this study composite brake block material was tested using a reduced-scale brake dynamometer. The size of the brake block specimen and the train wheel were reduced ¼ from original size.

2 Research Materials and Methods

The material tested was a railroad composite brake block with resin phenolic matrix (Fig. 1a) which had a size of 320 × 80 × 50 (mm). The brake block material was cut into test specimens with a size of 80 × 20 × 13 (mm) or reduced ¼ from the original size as shown in Fig. 1b. The brake block material that had been cut then was tested using the reduced-scale brake dynamometer as shown in Fig. 2.

This dynamometer was divided into several main parts, namely the wheel turning mechanism, the specimen suppression part, the friction force sensor and wheel temperature sensor. The 3 phase AC electric motor was used to rotate the wheel.

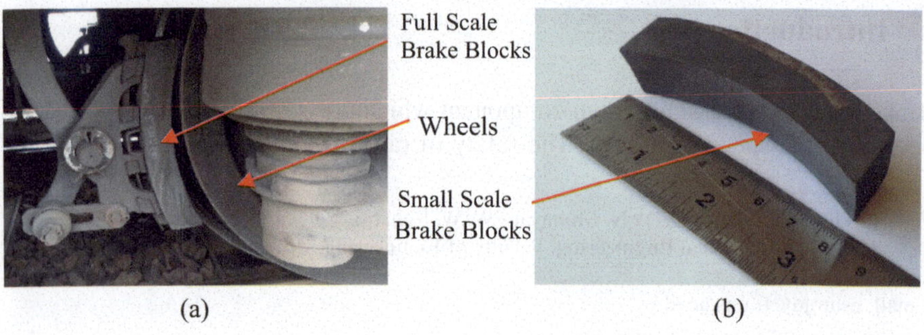

Fig. 1 Composite brake block materials: **a** full scale and **b** test specimen

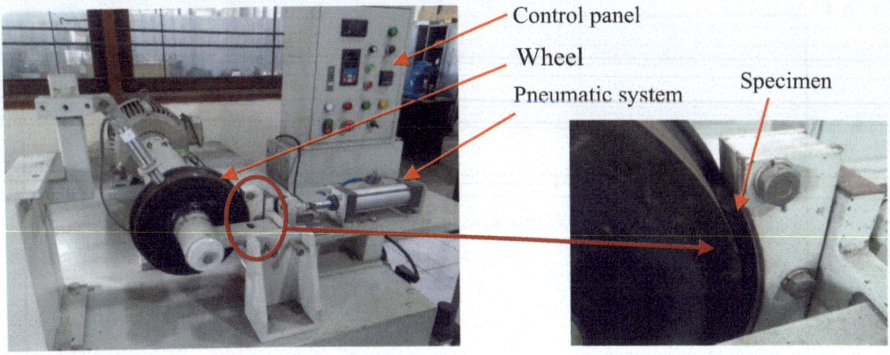

Fig. 2 A reduced-scale brake dynamometer

The rotation speed of the wheel was adjusted by using an inverter. The pneumatic system was mounted to adjust the magnitude of the contact force between the specimen and the wheel. A load cell is equipped on the dynamometer to measure friction force. Wheel temperature during friction testing was measured using an infrared thermometer.

By pressing the specimen onto the rotating wheel surface (Fig. 2) and measuring the friction force between the wheel and the specimen friction testing was performed. Friction measurements were carried out with variations in sliding speed 3, 6, 9, and 12 (m/s) and variations in load pressures in kgf (or contact pressure in MPa) 16 (0.10), 24 (0.15), 32 (0.20) and 40 (0.25). The friction coefficient is calculated using Eq. (1).

$$\mu(\text{friction coefficient}) = \text{friction force}/\text{normal force}. \quad (1)$$

3 Results and Discussion

The value of the relationship in each variation of speed and contact pressure with the overall friction coefficient can be seen in Figs. 3 and 4. Figure 3 shows the trend of changes in the friction coefficient as a function of sliding speed. The friction coefficient in each contact pressure variation decreased with increasing sliding speed. This phenomenon occurred due to the fact that shear strength of the brake block material on the surface decreased when the temperature increased [5]. The temperature rise for brake block surface was related to the heat generated due to friction between the wheel and the surface of the brake block. The faster the friction between the wheel and the surface of the brake block, the greater the amount of heat produced. Heat was partially lost in the environment and some of the heat was distributed between the

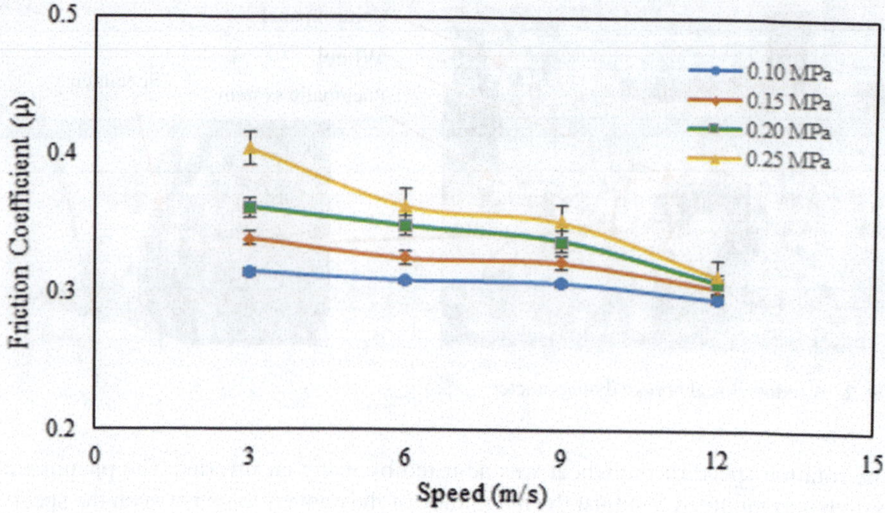

Fig. 3 The relationship of friction coefficient and sliding speed with various of contact pressure

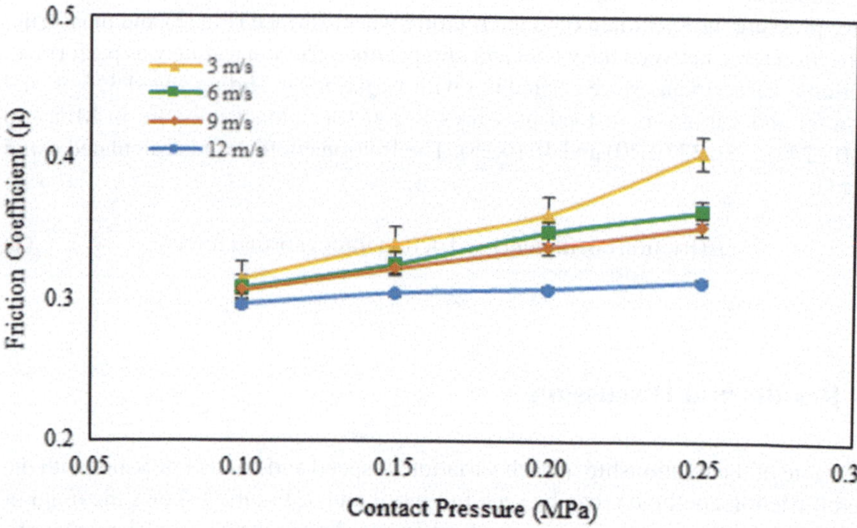

Fig. 4 The relationship of friction coefficient and contact pressure with various of sliding speed

wheel and the brake block, which lead to an increase in the contact surface temperature. The increase in temperature on the contact surface had an effect on decreasing the shear strength of the specimens on the contact surface so that the friction coefficient decreased. The highest friction coefficient value that was 0.312–0.404 was reached at the contact pressure of 0.25 MPa and the sliding speed of 3 m/s. The lowest friction coefficient value that was 0.296–0.314 was obtained at the contact pressure of 0.10 MPa and the sliding speed of 12 m/s.

Figure 4 shows the trend of changed in the value of the friction coefficient as a function of contact pressure. The friction coefficient in each sliding speed variation increased with increasing contact pressure. This was occurred because the micro-convex-apices (uneven surface) induce on increasing the actual contact area due to an increase contact pressure, then it caused an increase in the friction coefficient [6]. Increasing contact pressure caused normal stress and shear stress at contact zone increased. This condition generated increasing level of plastic deformation at the micro-convex-apices and then lead to increase the contact zone area at sliding interface. The highest friction coefficient value was reached when the speed was 3 m/s and the contact pressure was 0.25 MPa which was 0.314–0.404. The lowest friction coefficient value was produced at a speed of 12 m/s and when the contact pressure was 0.10 MPa which was 0.296–0.312.

Figure 5 was example of the graph of wheel temperature measurements during testing. The graph illustrated that temperature of the wheel surface increased during friction test performed. Wheel temperature after friction testing at various of sliding speed and contact pressure was displayed in Fig. 6. Figure 6 showed that the wheel temperature increased with increasing sliding speed and contact pressure. The increase in sliding speed and contact pressure caused an increased in heat at friction surface. This heat caused a rise in wheel temperature. It lead wheel temperature to rise.

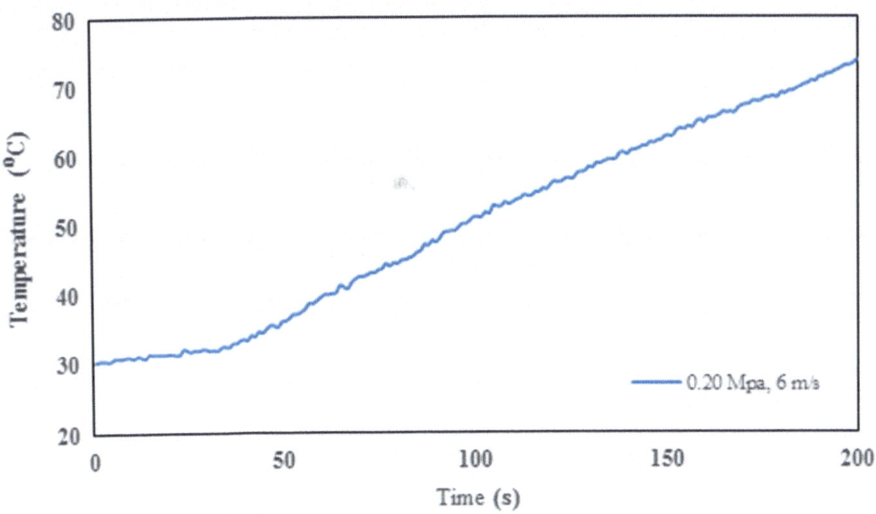

Fig. 5 Example of the results of temperature measurements with a load of 0.20 MPa at a sliding speed of 6 m/s

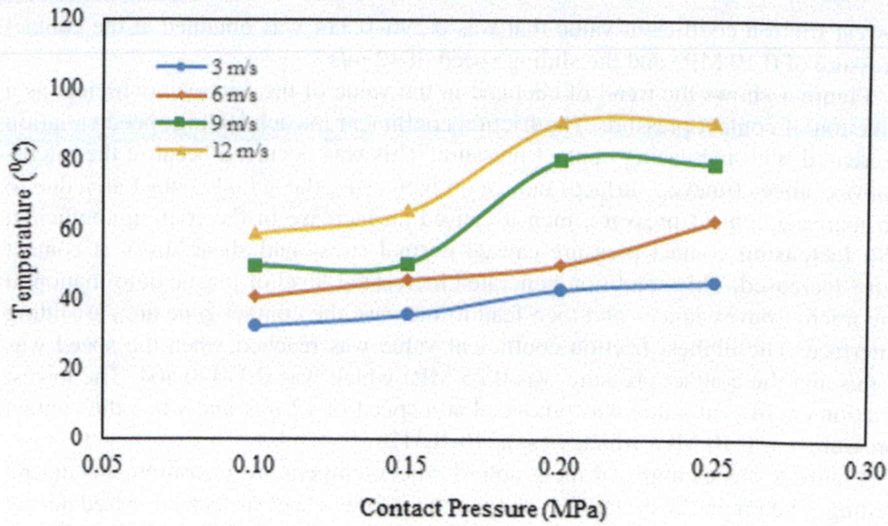

Fig. 6 The relationship of temperature and contact pressure with various of sliding speed

4 Conclusion

The friction coefficient of composite brake block was influenced by the friction testing parameters. The friction coefficient of composite brake blocks decreased with increasing sliding speed, but increased with increasing contact pressure. The friction coefficient of composite brake blocks addressed values that was vary from 0.296 to 0.404.

Acknowledgements The research was supported by the Ministry of Research, Technology, and Higher Education of the Republic of Indonesia through the Fundamental Research Grant with Contract No. 719/UNS27.21/PN/2019 and USAID through Sustainable Higher Education Research Alliances (SHERA) Program—Centre for Collaborative (CCR) National Center for Sustainable Transportation Technology (NCSTT) with Contract No. IIE00000078-ITB-1.

References

1. Vakkalagadda MRK, Srivastava DK, Mishra A, Racherla V (2015) Performance analyses of brake blocks used by Indian Railways. Wear 328:64–76
2. Verma PC, Alemani M, Gialanella S, Lutterotti L, Olofsson U, Straffelini G (2016) Wear debris from brake system materials: a multi-analytical characterization approach. Tribol Int 94:249–259
3. Pellizzari M, Cipolloni G (2017) Tribological behaviour of Cu based materials produced by mechanical milling/alloying and spark plasma sintering. Wear 376–377:958–967
4. Li J, Wongsa-Ngam J, Xu J, Shan DB, Guo B (2015) Wear resistance of an ultrafine-grained Cu-Zr alloy processed by equal-channel angular pressing. Wear 326–327:10–19

5. Blau PJ, Mclaughlin JC (2003) Effects of water films and sliding speed on the frictional behavior of truck disc brake materials. Tribol Int 36:709–715
6. Bao J, Li Z, Hu D, Yin Y, Liu T (2015) Frictional performance and temperature rise of a mining nonasbestos brake material during emergency braking. Adv Mater Sci Eng 2015:1–7

Sound Absorption of BCC Lattice Structures

Dg. H. Kassim, A. Putra, M. F. S. Che Hamid, and Mohd Rizal Alkahari

Abstract This study investigates the sound absorption characteristics of body centre cubic (BCC) lattice structure produced by additive manufacturing. Measurement of sound absorption coefficient was conducted by using impedance tube apparatus according to the two-microphone method described in ISO 10532-2 (2001). The thickness, strut diameter of lattice arm and air gap are the factors discussed on their effects on the absorption coefficient. The results can be seen to resemble the performance of a Helmholtz resonator where a peak absorption is present at particular frequency which corresponds the resonant frequency of the system. The peak amplitude and the frequency can be controlled by adjusting the size of the strut and the thickness of the structure.

Keywords BCC lattice · Sound absorber · Absorption coefficient

1 Introduction

Everywhere in the world, a developed area often suffers a serious noise pollution arise from various sources such as machineries, transportation or daily appliances [1, 2]. In recent years, the noise pollution become an important issue as it should be overcome and controlled in order to minimize the severe health risk emitted from it [3]. Abundance type sound absorbing material are available commercially to be use as an effective way to solve these problems. Porous sound absorbing materials are among the well-known to be used in room acoustic applications [4–6] due to the easy to be manufactured, easy to use and carry multiple possible design. Despite its broad spectrum, its technologies and materials to be use are often limits by the standard and manufacturing method. Moreover, porous absorber such as foam can

Dg. H. Kassim · A. Putra (✉) · M. F. S. C. Hamid · M. R. Alkahari
Fakulti Kejuruteraan Mekanikal, Universiti Teknikal Malaysia Melaka, Hang Tuah Jaya, 76100 Durian Tunggal, Melaka, Malaysia
e-mail: azma.putra@utem.edu.my

Dg. H. Kassim · A. Putra · M. R. Alkahari
Centre for Advanced Research on Energy, Universiti Teknikal Malaysia Melaka, Hang Tuah Jaya, 76100 Durian Tunggal, Melaka, Malaysia

U. Sabino et al. (eds.), *Proceedings of the 6th International Conference and Exhibition on Sustainable Energy and Advanced Materials*, Lecture Notes in Mechanical Engineering,
https://doi.org/10.1007/978-981-15-4481-1_8

give a negative impact to environment [7–9], resulting in extensive research to find acoustic absorber from different innovative materials [10–13].

This is where additive manufacturing becomes the way of solving those implications in producing sound absorber. Additive manufacturing is a process where a 3D computer aided model is transferred to a 3-dimensional object printed layer by layer through a deposited material or also known as 3D printing [14]. Additive manufacturing enables manufacturers to skip complex process on the new product development. It became reliable to industrial application due to the simple processing to produce a complex and reliable product [15]. It is also because of the rapid development and increasing affordability to certain corresponding field. This method is popular because it can produce a precise complex object.

A lattice structure is the suitable example to illustrate the ability of additive manufacturing in terms of producibility. It has attracted tremendous attention from researchers and engineers for multifunction design opportunities as well as their significant potential for a wide range of lightweight application [16]. There are approximately 14 distinct kinds of lattice structure exist. However, most metal and other solids in the nature have unit cell structure described as body centre cubic (BCC), face centred cubic (FCC) and hexagonal close packed (HCP). A single unit cell can be repeated to create a thicker structure to increase toughness or modification can be made from parent structure to achieve certain targeted properties [17].

Most of the previous studies performed on lattice structure focuses on the mechanical behaviour or energy absorption for strength focused applications [18–21]. However, there is limited study about how lattice structure can be used as a sound absorber. Therefore, the objective of the study is to investigate the simple body centre cubic (BCC) lattice structure acoustic properties as a potential sound absorber.

2 Methodology

2.1 Material and Sample Preparation

For this purpose of study, fused deposition modelling (FDM) was used to fabricate additively manufactured body centre cubic (BCC) lattice sample. Six samples in total were fabricated using Cube Pro Trio 3D printer. The material used was acrylonitrile butadiene styrene (ABS), a plastic polymer with increasing popularity in the engineering polymer field. Common ABS have 15–35% of acrylonitrile, 5–30% butadiene and 40–60% styrene. Although it has no specific melting point, the specifically extruded filament used in this study has a printing temperature of 230–240 °C with solid density of 1060 kg/m^3. This material is also known with flexibility and toughness. The surface can be smoothened through chemical treatment and more reliable when is subjected to external impacts compared to other raw materials whilst able to maintain its lightweight properties.

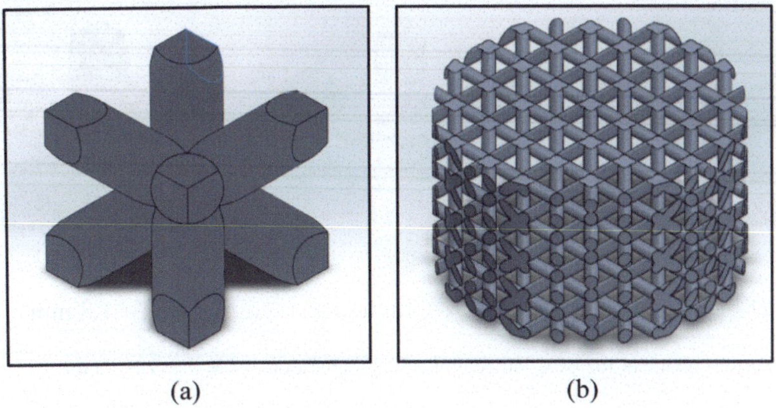

Fig. 1 **a** A BCC unit cell with cylindrical strut and **b** stacked structure of lattice sample

Body centre cubic (BCC) lattice were utilized as the basic unit cell for the fabricated sample. The angle of lattice strut was kept constant to minimize the effect of overhang during the printing process [22, 23]. Figure 1a shows a BCC unit cell and Fig. 1b shows repeated and stacked unit cells to form a cylindrical shape of 33.3 mm diameter. The structure was designed using computer aided design (CAD) software and was converted to 3D drawing format (typically a .STL format) before it can be printed.

The varied geometries were categorized by the different thickness, t (shown in Fig. 2) and strut diameter, D_S (shown in Fig. 3) followed by the different air gap, d applied during the sound absorption measurement. The details of the sample configurations and experiment are described in Tables 1 and 2 respectively.

Fig. 2 Example of the lattice structure with thickness of 15 and 20 mm

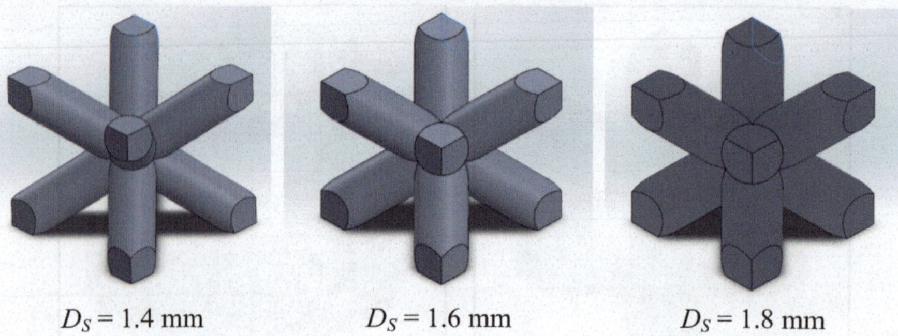

Fig. 3 Various diameter for BCC lattice unit cell configuration at x = y = z = 5 mm

Table 1 Geometry configuration for BCC lattice sample

Sample name	Thickness, t (mm)	Strut diameter, D_S (mm)
BCC1	15	1.6
BCC2	20	1.6
BCC3	30	1.6
BCC4	40	1.6
BCC5	40	1.4
BCC6	40	1.8

Table 2 Measurement conducted to assess the effect of sound absorption characteristics with different geometries

Measurement(s)	Sample(s)
Effect of thickness, t	BCC1, BCC2, BCC3 and BCC4
Effect of strut diameter, D_S	BCC3, BCC5 and BCC6
Effect of air gap, d	BCC6: no air gap, 10 and 20 mm

2.2 *Measurement*

Sound absorption coefficient of the 3D printed BCC lattice samples was experimentally measured by two microphone impedance tube method following the ISO 10534-2 standard [24]. Home-made aluminum impedance tube of 33.3 mm inner diameter was used under ambient temperature of 23 °C, sound speed of $c_0 = 343.9$ m/s, relative humidity of 60%, and air density of $\rho = 1.2$ kg/m^3. The frequency range used was between 500 and 6400 Hz. The in situ laboratory experimental setup and its schematic drawing are shown in Figs. 4 and 5 respectively.

Random sound signal was used for this particular measurement setup generated by the loudspeaker attached at the opposite direction from the BCC lattice sample. Signals from microphone 1, nearest to the loudspeaker located x_1 from the sample

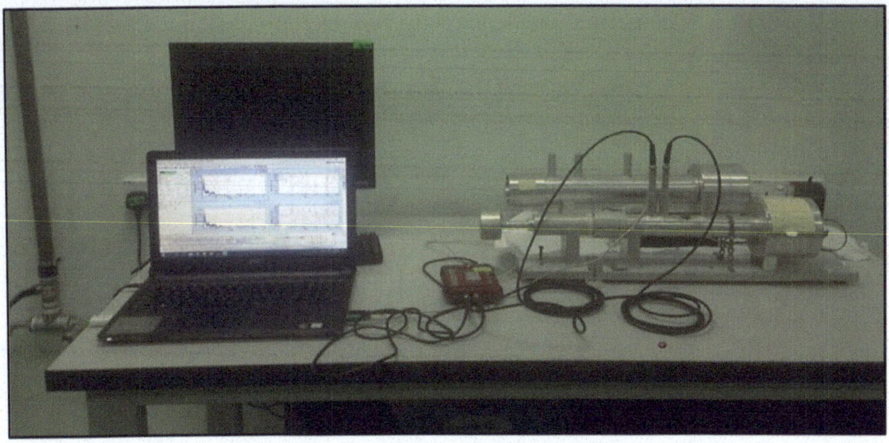

Fig. 4 In-situ two microphone impedance tube measurement setup

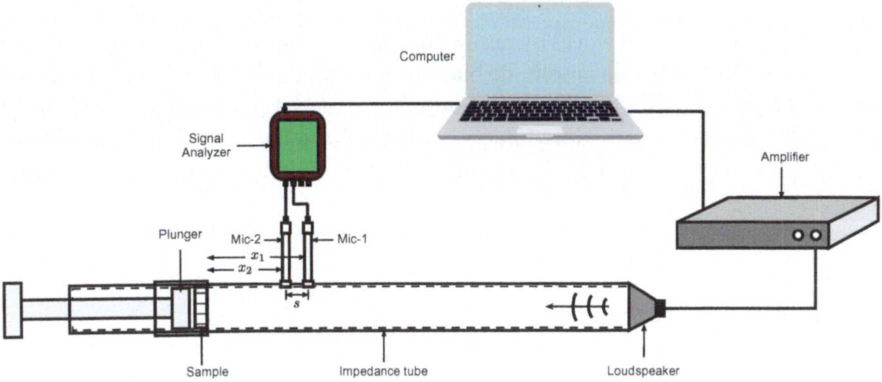

Fig. 5 Schematic diagram of sound absorption measurement setup

and microphone 2, nearest to the sample located at x_2, are used to calculate the incident and reflected wave transfer function, H_I and H_R [25–27]:

$$H_I = e^{-jk_0 s} \tag{1}$$

$$H_R = e^{jk_0 s} \tag{2}$$

where the distance of the two microphone is $s = x_1 - x_2$. The complex acoustic transfer functions, H_{12} is determined by:

$$H_{12} = \frac{p_2}{p_1} = \frac{e^{jk_0 x_2} + re^{-jk_0 x_2}}{e^{jk_0 x_1} + re^{-jk_0 x_1}} \tag{3}$$

where $k_0 = 2\pi f/c$. The normal incidence reflection factor, r is defined by [19, 23]:

$$r = \frac{H_{12} - H_I}{H_R - H_{12}} e^{2jk_0x_1} \tag{4}$$

The specific acoustic impedance, Z and normal incidence sound absorption coefficient, α were determined by:

$$Z = \left(\frac{1+r}{1-r}\right)\rho c_0 \tag{5}$$

$$\alpha = 1 - |r|^2 \tag{6}$$

3 Results and Discussions

Sound absorbing material works by turning mechanical energy of the sound wave into low heat thus preventing an amplitude buildup caused by reflective energy from hard surfaces of an enclosed space. A sound absorber material performance is classified by its efficiency to absorb sound energy. Material thickness, porosity and how it is positioned are among many factors that can manipulate the efficiency of sound absorber.

3.1 Effect of Thickness

Figure 6 shows the absorption coefficient of BCC lattice for sample BCC1, BCC2, BCC3 and BCC4 without an air gap behind the sample. The general pattern for the peak of absorption moves to the lower frequency with increasing thickness. However, there is obviously a decreasing bandwidth of absorption. Generally speaking, to mid low frequencies between 100 Hz and 2 kHz have higher wavelength [22] thus thicker material provides in better sound absorption in these frequencies explaining why thicker BCC lattice sample has high absorption peak [28, 29]. Thicker absorber also increases the resistance of flow as well as the dissipation of sound energy. Although the sound absorption seen at the frequencies between 2.5 and 3.5 kHz is predominantly low, it is still noticeable that the thicker BCC sample offers a better performance at frequency higher than 4 kHz.

Wrapping the sample with linen or cloth might reverse the performance of these lattice sample for the frequency above 2 kHz as presented by [3]. Using material such as rubber [30] or fire-retardant material are also said to produce better sound absorption [31]. However, for a micro perforated panel (MPP) increasing thickness will impair the performance due to the working principle of Helmholtz resonator.

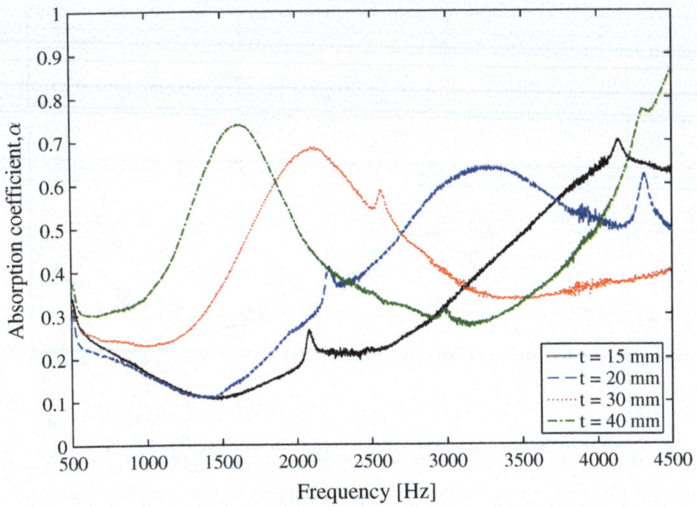

Fig. 6 Measured sound absorption coefficient of BCC1, BCC2, BCC3 and BCC4 lattice sample backed by rigid backing

Therefore, further careful investigation is needed for the BCC sound absorber by altering the strut diameter and air gap which are discussed in the following sections.

3.2 Effect of Strut Diameter

A different diameter of lattice strut is able to modify the peak of amplitude of sound absorption coefficient as shown in Fig. 7. The sample that has a greater diameter has the higher amplitude of absorption coefficient compared to other sample. The peak value of sound absorption coefficient is about 0.7 between the tested frequency which is around 1.5–2 kHz.

Larger diameter of strut will shrink the size of pores inside the structure and also the opening along the sound energy propagation. Smaller opening creates a higher viscosity which increases acoustic energy dissipation thus achieving higher sound absorption coefficient. However, if density is taken into consideration, a large strut diameter will be cost defective as more material has to be used to produce such structure.

3.3 Effect of Air Gap

Effect of the three selected BCC lattice sample backed by an air gap against the performance of sound absorption coefficient is shown in Fig. 8. Two significant

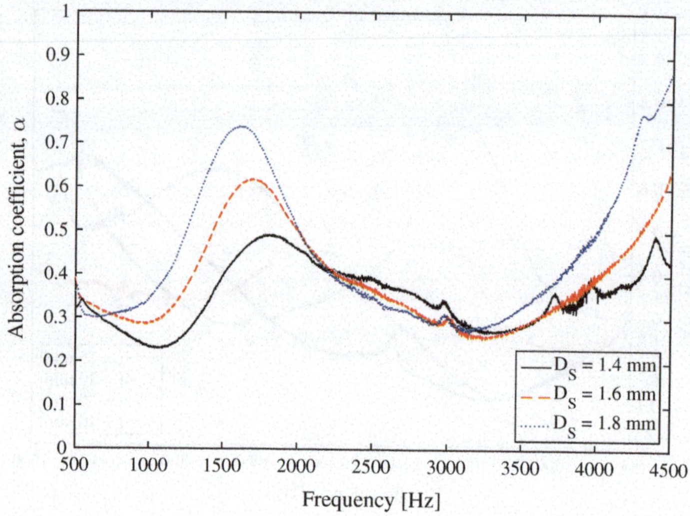

Fig. 7 Measured sound absorption coefficient of BCC3, BCC5 and BCC6 lattice sample with multiple strut diameters

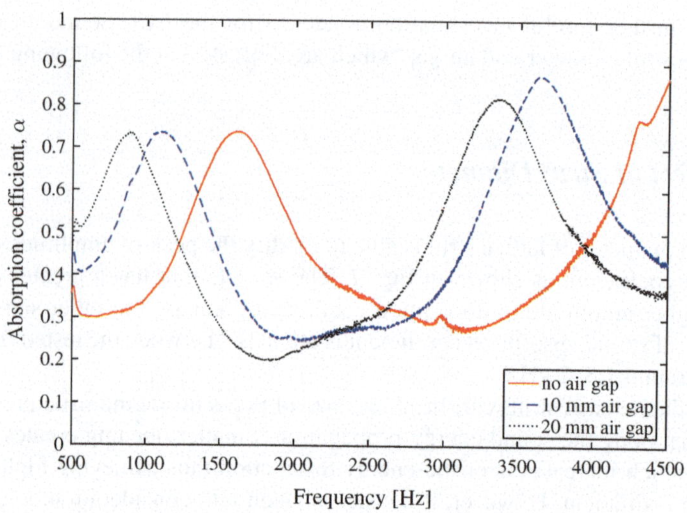

Fig. 8 Measured sound absorption coefficient of BCC6 lattice sample backed by different rigid backing

sound absorption peaks can be seen produced by all test samples. Increasing the air gap substantially shifts the absorption peak to the lower frequency. Sound absorption coefficient for the first peak is formed at frequency lower than 1.5 kHz with the value of 0.7 and is seen consistent even with increasing air gap. On the other hand, the second peak is formed of frequency of 3.5 kHz and above. This peak also experiences an increment with the expanded air gap. The second peak also exhibits a wider frequency bandwidth coverage compared to the former peaks.

Mass-spring resonance is created by the presence of air gap and thus the peak of absorption occurs when the stiffness of the air gap cancels the air mass of the holes in the sample. The presence of air gap layer can be a good alternative technique in order to improve the sound absorption coefficients. However, in harsh environment, unprotected and lowly maintain sound absorber, this can promote the accumulation of solid or fluid particle inside the absorber and the back layer. However, this may not be the case for lattice structure. For a thinner micro perforated panel (MPP), it was proven that the sound absorption characteristics did not affected by dust particle applied under controlled condition [32, 33].

4 Conclusion

The acoustic performance of body centre cubic (BCC) lattice structure has been studied and presented in this paper. Altering the thickness of lattice sample enhances the absorption mainly towards lower frequency, while larger strut diameter only enhances the peak of sound absorption without significant change in the frequency bandwidth. The introduction of air gap behind the sample shifts the peak frequency to lower frequency. Based on the results of this experiment, the lattice structures can be suggested to be the next possible sound absorber with careful control on the thickness, strut diameter and air gap.

Acknowledgements The authors Dg Hafizah binti Kassim, Azma Putra, Muhammad Fakrul Syahid Bin Che Hamid and Mohd Rizal Alkahari acknowledge Fakulti Kejuruteraan Mekanikal, UTeM for the conference registration fee.

References

1. Cao L, Fu Q, Si Y, Ding B, Yu J (2018) Porous materials for sound absorption. Compos Commun 10:25–35
2. Yang M, Chen S, Fu C, Sheng P (2017) Optimal sound-absorbing structures. Mater Horiz 4(4):673–680
3. Amares S, Sujatmika E, Hong TW, Durairaj R, Hamid HSHB (2017) A review: characteristics of noise absorption material. J Phys Conf Ser 908(1):012005
4. Paşayev N, Kocatepe S, Maraş N (2019) Investigation of sound absorption properties of nonwoven webs produced from chicken feather fibers. J Ind Text 48(10):1616–1635
5. Lu TJ, Hess A, Ashby MF (1999) Sound absorption in metallic foams. J Appl Phys 85(11):7528–7539
6. Lim ZY, Putra A, Nor MJM, Yaakob MY (2018) Sound absorption performance of natural kenaf fibres. Appl Acoust 130:107–114
7. Glé P, Gourdon E, Arnaud L (2011) Acoustical properties of materials made of vegetable particles with several scales of porosity. Appl Acoust 72(5):249–259

8. Asdrubali F, Schiavoni S, Horoshenkov KV (2012) A review of sustainable materials for acoustic applications. Build Acoust 19(4):283–311
9. Joshi SV, Drzal LT, Mohanty AK, Arora S (2004) Are natural fiber composites environmentally superior to glass fiber reinforced composites? Compos A Appl Sci Manuf 35(3):371–376
10. Álvarez-Láinez M, Rodríguez-Pérez MA, de Saja JA (2014) Acoustic absorption coefficient of open-cell polyolefin-based foams. Mater Lett 121:26–30
11. Koruk H, Genc G (2015) Investigation of the acoustic properties of bio luffa fiber and composite materials. Mater Lett 157:166–168
12. Ru J, Kong B, Liu Y, Wang X, Fan T, Zhang D (2015) Microstructure and sound absorption of porous copper prepared by resin curing and foaming method. Mater Lett 139:318–321
13. Sun P, Guo Z (2015) Preparation of steel slag porous sound-absorbing material using coal powder as pore former. J Environ Sci 36:67–75
14. Srinivas M, Babu BS (2017) A critical review on recent research methodologies in additive manufacturing. Mater Today Proc 4(8):9049–9059
15. Petrovic V, Vicente Haro Gonzalez J, Jordá Ferrando O, Delgado Gordillo J, Ramón Blasco Puchades J, Portolés Griñan L (2011) Additive layered manufacturing: sectors of industrial application shown through case studies. Int J Prod Res 49(4):1061–1079
16. Yan C, Hao L, Hussein A, Young P, Huang J, Zhu W (2015) Microstructure and mechanical properties of aluminium alloy cellular lattice structures manufactured by direct metal laser sintering. Mater Sci Eng A 628:238–246
17. Ibrahim Y, Li Z, Davies CM, Maharaj C, Dear JP, Hooper PA (2018) Acoustic resonance testing of additive manufactured lattice structures. Addit Manuf 24:566–576
18. Xiao Z, Yang Y, Xiao R, Bai Y, Song C, Wang D (2018) Evaluation of topology-optimized lattice structures manufactured via selective laser melting. Mater Des 143:27–37
19. Niu J, Choo HL, Sun W, Mok SH (2018) Analytical solution and experimental study of effective Young's modulus of selective laser melting-fabricated lattice structure with triangular unit cells. J Manuf Sci Eng 140(9):091008
20. Du Plessis A, Yadroitsava I, Yadroitsev I, le Roux SG, Blaine DC (2018) Numerical comparison of lattice unit cell designs for medical implants by additive manufacturing. Virtual Phys Prototyp 13(4):266–281
21. Shamvedi D, McCarthy OJ, O'Donoghue E, Danilenkoff C, O'Leary P, Raghavendra R (2018) 3D metal printed heat sinks with longitudinally varying lattice structure sizes using direct metal laser sintering. Virtual Phys Prototyp 13(4):301–310
22. Brackett D, Ashcroft I, Hague R (2011) Topology optimization for additive manufacturing. Proceedings of the solid free form fabrication symposium, pp. 348–362. Austin, TX
23. Cacace S, Cristiani E, Rocchi L (2017) A level set based method for fixing overhangs in 3D printing. Appl Math Model 44:446–455
24. Standard B (2001) Acoustics—determination of sound absorption coefficient and impedance in impedance tubes—part 2: transfer-function method. BS EN ISO, 10534-2
25. Chung JY, Blaser DA (1980) Transfer function method of measuring induct acoustic properties. I. Theory. J Acoust Soc Am 68(3):907–913
26. Chung JY, Blaser DA (1980) Transfer function method of measuring induct acoustic properties. II. Experiment. J Acoust Soc Am 68(3):914–921
27. Krüger J, Quickert M (1997) Determination of acoustic absorber parameters in impedance tubes. Appl Acoust 50(1):79–89
28. Azkorra Z, Pérez G, Coma J, Cabeza LF, Burés S, Álvaro JE, Erkoreka A, Urrestarazu M (2015) Evaluation of green walls as a passive acoustic insulation system for buildings. Appl Acoust 89:46–56
29. Adnan NQA, Rus AZM (2014) Sound absorption of laminated biopolymer foam and epoxy foam, vol 594. Trans Tech Publications, pp 291–295
30. Juliá E, Segura J, Nadal A, Gadea JM, Crespo JE (2013) Study of sound absorption properties of multilayer panels made from ground tyre rubbers. Fascicle Manag Technol Eng 1:147–150
31. Asdrubali F, D'Alessandro F, Schiavoni S (2008) Sound absorbing properties of materials made of rubber crumbs. J Acoust Soc Am 123(5):3037

32. Liu J, Herrin DW (2011) Effect of contamination on acoustic performance of microperforated panels 2011-01-1627. SAE Int J Passeng Cars Mech Syst 4:1156–1161
33. Herrin DW, Liu W, Hua X, Liu J (2017) A guide to the application of microperforated panel absorbers. Sound Vib 13

Application of Quality Function Deployment in Product Design and Development: Car Seat Case Study

Shafizal Mat, Mohd Farhe Hussin, Faiz Redza Ramli, Mohd Rizal Alkahari, Mohamad Ridzuan Jamli, Syahibudil Ikhwan Abdul Kudus, and Keith Case

Abstract Quality Function Deployment (QFD) plays a big role in product design and development. It is consisted of several techniques which are very useful and suitable to be applied in whatever field. This project focuses on how the QFD is implemented and applied in designing a car seat. Ergonomics study is important because it helps to analyze the human posture behavior. According to QFD, the process of designing a new product starts from customer's requirement. Then some of concepts are developed as the early picture resulting from the product design specification and customer's requirement. Other techniques were applied such as morphology chart, concept generation and concept screening. These techniques were then lead into concept selection using weighted scoring method. CAD was performed by using SolidWorks CAD software. An ergonomics analysis were also studied to see the final score of the design using RULA analysis. As the result from QFD process, a new ergonomics car seat design is developed.

Keywords Quality Function Deployment · Human posture · Ergonomics · Rapid upper limb assessment (RULA)

S. Mat (✉) · M. F. Hussin · F. R. Ramli · M. R. Alkahari · S. I. A. Kudus
Fakulti Kejuruteraan Mekanikal, Universiti Teknikal Malaysia Melaka, Hang Tuah Jaya, 76100 Durian Tunggal, Melaka, Malaysia
e-mail: shafizal@utem.edu.my

S. Mat · F. R. Ramli · M. R. Alkahari
Centre for Advanced Research on Energy, Universiti Teknikal Malaysia Melaka, Hang Tuah Jaya, 76100 Durian Tunggal, Melaka, Malaysia

M. R. Jamli
Fakulti Kejuruteraan Pembuatan, Universiti Teknikal Malaysia Melaka, Hang Tuah Jaya, 76100 Durian Tunggal, Melaka, Malaysia

S. I. A. Kudus
Fakulti Teknologi Kejuruteraan Mekanikal dan Pembuatan, Universiti Teknikal Malaysia Melaka, Hang Tuah Jaya, 76100 Durian Tunggal, Melaka, Malaysia

K. Case
School of Mechanical, Electrical and Manufacturing Engineering, Loughborough University, Loughborough, Leicestershire, UK

U. Sabino et al. (eds.), *Proceedings of the 6th International Conference and Exhibition on Sustainable Energy and Advanced Materials*, Lecture Notes in Mechanical Engineering,
https://doi.org/10.1007/978-981-15-4481-1_9

1 Introduction

To design a product well, a design teams needs to know what it is they are designing, and what the end-users will expect from it. Quality Function Deployment is a systematic approach to design based on a close awareness of customer desires, coupled with the integration of corporate functional groups. It consists of translating customer desires (for example, the ease of writing for a pen) into design characteristics (pen ink viscosity, pressure on ball-point) for each stage of the product development [11]. Ultimately the goal of QFD is to translate often subjective quality criteria into objective ones that can be quantified and measured and which can then be used to design and manufacture the product. It is a complimentary method for determining how and where priorities are to be assigned in product development. Quality Function Deployment was developed by Yoji Akao in Japan in 1966. By 1972 the power of the approach had been well demonstrated at the Mitsubishi Heavy Industries Kobe Shipyard [13] and in 1978 the first book on the subject was published in Japanese and then later translated into English in 1994. In Akao's words, QFD "is a method for developing a design quality aimed at satisfying the consumer and then translating the consumer's demand into design targets and major quality assurance points to be used throughout the production phase. QFD is a way to assure the design quality while the product is still in the design stage." As a very important side benefit he points out that, when appropriately applied, QFD has demonstrated the reduction of development time by one-half to one-third [1].

The 3 main goals in implementing QFD are:

1. Prioritize spoken and unspoken customer wants and needs.
2. Translate these needs into technical characteristics and specifications.
3. Build and deliver a quality product or service by focusing everybody toward customer satisfaction.

QFD uses some principles from Concurrent Engineering in that cross-functional teams are involved in all phases of product development. Each of the four phases in a QFD process uses a matrix to translate customer requirements from initial planning stages through production control. Each phase, or matrix, represents a more specific aspect of the product's requirements. Relationships between elements are evaluated for each phase. Only the most important aspects from each phase are deployed into the next matrix.

Phase 1, Product Planning: Building the House of Quality. Led by the marketing department, Phase 1, or product planning, is also called The House of Quality. Many organizations only get through this phase of a QFD process. Phase 1 documents customer requirements, warranty data, competitive opportunities, product measurements, competing product measures, and the technical ability of the organization to meet each customer requirement. Getting good data from the customer in Phase 1 is critical to the success of the entire QFD process. Phase 2, Product Design. This phase 2 is led by the engineering department. Product design requires creativity and innovative team ideas. Product concepts are created during this phase and part specifications are documented. Parts that are determined to be most important to meeting customer needs are then deployed into process planning, or Phase 3. Phase 3, Process Planning. Process planning comes next and is led by manufacturing engineering. During process planning, manufacturing processes are flowcharted and process parameters (or target values) are documented. Phase 4, Process Control. Finally, in production planning, performance indicators are created to monitor the production process, maintenance schedules, and skills training for operators. Also, in this phase decisions are made as to which process poses the most risk and controls are put in place to prevent failures. The quality assurance department in concert with manufacturing leads Phase 4.

Proper ergonomic design is necessary to prevent repetitive strain injuries, which can develop over time and can lead to long-term disability. Ergonomics (or human factors) is the scientific discipline concerned with the understanding of interactions among humans and other elements of a system, and the profession that applies theory, principles, data and methods to design in order to optimize human well-being and overall system performance. Ergonomics is employed to fulfill the two goals of health and productivity. It is relevant in the design of such things as safe furniture and easy-to-use interfaces to machines. Ergonomics is concerned with the "fit" between people and their technological tools and environments. It takes account of the user's capabilities and limitations in seeking to ensure that tasks, equipment, information and the environment suit each user. To assess the fit between a person and the used technology, ergonomists consider the job (activity) being done and the demands on the user; the equipment used (its size, shape, and how appropriate it is for the task), and the information used (how it is presented, accessed, and changed). Ergonomics draws on many disciplines in its study of humans and their environments including anthropometry, biomechanics, mechanical engineering, industrial engineering, industrial design, kinesiology, physiology and psychology.

1.1 Ergonomics Car Seat Design

A car seat is the chair used in automobiles. It is one of the most important components of vehicles and they are place where professional driver spend most of their time. Automotive seat, which are in contact with vehicle occupants, play an important role in improving the comfort and work environment of a driver and passengers. The improvement of automotive seating systems, particularly for the driver, has been the subject of intense interest for many years since a driver feels more fatigue than passengers. The paper describes a large variety of studies and up to date techniques developed for vehicle seats used by different type of transportation system such as trucks, cars, tractors, trains and aircrafts [2]. In designing a comfortable seat, it is important to understand the vibration environment to which individuals are exposed and how well they can tolerate this environment. Moreover, human sensitivity to low-frequency whole-body vibration (WBV) has pointed to ride quality as an important need in seat design [6].

The standard car seat is designed to support thighs, the buttocks, lower and upper back, and head support. The front driver and passenger seats of most vehicles have three main parts: the seat back (squab), seat base (cushion), and the headrest. These components are usually constructed from foam to provide comfort to the rider. When choosing this product, foam manufacturers must consider the most suitable foam for balancing comfort, support, safety, and recycling properties (ATASED 6). The cushioning agent is especially important when considering that moving cars can transmit vibrations near the human spine's resonant frequency of 3 Hz. The base can usually be moved forward and back on metal railings and may move up and down to adjust to different body types. This movement is accomplished either by manual latches or by electric levers. The standard base span is from the seated rear to just before the knees and the typical squab extends to the shoulders on the average height person. The head support is connected to the top of the squab by two metal circular tube shafts that are permanently attached to the head support and slide into two holes in the squab. The comfort and durability of a car seat depends on the quality and engineering of its four major components: backrest, cushion-spring platform, upholstery, and padding materials. Modern car seats (especially those used in higher-end sedans and SUVs) combine these four elements not only to create attractive silhouettes, but more importantly, to produce ergonomic and safe car seats.

Ergonomic car seats are great for vehicles used for long driving. Some very good car seats are pressurized to conform to the natural shape of every body type so that the passenger's back, neck, and hips are amply supported. When sitting in the car seat your posture is the most important factor when considering your comfort. The positions that drivers assume depend on their anthropometric characteristics, the range and type of adjustments available in the seat package, and each driver's preferred driving position (Ergonomics 1). Common available adjustments deal with concerns such as providing legroom, supplying back support, and giving head support. Available amenities include electric adjustments, choice in fabric covering, and temperature control. Important advancements in this area are of top concern to manufacturers, government officials and of course, consumers. An integral dimension to any seat design is the aspect of safety. As government regulations become more comprehensive, the design of the car seat must deal with more than just seat belts. Positioning of the occupant with respect to the steering wheel must be considered so injury is not caused by the steering wheel or air bag in a collision.

Nowadays, car manufacturers must try predicting car driver size and position to enhance safety in crashes. Occupant protection systems must find information about the driver's chosen seat position and the size of the driver especially if they are exceptionally large or small (Ergonomics 1). Engineers have gone through a series of tests to determine whether it is possible to determine driver position to boost safety. This process involves installing sensors in seats, utilizing specific CAE (computer assisted engineering) software for numerical evaluations, and experimental tools for laboratory test rigs and road tests (ATASED 5). Test results proved it possible to determine drivers' head and chest positions relative to potentially injurious elements such as the steering wheel and airbag; it is even possible to determine some physical dimensions of drivers [9]. Correct sitting posture at work is essential in helping to avoid neck and back pain. The chair should be comfortable and the controls easy to reach. Adjust the seat and back frequently to suit the task and maintain full back support. The pelvis should be tilted forwards enabling the spine to hold its natural 'S' shape. The workstation should suit the user with the height and position of the equipment being corrected for their stature. The challenge of designing for the human body is that it comes in so many different sizes and shapes. As a result, a design that may be comfortable for one person can be inappropriate for others. While most people believe it is relaxing, sitting is actually hard on the back because it transfers the full weight of the upper body onto the buttocks and thighs. Sitting, especially for long periods of time, can also cause increased pressure on the intervertebral discs—the springy, shock-absorbing parts of the spine.

The posture of the seated person is dependent on the design of the seat itself, individual sitting habits and the work to be performed. Seated postures are defined as the body position in which the weight of the body is transferred to a supporting area. The biomechanical considerations of seated postures include the spine, arms, and legs. The muscles at the back of the thighs influence the relative position of the spine and pelvis. The location and slope of the work area influence the position of the neck, shoulders, and upper extremities, when an individual is in a seated posture [12]. A comfortable driving posture is essential for ergonomic design for a driver. Driving is different with the seating on the chair. This is because when driver operate the vehicle, whole body are move same the vehicle that move. It can might be risk to the driver get the neck, back and shoulder pain when not adjust the seat properly before drive.

The height and inclination of the seat pan, combined with the position, shape, and inclination of the backrest influences the resulting seated posture. Low back stresses can be reduced with the use of back supports or seat backs. The most important factor in reducing low back stress is the inclination angle of the seat back itself. Leaning the back against an inclined backrest reduces the physical load on the intervertebral discs and the static strain of the back and shoulder muscles. This is due to the loads on the spine being greater when one is seated, as compared to when one is standing. When one is seated, the posterior elements of the spine place a greater load through the intervertebral discs.

The lumbar support should be placed in the lumbar region to achieve a more normal lordotic curvature when in the seated posture. In order to provide as much comfort as possible, the support should be adjustable in both height and size. The seat back width should allow users to be supported without arm interference. The shape should be convex from top to bottom to conform to the normal lordosis, and concave from side to side to conform to human anatomy and support the occupant in the seat. The tilt angle of the backrest should be adjustable. A backrest inclination of about 110° is considered an appropriate posture; however, greater inclination may be desirable by the user. The backrest tilt angle adjustment should be independent so that there is little or no effect on the front seat height or angle. Seats with high backrests are preferred since they provide both lower back and upper back support. To prevent back strain, it is also recommended that seats have lumbar support, because the lumber region is one of the most highly strained parts of the spine when sitting [12].

2 Methodology

2.1 Conceptual Design

Conceptual design is the very first phase of a design where drawings are the primary focus, which are comprised of simple plans and sections. These simple drawings should be able to lend themselves easily to more specific sets of plans.

The morphological chart is a method to generate ideas in an analytical and systematic manner [14]. Using the morphological chart, the product's purpose is divided in a set of (sub-) functions. For each of the (sub-) functions ideas are generated and combined into an overall solution. In a morphological chart (Fig. 1), possible components are listed based on their functions. Functions are listed in columns, and components are the means that realize the functions are listed in rows. The morphological method thus yields a matrix of functions and components that realize that function. Through careful selection and combination of a set of components, an idea comes about. This idea should be seen as a principle solution: a careful chosen combination of components that together form a conceptual solution [10].

The concept screening can be as simple as a checklist of criteria in form of questions that fall into 2 categories: 'must meet' and 'should meet' criteria. Must meet criteria are questions used to determine the viability of the opportunity. These criteria should be structured as closed-ended questions and are designed to provide go/no go decision points. Should meet criteria are often more specific. Pugh's Concept Selection matrix (Pugh method) is a valuable design tool for assessing design trade-off and sparking the creation of useful hybrid concepts. It is best to apply this method after the Voice of the Customer is captured. The Pugh matrix tool is used to facilitate a

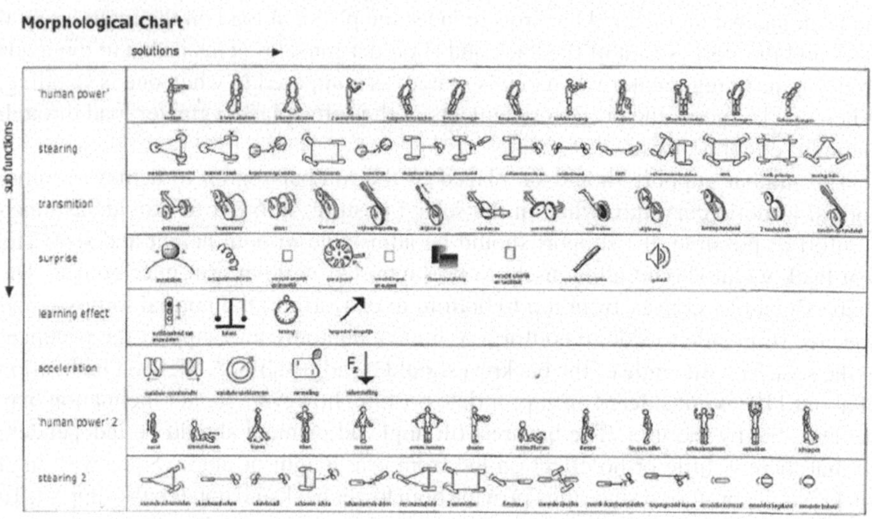

Fig. 1 Example of morphology chart [10]

team-based process for concept generation and selection. A reference concept or base concept is initially selected as the "best current concept". Several other concepts are evaluated according to their strengths and weaknesses against the reference concept. Usually the options are scored relative to the reference concept using a symbolic approach (one symbol for better than, another for neutral, and another for worse than). Comparison of the scores generated gives insight into the best alternative. The Pugh Method is a disciplined approach that helps create strong alternative concepts from weaker concepts. The advantage of this approach is it is extremely useful because it can help teams arrive at an optimum concept that may actually be a hybrid of the best of all the other concepts.

2.2 *Product Design Specification*

The product design specification is a listing of the critical parameters, specifications and requirements for the designing product. It is a statement of what the product should be and should do. This is a constantly evolving document. It is subject to change as the project progresses and as more information is learned. Detail is added as the design grows. The product design specification is driven by customer needs and the constraints imposed upon the system. Table 1 shows the product design specification (PDS) of the driver seat. There are nine criteria that take into consideration; performance, environment, maintenance, life in service, size, weight, material, aesthetics and ergonomics respectively. Performance of the seat is the main criteria. It should withstand about 90 kg of body weight. The seat can be adjusted upward and downward, and forward and backward according to driver's comfortability.

Table 1 Product design specification (PDS) for driver seat

No	Criteria	Parameter
1	Performance	It can support maximum body weight (100 kg). It can be adjustable upward and downward according to lumbar height and forward and backward according to driver's comfortability
2	Environment	It can withstand hot and cold temperature, vibration and shock loading of driver
3	Life in service	It can withstand for >5 years
4	Maintenance	No need special tools for maintenance and easy cleaning
5	Size	It should fit to present car interior
6	Weight	<15 kg
7	Material	Medium density flexible polymer foam
8	Aesthetics	Simple and fulfill needs and sporty look
9	Ergonomic	Seat will support body posture. Contour of the seat back according to driver's lumbar and less sharp edges

Table 2 Additional product design specification (PDS) of car seat

Item	Criteria	Advantage
Usage	• Easy to use • Easy to operate	• Saving time and energy
Cheap	• Low cost production • Use low cost material	• Affordable
Lifetime	• Can use for a long period of time	• No damaged • Saves the cost
Quality	• High quality with low cost material but durable and high quality • Consumer meet demand	• Use material with high quality
Market	• Easy to market • Faster process	• Can be used daily • Limited region problem solving

Second is environment criteria which is the seat withstand hot and cold temperature, vibration and shock loading of driver. Third the seat can withstand about more than 5 years or in other word, it is durable. Fourth, easy cleaning and no need special tools to maintain the seat. This mean the seat is user friendly and burden less. Fifth is about size. Size of the seat should fit to present car interior. This means the seat can be installed in any car. Sixth is weight of the seat. This seat has to be light and easy for installation. Maximum weight of the seat is less or equal to 15 kg. Seventh is about material of the seat and the best material that suit the seat is Medium density flexible polymer foam. Eighth, the seat has to be simple and fulfill customer's need and also sporty look. This parameter explains about aesthetics value. Lastly, according to ergonomics, the seat will support body posture. Contour of the seat back according to driver's lumbar and less sharp edges.

Table 2 shows the additional product design characteristics of car seat. This additional criteria is used as an indicator for development of the seat and to support the previous product design specification.

2.3 Concept Development

As shown in Fig. 2, first concept consist of complex design and shape. Therefore it is also consisted of complex curves section. The backrest is integrated with headrest which means there is restriction for user to adjust the height of the headrest. For advantage of this seat, it gives best support to whole driver's body. The second concept is basically an integrated design. It is look alike as the first concept. It is consisted of complex edges and shape. The headrest and the backrest are fully integrated and it same goes to the first concept. The advantage of this seat is it gives best support for the driver. In addition, this seat is suitable to be used for racing type car. The third concept is not consisted of complex design and shape. In addition there are only minimal curves and edges. The back rest and the headrest is separated which mean

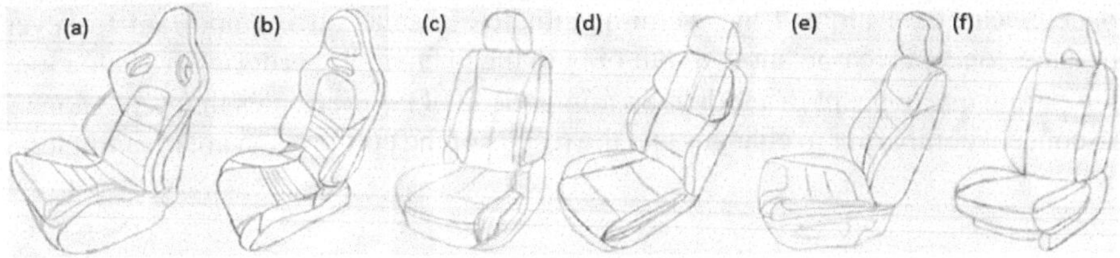

Fig. 2 **a** Concept 1, **b** concept 2, **c** concept 3, **d** concept 4, **e** concept 5 and **f** concept 6

the headrest can be adjusted up and down. It gives a good support for driver's body. The fourth concept of seat design also is not consisted of any complex design. It has minimal curves and edges but still give a better support for driver. Other the backrest and headrest are separated. All the curves and edges for the fifth concept are minimal and not too complex. This seat design is one of good design and shape. The backrest and the headrest are separated and this means the headrest can be adjusted up and down according driver's comfort. Other, it gives better support for the driver's body. The sixth concept has very minimal curves and edges but it still gives good support for driver's body. This seat design is one of the designs that consisted of separated backrest and headrest.

Design selection is the process of evaluating concepts with respects to customer needs and other criteria. By comparing the relative strengths and weakness of the concept, and selecting one or more concepts for further investigation, testing and development. The selection of an overall product concept will be focused from the beginning of the development process before selection of the best design is chosen. A design has been developed to find a design requirement by analyzing design sensitivity and by searching for the feasible design region. In addition, the best configuration concept that satisfies the design requirements has been identified using Quality Function Deployment and the Configuration Concept Selection Matrix. This Concept Selection Process can contribute to reduce the time and effort during early design stage [8].

There is having some criteria to make a selection of the product. There are has six step to selection process. The six-steps are:

1. Prepare the selection matrix based on customer needs.
2. Rate the concepts.
3. Rank the concepts.
4. Combine and improve the concepts.
5. Select one or more concepts to chosen.
6. Reflect on the results and the process.

The selection of the driver seat is related based on the specification that has been state and follow the ergonomics analysis. The selection design of the driver seat must follow the concept screening and concept scoring. The purpose of this stage is to narrow down the number of concepts quickly and to improve the concepts. In order to evaluate concepts effectively, some sort of criteria are needed against which concepts can be evaluated in a semi-quantitative manner. It is important to never compare concepts to one another, but only to the unchanging criteria for evaluation. Comparing one concept to another is a comparison of two variables that is as soon as the concept comparing to changes and the results of the comparison are meaningless.

2.4 Concept Screening and Scoring

In Table 3, the selection criteria are listed along of the screening matrix. The matrices focus attention on the customer needs and other decision criteria and on the product concepts for improvement and selection. These criteria are chosen based on customers needs. The selection criteria should be chosen to differentiate among the concepts. After careful consideration, the chosen concept become the benchmark or reference against which all concepts are rated.

For the score that gathered from the table, the concepts 3, 4, 5 and 6 are selected as a reference concept for the selection design. The concepts follow the customer needs criteria and will be continued for the next process.

Table 4 shows the concept scoring for all car seat design concepts. Concept scoring is used when increased resolution will better differentiate among competing concepts. In this stage, weight the relative importance of the selection criteria and focuses on comparisons with respect the criteria. The concepts scoring are determined by the weighted sum of the ratings. The concept which gets the highest score will proceed for the next stage. In this case the fifth concept got the highest score follow by third, sixth, fourth, first and second respectively.

Table 3 Concept screening method

Selection criteria		Concept variants						
		1	2	3	4	5	6	Ref
Ease of handling		–	0	+	+	+	+	0
Ease of use		0	0	0	0	+	+	0
Number readability		0	0	0	+	0	0	0
Ergonomics design		+	+	0	+	+	0	0
Load handling		+	0	0	0	+	0	0
Manufacturing ease		–	–	+	+	0	+	0
	Pluses	2	1	2	4	4	3	
	Same	2	4	4	2	2	3	
	Minus	2	1	–	–	–	–	
	Net	0	0	2	4	4	3	
	Rank	4	4	3	1	1	2	
	Continue?	No	No	Yes	Yes	Yes	Yes	

Table 4 Weighted rating method

		Concepts											
		1		2		3		4		5		6	
Selection criteria	Weight (%)	Rating	Weighted score	Rating	Weighted score	Rating	Weighted score	Rating	Weighted score	Rating	Weighted score	Rating	Weighted score
Ease of handling	15	2	0.3	2	0.3	3	0.45	3	0.45	3	0.45	3	0.45
Ease of use	15	2	0.3	2	0.3	4	0.6	3	0.45	4	0.6	3	0.45
Readability of settings	10	3	0.3	3	0.3	3	0.3	3	0.3	3	0.3	4	0.4
Ergonomics design	25	4	0.96	3	0.75	3	0.75	3	0.75	4	1	3	0.75
Ease of manufacture	20	2	0.4	2	0.4	3	0.6	2	0.4	3	0.6	3	0.6
Load handling	15	2	0.3	2	0.3	3	0.45	4	0.6	3	0.45	3	0.45
	Total score	2.56		2.35		3.15		2.95		3.4		3.1	
	Rank	5		6		2		4		1		3	
	Continue	No		No		No		No		Yes		No	

Concept 5 is chosen because of firstly this concept meets all the criteria and gained the highest total score after concept scoring and concept screening are performed. By comparing to the others, this concept is the most preferable design because it is the most ergonomically design regarding to all selection criteria in Table 4. Even though there are not too many supports but it can hold and support the whole driver's body. It is also can withstand amount of load. In addition, the backrest and the headrest are separated which mean the headrest can be adjusted according to driver comfortability.

3 Results and Discussion

Figure 3 shows the final seat design. This seat is consisted of four main parts which are seat back, seat base (cushion), side support and headrest. All the drawing processes are performed by using SolidWorks CAD software.

RULA analysis is done to find the ergonomics effect on driver which is given by scale number [4]. The RULA Method was developed and is particularly used to evaluate the risk to workers engaged sedentary tasks such as workers at video terminals or the risk from other tasks in which the operators sits or the risks for workplace in which the worker stands for a large part of the time. The analysis input data are body posture (head, trunk, and upper limbs), the force used, and the type of movements or actions performed, repetitively [7]. RULA analysis is used to analyze many facets of manikin posture based on a combination of automatically detected variables and user data [5]. By using data derived from the RULA equations, this analysis:

1. Considers multiple variables such as object weight, lifting distance, lowering distance, task frequency and action duration.

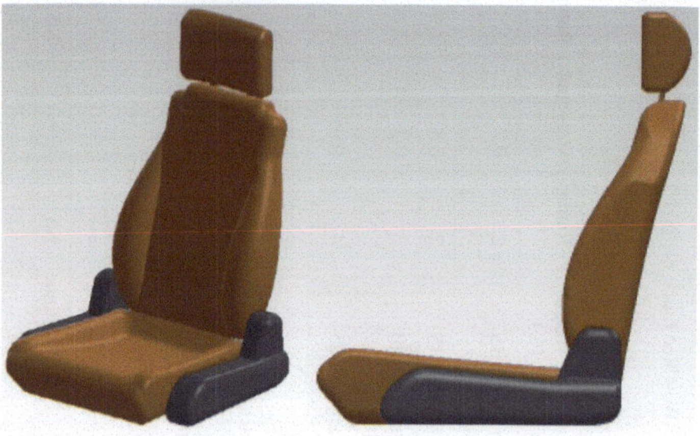

Fig. 3 New seat design

2. Gives you the option of adding task-specific variables such as whether the manikin is externally supported, if the manikin's arms are working across the midline of the body during a task, and whether the manikin's feet are balanced and well supported.
3. Creates a Summarized Report of the task.
4. Provides a quantified set of results noting whether the task and posture are acceptable, should be investigated further, should be investigated further but changed soon, or should be investigated further but changed immediately.

For this analysis, certain parts of manikin's body will be adjusted according to seat design and not all parts of the body are moved. This is because the body's part that undergo evaluation are back spine, leg, neck and head.

Table 5 shows the RULA action level and its description. The score is given as number scale from lowest score 1 to the highest score 7. Each action level has its own descriptions which explain about what are the consequences if the analysis meets the allocated score [15].

The final score of the final design is 2 as shown in Fig. 4. It indicates that posture is acceptable if it is not maintained or repeated for long periods. Regarding to redesigned process, the height of seat back is reduced and the seat back width also is reduced.

Table 5 RULA analysis score description [7]

Level	Score	Description
1	1 or 2	Acceptable posture if not maintained or repeated for long periods
2	3 or 4	Further investigation needed and may require changes
3	5 or 6	Investigation and changes required soon
4	>7	Investigation and changes required immediately

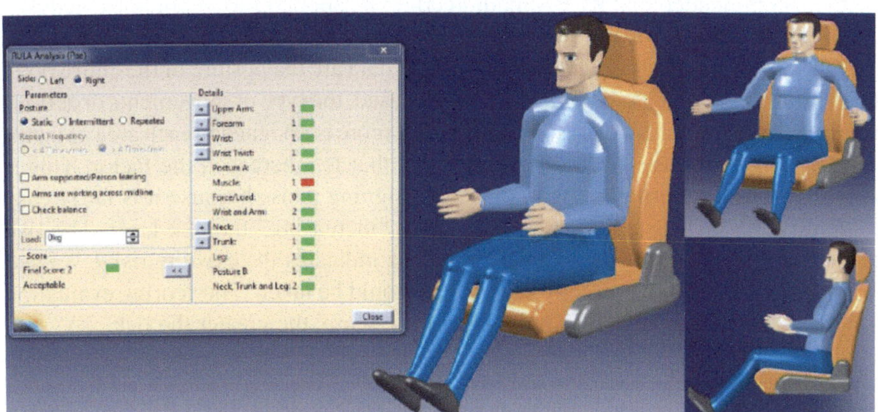

Fig. 4 RULA analysis of final design

As the result, the manikin's body is placed perfectly on the seat and the whole body is fully supported.

Material selection is an important step to determine a comfortable in term of ergonomics. Material selection is a step in the process of designing any physical object. In the context of product design, the main goal of material selection is to minimize cost while meeting product performance goals. Sitting on a hard seat cushion will adversely affect the pelvic bone. For the seat cushion material, elasticity as well as stiffness is required to provide comfort. Also, the material must not be too heavy, with reasonable service life duration and cost. Most importantly, it must provide support for a stable posture. The most optimum material that is identified is medium density flexible polymer foam.

4 Conclusion

QFD provides a better way of understanding of total design process which starts from customer's requirement until detail design process. All these steps are essential and for example the concept screening. At this stage, concepts that have been generated are filtered to see which the best to be developed is. Other objective that has been achieved is the ergonomics study. Ergonomics study is very important and it has great connection with QFD. This is because, when we develop a product, it has to be analyzed so that it does not harm the users and give benefit to the users. In other words, the product can be used for a long time. One of the analysis that was conducted is rapid upper limb assessment or RULA. This assessment is to analyze the human posture. According to this study, the assessment is about the posture of human sitting. After analysis is done, the seat is optimized and improvised based on the results that are collected.

It is recommended that REBA method could be applied for ergonomics analysis. The REBA Method [3] was developed to assess the risk of workers in health facilities and other service industries. The analysis input data are the posture of the entire body (head, trunk, upper and lower limbs), the force used, the type or movements or actions performed, repetitively and grasp conditions. A score is assigned to each area or factor examined that increases according to the risk that is inherent in the factor. This is particularly useful during the workplace redesigning phase because it is fairly easy to identify the factors that most contribute to "increasing" the risk level. The risk index that is obtained from the REBA analysis indicates the existing risk level and the urgency with which corrective measures should be taken. The scoring evaluation is same as RULA analysis which the higher the score, the greater the risk.

Acknowledgements The authors would like to thank the Centre for Research and Innovation Management (CRIM), Innovation and Sustainability in Machine Technologies Research Group (i-SMAT), and Fakulti Kejuruteraan Mekanikal, University Teknikal Malaysia Melaka (UTeM) for registration fees support to participate in ICE-SEAM2019.

References

1. Akao Y (1990) Quality function deployment: integrating customer requirements into product design. Productivity Press, Cambridge
2. Fai TC, Delbressine F, Rauterberg M (2007) Vehicle seat design: state of the art and recent development. In: Proceedings world engineering congress
3. Hignett S, McAtamney L (2000) Rapid entire body assessment (REBA). Appl Ergon 31(2):201–205
4. Mat S, Abdullah MA, Dullah AR, Shamsuddin SA, Hussin MF (2017) Car seat design using RULA analysis. In: Proceeding of mechanical engineering research day, pp 130–131
5. Mat S, Salim MA, Wan Mohamad WMF, Yusop MSM (2011) Design and ergonomics analysis of cutter machine locking device in aerospace industry. J Mater Sci Eng 5:145–154
6. Mayton AG et al (2007) Laboratory investigation of seat suspension performance during vibration testing. In: ASME 2006 international mechanical engineering congress and exposition. American Society of Mechanical Engineers Digital Collection
7. McAtamney L, Corlett EN (1993) RULA: a survey method for the investigation of work-related upper limb disorders. Appl Ergon 24(2):91–99
8. Park HU, Park MY, Lee SJ, Lee JW, Byun YH (2006) Development of requirement driven design concept selection process in aerospace system. In: International conference on computational science and its applications. Springer, Berlin, Heidelberg
9. Pheasant S (1998) Bodyspace: anthropometry, ergonomics and the design of work, 2nd edn. Taylor & Francis, London
10. Roozenburg NFM, Eekels J (1995) Product design: fundamentals and methods, vol 2. Wiley
11. Rosenthal SR (1992) Effective product design and development, how to cut lead time and increase customer satisfaction. Business One Irwin, Homewood
12. Saporta H (2000) Durable ergonomic seating for urban bus operators. Worksite redesign program. Oregon OSHA 417
13. Sullivan LP (1986) Quality function deployment, quality progress, pp 39–50
14. Yusop MSM, Mat S, Yusuff MA, Shamsuddin SA, Dullah AR (2018) Outboard boat engine lifter trolley design in fishing industries. In: Proceedings of mechanical engineering research day 2018, pp 108–110
15. Yusop M, Mat S, Ramli FR, Dullah AR, Khalil SN (2018) Design of welding armrest based on ergonomics analysis: case study at educational institution in Johor Bahru, Malaysia. ARPN J Eng Appl Sci 13(1):309–313

Structural Assessment Review of Type-C Independent Tank in LNG Bunkering Ship

Teguh Muttaqie, Seung Geon Lee, Sang-Rai Cho, and Jung Min Sohn

Abstract In recent years, Liquefied Natural Gas (LNG) becomes one of the most suitable solutions to fulfil world primary eco-friendly energy demand. Consequently, bunkering ship is designed to charge liquefied gas with the increasing need for natural gas. Due to the extremely dangerous and cryogenic cargo, the design of LNG bunkering ship requires very high-level structural design/analysis technologies rather than conventional ships. Therefore, this paper describes a general procedure of the structural safety assessment for the independent type C tank of LNG bunkering ship. It aims to present a complete procedure for strength evaluation of independent type C tank including hull strength assessment.

Keywords Structural assessment review · FE analysis · LNG type C tank · LNG bunkering ship

1 Introduction

Liquefied Natural Gas (LNG) is clean energy thus a better alternative to conventional oil-based fuels. The global reserves remain huge, it is estimated that the global gas reserves will last us at least another 200 years [1]. In recent years, the market of small and medium LNG carriers have been growing rapidly for domestic trade in South-East Asian countries such as Indonesia. One of the more promising solutions to meet

---

T. Muttaqie · S. G. Lee · J. M. Sohn (✉)
Department of Naval Architecture and Marine System Engineering, Pukyong National University, Busan 48513, Republic of Korea
e-mail: jminz@pknu.ac.kr

T. Muttaqie
Agency for the Assessment and Application of Technology (BPPT), Jakarta 10340, Indonesia

S.-R. Cho
School of Naval Architecture and Ocean Engineering, University of Ulsan, Ulsan, Korea

J. M. Sohn
Interdisciplinary Program Marine Convergence Design, Pukyong National University, Busan 48513, Republic of Korea

U. Sabino et al. (eds.), *Proceedings of the 6th International Conference and Exhibition on Sustainable Energy and Advanced Materials*, Lecture Notes in Mechanical Engineering,
https://doi.org/10.1007/978-981-15-4481-1_10

this growing need is to use the LNG bunkering ship to support LNG infrastructure from Shore-to-Ship, and/or Ship-to-Ship.

The LNG is insulated at cryogenic temperature and slightly pressurised above atmospheric pressure. The boil-off gas is used as fuel in the ship boilers. LNG tank is a pressure container of LNG, according to the "International Carriage of liquefied gases in bulk ship construction and equipment rules" [2], LNG tank can be divided into 4 categories, Integral tanks, Membrane tanks, Semi-membrane tanks and Independent tanks. Membrane tanks are widely used in the construction of large LNG ships which the tank is directly fitted into the volume inner hull. However, sloshing phenomena can be possibly damaging to the insulation systems inside the tank [3, 4]. Other types are the independent tanks type A, and B, including the prismatic (SPB) and spherical (Moss). These type can minimize the sloshing loads due to a lot of internal swash member inside [5]. For the LNG bunkering ship, C-type Independent has more advantages. Type C independent tanks are pressure vessels which are designed for pressures greater than 2 bar and are cylindrical or spherical in shape. Type C tanks can be designed for much higher vapor space pressures than Type A and Type B tanks. A secondary barrier is also not required for Type C tanks.

This paper deals with the structural design of the type C independent cargo tanks and its design of the LNG bunkering ship. Figure 1 presents the flowchart of the structural assessment review of the Type-C LNG bunkering ship. To validate the assessment, acceptance criteria using the IGC code and Korean Register [6] will be used. Verification of the shell thickness of the type-C tank also performed using ABAQUS FEA [7].

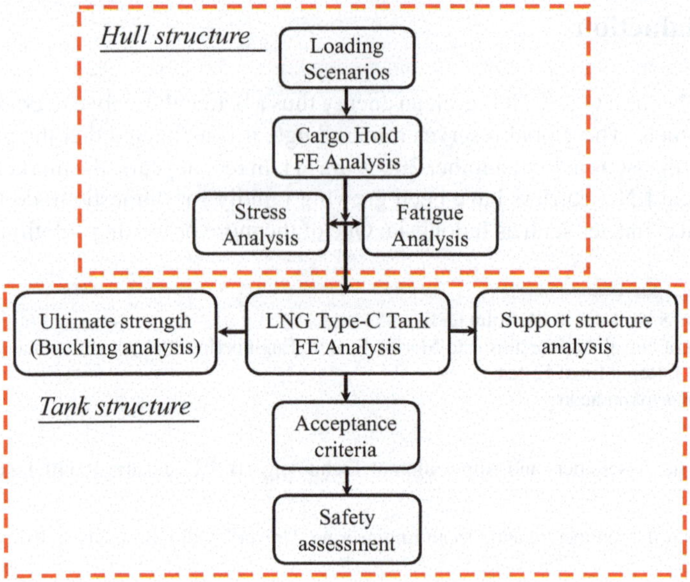

Fig. 1 Flow chart of structural safety assessment of type-C LNG tank of the bunkering ship

2 Strength Evaluation of Type-C Independent Tank

The regulation used in this work is the Korean Register international code for the construction and equipment of ships carrying liquefied gases in bulk (IGC Code) where it originates from the International Maritime Organisation (IMO) within IGC Code. The detailed design parameter of the bunkering ship and the LNG cylindrical tank are shown in Table 1.

Figure 2 shows the general arrangement of a 500 cbm bunkering ship which was used as the subject of this study. The ship is characterized by typical Ulstein X-bow and 23.25 m length of cargo hold where the tank was located. Overall the ship has 45.65 m length and the full load draught is 2.5 m.

2.1 *Cryogenic Material*

It is noted that the material that was considered in this study is specialized to be used for cryogenic applications. Recently, an amendment to the IGC codes has been made to include high manganese austenitic steel for cryogenic service. Therefore, the high

Table 1 Design parameter of 500 cbm bunkering ship and the LNG tank

Ships parameter		Tank parameter	
Length of ship	45.65 m	Diameter	5.55 m
Breadth	12.40 m	Length	21.75 m
Depth	4.50 m	LCG from midship	1.55 m
Draft	2.50 m	TCG from midship	0.00 m
Block coefficient	0.74	VCG from base	4.95 m
Service speed	8.00 knots	Material	High manganese
		Yield strength	400 MPa
		Tensile strength	660 MPa

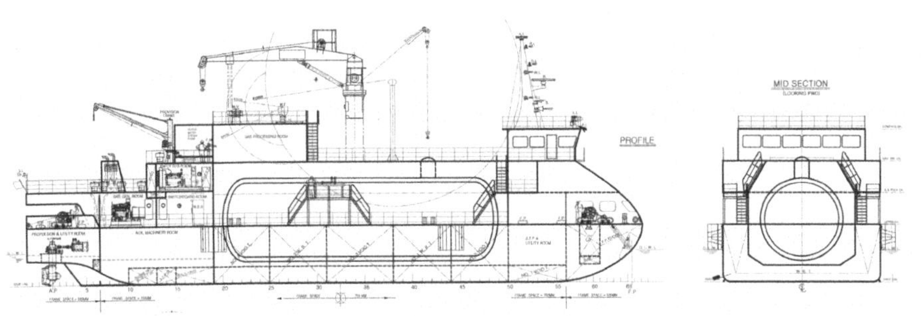

Fig. 2 Schematic illustration of 500 cbm bunkering ship

Table 2 Chemical composition of high manganese steel in weight percentage

C	Si	Mn	P	S	B	N	Nb + Ti + V
0.2–0.6	0.1–1.0	22.0–26.0	<0.03	<0.01	0–0.01	0–0.15	0–0.3

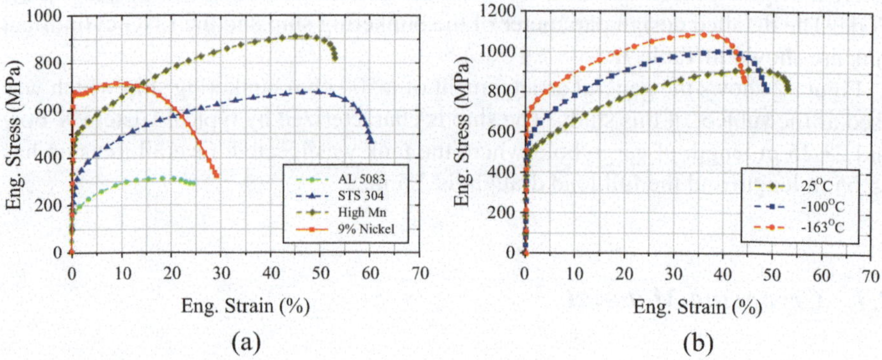

Fig. 3 The tensile test result, **a** results at room temperature for various material, **b** high Mn under cryogenic condition [9]

manganese austenitic steel was chosen considering its high yield strength under the cryogenic environment. Table 2 shows the weight percentage of the material composition of high manganese steel according to Korean Standard (KS-D-3031) high manganese austenitic steel plates for pressure vessels for low-temperature service. Figure 3 shows the yield strength and ultimate tensile strength of base material High Mn steel compared with other cryogenic materials, the variation to temperature was also given. The yield and tensile value which shown in Table 1 are the properties standard of High Mn "as-welded condition" [8].

From the above results, it was discovered that the high manganese material can sustain high elongation of 80% with the highest tensile strength compared to other cryogenic material. Therefore, the application of high manganese can be submitted for cargo tank and fuel tank for the carriage of LNG.

2.2 *Design Steam Pressure* (P_0) *and Accelerated Liquid Pressure* (P_{gd})

According to the IGC rule [2, 6], type-C Independent tank should comply to the pressure vessel standard, and the design stem pressure is larger than the value calculated by Eq. (1), as follows

$$P_0 = 0.2 + AC(\rho r)^{1.5} \quad (1)$$

$$A = 0.00185\left(\frac{\sigma_m}{\Delta\sigma_A}\right)^2 \quad (2)$$

Here, P_0 is the design steam pressure, C is a characteristic tank dimension which has to be taken as the greatest value from the size of tank height h, $0.75b$ width in transverse ship direction or $0.45l$ length of the tank in ship's longitudinal direction; ρr is the relative density of the LNG to water at design temperature (0.5). σ_m is the allowable stress of material (MPa) which has to be smaller than two following value, R_m/A and R_e/B. R_m is the specified minimum tensile strength of the material, 660 MPa; R_e is the specified minimum yield strength of material, 400 MPa; A and B is the fitness value of manganese steel, 3 and 5, respectively. $\Delta\sigma_A$ is the allowable dynamic membrane stress corresponds to double amplitude at probability level of 10^{-8} which equals to 55 MPa for ferritic-perlitic, martensitic and austenitic steel. It is calculated from Eqs. (1) and (2). The design steam pressure is 0.29 MPa.

The internal design pressure, P_{eq} is the sum of steam pressure P_0 and internal accelerated liquid pressure, P_{gd}, where it is created as a result of acceleration on the centre of gravity of the cargo due to the motions of the ship. This explained parameter should be calculated as follows:

$$P_{gd} = \alpha_\beta Z_\beta\left(\frac{\rho}{1.02\times 10^5}\right) \quad (3)$$

However, in the calculation of P_{gd} the effects of liquid sloshing are not included. From Eq. (3), The internal liquid pressure from the resultant dimensionless acceleration was produced by gravity and motion in arbitrary direction, α_β, and liquid height at arbitrary direction Z_β (m); where ρ is the density of the cargo at design temperature.

In order to solve Eq. (3), guidance formulae for acceleration components are needed. In the KR rules the acceleration components which corresponds to the probability of exceedance of 10^{-8} in the North Atlantic are given. Values of α_x, α_y, α_z depend on the ship and tank dimension. The value of α_x are considered as longitudinal acceleration produced by surging, pitching motion and gravity component of pitching; α_y the transverse motion produced by swaying, yawing, rolling, and gravity component of rolling; α_z vertical motion acceleration by heave, pitch and possibly rolling. These acceleration components can be calculated as follows,

$$\alpha_x = \pm\alpha_0\sqrt{0.06 + A^2 - 0.25A} \quad (4)$$

$$\alpha_y = \pm\alpha_0\sqrt{0.6 + 2.5\left(\frac{x}{L_0} + 0.05\right)^2 + K\left(1 + 0.6K\frac{z}{B}\right)^2} \quad (5)$$

$$\alpha_z = \pm\alpha_0\sqrt{1 + \left(5.3 - \frac{45}{L_0}\right)^2\left(\frac{x}{L_0} + 0.05\right)^2\left(\frac{0.6}{C_b}\right)^{1.5} + \left(\frac{0.6yK^{1.5}}{B}\right)^2} \quad (6)$$

Table 3 Results of dimensionless acceleration

Acceleration components	
Vertical acceleration, α_z	0.73
Transverse acceleration, α_y	0.95
Longitudinal acceleration, α_x	0.46

where

$$\alpha_0 = 0.2\frac{V}{\sqrt{L_0}} + \frac{34 - \left(\frac{600}{L_0}\right)}{L_0} \tag{7}$$

$$A = \left(0.7 - \frac{L_0}{1200} + 5\frac{z}{L_0}\right)\left(\frac{0.6}{C_b}\right) \tag{8}$$

x is the longitudinal distance (m) from amidships to centre of gravity of the tank, (+) indicates forward to amidships, (−) is located after of amidships, y is transverse distance (m) from centerline to the centre of gravity of the tank, z is vertical distance (m) from the ship's actual waterline to the centre of gravity of tank, where (+) above and (−) below the waterline. L_0 is the length of the ship (m), B is the greatest moulded breadth of the ship (m), K is constantly equal to 1.0, V is the service speed (knot), C_b is the block coefficient. Table 3 contains results of dimensionless acceleration.

Having finished with the considered case of the dimensionless acceleration value, the resultant acceleration can be acquired to estimate the internal accelerated liquid pressure (P_{gd}). The values of α_β can be determined according to Fig. 2, where β is the arbitrary degree. Following from Ref. [10] the corresponding parameters of the resultant acceleration of Fig. 4 are obtained. For the longitudinal direction β and α_β are to be determined with α_x substituted for α_y.

$$\beta\text{max} = \arctan\frac{\alpha_y}{\sqrt{(1 - \alpha_z 2)}} \tag{9}$$

$$\alpha_\beta = \frac{\cos\beta\alpha_y^2 + \alpha_z\alpha_y\sqrt{\cos^2\beta\alpha_y^2 + \sin^2\beta\alpha_z^2 - \sin^2\beta}}{\cos^2\beta\alpha_y^2 + \sin^2\beta\alpha_z^2} \tag{10}$$

2.3 *Summaries of the Calculated Design Pressure and Scantling*

The main structural member of the LNG type C tanks consists of the cylindrical shells and dished end. The initial scantling calculation can be determined using boiler and pressure vessels rule [11] as given below,

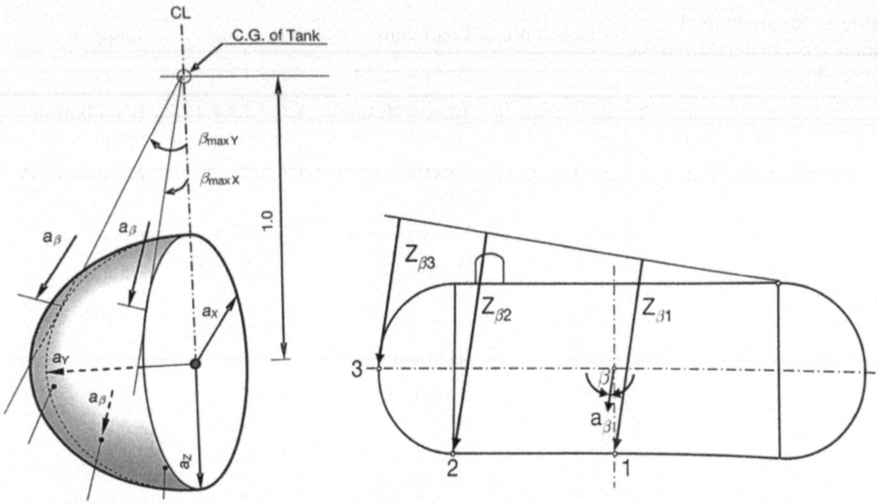

Fig. 4 Determination of the acceleration and liquid height [6]

$$t_{shell} = \frac{PD}{2fJ - 1.2P} + 1 \quad (11)$$

$$t_{sphere} = \frac{PR}{2fJ - 0.2P} + 1 \quad (12)$$

where P is design pressure (MPa), f is the allowable membrane stress (MPa), J is the welding efficiency (0.45–1.00). However, the value of safety factor shall be considered as the thickness correction factor, and commonly taken as 1.5–2.0. As discussed in the previous subsection, the calculated parameters are summarized in Table 4.

Table 4 Summaries of the calculated parameter

Design pressure parameter (MPa)	
Design steam pressure, P_0	0.293
Accelerated liquid pressure, P_{gd}	0.040
Total internal design pressure, $P_{eq} = P_0 + P_{gd}$	0.333
Design ext pressure, P_e	0.045
Total ext pressure, $P_t = P_{eq} - P_e$	0.288
Cylindrical thickness after correction, t_{shell}	14 mm
Spherical dished-end thickness after correction, t_{sphere}	20 mm

Table 5 Required load parameters obtained from the IGC code

Load no.	Load item	IGC code	Input
1.	Body weight	Ch.4.12.1	1.0 G
2.	Temperature	Ch.4.13.4.1	LNG boiling temp
3.	External pressure	Ch.4.13.3	$P_{ext\ total} = P_0 + P_{gd} - 0.045$
4.	Vapor pressure	Ch.4.23.1.2	P_o
5.	Hydrostatic pressure	Ch.4.23.6.1	1.5 P_o
6.	Vertical acceleration	Ch.4.14.1.2.1	αz
7.	Transverse acceleration	Ch.4.14.1.2.2	αy
8.	Longitudinal acceleration	Ch.4.14.1.2.3	αx
9.	Collision load	Ch.4.15.1	(+) 0.5 G, x-direction
			(−) 0.25 G, x-direction
10.	Load due to flooding	Ch.4.15.2	$P_{hyd} = \rho$ gh
11.	Static heel loads (0–30°)	Ch.4.13.9	Same as load no. 1–10

3 Load Parameter

Due to their special characteristic of LNG tank, e.g., high pressures and low temperature, special attention needs to be made. There are several loads which need to be considered in accordance with IGC Code, such as permanent, functional, environmental, and accidental loads. Table 5 shows required some of the load parameters for evaluating the local strength of the Type C tank. However, in this study only the first four numerical results presented in the forthcoming section.

4 Finite Element Model

For the structural strength evaluation purpose, the finite element model has been constructed by considering various loading conditions. The FE model consists of a full length of cargo hold section, the LNG tank and its subdivision (swash bulkhead, ring stiffeners, ellipsoidal head, bottom, and saddle support structure).

The constraints are applied in defining the saddle to tank connection. Two saddle-support structures were used. The fixed saddle is applied at the front where the sliding connection on the rear. Figures 5 and 6 show the finite element model of 500 cbm bunkering ship corresponding to their thickness (mm). The FE model is generated

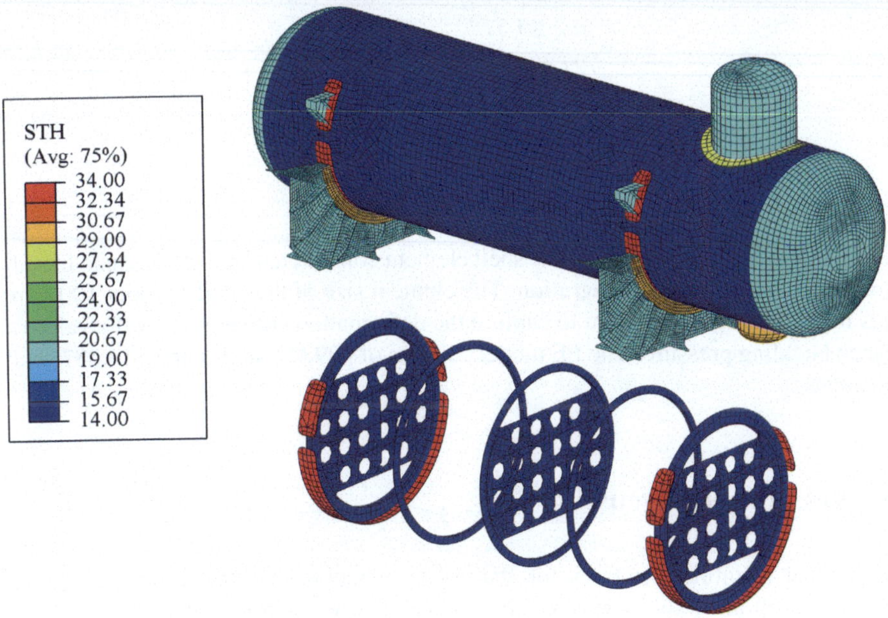

Fig. 5 A FE model for LNG type C tank

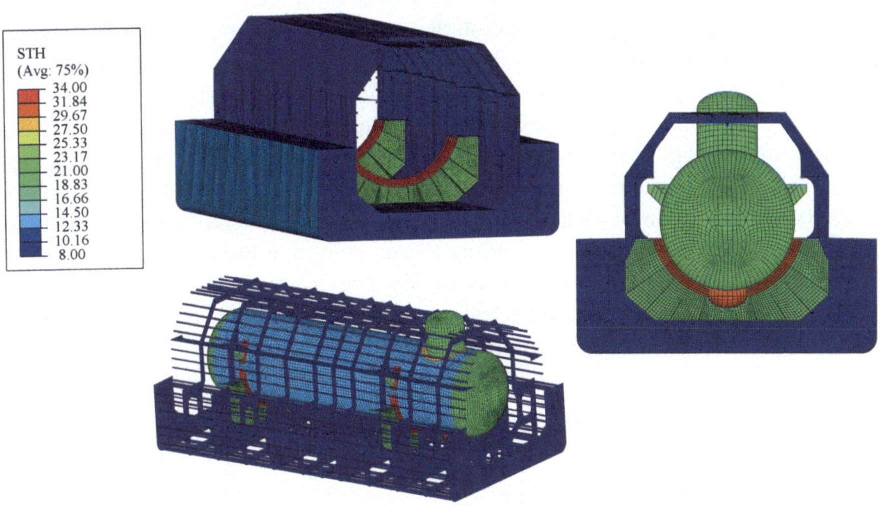

Fig. 6 A global FE model consisted of the hull and cylindrical tank

Table 6 Strength acceptance criteria

Strength criteria
$f = 220$ MPa
$\sigma_L \leq 1.5f$
$\sigma_b \leq 1.5f$
$\sigma_L + \sigma_b \leq 1.5f$
$\sigma_m + \sigma_b \leq 1.5f$
$\sigma_m + \sigma_b + \sigma_g \leq 3.0f$
$\sigma_L + \sigma_b + \sigma_g \leq 3.0f$

using S4R type shell elements. The shell element constructed by 4-node, quad literal-dominated with reduced integration. The element size of the shell is about 100 mm. this mesh density is sufficient to capture the deformation converging by the value of Eigen buckling pressure. The FE model consists of 309,625 nodes and 316,287 shell elements.

5 Results and Discussion

Equivalent membrane stresses for shell/plate elements from FE results are to be checked with allowable stresses. The allowable stresses limit related to the base material is applicable for the main structure and the LNG tank. The results are mainly used for checking yielding of the hull structure and the interaction between the hull and the saddle supports. The details of the strength limit are given in Table 6; where σ_m equivalent to primary general membrane stress; σ_L is primary local membrane stress; σ_b equivalent to primary bending stress; σ_g equivalent to secondary stress.

Results from the analyses covered the loads for structure-weight, external pressure, acceleration, etc., as presented in Table 5. The sample FE results from load case 1–4 presented in Fig. 7, the yielding check was done for all components including the tank shell, hull structure, and two saddle supports. The maximum stress occurs at the fixed connection of saddle and tank. In this case, special attention needs to be made in this area. However according to the strength criteria, the local primary membrane stress and the bending stress are less than 1.5 times the allowable stress. It can be concluded that the current design of LNG Type C tank and its subdivision structures fulfilled the strength requirement in the IGC Code.

6 Conclusions

In this study, the methodology and procedure of strength evaluation on the LNG Type C independent tank of bunkering ship have been reviewed. The assessment includes

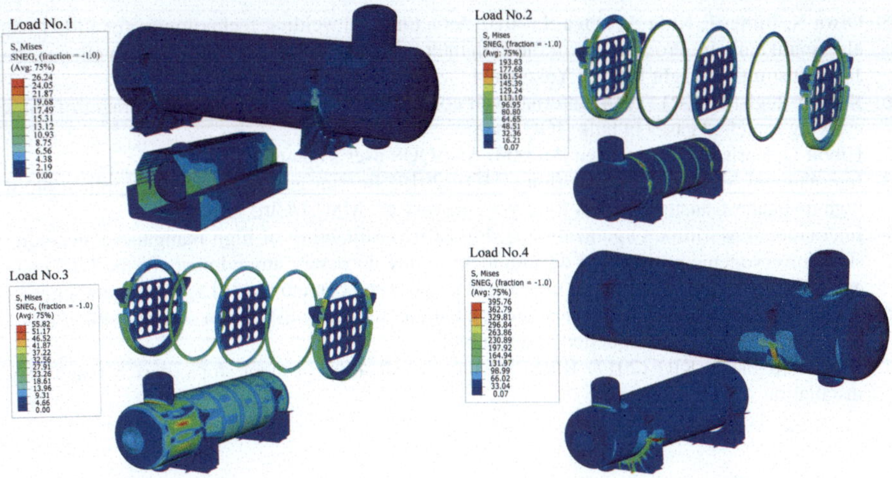

Fig. 7 Stress distribution from load case no. 1–4

the global cargo hold structure and the tank itself. The presented example of ship parameter and tank illustrates the application of 500 cbm bunkering ship.

The procedure provided in this study, evaluating numerically the strength limit according to the required assessment from the IGC Code. The FE results so far have demonstrated that the yielding requirements are verified for the tank and its support structures. FE results show that the critical pressure is less than 1.5 times the allowable stress. Therefore the case study which has presented may be useful in designing similar structure.

Acknowledgements This work is a combined result of "Development of LNG Bunkering System for Coastal Trading LNG-fueled Ships (PMS4150)" and "Development of LNG Bunkering Operation Technologies based on Operation System and Risk Assessment" supported by the Ministry of Ocean and Fisheries in South Korea.

References

1. International Gas Union (IGU) (2018) 2018 world LNG report, 2018 edn. In: 27th world gas conference
2. International Maritime Organization (IMO) (2016) Resolution MSC.370 (93) international code for the construction and equipment of ships carrying liquefied gases in bulk, IGC code IMO, 2016 edn
3. American Bureau of Shipping (ABS) (2006) Guidance notes on strength assessment of membrane-type LNG containment systems under sloshing loads
4. Wang B, Kim JW (2007) Strength evaluation of LNG containment system considering fluid-structure interaction under sloshing impact pressure. In: Proceedings of the 26th international conference on ocean, offshore and Arctic engineering OMAE. ASME, San Diego

5. Oiwa N, Iijima T, Kida A, Ohga S (2010) Structure and welding technologies for SPB LNG aluminum tank. In: Proceedings of the 12th international conference on aluminium alloys. The Japan Institute of Light Metals, Yokohama
6. Korean Register (KR) (2014) International code for the construction and equipment of ships carrying liquefied gases in bulk (IGC code)
7. Hibbit D, Karlsson B, Sorensen P (2000) ABAQUS user's manual. USA
8. International Maritime Organization (IMO) (2019) Interim guidelines on the application of high manganese austenitic steel for cryogenic service, MSC.1/Circ.1599
9. International Maritime Organization (IMO) (2016) Suitability of high manganese austenitic steel for cryogenic service and development of any necessary amendments to the IGC code and IGF code, sub-committee on carriage of cargoes and containers, 3rd session agenda item 8
10. Llyod Register (LR) (2013) Rules and regulations for the construction and classification of ships for the carriage of liquefied gases in bulk
11. Korean Register (KR) (2017) Rules for the classification of steel ships, part 5 machinery installation

Gas Dispersion Analysis on the Open Deck Fuel Storage Configuration of the LNG-Fueled Ship

Haris Nubli, Aditya Rio Prabowo, and Jung Min Sohn

Abstract The LNG (Liquefied Natural Gas) fueled ship is one of the solution to solve the emission of SO_x and NO_x in conventional maritime fuel. Despite the advantages, the LNG fuel is more dangerous due to their low flash point. The leakage of gas fuel is one of the threat of LNG-fueled vessel, due to its explosive characteristic. This study aims to establish the exclusion zone due to LNG dispersion. The analysis contained the potential leakage and gas dispersion along to the exposed area. A verified Computational Fluid Dynamics (CFD) analysis based on the time accuracy assumption using KFX software was performed to simulate the dynamic behavior of the vapor dispersion. The scenarios were categorized by the leak size, leak location, and mass flow rate. The required environment parameters, e.g., wind speed and wind directions were also considered. It is found that the behavior of dispersed vapor was strongly depending on the leak location and mass flow rate. Surprisingly, the environment parameter has slightly affected the distance of the vapor dispersion. According to these findings, to deliver risk mitigation in a gas leakage accident, the exclusion zone was provided.

Keywords LNG-fueled ship · Gas dispersion analysis · Computational fluid dynamics · Exclusion zone

H. Nubli · J. M. Sohn (✉)
Interdisciplinary Program of Marine Convergence Design, Pukyong National University, Busan 48513, Republic of Korea
e-mail: jminz@pknu.ac.kr

A. R. Prabowo
Department of Mechanical Engineering, Universitas Sebelas Maret, Surakarta 57126, Indonesia

J. M. Sohn
Department of Naval Architecture and Marine System Engineering, Pukyong National University, Busan 48513, Republic of Korea

U. Sabino et al. (eds.), *Proceedings of the 6th International Conference and Exhibition on Sustainable Energy and Advanced Materials*, Lecture Notes in Mechanical Engineering,
https://doi.org/10.1007/978-981-15-4481-1_11

1 Introduction

In the energy sector market, LNG is one of the fastest-growing commodity. The growth in LNG demand has been about 8% per year over the past decade [1]. According to the International Energy Agency in 2017, the forecast of LNG consumption in the marine sector will reach 47 million tons per annum in 2040. Due to the high demand for LNG fuel, assumed the demand for new building ships will reach about 4000 of LNG fueled ships in 2030 [2].

LNG (liquefied natural gas) is an alternative fuel for shipping industries, instead of the heavy fuel oil and marine diesel oil. LNG fuel has cleaner exhaust gas, where free from SO_x and PM (particulate matter) [3]. The policies of environmental protection become strict. Since the International Maritime Organization issued the Regulations 19–23 in MARPOL Annex VI, allowed the emission of NO_x amount per engine power unit only 75% for Tier II standard and also in the ECA (emission control area) was mandatory to use LNG fuel for Tier III standard in North America and Europe [4]. Therefore, to respond to the clean energy challenge, the LNG-fueled ship is one of the solutions for shipping industries.

To ensure the safety of the personnel, it is necessary to investigate the risk of fuel gas supply system for LNG fueled ship. It recommends to establish the safety zone to restrict the personnel into the hazardous area. The safety zone is determined as the result of the calculation of a gas dispersion distance [5].

There are several previous investigations of exclusion zone for natural gas dispersion. An investigation of the natural gas leakage of LNG-fueled ship was performed by Li et al. [6]. They were distinguished some events according the type of the equipment. The numerical analysis of natural gas dispersion was performed using CFD method. The results of gas dispersion can used to determine the dangerous zone. Besides the environmental conditions, the geometric shape and size also affecting a dispersed gas. This examination was brought by Park et al. [7]. Where the ship to ship bunkering activities was brought with various models of the ship types. Also, the CFD simulation was used as the numerical method of natural gas dispersion. They were determines the safety zone from the gas dispersion results.

In this context, the objective of this study is to develop the method of the natural gas dispersion. A hypothetical LNG fueled ship was used in this investigation, with the open deck configuration of FGSS (fuel gas supply system) and LNG fuel storage. It is also considering the probability of equipment failure frequencies which represent as leak size and mass flow rate of gas leakage. The environmental conditions also considered e.g. wind direction and wind speed. The validated CFD (computational fluid dynamics) based software Kameleon FireEx was used as a numerical tools for dispersion analysis. Thus, the exclusion zone can determined.

2 Methodology and Data Arrangement

It is an important step to collect the data which was affected to the natural gas dispersion. The collected data was used for setting a scenario and performs a random sampling method for data selections. The selected data was used as the variable for CFD simulations. Then, the results can be plotted to determine the exclusion zone. The framework of this study is shown in Fig. 1.

2.1 Initial Leak Frequencies

Each part or equipment on the system have the different characteristic to fail, it can be mentioned as the initial leak frequencies. The initial leak frequencies represented as a probabilistic frequency of equipment to leak per years. The historical data of initial leak frequencies were collected from the UK HSE for "Failure Rate and Event Data for use within Land Use Planning, Risk Assessments" and Baker Risk's risk analysis database is shown in Table 1 [8, 9].

Fig. 1 A study framework of the natural gas dispersion analysis

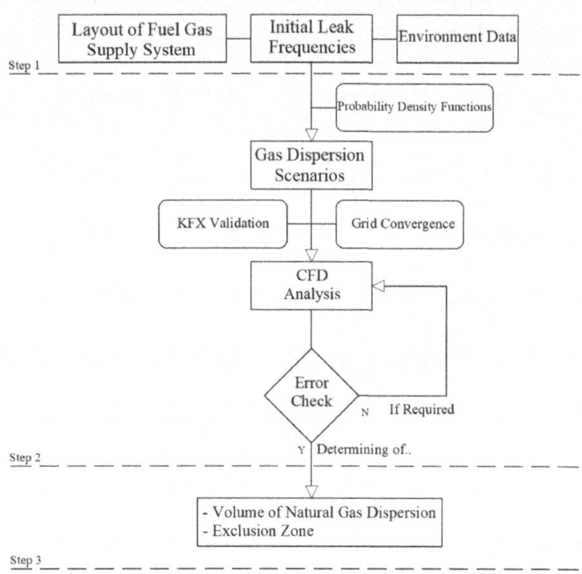

Table 1 The initial leak frequency each equipment [8, 9]

Equipment	Initial leak frequency			
	Vapor release	Minor failure	Major failure	Catastrophic
Atmospheric tank	5.E−03	1.E−03	1.E−04	2.E−05
Heat exchanger	1.E−03	2.E−04	4.E−05	2.E−05
LNG vessel	4.E−05	3.E−06	1.E−06	5.E−08
Pressure vessel	2.E−03	3.E−03	1.E−04	5.E−06
Compressor	5.E−02	3.E−03	0.E+00	5.E−04
Vaporizer	1.E−03	2.E−04	4.E−05	2.E−05
Pipe	$3E{-}6/D_p \times (D_h/25.4)$; D_p = pipe diameter, D_h = hole diameter (mm)			

2.2 Layout and Leak Locations

The hypothetical LNG fueled ship of the Oil Tanker was used in this study. The fuel gas supply system was located on the main deck, in front of the superstructures. The ship model is shown in Fig. 2a.

The fuel gas supply system in this study is featured with a re-liquefaction system for boil-off gas instead of burning it off during the voyage [10]. Thus, the FGSS branches off into the direct flow, which converted to be a vapor in a vaporizer before consumed by the engines. The boil-off gas should pass the condenser to turn into liquid again and flow to the engine through the flash tank [11]. The FGSS layout detail is shown in Fig. 2b.

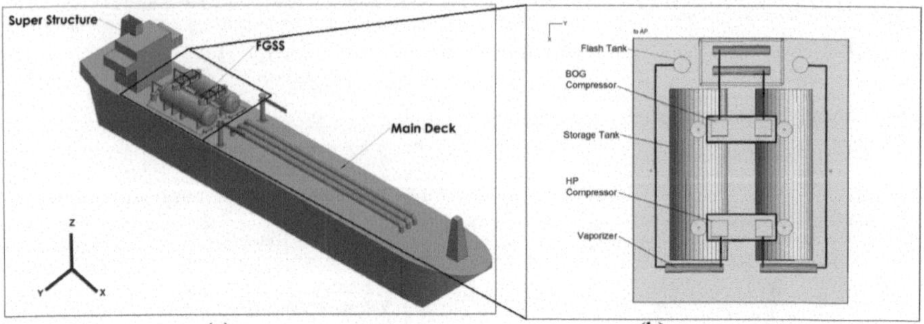

Fig. 2 A 3D model of LNG fueled ship (**a**), layout of FGSS (**b**)

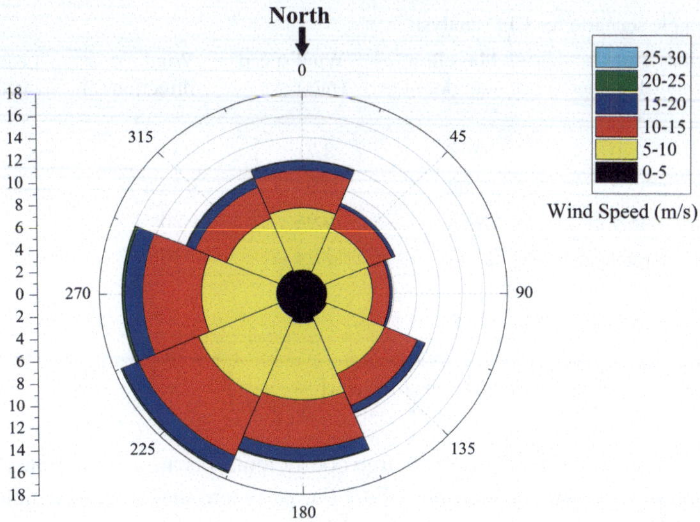

Fig. 3 The wind-rose, located in North Sea [12]

2.3 *Environment Conditions*

The effects of wind were considered in this study. The European North Sea area was selected as the location of the gas leak accident. The wind data consisted of the wind direction and wind speed, the obtained wind-rose is shown in Fig. 3.

2.4 *Exclusion Zone*

The area covered with natural gas cloud defined as the exclusion zone. For safety purpose, the exclusion zone was defined as exposed to the half of LFL (lower flammability limit) above [13]. According to DNV-GL rules, the gas detection system shall be activated at 30% of LFL [14]. Thus, the volume fraction contour consisted of 30%LFL, 50%LFL, LFL, and the UFL (upper flammability limit). As given the natural gas LFL is 5% concentration of natural gas per total volume of air, while UFL is 15% [15].

3 Selected Gas Dispersion Scenario

Initial leak frequency data for the pipe, it only considers the hole diameter. Then the mass flow rate can be predicted by a function which related to hole size and fluid in compression. The function is shown in Eq. (1) [16],

Table 2 Sample scenario for CFD analysis

No. scenario	Hole diameter (mm)	Mass flow rate (kg/s)	Wind speed (m/s)	Wind direction (°)	Leak direction
1	6.0	0.17	8.71	237	X
2	39.3	7.09	8.9	185.8	−Y
3	45.9	9.68	9.96	20	−X
4	69.9	22.43	12.24	159	Y

$$\dot{m} = C_d A \sqrt{\gamma \rho_0 P_0 \left(\frac{2}{\gamma + 1}\right)^{\frac{\gamma+1}{\gamma-1}}} \quad (1)$$

where, $\dot{m}$ is mass flow rate, C_d is discharge coefficient, A is area of the hole diameter, P_0 is a discharge pressure which has 3E+7 Pa, ρ_0 is real gas density which defined 511.01 kg/m^3, and γ is heat capacity ratio of gases which has 1.32.

For considering the probability of the event. The total selected scenario was 100 number. However in this study, there were 4 selected scenario which included all leak direction.

The selected parameter is shown on Table 2.

4 Results and Discussion

4.1 Grid Convergence

The grid convergence test was performed to determine the proper size of the grid. The input value for grid convergence test has same for each grid number. The hole diameter is 15.7 mm, for mass flow rate is 1.13 kg/m^3, for wind direction and wind speed has 197.7° and 6.8 m/s respectively. The absolute velocity captured after performing a CFD simulation, this value used as a comparison for each grid size. The grid sizes ranged from 2E+5 to 1E+6 cells. The increasing grid size shows that the absolute velocity becomes constant in 4E+5 cells. However, 6E+5 cells were used due to this cell number had a difference percentage of 0.11 compared with the finest grid. The grid size comparison is shown in Table 3.

4.2 Vapor Dispersion

The vapor dispersion distance was measured with 50%LFL, consisted of length (a), width (b) and height (c). With this three-line, the area of exclusion zone defined as an

Table 3 Result of grid convergence test

Grid number	Absolute velocity	Difference (%)
2.00E+05	0.5800	4.93
4.00E+05	0.5523	0.16
6.00E+05	0.5520	0.11
8.00E+05	0.5524	0.18
1.00E+06	0.5514	–

equivalent ellipse in the XY plane, XZ plane and YZ plane. The distance and area of vapor dispersion was shown in Table 4. The vapor cloud movement affected by the wind speed and direction, also it was depending on the leak direction each scenario. The contour plot of natural gas dispersion was shown in Figs. 4, 5, 6 and 7, the color of the contour has a value of the mass fraction of natural gas in percent of LFL. It has been 30%LFL, 50%LFL, LFL, and UFL or 1.5%, 2.5%, 5%, and 15% respectively in the percentage of total volume.

The CFD analysis has performed with the steady-state flow. The simulation stopped until the gas volume, velocity and fuel balance plot had a constant value.

Table 4 Defined exclusion zone for total scenario

No. scenario	Distance of gas cloud (m)			Area of gas cloud (m^2)		
	Length (a)	Width (b)	Height (c)	XY area	XZ area	YZ area
1	4.67	0.69	0.26	10.17	3.83	0.57
2	26.43	16.19	8.76	1343.92	726.59	445.19
3	12.07	7.64	10.56	289.50	400.01	253.32
4	38.99	84.67	11.73	10,365.28	1435.81	3117.90

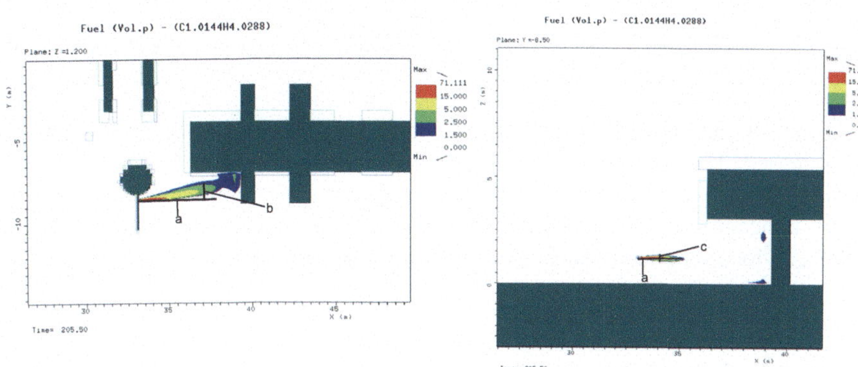

(a) (b)

Fig. 4 The contour plot of XY plane view of scenario 1 (**a**) and contour plot of XZ plane view (**b**)

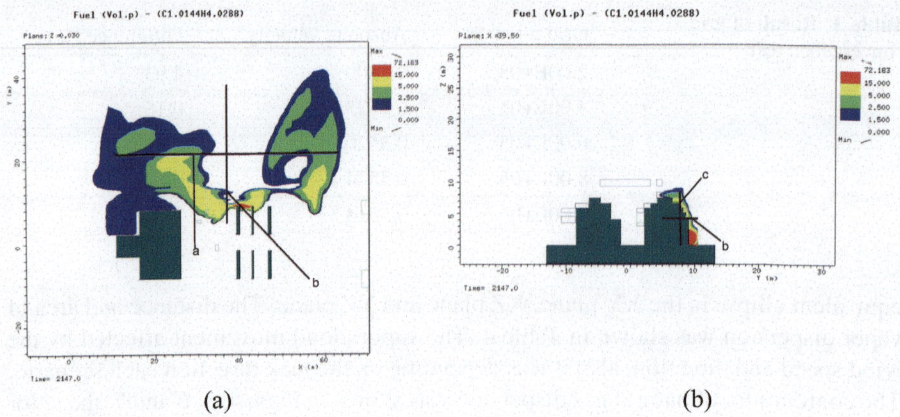

Fig. 5 The contour plot of XY plane view of scenario 2 (**a**), and contour plot of YZ plane view (**b**)

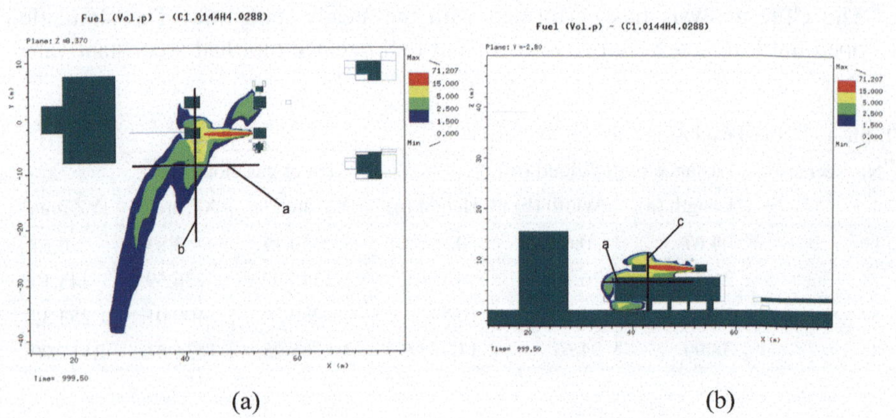

Fig. 6 The contour plot of XY plane view of scenario 3 (**a**), and contour plot of XZ plane view (**b**)

Therefore, the amount of the natural gas vapor volume was not depending on the time-released. The volume of releasing vapor was shown in Table 5.

The CFD software captured the volume of vapor released within the range of LFL to UFL concentration. The accumulation of vapor cloud was significantly affected by the leak location, due to each location has a different obstacle.

5 Conclusion

This paper propose a method of natural gas dispersion on the FGSS of LNG fueled ship and setting the exclusion zone. The layout of FGSS, initial leak frequencies,

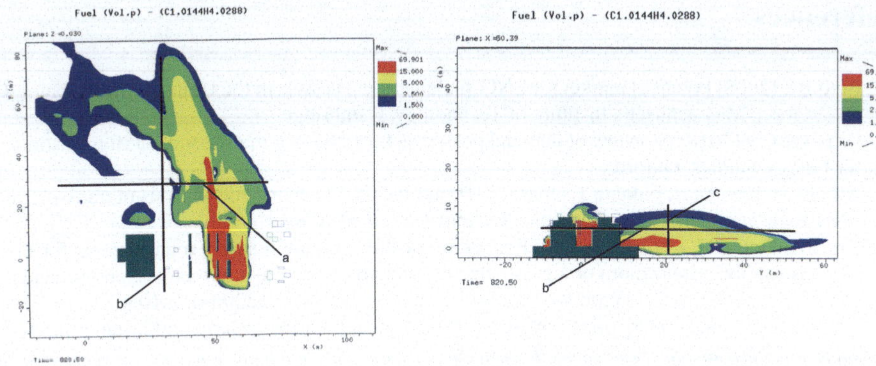

(a) (b)

Fig. 7 The contour plot of XY plane view of scenario 4 (**a**), and contour plot of YZ plane view (**b**)

Table 5 Volume of released vapor

No. scenario	Leak coordinates (m)			Vapor volume at LFL (m^3)
	X	Y	Z	
1	33.10	−8.50	1.20	0.26
2	39.50	10.40	0.0	872.40
3	52.12	−2.80	8.37	318.01
4	50.39	−10.40	0.03	4749.56

and environment data are the main parameter to establish each scenario. Those data can determine the leak locations, rate of leak flows, wind speed and wind direction respectively. Prior to performing the CFD analysis, the grid convergence test was performed to obtain the reasonable results. In this study, the 6.0E+5 of grid size was used due to considering the simulation time. The difference between medium and fine grid size were quite small.

The vapor dispersion has different behavior due to effects of the relation of those parameters. Scenario 2 has a larger vapor volume than scenario 3, however the mass flow rate of scenario 3 is higher than scenario 2. It is due to scenario 3 located on the $Z = 8.37$ m, the highest leak location where there is no obstacle. Thus, the vapor cloud was not accumulated.

Acknowledgements The first author gratefully acknowledge the financial support and publishing with the grant from BK21 plus MADEC Human Research Development Group, South Korea. This work was supported by the National Research Foundation of Korea (NRF) grant funded by the Korea government (MSIT) (No. NRF-2018R1C1B5085449).

References

1. Brett BC (2008) Potential market for LNG-fueled marine vessels in the United States. Doctoral dissertation, Massachusetts Institute of Technology, Cambridge
2. Le Fevre CN (2018) A review of demand prospects for LNG as a marine fuel. Oxford Institute for Energy Studies, Oxford
3. Adachi M, Kosaka H, Fukuda T, Ohashi S, Harumi K (2014) Economic analysis of trans-ocean LNG-fueled container ship. J Mar Sci Technol 19(4):470–478
4. Kim JH, Doh DH, Choi BC (2018) Evaluation of the ventilation safety requirements for the fuel gas supply system room of a gas-fueled vessel: simulated leaks of methane and propane. J Mech Sci Technol 32(11):5521–5532. https://doi.org/10.1007/s12206-018-1050-7
5. International Standard Organization (2015) 18683: guidelines for systems and installations for supply of LNG as fuel to ships. ISO, Geneva
6. Li XJ, Zhou RP, Konovessis D (2016) CFD analysis of natural gas dispersion in engine room space based on multi-factor coupling. Ocean Eng 111:524–532. https://doi.org/10.1016/j.oceaneng.2015.11.018
7. Park S, Jeong B, Yoon JY, Paik JK (2018) A study on factors affecting the safety zone in ship-to-ship LNG bunkering. Ships Offshore Struct 13:312–321. https://doi.org/10.1080/17445302.2018.1461055
8. Health and Safety Executive (2017) Failure rate and event data for use within land use planning risk assessments. UK HSE, Bootle
9. Moosemiller M (2011) Development of algorithms for predicting ignition probabilities and explosion frequencies. J Loss Prev Process Ind 24(3):259–265. https://doi.org/10.1016/j.jlp.2011.01.012
10. Shin Y, Lee YP (2009) Design of a boil-off natural gas re-liquefaction control system for LNG carriers. Appl Energy 86(1):37–44. https://doi.org/10.1016/j.apenergy.2008.03.019
11. MAN BandW (2014) Project guide: MAN BandW S70ME-C8.2-GI-TII. MAN Diesel and Turbo, Copenhagen
12. Meteoblue. https://www.meteoblue.com/en/products/historyplus/windrose/61.36N-3.24E0_UTC. Accessed 22 Aug 2019
13. Fu S, Yan X, Zhang D, Li C, Zio E (2016) Framework for the quantitative assessment of the risk of leakage from LNG-fueled vessels by an event tree-CFD. J Loss Prev Process Ind 43:42–52. https://doi.org/10.1016/j.jlp.2016.04.008
14. Veritas DN, Lloyd G (2018) Rules for classification: ships, part 5, chap 7. In: Liquefied gas tankers. DNV-GL, Høvik
15. Cashdollar KL, Zlochower IA, Green GM, Thomas RA, Hertzberg M (2000) Flammability of methane, propane, and hydrogen gases. J Loss Prev Process Ind 13(3–5):327–340
16. CCPS (1999) Guidelines for chemical process quantitative risk analysis, 2nd edn. Wiley, New York

Rheological Properties of Magnetorheological Elastomer Using Cobalt Powder as Filler

Afiq Azri Zainudin, Kamal Hafiz Khalid, Siti Aishah Abdul Aziz, Saiful Amri Mazlan, Nur Azmah Nordin, Hafizal Yahaya, and Abdul Yasser Abd Fatah

Abstract Carbonyl iron particles (CIP) have been widely used as magnetic particles in magnetorheological elastomer (MRE) fabrication. This kind of magnetic particle exhibits high magnetic saturation however low conductivity. Therefore, in this study, cobalt particle has been introduced as filler in MRE fabrication. It is well known that cobalt offered dual properties which are magnetic and electrical properties. The properties offered by the cobalt are believed to enhance the magnetic and rheological properties of the isotropic MRE. Therefore, the isotropic MRE was fabricated using silicone rubber (SR) as the matrix element interspersed with 53 wt% of cobalt powder. The effect of magnetic was experimentally investigated in this study using Vibrating Sample Magnetometer (VSM). Meanwhile, the rheological properties related to the frequency and current sweep were examined using rheometer Anton Paar in the

A. A. Zainudin · K. H. Khalid · S. A. A. Aziz · S. A. Mazlan (✉) · N. A. Nordin · H. Yahaya
Engineering Materials and Structures (eMast) iKohza, Malaysia-Japan International Institute of Technology (MJIIT), Universiti Teknologi Malaysia, Jalan Sultan Yahya Petra, 54100 Kuala Lumpur, Malaysia
e-mail: amri.kl@utm.my

A. A. Zainudin
e-mail: afiqzn@gmail.com

K. H. Khalid
e-mail: kamalh859@gmail.com

S. A. A. Aziz
e-mail: aishah118@gmail.com

N. A. Nordin
e-mail: nurazmah.nordin@utm.my

H. Yahaya
e-mail: hafizal.kl@utm.my

S. A. A. Aziz · S. A. Mazlan
Advanced Vehicle System (AVS) Research Group, Universiti Teknologi Malaysia, Jalan Sultan Yahya Petra, 54100 Kuala Lumpur, Malaysia

A. Y. A. Fatah
Razak Faculty of Technology and Informatics, Universiti Teknologi Malaysia, Jalan Sultan Yahya Petra, 54100 Kuala Lumpur, Malaysia
e-mail: yasser.kl@utm.my

U. Sabino et al. (eds.), *Proceedings of the 6th International Conference and Exhibition on Sustainable Energy and Advanced Materials*, Lecture Notes in Mechanical Engineering,
https://doi.org/10.1007/978-981-15-4481-1_12

absence and presence of the magnetic field. Moreover, the magnetic properties of MRE with cobalt exhibit higher magnetic saturation up to 78.74 emu/g as compared to conventional MRE. In the meantime, the storage modulus of MRE with cobalt depicted an enhancement of field-dependent modulus at all magnetic field applied. Considering that the introduced cobalt as a filler and the obtained results are satisfactory, the introduced study might open new avenues for the cobalt to be used as filler in the MRE fabrication.

Keywords Cobalt · Magnetorheological elastomer · Rheological properties · Magnetic properties

1 Introduction

Magnetorheological (MR) materials are known as smart materials because of their rheological properties can be converted continuously, rapidly and reversibly by applying external magnetic field [1, 2]. Generally, there are four types of MR materials including MR fluid (MRF), MR elastomer (MRE), MR grease (MRG) and MR foam [3]. The most common MR material is MRF which consists of magnetic particles suspended in a non-magnetic fluid, which was discovered by Jacob Rabinow in 1948 [4]. However, MRF has a few drawbacks such as sedimentation and leaking problems that limit its application [5]. Thus, these problems have been solved by switching the fluid matrix to a solid matrix which is known as MRE [6]. MRE commonly consists of rubber as a matrix, and magnetic particles [7]. Many elastomer matrix has been explored and used in the MRE fabrication by previous researchers such as silicone rubber [8] and natural rubber [9]. Silicone rubber has been widely used as the matrix in MRE due to easy fabricated from liquid precursors [10]. Meanwhile, iron and carbonyl iron particles (CIPs) have been generally used as magnetic particles by researchers due to its high saturated magnetization, high permeability, and low remnant. The magnetic particle size used in MRE usually in micrometers specifically between 1 and 10 μm because smaller particles provide a more effective area of interfacial friction involving the particle and matrix. MRE can be classified into two categories, which are isotropic and anisotropic MREs according to the orientation of the particles inside the matrix.

The magnetic particles are randomly dispersed in the isotropic MRE, while in the anisotropic MRE, the magnetic particles are aligned and formed chain-like structures [11]. The isotropic MREs are cured in the presence of a magnetic field while the matrix materials are still in liquid precursors, thus the magnetic particles can move along the direction of the magnetic field. MRE has been used in tunable automotive mounts, vibration absorber and auto-mobile suspensions [12]. Besides CIPs, cobalt also is another suitable magnetic particle that can be used in MRE. In addition, cobalt also exhibits dual properties; magnetic and electrical properties. The research regarding MRE with cobalt is considered rare. Previous research by Tong et al. [13] studied the shape of the cobalt in MRE. In his studies, an anisotropic MRE with two types of shape namely flower-like and spherical shape were fabricated and analyzed. The result showed that the flower-like shape cobalt particles exhibit higher performance of storage modulus up to 0.8 MPa for the frequency sweep test as compared to the spherical shape. However, the rheological and magnetic properties of the previous study are not thoroughly examined. Therefore, this study aims to fabricate and investigate the rheological and magnetic properties of isotropic MRE based cobalt. The magnetic and rheological properties of MRE with cobalt will be examined and discussed in detail.

2 Experimental Procedures

2.1 *Materials*

Cobalt powder contains a 99.8% trace metal basis with an average of 2 μm diameter, obtained from Sigma-Aldrich, USA are used as a magnetic particle. Silicone rubber (SR) NS 625 A is used as the matrix in the MRE fabrication together with the curing agent NS 625 B which both purchased from Nippon Steel Co, Japan.

2.2 *Procedures for Preparing MRE Based Silicone Rubber*

A single MRE was prepared through a mixing process. First, 47 wt% of SR was mixed with 53 wt% of cobalt powder using a mechanical stirrer at 280 rpm until the mixture was visually homogenous. Then, 2% of the curing agent from the total weight percent was added, and the mixture was continuously stirred for another 1 min. The mixture was then poured into the steel mold and cured for 2 h.

2.3 *Magnetic Properties Test*

The magnetic properties of the MRE were investigated using Vibrating sample magnetometer (VSM); MicroSense, USA at room temperature. The Teflon tape was used to secure the MRE sample with 0.07 g that mounted in the VSM sample-holder which will continuously vibrate during the analysis. The magnetic test was subjected to a maximum magnetic field of 1.5 T.

2.4 *Rheological Properties Test*

The rheological test was performed using a rotational rheometer (Physica MCR 302, Anton Paar Company, Austria). The sample used is subjected to a 20 mm diameter and 1 mm thickness, respectively. The sample was placed between a rotating disk and a parallel base plate (PP20/MRD/1T). An oscillatory shear test method was applied for all rheological tests, and the measuring temperature was kept constant at 25 °C. For the frequency sweep test, the strain was kept constant at 0.01% and continues frequency of 0.1–100 Hz. The frequency sweep test was then subjected to the current applied varied from 0 to 5 A. On the other hand, the current sweep test was carried out to analyze the effect of the magnetic field of the MRE samples from 0 to 5 A.

3 Results and Discussion

3.1 *Vibrating Sample Magnetometer*

The magnetic properties of cobalt have been investigated by the VSM. Figure 1 shows the hysteresis loops of the MRE sample that have been measured in the field up to 15,000 Oe. The important magnetic parameters related to magnetic saturation M_S, coercivity H_C, and retentivity magnetization, M_R at room temperature of 25 °C were examined. Table 1 shows a summary of the magnetic performance for the MRE sample prepared in this study.

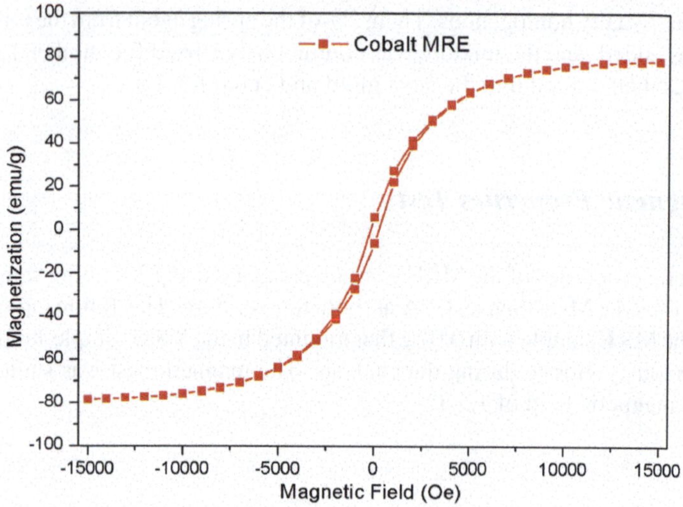

Fig. 1 Magnetization curves at room temperature

Table 1 Magnetic characteristics of MRE sample

Sample	M_S, emu/g	M_R, emu/g	H_C, Oe
Isotropic MRE	78.74	6.09	215.47

Table 1 depicted that the MRE with cobalt as magnetic particles exhibit a magnetic saturation of 78.74 emu/g. The magnetic remanence of the MRE with cobalt is 6.09 emu/g and the coercivity of the isotropic MRE samples is 215.47. This result is parallel with the previous study that showed the MRE with cobalt exhibit increment of remanence which can be seen from Fig. 1 [14].

3.2 *Effect of Frequency*

The results obtained from the frequency sweep test at 0, 2, and 4 A of applied current are presented in Fig. 2. Based on Fig. 2, the storage modulus increases parallel with the increment of frequency and magnetic field. Table 2 depicted the initial and maximum storage modulus of isotropic MRE with cobalt at different applied currents.

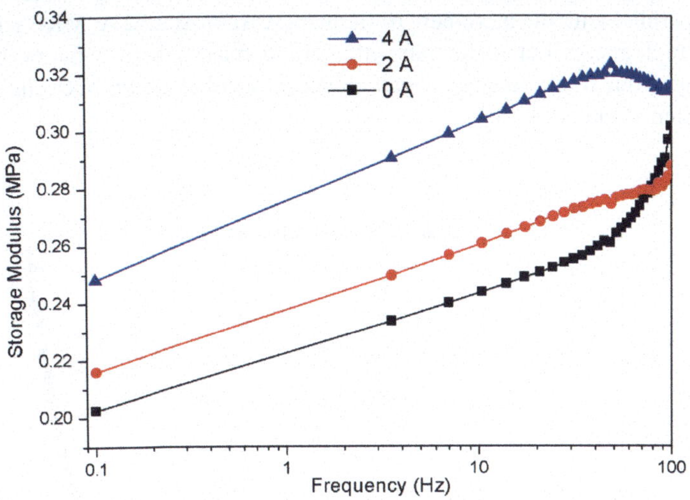

Fig. 2 Frequency sweep test of MRE with cobalt at off- and on-state conditions

Table 2 Initial and maximum storage modulus of MRE at different applied current

Current (A)	Initial storage modulus	Maximum storage modulus
0	0.2025	0.3016
2	0.2162	0.2878
4	0.2482	0.3154

This is because the high shear stress was formed at the surface and in the MRE samples particularly at high excitation frequency during shear deformation [15]. Consequently, the samples were stiffened and resulted in increasing of storage modulus. However, when the frequency exceeds a critical value, they are in glassy states and the storage modulus grows rapidly with increasing frequency [13].

3.3 *Effect of Loss Factor*

Tan δ or known as loss factor which represents the damping property of MRE is one of an important parameter in characterizing the damping capacity of MRE materials. Tan δ is the ratio of the loss modulus to the storage modulus. The loss factor is affected by small groups and chains and molecular movement in the polymer structure. Thus, the higher loss factor showed higher molecular mobility. As shown in Fig. 3, cobalt which acts as magnetic particles in the MRE fabrication exhibit an increment as the frequency increased. At the same time, the loss factor increased with an increment of the magnetic field.

As a filler in the silicone matrix, cobalt particles might lead to difficulty in matrix chain mobility throughout the frequency amplitude. Based on Fig. 3, the loss factor increases parallel with the increment of frequency. At on-state 2 A, MRE with cobalt dissipates high energy but as the magnetic field increases, the loss factor decreases. This is maybe due to the strong interaction between cobalt and rubber chains which lead to stiffened material [16].

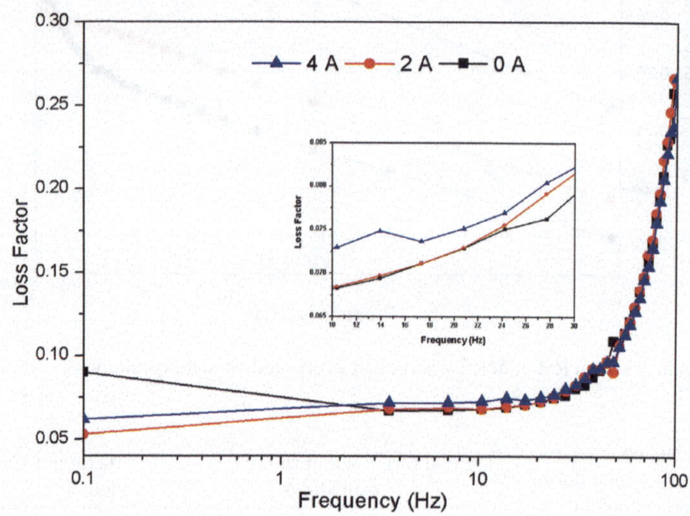

Fig. 3 Loss factor against frequency at various magnetic fields

3.4 *Effect of Magnetic Field*

The viscoelastic properties such as storage modulus and MR effect are the parameter to evaluate the MRE performance. Storage modulus is defined as that the strain energy is temporarily stored during deformation and can be recovered later. The ratio of magneto-induced modulus and the initial modulus is simplified as MR effect. Below is the formula for MR effect

$$G_{MRE}(\%) = \frac{\Delta G}{G_0} \times 100\% \qquad (1)$$

where G_0 is the initial storage modulus, and G_{max} is the maximum storage modulus at the highest magnetic flux density. Figure 4 depicts the correlation between magneto-induced storage and samples at various magnetic flux densities. The storage modulus the sample shows an increasing trend with the magnetic field. The initial storage for MRE with cobalt at 0 A is 0.20 MPa. Meanwhile, the initial storage modulus of 2 and 4 A are 0.21 and 0.25 MPa. This is due to the magnetic particles inside the interparticle interaction become strong resulting in higher storage modulus.

Meanwhile, the MR effect for isotropic MRE with cobalt is 27.78% as compare to MRE based graphite which is 7.2% [17]. This might be due to the high magnetic saturation offered by the cobalt powder as compared to the carbonyl iron particles. The more particles will make the MRE becomes brittle, thus decreasing the MR effect. Table 3 depicted the initial storage modulus, G_0, the absolute MR effect, ΔG and relative MR effect of MRE with cobalt.

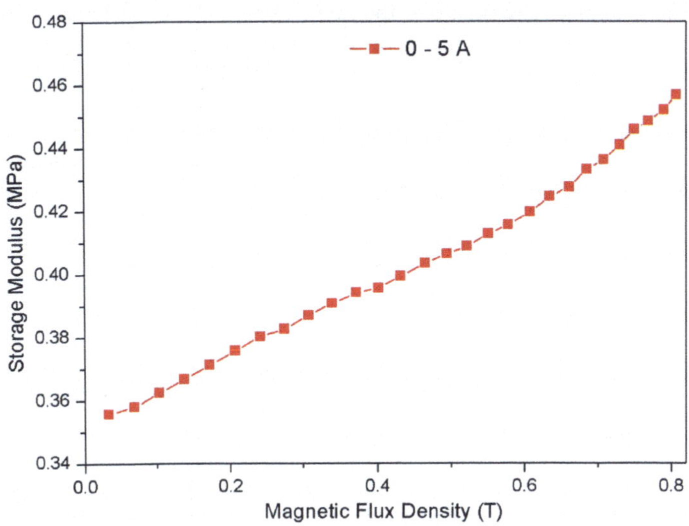

Fig. 4 Current sweep test

Table 3 The initial storage modulus (G_0), the magnetically induced modulus (ΔG), and the MR effect

Sample	G_0 (MPa)	ΔG (MPa)	MR effect (%)
Isotropic cobalt MRE	0.36	0.10	27.78
Isotropic graphite MRE (Shabdin et al. [17])	0.58	0.04	7.2

As shown in Table 3, the initial storage modulus of isotropic MRE with cobalt is lower than the MRE with graphite which can be related to the stiffer of the materials; the higher the modulus, the stiffer the samples. Due to the softer samples of MRE with cobalt, the MR effect of the MRE with cobalt is 74% higher as compared to the MRE with Graphite. This is believed due to the softer samples that allowed the higher "movement" of the cobalt particles in the MRE during the magnetic field applied.

4 Conclusion

In this study, isotropic MRE with cobalt was successfully prepared through a mixing process. The experimental test was undertaken using a rotational rheometer to investigate the rheological properties. The results have shown that the MRE exhibits a magnetic saturation of 78.74 emu/g. Besides, the significant increment of storage modulus of MRE was also observed under various magnetic flux densities from 0 to 0.8 T at room temperature. The result has shown that the MR effect of MRE with cobalt is 27.78%, which is almost 74% higher than the previous report by another researcher. It is finally remarked that the investigation of other characteristics, such as conductivity and morphological analysis will be undertaken as the next phase of this study.

Acknowledgements This study was financially supported by the Universiti Teknologi Malaysia under Transdisciplinary Research Grant (Vot No: 07G13) and Professional Development Research University grant (Vot No: 04E02).

References

1. Ubaidillah, Sutrisno J, Purwanto A, Mazlan SA (2015) Recent progress on magnetorheological solids: materials, fabrication, testing, and applications. Adv Eng Mater 17(5):563–597
2. Margida AJ, Weiss KD, Carlson JD (1996) Magnetorheological materials particles based on iron alloy particle. Int J Mod Phys B 10(23 and 24):3335–3341
3. Carlson JD, Jolly MR (2000) MR fluid, foam and elastomer devices. Mechatronics 10(4):15
4. Rabinow J (1948) The magnetic fluid clutch. Trans Am Inst Electr Eng 1308–1315
5. Ngatu GT, Wereley NM, Karli JO, Bell RC (2008) Dimorphic magnetorheological fluids: exploiting partial substitution of microspheres by nanowires. Smart Mater Struct 17(4):8
6. Portillo MA, Iglesias GR (2017) Magnetic nanoparticles as a redispersing additive in magnetorheological fluid. J Nanomater 2017

7. Ahmad Khairi MH et al (2017) The field-dependent complex modulus of magnetorheological elastomers consisting of sucrose acetate isobutyrate ester. J Intell Mater Syst Struct 28(14):1993–2004
8. Lötters JC, Olthuis W, Veltink PH, Bergveld P (1997) The mechanical properties of the rubber elastic polymer polydimethylsiloxane for sensor applications. J Micromech Microeng 7(3):145–147
9. Wahab NAA et al (2016) Fabrication and investigation on field-dependent properties of natural rubber based magneto-rheological elastomer isolator. Smart Mater Struct 25(10):1–11
10. Chen L, Gong XL, Jiang WQ, Yao JJ, Deng HX, Li WH (2007) Investigation on magnetorheological elastomers based on natural rubber. J Mater Sci 42(14):5483–5489
11. Lu X et al (2012) Mechanical and structural investigation of isotropic and anisotropic thermoplastic magnetorheological elastomer composites based on poly(styrene-b-ethylene-co-butylene-b-styrene) (SEBS). Rheol Acta 51(1):37–50
12. Fu J, Yu M, Dong XM, Zhu LX (2013) Magnetorheological elastomer and its application on impact buffer. J Phys Conf Ser 412(1)
13. Tong Y, Dong X, Qi M (2018) Improved tunable range of the field-induced storage modulus by using flower-like particles as the active phase of magnetorheological elastomers. Soft Matter 14(18):3504–3509
14. Tong Y, Dong X, Qi M (2019) Payne effect and damping properties of flower-like cobalt particles-based magnetorheological elastomers. Compos Commun 15:120–128
15. Rahman NANA, Mazlan SA, Aziz SAA, Nordin NA (2018) Magnetorheological elastomer silicone-based containing corroded carbonyl iron particles. Mater Sci Eng 772:51–55
16. Marcovich NE, Villar MA (2003) Thermal and mechanical characterization of linear low-density polyethylene/wood flour composites. J Appl Polym Sci 90(10):2775–2784
17. Shabdin MK et al (2019) Material characterizations of Gr-based magnetorheological elastomer for possible sensor applications: rheological and resistivity properties. Mater (Basel) 12(3):391

Optimization of Compression Molding Parameters for Pineapple Leaf Fiber Reinforced Polypropylene Composites Using Taguchi Method

Mohd Zulkefli Selamat, Ayu Natasya Kasim, Sivakumar Dhar Malingam, and Mohd Ahadlin Mohd Daud

Abstract The appropriate compression molding parameters of the fabrication process of fiber reinforced polymer (FRP) composites materials are important to ensure the quality of composite samples. Therefore, this paper presents the determination of optimal compression molding parameters for pineapple leaf fiber (PLF)/polypropylene (PP) composite to get larger values of tensile strength through Taguchi method. The Taguchi's L9 orthogonal array has been used as a design of experiment (DOE) while the tensile strength is assumed to be quality characteristic (responses). In order to determine optimal compression molding parameter, four parameters of compression molding, which were temperature, pressure, pre heat duration and compression duration have been analyzed. The nine samples of the composites were fabricated with 60 wt% fiber loading and 30 mm fiber length with random orientation. From the analysis of tensile test result, it was determined that the optimum compression molding parameters for PLF/PP composite which are 30 kg/cm^2 for pressure, 175 °C for temperature, 6 and 4 min to pre heat and compression duration respectively and give the higher tensile strength of 34.57 MPa. From this result, it is concluded that the 60/40 wt% (PLF/PP) composite ratio with 30 mm fiber length that were fabricated with optimal compression molding parameters is the best combination to get high value of tensile strength. There are only slight differences between the predicted and experimental values, where the experimental tensile strength only decreased by 1.9% compared to the predicted value.

Keywords Pineapple leaf fiber (PLF) · Polypropylene (PP) · Biodegradable fiber reinforced plastics (BFRP) · Taguchi method

---

M. Z. Selamat (✉) · A. N. Kasim · S. D. Malingam · M. A. M. Daud
Fakulti Kejuruteraan Mekanikal, Universiti Teknikal Malaysia Melaka, Hang Tuah Jaya, 76100 Durian Tunggal, Melaka, Malaysia
e-mail: zulkeflis@utem.edu.my

Centre for Advanced Research on Energy, Universiti Teknikal Malaysia Melaka, Hang Tuah Jaya, 76100 Durian Tunggal, Melaka, Malaysia

U. Sabino et al. (eds.), *Proceedings of the 6th International Conference and Exhibition on Sustainable Energy and Advanced Materials*, Lecture Notes in Mechanical Engineering,
https://doi.org/10.1007/978-981-15-4481-1_13

1 Introduction

Nowadays, the demand for the use of fiber reinforced polymer (FRP) composites material has gained popularity in the engineering field. Most of the current FRP composites that are used in industry are made from synthetic fiber material such as aramid, carbon, boron, basalt and fiber glass. However, these kinds of fibers are harmful to the environment and human health because they enhanced with chemicals toxicants such as mercury and carbon monoxide, organic compounds and methyl alcohol. Therefore, for those who are continuously exposed to hazardous work environment involving synthetic fiber material, they may experience upper respiratory tract, acute irritation of the skin and eyes, as well as having a high risk to get lung related diseases. Additionally, the production of synthetic fiber causes air pollution that contributes to global warming and the depletion of petroleum resources. Besides that, the product from synthetic fiber does not degrade and these results in the increase of environmental pollution and threaten humanity, wildlife and destroys nature significantly [1–3].

In order to overcome this problem, researchers contribute to protect green earth by replacing synthetic fiber with natural fiber in the manufacturing of FRP composite. Development of biodegradable fiber reinforced polymer (BFRP) composite will help to reduce ecological concerns on the consumption of petroleum base resources, environmental pollution and waste disposal problems. Additionally, the use of natural fiber composites offers a combination of excellent mechanical properties, dielectric properties, and environmental advantages such as low density, low cost, reasonable specific strength, high toughness, being recyclable, biodegradable, renewable and carbon dioxide (CO_2) neutral [4–14]. Biodegradable Fiber Reinforce Polymer (BFRP), also known as natural fiber composite is becoming an important research topic in this decade as a replacement of conventional material, especially in the automotive, marine, packaging, furniture and building construction industries [14–16].

There are a few examples of natural fibres that can be extracted from a plant such as pineapple leaf, palm leaf, banana leaf, hemp, kenaf, bamboo and coconut shell fibre. From all-natural fibres, pineapple leaf fibre (PLF) seems to have the highest cellulose content which makes the fibres can produce good mechanical properties. PLF is one of the most attractive candidates as a strengthening natural fiber. It has several advantages such as small environmental load because it grows rapidly and takes only several months to grow. Thus, it is easy to regenerate after harvesting, and the PLF has relatively high strength compared with other natural fibers such as kenaf, jute and cotton [15]. Additionally, the PLF is a typical bio-resource which has not been fully utilized. Currently, only the fruit of the pineapple is used while the leaf is burnt or thrown away, thus causing pollution and wastage of a potential source of natural fiber [16, 17]. Hence, the use of PLF in the BFRP composite materials industry can reduce environmental pollution, waste disposal problems and ecological concern especially in Malaysia.

Recently, plenty of research focused on the characterization of mechanical and physical properties of PLF reinforced polymer composite [4, 5, 11, 16, 18]. But literature concerning fabrication process is few. The appropriate parameters in the composite fabrication process are vital to ensure the quality of the composites. One of the fabrication techniques in manufacturing of this composite is through the hot compression molding. Where, the composite samples which have been fabricated with the compound of the composite materials were placed in a mold with the specified dimension, and then the mold was pressed in hot press machine [4, 5, 18].

In this study, an alternative approach based on Taguchi method is used to minimize the cost, experimental time and to obtain an optimal parametric design for various parameters and their interactions [11, 19, 20]. Montgomery [21] mentioned that the objective of the parameter design is to optimize the settings of the process parameter values for enhancing performance characteristics and to distinguish the product parameter values under the optimal process parameter values. Therefore, the parameters design is the key step in the Taguchi method to achieve high quality products without increasing cost. The objective of the present work is to determine the optimal compression molding parameters for PLF/PP composite to get larger value of tensile strength.

2 Methodology

PLF: Fresh pineapple leaves from Josapine cultivar was collected from cultivation areas in Pontian, Johor Malaysia. Decortication machine was used to extract the fiber from the fresh leaves. Alkaline treatment using sodium hydroxide (NaOH) was conducted with the purpose to produce high quality fibers. Modifying natural fibers with alkali has greatly improved tensile properties, stabilizes the molecular orientation, removed all the impurities and moisture content, and improved the adhesion between hydrophilic PLF and hydrophobic PP. After alkaline treatment, the available long fibers were cut into approximately 30 mm length as shown in Fig. 1a.

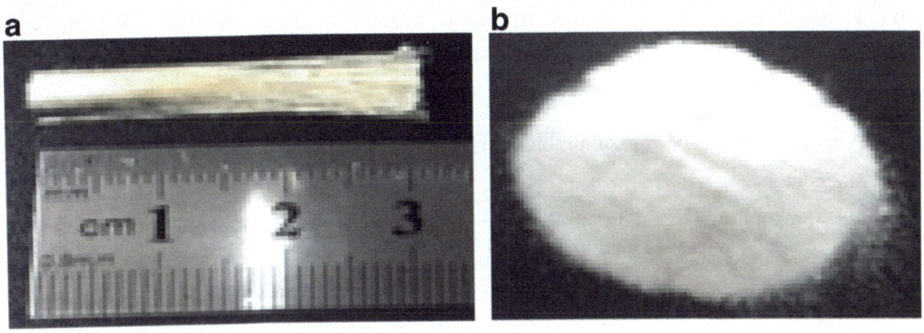

Fig. 1 **a** PLF (approximately 30 mm), **b** polypropylene (250 μm)

Table 1 Parameters outline

Parameters	Values
Temperature (°C)	170–180
Pressure (kg/cm^2)	20–35
Pre heat duration (min)	4–7
Compress duration (min)	3–6

2.1 *Determination of Compression Molding Parameter for Virgin PP Powder*

In order to determine the appropriate compression molding parameters, virgin PP was compressed with different parameters according to their own properties. The melting point of PP is around 165–170 °C. In natural context, it has a heat transfer during the process. Therefore, the selected appropriate value of temperature was from 170 °C until 180 °C. Pressure is a crucial parameter for the hot compression molding. A suitable pressure needs to be determined to make sure the PP is compact and does not overflow from the mold. The selected value of pressure was from 25 to 35 kg/cm^2. Estimating the correct duration for the pre heat and compression also plays an important role in ensuring sufficient heat transfer from the hot press machine to the mold. Table 1 shows the parameters outline that was used to determine the appropriate compression molding parameters.

2.2 *Optimization Compression Molding for PLF/PP Composite Parameter Through Taguchi Method*

The appropriate compression molding parameters for PLF/PP composite fabrication process need to be determined to ensure the quality of the composite samples. The three best parameters from virgin PP powder were used as a guideline. The four factors were determined which were temperature (°C), load (kg/cm^2), pre heat duration (m) and compression duration (m), with three levels taken for each of factors as shown in Table 2. Taguchi method was used to determine the optimal compression

Table 2 Four factors of compression molding parameter at three different levels

Levels	Factors			
	Temperature (°C)	Load (kg/cm^2)	Pre heat duration (m)	Compress duration (m)
1	170	25	4	3
2	175	30	5	4
3	180	35	6	5

molding parameter, minimize the experimental time and cost. The set of experiment for the four different compression molding parameters with three different levels entirely gives large set of experiments. To minimize the set of experiments the Taguchi method was applied with notation of L9 (43) orthogonal arrays with four factors and three levels. In this study, L9 (43) orthogonal arrays with four factors and three levels was selected by using Minitab statistical software before sampling, testing and examination of the outcomes. In total, the number of PLF/PP composite samples needed to be fabricated based on this setup is 9. The fabrication of PLF/PP composite for each sample follows the respective parameter setting as specified in Table 3.

The composite samples were fabricated by a manual mixing technique and were compression molded to form a sheet of composites with 3 mm thickness. The composition of the PLF/PP composite by weight percentage (wt%) and the fiber length was fixed at 60/40 and 30 mm respectively. Fabrication of the composite samples begins by carefully placing the compound of the PLF and PP in a mold with random orientation of the fibers as shown in Fig. 2. Then, the mold was placed in hot

Table 3 DOE with L9 (4^3) orthogonal arrays

No. of samples	Temperature (°C)	Pressure (kg/cm^2)	Pre heat duration (m)	Compress duration (m)
1	170	25	4	3
2	170	30	5	4
3	170	35	6	5
4	175	25	5	5
5	175	30	6	3
6	175	35	4	4
7	180	25	6	4
8	180	30	4	5
9	180	35	5	3

Fig. 2 Random orientation of fiber

press machine and pressed to form a sheet of composite follows the perspective compression molding parameter setting as specified in Table 3. Following this, the mold undergoes the cooling process for 30 min and the final sample of the composite sheet is removed from the mold. Table 4 shows samples of PLF/PP composite for 9 different compression molding parameters. After the completion of the compression molding process, tensile test was conducted to determine the tensile properties as characteristic or response.

2.3 *Tensile Test*

Tensile test is performed according to ASTM D 3039: Standard Test Method for Tensile Properties of Polymer Matrix Composite Materials. The specimens having dimension of 140 mm length, 13 mm width and 3 mm thickness were tested using Instron Universal Testing Machine (Model 5585H) controlled by Bluehill 2 software with a 1 kN load test and operated at constant head-speed tests of 2 min/mm. The specimen was fixed in the grips of the testing machine and load was applied hydraulically till rupture occurs. Four specimens are prepared and tested for each respective parameter setting.

3 Results and Discussion

3.1 *Determination of Compression Molding Parameter for Virgin PP Powder*

The parameters evaluation was based on the visual observation of the product of each parameter has been used. The parameter that could produce the best appearance, or in which the yielded PP that are completely melted and does not overflow from mold, was used for the next fabricate of PLF/PP composite samples [4, 22]. Table 5 shows the three best samples from the three different parameters outline that was used. Based on visual observation, it was identified that the three samples have given the good condition of the samples. Which are parameters number 1 and 2 shows the sheet of PP which had the yielded PP that are incompletely melted, however the appearance still acceptable because the melting PP is not overflow from the mold. Meanwhile parameter number 3, it showed the best appearance, or in which the yielded PP that are completely melted and does not overflow from mold. So that the set-in parameters 3 was chosen as an appropriate parameter and had been used for next fabricate of PLF/PP composite samples.

Table 4 Samples of PLF/PP composite for each parameter

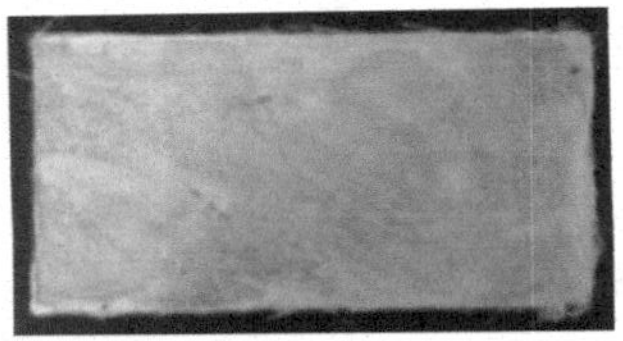 Sample 1	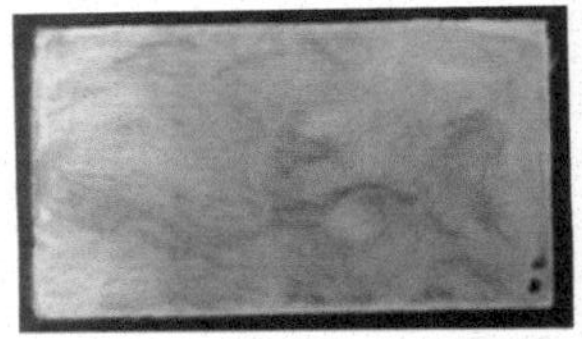 Sample 2	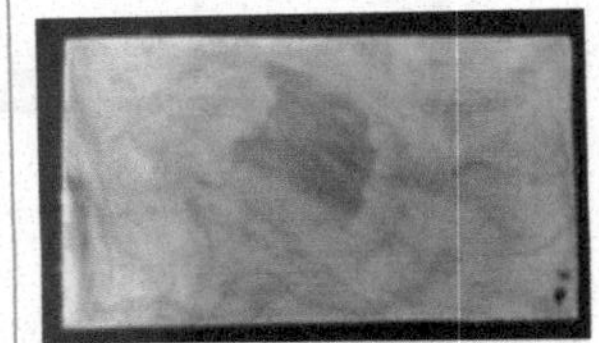 Sample 3
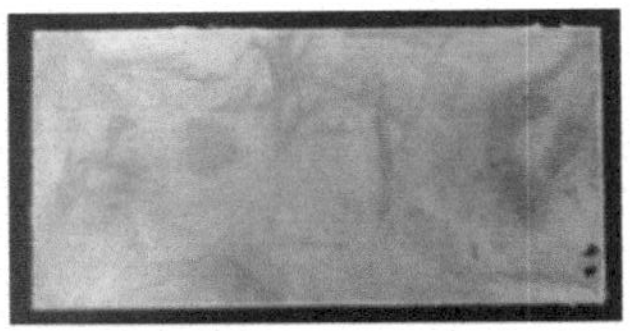 Sample 4	 Sample 5	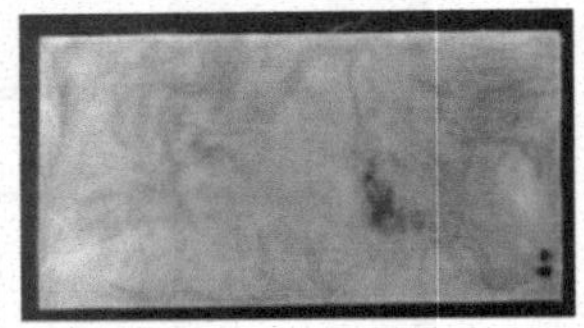 Sample 6
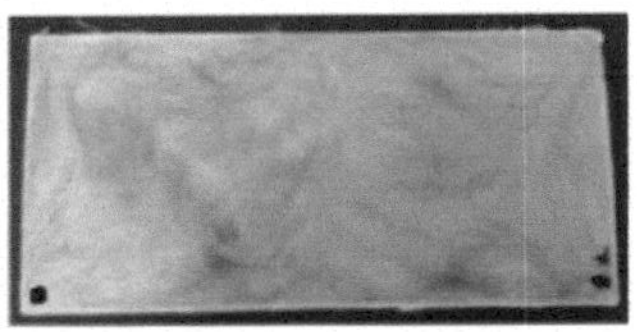 Sample 7	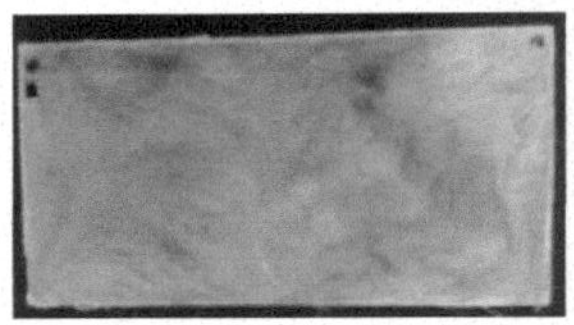 Sample 8	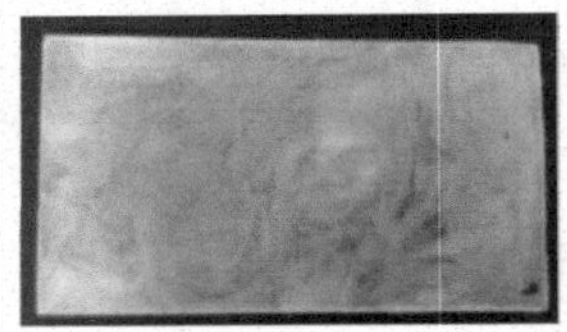 Sample 9

Table 5 Conditions of the three best of the samples (reorganize)

Parameters	1	2	3
Temperature (°C)	170	175	180
Pressure (kg/cm^2)	25	30	35
Preheat duration (m)	4	5	6
Compress duration (m)	3	4	5
Result			

3.2 *Optimization Compression Molding Parameter of PLF/PP Composite Through Taguchi Method*

Table 6 shows the actual result of tensile strength (TS) with their computed mean. Additionally, Table 6 shows the respond table for means for each level of tensile strength. This data was then plotted as shown in Fig. 3 which shows that the means for larger is better for tensile strength and the significant interaction. Table 7, pressure (kg/cm^2) is a dominant parameter on the tensile strength followed by pre heat duration (m), compress duration (m) and temperature (°C).

Table 6 Means of different responses

No. of samples	TS A (MPa)	TS B (MPa)	TS C (MPa)	TS D (MPa)	Mean	P mean
1	13.10	4.73	5.86	14.23	9.4800	35.8225
2	22.97	30.83	31.08	25.79	27.6675	
3	28.73	26.75	18.46	18.96	23.2250	
4	35.10	21.94	13.57	12.04	20.6625	
5	32.82	23.41	24.07	37.90	29.5500	
6	27.27	14.71	19.74	19.83	20.3875	
7	35.16	17.98	25.97	25.45	26.1400	
8	20.68	26.15	37.27	28.95	28.2625	
9	6.39	16.87	24.01	18.12	16.3475	

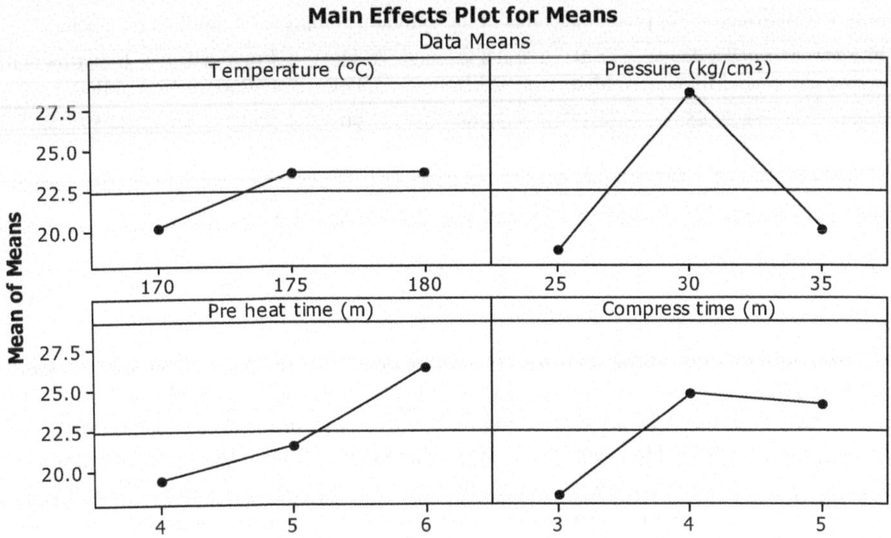

Fig. 3 Main effects plot for means

Table 7 Response table for means (larger is better)

Level	Temperature (°C)	Pressure (kg/cm^2)	Pre heat duration (m)	Compress duration (m)
1	20.12	18.76	19.38	18.46
2	23.53	28.49	21.56	24.73
3	23.58	19.99	26.31	24.05
Delta	3.46	9.73	6.93	6.27
Rank	4	1	2	3

Based on the evidence from Table 8 and Fig. 3, the prediction compression molding parameters for short PLF/PP composite which are 30 kg/cm^2 for pressure, 175 °C for temperature, 6 min for pre heat duration and 4 min for compress duration.

Once the prediction of compression molding parameters has been determined, the final step for this phase is to verify the optimal level of the compression molding parameters [23]. Whereby, composite sample was fabricated with the same composition of PLF/PP which is 40 wt% PLF and 60 wt% PP by using the prediction compression molding parameters. After the completion of the compression molding process, tensile test was conducted to determine the tensile properties. The estimated

Table 8 Factor level for predictions

Temperature (°C)	Pressure (kg/cm^2)	Pre heat duration (m)	Compress duration (m)
175	30	6	4

Table 9 Confirmation for prediction value of the optimum compression molding parameters

PLF/PP composite	TS A (MPa)	TS B (MPa)	TS C (MPa)	TS D (MPa)	Predicted (MPa)	Experimental (MPa)
40/60	32.45	34.33	33.60	37.90	35.82	34.57

mean for the tensile strength was calculated by using Eq. (1).

$$\overline{X} = \frac{\sum X}{n} \tag{1}$$

The result of the confirmation experiment utilizing the prediction compression molding parameters of tensile strength is shown in Table 9. Based on the results, it is shown that there is good agreement between the predicted and experimental mean value [24, 25]. There are only slight differences where the experimental tensile strength was only 3.94% lower compared to the predicted value. Even though the tensile strength from confirmatory experimental was lower as compared to the predicted values, however, this confirmatory value was still higher as compared to the means of nine samples of different parameters as shown in Table 6. Therefore, the prediction compression molding parameters were accepted as the optimal compression molding parameter for PLF/PP composite which is 30 kg/cm^2 for pressure, 175 °C for temperature, 6 min for pre heat duration and 4 min for the compress duration with 34.57 MPa.

4 Conclusion

The result of the study revealed that the pressure during compression molding appears to be the most dominant factor in influencing the tensile strength of the PLF/PP composite. In addition, the effects from pre heat duration, compress duration and temperatures are significant and need to be considered in the compression molding process of PLF/PP composite. Based on the result of Taguchi analysis, the optimum compression molding parameters for the tensile strength was 30 kg/cm^2 pressure, 175 °C temperature, 6 min pre heat time and 4 min compress time and give the tensile strength of 34.57 MPa. Determination of the effective fiber length and fiber loading of PLF/PP composite to get better with larger values of tensile strength will be the focus of our future work.

Acknowledgements The authors would like to thank the Malaysia Ministry of Higher Education, Malaysia Ministry of Science, Technology and Innovation for sponsoring this work under Grant FRGS/2/2014/SG06/FKM/02/F00237 and Advanced Material Research Group (A-MAT), Faculty Mechanical Engineering, University Teknikal Malaysia Melaka (UTeM) for financially sponsoring, facilities and gratefully knowledge during this research.

References

1. Roslan SAH, Hassan MZ, Rasid ZA, Zaki SA, Daud Y, Aziz S, Sarip S, Ismail Z (2015) Mechanical properties of bamboo reinforced epoxy sandwich structure composites. Int J Autom Mech Eng 12:2882–2892
2. Kengkhetkit N, Amornsakchai T (2014) A new approach to "greening" plastic composites using pineapple leaf waste for performance and cost effectiveness. Mater Des 55:292–299
3. Zaimy AGMK, Zafiah A, Rus M, Ab Latif N, Nurulsaidatulsyida S (2013) Mechanical and thermal properties of waste bio-polymer compound by hot compression molding technique. J Mech Eng Sci 5:582–591
4. Selamat MZ, Razi M, Kasim AN, Sivakumar DM, Azma P, Daud MAM, Yuhazri Y (2016) Mechanical properties of starch composite reinforced by pineapple leaf fiber (PLF) from josapine cultivar. ARPN J Eng Appl Sci 11(16):9783–9788
5. Kasim AN, Selamat MZ, Aznan N, Sahadan S, Daud MAM, Salleh S, Jumaidin R (2015) Effect of pineapple leaf fiber loading on the properties of pineapple leaf fiber–polypropylene composite. J Teknol 77(21):117–123
6. Selamat MZ, Kasim AN, Shamsudin SA, Mohd Daud MA, Dhar S (2014) Effect of bamboo fibre length on the mechanical properties of bamboo fibre/polypropylene composite. In: 8th MUCET, Melaka, Malaysia
7. Dhal JP, Mishra SC (2012) Processing and properties of natural fiber-reinforced polymer composite. J Mater 2013:1–6
8. Vinod B, Sudev L (2013) Effect of fiber orientation on the flexural properties of PALF reinforced bisphenol composites. Int J Sci Eng Appl 2:166–169
9. Phong NT, Fujii T, Chuong B, Okubo K (2012) Study on how to effectively extract BFs from raw bamboo and waste water treatment. Mater Sci Res 1:144–155
10. Abdul Khalil HPS, Bhat IUH, Jawaid M, Zaidon D, Hermawan A, Hadi YS (2012) Bamboo fiber reinforced biocomposites: a review. Mater Des 42:353–368
11. Raghavendra G, Acharya SK, Deo CR, Mishra P (2012) Fabrication-modelling and analysis on tribological performance of natural composites using Taguchi approach. Procedia Eng 38:2635–2644
12. Jeyanthi S, Janci Rani J (2012) Influence of natural long fiber in mechanical, thermal and recycling properties of thermoplastic composites in automotive components. Int J Phys Sci 7:5765–5771
13. Chandramohan D, Marimuthu K (2011) Tensile and hardness tests on natural fiber reinforced polymer composite material. Int J Adv Eng Sci Technol 6:97–104
14. Alves C, Silva A, Reis L, Freitas M, Rodrigues L (2010) Ecodesign of automotive components making use of natural jute fiber composites. J Clean Prod 18:313–327
15. Jawaid MHPS, Khalil HA (2011) Cellulosic/synthetic fibre reinforced polymer hybrid composites: a review. Carbohyd Polym 86(1):1–18
16. Mohamed AR, Sapuan SM, Khalina A (2010) Selected properties of hand-laid and compression molded vinyl ester and pineapple leaf fiber (PALF) reinforced vinyl ester composites. Int J Mech Mater Eng 5(1):68–73
17. Mohamed A, Sapuan S, Shahjahan M, Khalina A (2009) Characterization of pineapple leaf fibers from selected Malaysian cultivars. J Food Agric Environ 7:235–240
18. Kasim AN, Selamat MZ, Daud MAM, Yaakob MY, Putra A, Sivakumar D (2016) Mechanical properties of polypropylene composites reinforced with alkaline treated pineapple leaf fibre from Josapine cultivar. Int J Autom Mech Eng 13:3157–3167
19. Nalbant M, Gökkaya H, Sur G (2007) Application of Taguchi method in the optimization of cutting parameters for surface roughness in turning. Mater Des 28(4):1379–1385
20. Ghani JA, Choudhury IA, Hassan HH (2004) Application of Taguchi method in the optimization of end milling parameters. J Mater Process Technol 145(1):84–92
21. Montgomery DC (1997) Design and analysis of experiments, 4th edn. Wiley, New York
22. Li X, Tabil LG, Panigrahi S (2007) Chemical treatments of natural fiber for use in natural fiber-reinforced composites: a review. J Polym Environ 15:25–33

23. Suherman H, Bung U (2015) Optimization of moulding parameters on the electrical conductivity of carbon black/graphite/epoxy composite for bipolar plate using the Taguchi method. Adv Mater Res 1119:201–206
24. Selamat MZ, Sahari J, Muchtar A, Muhamad N (2011) Simultaneous optimization for multiple responses on the compression moulding parameters of composite graphite—polypropylene using taguchi method. Key Eng Mater 471–472:361–366
25. Selamat MZ, Tahir MSZ, Kasim AN, Dharmalingam S, Putra A, Yaakob MY, Daud MAM (2018) Effect of starch sizes particle as binder on short pineapple leaf fiber composite mechanical properties. MATEC Web Conf 150:04008

Interleaved Carbon Fibre Composites with Shape Memory Capability for Use in Hinge Deployment

Dharu Feby Smaradhana and Budi Santoso

Abstract The use of solar panels for generating power of satellite demands deployable structures which can be folded before launch and be deployed while in space. A hinge is a simple deployable structure that can be used as a driving device for deploying solar panels. Hinge made of polystyrene-interleaved carbon fibre composites having shape memory capability was proposed and investigated. It has been shown that the interleaved composites are able to be deformed from flat shape to U-shape at elevated temperature (re-shaping stage) and then able to maintain its deformed shape at room temperature. Moreover, the composites have ability to recover to its original shape once re-heated in an unconstrained state (shape recovery stage). The concept was conducted by attaching two panels on both sides of the interleaved composites (hinge) during those stages to see the shape memory capability. The outputs of this work are the spring back angle after the re-shaping stage and the final angle after shape recovery stage. The predictive analysis using simple beam theory is compared to the experimental results. The results showed the possibility of interleaved carbon fibre composites to be used in deployable structures.

Keywords Deployable structures · Hinge · Shape memory composites · Carbon fibres · Interleaves

1 Introduction

Nowadays, the main energy of satellites is generated from solar arrays. Thus, developing an efficient mechanism for deployment of solar arrays is important. The use of either control mechanism [1–3] or shape memory alloys [4–6] is complex and increases the weight of the structures. Shape memory composites (SMCs) are expected to be the alternative materials to overcome those problems because SMCs provide high stiffness- and strength-to-weight ratios and high flexural strains without damage [7].

D. F. Smaradhana (✉) · B. Santoso
Mechanical Engineering Department, Universitas Sebelas Maret, Surakarta, Indonesia
e-mail: dharufs@gmail.com

U. Sabino et al. (eds.), *Proceedings of the 6th International Conference and Exhibition on Sustainable Energy and Advanced Materials*, Lecture Notes in Mechanical Engineering,
https://doi.org/10.1007/978-981-15-4481-1_14

The manufacturing of SMCs is conducted through modification of the interface of the polymer composites. The interface can be modified by giving coating to the fibres [8–11] or placing interleaves in between the adjacent plies of laminated composites [12–14]. The modification of interface was initially conducted to improve the mechanical properties of composites, especially its toughness [15–17], but recent authors have investigated this to create functional composites. Tridech et al. [18] developed and investigated carbon fibre reinforced polymer composites (CFRP) having ability to control the stiffness. The authors used polyacrylamide (PAAm) which was coated onto carbon fibres via electrocoating process before impregnated into the resin. Heating PAAm above its glass temperatures (Tg) caused the PAAm to soften and allowed carbon fibres to slide within the matrix. This made CFRP become flexible and lose the flexural stiffness up to 88%.

Robinson et al. [19] developed CFRP having a shape memory effect without the use of shape memory constituents such as shape memory polymers by modifying the interface of laminated composites. An interleaf made of polystyrene (PS) (thermoplastic) was placed in between the adjacent plies to allow the laminae to slide relatively to each other once deformed in a desired shape at high temperature, above the Tg (glass transition temperature) of the interleaves but below the Tg of epoxy (matrix of the composite). Afterwards, the composite was able to maintain the deformation at room temperature. Once re-heated in an unconstrained state, the composite was almost fully recovered to its original shape. The deployment of box had been investigated and showed a promising result.

In this paper, carbon fibre shape memory composites developed from the previous work [19] was manufactured and investigated to deploy composite panels which mimic the process of solar panels deployment on satellite. Simple beam theory calculation was used to get the predictive spring back angle after reshaping stage and final angle after shape recovery stage in order to compare to the experimental results.

2 Investigation of Test Specimens

2.1 *Materials*

CFRP plies (T300-914C) produced by Hexcel, UK, were used in this work. The interleaf materials used were polystyrene (PS) (ST316310) bought from Goodfellow, UK, in the form of pellets. These pellets were then manufactured into a film using heated press. For analysis of shape memory effect, Tg of 914 epoxy and PS is 180 °C and 100 °C respectively.

2.2 Preliminary Design and Prediction

Design There are two types of specimens in this work: specimen having polystyrene films (PS) and CFRP plies in the stacking sequences (see Fig. 1), termed hinge, as the actuator to deploy panels, and specimen only consisting of CFRP plies with quasi-isotropic arrangement ($[0°/45°/-45°/90°]_s$) without the addition of PS films, termed solid panel, which functioned as the materials which were deployed. There are two stages in this study to understand shape memory behaviour of hinge specimens. First stage is reshaping process and second is shape recovery process, as shown in Fig. 2. The design of full specimens is shown in Fig. 3 with two panels attached on the left and right sides of the hinge. The deployment is conducted from 180° bend specimen to flat specimen.

Minimum reshaping radius Once the specimen is bent, there will be a compressive load on one side of the layer and a tensile load on another. Therefore, minimum reshaping radius of the hinge must be considered to prevent fibre failure. The strain

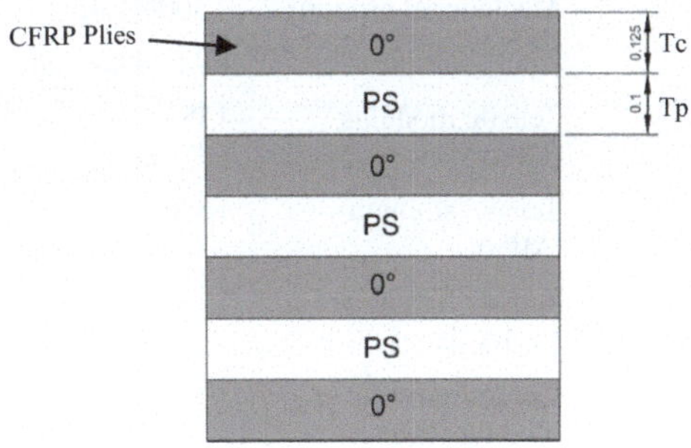

Fig. 1 Stacking sequences of hinge specimen

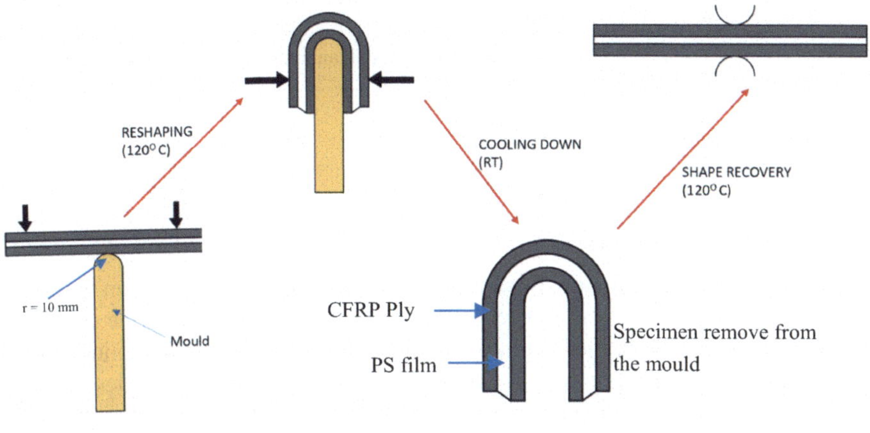

Fig. 2 Shape memory concept

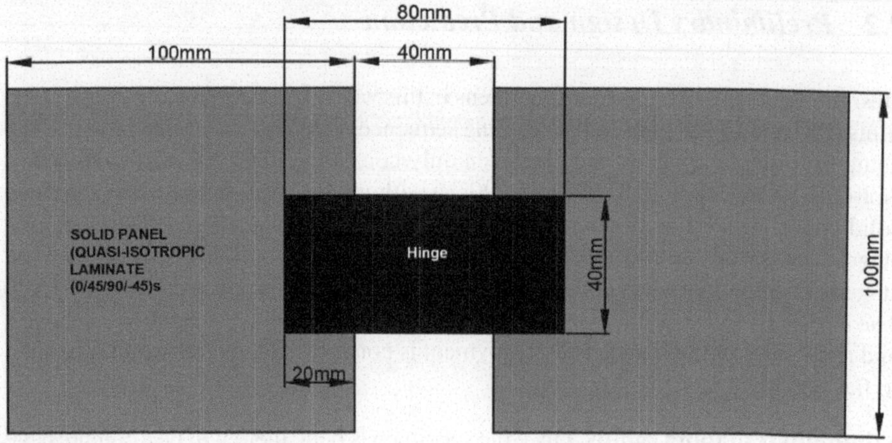

Fig. 3 Full design of specimens

Table 1 T300/914C properties (provided by Hexcel datasheet)

Mechanical property	T300/914C	
	Fibre direction	Transverse direction
Tensile modulus (GPa)	135	8.5
Tensile strength (MPa)	1650	79
Tensile failure strain ($\times 10^{-3}$)	12.2	9.3
Compressive modulus (GPa)	**	**
Compressive strength (MPa)	1350	230
Compressive failure strain ($\times 10^{-3}$)	10.0	27.1

**The data is not available and not considered in this study

due to bending can be determined using Eq. (1).

$$\varepsilon = \frac{tc}{2R} \tag{1}$$

ε, tc, R in Eq. (1) are strain, ply thickness and radius respectively; 1/R can be called curvature. The thickness of plies used in this work is 0.125 mm. From the equation, the smaller radius, the higher strain is obtained. By inputting the maximum strain of the plies used, CFRP prepregs T300/914C, in the mechanical properties given in Table 1, the minimum reshaping radius can be determined. Only will fibre direction

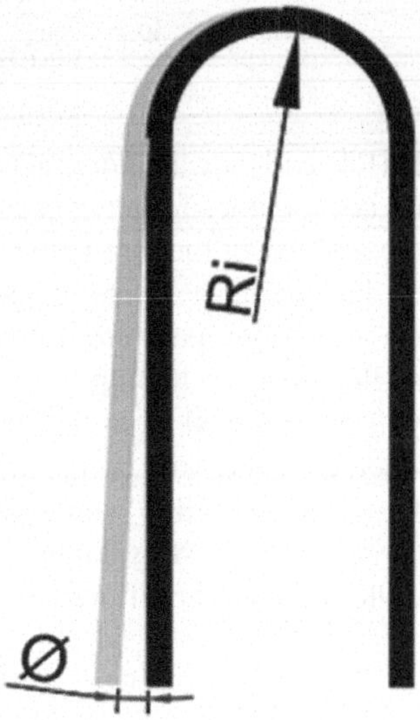

Fig. 4 Spring back angle of reshaped hinge

be used for hinge in this work. Thus, the minimum reshaping radius is 5.12 mm, while the inner hinge radius in this study was 10 mm.

Spring back after reshaping stages After reshaping, the hinge is released at the room temperature. Moreover, it will show a small amount of deflection to the intended shape, called spring back (see Fig. 4). According to Robinson et al. [19], the magnitude of spring-back angle (Ø) can be predicted by calculating the moments needed to reshape the hinge. The moments of lamina can be estimated using Eq. (2).

$$Mi = \frac{Eiwtc^3}{12Ri} \tag{2}$$

Mi is the moment needed to result a change of curvature (1/Ri) for a given lamina i. In this work, if the hinge is expected to change from flat shape to U shape with 10 mm radius which means that the curvature of this specimen has changed from zero to 1/10 mm^{-1}. Furthermore, Ei is the young's modulus of the lamina; the magnitude depends on the fibre direction (see Table 1); in this work, the value of fibre direction is used, while w is the width of the specimen.

The moments of the laminate (MT) can be determined by summing Mi for all CFRP plies as shown in Eq. (3).

$$MT = \sum_{\text{For all CFRP}} Mi \tag{3}$$

The moments of the laminate (MT) are applied in the opposite direction to the reshaped specimen which will cause the spring-back. The relation between the moments (MT) and the spring-back angle is presented in Eq. (4).

$$\text{Ø} = \frac{1}{(EI)RT} \int_{\text{Over entire midline of the specimen}} MT\, ds \quad (4)$$

EI(RT) and s are the flexural modulus of the laminate at room temperature determined by beam theory for a beam section of multiple materials [20] and the distance along the midline of the laminate respectively.

The spring-back angle also can be estimated using the ratio of flexural stiffness at elevated temperature to that at room temperature multiplied by the change of the angle during reshaping process as expressed in Eq. (5) [19]. In this work, the hinge will be reshaped from flat to be U shape in which the change of the angle is 180°. Therefore, there are two predicted results for the spring back angle of reshaped hinge from Eqs. (4) and (5) which were compared to the experimental results. Table 2 gives the predicted results using Eqs. (5) and (4); the calculation of both equations only indicated a slight difference.

$$\text{Ø} = \frac{Flexural\ stiffness\ at\ elevated\ temparture}{Flexural\ stiffness\ at\ room\ temperature} \times 180° \quad (5)$$

Shape recovery angle The hinge lifts two panels in the recovery process (see Fig. 5a), so that the weight of the panels influences the recovered shape (see Fig. 5b). Recovered shape can be predicted by estimating the moments acting in the opposite

Table 2 Predicted results of spring back and recovered angle

	Spring back angle (flat to 180° bend specimen) (°)		Recovered angle (180° bend to flat specimen) Equation (8) (°)
	Equation (4)	Equation (5)	
Hinge (cfrp/ps)	1.62	1.35	5.51

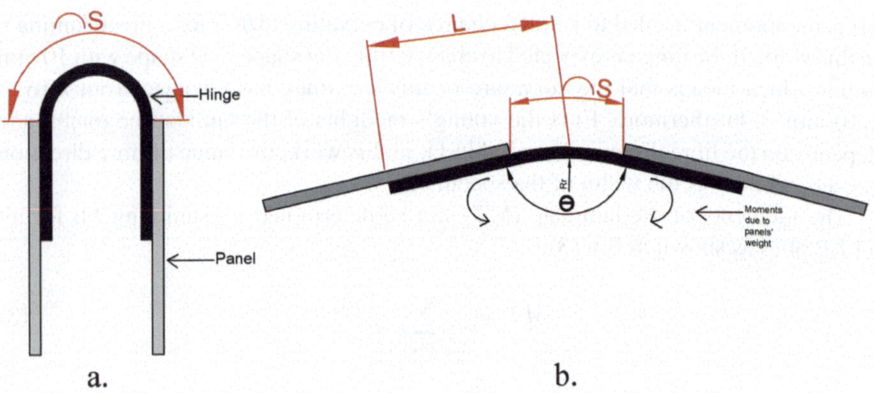

Fig. 5 Effect of solid panels weight on shape recovery angle **a** before shape recovery, **b** after shape recovery

direction to the recovery force due to the weight of the panels. The moments can be calculated as expressed in Eq. (6).

$$\mathrm{Mw} = \mathrm{P} \times \mathrm{L} \tag{6}$$

Mw is the moments due to weight; L is length of the moment due to weight of a panel which is the length from the centre of the hinge to the centre of solid panel, and P is the mass of one panel. As the recovery process occurs at the elevated temperature, recovery radius (Rr) (see Fig. 5b) can be obtained by estimating the relation between flexural modulus of the hinge at high temperature and the moment because of the weight of the panel, as presented in Eq. (7).

$$Rr = \frac{(EI)HT}{Mw} \tag{7}$$

After obtaining the magnitude of Rr, the angle of the recovered shape (θ) (see Fig. 5b) can be predicted using Eq. (8).

$$\theta = \frac{S}{Rr} \tag{8}$$

S is the half of the arc length of the hinge (see Fig. 5) which is assumed that its magnitude does not change from reshaped form to recovered form. In addition, the results need to be converted from radian to degree. Table 2 represents the predictive result of recovery angle using Eq. (8).

2.3 *Experimental Investigation*

Manufacturing The hinge laminate was laid up the same as the stacking sequences shown in Fig. 1, while two quasi-isotropic laminates with stacking sequences of $[0°/45°/-45°/90°]_s$ were manufactured as solid panels. The dimension of the laminates was 120 mm × 120 mm both for hinge and solid panels. Afterwards, the plies were placed on the plate and covered with peel ply, breather and vacuum bag respectively for curing preparation. The curing was conducted using autoclave with temperature of 175 °C and pressure of 100 Psi. The laminates were held at that temperature for an hour and then cooled down with the same pressure. After curing stage, the laminates were cut into the specific dimension: 80 mm × 40 mm (the longer dimension in the 0° direction of laminate) and 100 mm × 100 mm for solid panels.

Re-shaping stage The hinge was bent from flat to 180° bend specimen with radius of 10 mm. A male mould made of wood was used to reshape the hinge. The mould should be wrapped by a release film. Afterwards, the flat specimen was placed on the mould (Fig. 2). The specimen and the mould were then covered with a breather cloth and sealed with a vacuum bag before putting in the oven to be heated to 120 °C. The specimen was held at that temperature for 15 min before deformed manually towards the mould using hands. The vacuum pump was then switched on to hold the new shape. The specimen was cooled down at the room temperature with vacuum applied. After reshaping process, the deformed specimen was scanned to measure the spring back angle.

Shape recovery stage The recovery process was performed in an environmental chamber (Instron) at the temperature of 120 °C by clamping the hinge using metal rods. A Pentax camera was used to film the process which was stopped after the specimen has been fully recovered or after 30 min. For the full-assembly specimen (see Fig. 6), an adhesive (P-2, TML, Japan) was used to bond the reshaped hinge and the solid panels together before shape recovery stage. The aim was to investigate the deployment of this configuration. At last, the shape recovery angle was investigated using a scanner to measure the magnitude of the enclosed angle. The video frames of full-assembly specimen are shown in Fig. 7.

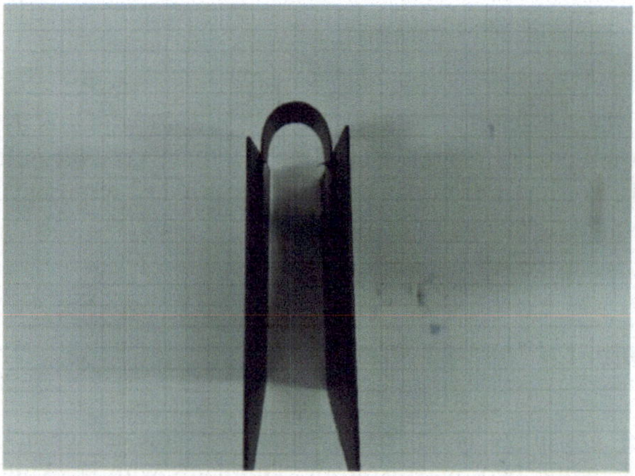

Fig. 6 Full-assembly specimen before shape recovery stage

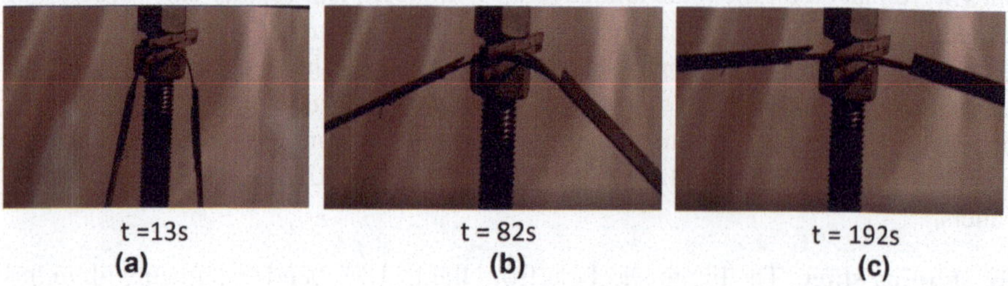

Fig. 7 Video frames of specimen during shape recovery stage

Optical microscope inspection Optical microscope was used to investigate the behaviour of cfrp and polystyrene after recovery stages. The hinge was cut into small pieces (20 mm × 20 mm), mounted into epoxy resin (Epoxycure, Buehler, UK) in a small pot and then cured for 8 h at room temperature. A grinder-polisher machine (Metaser/Metapol, Buehler, UK) was used to polish the samples. There were five sandpapers (SICP320, 600, 1000, 2500 and 4000) and diamond-based dispersions (6, 3 and 0.06 μm) used in the polishing process. The polished samples were then examined using an optical microscope (BX51 M/DP70, Olympus, Japan). The result of polished cross-section is shown in Fig. 9.

3 Result and Discussion

The values of experimental spring back angle of the hinge in full-assembly specimen are almost equal to the predicted values in Table 2, though the predicted values are slightly lower by 0.2° and 0.5° for Eqs. (4) and (5) respectively.

The specimen was successfully deployed to be flat shape within 192 s as shown in Fig. 7. Figure 8 shows the photograph of the specimen after recovery stage. It is clearly seen that the specimen did not fully return to original shape; there was a small angle in the hinge as large as 7.84°. This angle is expected due to the weight of the solid panels attached on the sides of hinge which has been predicted using Eq. (8). However, this final recovered angle was higher than the prediction (see Table 2) which only constitutes 5.51°. This might be caused by the damages in interleaf films as shown in Fig. 9. There seems to be two types of damages from the micrograph: void and delamination of PS film. The later proved that the bonding between CFRP ply and PS interleaf is weak. This was not predicted in the predictive analysis and was likely to reduce the ability of hinge to recover to its original shape.

The drawback of this configuration is that the solid panels should be manually attached to the hinge using an adhesive after the hinge was reshaped. The joining

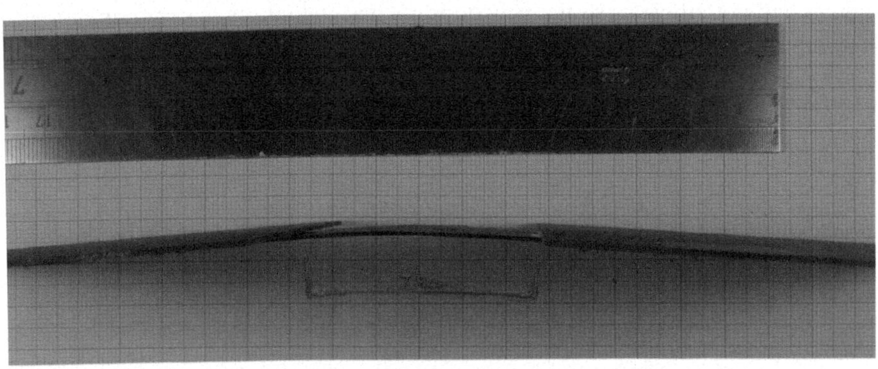

Fig. 8 Specimen after shape recovery stage

Fig. 9 Micrograph from optical microscope of polished cross-section of hinge specimen after shape recovery stage

process was difficult, and misalignment can also occur in joining process which can cause failure in shape recovery process.

4 Conclusion

It has been shown that interleaved carbon fibre composites (hinge) can be deformed at elevated temperature and able to maintain the deformed shape at room temperature with only small amount of spring back angle which has been predicted using a simple calculation. A successful panels deployment has been shown during shape recovery stage, yet it did not fully return to its original shape due to the weight of solid panels attached on the hinge. Though predictive analysis has been conducted to estimate the effect of the weight of solid panels, the final angle in experimental analysis was bigger than that in predictive analysis which was caused by the damages in interleaf films as shown in micrograph of polished cross section of the hinge after recovery stage. This damage was likely to reduce the energy required in shape recovery process and might be caused by the weak bonding between CFRP ply and polystyrene film. Furthermore, the use of adhesive in bonding reshaped hinge and solid panels was difficult and not efficient, so the new design with integral hinge configuration will be developed in further research.

Acknowledgements We gratefully acknowledge the financial support for Dharu Smaradhana provided by LPDP. We thank Prof. Paul Robinson from The Composite Centre, Imperial College London, for guidance, all facilities and specimens in carrying out this research.

References

1. Tsuda Y, Mori O, Funase R, Sawada H, Yamamoto T, Saiki T, Endo T, Kawaguchi J (2011) Flight status of IKAROS deep space solar sail demonstrator. Acta Astronaut 69:833–840

2. Wie B, Furumoto N, Banerjee AK, Barba PM (1986) Modeling and simulation of spacecraft solar array deployment. J Guid Control Dyn 9:593–598
3. Ning-Cong X, Yan-Feng L, Hongzhong H (2009) Reliability analysis method of deployment mechanism of a satellite solar arrays. J Astronaut 4:65
4. Van Humbeeck J (2001) Shape memory alloys: a material and a technology. Adv Eng Mater 3:837–850
5. Jacobs S, Coconnier C, DiMaio D, Scarpa F, Toso M, Martinez J (2012) Deployable auxetic shape memory alloy cellular antenna demonstrator: design, manufacturing and modal testing. Smart Mater Struct 21:75013
6. Godard OJ, Lagoudas MZ, Lagoudas DC (2003) Design of space systems using shape memory alloys. In: Smart structures and materials 2003: smart structures and integrated systems. International Society for Optics and Photonics, pp 545–559
7. Holden MS, Wadhams TP, Parker RA (2006) Report documentation page, 298. New York
8. Reifsnider KL (1994) Modelling of the interphase in polymer-matrix composite material systems. Composites 25(7):461–469
9. Roy AK (1999) Engineered interfaces in fiber reinforced composites
10. Labronici M, Ishida H (1994) Toughening composites by fiber coating : a review
11. Thomason JL, Adzima LJ (2001) Sizing up the interphase: an insider's guide to the science of sizing. Compos A Appl Sci Manuf 32(3–4):313–321
12. Chan W (1991) Design approaches for edge delamination resistance in laminated composites. J Compos Technol Res 91–96
13. Saghafi H, Ghaffarian SR, Salimi-Majd D, Saghafi HA (2017) Investigation of interleaf sequence effects on impact delamination of nano-modified woven composite laminates using cohesive zone model. Compos Struct 166:49–56
14. Maples HA, Smith O, Burgstaller C, Robinson P, Bismarck A (2016) Improving the ply/interleaf interface in carbon fibre reinforced composites with variable stiffness. Compos Sci Technol 128:185–192
15. Chen SF, Jang BZ (1991) Fracture behaviour of interleaved fiber-resin composites. Compos Sci Technol 41(1):77–97
16. Kim J, Mai YW (1991) Effects of interfacial coating and temperature on the fracture behaviours of unidirectional Kevlar and carbon fibre reinforced epoxy resin composites. J Mater Sci 26(17):4702–4720
17. Kim JK, Mai YW (1991) High strength, high fracture toughness fibre composites with interface control—a review. Compos Sci Technol 41(4):333–378
18. Tridech C, Maples HA, Robinson P, Bismarck A (2013) High performance composites with active stiffness control. ACS Appl Mater Interfaces 5(18):9111–9119
19. Robinson P, Bismarck A, Zhang B, Maples HA (2017) Deployable, shape memory carbon fibre composites without shape memory constituents. Compos Sci Technol 145:96–104. https://doi.org/10.1016/j.compscitech.2017.02.024
20. Jones RM (1999) Mechanics of composite materials. Taylor & Francis, London

Rheological Properties of Mg Substituted Cobalt Nickel Ferrite Nanoparticles as an Additive in Magnetorheological Elastomer

Siti Aishah Abdul Aziz, Mohd Syafiq Abdull Aziz, Muhammad Kashfi Shabdin, Saiful Amri Mazlan, Nur Azmah Nordin, Hafizal Yahaya, and Rizuan Mohd Rosnan

Abstract Additive has been used widely in magnetorheological elastomer (MRE) fabrication in order to enhance the magnetic, electrical and rheological properties. In this study, the ferrite nanoparticles namely Magnesium (Mg) substituted Cobalt Nickel Ferrite is introduced as an additive in order to enhance the magnetic and rheological properties of MRE. The conventional co-precipitation method is used to synthesize the Mg substituted Cobalt Nickel Ferrite nanoparticles. The 1 wt% of spherical Mg substituted Cobalt Nickel Ferrite nanoparticles with a range size of 50 nm are then sonicated using ultrasonication before mixing with 70 wt% carbonyl iron particles (CIPs) and silicon-rubber (SR) as a matrix. Two prepared samples of MRE with and without Mg substituted Cobalt Nickel Ferrite nanoparticles are characterized using Vibrating Sample Magnetometer (VSM). Meanwhile, the rheological properties related to the frequency and magnetorheological (MR) effect in off- and on-state condition are determined by using rotational rheometer. The result depicted that the magnetic saturation of MRE with Mg substituted Cobalt Nickel Ferrite nanoparticles increased by 2%. Meanwhile, the maximum storage modulus of MRE with Mg substituted Cobalt Nickel Ferrite nanoparticles enhanced up to 13% as compared to conventional MRE. In the case of energy dissipation, the MRE + S1, exhibit higher energy dissipation as compared to conventional MRE. In the meantime, the relative MR effect of MRE with 1 wt% Mg substituted Cobalt Nickel Ferrite nanoparticles can reach up to 215%, as compared to conventional MRE. The enhancement of magnetic and rheological properties of MRE with Mg substituted Cobalt Nickel Ferrite nanoparticles suggest that the nanoparticles additive fill the

---

S. A. A. Aziz · M. S. A. Aziz · S. A. Mazlan (✉) · N. A. Nordin · H. Yahaya
Engineering Materials and Structures (eMast) iKohza, Malaysia-Japan International Institute of Technology (MJIIT), Universiti Teknologi Malaysia, Jalan Sultan Yahya Petra, 54100 Kuala Lumpur, Malaysia
e-mail: amri.kl@utm.my

M. K. Shabdin · S. A. Mazlan
Advanced Vehicle System (AVS) Research Group, Universiti Teknologi Malaysia, Jalan Sultan Yahya Petra, 54100 Kuala Lumpur, Malaysia

R. M. Rosnan
Jeol (Malaysia) Sdn. Bhd., 508, Block A, Kelana Business Center 97, Jalan SS7/2, 47301 Kelana Jaya, Petaling Jaya, Malaysia

U. Sabino et al. (eds.), *Proceedings of the 6th International Conference and Exhibition on Sustainable Energy and Advanced Materials*, Lecture Notes in Mechanical Engineering,
https://doi.org/10.1007/978-981-15-4481-1_15

void and improved the interaction between CIPs resulted in increment of storage modulus.

Keywords Magnetorheological elastomer · Additive · Ferrite nanoparticles · Magnetic · Rheological properties

1 Introduction

Magnetorheological elastomer (MRE) is a smart material which consists of micron size iron particles and matrix; generally, rubber [1]. An MRE can change their rheological and mechanical properties reversibly and continuously in the presence of an external magnetic field [2, 3]. The MRE is beneficial to a lot of potential intelligent systems and structures including dampers, isolators, sensors and shock absorbers systems [4, 5]. MRE can be classified into two types which depends on the curing condition; anisotropic and isotropic. During the absence of magnetic field in curing condition, the magnetic particles are randomly dispersed which known as isotropic MRE. Meanwhile, in the presence of magnetic field during curing condition, the magnetic particle that been dispersed in rubber medium will form a chain-like structure in which known as anisotropic MRE. MRE offered an outstanding properties as compared to other magnetorheological materials which is magnetorheological fluid (MRF) as an MRE materials doesn't have to deal with sedimentation and leakage problem [6]. However, MRE has been reported to exhibit low MR effect, mechanical and damping properties which restricted the use of MRE in further application.

Therefore, in order to overcome these problems, previous researchers have introduced an additive in order to overcome these kinds of problems. To date, few types of additive have been used in MRE fabrication such as plasticizer [1, 7], nanoparticle [8–10] and processing aids [11]. At such, Khairi et al. [1] reported that the used of the sucrose acetate isobutyrate (SAIB) ester as plasticizer did soften the matrix and enhanced the MR effect up to 77% due to the decrement of zero-field modulus. Meanwhile, Chen et al. [10] reported that by using carbon black as additives, the well bonding between magnetic particles were observed due to the magnetic particles embedded in and form better interaction with the matrix. Thus, as a result, the MR effect did increase up to 104% with higher concentration of carbon black. In the meantime, the mechanical properties of MRE with carbon black also showed an enhancement of tensile strength due to the carbon black did occupy the void and distribute homogenous stress distribution in the matrix. In other study, Aziz et al. [8] fabricated different concentration of carboxyl multiwall carbon nanotubes (COOH-MWCNTs) as an additives in the MRE fabrication and the author found that the MRE with 1 wt% COOH-MWCNTs depicted the highest MR effect of 17.5% due to homogenous dispersion of COOH-MWCNTs in the MRE.

Recently, nanosized ferrite particles has been utilized as an additive in order to overcome the sedimentation that occurred in MR fluid [12]. In this study, the author found that the sediment stability and rheology properties did improved with

0.5 wt% Zn doped ferrite. Furthermore, Hajalilou et al. [13] claimed that cobalt ferrite, $CoFe_2O_4$ nanoparticles has an outstanding properties due to high cubic magneto crystalline anisotropy, moderate saturation magnetization (M*s*) approximately at 52 emu/g, high coercivity (H*c*), strong anisotropy along with good mechanical hardness and chemical stability. However, the use of ferrite nanoparticles has never been reported in the MRE fabrication. It is believed that this kind of additive might altered the interparticle interaction thus improved the modulus and damping properties of MRE. Therefore, in this study, 1.0 wt% of Mg substituted Cobalt Nickel Ferrite nanoparticles is introduced as an additive in MRE fabrication and their rheological properties were examined using a rheometer.

2 Methodology

2.1 Samples Fabrication

In this study, two types of silicone based-MRE samples were prepared through mixing process. 70 wt% of carbonyl iron particles (CIPs) were mixed with 29 wt% of silicon rubber (SR) (NS 625 A) by using mechanical stirrer, Multimix High Speed Dispersed (HSD) approximately for 10 min until the mixture was visually homogenous. Meanwhile, the 1 wt% of Mg substituted cobalt nickel ferrite nanoparticles, $Co_{0.5}Ni_{0.1}Mg_{0.4}Fe_2O_4$ (after this known as S1) with the size of approximately 50 nm were sonicated through ultrasonicator for 10 min. The S1 were then mix thoroughly in the mixing and mixed for another 10 min. Then, 2 wt% of curing agents (NS 625 B) was added and the sample were continuously mixed for another 1 min. Later, the mixture was poured into the steel mould (cylinder shape) and cured in the absence of magnetic field for 24 h.

2.2 Samples Fabrication

The magnetic properties of the MRE samples were measured by using Vibrating Samples Magnetometer (VSM), Microsense FCM10, USA. The MRE samples were subjected to the measurement of magnetic field in the range of −15 k to +15 k Gauss, respectively. Meanwhile, the rheological properties of the MRE samples related to frequency sweep and MR effect were observed under various magnetic fields using an oscillation mode via rheometer (Physica MCR 302, Anton Paar, Austria). The frequency sweep test was carried out under the influence of magnetic field from 0 to 5 A at a constant strain of 0.01% by changing the frequency from 0.1 until 100 Hz. Meanwhile, in order to observe the MR effect, the MRE samples were evaluated under the magnetic varied from 0 to 5 A at a constant shear strain and frequency; 0.01% and 1 Hz, respectively.

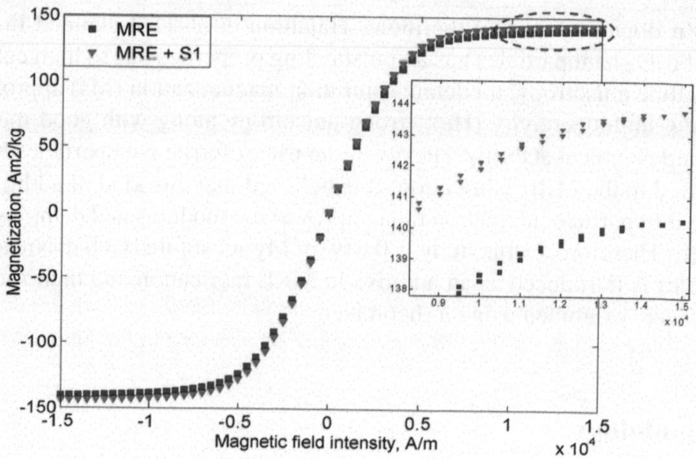

Fig. 1 Hysteresis loop for all MRE samples

Table 1 Summary of magnetic properties of MRE samples

Samples	M_S, emu/g	M_R, emu/g	H_C, Oe
MRE	140.3	0.32	8.81
MRE + S1	143.7	0.59	15.64

3 Results and Discussion

3.1 Magnetic Properties

Figure 1 depicted the magnetic properties related to saturation, remanence and coercivity of MRE samples. The result revealed that the MRE + S1 has the highest magnetic saturation, Ms, 144 emu/g while the conventional MRE exhibit 139 emu/g. The summary of magnetic properties of the MRE samples are shown in Table 1.

The improvement of magnetic saturation in MRE + S1 might be due to the Mg substituted cobalt nickel ferrite nanoparticles filled the inner void between the magnetic particles thus resulted in enhancement of particle interaction. However, the remanence and coercivity of MRE + S1 is still higher as compared to the conventional MRE.

3.2 Frequency Sweep

Figure 2 represent the storage modulus of MRE samples with and without S1 at off- and on-state condition. Both samples exhibit increments of storage modulus with the increment of frequency parallel with the increment of magnetic field. At

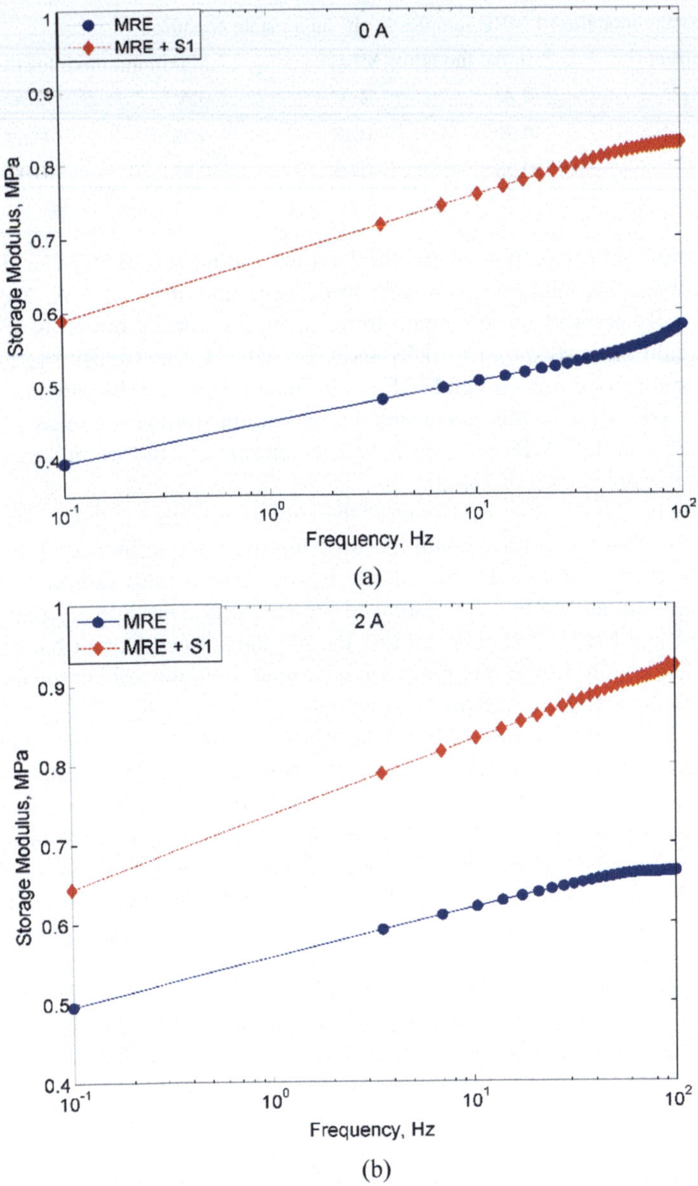

Fig. 2 Storage modulus versus frequency at off-state (0 A) and on-state (2 A) condition

Table 2 Storage modulus of MRE samples at off- an on-state condition

Sample/current	Initial modulus (MPa)		Maximum modulus (MPa)	
	0 A	2 A	0 A	2 A
MRE	0.39	0.48	0.58	0.67
MRE + S1	0.58	0.65	0.82	0.92

such, conventional MRE showed an initial storage modulus of 0.39 MPa at off-state condition, while increased to 0.48 MPa at on-state condition of 2 A. Meanwhile, the MRE + S1 showed an increment trend of initial storage modulus at off- and on-state condition correspond to 0.58 and 0.65 MPa. Concurrently the maximum storage modulus for conventional MRE at off- and on-state condition were observed at 0.58 and 0.67 MPa. In the meantime, the maximum storage modulus of MRE + S1 were 0.82 and 0.92 MPa at 0 and 2 A. The summary of the storage modulus for all the samples are shown in Table 2.

As stated in Table 2, the increment of the storage modulus is due to the "movement" of CIPs that try to form chain like structure with the influence of an external magnetic field thus strengthen the particle interaction for both samples. However, in other situation, an MRE + S1 displayed higher storage modulus as compared to the conventional MRE. It is believed that the Mg substituted Cobalt Nickel Ferrite Nanoparticles which exhibit magnetic behavior tend to fill the void in the matrix and CIPs thus enhance the interparticle interaction.

The energy dissipated during the shear testing is related to the loss modulus of the materials. Commonly, in MRE or any composite materials, both matrix-filler and filler-filler as well as the interfacial interaction contribute to these losses. Figure 3 depicted the loss modulus of MRE samples at off- and on-state condition. As can be seen from Fig. 3a–b, conventional MRE samples exhibit low loss modulus with increasing frequency as compared to MRE + S1. At off-state condition, the conventional MRE shown an initial loss modulus of 0.030 MPa and slightly increased up to 0.031 MPa at on-state condition of 2 A. Moreover, at maximum frequency of 100 Hz, the loss modulus of conventional MRE at off-state condition reflect 0.089 MPa and increased up to 0.094 MPa at 2 A. Meanwhile, for the MRE + S1, the initial modulus at off- and on-state condition are 0.052 and 0.049 MPa, respectively. At higher frequency of 100 Hz, the loss modulus increased up to 0.111 and 0.113 MPa at off- and on-state condition. Generally, during the dynamic mechanical testing, the increment of matrix-filler or filler-filler bonding resulted in the increment value of loss modulus. The introduction of S1 in the MRE is assumed to enhance the particle interaction as the nanoparticle tend to fill the CIPs void thus enhanced the particle interaction in the matrix. Therefore, more energy dissipation occurred during the shear test. In addition, with the influence of magnetic field at on-state condition of 2 A, the CIPs and Mg substituted Cobalt Nickel Ferrite Nanoparticles which exhibit magnetic behavior, tend to experience better interparticle interaction during the shear test consequently resulted in higher energy dissipation.

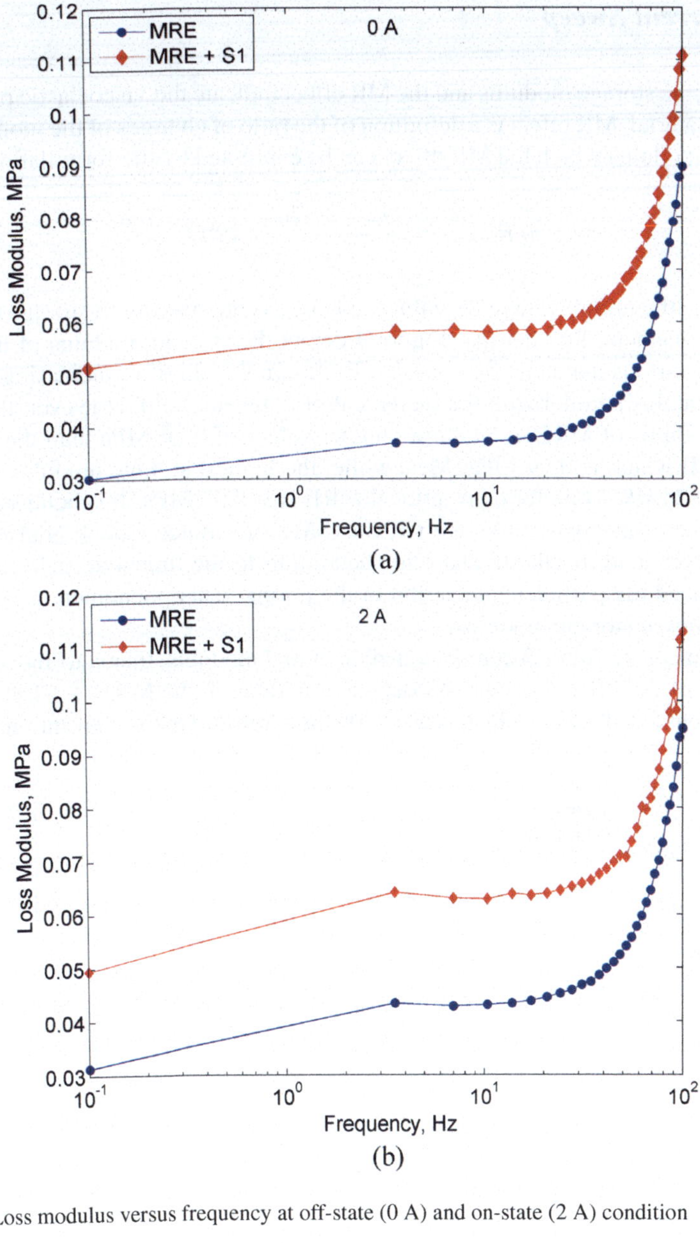

Fig. 3 Loss modulus versus frequency at off-state (0 A) and on-state (2 A) condition

3.3 Current Sweep

Generally, the storage modulus and the MR effect indicate the viscoelastic properties of MRE material. MR effect is a definition of the ratio of changes of the modulus and the initial modulus which the MR effect can be expressed by the formulation below;

$$MR_{effect} = \frac{G_{max} - G_0}{G_0} \times 100\% \quad (1)$$

where G_0 is the initial storage modulus, and G_{max} is the maximum storage modulus at highest magnetic flux density. Figure 4 depict the storage modulus of the MRE samples at various magnetic flux density. Both samples show an increasing trend of storage modulus parallel with the increment of magnetic field. However, the initial storage modulus of MRE + S1 shows higher values of 0.26 MPa than the conventional MRE which is 0.19 MPa. Meanwhile, the highest storage modulus of MRE + S1 is 0.82 MPa while for conventional MRE is at 0.53 MPa. It is believed during the presence of magnetic field, the CIPs and Mg substituted Cobalt Nickel Ferrite Nanoparticles tend to vibrate and align according to the magnetic field direction, thus enhanced the particles interaction in the matrix. This phenomenon lead to an enhancement of storage modulus.

Simultaneously, Table 3 demonstrated the initial modulus, magneto-induced, and MR effect of the MRE samples. As depicted in Table 3, the MRE + S1 exhibit an enhancement of MR effect which is up to 20% as compared to the conventional MRE.

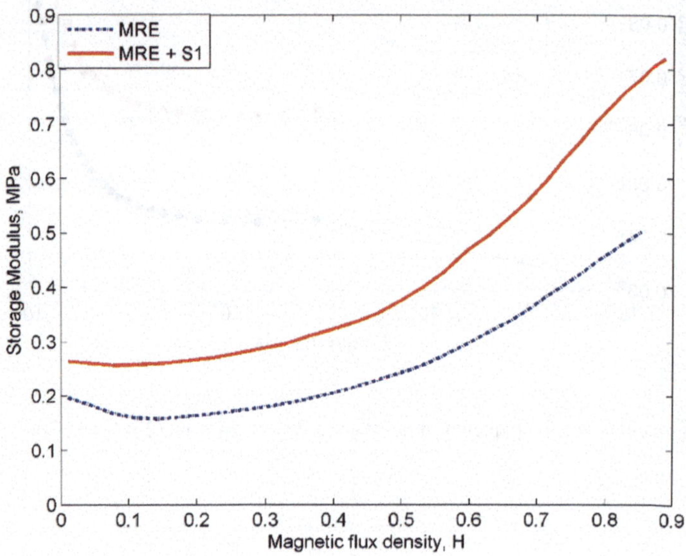

Fig. 4 MR effect for MRE samples

Table 3 The initial modulus, magneto-induced modulus and MR effect of the MRE samples

Sample	G_0 (MPa)	ΔG (MPa)	MR effect (%)
MRE	0.19	0.34	178.95
MRE + S1	0.26	0.56	215.38

It is also observed that the MRE + S1 is stiffer as compared to the conventional MRE as the initial modulus is higher.

4 Conclusion

In this study, the effects of MRE with and without the presence of Mg substituted cobalt nickel ferrite nanoparticles as an additive were experimentally studied. The magnetic and rheological properties of two different samples of MRE were analyzed and compared. The MRE with Mg substituted Cobalt Nickel Ferrite Nanoparticles as an additive possess high magnetic saturation of 143.7 emu/g in which is 2% higher than the conventional MRE. In the meantime, the ferrite nanoparticles did occupy the void between the CIP, resulted in an enhancement of the interparticle chains in the matrix thus enhanced the storage modulus up to 42%. Meanwhile, the initial loss modulus of MRE + S1 exhibit higher energy dissipation in the range of 58–73% in the absence and presence of magnetic field. Despite that, the improvement of the interaction between CIPs and Mg substituted Cobalt Nickel Ferrite Nanoparticles lead to high MR effect as high as 215%, respectively. This finding revealed that with the addition of small amount of ferrite nanoparticles did altered and changed the magnetic and rheological properties of MREs.

Acknowledgements This study was financially supported by the Universiti Teknologi Malaysia under Transdisciplinary Research Grant (Vot No: 06G77) and Professional Development Research University grant (Vot No: 04E02).

References

1. Khairi MHA, Mazlan SA, Ubaidillah, Ku Ahmad KZ, Choi S-B, Abdul Aziz SA, Yunus NA (2017) The field-dependent complex modulus of magnetorheological elastomers consisting of sucrose acetate isobutyrate ester. J Intell Mater Syst Struct 28:1993–2004
2. Kramarenko EY, Chertovich AV, Stepanov GV, Semisalova AS, Makarova LA, Perov NS, Khokhlov AR (2015) Magnetic and viscoelastic response of elastomers with hard magnetic filler. Smart Mater Struct 24:035002
3. Sorokin VV, Stepanov GV, Shamonin M, Monkman GJ, Kramarenko EY (2017) Magnetorheological behavior of magnetoactive elastomers filled with bimodal iron and magnetite particles. Smart Mater Struct 26:035019
4. Ahamed R, Choi S-B, Ferdaus MM (2018) A state of art on magneto-rheological materials and their potential applications. J Intell Mater Syst Struct 29:2051–2095

5. Wahab NAA, Mazlan SA, Ubaidillah, Kamaruddin S, Ismail NIN, Choi SB, Sharif AHR (2016) Fabrication and investigation on field-dependent properties of natural rubber based magneto-rheological elastomer isolator. Smart Mater Struct 25:107002
6. Yang Z, Qin C, Rao Z, Ta N, Gong X (2014) Design and analyses of axial semi-active dynamic vibration absorbers based on magnetorheological elastomers. J Intell Mater Syst Struct 25:2199–2207
7. Mitsumata T, Kimura Y, Yamauchi T, Kanauchi S, Kawai M, Tamesue S (2015) Effect of plasticizer on the magnetoelastic behavior for magnetic polyurethane elastomers. Chem Lett 44:177–178
8. Aziz SAA, Ubaidillah, Mazlan SA, Ismail NIN, Choi S (2018) Implementation of functionalized multiwall carbon nanotubes on magnetorheological elastomer. J Mater Sci 53:10122–10134
9. Poojary UR, Hegde S, Gangadharan KV (2018) Experimental investigation on the effect of carbon nanotube additive on the field-induced viscoelastic properties of magnetorheological elastomer. J Mater Sci 53:4229–4241
10. Chen L, Gong XL, Li WH (2008) Effect of carbon black on the mechanical performances of magnetorheological elastomers. Polym Test 27:340–345
11. Aziz SAA, Mazlan SA, Ismail NIN, Ubaidillah, Choi SB, Nordin NA, Mohamad N (2018) A comparative assessment of different dispersing aids in enhancing magnetorheological elastomer properties. Smart Mater Struct 27:1–30
12. Han JK, Lee JY, Choi HJ (2019) Rheological effect of Zn-doped ferrite nanoparticle additive with enhanced magnetism on micro-spherical carbonyl iron based magnetorheological suspension. Colloids Surf A Physicochem Eng Asp 571:168–173
13. Hajalilou A, Mazlan SA, Abbasi M, Lavvafi H (2016) Fabrication of spherical $CoFe_2O_4$ nanoparticles via sol-gel and hydrothermal methods and investigation of their magnetorheological characteristics. RSC Adv 6:89510–89522

Rheological Behavior of Graphite Induced Anisotropic Magnetorheological Elastomer

Muhammad Kashfi Shabdin, Mohd Azizi Abdul Rahman, Saiful Amri Mazlan, Siti Aishah Abdul Aziz, and Nurhazimah Nazmi

Abstract The rheological properties of magnetorheological elastomer (MRE) with graphite (Gr) was evaluated in the present research. Two types of MRE samples namely anisotropic MRE and GrMRE were fabricated with Gr as a filler during MRE fabrication. The effect of Gr as filler on the rheological properties of MRE was experimentally investigated and compared with those of conventional MREs. Rheological properties under different magnetic fields at zero-field modulus and on-state of 4 A were evaluated under oscillatory rheometer test. Moreover, the effect of curing conditions of GrMRE towards the MR effect was thoroughly examined. It was observed that with the addition of Gr, the rheological properties such as the storage modulus are improved, particularly at low strain amplitudes which showed that the initial storage modulus increase at 48% for both off- and on-state conditions when compared between GrMRE and reference sample without Gr while less dissipation of energy detected in both sample as the loss modulus does not change so much. In addition, the values of MR effect were found to increase in conventional MRE at 32.8% while GrMRE showed 28.3% of MR effect, in which, this kind of finding could be practically used in the fabrication process of potential sensing devices.

Keywords Magnetorheological elastomer · Graphite · Resistance

1 Introduction

Researchers interested in intelligent materials such as magnetorheological elastomer (MRE) has shown a fast augmentation. This acceleration is because this kind of materials can reversibly response to the external magnetic field exerted on them. Besides,

M. K. Shabdin · M. A. A. Rahman (✉) · N. Nazmi
Advanced Vehicle System Research Group, Universiti Teknologi Malaysia, Jalan Sultan Yahya Petra, 54100 Kuala Lumpur, Malaysia
e-mail: azizi.kl@utm.my

S. A. Mazlan · S. A. A. Aziz
Engineering Materials and Structures (e-MAST), Malaysia Japan International Institute of Technology, Universiti Teknologi Malaysia, Jalan Sultan Yahya Petra, 54100 Kuala Lumpur, Malaysia

U. Sabino et al. (eds.), *Proceedings of the 6th International Conference and Exhibition on Sustainable Energy and Advanced Materials*, Lecture Notes in Mechanical Engineering,
https://doi.org/10.1007/978-981-15-4481-1_16

other properties such as electric, mechanical, stress, pH, moisture, temperature and so on could also alter the feedback of these smart materials. Magnetorheological elastomer (MRE) generally comprises of micron-size magnetizable particles, matrix, and fillers [1]. Carbonyl iron particles (CIP) are most frequently used as the magnetic particles owing to its high saturation magnetization of metallic elements as well as high permeability and low remnant magnetization. The particles are dispersed in various types of rubber matrixes such as natural rubber or silicone rubber [2]. As a potential approach to enhancing the properties of MRE, fillers such as carbon black [3], multiwall carbon nanotubes [4, 5], and nanowires [6] have been used in the previous studies. Apart from having the magnetic particles embedded in the matrix, a few researchers have utilized other types of particles to further enhance the mechanical and chemical or electrical properties of MREs such as nickel and graphite [7]. The introduction of graphite powder is likely to generate a new electrical conductivity property while maintaining the existing rheological properties of MRE. The feasibility study on the microstructure and rheological properties of GrMRE has been conducted by Tian et al. [8]. The authors used 10 g of CIPs, 3 g of silicone oil, and 3 g of silicone rubber with various weight fractions of graphite (0–5 g) for fabricating the MRE. Two types of GrMRE were prepared by varying the conditions between isotropic and anisotropic during the curing process; it was observed that the particles were randomly distributed in the isotropic condition, while they were uniformly arrayed as chains in the anisotropic condition. This difference in particle structure contributed toward the increment in the initial storage and loss moduli under the anisotropic condition. In addition, it has been observed that a higher volume fraction of the graphite filler can enhance the MR effect, resulting in higher yield stress under both the conditions. However, the study was limited to the investigation of the rheological properties at low graphite composition; where this study utilized higher graphite weight percentage to investigate the effect of align particles to various magnetic field modulus. Higher dynamic mechanical properties could be a good indication for the realization of future sensing application.

2 Experimental Methods

2.1 Materials

The CIP used was obtained from BASF, Germany and has an average size of 6 μm. The room temperature vulcanization (RTV) Silicone Rubber (SR), NS625, and Silicone oil (SO) DC200 (100cs) were purchased from Nippon Steel Co, Japan, whereas the graphite powder, Gr (code: 8169-00) of average size 16 μm and density 1.8 g cm^{-3} was obtained from R&M Chemicals, Evergreen Engineering and Resources Co., Malaysia.

2.2 GrMRE Fabrication

A similar method to the common polymer process was used in the fabrication of Gr–MRE sample [5]. A series of GrMRE samples were synthesized with 12.5 g CIP, 22.5 g SR, 20 g Gr, and 5 g SO. Initially, the CIP was mixed with SR and SO in a beaker at room temperature of 25 °C. The GrMRE fabrication includes sonication process which was performed for the Gr particles to diminish the weak Van der Waals interaction before mixing the powder with CIP, SR, and SO. The mixture was then stirred for 10 min at 280 rpm to obtain a homogenous mixture. Then, graphite powder was added to the mixture and stirred for another 10 min. Then, the curing agent was added to the mixture to let the sample cure. The curing process followed the procedure provided by the Nippon Steel technical bulletin which described the 2% of curing agents for overall wt% of silicon rubber used. The curing process happens in an onstate condition with the presence of a magnetic field. The mixture was then transferred into the mould of 60 mm diameter and 1 mm thickness. Lastly, the mixture was placed in the mould for curing and left for approximately 1 h.

2.3 Material Characterization

The sample undergo microstructure observation using the Low Vacuum Scanning Electron Microscope (LVSEM): JSM–IT300, JEOL, Japan. On the other hand, the rheological properties of the Gr–MRE samples were measured using an oscillation rheometer (Physica MCR 302, Anton Paar, Germany). The sample was exposed to a magnetic field in the rheometer, generated by a magnetorheological device (MRD 170). The circular sample at approximately ~1 mm, was then fixed in the rheometer cavity plate. A shear strain in the range 0.001–25% was applied on the sample during the experimental investigation by using a pp20 rotary disc 20 mm width and 1 mm thickness.

2.4 Rheological Characterization

Strain and current amplitude test were executed to determine the dynamic performance of the GrMRE samples. Experimental data were gathered with various amplitudes of oscillation at zero and applied magnetic field strengths. In the sweep strain test, the storage modulus was obtained by changing the strain from 0.001 to 25% at various magnetic fields by providing different amperes at 0 and 4 A, which are equivalent to 0 and 0.69 T, respectively and the condition of the consistent frequency of 1 Hz at 25 °C. Meanwhile, the sweep current test was executed by delivering continuous current inputs of 0 and 4 A at a consistent strain of 0.01% (in the linear viscoelastic region) and a consistent frequency of 1 Hz at 25 °C. The MR effect formula is expressed as follows [9]:

$$\text{MR effect} = \frac{G_{max} - G_0}{G_0} \times 100\%,$$

where G_{max} is the maximum modulus, and G_0 is the initial modulus.

3 Results and Discussion

3.1 LVSEM Characterization and Rheological Properties

Figure 1 shows the CIPs alignment in the LVSEM micrograph. The alignment follows the direction of magnetic field density applied across the sample during the curing process. Subsequently, rheological data sets were obtained in the various amplitude of oscillation in the presence of magnetic fields intensity. Swept strain test was carried out by differentiating strain from 0.001 to 25% at zero magnetic fields and at the presence of the magnetic field, which is off state and onstate, respectively. The analysis of the testing showed the storage modulus, loss modulus and loss factor. The experiment was conducted at a constant frequency of 1 Hz at room temperature which is 25 °C. Meanwhile, current swept test was performed by providing stable current from 0 to 4 A at a steady strain of 0.001% to ensure it is in the linear viscoelastic region and at a consistent frequency of 1 Hz at 25 °C.

Effect of anisotropic GrMRE on the storage modulus Figure 2 shows storage modulus versus strain amplitude at off-state (a) and onstate (b) conditions. In both

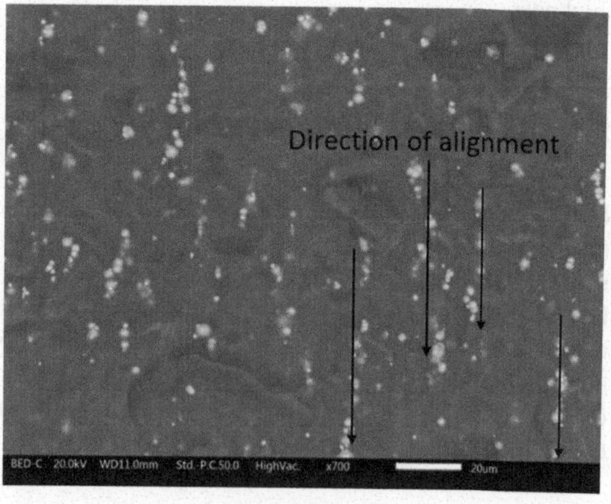

Fig. 1 Micrograph showing the direction of CIPs alignment

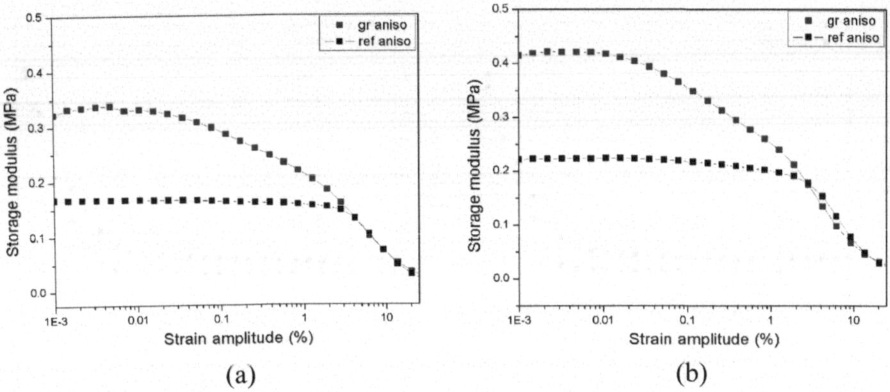

Fig. 2 Storage modulus against strain amplitude at **a** off-state, and **b** on-state conditions of anisotropic MRE (ref) and GrMRE samples

cases, the storage modulus inclination is decreasing with increasing strain amplitude and it can be observed that the initial modulus of MRE filled with graphite is higher when compared to the conventional MRE. For MRE, it goes down smoothly within 1% and starting to plummet significantly over 5% strain, with which within 1% strain, the trend shows approximately linear relationship with the strain. Subsequently, GrMRE shows earlier drop at 0.01% strain and meet the MRE dropline at 5% strain. It could also be noted that at onstate condition, the initial storage modulus of the samples is higher when compared to the off-state condition.

It is known that under an oscillating force, either elastic or viscous properties of materials affects strain amplitude [10]. The recoverable strain energy in the deformed sample could be assessed by the storage modulus of rubber. Graphite addition consequently enhanced the storage modulus of the MRE, which means that the composite is stiffer in the presence of the filler triggered by the stronger matrix-filler interaction. This non-linear property of filled rubber could be explained as the Payne effect and could be described by the presence of a filler structure in the rubber carrier matrix above the maximum permeability [11].

Effect of GrMRE on loss factor Figure 3 shows the variation of loss factor or tan δ of MRE with and without graphite addition against strain amplitude. Loss factor is affected by the movement of the molecule chains within the morphology of the polymer which indicates that as tan δ increases, there is an increment in the molecular mobility occurs in the composite. Based on the figure, the change in loss factor could not be clearly distinguished between off-state and onstate conditions. Both conditions, however, show that MRE with graphite has a higher loss factor when compared to the conventional MRE and the change could be determined from 5 to 25% strain. Besides, there is approximately linear relationship between loss factor and the strain at 0.001–1% strain. This phenomenon probably due to the higher degree

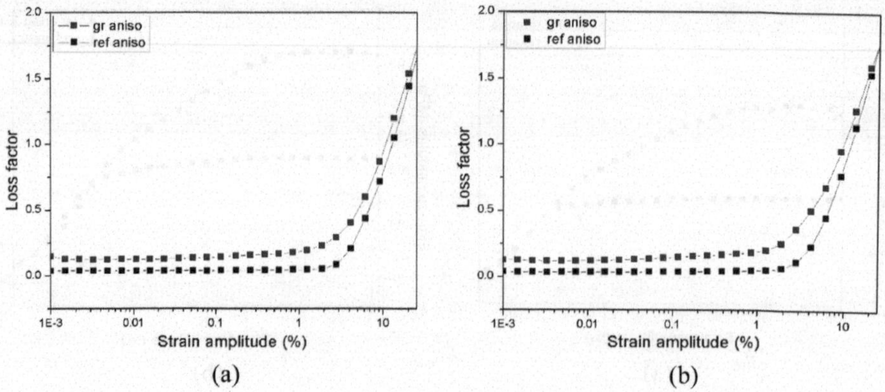

Fig. 3 Loss factor versus strain for **a** off-state, **b** on-state of all MRE samples

of different size particle mobility in GrMRE compared to MRE. Besides, particles agglomeration caused the sample to be brittle which could also lead to higher loss factor.

Effect of GrMRE on MR effect Figure 4 demonstrates storage modulus against magnetic field intensities at a consistent strain amplitude of 0.01%. The MR effect of every sample was obtained from the ratio of storage modulus in Fig. 4 which was determined from the variation of applied magnetic field to the MRE samples. It could be observed that the MR effect of graphite filled MRE experiencing lower percentage than conventional MRE at 28.3% and 32.8%, respectively. This results somehow agree with research done by Tian et al. [8] which stated that the growth of graphite weight fraction brings about decrement in the MR effect. The author also reported the MR effect value of 2.4 for the anisotropic graphite containing MRE by

Fig. 4 Storage modulus against magnetic field intensity for both GrMRE and MRE samples

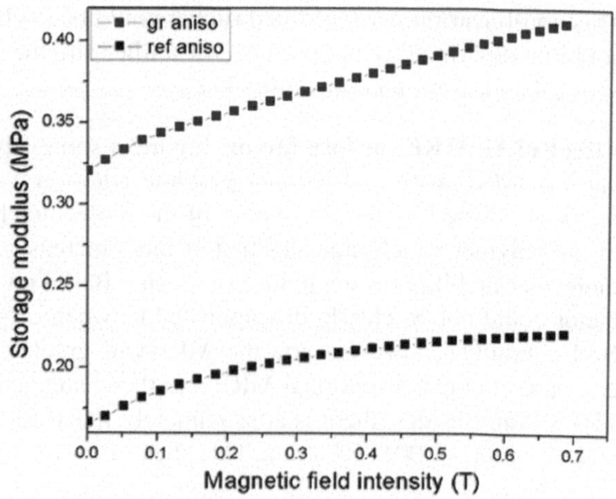

using 23.81 wt% graphite filler. This phenomenon occurs because of the contribution of graphite powder to the initial stiffness of the sample, which causes MR effect to exert less effect on the GrMRE compared to the MRE.

4 Conclusion

MREs filled with graphite has been well fabricated and analyzed. Conclusively, the dynamic mechanical evaluation depicted that the initial stiffness of the GrMRE increased with graphite addition of 33 wt%. Analogous work on rheological behavior has displayed that graphite-based MRE shows better MR performance in term of initial storage modulus at low strain amplitude for both off- and on-state even though the MR effect is lower in anisotropic graphite filled MRE. This finding could be further investigated to utilize the attributes of GrMRE for sensing application.

Acknowledgements The authors gratefully acknowledge the financial support of the Ministry of Education in Malaysia and Universiti Teknologi Malaysia under grant vote no: TDR (06G16) and PDRU (04E02).

References

1. Gong XL, Zhang XZ, Zhang PQ (2005) Fabrication and characterization of isotropic magnetorheological elastomers. Polym Test 24(5):669–676
2. Ubaidillah, Sutrisno J, Purwanto A, Mazlan SA (2015) Recent progress on magnetorheological solids: materials, fabrication, testing, and applications. Adv Eng Mater 17(5):563–597
3. Abdullateef AA et al (2012) Natural rubber nanocomposites with functionalized carbon nanotubes: mechanical, dynamic mechanical, and morphology studies. J Appl Polym Sci 125:E76–E84
4. Li R, Sun LZ (2011) Dynamic mechanical behavior of magnetorheological nanocomposites filled with carbon nanotubes. Appl Phys Lett 99(13):131912
5. Aziz SAA et al (2016) Effects of multiwall carbon nanotubes on viscoelastic properties of magnetorheological elastomers. Smart Mater Struct 25(7):077001
6. Antonel PS, Oliveira CLP, Jorge GA, Perez OE, Leyva AG, Negri RM (2015) Synthesis and characterization of $CoFe_2O_4$ magnetic nanotubes, nanorods and nanowires. Formation of magnetic structured elastomers by magnetic field-induced alignment of $CoFe_2O_4$ nanorods. J Nanopart Res 17(7)
7. Zou H, Zhang L, Tian M, Wu S, Zhao S (2010) Study on the structure and properties of conductive silicone rubber filled with nickel-coated graphite. J Appl Polym Sci 115(5):2710–2717
8. Tian TF, Li WH, Alici G, Du H, Deng YM (2011) Microstructure and magnetorheology of graphite-based MR elastomers. Rheol Acta 50(9–10):825–836
9. Li WH, Zhang XZ, Du H (2013) Magnetorheological elastomers and their applications. Adv Elastom 12:357–374

10. Sui G, Zhong WH, Yang XP, Yu YH, Zhao SH (2008) Preparation and properties of natural rubber composites reinforced with pretreated carbon nanotubes. Polym Adv Technol 19(11)
11. Ponnamma D, Sadasivuni KK, Grohens Y, Guo Q, Thomas S (2014) Carbon nanotube based elastomer composites—an approach towards multifunctional materials. J Mater Chem C 2(40):8446–8485

Intrinsic Apparent Viscosity and Rheological Properties of Magnetorheological Grease with Dilution Oils

N. Mohamad, M. A. Rosli, Siti Aishah Abdul Aziz, Saiful Amri Mazlan, Ubaidillah, Nur Azmah Nordin, Hafizal Yahaya, and Abdul Yasser Abd Fatah

Abstract In this study, the principal rheological properties of the magnetorheological grease (MRG) is experimentally investigated with various types and percentages of dilution oils. The stability in terms of oil separation and rheological properties of the natural plant-based dilution oil; castor oil was compared with petroleum-based; kerosene and hydraulic oils. Several samples of MRGs with different types and ratios of dilution oils were prepared by mixing spherical carbonyl iron (CI) particles and grease using mechanical stirrer. The rheological test was performed under rotational mode of shear rheometer by changing the magnetic fields from 0 to 0.7 T at room temperature condition. The results showed that the apparent viscosity of the dilution oils of kerosene has the lowest apparent viscosity followed by castor and hydraulic oils. It has been observed that by addition of dilution oils more than 10 wt% has reduced the stability of the samples. It is also noted that the increment of dilution oils percentages in MRGs have insignificant effect towards linear viscoelastic (LVE) region. This fact indicated that the improvement of the rheological properties was dependent on the initial viscosity as well as optimum percentage of the dilution oils utilized in the MRG suspension without visibility of the oil separation. In addition, the comparison of the MRG performance in terms of yield stress between different types and percentages of dilution oils was evaluated and discussed in detail.

Keywords Magnetorheological grease · Additive · Dilution oil · Rheological properties · Apparent viscosity · Oil separation · Yield stress

N. Mohamad · M. A. Rosli · S. A. A. Aziz · S. A. Mazlan (✉) · N. A. Nordin · H. Yahaya
Engineering Materials and Structures (eMast) iKohza, Malaysia-Japan International Institute of Technology (MJIIT), Universiti Teknologi Malaysia, Jalan Sultan Yahya Petra, 54100 Kuala Lumpur, Malaysia
e-mail: amri.kl@utm.my

Ubaidillah
Mechanical Engineering Department, Faculty of Engineering, Universitas Sebelas Maret, Jl. Ir. Sutami 36 A, Kentingan, Surakarta 57126, Central Java, Indonesia

A. Y. A. Fatah
Razak Faculty of Technology and Informatics, Universiti Teknologi Malaysia, Jalan Sultan Yahya Petra, 54100 Kuala Lumpur, Malaysia

U. Sabino et al. (eds.), *Proceedings of the 6th International Conference and Exhibition on Sustainable Energy and Advanced Materials*, Lecture Notes in Mechanical Engineering,
https://doi.org/10.1007/978-981-15-4481-1_17

1 Introduction

Magnetorheological grease (MRG) is classified as a smart material due to the ability to alter its rheological properties rapidly, continuously and reversibly with the presence of magnetic field. The magnetic particles in the MRG will be induced and formed chain-like structures parallel to the magnetic fields direction once subjected to the applied field [1]. This phenomenon has transformed the rheological properties of MRGs including apparent viscosity, storage modulus and yield stress by altering the gel-like structures into solid-like states within milliseconds [2–5]. This behavior also has tremendously improved the stability of the magnetic particles suspend in the grease medium and no sedimentation is observed compared to conventional magnetorheological (MR) fluid [5]. Typically, the sedimentation occurred in the MR fluid due to the difference in viscosities between magnetic particles and hydrocarbon oils which react as suspending medium [6]. In contrast, grease which possessing thixotropic behavior able to suspend the magnetic particles for a long time owing to its thick viscosity that hindered the magnetic particles from sediment [7–10]. On top of that, the advantages such as self-sealing ability which able to prevent the leakage problem in device has made the MRG as an option to replace the conventional MR fluid [5].

Besides, MRG is preferable to be utilized in the MR devices such as MR damper, MR brake and MR clutch as studied by [8–10]. Gordaninejad et al. [8] has conducted a research on the MR damper. The result showed that the usage of the MRG in the MR damper has better controllability compared to the MR fluid without presence of sedimentation. In another study, the performance of the MRG in the MR clutch has been studied by Kavlicoglu et al. [10]. It was observed that the MRG manage to increase the torque capacity up to 75% in which appropriate for torque transfer applications. Apart of the study on the MRG performances, previous researchers had investigated extensively on the samples preparation and field-dependent rheological properties of MRGs [2, 6, 10, 11]. In 2007, Sahin et al. [3] has examined the rheological properties of the MRG in terms of apparent viscosity. The results depicted that the MRG had experienced shear-thinning behavior where the apparent viscosity decrease parallel with the increment of shear rates. The study has been continued by Mohamad et al. [11] on the effect of magnetic particles fractions under influenced of magnetic fields. From the study, it was found that the MRG with 70 wt% of magnetic particles had the highest yield stress (52.7 kPa) and MR effect (952.38%). In addition, the magnetic particles shape which has also influenced the rheological performance of the MRG has been investigated by [7]. The study concluded that the plate-like shape of magnetic particles required lower fractions to achieved high yield stress and MR effect compared to the spherical shape. This proved that the magnetic particles have triggered the rheological performance of the MRG. However, by utilizing the high fractions of magnetic particles in the grease has led to the higher off-state (no magnetic field) viscosity which is less considered by other researchers. Thus, Kim et al. [12] had investigated the effect of addition additives in the MRG towards rheological properties. It can be summarized that the addition of 5% of kerosene oil in the MRG has decrease the apparent viscosity as well as yield stress significantly. In addition, better distribution of magnetic particles is obtained.

In fact, the study of the dilution oil as an additive in the MRG has been studied by [12], however, the knowledge on several issues such as compatibility of the dilution oil towards MRG behavior and chain formation is still lack. Consequently, the objective of this study is to investigate the influence of different types and percentages of dilution oils towards MRG performance, to characterize and analyse the performance of MRG with three different types of dispersing oils as the additive. In order to achieve this research goal, a series of MRG samples were prepared using different types of dilution oils; castor, kerosene and hydraulic oils. The dilution oils percentages were varied in between 5 and 15% with the total solid content of 70 wt%. Results in terms of oil separation and rheological properties such as apparent viscosity, shear stress and yield stress are presented and discussed accordingly.

2 Methodology

2.1 *Sample Preparation*

Several samples of MRGs were prepared by mixing the carbonyl iron (CI) particles, grease and dilution oils using mechanical stirrer. Soft magnetic CI particles (OM grade, BASF, Germany) in spherical shape with diameter of 5 μm and density of 7.874 g/cm^3 was used as purchased. Meanwhile, the NPC Highrex HD-3 Grease was chosen as the suspending medium. The standard grade of the grease is NLGI 3.

In this study, three types of dilution oils; castor, kerosene and hydraulic oils were selected as dilution additive to reduce the initial apparent viscosity of the suspension. The density of the castor, kerosene and hydraulic are 0.96, 0.81 and 0.88 g/cm^3, respectively. As a first step, the CI particles with 70 wt% was dispersed in the grease medium using mechanical stirrer. Subsequently, 5–15% of the dilution oils in the volume were added to the mixture and continuously stirred for 1 h until homogenous suspension is obtained. All the samples preparation was conducted in the room temperature. The same procedure was repeated for MRG without dilution oils as reference. The details of the MRG compositions are shown in Table 1.

2.2 *Samples Characterization*

The rheological properties of MRG were characterized by using a commercial shear-rheometer (MCR302, Anton Paar, Austria). 1 mL of MRG samples was placed in between two parallel plates with 1 mm of constant gap distance. The rheological measurements were performed under rotational mode which comprising of magnetic flux density and shear rates sweep tests. The measurement on magnetic flux density

Table 1 Compositions of MRG with different types and percentages of dilution oils

Samples	CI particles (wt%)	Grease (wt%)	Dilution oils (wt%)		
			Castor	Kerosene	Hydraulic
MRG1	70	30			
MRG2	70	25	5		
MRG3	70	20	10		
MRG4	70	15	15		
MRG5	70	25		5	
MRG6	70	20		10	
MRG7	70	15		15	
MRG8	70	25			5
MRG9	70	20			10
MRG10	70	15			15

sweep test was conducted by varied the magnetic flux density by manipulating the coil applied current with constant shear rate of 400 s^{-1}. Meanwhile, the shear rates sweep test was examined by manipulating shear rates from 0 to 2000 s^{-1} with the increment of current in the interval of 1 A. Furthermore, the viscoelastic properties of MRGs samples were measured under oscillation mode with variation of strains from 0.0001 to 10%. The apparent viscosity, shear stress, storage modulus and yield stress of the MRGs samples were analyzed and recorded. Besides, the MRGs samples were placed in cylindrical tubes for the stability test. The time for oil separation to be occurred in the MRGs was observed and recorded for 48 h. All the experiments were performed under room temperature condition.

3 Results and Discussion

The effect of different types and percentages of dilution oils under influence of magnetic fields have been investigated. Figure 1 shows the apparent viscosity of the MRGs are increased along with the increasing of magnetic fields. Nevertheless, the initial apparent viscosity of the MRGs are found decreased as the percentages of dilution oils increased.

This phenomenon proved that the formation of chain structures between CI particles have been weakened by the presence of dilution oils. The addition of dilution oils which react as a lubricant in the MRGs had improved the distribution of the CI particles in suspending medium during off-state condition [13, 14]. As a result, the void distance between CI particles has also increased. In addition, the CI particles seem to be less attached to the grease medium owing to the dilution oils may facing slipping effect during shearing process [15]. Thus, stronger magnetic field induced

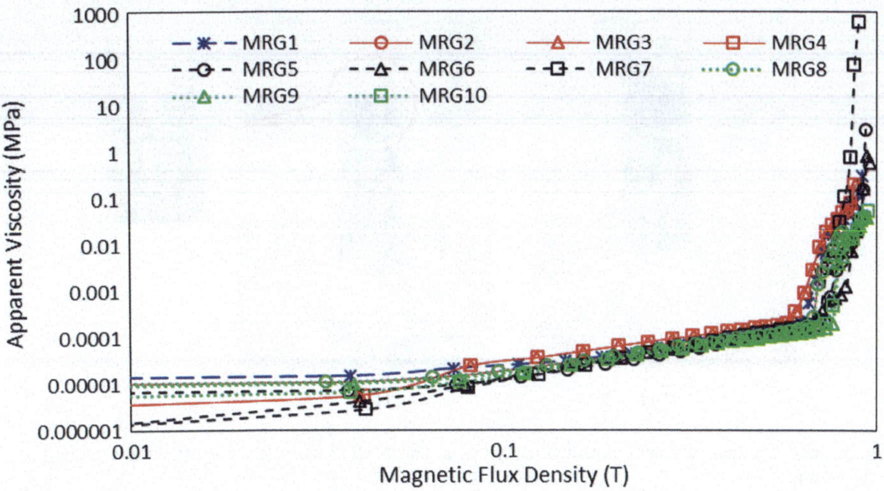

Fig. 1 Apparent viscosity of MRG with different types of dilutions oils and percentages as function of magnetic flux density

is required for the CI particles to move and form stronger chains alignment against shearing factor.

By referring to Fig. 2, the shear stress for MRG1, MRG2, MRG5 and MRG8 samples are increased parallel with increasing of shear rates. With the presence of the dilution oil, the shear stress will be reduced. This reduction occurred for both

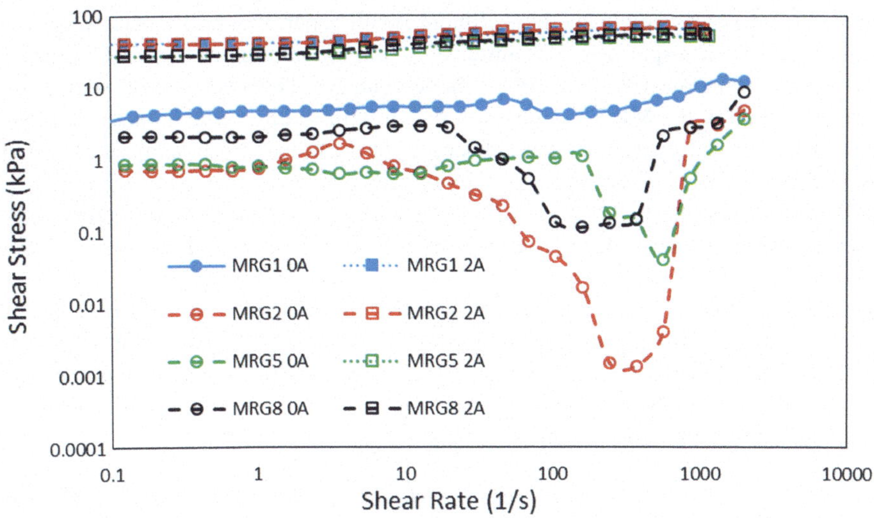

Fig. 2 Comparison of off- and on-state conditions of shear stress for MRG with different types of dilution oils for 5% as function of shear rates

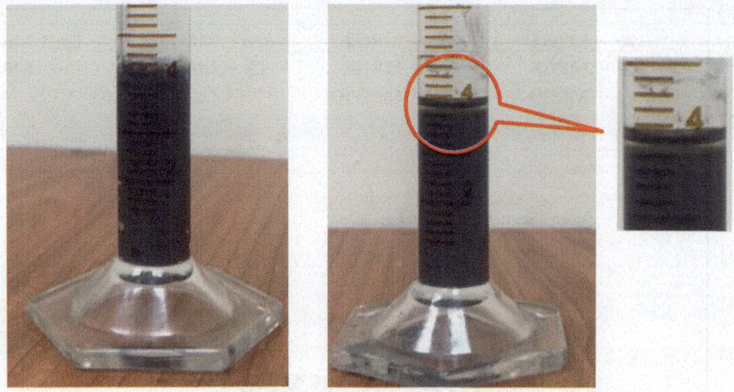

(a) (b)

Fig. 3 Stability test of MRG samples for 48 h. **a** The MRG7 sample right after been placed and **b** after 24 h

off-state and on-state condition. The CI particles in the grease medium which are less restricted to move have been facilitated a better chain formation with the presence of dilution oils on on-state condition [12].

The stability test of MRG samples are depicted in Fig. 3. Each of samples were transferred into cylindrical tubes and placed in the static table for observation. The stability rate can be calculated using Eq. (1) as follows

$$Stability\, rate\,(\%) = Q/P \times 100 \tag{1}$$

where Q denotes as height of the sediment or oil separation and P represents the total height of the suspension. It is anticipated that the stability of the MRG samples will reduce by increasing of dilution oils percentages. However, the oil separation only been seen in the MRG7 sample after 24 h with stability rate reduction of 5%. Meanwhile, no oil separations were observed for the other MRGs samples although after 48 h.

This confirms that the stability of the suspension dependent on the percentages and types of dilution oils. Kerosene oil which has lower viscosity is easily undiluted from the suspension and form a layer on top of grease and CI particles. This indicates that the kerosene oil is less compatible with the NPC Highrex HD-3 Grease which has been used as the suspending medium in this study. Therefore, the structuration formation between oil and grease entanglement fibrous structure has not taken place.

The MRG's samples performance in terms of yield stress was calculated using Bingham plastic regression as shown in Eq. (2);

$$\tau = \tau_y + \eta\dot{\gamma} \text{ for } |\tau| \geq \tau_y, \dot{\gamma} > 1 \tag{2}$$

The value of yield stress for MRG with different types and percentages of dilution oils was tabulated in Table 2. It shows that the yield stress efficiency of all MRG

Table 2 The value of yield stress with variation of magnetic fields

Samples	Yield stress (kPa)				
	0 T (0 A)	0.24 T (1 A)	0.45 T (2 A)	0.64 T (3 A)	0.84 T (4 A)
MRG1	4.41	14.51	42.14	65.25	78.25
MRG2	0.78	15.19	41.00	57.23	73.23
MRG3	0.44	13.33	35.15	50.00	69.60
MRG4	0.35	9.75	34.43	51.70	69.33
MRG5	0.72	10.55	25.71	40.85	77.50
MRG6	0.79	12.60	35.63	62.08	84.71
MRG7	0.21	7.00	23.50	43.04	70.76
MRG8	2.04	6.83	27.90	56.12	77.00
MRG9	1.68	7.19	35.64	57.20	78.74
MRG10	1.04	10.92	42.26	63.92	79.87

samples were improved by increasing the magnetic field. Meanwhile, the yield stress is reduced by increasing percentages of dilution oils during off-state condition. Generally, MRG with presence of dilution oils have lower yield stress values compared to the reference sample; MRG1 at off-state condition. The addition of dilution oil in the suspension has alter the initial viscosity of the grease which leads to the properties modifications and directly influence the performance of the MRG. The presence of dilution oil as lubricants has improved the dispersion of CI particles in the grease medium in which better alignment of chain structures has been achieved. Thus, higher yield stress is obtained by increasing the percentages of dilution oils with the increment of magnetic field. Moreover, the performance of the MRG suspensions are also dependent on the compatibility of the dilution oils towards grease medium used in this study. The hydraulic dilution oil showed the well compatibility and interaction with grease medium followed castor and kerosene oils.

Figure 4a–c shows the effect of shear strain towards storage modulus of the MRGs with different types and percentages of oils under off- and on-state condition. The most important parameter to be determined in this oscillation mode test is linear viscoelastic (LVE) region which indicating the breaking point of the materials under device operation. The addition of different types and percentages of oils on MRGs show insignificant effect towards LVE region; 0.002%. Besides that, an improvement of the LVE region is observed for all MRGs samples by applying magnetic field from 0.002 to 0.03%.

This phenomenon proves that the chain structure between the CI particles are not destroyed until the shear strain exceeded the LVE region [16]. The presence of magnetic fields has formed thick and strong chain structures which indirectly enhanced the viscoelastic properties of the MRGs samples. Moreover, the decrement of storage modulus over shear strain indicated that the MRGs samples have also experiencing the Payne effect. In this study, the dependence of dilution oils towards the MRGs is less noticed on the viscoelastic properties, however, obviously observed on the

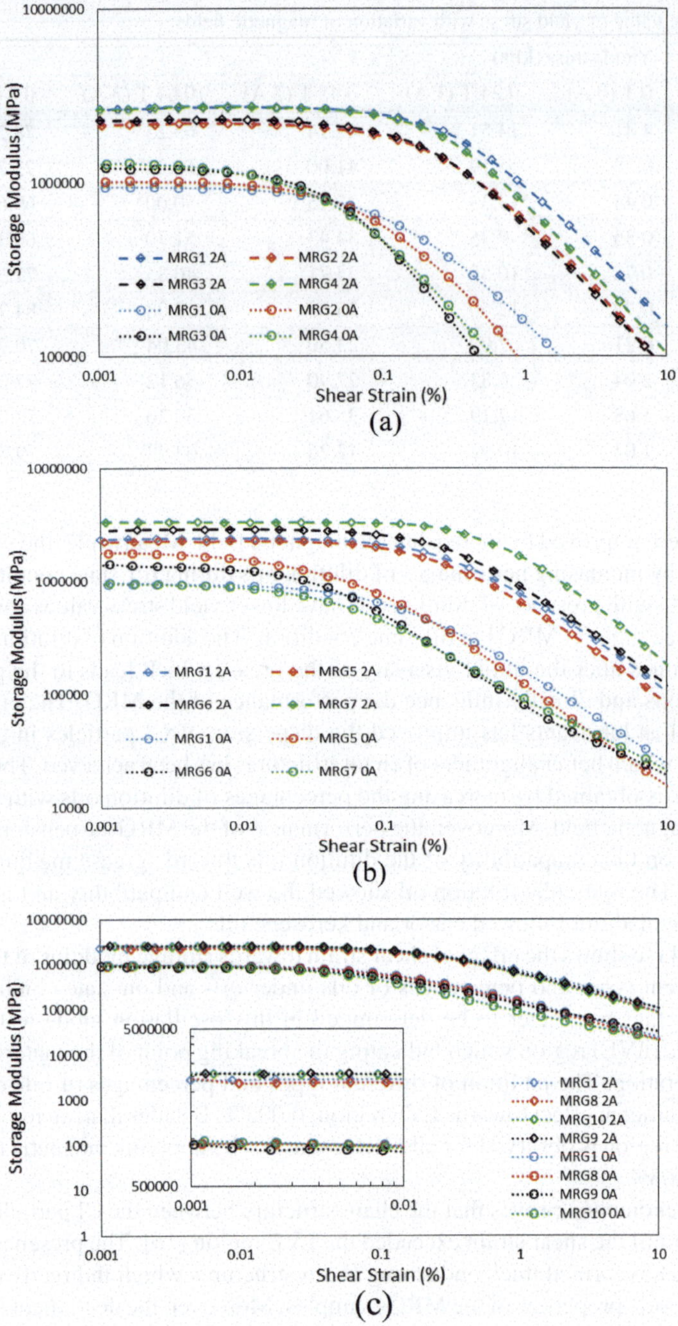

Fig. 4 Comparison of storage modulus for MRGs as function of shear strain under off- and on-state conditions. **a** MRGs with castor oil, **b** MRGs with kerosene oil and **c** MRGs with hydraulic oil

stability and viscosity of the samples. In addition, the usage of thick grease as suspending medium has increased the restriction on the CI particles movements. Thus, addition of the dilution oils in small amount has not affected the grease properties. Consequently, more energy able to be stored and reduced the energy dissipation from the MRGs samples.

4 Conclusion

In this work, the influence of different types and percentages of dilution oils in MRG towards rheological properties have been investigated. The importance of this work is to study the behavior of the suspension on apparent viscosity, yield stress as well as stability with the presence of dilution oils. Thus, ten samples were prepared with different types and percentages of dilution oils. The dilution oils utilized in this study were castor, kerosene and hydraulic oils, while the percentages variate from 5 to 15% from total weight of suspension. The rheological properties of MRGs were characterized using commercial rheometer. The apparent viscosity including yield stress of the MRGs samples were reduced with the increment on dilution oils percentages. The dilute suspension of MRGs was observed for kerosene oil followed by castor oil, hydraulic oil and reference sample. Besides, the oil separation has been appeared in the MRG with 15% of kerosene oil in 24 h observation compared to the other samples. This is happened owing to the incompatible structuration formation between oil and grease fibrous structure. In the meantime, the addition of dilution oils percentages between 5 and 15% shows insignificant effect on the LVE region for both conditions. Finally, it can be summarized that the rheological properties of MRG is dependent on the types and amount of the dilution oil added to the suspension. It is remarked that a comprehensive investigation on the viscoelastic and physical properties of the MRG with dilution oils will be undertaken as a second phase of this work.

Acknowledgements This study was financially supported by the Universiti Teknologi Malaysia under Transdisciplinary Research Grant (Vot No: 06G77) and Professional Development Research University grant (Vot No: 04E02). The research is also partially funded by USAID through Sustainable Higher Education Research Alliances (SHERA) Program—Center for Collaborative (CCR) National Center for Sustainable Transportation Technology (NCSTT) with Contract No. IIE00000078-ITB-1, as well as Universitas Sebelas Maret (UNS) through Hibah Kolaborasi Internasional 2019.

References

1. Jolly MR, Carlson JD, Munoz BC (1996) A model of the behaviour of magnetorheological materials. Smart Mater Struct 5:607–614
2. Park BO, Park BJ, Hato MJ, Choi HJ (2011) Soft magnetic carbonyl iron microsphere dispersed in grease and its rheological characteristics under magnetic field. Colloid Polym Sci 289:381–386
3. Sahin H, Wang X, Gordaninejad F (2009) Temperature dependence of magneto-rheological materials. J Intell Mater Syst Struct 20:2215–2222
4. Fu LJ, Liao CR, Ma JZ (2011) Test method for rheological behaviors of magnetorheological grease based on coaxial cylinder shear mode and rheometer. Adv Mater Res 291–294:1929–1934
5. Sukhwani VK, Hirani H (2008) A comparative study of magnetorheological-fluid-brake and magnetorheological-grease-brake. Tribol Online 3:31–35
6. Shah K, Phu DX, Seong M-S, Upadhyay RV, Choi S-B (2014) A low sedimentation magnetorheological fluid based on plate-like iron particles, and verification using a damper test. Smart Mater Struct 23:027001
7. Mohamad N et al (2018) A comparative work on the magnetic field-dependent properties of plate-like and spherical iron particle-based magnetorheological grease. PLoS ONE 13:e0191795
8. Gordaninejad F, Miller M, Wang X, Sahin H, Fuchs A (2007) Study of a magneto-rheological grease (MRG) damper. In: Matsuzaki Y, Ahmadian M, Leo DJ (eds) Proceedings of SPIE, vol 6525, pp 65250G–65250G–6
9. Sugiyama S, Sakurai T, Morishita S (2013) Vibration control of a structure using magneto-rheological grease damper. Front Mech Eng 8:261–267
10. Kavlicoglu BM, Gordaninejad F, Wang X (2013) Study of a magnetorheological grease clutch. Smart Mater Struct 22:125030
11. Mohamad N, Mazlan SA, Ubaidillah, Choi SB, Nordin MFM (2016) The field-dependent rheological properties of magnetorheological grease based on carbonyl-iron-particles. Smart Mater Struct 25
12. Kim J, Ko J, Liu Y, Kim I, Choi H (2012) Effect of medium oil on magnetorheology of soft magnetic carbonyl iron particles. 170:5246
13. Leong SAN, Mazlan SA, Mohamad N, Aziz SAA, Ubaidillah (2016) An overview of nanoparticles utilization in magnetorheological materials. AIP Conf Proc 1710:020002
14. Aziz SAA et al (2018) A comparative assessment of different dispersing aids in enhancing magnetorheological elastomer properties. Smart Mater Struct 27
15. de Vicente J, López-López MT, Durán JDG, González-Caballero F (2004) Shear flow behavior of confined magnetorheological fluids at low magnetic field strengths. Rheol Acta 44:94–103
16. Gong X et al (2012) The investigation on the nonlinearity of plasticine-like magnetorheological material under oscillatory shear rheometry. J Rheol (N Y N Y) 56:1375

Effect of Different Curing Conditions on the Morphological and Rheological Properties of Rigid Magnetorheological Foam

Noor Sahirah Muhazeli, Siti Maisarah Abd Aziz, Nur Azmah Nordin, Saiful Amri Mazlan, Siti Aishah Abdul Aziz, and Hafizal Yahaya

Abstract The study aims to investigate the effect of curing conditions on the magnetorheological (MR) foam that consists of carbonyl iron particles (CIPs) in the matrix of polyurethane (PU) foam. MR foams with 35 wt% of CIPs are fabricated via in situ method in two different curing conditions, particularly in the absence and presence of magnetic field known as isotropic and anisotropic conditions, respectively. The morphology between isotropic and anisotropic MR foams is compared, also with the virgin foam as control sample and correlation towards the changes of rheological properties are analyzed. The rheological properties of MR foams related to MR effect and storage modulus, under sweep frequencies at various magnetic fields, are further examined via rheometer. The results display an oriented structure of CIPs between the struts of MR foam after curing in anisotropic condition, whereas the random distribution of CIPs has been observed in the isotropic MR foam, and no CIPs identified in the struts of virgin foam. Meanwhile, in the rheological test, it is revealed that the anisotropic MR foam resulted in the highest MR effect, about 89% compared to the isotropic MR foam that possesses only 10%. The increment of storage modulus towards the on-state and off-state frequencies swept also slightly different for each curing conditions which revealed that the anisotropic MR foam obtained the highest value of storage modulus which is ~1.33 MPa compared to the isotropic MR foam and the virgin foam, that only presented at ~1.14 MPa and ~0.84 MPa, respectively. The results from this study are believed to contribute the body of knowledge in producing various dynamic performances of MR foams with controllable structures and properties, for different areas and applications.

N. S. Muhazeli · S. M. A. Aziz · N. A. Nordin (✉) · S. A. A. Aziz · H. Yahaya
Engineering Materials and Structures (eMast) Research Laboratory, Malaysia-Japan International Institute of Technology (MJIIT), Universiti Teknologi Malaysia, Jalan Sultan Yahya Petra, 54100 Kuala Lumpur, Malaysia
e-mail: nurazmah.nordin@utm.my

S. A. Mazlan
Advanced Vehicles Systems (AVS) Research Laboratory, Malaysia-Japan International Institute of Technology, Universiti Teknologi Malaysia, Jalan Sultan Yahya Petra, 54100 Kuala Lumpur, Malaysia

U. Sabino et al. (eds.), *Proceedings of the 6th International Conference and Exhibition on Sustainable Energy and Advanced Materials*, Lecture Notes in Mechanical Engineering,
https://doi.org/10.1007/978-981-15-4481-1_18

Keywords Magnetorheological (MR) foam · Isotropic · Anisotropic · Morphological · Rheological · MR effect

1 Introduction

Magnetorheological (MR) materials are categorized as one of the smart materials which acts continuously, rapidly and reversibly change in its rheological properties by presence of magnetic field [1]. MR foam is classified as one of the MR materials, comprised of micron-sized magnetically permeable particles that are embedded in the absorbent matrix of foam [2, 3]. In addition, MR foam has various interesting advantages such as controllable, lightweight, excellent sound and shocks absorbing properties and exhibits better insulation performance [3–5]. Previously, the area of applications of conventional composite foam was generally commercialized in construction industries and appliance materials insulation [6]. Recently, the innovative research on developed MR foam has led this materials to be potentiality used in acoustic industry as it contribute great performance towards the acoustic absorption properties, especially in the low-frequency region of sound absorption that was difficult to be achieved by conventional composite foam [4, 7]. It is believed that the addition of CIPs in the rigid foam could enhance the passive foam to become an active one in order to absorb both in low and high frequency ranges of sound and noises along with other properties that can be tuned controllably by manipulating induced magnetic field.

MR foam can be fabricated via two different curing conditions such as; isotropic and anisotropic. Isotropic curing condition can be described as the fabrication of MR foam with randomly dispersed magnetic particles (CIPs), forming in the absorbent matrix without the presence of the magnetic field [8]. Meanwhile, anisotropic curing conditions can be described as the alignment of the magnetic particles (CIPs) along the magnetic flux direction during curing and foaming process [9]. According to Sorrentino et al. [10], anisotropic curing condition of MR foam resulted in the strong enhancement of stress-strain behavior compared to isotropic curing condition due to reordering of micro-particles along the magnetic field. Plus, a study by Gong et al. [2] had notably proved that anisotropic MR foam has a higher value of acoustic absorption coefficient and wider frequency range of acoustic absorption coefficient compared to isotropic MR foam. In the meantime, the mechanical properties related to mechanical stiffness and strength also increased in the anisotropic MR foam as compared to isotropic MR foam [10–12].

However, the study of MR foam is still in the early stage, while the detail of mechanism and the correlation between modifications structure of the rigid type of MR foams towards the changes of rheological properties are still limited and incomplete. In order to achieve this target, rigid MR foam samples are fabricated via different curing conditions, particularly isotropic and anisotropic. The MR foam samples consist of; CIPs as the reinforcement filler, and polyol and isocyanates as the main matrix fabricated via an in situ method of fabrication. In this study also, the

Table 1 Composition of MR foam with 35 wt% CIPs

Sample	Total weight of polyol and isocyanates (g) (1:1)	Quantity of CIPs (g)	Curing conditions	Remark
A	20	0	Isotropic	The control sample (virgin foam)
B	20	7	Isotropic	MR foam
C	20	7	Anisotropic	MR foam

investigation on the effect of different curing conditions; isotropic and anisotropic on the morphological properties and rheological performance of MR foams are further evaluated. This novel feature is important to be analyzed in order to evaluate the correlation between the CIP's distribution and the foam pores as this would contribute to the final rheological properties of rigid MR foams. The details of the results are presented in this paper.

2 Materials and Experimental Method

2.1 *Materials*

MR foam was prepared by using two main ingredients which were; (1) polyurethane (PU) foam component that consist of polyether polyols (PP) and polymethylene polyphenyl polyisocyanates (PPP) agent and (2) CIPs as the reinforcement filler that was obtained from CK Materials Laboratory Co. Ltd., Seoul, Korea. The control sample (virgin foam) and MR foams samples (35 wt% of CIPs) with and without the appliance of the magnetic field applied (isotropic and anisotropic), were fabricated in this study. The detail information on the total weight of PU foam components, the quantity of CIPs and the type of curing conditions were tabulated in Table 1.

2.2 *Fabrication of MR Foam*

Isotropic and anisotropic MR foam samples were prepared via in situ fabrication by using a mechanical stirrer at a speed of 350 rpm, at room condition. At first, polyol was stirred for five minutes until the white color of the mixed polyol has appeared. Then, CIPs were added and mixed with the polyol for five minutes. Next, isocyanates were poured into the mixture and continuously stirred for 10 s. Then, the whole mixture was poured immediately into a mold and it was left for curing; no magnetic field applied for an isotropic sample during condition and for anisotropic sample, the magnetic field of 0.2 A was used during the curing process. After that, it was left at room temperature for 24 h. As a contrast, the virgin foam, without the addition of CIPs was prepared.

2.3 *Morphological Characterization*

To observe the dispersion of CIPs in the matrix of MR foam, Low Vacuum-Scanning Electron Microscope (LV-SEM), JEOL JSM-IT300 was used in this study. The purpose of morphological characteristics was to observe the distribution of the CIPs in the foam structures, after undergoing two different curing conditions; isotropic and anisotropic. The MR foams samples were observed at an accelerating voltage of 5 kV, with the capability of 950 times of magnification (20 μm).

2.4 *Rheological Properties Test*

The rheological tests were performed using a rotational rheometer (Physica MCR 302, Anton Paar Company, Germany). An oscillatory shear test method had been applied for the rheological testing and the measuring temperature is fixed at 25 °C, which was controlled by the temperature control device (Viscotherm VT2, Anton Paar, Germany). The samples used for rheological tests have a diameter and thickness of 20 mm and 1 mm, respectively. The circular sample was placed between a rotating disk and a parallel base plate (PP20/MRD/TI). The current sweep test and frequency sweep testing are carried out to analyze the effect of the magnetic field towards the rheological properties of MR foam. For the current sweep test, the frequency was kept constant, at 0.0001% and 1 Hz, and the applied of magnetic field was varied from 0 to 0.8 T. While, the frequency sweep test was performed by varying the frequency from 1 Hz until 100 Hz and varied current at 0 A (off state) and 2 A (on state), at a constant strain of 0.0001%.

3 Result and Discussion

3.1 *Morphological Characterization*

The LV SEM images of virgin foam and MR foams fabricated via different curing conditions; isotropic and anisotropic condition are presented in Fig. 1.

Figure 1a showed the morphological micrographs of virgin foam and it can observed that the strut of the virgin foam pores was vacant (no CIPs) compared to Fig. 1a, b. In addition, in Fig. 1b which represent the isotropic MR foam, it was

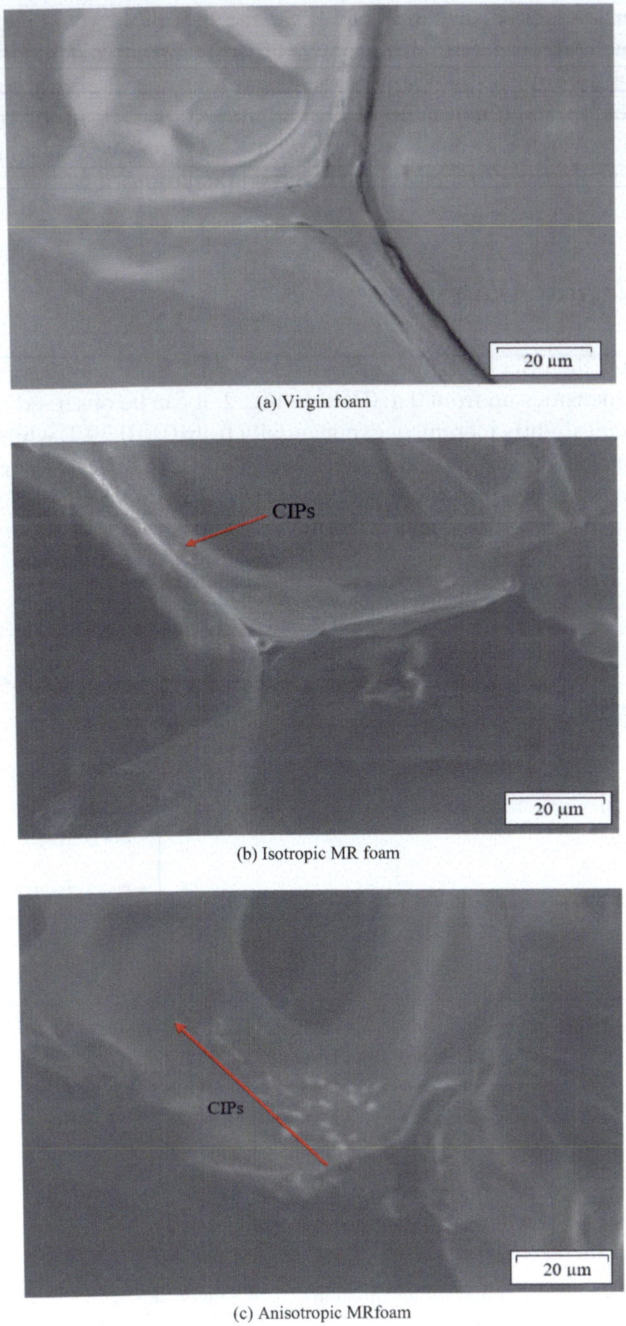

(a) Virgin foam

(b) Isotropic MR foam

(c) Anisotropic MRfoam

Fig. 1 LV SEM images of MR foams

clearly shown a random distribution of CIPs in the strut of the foam pores compared to Fig. 1c; which represent the anisotropic MR foam sample that depicts an alignment of the CIPs in the strut of the foam pores. Thus, from the closed-up images of LV SEM, it can be observed that there was an alignment of CIPs in the matrix of MR foam structures which was cured in anisotropic condition due to the application of magnetic field compared to the MR foam cured in isotropic condition.

3.2 *Rheological Analysis*

Figure 2 shows the effect of storage modulus on MR foam samples as a result of magnetic flux densities up from 0 to 0.8 T. In Fig. 2, it can be observed that isotropic MR foam exhibit slightly increment exponentially from 0 to 0.39 T, while anisotropic MR foam depicted a significant increment from 0 to 0.56 T. Besides, in term of the of storage modulus, both isotropic and anisotropic MR foam showed a difference of 0.54 MPa and 2.68 MPa (Table 2) in the increment values of storage modulus,

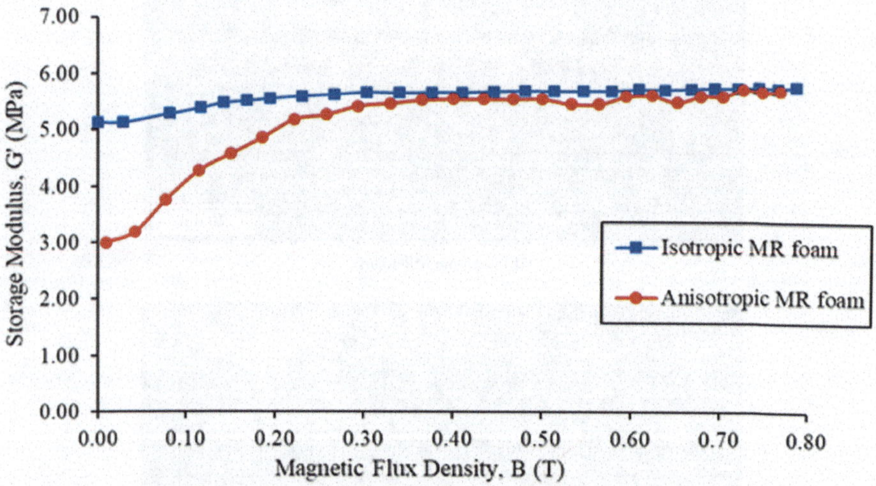

Fig. 2 Effect of magnetic fields on the storage modulus of MR foams with 35 wt% CIPs in isotropic and anisotropic curing conditions

Table 2 MR effect of the MR foams at two curing conditions

MR foam sample (35 wt% CIPs)	Viscoelastic properties of rigid MR foams		
	Initial storage modulus, G'_0 (MPa)	Maximum storage modulus, G'_{max} (MPa)	MR effect (%)
Isotropic	5.13	5.67	10
Anisotropic	3.00	5.68	89

respectively. Thus, it is observed that the range value of the storage modulus of anisotropic MR foam was higher compared to isotropic MR foam, even though the isotropic MR foam sample was obviously stiffer compared to anisotropic MR foam. This phenomenon is believed due to the pore's growth factor that is inhibited by the higher density and hardness of CIPs in isotropic MR foam [4]. Meanwhile, compared to anisotropic MR foam, the magnetic field that was applied has decreased the obstruction of higher density and hardness of CIPs and led to the higher range value of storage modulus which provided significant improvement towards MR foam material characterization.

Moreover, the average values of the MR effect in terms of percentage (%) were calculated and presented in Table 2. From the plotted graph in Fig. 2 and Table 2, it shown that the MR foam fabricated in anisotropic curing condition has superior MR effect compared to isotropic MR foam which exhibits particularly at 89% and 10% of the MR effect average values, respectively. This phenomenon can be explained by the incorporation of CIPs and the application of the magnetic field that was applied during the curing condition towards the MR foam development during the foaming process. Also, the presence of the magnetic field during the fabrication of anisotropic MR foam attributed to the increase in the particle interaction of CIPs that leads to stronger filler-matrix interactions [13]. This also agrees well with the finding from the previous research in which the anisotropic MR foam showed a higher MR effect compared to isotropic MR foam [9].

Generally, the motion of the polymer molecular chain is frequency dependence, hence, in this study, the frequency dependence of MR foam cured with and without the magnetic field was evaluated. Figure 3 presents the changes of MR foam storage

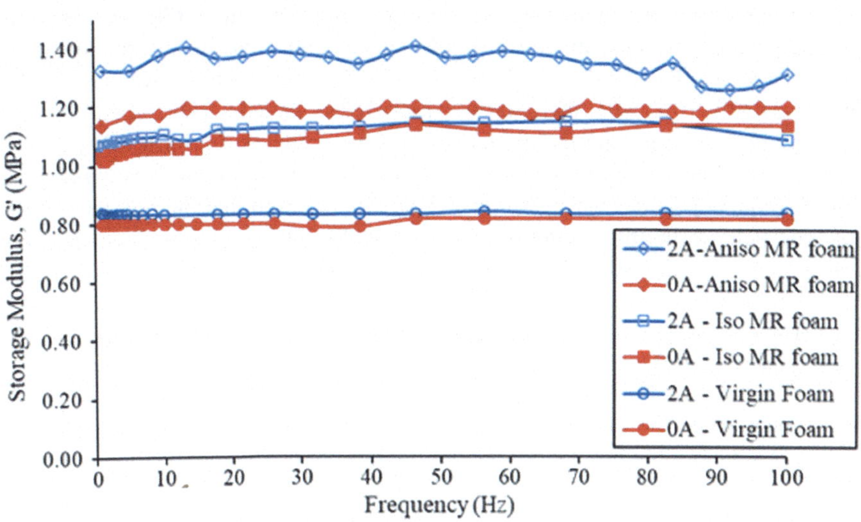

Fig. 3 Effect of frequencies on the storage modulus of virgin foam and MR foam with 35 wt% CIPs at 0 and 2 A applied current

modulus, correspond to frequencies at off state (0 A) and on state (2 A) conditions. The graph revealed that there is an increment in the initial storage modulus as the frequencies and the currents increased. Moreover, at on-state condition (2 A), the anisotropic MR foam indicated the highest initial storage modulus at 1.33 MPa compared to isotropic MR foam and virgin foam which at 1.14 and 0.84 MPa. A similar pattern is also observed in off state (0 A) condition in which anisotropic MR exhibit the highest storage modulus at 1.18 MPa compared to isotropic MR foam and virgin foam which at 1.02 and 0.79 MPa. This result agreed well with the finding of previous studies where the storage modulus increases when there was an increase in the current applied on the MR samples [2]. As the current increase, the values of storage modulus of virgin foam and MR foam curing in anisotropic and isotropic also increased. However, in terms of frequency, it was shown that the storage modulus of virgin foam and MR foam was independent of the frequency applied. When there is an increase in the frequency values, there is no significant effect on the values of the storage modulus of the virgin foam and the MR foam samples.

4 Conclusion

In this study, the effects of two curing conditions particularly isotropic and anisotropic curing conditions on MR foam are investigated. The microstructure and the rheological properties were characterized in details. The CIPs formed an aligned structure in the struts of the MR foam after magnetic field was applied during the curing condition, compared to the MR foam that was cured without the presence of CIPs that were randomly distributed and virgin foam which was no CIPs was generally identified in the struts of foam pores. Besides, it was found that anisotropic MR foam depicted a high value of MR effect which was 89% compared to isotropic MR foam which was only 10%. MR foam that was prepared in an anisotropic condition also indicated higher initial storage modulus in the frequency sweep test compared to isotropic MR foam and virgin foam. However, the storage modulus was independent of frequency values was observed. Thus, it is believed that the presence of the magnetic field during the curing of MR foam had influenced the performance and corresponding on the MR effect of the materials. Due to the unique properties of MR foam, this material can be potentially used as the sound insulation and absorption parts which can outspread the interesting materials applications, especially in vibration and acoustic absorption.

Acknowledgements This study was financially supported by the Universiti Teknologi Malaysia under Trans-disciplinary Research Grant (Vot No: 07G13) and Fundamental Research Grant Scheme (Vot No: 5F001).

References

1. Ubaidillah, Sutrisno J, Purwanto A, Mazlan SA (2014) Recent progress on magnetorheological solids: materials, fabrication, testing, and applications. Adv Eng Mater 17:563–597
2. Gong Q, Wu J, Gong X, Fan Y, Xia H (2013) Smart polyurethane foam with magnetic field controlled modulus and anisotropic compression property. RSC Adv 3:3241–3248
3. Carlson JD, Jolly MR (2000) MR fluid, foam and elastomer devices. Mechatronics 10:555–569
4. Muhazeli NS, Nordin NA, Mazlan SA, Rizuan N, Choi S (2019) Characterization of morphological and rheological properties of rigid magnetorheological foams via in situ fabrication method. J Mater Sci 54:13821–13833
5. Nski I (2010) Acoustic absorption of foams coated with MR fluid under the influence of magnetic field. Intell Mater Syst Struct 21:125–131
6. Verdolotti L, Di Caprio MR, Lavorgna M, Buonocore GG (2017) Polyurethane nanocomposite foams: correlation between nanofillers, porous morphology, and structural and functional properties. In: Polyurethane polymers: composites and nanocomposites. Elsevier, pp 277–309
7. Wu J, Gong X, Fan Y, Xia H (2010) Anisotropic polyurethane magnetorheological elastomer prepared through in situ polycondensation under a magnetic field. Smart Mater Struct 19:105007–105013
8. Weng L, Wang T, Ju P-H, Liu L-Z (2018) Preparation and properties of polyimide/Fe_3O_4 composite foams. Pigm Resin Technol 47:173–179
9. Sorrentino L, Aurilia M, Forte G, Iannace S (2008) Composite polymeric foams produced by using magnetic field. Adv Sci Technol 54:123–126
10. Sorrentino L, Aurilia M, Cafiero L, Iannace S (2011) Nanocomposite foams from high-performance thermoplastics. J Appl Polym Sci 122:3701–3710
11. D'Auria M, Davino D, Pantani R, Sorrentino L (2016) Polymeric foam-ferromagnet composites as smart lightweight materials. Smart Mater Struct 25:055014–055027
12. Davino D, Mei P, Sorrentino L, Visone C (2012) Polymeric composite foams with properties controlled by the magnetic field. IEEE Trans Magn 48:3043–3046
13. Niedermeier W, Luginsland H, Fro J (2005) The effect of filler–filler and filler–elastomer interaction on rubber reinforcement. Compos Part A Appl Sci Manuf 36:449–460

Mini Review on Effect of Coatings on the Performance of Magnetorheological Materials

S. K. Mohd. Jamari, U. Ubaidillah, Siti Aishah Abdul Aziz, Nur Azmah Nordin, A. Fajrin, and Saiful Amri Mazlan

Abstract Magnetorheological materials have attracted a great deal of interests nowadays due to its controllable mechanical properties upon the application of external magnetic field. Its ability to change its rheological properties in a split second has found its way in the applications that require absorption and isolation of vibration and noise. However, the problems with oxidation, sedimentation and aggregation of the magnetic particles hinder the optimum performance that can be utilised with this smart material. This includes the reduced performance of yield stress, shear stress, shear modulus and storage modulus and over a long operational period, will affect its magnetisation properties. Hence, there is a need to protect the magnetic particles with coating layer which can overcome these drawbacks. The main focus of this work is to present an overview on the aforementioned problems in MR materials that can be controlled by applying protective coating on the magnetic particles. Several works have reported the enhancement of performances such as oxidation resistance, interface between particles and the carrier medium as well as sedimentation stability by introducing coated magnetic particles in the MR materials.

Keywords Magnetorheological · Coating · Oxidation · Rheology · Magnetic particles

1 Introduction

Smart materials can be defined as materials that possess designed properties which can be controlled by applying external stimuli such as thermal, pH, electric field,

S. K. Mohd. Jamari · S. A. A. Aziz · N. A. Nordin (✉) · S. A. Mazlan
Malaysia-Japan International Institute of Technology, Universiti Teknologi Malaysia, Jalan Sultan Yahya Petra, Kampung Datuk Keramat, 54100 Kuala Lumpur, Malaysia
e-mail: nurazmah.nordin@utm.my

U. Ubaidillah · A. Fajrin
Mechanical Engineering Department, Faculty of Engineering, Universitas Sebelas Maret, Jalan Ir. Sutami 36A, Kentingan, Surakarta 57126, Central Java, Indonesia

U. Ubaidillah
National Center for Sustainable Transportation Technology (NCSTT), Bandung, Indonesia

U. Sabino et al. (eds.), *Proceedings of the 6th International Conference and Exhibition on Sustainable Energy and Advanced Materials*, Lecture Notes in Mechanical Engineering,
https://doi.org/10.1007/978-981-15-4481-1_19

magnetic field, moisture, light, wavelength etc. These stimuli-responsive materials can change their behaviour in terms of physical and mechanical properties (e.g. shape, size, and viscosity, thermal, and optical properties) rapidly and interchangeably.

One of the smart materials that have caught the researchers' interest world-wide nowadays is magnetorheological (MR) material. Also known as magneto-active or magneto-sensitive materials, MR materials have alterable rheological properties relative to the magnitude of external magnetic field applied on it. This reversible process can continuously be accomplished in a split second [1–8]. MR material consists of magnetic particles suspended in a carrier medium such as fluid, grease, foam, elastomer and recently, plastomer, which then can be categorised as MR fluid, MR grease, MR foam, MR elastomer and MR plastomer, respectively. The types of carrier media used in MR materials have been summarised in Table 1.

MR fluid is the first MR material that was developed with the idea of controlling the rheological properties of the material by tailoring the applied external magnetic field. The magnetic particles are dispersed in a non-colloidal suspension which give excellent apparent yield stress. This enables MR fluid to become a good prospect in the applications that require active control of vibration and transmission of torque [5]. The material possesses Bingham fluid characteristics under the application of the magnetic field. MR grease/gel was then introduced to reduce the severe sedimentation in MR fluid due to higher viscosity of its carrier medium compared to that of MR fluid. Its carrier medium that has state between elastic solid and liquid, behaves like solid gel at room temperature but acts like fluid upon increasing of temperature above the phase change point [5]. Later, MR elastomer was established with the aim

Table 1 Summary of some of carrier media used in MR materials

Types of MR material	Carrier medium	References
MR fluid	Silicone oil	[9–12]
	Hydrocarbon oil	[13, 14]
	Honey	[15]
MR grease	Commercial grease	[16, 17]
	Modified silicone oil	[18]
	Polyurethane + castor oil	[19]
MR foam	Polyurethane	[20–22]
	Silicone PDMS	[23]
MR elastomer	Silicone rubber	[24–26]
	Natural rubber	[27–29]
	Styrene-ethylene-butylene-styrene (SEBS)	[7]
	PDMS	[30, 31]
	Polyurethane	[32]
MR plastomer	Polypropylene glycol + toluene diisocyanate	[33, 34]
	Paraffin wax + petroleum jelly	[35]

of locking the magnetic particles in the matrix of medium carrier and eventually diminish the accumulation problem of magnetic particles that occur in MR fluid and MR grease/gel.

Due to their field-responsive behaviour, MR materials have wider its applications in automotive industry (e.g. suspension system, dampers), noise reduction and soundproofing in acoustic absorption, vibration attenuation in seismic isolators, electromagnetic shielding and smart sensing [7]. Such applications are possible for MR materials due to behaviour of the magnetic particles embedded/suspended in MR materials. During the absence of external magnetic field (off-state), the magnetic particles are dispersed throughout the medium carrier. Once an external magnetic field are applied on the sample (on-state), the magnetic particles are polarised according to the direction of the magnetic. The formation of chain-like structure due to the applied magnetic field causes the flow of the carrier medium to be restricted, thus increasing the viscosity or stiffness of the medium [4]. The mechanism of the MR materials can be simplified as shown in Fig. 1. The most commonly used magnetic particles is carbonyl iron particles (CIP) while increased interest in other magnetic particles such as reduced iron [36], manganese ferrite [12], cobalt ferrite [37] and neodymium iron boron [34] has been observed. CIP has been widely used in MR materials due to its high permeability, low remnant magnetisation and high saturation magnetisation [10, 30].

The applied external magnetic field can also contribute another factor to the MR materials; the curing process of MR foam, MRE and MRP. This process either with or without presence of magnetic field is called as isotropic and anisotropic curing process, respectively. In isotropic condition, with the absence of external magnetic field during fabrication process will allow the magnetic particles to randomly distribute in matrix phase of MR materials. While in the anisotropic condition, the distribution of the particles is dependent on the direction of magnetic field and thus,

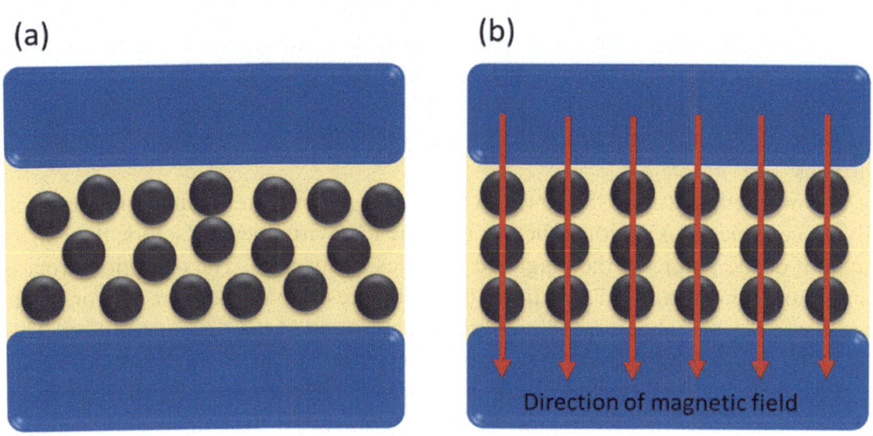

Fig. 1 The mechanism of MR materials during off-set (**a**) and on-set (**b**)

particle chains are obtained [24]. This will be resulted in both mechanical and magnetic anisotropy, which means the cured material possesses different properties along different axes [38], hence improving the mechanical properties such as dynamic stiffness and damping of the composite material [39, 40] as well as higher MR effect compared to isotropic composites [41]. It is also found that the loss and the storage modulus of the bare material can be enhanced with anisotropic MR materials compared to isotropic one which exhibited a reduction in both parameters [42]. Sapouna et al. have developed a novel isotropic/anisotropic MRE composite which exhibited higher tangent of the loss angle than isotropic MRE by keeping the stiffness lower than the anisotropic without reducing its MR effect [43].

2 Problems Related to MR Materials

Despite the improvement of the mechanical properties and the advantages that MR material possesses over conventional dampers and vibration isolators, it is far from being a perfect material. Serious sedimentation of magnetic particles in MRF have been widely discussed. High polydispersity or mismatch of densities between the magnetic particles and the carrier fluid can cause the particles to settle within a short period of time [10, 44, 45]. The sedimentation stability in MRF can also be increased by adding fillers in the carrier fluid. These fillers may come in the form of nano-particles such as organic clay, fumed silica, carbon nanotubes and graphite micro-particles which can form space particulate network [3, 44] as well as surfactants. However, the increased viscosity of the fluid can be another problem in some application during the off-state condition [44].

The sedimentation also can initiate other problem; the aggregation of the magnetic particles, which can cause wear and abrasion due to friction among the particles [46, 47]. Hu et al. and Zhang et al. found that the higher the magnetic field applied on MRF, the higher the coefficient of friction as well as the wear volume when tested under fretting conditions [48, 49]. This not only could occur in MRF, but also in other MR materials as the continuous process of on- and off-state may also contribute to the wear problem, especially in MR materials with high percentage of magnetic particles.

The magnetic particles dispersed in the carrier medium in MR materials are usually susceptible to oxidation, whether it is due to environmental factors, the carrier medium or the nature of the magnetic particles themselves. The oxidation process can be undergone in "wet" condition (with the presence of moisture and/or acidic medium) or in the absence of humidity, which is called the "dry" condition. This "dry" oxidation or also known as thermo-oxidation is caused by the combination of other oxidants such as oxygen and high heat [50]. In the case of MRE, although the magnetic particles are not exposed to air, reactive species will be able to diffuse through the elastomeric matrices over time, thus, alter the properties of the elastomer as well as the bonding between the matrices and the particles [30, 51]. This can affect the performance of the MR material itself, i.e. iron oxides have relatively lower saturation magnetisation, M_s compared to carbonyl iron [9]. Plus, the off-state viscosity of the media can be undesirably increased due to the oxidation products after a long term operation [52]. The iron oxide layer on the surface of the CIP in MRE can further degrade the polymeric elastomer due to large amount of oxygen that have been integrated into the matrices [30]. Lokander et al. reported that the incorporation of iron particles in the rubber matrices causes faster formation of oxidised skin of the elastomer as well as faster degradation of the sample. The higher amount of the iron particles embedded, the lower the oxidative stability. In the most probable theory explained by the team, this is due to the catalytic effect of the iron ions on the decomposition of the hydroperoxides [29]. The hygrothermal effect (the effect due to the movement of heat and moisture) experienced by the MR materials may also reduce the strength of the bonds between the magnetic particles and the carrier medium itself, especially for MRE and MR foam.

Therefore, extensive studies have been carried out to reduce the drawbacks mentioned before, hence the idea of coating the particles has been proposed. Coated magnetic particles have been proved to solve the sedimentation problem in MRF and thus eliminate the aggregation problem.

3 Several Works on Coated Magnetic Particles for MR Material Applications

Park et al. demonstrated that the dispersion stability in their MRF that consists of lubricant oil as the medium carrier has been increased by replacing PMMA-coated CIP with the pristine CIP [46]. The sedimentation stability was improved due to the particles having lower density compared to the uncoated ones while the higher re-dispersion of the particles after caking test was credited to the surface property of the cross-linked PMMA such as the electrostatic repulsion and the steric repulsion of the polymer.

The dispersion and sedimentation stability improvement of the MRF have also been proved by Tae et al. [47] with the grafting of polyaniline (PANI) on the CIP. They found that the rate of sedimentation of the PANI-coated CIP is lower than that of pure CIP which suggesting that it has better dispersity. This enhancement is said to be attributed to the reduction of density mismatch between the particles and the carrier fluid as well as the shape of the coated particles where the slightly non-spherical defects and the rough surface of the coating increases the surface area of the particles and thus decreases the density of the particles.

Fuchs et al. found that fluorinated styrene coated iron particles in MRE has superior mechanical properties after being exposed to oxidation test for 72 h compared to that of uncoated one [53]. This was proved as the coated iron particles MRE required only 3% of force to achieve 20% strain compared to 17% of that of uncoated samples after the oxidation test.

An accelerated heat chamber is used by Behrooz et al. to investigate the behaviour of pure elastomer, uncoated CIP MRE and coated CIP MRE under oxidative environment [51]. It was found that while the oxidation affects the elastomer matrices of all three samples, the coated CIP MRE demonstrated the lowest decrease of effective shear modulus by 2.3 times compared to nearly 4 times of uncoated CIP MRE and 3 times of pure elastomer. This shows that coated particles in MRE preserves the stiffness of MRE in oxidative environment.

A study by Yu et al. showed that polyaniline-coated CIP were capable to improve the interfacial interactions between the CIP and polyurethane/epoxy matrix of MRE due to the covalent bonds between these two component [54]. It was claimed that due to this improved property, the coated CIP MRE displayed higher storage modulus, lower loss factor and smaller magnitude of Payne effect.

Cvek et al. found that MRF with poly(glycidyl methacrylate) (PGMA) coated CIP has improved thermo-oxidation stability and increase the oxidation temperature of the particles. They also noted that there is only slight enhancement of thermo-oxidation when the molecular weight of PGMA was doubled. The sedimentation stability has also significantly enhanced due to the decreasing of bulk density with improved particles-silicone oil interaction due to the PGMA chains presence [44]. In another work, also done by Cvek et al. the same polymer chain showed excellent resistance to acid-corrosion. An interesting added point to this PGMA polymer chain is the absence of toxicity which makes it a promising material for drug delivery biomedical applications [55].

The declining of damping property and MR effect has also been observed by Malecki et al. in their MRE containing silica-coated CIP. The team suggested that the better adhesion of CIP to the matrix and the greater distances between the CIP which causes weaker magnetic interactions are the causes of such behaviour of the MRE [7].

A custom-designed MR brake was used by Nguyen et al. to evaluate the performance of silica-coated CIP of MRF in MR brake system [45]. It was found that the coated CIP MRF with lower density and higher thermal and chemical stability exhibited better tracking accuracy compared to the MRF that consists of pristine CIP [45].

Despite the enhancement of several properties in MR, there is also a disadvantage of having coated magnetic particles in MR samples. While magnetisation properties of the samples are crucial in MR materials, not all works have discussed on this, and if there is any, almost all works reported a decline of magnetisation, mainly the magnetic saturation, in MR samples with coated particles [7, 10, 25, 30, 44, 46, 47, 54–56]. Theoretically and logically, non-magnetic layer surrounding a magnetic particles certainly decreases the magnetic properties of the material. To the best of our knowledge, there is no work reported that is able to maintain or increase the magnetic saturation of MR materials while containing coated magnetic particles in them.

4 Summary and Outlook

This overview has attempted to cover the problems related to MR materials that are successfully being dealt with applying coating layer on the magnetic particles as well as some recent works that have been successfully overcame the aforementioned problems. However, the decrease in magnetic properties that are vital in the operation of MR materials has not been able to be overcome yet, therefore there are potentials remain to be made in the future in the development of protective coatings on magnetorheological materials.

References

1. Stenberg B, Lokander M, Reitberger T (2004) Magnetorheological elastomers—possibilities and limitations. Annu Trans Nordic Rheol Soc 12:163–170
2. Li W, Zhang X, Du H (2013) Magnetorheological elastomers and their applications. In: Visakh PM, Thomas S, Chandra AK, Mathew AP (eds) Advances in elastomers I: blends and interpenetrating networks. Springer, Berlin, pp 357–374
3. Hu B, Fuchs A, Huseyin S, Gordaninejad F, Evrensel C (2006) Atom transfer radical polymerized MR fluids. Polymer (Guildf) 47:7653–7663
4. Carlson JD, Jolly MR (2000) MR fluid, foam and elastomer devices. Mechatronics 10:555–569
5. Leong SAN et al (2016) An overview of nanoparticles utilization in magnetorheological materials. AIP Conf Proc 1710:020002
6. Mazlan SA, Ekreem NB, Olabi AG (2008) An investigation of the behaviour of magnetorheological fluids in compression mode. J Mater Process Technol 201:780–785
7. Malecki P, Krolewicz M, Krzak J, Kaleta J, Piglowski J (2015) Dynamic mechanical analysis of magnetorheological composites containing silica-coated carbonyl iron powder. J Intell Mater Syst Struct 26(14):1899–1905
8. Ubaidillah, Sutrisno J, Purwanto A, Mazlan SA (2014) Recent progress on magnetorheological solids: materials, fabrication, testing, and applications. Adv Eng Mater 1–35
9. Plachy T, Kutalkova E, Sedlacik M, Vesel A, Masar M (2018) Impact of corrosion process of carbonyl iron particles on magnetorheological behavior of their suspensions. J Ind Eng Chem 66:362–369
10. Zhang P, Zhen Y, Jin H, Lee C (2018) Tribological and rheological tests of core-shell typed carbonyl iron/polystyrene particle-based magnetorheological fluid. J Ind Eng Chem 68:342–349
11. Lee JH, Choi HJ (2018) Synthesis of core-shell formed carbonyl iron/polydiphenylamine particles and their rheological response under applied magnetic fields. Colloid Polym Sci 296:1857–1865
12. Wang G, Ma Y, Tong Y, Dong X (2017) Development of manganese ferrite/graphene oxide nanocomposites for magnetorheological fluid with enhanced sedimentation stability. J Ind Eng Chem 48:142–150
13. Bramantya MA, Sawada T (2011) The influence of magnetic field swept rate on the ultrasonic propagation velocity of magnetorheological fluids. J Magn Magn Mater 323:1330–1333
14. Lanzetta M, Iagnemma K (2013) Gripping by controllable wet adhesion using a magnetorheological fluid. CIRP Ann Manuf Technol 62(1):21–25
15. Bica I, Anitas EM (2018) Magnetic field intensity effect on electrical conductivity of magnetorheological biosuspensions based on honey, turmeric and carbonyl iron. J Ind Eng Chem 64:276–283

16. Mohamad N, Mazlan SA, Ubaidillah (2016) Effect of carbonyl iron particles composition on the physical characteristics of MR grease. AIP Conf Proc 1717:040027
17. Gordaninejad F, Miller M, Wang X, Sahin H, Fuchs A (2007) Study of a magneto-rheological grease (MRG) damper. In: Active and passive smart structures and integrated systems, vol 6525, pp 1–6
18. Zheng J, Li Y, Wang J, Shiju E, Li X (2018) Accelerated thermal aging of grease-based magnetorheological fluids and their lifetime prediction. Mater Res Express 5:085702
19. Yang P, Yu M, Luo H, Fu J, Qu H, Xie Y (2017) Improved rheological properties of dimorphic magnetorheological gels based on flower-like carbonyl iron particles. Appl Surf Sci 416:772–780
20. Sung G, Wan J, Hyeun J (2016) Fabrication of polyurethane composite foams with magnesium hydroxide filler for improved sound absorption. J Ind Eng Chem 44:99–104
21. Bandarian M, Shojaei A, Rashidi AM (2011) Thermal, mechanical and acoustic damping properties of flexible open-cell polyurethane/multi-walled carbon nanotube foams: effect of surface functionality of nanotubes. Polym Int 60:475–482
22. Baferani AH, Katbab AA, Ohadi AR (2017) The role of sonication time upon acoustic wave absorption efficiency, microstructure, and viscoelastic behavior of flexible polyurethane/CNT nanocomposite foam. Eur Polym J 90:383–391
23. Makarova LA et al (2019) Magnetorheological foams for multiferroic applications. J Magn Magn Mater 485:413–418
24. Agirre-Olabide I, Kuzhir P, Elejabarrieta MJ (2018) Linear magneto-viscoelastic model based on magnetic permeability components for anisotropic magnetorheological elastomers. J Magn Magn Mater 446:155–161
25. Seung HK, Ji SA, So YC, Kyoung HC, Hyoung JC (2019) Poly(glycidyl methacrylate) coated soft-magnetic carbonyl iron/silicone rubber composite elastomer and its magnetorheology. Macromol Res
26. Guan X, Dong X, Ou J (2008) Magnetostrictive effect of magnetorheological elastomer. J Magn Magn Mater 320:158–163
27. Abdul Aziz SA et al (2018) Effects of multiwall carbon nanotubes on viscoelastic properties of magnetorheological elastomer. Smart Mater Struct 25:077001
28. Yunus NA et al (2016) Rheological properties of isotropic magnetorheological elastomers featuring epoxidised natural rubber. Smart Mater Struct 25:107001
29. Lokander M, Reitberger T, Stenberg B (2004) Oxidation of natural rubber-based magnetorheological elastomers. Polym Degrad Stab 86(3):467–471
30. Cvek M, Mrlik M, Ilcikova M, Mosnacek J, Munster L, Pavliner V (2017) Synthesis of silicone elastomers containing silyl-based polymer-grafted carbonyl iron particles: an efficient way to improve magnetorheological, damping, and sensing performances. Macromolecules 50(5):2189–2200
31. Yao J, Sun Y, Wang Y, Fu Q, Xiong Z, Liu Y (2018) Magnet-induced aligning magnetorheological elastomer based on ultra-soft matrix. Compos Sci Technol 162:170–179
32. Wu J, Gong X, Chen L, Xia H, Hu Z (2009) Preparation and characterization of isotropic polyurethane magnetorheological elastomer through in situ polymerization. J Appl Polym Sci 114:901–910
33. Xu J et al (2018) The dynamic mechanical properties of magnetorheological plastomers under high strain rate. Compos Sci Technol 159:50–58
34. Zhao W, Pang H, Gong X (2017) A novel magnetorheological plastomer filled with NdFeB particles: preparation, characterization and magnetic-mechanic coupling properties. Ind Eng Chem Res 56(31):8857–8863
35. Xuan S, Zhang Y, Zhou Y, Gong X (2012) Magnetic plasticine: a versatile magnetorheological material. J Mater Chem 22:13395–13400
36. Pu H, Jiang F, Wang Y, Yan B (2010) Soft magnetic composite particles of reduced iron coated with poly (p-xylylene) via chemical vapor deposition polymerization. Colloids Surf A Physicochem Eng Asp 361(1–3):62–65

37. Wang G et al (2015) Facile synthesis and magnetorheological properties of superparamagnetic $CoFe_2O_4$/GO nanocomposites. Appl Surf Sci 357:2131–2135
38. Schubert G, Harrison P (2016) Magnetic induction measurements and identification of the permeability of magneto-rheological elastomers using finite element simulations. J Magn Magn Mater 404:205–214
39. Sohoni GB, Mark JE (1987) Anisotropic reinforcement in elastomers containing magnetic filler particles. J Appl Polym Sci 34(8):2853–2859
40. Tian T, Nakano M (2018) Fabrication and characterisation of anisotropic magnetorheological elastomer with 45° iron particle alignment at various silicone oil concentrations. J Intell Mater Syst Struct 29(2):151–159
41. Kukla M, Górecki J, Malujda I, Tala K, Tarkowski P (2017) The determination of mechanical properties of magnetorheological elastomers (MREs). Procedia Eng 177:324–330
42. Puente-Córdova JG, Reyes-Melo ME, Palacios-Pineda LM, Martínez-Perales IA, Martínez-Romero O, Elías-Zúñiga A (2018) Fabrication and characterization of isotropic and anisotropic magnetorheological elastomers, based on silicone rubber and carbonyl iron microparticles. Polymers (Basel) 10(12):1343
43. Sapouna K, Xiong YP, Shenoi RA (2017) Dynamic mechanical properties of isotropic/anisotropic silicon magnetorheological elastomer composites. Smart Mater Struct 26(11):115010
44. Cvek M et al (2015) A facile controllable coating of carbonyl iron particles with poly (glycidyl methacrylate): a tool for adjusting MR response and stability properties. J Mater Chem C 3:4646–4656
45. Nguyen P, Do X, Jeon J, Choi S, Liu YD, Choi HJ (2014) Brake performance of core–shell structured carbonyl iron/silica based magnetorheological suspension. J Magn Magn Mater 367:69–74
46. Park BJ, Kim MS, Choi HJ (2009) Fabrication and magnetorheological property of core/shell structured magnetic composite particle encapsulated with cross-linked poly (methyl methacrylate). Mater Lett 63(24–25):2178–2180
47. Tae HM, Hyoung JC, Kim N-H, Park K, You C-Y (2017) Effects of surface treatment on magnetic carbonyl iron/polyaniline microspheres and their magnetorheological study. Colloids Surf A 531:48–55
48. Hu ZD, Yan H, Qiu HZ, Zhang P, Liu Q (2012) Friction and wear of magnetorheological fluid under magnetic field. Wear 278–279:48–52
49. Zhang P, Lee KH, Lee CH (2017) Fretting friction and wear characteristics of magnetorheological fluid under different magnetic field strengths. J Magn Magn Mater 421:13–18
50. Jamari SKM (2015) Corrosion coatings using conducting polymer and green corrosion inhibitors. University of Malaya
51. Behrooz M, Sutrisno J, Zhang L, Fuchs A, Gordaninejad F (2015) Behavior of magnetorheological elastomers with coated particles. Smart Mater Struct 24(3):35026
52. Roupec J, Mazurek I (2011) Stability of magnetorheological effect during long term operation. In: Jablonski R, Brezina T (eds) Mechatronics. Springer, Berlin, Heidelberg
53. Fuchs A, Sutrisno J, Gordaninejad F, Caglar MB, Yanming L (2010) Surface polymerization of iron particles for magnetorheological elastomers. J Appl Polym Sci 117:934–942
54. Yu M, Qi S, Fu J, Zhu M, Chen D (2017) Understanding the reinforcing behaviors of polyaniline-modified carbonyl iron particles in magnetorheological elastomer based on polyurethane/epoxy resin IPNs matrix. Compos Sci Technol 139:36–46
55. Cvek M et al (2015) The chemical stability and cytotoxicity of carbonyl iron particles grafted with poly (glycidyl methacrylate) and the magnetorheological activity of their suspensions. RSC Adv 5:72816–72824
56. Fan M, He Z, Pang H (2013) Microwave absorption enhancement of CIP/PANI composites. Synth Met 166:1–6

Cartographer Local SLAM Optimization Using Multistage Distance Scan Scheduler

Abdurahman Dwijotomo, Mohd Azizi Abdul Rahman, Mohd Hatta Mohammed Ariff, and Hairi Zamzuri

Abstract This paper presents the utilization of Google's simultaneous localization and mapping (SLAM) called Cartographer, and improvement of the existing processing speed using multistage distance scheduler. The presented approach optimizes the Local SLAM part in Cartographer to correct local pose based from Ceres scan matcher by integrating scheduling software, which controls the distance of light detection and ranging (LiDAR) sensor and scan matcher's search window size. In preceding work, the multistage distance scheduler was successfully tested in the actual vehicle to map the road in real-time. Multistage distance scheduler means that local pose correction is done by limiting the distance scan of LiDAR and search window with the help of scheduling algorithm. The scheduling algorithm manages the SLAM to swap between small scan size (25 m) and large scan size (60 m) LiDAR at a fixed time during map data collection; thus it can improve performance speed efficiently better than full-sized LiDAR while maintaining the accuracy of full distance LiDAR. By swapping the scan distance of sensor between small and long-range scan, and adaptively limit search size of scan matcher to handle difference scan size, it can improve pose generation performance time around 15% as opposed against fixed scan distance 60 m while maintaining similar pose accuracy and large map size.

Keywords Cartographer · LiDAR sensor · SLAM · Multi-level

1 Introduction

Cartographer SLAM is one of Simultaneous Localization and Mapping (SLAM) methods developed by Google, which integrates compatibility with various sensor devices [1]. It was developed with the focus for portable mapping devices which work

A. Dwijotomo · M. A. A. Rahman (✉) · M. H. M. Ariff · H. Zamzuri
Advanced Vehicle System Research Group, Universiti Teknologi Malaysia, Jalan Sultan Yahya Petra, 54100 Kuala Lumpur, Malaysia
e-mail: azizi.kl@utm.my

H. Zamzuri
Emoovit Technology Sdn. Bhd., Level 1, Futurise Centre, Persiaran Apec, 63000 Cyberjaya, Selangor, Malaysia

U. Sabino et al. (eds.), *Proceedings of the 6th International Conference and Exhibition on Sustainable Energy and Advanced Materials*, Lecture Notes in Mechanical Engineering,
https://doi.org/10.1007/978-981-15-4481-1_20

anywhere. The source codes have been made open source since 2016 and further improved with a wide-open source community support. At the earliest development, the Cartographer uses LiDAR sensor and IMU to perform trajectory calculation and building the map, however as the time progressed it has been constantly improved with the integration of various sensor devices including Odometry, IMU, GPS, etc. This makes the Cartographer known as the only SLAM methods with the broadest range of sensors utilized.

SLAM is an algorithm to compute trajectory and generate maps based on the surrounding environment data. The SLAM methods are not new research and are not focus on this paper. However, this paper contributes to improving current Cartographer SLAM algorithm by reducing the computational load with the usage of multistage distance scheduler.

2 Related Work

SLAM is an algorithm to perform pose calculation based on the data of the map environment generated and vice versa. It is a critical component in autonomous vehicle navigation to provide map awareness and accurate pose estimation. The methods require sensor which can sense the environment for a map and pose generation with some additional odometry sensors to improve accuracy [2, 3]. For environment sensing mostly it uses either cameras or Light Radar (LiDAR) which later record data of object distance in a point cloud format. The maps are generated through the accumulation data of LiDAR or camera during vehicle/robot motion. To perform pose calculation, SLAM mainly utilizes scan matcher on the environment sensor against the map to locate translational and rotational errors during motion, which in turn can be used to compute pose.

Reliable real-time process in SLAM is hard to achieve. Most of the SLAM techniques will drop performance as it travels more distances [4, 5]. This is related to the map accumulation data and the usage of scan matcher. In SLAM, there are two usages of scan matcher; firstly, it is for pose estimation using a comparison of current data and previous data during motion. The second one is a pose correction by comparing current data sensor against the overall accumulated map generated. The scan matcher in pose estimation can only provide local maxima, and it will accumulate errors resulting in drift after some time.

Meanwhile, the pose correction process using global map scan matcher can obtain global maxima result which it tries to reduce the drift error. However, the bigger the size of the map, the slower it becomes because the scan matcher tries to match the sensor data against every possible combination on the map. In autonomous, the slower the navigation processing speed can result in loss of important information which leads to non-smooth vehicle maneuver, a slow response time that can lead to an accident, etc. Thus every technique to improve processing speed is important.

There are several techniques introduced to improve speed performance like the use of parallel programming using graphics processing units or known as GPUs or multi-threaded execution to boost the speed [6–9]. Normally, most of the scan matchers are performed with serial CPU computation. GPU is a specialized processor which handles the multimedia process and has a great parallel computational power. Because maps or environment data are similar to multimedia images and scan matcher techniques mostly process massive matrix array numbers, a GPU computation can significantly improve the time performance when in use. Even though it is efficient, it must be noted that the process requires additional costs for hardware, and not every SLAM algorithm is compatible with parallel executions.

Another technique to improve scan matcher processing speed is done with the usage of the grid-based map [4]. In 3D SLAM, the map of point clouds can be simplified using a method called voxel grid mapping. Voxel is a collection of the aligned boxes in 3D which hold information of down-sampled 3D point clouds. As the map becomes less complex, the scan matcher can process data faster, but the accuracy will also be less dependable. This also applies in 2D based SLAM [10]. A grid 2D map has been used to increase computational power similarly.

Outside grid-based maps, there is a feature-based map (e.g., a topography map) which saves some crucial features from sensor scans such as edge lines, planar surfaces, etc. [11, 12]. Rather than using scan matcher approach, the pose is obtained with a similar scan matcher technique by associating the obtained features against the saved topography map and then interpolate the corrected pose. The pose generation is faster than a normal map scan matcher technique since it skips other information outside referred features. However, if the environment lacks the referred specific features on the map, it can result in wrong pose computation.

Another technique to improve performance is the usage of multi-resolution based grid map like Google's Cartographer [1, 13]. In Cartographer, map grids with lower resolution are utilized to obtain the initial pose with scan matcher. Later, the initial pose accuracy is increased further with the usage of scan matcher in environment data or map with higher resolutions, which included in the range of previous low-resolution range scan results. This method is called Depth First Search Branch and Bound, where every computation is connected with previous computation states. The use of scan matcher in higher resolution maps with known initial pose can eliminate unnecessary computation of scan matcher when we try to compute possible rotation and translation ever during motion.

In this paper, we use a method inspired by the multigrid resolution map from the Cartographer. Instead of using multi-resolution maps, it utilizes scheduler to control the distance focus of the LiDAR sensor in Cartographer. In SLAM, it is known that accuracy and map distance sizes which sampled from the sensor will be increased proportionally. To create the map with full size; however, it is not necessary to use full sensor data every time. The scheduler only needs to use full sensor data at a fixed frequent time to build a full map. Thus, it can maintain accuracy and improve computational power.

3 System Overview

The work presented in this study is based on the modified Google's Cartographer. Cartographer is a SLAM method which focuses on real-time performance and portability. It uses grid maps based representation with flexible resolution and sensor choices. The Cartographer is divided into two processes. The first one is called Local SLAM, in which the method uses a Ceres scan matcher on submap (i.e., a local map) to get the estimated pose and orientation of the vehicle. The other process is known as Global SLAM to optimize the pose by utilizing overall maps generated (i.e., a global map) for pose correction. Some modifications are done with Cartographer's Local SLAM to integrate the multi-distance scheduler to control the sensor distances, and scan matcher's search windows during the map and pose generation.

3.1 Distance Scheduler in LiDAR Scans

The use of multi-distance scheduler is inspired by the usage of grid-based SLAM, which uses a multistage resolution based grid to perform pose optimization [1]. In grid-based SLAM, any hill climbing based approach faces a risk of getting stuck in local minima. As the presented approach is based on gradient ascent, it is also potentially prone to local minima issue. The problem is mitigated with a multistage resolution map representation similar to pyramid approaches in computer vision. Instead of using a single map, it uses multiple map levels down sampled from the actual grid map. This generative approach ensures the map is consistent across the scales while at the same time avoiding costly down sampling operations. The scan alignment process and pose estimation in grid-based SLAM is started with the lowest resolution map first, with the resulting pose getting used as a starting estimate for the next level.

The distance scheduler approach is done by limiting the distance of LiDAR scan which uses to generate point cloud (H) from LiDAR sensor, where H contains an array of object distance point $p\_point_{x,y,z}$ in sensor local coordinate. It is commonly known in SLAM that the size of the map not only increasing the pose accuracy but also increasing the computational burden. The idea of Multi distance usage is to increase the processing speed by using a smaller scan to generate map while maintaining the accuracy similar to a full-sized scan. This is done with the help of scheduling algorithm which swaps between small and full distance scan map at a period of interval. Table 1 shows the algorithm of the scheduler to control sensor scan distance.

Table 1 Pseudocode for multistage distance scheduler

```
Function Multi Distance Scheduler (PointsCloud_H, Period_ratio_on,
Max_Distances, Min_Distances)
  if (step scheduler == Period_ratio_time_on)
    for all p_point in PointsCloud do
      if p_point > Max_Distances
          Null p_point in PointClouds data
      end if
    end for
  end if
  if (step scheduler != Period_ratio_time_on)
    for all p_point in PointsCloud do
      if p_point > Min_Distances
          Null p_point in PointClouds data
      end if
    end for
  end if
   Step scheduler = Step scheduler + current_time
  return PointsCloud_H
  end function
```

3.2 Map Construction

The process to generate a map is divided into two. Firstly, the submap which is a map representation based on local coordinate. The other one is a global map which constructed by combining the accumulated submap which then optimized using a loop closing algorithm described in Sect. 3.4. The submap generation is an iterative process by comparing LiDAR scan coordinate with a submap coordinate frame. The pose transformation of the scan frame ξ in the submap is represented as T_ξ and is defined in (1):

$$T_\xi p = \underbrace{\begin{pmatrix} \cos\xi\theta & -\sin\xi\theta \\ \sin\xi\theta & \cos\xi\theta \end{pmatrix} p}_{rotation} + \underbrace{\begin{pmatrix} \xi x \\ \xi y \end{pmatrix}}_{translation} \tag{1}$$

Submap is represented in the probability grid form. In order for an object to be inserted in the map, it uses miss and hit probability algorithm on the scan. Multiple consecutive scans are necessary for this process. For every hit of LiDAR data scans, it is inserted to grid point associated with each pixel that intersects. For every miss, the data scans are inserted into the grid associated with the pixel excluding grid points which already have the hit set. Every grid has associated probability hit p_{hit} also, their associated $Weight_{map\_size}$ is based on scheduler step whatever use the small scan or large scan to update the map in Eqs. (2, 3). Be noted that the $Weight_{map\_size}$ in large sensor scans must be larger than smaller scan because it contains more feature information.

$$odds(p) = \frac{p}{1 - p'} * Weight_{map\_size} \tag{2}$$

$$M_{new}(x) = clamp(odds - 1(odds(M_{old}(x)) \cdot odds(p_{hit})) \tag{3}$$

The similar equation is also used to a probability miss weight calculation p_{miss}.

3.3 Ceres Scan Matching

The scan matching is a process to compare a scan into the submap and to find relative translation and rotation differences. The scan pose is obtained using the relationship between these variables by utilizing Ceres based scan matcher. The scan matcher is also responsible for finding a scan pose that maximizes the probabilities at scan point in the submap. The problem is described as nonlinear least squares. In the process of scan points in the submap, due to the usage of multi distances scan scheduler, it is unnecessary to use a larger search window (W) to match the scan into the submap. The small LiDAR scan requires a smaller search window to increase processing speed.

When scan uses long LiDAR distance:

$$\underset{\xi \in Wmax}{argmin} \sum_{k=1}^{K} \left(1 - M_{smooth}\left(T_{\xi} h_k\right)\right)^2 \tag{4}$$

When scan uses smaller LiDAR distance:

$$\underset{\xi \in Wmin}{argmin} \sum_{k=1}^{K} \left(1 - M_{smooth}\left(T_{\xi} h_k\right)\right)^2 \tag{5}$$

where h_k is a scan frame transformed by T_ξ to the submap frame. The function M_{smooth} is a bi-cubic interpolation smooth filter for probability values of submap.

3.4 Closing Loops Using Global SLAM

During the process of Local SLAM, as the scan is matched against the previous scan, it slowly accumulates errors. The error drift will grow bigger as the travel distances become longer. In order to reduce this error, the pose obtained from the Local SLAM needs to be improved further using a method called Global SLAM. The approach uses a Sparse Pose Adjustment (SPA) technique [14]. A scan matcher is used on a broader range map, which is the overall accumulated submap (global map) for pose correction.

Optimization problem

In an optimization problem, loop closing detection to reduce drift errors in SLAM is formulated as nonlinear squares. Every few seconds, Ceres scan matcher is utilized to compute the SPA as defined in Eq. (6):

$$\underset{P^M P^S}{argmin} \frac{1}{2} \sum_{ij} \rho\left(E^2\left(\xi_i^m, \xi_j^s; \sum_{ij}, \xi_{ij}\right)\right)^2 \tag{6}$$

where the submap poses $P^M = \{\xi_i^m\}_{i=1,\ldots,m}$ and the scan pose $P^S = \{\xi_i^m\}_{j=1,\ldots,n}$ in the world are optimized with some constraints. These constraints are in the form of the relative pose ξ_{ij} and their covariance matrices $\sum_{ij}$. For a pair of submap i and j, the pose ξ_{ij} describes the coordinate of the submap frame where the scan was matched. The residual E for such a constraint is computed by:

$$E^2\left(\xi_i^m, \xi_j^s; \sum_{ij}, \xi_{ij}\right) = e\left(\xi_i^m, \xi_j^s; \xi_{ij}\right)^T \sum_{ij}^{-1} e\left(\xi_i^m, \xi_j^s; \xi_{ij}\right), \tag{7}$$

$$e\left(\xi_i^m, \xi_j^s; \xi_{ij}\right) = \xi_{ij} - \begin{pmatrix} R_{\xi_i^m}^{-1}\left(t_{\xi_i^m} - t_{\xi_j^s}\right) \\ \xi_{i;}^m - \xi_j^s \end{pmatrix} \tag{8}$$

ρ loss function using *Huber loss* is utilized to reduce the influence of outliers.

Branch and bound scan matching (BBS)

The algorithm to find matched scan ξ^* can be described in equation Branch and Bound Scan Matching (BBS) (9). Where W is the search window and $M_{nearest}$ is map extended to nearest grid point to corresponding pixel.

$$\xi^* = \underset{\xi \in W}{argmax} \sum_{k=1}^{K} M_{nearest}\left(T_\xi h_\xi\right) \tag{9}$$

The algorithm of scan matching try to compute the matched scan with the best score inside search window W within finite step r for translation and step δ_θ for rotation (10).

$$w_x = \frac{Wx}{r},\ w_y = \frac{Wy}{r},\ w_\theta = \frac{W_\theta}{\delta_\theta} \tag{10}$$

Table 2 describe the native algorithm of branch and bound scan matcher.

However, finding a matched submap in overall accumulated global map requires a large search window, the native matching methods are slow due to iteration within a search window radius. To boost performance efficiency, the Cartographer uses a method called Depth First Search Branch and Bound Scan Matching (DFS BBS) to reduce computational burdens. This is done by computing good upper bound (c) score with the help of multi-resolution grid maps. The existing submap is downscaled

Table 2 Branch and bound scan matcher

```
- ∞ → best-score
for jx =-W_x until W_x do
   for jy = -W_y until W_y do
      for jθ = -W_θ until W_θ do
         score ← Σ_{k=1}^{K} M_nearest (T_ξ h_ξ) h_k
      if score > best_score then
         ξ_0 + (rj_x, rj_y, δ_θ j_θ) → match
         score → best_score
      end if
     end for
   end for
end for
return best-score and match
```

to several multi low-resolution maps for instances, 1/4 map size, 2/4 map size, 3/4 map size, etc., and then be used as scan matcher inputs. With downscaled submap, the scan matcher can compute faster because it contains fewer pixels at the cost of accuracy. The maps with lowest resolutions are used first for scan matcher inputs. The results of scan matcher with lower resolution are used as preliminary inputs for the next iteration of scan matcher in higher resolution maps. The upper bound values found in previous scan matcher can be used as references for the next iteration input with stricter search windows (i.e., more limited angular step and radius range). This process is repeated on the map on full resolutions. The algorithm for DFS BSS is described in Table 3.

Table 3 DFS-BBS

```
score_thresold -> best_score
Memorize and compute a score for every element in C_0.
Initialize a stack C with C_0 sorted by score, the maximum score at the
top.
while C is not null do
   Pop C from the stack C
   if score(c) > best_score  then
      if C is leaf node then
         ξ_c -> match
         score(c) -> best_score
      else
      Branch: split c into nodes C_c
      Compute and memorize a score for each element in C_c
      Insert C_c onto the stack C, sorted by score, the highest score last
      end if
   end if
end while
return match and best_score when set
```

4 Experiment Result

In this section, we presented some experiments of our SLAM methods computed using VLP-16 LiDAR sensor together with the inertial measurement unit modeled Spatial IMU. The test is conducted in an actual environment using a vehicle equipped with the LiDAR sensor. The computing specification uses Intel Core i5 with a dual-core processor of 2.4 GHz. Figure 1 shows the setup of the system for SLAM.

The Cartographer utilizes the Robot Operating System (ROS) available in Linux. For performance evaluation, the multi-distance scheduler implementations are analyzed from the trajectory and map generated. We performed a loop closing run, where we start the SLAM run, and it ends in the same starting position after traveling for some distances. Thus, we can know the drift error affected when compared to the standard Cartographer SLAM against Cartographer with a multi-distance scheduler by measuring position difference between map starting position and the end pose. For the first test, it was conducted at commercial areas in Cyberjaya, as shown in Fig. 2.

The tests are done in 30 km/h speed limit to reduce LiDAR sensor distortion in high speed motion [15]. The travel distances of tests are 300, 400, 1000 m simultaneously. For evaluation of multi-distance scheduler, we compared it against the standard Cartographer method using fixed 25 and 60 m LiDAR scan distances, while the scheduler algorithm used 25/60 m swappable variable distance. Figure 3 shows the results of the trajectories of the experiment conducted.

As can be seen, Fig. 3a–c. have a similar pattern. For 25 m LiDAR scan distance, it has larger drifts due to smaller data sampled from the map. The drifts from 25 m scan in the map for 3a, 3b, 3c are around 15 m, 25 m, and 15 m respectively from starting position compared to end position. The results differ between maps, and the most

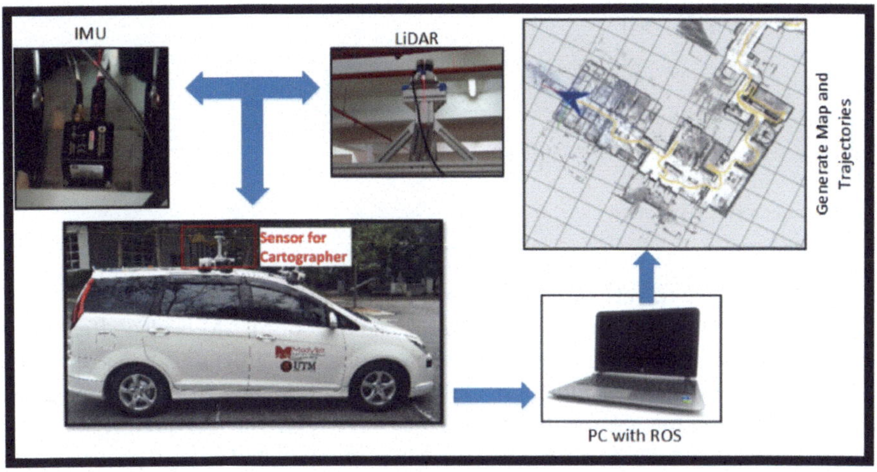

Fig. 1 SLAM system configuration

Fig. 2 Parking area (left); road area 1 (mid); road area 2 (right)

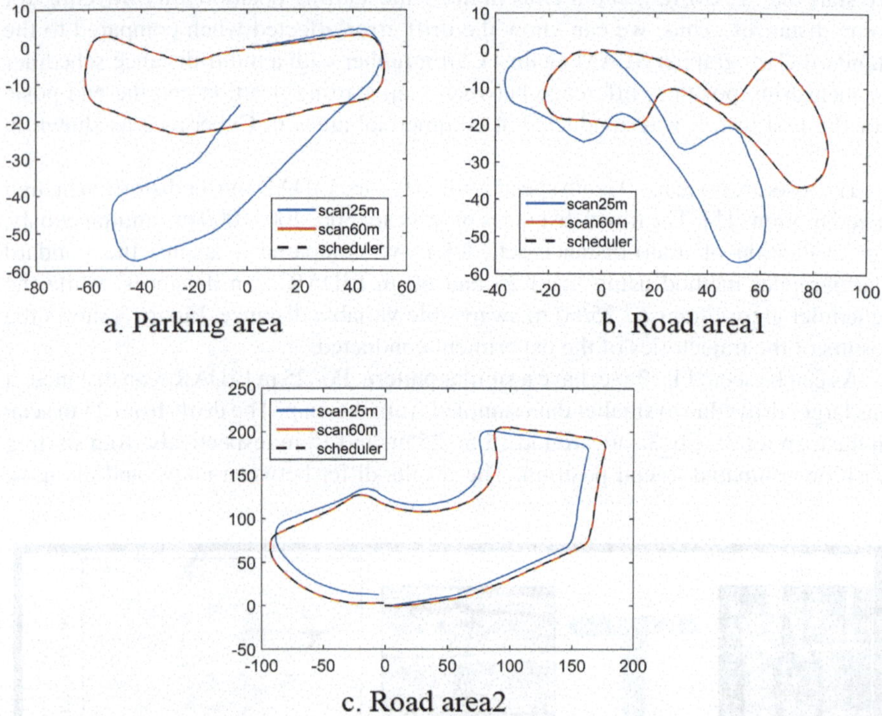

Fig. 3 Trajectories generation

prolonged distance travels should have more substantial drift compared to smaller distance travel. However, this is not the case. The drift of SLAM is also affected by features of the map. The features can be such as physical structures, poles, etc. The more features of the map sampled, the less drift error obtained. The road area 2 has most building structures while the parking area has lesser. The road area 1, however, has enormous drifts due to plenty of vegetation in the map. Cartographer uses 3D mapping compressed in 2D grid space to create a map and compute pose with it. Vegetation, when compressed in 2D space, will not create solid substance because the leaves in the tree are hard to be sampled by the LiDAR. Therefore, trajectories on the road area 1 are bigger than others. The trajectories of the scheduler with 25/60 m, when compared to a fixed 60 m, are almost identical. The drifts are lesser around 30 cm, 1 m, and 3 m for both setups with a variance of points between 10 cm point difference between the adaptive scheduler and a fixed 60 m distance. For map generation, it can be seen in Fig. 4.

Figure 4 shows that the maps generated from Cartographer differ between 25 m fixed scan, 60 m fixed scan, and the multi-distance scheduler. The 25 m fixed distance has a smaller map but larger drifts. The 60 m fixed scan distance has the largest size with the smallest drift errors. Meanwhile, the multi-distance scheduler maintains small drift as seen in the map, but it also generates a similar-sized map comparable to a fixed 60 m scan configuration due to the SLAM utilize sensor large scan size (60 m) even though it's not consistent due to scheduler. It must be noted that the computation load still increases linearly with distance travel because the proposed scheduler method does not optimize Global SLAM. The current improvement is made in Local SLAM part of Cartographer.

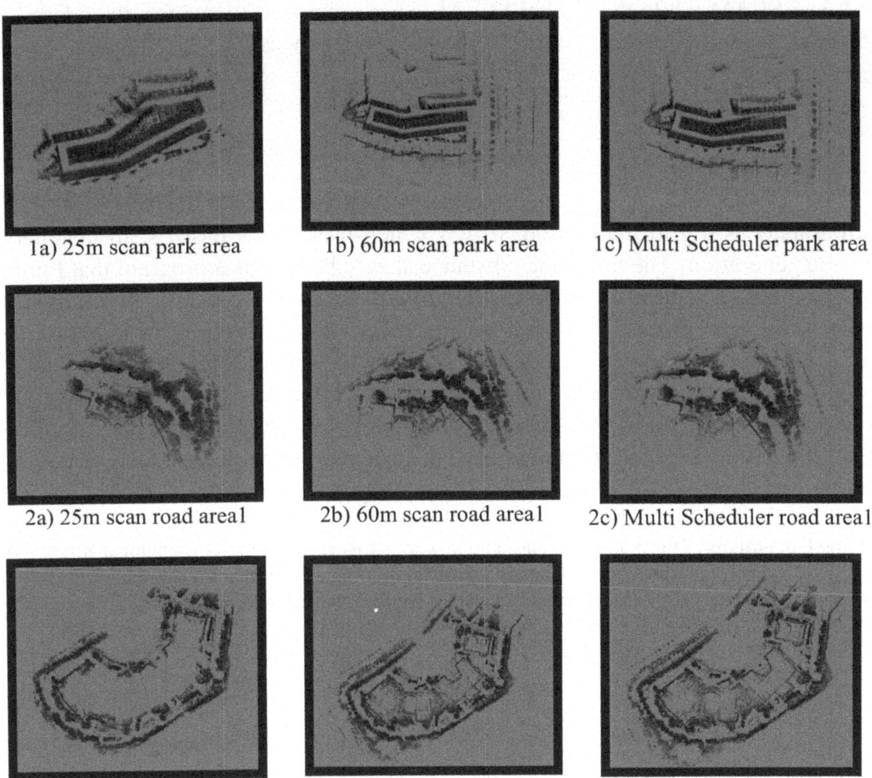

1a) 25m scan park area
1b) 60m scan park area
1c) Multi Scheduler park area
2a) 25m scan road area1
2b) 60m scan road area1
2c) Multi Scheduler road area1
3a) 25m road area2
3b) 60m scan road2 area
3c) Multi Scheduler road2 area

Fig. 4 Maps generation from cartographer

Table 4 Average pose generation time

Lidar draw distances	Elapsed time for pose generation (map 1) (ms)	Elapsed time for pose generation (map 2) (ms)	Elapsed time for pose generation (map 3) (ms)
25 m	36	36	41
Scheduler 25/60	52	52	61
60 m	60	61	70

Further study is necessary to reduce the impact of map size by improving Global SLAM part. Nevertheless, the Local SLAM improvement using scheduler is effective to reduce the load by up to 15%. Table 4 shows the computation performance comparison.

The results of pose generation time have expected time result. The fixed 25 m sensor scans are faster than any other setup followed by the scheduler and lastly a 60 m fixed sensor scan. The computation time also increases linearly following travel distances. This is due to the multi distances scheduler only implemented on the Local SLAM while the Global SLAM is not optimized to support the adaptive scheduler. The more distance it travels, the slower it processes.

5 Conclusion

Cartographer is a grid map based SLAM which uses Ceres based scan matchers for pose generation. The multistage distance scan scheduler is a program that limits sensor ranges to reduce the computational burden of scan matcher. By swapping the scan distance of sensor between smaller range (25 m) and longer range (60 m) and adaptively limit search size of scan matcher to handle difference scan size, it can improve the pose generation performance time around 15% as opposed against fixed scan distance 60 m. Meanwhile, it is also able to produce similar accuracy versus fixed scan distance 60 m in term of pose generation with the added benefit of maintaining large map size.

Acknowledgements This material is based upon work supported by the i-Drive team at Advanced Vehicle System Research Group, Malaysia Japan International Institute of Technology (MJIIT). This work is funded by the Ministry of Education Malaysia and Universiti Teknologi Malaysia, under VOT 06G16. The author also would like to acknowledge Emoovit Technology Sdn. Bhd., for their knowledge sharing and suggestions to improve researches quality.

References

1. Hess W, Kohler D, Rapp H, Andor D (2016) Real-time loop closure in 2D LIDAR SLAM. In: IEEE international conference on robotics and automation (ICRA), Stockholm
2. Khairuddin AR, Talib MS, Haron H (2016) Review on simultaneous localization and mapping (SLAM). In: IEEE international conference on control system, computing and engineering (ICCSCE), George Town
3. Krinkin K, Filatov A, Filatov AY, Huletski A, Kartashov D (2018) Evaluation of Modern Laser Based Indoor SLAM Algorithms. In: Conference of open innovations association (FRUCT), Jyvaskyla
4. Tiar R, Lakrouf M, Azouaoui O (2015) FAST ICP-SLAM for a bi-steerable mobile robot in large environments. In: IEEE international workshop of electronics, control, measurement, Liberec
5. Bahreinian SF, Palhang M, Taban MR (2016) Investigation of RMF-SLAM and AMF-SLAM in closed loop and open loop paths. In: International conference of signal processing and intelligent systems (ICSPIS), Tehran
6. Lee D, Kim H, Myung H (2012) GPU-based real-time RGB-D 3D SLAM. In: International conference on ubiquitous robots and ambient intelligence (URAI), Daejeon
7. Ratter A, Sammut C, McGill M (2013) GPU accelerated graph SLAM and occupancy voxel based ICP for encoder-free mobile robots. In: IEEE/RSJ international conference on intelligent robots and systems, Tokyo
8. Zhang H, Martin F (2013) CUDA accelerated robot localization and mapping. In: IEEE conference on technologies for practical robot applications (TePRA), Woburn
9. Song J, Wang J, Zhao L, Huang S, Dissanayake G (2018) MIS-SLAM: real-time large-scale dense deformable SLAM system in minimal invasive surgery based on heterogeneous computing. IEEE Robot Autom Lett 3(4):4068–4075
10. Kohlbrecher S, Stryk OV, Meyer J, Klingauf U (2011) A flexible and scalable SLAM system with full 3D motion estimation. In: IEEE international symposium on safety, security, and rescue robotics, Kyoto
11. Zhang J, Singh S (2014) LOAM: Lidar odometry and mapping in real-time. In: Robotics: science and systems conference, Pittsburgh
12. Mur-Artal R, Tardós JD (2017) ORB-SLAM2: an open-source SLAM system for monocular, stereo, and RGB-D cameras. IEEE Trans Rob 33(5):1255–1262
13. Greene WN, Ok K, Lommel P, Roy N (2016) Multi-level mapping: real-time dense monocular SLAM. In: Multi-level mapping: real-time dense monocular SLAM, Stockholm
14. Konolige K, Grisetti G, Kümmerle R, Burgard W, Limketkai B, Vincent R (2010) Sparse pose adjustment for 2D mapping. In: IROS, Taipei
15. Hong S, Ko H, Kim J (2010) VICP: velocity updating iterative closest point algorithm. In: IEEE international conference on robotics and automation, Anchorage

Effect of Corroded Plate-Like Iron Particles on the Rheological Properties of Magnetorheological Elastomer

Nurul Liyana Burhannuddin, Nur Nabila Balqis Zolkifli, Nur Azmah Nordin, Siti Aishah Abdul Aziz, Saiful Amri Mazlan, and Hafizal Yahaya

Abstract Prior studies have shown that the storage modulus and MR effect of MR materials were mainly determined by the performance of magnetic particles. However, there is still doubt for its utilization in a long-term operation under various external loads and conditions. In MREs, even though the magnetic particles are embedded in an elastomeric matrix and not exposed to the air, corrosion agents however such as moisture and oxygen can penetrate the structure phase and alter their properties. This corrosion phenomenon can significantly affect the performance of MRE devices. In this work, the effect of corroded plate-like carbonyl iron particles (CIPs) on the properties of magnetorheological elastomer (MRE) is investigated. The CIPs are corroded via accelerated corrosion test in various concentrations of diluted hydrochloric acid; particularly 0.5, 1.0 and 1.5 vol.% HCl. The morphology of non-corroded and corroded plate-like CIPs are characterized via SEM prior to fabrication of MREs. The rheological properties of MREs are tested using a rheometer under varying applied currents and frequencies. The results from SEM images show that the sample of non-corroded plate-like CIPs has a smooth surface structure while the corroded plate-like CIPs looked rougher on its surface structure. The storage modulus of MRE is increased with the increased of frequencies but decreased

N. L. Burhannuddin · N. N. B. Zolkifli · N. A. Nordin (✉) · S. A. A. Aziz · S. A. Mazlan · H. Yahaya
Engineering Materials and Structures (eMast) iKohza, Malaysia-Japan International Institute of Technology (MJIIT), Universiti Teknologi Malaysia, Jalan Sultan Yahya Petra, 54100 Kuala Lumpur, Malaysia
e-mail: nurazmah.nordin@utm.my

N. L. Burhannuddin
e-mail: nuruliyana.13@gmail.com

N. N. B. Zolkifli
e-mail: nabilanaila96@gmail.com

S. A. A. Aziz
e-mail: aishah118@gmail.com

S. A. Mazlan
e-mail: amri.kl@utm.my

H. Yahaya
e-mail: hafizal.kl@utm.my

U. Sabino et al. (eds.), *Proceedings of the 6th International Conference and Exhibition on Sustainable Energy and Advanced Materials*, Lecture Notes in Mechanical Engineering,
https://doi.org/10.1007/978-981-15-4481-1_21

with MRE containing higher corrosion rates. Similarly, the MR effect are decreased from 90.6% for MRE containing of non-corroded plate-like CIP, to 87%, 81.8%, and 71.8%, respectively with increasing of corrosion rate plate-like CIPs in MRE. The obtained results demonstrate that higher purity of the CIPs, the greater the impact on storage modulus and MR effect of MRE. Hence, significant finding of this research is related to the purity of CIPs that play an important role to ensure the continuous reliable performance of MRE devices for long-term applications.

Keywords Magnetorheological elastomer · Carbonyl iron particle · Plate-like · Corrosion · Rheological properties

1 Introduction

Magnetorheological (MR) materials are classified under class of smart materials due to its rheological properties that can be actively changed correspond to applied magnetic field [1, 2]. A solid-carrier matrix, normally a viscoelastic polymer material that consists of magnetically permeable particles is known as magnetorheological elastomers (MRE). Typically, the silicone rubber (SR) is commonly utilized as a matrix phase as it is generally soft and deformable at room temperature. The stiffness of MREs will continuously vary under magnetic fields due to feature of MRE that is controllable modulus [3]. Thus, it is potential to be widely used in various applications such as in automotive bushings and engine mounts [4], besides could act as actuator and vibration absorber as well [3]. MREs also use in medical devices, particularly for MR brake that has been applied into artificial prosthetic knee for rehabilitation [5, 6]. Others application that utilize MRE in the devices are tyre pressure control, variable spring rate, propeller shaft absorber and remote controllable attachment [6].

Nonetheless, characteristics of magnetic particles such as size/shape, distribution in the matrix and concentration of magnetic particles are mainly affected the overall performance of the MRE [7]. In fact, carbonyl iron (CIP) with micron size is widely used as magnetic particles in MR material and frequently used shape is spherical. In recent years, there has been an increasing interest in studies on the particle shape's influence on the rheological properties of the plate-like CIPs. The possibility outcomes of using the plate-like CIPs instead of conventional spherical CIPs is closely related to storage modulus and MR effect of MRE [8]. Hapipi et al. [9], investigated the MRE in terms of rheological properties of the CIPs on the basis of two different shapes: spherical and plate-like, respectively. They reported that the MRE, which embedded with plate-like iron particles, exhibited higher dynamic modulus than MRE with spherical shape at a same magnetic field strength. In addition, they proved the MRE with plate-like shaped enhanced MR effect as well.

However, despite its utilization in a long-term operation under various external loads and conditions, the rheological properties of MRE will be changed. It is believed that the MR material has been deteriorated after exposing to long duration under various conditions. This would cause formation of rust or corrosion layer on the surface of iron particles in the MRE after the material being exposed to the contaminated gases and oxygen in the air, in the repeated cycles of MR systems. The corrosion of iron particle in MRE will cause the material have to low ductility, brittle phase and easy to crack. Subsequently, it would weaken the performance of the material [10]. However, few writers have been put an interest to investigate the corroded CIPs in MR materials. For instance, Han et al. [11] had focused on the effect of corrosion in the magnetorheological fluid (MRF) and they found that the corrosion effect of magnetic particles was able to decrease the properties of MRF. Moreover, in the MRF also contain CIPs as magnetic particle that can corroded in presence of air and water due to oxidation and can cause the surface roughness increase [12].

However, too little attention has been focused on the effect of corroded CIPs in MREs and the resultant performance due to the corrosion phenomenon. Moreover, the study on the effect of the corroded plate-like CIPs has not been reported as far as the authors' knowledge. The aim of this research is to experimentally investigate the effect of corroded plate-like CIPs on the rheological properties of MRE. In order to achieve this goal, various corrosion rates are performed by controlling the immersion time of CIPs in various concentration of diluted HCl. The change of plate-like CIP's surface structure will be characterized via scanning electron microscope (SEM) and correlation between corroded plate-like CIP and the corresponding rheological properties of MREs will be further analysed.

2 Measurement Methods

2.1 *Raw Materials*

Spherical carbonyl iron particles (CIPs) were supplied by CK Materials Lab Co., Ltd. with their average particle size between 3–5 μm. Silicone rubber (SR) from Nippon Steel (NS 625-A) were used as a matrix phase together with NS 625 B as curing agent. Hydrochloric acid (HCL) diluted with water were prepared for the accelerated corrosion process.

2.2 *Preparation of Plate-Like CIPs*

In this work, the plate-like CIPs were produced from spherical through a mechanical alloying process using a high-energy rotary ball mill machine from Tencan Company (QM-5 model) at rotational speed 180 r/min for 40 h using zirconia ball with ball-to-powder weight ratio of 20:1. In order to overcome particles adhesion and improve the efficiency plate-like particles, pure ethanol is added as process controlling agent [13].

2.3 Accelerated Corrosion Test

At first, 0.5 vol. % of hydrochloric acid (HCI), about 0.5 ml was diluted in 99.5 ml of distilled water at room temperature. Then, 40 g of plate-like CIPs was slowly poured into the acid solution and stirred continuously for one minute. Later, the mixture was left for 30 min. Then, the corroded of plate-like CIPs was separated from the diluted HCI by using a magnet and was washed using ethanol few times in order to remove the residue of water. Afterwards, the corroded CIP was undergone a drying process, which was divided into three stages; first, a sonication or ultrasonic processing was performed to avoid the corroded particles become de-agglomeration. Secondly, the solution was ready for filtration by using a vacuum pump prior to sieving in final stage and getting a uniform size of particles. The same steps of accelerated corrosion test were repeated for 1.0 and 1.5 vol.% of HCl, respectively.

2.4 MRE Fabrication

Four types of MRE were prepared through a mixing process. 40wt% silicon rubber (SR) was mixed with 60wt% of non-corroded PL CIPs using a mechanical stirrer at 280 rpm for 10 min until the mixture is visually homogenous. Then, the curing agent was added, and the mixture was stirred for another 1 min. The mixture then were poured into the steel mould and cured in the absence of magnetic field for 24 h. The process was repeated for fabrication of MREs with corroded plate-like CIPs that was immersed in three different concentrations of HCL (0.5, 1.0, 1.5 vol. % HCl) individually.

2.5 Characterization

The surface morphologies of the plate-like CIPs was observed using a scanning electron microscope (SEM, JEOL-IT300) to evaluate the microstructure of plate-like CIPs before and after corrosion process. The test was conducted with the application of an accelerating voltage 5 kV with the magnification 2500× for surface morphology.

The rheological properties of MRE samples were measured by using a rheometer (Model MCR 302, Anton Paar). The samples used for rheological test had a diameter and thickness of 20 mm and 1 mm, respectively. The sample was placed between a rotating disk and a parallel base plate (PP20/MRD/TI). Two kinds of testing were involved to measure the dynamic properties such as storage modulus and MR effect. At first, the frequency sweep test which condition of strain amplitude was set at 0.002% and the frequency was set from 0.1 to 100 Hz with the applied current was varied from 0 to 4 A. Secondly, the magnetic field sweep test was measured in condition of strain amplitude and frequency were fixed at 0.002%, and 1 Hz, respectively with various of applied from 0 to 4 A.

3 Results and Discussion

3.1 Morphology of Plate-Like CIPs

The microstructure of plate-like CIPs and corroded plate-like CIPs are observed by using SEM. Figure 1a shows the morphology of the CIPs after ball milling process that caused changed of CIP from spherical to plate-like (flattened) shape. As shown in this figure, the non-corroded PL CIPs has smooth surfaces. The result obtained that ball milling process was effective to help in adjustment of shape of particles into plate-like shaped [14]. However, after the accelerated corrosion test was done on the CIP with 1.0 vol.% HCl, the plate-like shapes has shown to be corroded as presented in Fig. 1b. It can be seen clearly that the corroded plate-like CIPs have more rougher surfaces with white dots have been formed on the surface of the plate-like CIPs and it is indicated that the corrosive layer on the surface of oxidized magnetic particles has been formed on the particle's surfaces due to corrosion process.

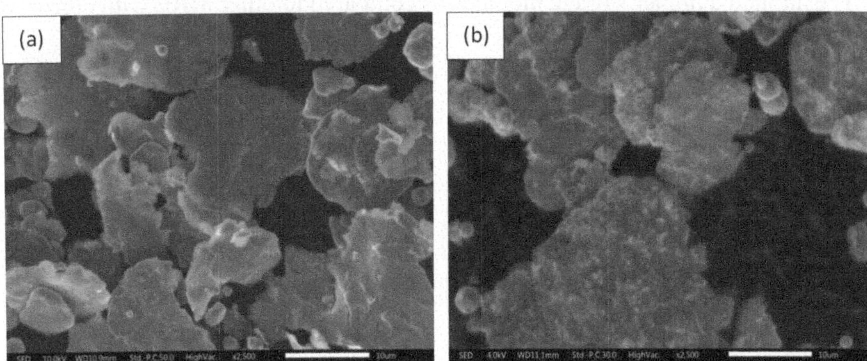

Fig. 1 SEM images of plate-like CIPs. **a** Non-corroded and **b** corroded with 1.0 vol.% HCl

3.2 *Rheological Properties*

3.2.1 Frequency Sweep Test

The frequency sweep test becomes necessary to apply in order to evaluate the dynamic stiffness of MRE. The obtained results of storage modulus of MRE samples at different magnetic field strength with various applied currents of 0, 1, 2, 3 and 4 A are measured and presented in Fig. 2.

It could be observed that the storage modulus for all samples of MREs including non-corroded and corroded CIPs had increased as the applied current and the frequency increased. To be specific, as shown in Fig. 2a, the initial storage of MRE with non-corroded CIPs sample has increased with the increasing of applied currents. In the same way, similar trends of initial storage modulus of MRE with corroded CIPs samples are observed in Fig. 2b–d. However, for MRE with corroded CIPs samples, the increasing of concentration of HCl will be resulted in the decreasing of initial storage modulus. For further explanation, the relationship between storage modulus and frequency measured at off-state (0 A) and on-state condition (3 A) of applied currents as presented in Fig. 3.

As shown in this figure, the initial of storage modulus of MRE samples is gradually increased with increasing of frequencies at both off- and on-state condition. In the absence of external magnetic field, the MRE with non-corroded CIPs sample is exhibited a higher storage modulus as compared to other MRE with corroded

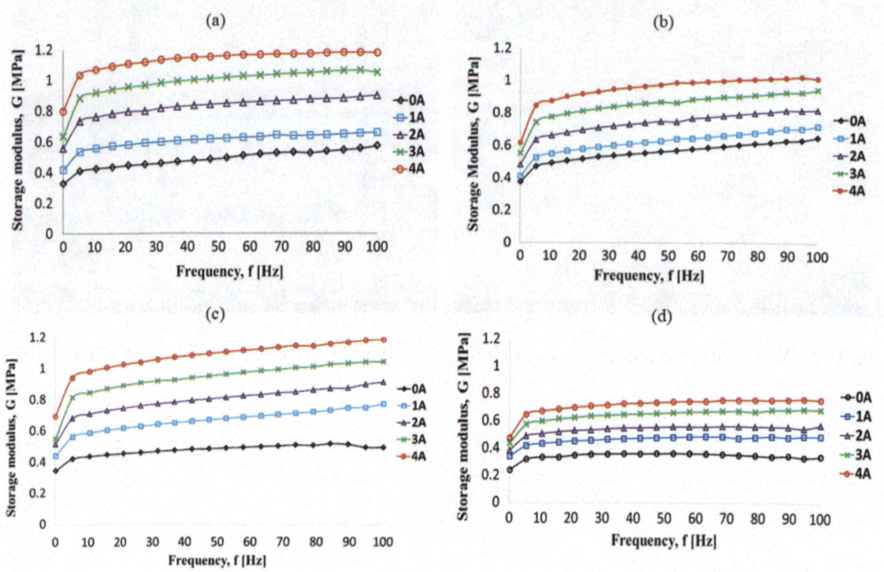

Fig. 2 Frequency dependency of storage modulus of MRE samples at various currents; MRE with **a** non-corroded CIPs, **b** 0.5% HCI, **c** 1.0% HCI and **d** 1.5% HCI of corroded CIPs

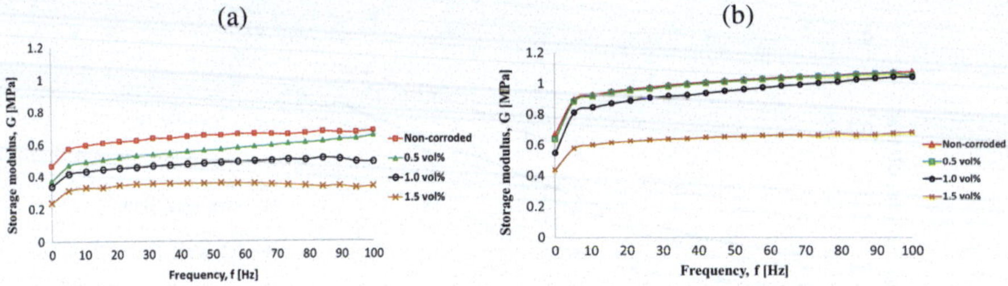

Fig. 3 Frequency dependency of storage modulus of MRE at **a** off-state (0A) and **b** on-state (3A) conditions

CIPs sample. It is noted that corrosive layer appeared on the corroded CIPs was mostly brittle and this phenomenon has led to decrease the dynamic properties such as storage modulus. Nonetheless, when the external magnetic field was applied, the chain structure formation between magnetic particles and matrix in MRE samples becomes significant effect on storage modulus [15]. As can be seen from the Fig. 3b, the MRE with non-corroded CIPs is more sensitive to the testing frequency as compared to other MRE with corroded CIPs samples. This sensitivity aspect makes the interaction of the non-corroded CIPs with regards to their higher purity had resulted in strong bonding between CIPs within the matrix. Despite this, the storage modulus of MRE with corroded CIPs sample decreases as the corrosion rate is increased. It is due to less purity of CIPs which has led to poor bonding between the CIPs in the MRE. This suggests that purity of CIPs plays an important key in order to get a better matrix-particle interaction and stronger bonding, having will lead to ease magnetization.

3.2.2 Magnetic Field Sweep Test

The key parameter for calculating the relative change of modulus is known as MR effect, the ratio of magneto-induced modulus and the initial modulus which can be expressed by the following equation:

$$MR\,effect(\%) = \frac{(G_{\max} - G_0)}{G_0} \times 100\%$$

where G_0 is the initial storage modulus, and $G_{\max}$ is the maximum storage modulus at highest magnetic flux density

The MR effect of MREs was investigated under the current sweep test with varied applied currents while shear strain and frequency were kept constant, at 0.002% and 1 Hz, respectively. The relationship between the storage modulus of MRE samples at various magnetic flux densities are measured and plotted in Fig. 4. From this figure, there is a clear trend increasing of the storage modulus of all MREs the increasing of magnetic field strength.

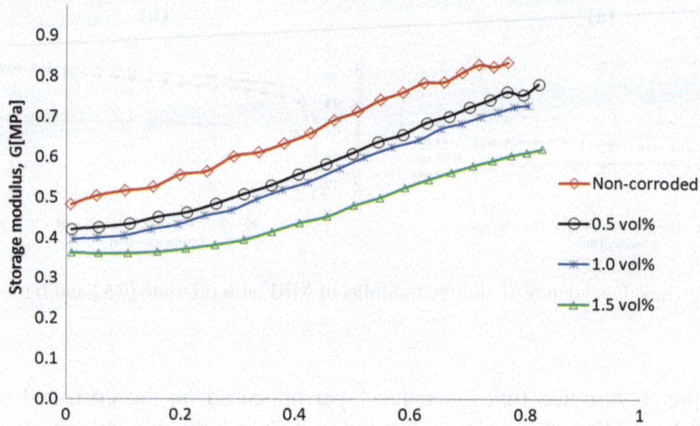

Fig. 4 Transition of storage modulus of MREs correspond to magnetic flux densities for MREs contained non-corroded and corroded CIPs

The initial modulus, magneto-induced modulus and MR effect at different magnetic field strengths for all samples were summarized in Table 1. As referred to the table, the non-corroded CIPs based MRE has the highest initial modulus compared to other samples, about 0.48 MPa, followed by the MRE with corroded PL CIPs of 0.5, 1.0 and 1.5 vol.% HCl which are 0.42, 0.40, and 0.36 MPa, respectively. The changes in the storage modulus, ΔG showed a decrement trend with the increment of corroded PL CIPs correspond to higher HCl concentrations. It is due to the increment of corrosion rate of CIPs that finally would lower the interaction between CIPs in MRE. In fact, the MRE with non-corroded CIPs sample has the largest MR effect, about 90.63%, followed by the MREs with corroded CIPs of 0.5, 1.0 and 1.5% correspond to 87.0%, 81.84% and 71.77% MR effect, respectively.

Following the reduction of MR effect, a significant of purity of CIPs content has an important influence on the final performance of the MRE rheological property which has been confirmed with the micrograph analysis mentioned in Fig. 1b. The existence layer occurred in the MRE samples with the presence of HCl is believed

Table 1 The initial storage modulus (G_0), the magnetically absolute modulus (ΔG) and the MR effect of various MRE samples

MRE sample	Initial modulus, G_o (MPa)	Absolute modulus, ΔG (MPa)	Average MR effect (%)
Non-corroded CIP	0.48	0.35	90.63
CIP + Diluted 0.5% HCl	0.42	0.36	87.00
CIP + Diluted 1.0% HCl	0.40	0.33	81.84
CIP + Diluted 1.5% HCl	0.36	0.26	71.77

to have an adverse effect on the rheological properties particularly the MR effect. In the meantime, with the existence of corrosive layer of the corroded CIPs due to the HCl effect, it is assumed that this layer might hinder the movement/mobility of CIPs particle during the influence of magnetic field, thus decreased the MR effect.

4 Conclusion

The effect of corrosion rate of PL CIPs on the rheological properties of the MREs by immersion of CIPs with various concentration of hydrochloric acid (HCI) on MRE silicone-based was investigated. The PL CIPs was successfully prepared from SL CIPs by using ball milling method prior to undergo the accelerated corrosion test. The PL CIPs were forcibly corroded by performing the accelerated corrosion test which chemically reacted using diluted hydrochloric acid (HCI) with various concentrations. The pure PL CIPs (non-corroded) were used as a reference. The morphologies was performed to investigate the changes of PL CIPs after corrosion process. The result of non-corroded PL CIPs was compared with corroded CIP with immersion of 1.0 vol.% HCI which result shown the corroded PL CIPs have rougher surfaces with white dots have been formed on the surface which resulting in oxidized layer form due to corrosion process. The relative changes in MR effects are observed via frequency and magnetic field sweep test using a rotational rheometer. The observed of field-dependent modulus of MRE with non-corroded PL CIPs was found to be higher from other MRE with corroded PL CIPs samples. It is believed that corrosion effect may deteriorate the material's characteristics by chemical interaction with the environment. In fact, corroded layer on the surface structure of materials is known as hard and brittle phase and this would weaken the initial properties of the material. In conclusion, the study of corrosion phenomenon on the magnetic particles of PL CIPs has become our concern and important influence in improving and controlling the purity of PL CIPs to be more reliable in long term use towards the MRE applications.

Acknowledgements This study was financially supported by Universiti Teknologi Malaysia under Trans-disciplinary Research Grant (Vot No: 07G13) and Fundamental Research Grant Scheme (Vot No: 5F001). The research is also acknowledged and partially supported from Microscopy Lab Malaysia-Japan International Institute Technology (MJIIT), University Teknologi Malaysia (UTM).

References

1. Boczkowska A, Awietjan S (2012) Microstructure and properties of magnetorheological elastomers. In: Advanced elastomers technology, properties and applications, pp 148–180. https://doi.org/10.5772/50430
2. Ubaidillah, Mazlan SA, Sutrisno J, Zamzuri H (2014) Potential applications of magnetorheological elastomers. Appl Mech Mater 663:695–699. https://doi.org/10.4028/www.scientific.net/AMM.663.695

3. Ubaidillah, Hadi S, Harjana, Hairuddin K, Mazlan SA (2018) Design and fabrication of magnetorheological elastomer vibration isolator. J Mech Eng 5:192–210
4. Li W, Zhang X (2008) Research and applications of MR elastomers. Recent Patents Mech Eng 1:161–166. https://doi.org/10.2174/2212797610801030161
5. Jonsdottir F, Gudmundsson FB, Gudmundsson KH, Lecomte C (2015) Preparation and characterization of a prototype magnetorheological elastomer for application in prosthetic devices. In: Thematic conference on smart structures and materials
6. Ubaidillah, Sutrisno J, Purwanto A, Mazlan SA (2015) Recent progress on magnetorheological solids: Materials, fabrication, testing, and applications. Adv Eng Mater 17:563–597. https://doi.org/10.1002/adem.201400258
7. Song HJ, Wereley NM, Bell RC, Planinsek JL, Filer JA (2009) Field dependent response of magnetorheological elastomers utilizing spherical Fe particles versus Fe nanowires. J Phys: Conf Ser 149:012097. https://doi.org/10.1088/1742-6596/149/1/012097
8. Shah K, Choi S (2015) The in fluence of particle size on the rheological properties of plate-like iron particle based magnetorheological fluids. Smart Mater Struct 24:8. https://doi.org/10.1088/0964-1726/24/1/015004
9. Hapipi N, Aishah S, Aziz A, Amri S, Bok S, Mohamad N, Hana M, Khairi A, Yasser A, Fatah A, Universiti M, Sultan J, Petra Y, Lumpur K (2019) Results in physics the field-dependent rheological properties of plate-like carbonyl iron particle-based magnetorheological elastomers. Results Phys 12:2146–2154. https://doi.org/10.1016/j.rinp.2019.02.045
10. Sunkara SR, Root TW, Ulicny JC, Klingenberg DJ (2009) Iron oxidation and its impact on MR behavior. J Phys Conf Ser 149:012081. https://doi.org/10.1088/1742-6596/149/1/012081
11. Han YM, Kim S, Park YD, Kang JW, Choi SB (2015) Field-dependent characteristics of magnetorheological fluids containing corroded iron particles. Smart Mater Struct 24:115016. https://doi.org/10.1088/0964-1726/24/11/115016
12. Miao C, Shen R, Wang M, Shafrir SN, Yang H, Jacobs SD (2011) Rheology of aqueous magnetorheological fluid using dual oxide-coated carbonyl iron particles. J Am Ceram Soc 94:2386–2392. https://doi.org/10.1111/j.1551-2916.2011.04423.x
13. Shilan ST, Mazlan SA, Ido Y, Hajalilou A, Jeyadevan B, Choi S, Yunus NA (2016) A comparison of field-dependent rheological properties between spherical and plate-like carbonyl iron particles-based magneto-rheological fluids. Smart Mater Struct 25:1–9. https://doi.org/10.1088/0964-1726/25/9/095025
14. Zheng D, Liu T, Zhou L, Xu Y (2016) Electromagnetic absorbing property of the flaky carbonyl iron particles by chemical corrosion process. J Magn Magn Mater 419:119–124. https://doi.org/10.1016/j.jmmm.2016.06.008
15. Mohamad N, Mazlan SA, Ubaidillah, Choi S, Abdul Aziz SA (2019) The field-dependent viscoelastic and transient responses of plate-like carbonyl iron particle based magnetorheological greases. Intell Mater Syst Struct 1–10. https://doi.org/10.1177/1045389x19828504

Optimization of Mechanical Properties of Unsaturated Polyester Composites Reinforced by Microcrystalline Cellulose Various Treatments Using the Taguchi Method

Sakuri Sakuri, Eko Surojo, and Dody Ariawan

Abstract This research used Taguchi method to determine the most influencing processing parameters of stress properties. The composite is made of unsaturated polyester reinforced by microcrystalline cellulose. The research processing parameters includes: stirring time, stirring speed, the addition of Microcrystalline cellulose (MCC) volume fraction, and temperature change. Taguchi experiment used orthogonal array $L_9(3)^4$ design concerning 4 influencing parameters and 9 combinations. The DOE (Design of Experiment) and signal to noise ratio (S/N) result shows the influence of each level shown by S/N ratio graph. The testing result shows that the best tensile stress was obtained at 10 min stirring time, 250 rpm stirring speed, 5% MCC addition of value fraction, and 40 °C treatment temperature. Meanwhile the maximum flexural strength was obtained at 30 min stirring time, 250 rpm stirring speed, 5% MCC addition of volume fraction, and 24 °C treatment temperature. Anova test results showed that the addition of 5% MCC volume fraction had the highest influence with a contribution of 55.1% and the lowest turntable time of 7.81% against tensile strength. Flexural structure is influenced by the addition of volume fraction of 41.2% and the lowest turnover time of 7.72%. Respectively tensile highest structure 49.4 and flexural 100.86 MPa.

Keywords Microcrystalline cellulose · Unsaturated polyester · Taguchi · S/N ratio

1 Introduction

The ability of renewable natural resource is being increased for advance material development. Microcrystalline Cellulose (MCC) as an ingredient taken from plant was developed as composite booster. It started in 2004 by mixing MCC into PVA (Polyvinyl Alcohol) to enhance the tensile strength and modulus [1]. Polypropylene

S. Sakuri · E. Surojo · D. Ariawan (✉)
Department of Mechanical Engineering, Faculty of Engineering, Universitas Sebelas Maret, Jl. Ir. Sutami no. 36A Kentingan, Surakarta 57126, Indonesia
e-mail: dodyariawan@staff.uns.ac.id

S. Sakuri
Department of Mechanical Engineering, STT Wiworotomo Purwokerto, Purwokerto, Indonesia

U. Sabino et al. (eds.), *Proceedings of the 6th International Conference and Exhibition on Sustainable Energy and Advanced Materials*, Lecture Notes in Mechanical Engineering,
https://doi.org/10.1007/978-981-15-4481-1_22

and MCC was made as a composite using roller at temperature 200 °C to produce composite with tensile and bending strength increase to 4.4 times [2]. Nanocrystalline cellulose (NCC) was tested by Alper Kiziltas. NCC, polyethylene terephthalate (PET), and Polytrimethylene terephthalate (PTT) was heated until 2300 °C. Blending and injection molding was done to increase the tensile strength of the composite [3].

In make MCC composite, can be mixed in thermosetting by combining nanowhiskers cellulose and epoxy [4]. Cellulose can be evenly mixed by curing method. The result shows the increase of mechanical properties and modulus. To make composite, wood ash and MCC were mixed with Unsaturated polyester by mixing and casting method [5]. NCC and MCC was used as an upholstery composite of art equipments with weight concentration 5.10 and 30 wt% respectively. Both are able to enhance the modulus elasticity, decrease the thermal expansion, and enhance the creep strength [6].

Some treatments were done to increase the tensile and flexural. MCC, PLA (polylactic acid), and Montmorillonite were mixed by stirring them at 60 °C [7], MCC was stirred in 600 rpm and degassing [8], screw extruder 50 rpm [9], and mixed with high speed agitation at temperature 100 °C for 10 min. [10]. Thus, the tensile and flexural increase can be done by the addition of MCC fraction and various other treatments. This research is also needed to find out whether the treatment in Microcrystalline cellulose processing on composites will have a significant influence in increasing the tensile and bending strength.

The research was started by mixing MCC powder with pada unsaturated polyester. The mixing process used steerer to determine the rotational speed and the temperature. Mixing time (stopwatch) and volume fraction (scale) were based on the material density. This research will use the parameter effect on the mechanical properties by using DOE. Taguchi method was used to determine the optimum value of tensile and bending strength by varying the process and the volume adding.

2 Experiment

2.1 Material and Process

Microcrystalline Cellulose was obtained from Sigma Aldrich with Serial Number 310697, size 20 micron pH 5–7, and density 0.5 g/ml (250 °C) density. Unsaturated and Catalyst (methyl ethyl ketone) were obtained from Justus Kimia Raya Semarang Indonesia. The mintage process used hand lay up method. Composite was left in room temperature for 60 min. Then, it was put into microwave with temperature 600 °C for 120 min. After that, before being specimen, it was left in room temperature for 24 h. Tensile test used UTM (Universal Tensile Machine) ASTM D3039 and bending test used ASTM D790.

Table 1 Parameters and design level

Parameter	Code	Unit	Level 1	Level 2	Level 3
A: Time	t	Minute	10	20	30
B: Rotational speed	S	rpm	150	250	350
C: Volume fraction	V	Percent	1	2	5
D: Temperature	T	Celsius	40	60	80

2.2 *Experimental Design and Procedure*

Taguchi is a general approach that is applied to optimize the design's parameter. It is based on Orthogonal Array (OA) providing regressive variation to obtain a significant value of an experiment to produce the optimum parameter. Taguchi often applies S/N analysis to obtain the optimum prediction result and Anova analysis to obtain the influence and contribution value. S/N (Signal to noise) uses *Larger Is Better* setting [11].

Various MCC processing need appropriate processing and composition mixing. Thus, the research will review MCC mixing level 3 and 4 with orthogonal $L_9(3^4)$ factor based on Taguchi to determine the optimum tensile strength and flexural. The parameter and design level is shown in Table 1.

3 Results and Discussion

3.1 *Signal to Noise (S/N)*

The data was analyzed by Weibull distribution and obtained Anderson Darling (AD) value = 0.359 (less than 1). It is inferred that the data was well-spread. Signal to noise was calculated on each controlling factor to minimize the deviation process and the tensile and bending stress variations. S/N ratio was formulated by the researchers, and they chose the highest value to optimize the quality of the experiment. In this research, the more the respond, the better the result [12]. The formulation is:

$$S/N = -10 \log\left(\frac{1}{n}\sum_{i=1}^{n}\frac{1}{Y_i^2}\right)$$

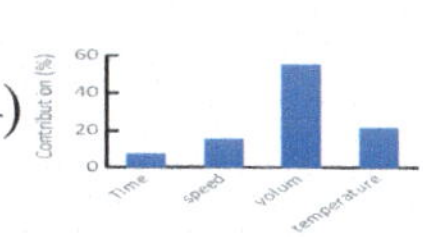

where *n* tests the test recurrence, and Y_i is the result. The result of tensile and flexural test referring to orthogonal array processed using Minitab 18 for Taguchi DOE (Design of Experiment) Taguchi can be seen in Table 2.

Table 2 Result S/N ratio

Test	t	S	V	T	S/N ratio	
					Tensile	Flexural
1	10	150	1	40	33.12	38.88
2	10	250	2	60	33.45	38.83
3	10	350	5	80	33.43	38.93
4	20	150	2	80	33.06	38.67
5	20	250	5	40	33.40	39.5
6	20	350	1	60	33.25	37.88
7	30	150	5	60	33.47	39.13
8	30	250	1	80	33.19	38.79
9	30	350	2	40	33.27	38.76

S/N ratio calculation is used to figure out the most influencing factor on the test result. Its response affects the factors/level and mean. Table using Larger is Better S/N setting shows that the addition of volume fraction has the highest effect on the increase of the tensile stress. The next factors are temperature, rotational speed, and rotation time. The result of main effects plot for S/N ratio can be seen in Fig. 1.

Figure 1 shows the main effect plot dan S/N ratio describing the best treatment to obtain the maximum tensile strength, with 10 min stirring time, 250 rpm stirring speed, 5% MCC addition, and temperature 40 °C. Meanwhile, the maximum flexural strength will be reached in 30 min stirring time, 250 rpm stirring speed, 5% MCC addition, and temperature 24 °C. Table 3 shows the influence and contribution level of each factor on the tensile strength.

Table 4 shown the determined of the influence and the contribution of each factor on the flexural strength. Figure 2 shown that this Anova test result shows the contribution level of each treatment to obtain the maximum tensile stress. It started

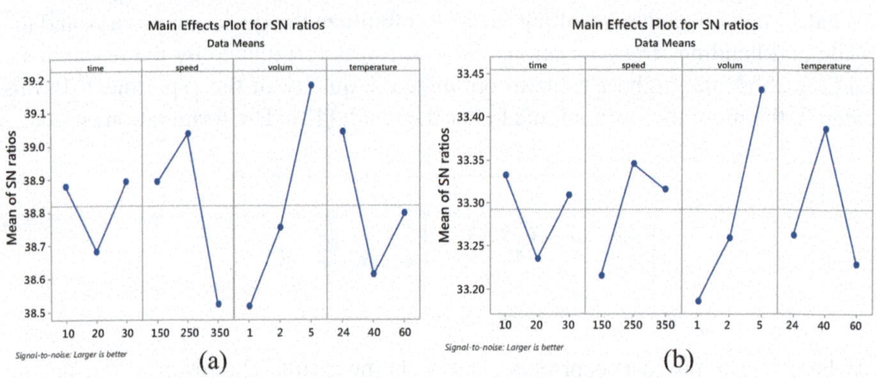

Fig. 1 Main effect plot for S/N ratios. **a** Tensile stress value, **b** flexural stress value

Table 3 Analysis of variance

Source	DF	Seq. SS	Contribution	Adj. SS	Adj. MS	F-value	*P*-value
Time	2	0.3994	7.81%	0.3994	0.1997	*	*
Speed	2	0.8077	15.80%	0.8077	0.4039	*	*
Volume	2	2.8166	55.10%	2.8166	1.4083	*	*
Temperature	2	1.0880	21.28%	1.0880	0.5440	*	*
Error	0	*	*	*	*		
Total	8	5.1117	100.00%				

Table 4 Analysis of variance

Source	DF	Seq. SS	Contribution	Adj. SS	Adj. MS	F-value	*P*-value
Time	2	11.92	7.72%	11.92	5.99	*	*
Speed	2	34.09	22.09%	34.09	17.06	*	*
Volume	2	63.93	41.42%	63.93	31.96	*	*
Temperature	2	44.42	28.78%	44.42	22.21	*	*
Error	0	*	*	*	*		
Total	8	154.36	100.00%				

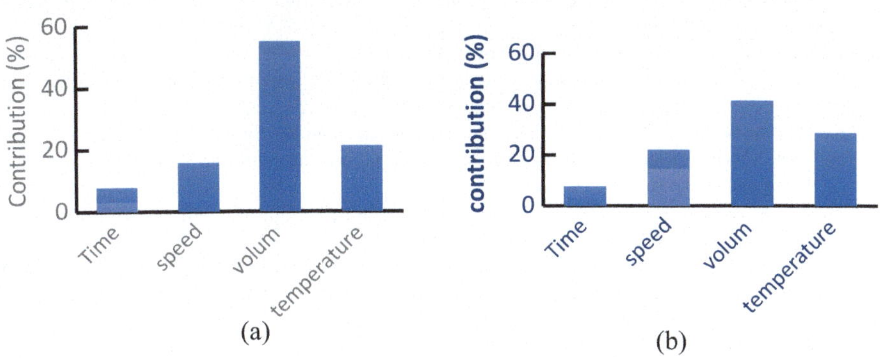

Fig. 2 The factors' contribution. **a** Tensile strength, **b** flexural strength

with the 58.67% addition of volume fraction and continued with 28.54% temperature, stirring time, and stirring speed. The contribution of each factor in flexural test shows that volume adding, temperature, and stirring time give 41.42%, 28.78%, and 7.72% contribution respectively.

Compared to the previous research [13], unsaturated with no treatment showed 29.37 MPa of tensile strength and 75.6 MPa of flexural strength. It means that there is a 20.03 MPa 68.19% tensile strength increase and 25.26 MPa 33.41% bending strength increase. In this treatment and method there is an increase in MCC percentage, which in the study [14] an increase of 20% MCC only produces a tensile strength up to 24–36.6 MPa. Whereas in this study only 5% MCC produced 29.37 MPa of Tensile strength.

4 Conclusion

Based on the analysis result and the calculation, using Taguchi method with S/N ratio analysis and Analysis of Variance, the best treatment is 5% MCC adding, 250 rpm stirring speed, 10 min stirring time, and 40 °C temperature. The highest contribution with 5% MCC adding has the best effect by providing 55.10% contribution for tensile strength and above 41.42% for flexural strength. Temperature and speed also give a quite significant contribution.

References

1. Mathew AP, Oksman K, Sain, M (2005). Mechanical properties of biodegradable composites from poly lactic acid (PLA) and microcrystalline cellulose (MCC). Journal of Applied Polymer Science 97(5):2014–2025
2. Treimanis A, Laka MARIANNA, Chernyavskaya S, Ganster JOHANNES, Erdmann J, Ziegler L, Birska I (2016) Microcrystalline cellulose fillers for use in hybrid composites with polyethylene and lignin. Cell. Chem. Technol 50(1):117–125
3. Kiziltas A, Gardner DJ, Han Y, Yang, HS (2014) Mechanical properties of microcrystalline cellulose (MCC) filled engineering thermoplastic composites. Journal of Polymers and the Environment 22(3):365–372
4. Tang L, Weder C (2010) Cellulose wisker/epoxy resin nanocomposite ACS Publication. Appl Mater Interfaces. https://doi.org/10.1021/am900830h
5. Karakus K, İlkay ATAR, BAŞBOĞA İH, Bozkurt F, MENGELOĞLU F (2017) Wood ash and microcrystalline cellulose (MCC) filled unsaturated polyester composites. Kastamonu Üniversitesi Orman Fakültesi Dergisi 17(2):282–289
6. Cataldi A, Dorigato A, Deflorian F, Pegoretti A (2014) Effect of the water sorption on the mechanical response of microcrystalline cellulose-based composites for art protection and restoration. Journal of Applied Polymer Science 131(18)
7. Arjmandi R, Hassan A, Haafiz MM, Zakaria Z, Inuwa IM (2014) Characterization of poly-lactic acid/microcrystalline cellulose/montmorillonite hybrid composites. Malaysian J. Anal. Sci 18:642–650
8. Dhogan N, Mchugh TH (2007) Effect of microcrystalline cellulose on functional properties of hydroxy propyl methyl cellulose microcomposite film. J Food Sci 72
9. Boran, S (2016) Biological properties of nutshell and microcrystalline cellulose filled density polyethylene composite. Arch Mater Sci Eng 78(2):71
10. Wang B, Zhang HR, Huang C, Chen XF, Shi SL, Xiong L, Luo J, Chen XD (2015) Polyvinyl chloride/attapulgite/micro-crystalline cellulose (MCC) composites preparation and analysis of the role of MCC as a compatibilizer. BioResources 10(4):7693–7703
11. Kumar Y, Singh H (2015) Multi response in dry turning process using Taguchi approach and utility concept. Proc Mater Sci 5:2142–2151
12. Roy RK (1990) A primer on the Taguchi method. Van Nostrand Reinhold, NY, pp 100–154

13. Sakuri S, Ariawan D, Surojo E (2019) Mechanical properties of microcrystalline cellulose filled unsaturated polyester blend composite with various treatment. In: AIP production AIP conference proceedings, vol 2097, pp 030065. http://doi.org/10.1063/1.5098240
14. Kiziltas A, Gardner DJ, Han Y, Yang, HS (2010) Determining the mechanical properties of microcrystalline cellulose (MCC)-filled PET-PTT blend composites. Wood and Fiber Science 42(2):165–176

Effect of High Sintering Temperature on the Cobalt Ferrite Synthesized Via Co-precipitation Method

Siti Maisarah Ahmad Tarmizi, Muhammad Amin Zamri, Nur Azmah Nordin, Rizuan Mohd Rosnan, Saiful Amri Mazlan, Hafizal Yahaya, and Siti Aishah Abdul Aziz

Abstract Magnetic particle is one of the main elements used in magnetorheological (MR) materials. In this study, the magnetic cobalt ferrite nanoparticles are successfully synthesized via co-precipitation method at different sintering temperatures. The nanoparticles are prepared in neutral condition (pH 7) at different temperatures of 900, 1000 °C for 8 h, individually. The properties of cobalt ferrite related to phase analysis, microstructure and magnetic properties were characterized by particle size analyzer (PSA), field-emission scanning electron microscopy (FESEM), X-ray diffraction (XRD), and vibrating sample magnetometer (VSM). The result showed that the size of cobalt ferrite powders are 16.72, 17.76 and 37.29 nm correspond to different temperatures of 900, 1000 and 1100 °C, respectively. This indicated that the higher sintering temperature induced to the synthesized cobalt ferrite, greater

S. M. A. Tarmizi · M. A. Zamri · N. A. Nordin (✉) · S. A. Mazlan · H. Yahaya · S. A. A. Aziz
Engineering Materials and Structures (eMast) iKohza, Malaysia-Japan International Institute of Technology (MJIIT), Universiti Teknologi Malaysia, Jalan Sultan Yahya Petra, 54100 Kuala Lumpur, Malaysia
e-mail: nurazmah.nordin@utm.my

S. M. A. Tarmizi
e-mail: maisarahtarmizi94@gmail.com

M. A. Zamri
e-mail: aminzamri96@gmail.com

S. A. Mazlan
e-mail: amri.kl@utm.my

H. Yahaya
e-mail: hafizal.kl@utm.my

S. A. A. Aziz
e-mail: aishah118@gmail.com

R. M. Rosnan
Jeol (Malaysia) Sdn. Bhd., 508, Block A, Kelana Business Center 97, Jalan SS7/2, Kelana Jaya, 47301 Petaling Jaya, Malaysia
e-mail: rizuanmr@jeolmal.com

S. A. Mazlan
Advanced Vehicle System (AVS) Research Group, Universiti Teknologi Malaysia, Jalan Sultan Yahya Petra, 54100 Kuala Lumpur, Malaysia

U. Sabino et al. (eds.), *Proceedings of the 6th International Conference and Exhibition on Sustainable Energy and Advanced Materials*, Lecture Notes in Mechanical Engineering, https://doi.org/10.1007/978-981-15-4481-1_23

size of the nanoparticles will be obtained. Meanwhile, the micrograph of cobalt ferrite powders poses a polygonal shape with all exhibit the face centered cubic (FCC) structure. In the meantime, the magnetic saturation, M_s increased with the size of nanoparticles increased from 28.95 to 38.85 emu/g, simultaneously correspond to increasing in the sintering temperatures. The results underlined that the sintering temperature did affect and alter the size of $CoFe_2O_4$ nanoparticles that resulted in magnetic properties enhancement.

Keywords Magnetic nanoparticles · Cobalt ferrite · Co-precipitation · Sintering temperature

1 Introduction

Recently, addition of magnetic particles has been widely used in polymer based materials such as elastomer, grease, plastomer and foam. Known as magnetorheological (MR) materials, these smart materials are actively respond to applied magnetic field. With presence of magnetic particles, the mechanical properties of the polymer materials can be enhanced by alignment of the chain-like particles parallel with the field thus generate MR effect. Indeed, previous studies have reported the use of carbonyl iron particles (CIPs) in most MR materials [1–3]. Magnetic particles in powder form is more suitable than a bulk shape in the application of MR materials due to easily being dispersed and would be uniformly distributed in the carrier medium of MR materials.

Alternative for CIPs, iron based particles such as ferrites (Fe_2O_4), magnetite (Fe_3O_4), maghemite (γ-Fe_2O_3) and hematite (α-Fe_2O_3) can also be used as filler in the MR materials. Despite high magnetic saturation and good mechanical strength, this particles is pricey and industrial manufactured. Cobalt ferrite or $CoFe_2O_4$ intrigued the interest amongst researchers to explore more because of their promising chemical and physical properties such as high coercivity, moderate saturation magnetization (80 emu/g), high mechanical hardness, exceptional chemical stability and high Curie temperature (520 °C) [1–5]. The morphology and properties of cobalt ferrite can be varied depending on the type of methods and parameters used during the synthesis process of the particles such as pH, drying temperature and sintering temperature. Currently, various methods have been applied to synthesis cobalt ferrite particles. For example, co-precipitation [5–8], microemulsion [9], sol-gel technique [10, 11] and polymer pyrolysis [12]. Among these methods, co-precipitation is more favorable method to prepare the cobalt ferrite particles since can be synthesized in bulk scale, low cost, high crystallinity, nontoxic precursors and good texture of the particles. It also yields homogeneity of the final product.

Meanwhile, one of the parameter that mostly been focused during synthesis cobalt ferrite powder is sintering temperature. Based on previous studies, Kurian et al was carried out with comparison method; sol gel and co-precipitation [4]. Then, the powder was sintered at 450 and 750 °C for 4 h, individually. For co-precipitation method, at 450 °C the size of cobalt ferrite was 14.45 nm and increase to 35.09 nm for 750 °C of sintering temperature. It stated that co-precipitation method indicated better distribution of homogenous nanoparticles compared with sol gel method. In 2018, synthesized cobalt ferrite nanoparticles using co-precipitation method was done by varied the sintering temperature from room temperature to 100 °C [6] It depicts that the magnetic saturation of cobalt ferrite increase as the size of the particles increase. However, various methods bring challenge and curiosity to the scientist on how to obtain the desirable characteristics and properties for best performance of cobalt ferrite particles for some applications In this study, the effect of high sintering temperature on the synthesized cobalt-ferrite particles has been focused and correlation towards the changed of magnetic properties of the particles analyzed. The cobalt-ferrite particles produced via co-precipitation method.

2 Experimental

2.1 Synthesis Powder

Cobalt ferrite powders were synthesized by a co-precipitation method in a neutral condition (pH 7). Iron (III) nitrate nanohydrate, $Fe(NO_3)_3 \cdot 9H_2O$ and cobalt (II) acetate tetrahydrate, $Co(CH_3COO)_2 \cdot 4H_2O$ would be used as starting raw materials, obtained from Qrec brand. These two chemicals were mixed, together with distilled water that acted as solvent for the solution. The solution was heated at 100 °C for two hours and stirred with magnetic stirrer constantly with 200 rpm. Then, sodium hydroxide pellets that act as extracting agent were added into the solution to extract the synthesized powder. Two separated products were observed in the solution where the cobalt ferrite powder was sediment respective to higher density than the overall solution. The precipitate was washed once a day with distilled water until the pH of the solution reached the neutral condition of pH 7. Once the neutral pH precipitate obtained, the powder was dried in an oven to remove the excess water content in the powder, at 200 °C for 24 h. Next, the powder was sintered at three temperatures, particularly 900, 1000 and 1100 °C to make the material more compact and obtained the pure phase of single-phase spinel structures of the compound particles.

2.2 Characterization of Powder

Cobalt ferrite powder undergoes various characterizations, in terms of structural characterization, morphological, chemical composition, particles size distribution and magnetic properties correspond to different sintering temperatures. X-ray diffraction was carried out by using X-ray diffractometer (Empyrean, Pan analytical) with

Cu K_α radiation of k = 0.154 nm. The lattice parameter, a was calculated with formula; $a = d\sqrt{(h^2 + k^2 + l^2)}$ where h, k, and l is the peak of diffraction plane. In consideration of spinel structure that contains 8 molecules, X-ray density (d_x) was obtained with relation $d_x = 8\,M/Na^3$ where M is molecular weight, N is Avogadro number (6.022×10^{23} mol^{-1}) and a is the lattice parameter.

The morphological of powder was characterized via Field-Emission Scanning Electron Microscopy (FESEM), brand JEOL with model of JSM-7800F. The FESEM also is equipped with energy dispersive X-ray (EDX) that functioned to analyze the chemical composition individually. Meanwhile, magnetic properties of the cobalt ferrite powders were examined via vibrating sample magnetometer (VSM) with quartz powder cup as holder, brand MicroSense USA. In order to determine the average size of the cobalt ferrite powders, particles size analyser (PSA) was used and the PSA used is SALD 2300 with a measurement range of 17–2500 μm.

3 Results and Discussions

3.1 Particle Size Distribution

The particle size of cobalt ferrite that undergo sintering temperatures of 900–1100 °C was analyzed via PSA. In particular, for 900 °C, the average particle's size is noted around 16.71 nm. For 1000 °C, the particle' size was 17.75 nm while, for the highest sintering temperature of 1100 °C, 37.30 nm was the size of the particles.

As shown in Fig. 1, the average size of the cobalt ferrite powder for 900, 1000 and 1100 °C are 16.71, 17.75 and 37.30 nm, respectively, indicating that the increment of the sintering temperatures has increased the particle size of cobalt ferrite powder. However, the particle' size at 1100 °C sintering temperature shows a significant different compared to size at 900 and 1000 °C sintering temperatures. In fact, the

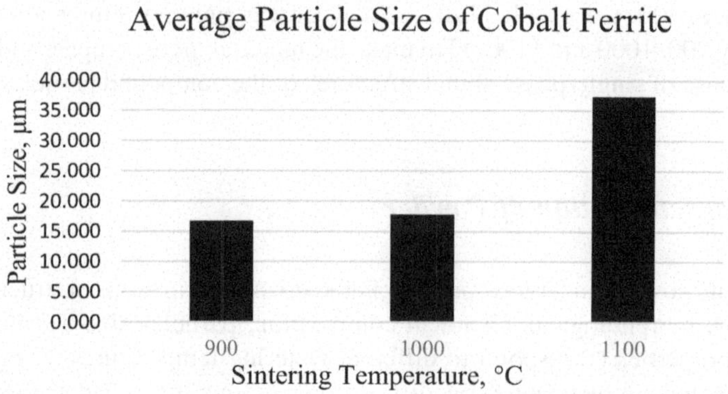

Fig. 1 The average size of cobalt ferrite powder for three sintering temperatures

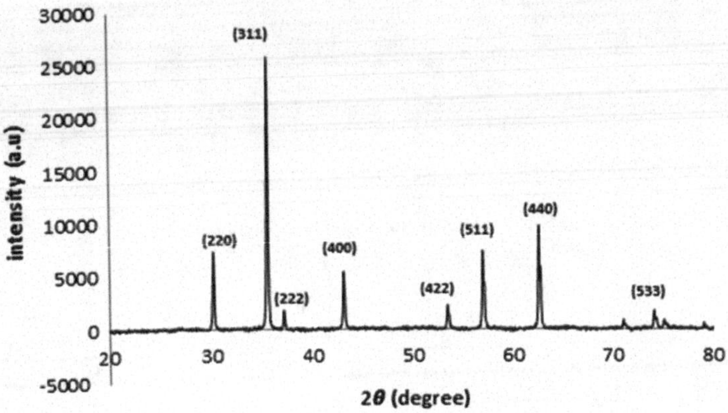

Fig. 2 XRD patterns for the cobalt ferrite powder sintered at 1100 °C

difference is about 20 μm respective to value at 900 and 1000 °C sintering temperatures. Sintering temperature is known a parameter that could enlarge the grain size of particle/structure and as a result, the grain would grow and affect the particle' size. Higher sintering temperature, the bigger the particle's size would be obtained [13, 14]. In fact, the sintering temperature that is lower than 1000 °C has no significant effect on the particle size distribution as shown is the figure and as was reported by previous studies [4, 7, 8].

3.2 *Structural Characterization*

Figure 2 presents the XRD pattern of cobalt ferrite powder that sintered at 1100 °C. As can be seen, the diffraction peaks obtained for the plane (220), (311), (400), (422), (511), (440) and (533) point out the formation of single phase spinel structure of cobalt ferrite. The sharp peak of the planes indicates crystallinity structure of the cobalt ferrite powder. The value of lattice parameter and X-ray density obtained for the cobalt ferrite powder from the calculations are 8.368 Å and 5.318 g/cc, respectively.

3.3 *Morphological Characterization*

Figure 3 illustrates the morphology of cobalt ferrite powder obtained from FESEM at a magnification of 100,000×. It can be seen the powder exhibit irregular polygonal shape, most in octahedron shape of particles. However, the cobalt ferrite powders are shown to be agglomerated as in Fig. 3.

Fig. 3 FESEM images of cobalt ferrite powder of ×100,000 magnifications

In order to clarify the chemical composition of the observed powder, EDX analysis was carried out. Figure 4 illustrates the EDX mapping analysis on the interest area that contained of cobalt ferrite powder. In particular, the elemental mapping shows that the scan cobalt ferrite powder is composed of Co, Fe and O, which were presented in yellow, green and red colours, respectively. From Table 1, it presents the mass and the percentage of the element in the cobalt ferrite powder. Meanwhile, the atomic percentage of each element shows that the element of Fe is twice than the Co while element O is fourth times the element Co and twice than the element of Fe that tally with the chemical formula of cobalt ferrite which is $CoFe_2O_4$. This data can predict that the chemical compound of the powder is around cobalt ferrite powder and this result can be correlated with the XRD result shown before.

3.4 Magnetic Properties

Figure 5 presents the hysteresis loop of Cobalt ferrite powders at three sintering temperatures. It can be noted that the magnetic saturation, M_s at the temperatures of 900, 1000 and 1100 °C are 73.43, 80.07 and 93.07 emu/g, respectively. In brief, increasing the sintering temperature has increased the M_s value of the cobalt ferrite

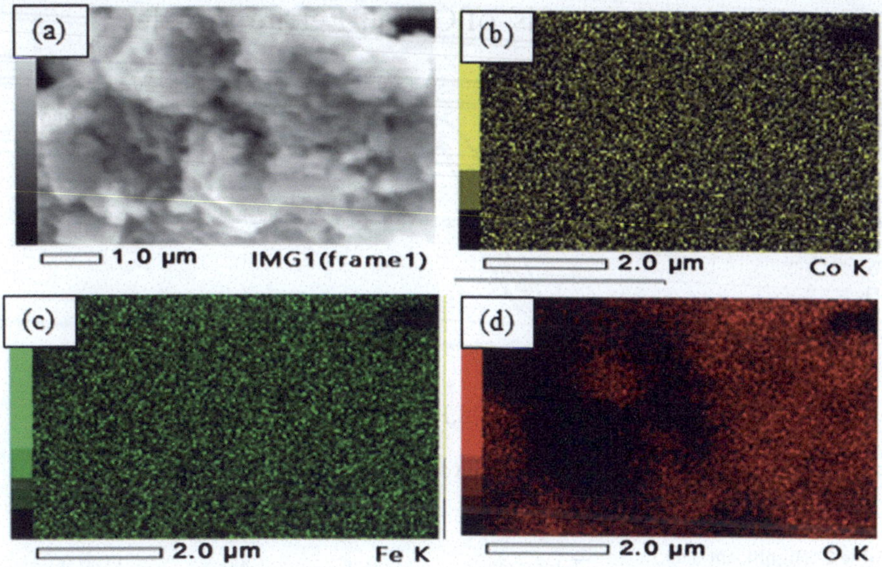

Fig. 4 Elemental mapping the Cobalt ferrite powders; **a** interest area, **b** presence of Co element, **c** element Fe and **d** element O

Table 1 Mass and atomic percentage of elements in Cobalt ferrite powders

Element	Mass%	Atom%
Co	15.55	12.15
Fe	29.53	26.47
O	54.92	61.38
Total	**100.00**	**100.00**

Bold represent the subject of the variables and the end of the results

powders. This result might be due to the increasing of particle's size of the particles as increasing the sintering temperatures. Correspondingly, the remanence, M_r value at the temperature of 900 °C is about 10.48 emu/g, followed by 11.91 emu/g for 1000 °C and the highest M_r is at 1100 °C with 28.56 emu/g. Meanwhile, the coercivity, H_c of the Cobalt ferrite powders is 157.70 Oe for 900 °C sintering temperature, 169.27 Oe for 1000 °C and the highest H_c value is at 1100 °C with 564.55 Oe. All the results have the similar trend as increasing the sintering temperature that has enhanced the magnetic properties, respectively (Table 2).

Three elements in the hysteresis graph, particularly M_s, M_r and H_c values are shown to be increased with the increased of sintering temperatures. This trend also was reported in previous study [4, 6, 15]. It is believed that when the sintering temperature increased, the particles would receive more energy and enlarge the grain size of the particles [16]. It would affect the small grains of particles to be merged,

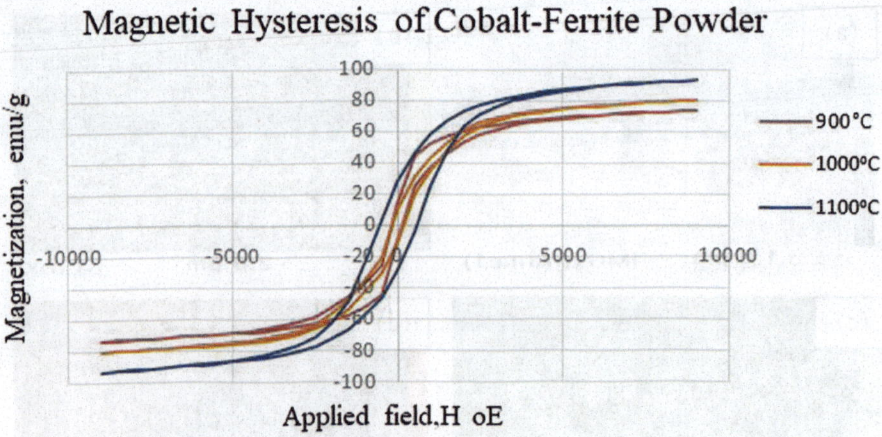

Fig. 5 Magnetic hysteresis of cobalt ferrite powder at different sintering temperatures

Table 2 Significant data from magnetic hysteresis

Temperature	900 °C	1000 °C	1100 °C
M_s (emu/g)	73.43	80.07	93.07
M_r (emu/g)	10.48	11.91	28.56
H_c (Oe)	157.50	169.27	564.55

became bigger and consequently the resultant particle's size will be increased. The phenomenon also may help to re-orientate the spin (in the field direction) and domains to magnetic align [17]. Therefore, the magnetic value is significantly increased. The magnetic properties cobalt ferrite powders sintered at 1100 °C are drastically increased compared to the other sintering temperatures (900 and 1000 °C) due to significant improved in particle's size as reported in Fig. 5. Therefore, it can be concluded that the sintering temperature has affected the physical and magnetic properties of cobalt ferrite powders especially for the temperature more than 1000 °C.

4 Conclusion

The synthetization of cobalt ferrite powder via co-precipitation method and being sintered at three different temperatures were achieved. In brief, increased the sintering temperature has increased the physical and magnetic properties of cobalt ferrite powder. The characterization of the cobalt ferrite powder shows that the magnetic saturation, M_s of the powder at 900, 1000 and 1100 °C are 73.43, 80.07 and 93.07 emu/g, respectively indicating that the increment trend of M_s as increasing the sintering temperature. Perhaps, the cobalt ferrite powder have a great potential as replacement and alternative for CIPs in some applications that required properties moderate magnetic saturation.

Acknowledgements This study was financially supported by the Universiti Teknologi Malaysia under Trans-disciplinary Research Grant (Vot No: 07G13) and Fundamental Research Grant Scheme (Vot No: 5F001). The authors acknowledge and thank MJIIT Microscopy Laboratory for assistance for this study.

References

1. Rahman NANA, Mazlan SA, Aziz SAA., Nordin NA, Ubaidillah R, Hapipi N (2018) Magnetorheological elastomer silicone-based containing corroded carbonyl iron particles. Key Eng Mater 772:51–55
2. Ubaidillah R, Sutrisno J, Purwanto A, Mazlan SA (2015) Recent progress on magnetorheological solids: materials, fabrication, testing, and applications. Adv Eng Mater 17(5):563–597
3. Wahid SA, Ismail I, Aid S, Rahim MSA (2014) Magneto-rheological defects and failures: a review. IOP Conf Ser: Mater Sci Eng 114
4. Kurian M, Thankachan S, Nair DS, Aswathy EK, Babu A, Thomas A, Binu Krishna KT (2015) Structural, magnetic, and acidic properties of cobalt ferrite nanoparticles synthesised by wet chemical methods. J Adv Ceram 4(3):199–205
5. Sagadevan S, Podder J, Das I (2017) Synthesis and characterization of cobalt ferrite ($CoFe_2O_4$) nanoparticles prepared by hydrothermal method. In: Recent trends in materials science and applications, pp 145–152
6. Stein CR, Bezerra MTS, Holanda GHA, André-Filho J, Morais PC (2018) Structural and magnetic properties of cobalt ferrite nanoparticles synthesized by co-precipitation at increasing temperatures. AIP Adv 8(5)
7. Ristic M, Krehula S, Reissner M, Jean M, Hannoyer B, Musić S (2017) Synthesis and properties of precipitated cobalt ferrite nanoparticles. J Mol Struct 1140:32–38
8. Olusegun SJ, Freitas ETF, Lara LRS, Stumpf HO, Mohallem NDS (2019) Effect of drying process and calcination on the structural and magnetic properties of cobalt ferrite. Ceram Int 45(7):8734–8743
9. Liu C, Rondinone AJ, Zhang ZJ (2000) Synthesis of magnetic spinel ferrite $CoFe_2O_4$ nanoparticles from ferric salt and characterization of the size-dependent superparamagnetic properties. Pure Appl Chem 72(1–2):37–45
10. Hunyek A, Sirisathitkul C, Harding P, Harding DJ (2012) Structural and magnetic properties of cobalt ferrites synthesized using sol-gel techniques. Mater Sci-Poland 30(3):278–281
11. Shinde AB (2013) Structural and electrical properties of cobalt ferrite nanoparticles. Int J Innov Technol Explor Eng 3(4)
12. Swatsitang E, Phokha S, Hunpratub S, Usher B, Bootchanont A, Maensiri S, Chindaprasirt P (2016) Characterization and magnetic properties of cobalt ferrite nanoparticles. J Alloy Compd 664:792–797
13. Xavier S, Thankachan S, Jacob BP, Mohamed EM (2013) Effect of sintering temperature on the structural and magnetic properties of cobalt ferrite nanoparticles. Nanosyst: Phys Chem Math 4(3):430–437
14. Syamimi NF, Amin Matori K, Lim WF, Abdul Aziz S, Mohd Zaid MH (2014) Effect of sintering temperature on structural and morphological properties of europium (III) oxide doped willemite. J Spectrosc 1–8
15. Kumar L, Kar M (2011) Effect of annealing temperature and preparation condition on magnetic anisotropy in nanocrystalline cobalt ferrite. IEEE Trans Magn 47(10):3645–3648

16. Bjørk R, Tikare V, Frandsen HL, Pryds N, Blendell J (2013) The effect of particle size distributions on the microstructural evolution during sintering. J Am Ceram Soc 96(1):103–110
17. Issa B, Obaidat IM, Albiss BA, Haik Y (2013) Magnetic nanoparticles: surface effects and properties related to biomedicine applications. Int J Mol Sci 14(11):21266–21305

The Straight Blade Application to Increasing the Performance of the Savonius Water Turbine (Simulation Study)

Ahmad Irham Rahimi, Dhimas Cahyo Anindito, Dominicus Danardono, and Syamsul Hadi

Abstract Decreasing fossil energy sources encourages people to switch from fossil energy to renewable energy because of energy as an essential aspect of life. One of the current efforts to utilize water resources as renewable energy is the Savonius water turbine development. Conventional Savonius water turbine has a simple construction and application. However, the turbine produces relatively small power that requires further assessment to improve its performance. The purpose of this simulation study is to determine the effect of the placement angle of straight blade on the performance of Savonius water turbines on the horizontal axis. Four placement angles of the straight blade at two variations of the number of blades were investigated. This research was conducted by designing a simulation model using software ANSYS 15 with CFX Solver. The water velocity of 1.5 m/s was applied to the turbine with a diameter of 164 mm and a height of 82 mm. The results present torque distribution, flow velocity, and power coefficient (C_p). Based on research, Savonius turbines with placement angle of 85° has the best power coefficient with a C_p value of 0.086 on the Tip Speed Ratio of 0.7. This study was able to increase the power coefficient by 26% with the addition of a straight blade.

Keywords Straight blade · Renewable energy · Savonius water turbine · Power coefficient · Hydropower

1 Introduction

Energy is an important aspect of life. The decline in fossil energy sources has encouraged humans to turn to renewable energy. Fossil power plants have supplied most of Indonesia's electricity needs. In 2017 all energy needs in Indonesia depend on

A. I. Rahimi · D. Danardono · S. Hadi (✉)
Department of Mechanical Engineering, Universitas Sebelas Maret, Jl. Ir. Sutami 36A, Surakarta 57126, Indonesia
e-mail: Syamsulhadi@ft.uns.ac.id

D. C. Anindito
Graduate School of Mechanical Engineering, Universitas Sebelas Maret, Jl. Ir. Sutami 36A, Surakarta 57126, Indonesia

U. Sabino et al. (eds.), *Proceedings of the 6th International Conference and Exhibition on Sustainable Energy and Advanced Materials*, Lecture Notes in Mechanical Engineering,
https://doi.org/10.1007/978-981-15-4481-1_24

fossil energy around 76.3% [1]. Reviewing the data, the government has launched various programs to reduce the use of fossil energy in Indonesia. One of the programs launched by the Indonesian government is the Green Building Program. Green building program is a solution that can improve the efficiency of a building by utilizing resources. So that this can reduce the negative impact of environmental contamination [2]. Rainwater and household waste in high rise buildings that are usually only wasted have great potential to be used as electricity generation. This is supported by the climatic conditions of Indonesia, which have an average rainfall intensity of 1871.03 mm [3]. Rainwater that has flowed into one pipe can be used for electricity generation with a Rain Water Harvesting system.

The various method has been carried out to design water turbines to produce maximum power efficiency [4]. One interesting study has been applied to wind turbines for water, such as the Savonius turbine [5]. Savonius turbine has an initial torque able to spin with low fluid speed. Some advantages from Savonius turbines are rotation independent of the direction of fluid flow, good initial characteristics, easy to design, and vibration load will be slightly on the supporting structure [6]. However, the Savonius turbine still has a relatively low efficiency [7]. Research on the Savonius turbine usually use wind movement sources but Savonius turbine used water sources to increase the performance by 61.32% compared to wind sources [5].

Various research were carried out to improve the efficiency of the Savonius turbine by modifying the shape of the turbine. Researchers have modified the Savonius turbine by adjusting the angle of the blade curvature (ψ). This research was conducted by comparing the angle of the blade curvature of 110°, 120°, 130° and 140°. The Savonius turbine with the best performance is produced by the angle of the blade curvature of 120° with a coefficient of power (C_p) value of 0.23 at TSR = 1.075 and electrical power of 39.15 Watt [8]. Other research also was conducted by comparing the number of blades starting from 2, 3, 4, 5 and 6. The Savonius turbine with the number of blades 3 has the best performance, which can produce a coefficient of power (C_p) value of 0.23 at TSR = 1, 7, and electrical power of 34,834 W [9].

The turbine blade design for the Savonius has studied by experiments and simulations. Altan et al. have researched comparing Savonius turbines with the addition of a straight blade to conventional Savonius turbines. The best performance is owned by turbine design with Aspect Ratio (AR) = 0.7, Overlap Ratio (OR) = 0, and end blade parameters (D_0/D) = 1.1 with power coefficient value (C_p) 0.21 at TSR = 0.69 [10]. The best performance possessed by the Savonius turbine with the addition of a straight blade can increase the power coefficient by 20% compared to conventional Savonius turbines [11].

Savonius turbines can still be developed further until now. One of the development efforts made by the Savonius turbine is the modification of the turbine geometry to increase the efficiency of the Pico Hydro. Therefore based on the literature, further development is needed for the addition of a straight blade using a simulation method to improve the performance of Savonius water turbines.

2 Savonius Rotor

One type of vertical axis wind turbine, which is a drag type that can be used in low-velocity wind fluids is Savonius wind turbines. This turbine is a very simple shape formed from two semicircular blades. The simple shape of the Savonius turbine is also easily modified its geometric design. Savonius rotors can also be applied in two sources of movement with water or wind. Savonius rotor design that is developed will be able to rotate from all directions of fluid flow [12]. So this savonius rotor design is ideally made for areas with very turbulent flow.

2.1 Principal Savonius Water Turbine Rotor

Savonius water turbine rotor is a turbine that uses water energy as its driving force to produce electrical energy by rotating the generator due to the drag force caused by water pressure that hits the turbine blade. Based on aerodynamic principles, Savonius rotors utilize the drag force when extracting fluid energy from fluid flow through the rotor blades. So the concave surface has resistance coefficient higher than the convex surface. Therefore, the semi-cylindrical concave surface that has passed through the fluid will provide a higher drag than the other side so that the rotor rotates. The working principle of the Savonius water turbine is that water from the upper reservoir is channeled through pipes to the turbine blades, which then spin and produce electrical energy from the rotor rotation [13].

2.2 Rotor Performance Parameter Equation

Several parameters are needed to determine the performance of a savonius rotor in this research. Research results that have been carried out with the simulation method will be processed using equations to calculate the rotor performance parameters. The parameters of Savonius rotor performance is as follows:
Power:

$$\mathrm{P_{Available}} = 0.5 \times \rho \times \mathrm{A} \times \mathrm{V}^3 \tag{1}$$

$$P_{Rotor} = T \times \omega \tag{2}$$

Tip Speed Ratio (TSR) is the tangential velocity at the tip of the rotor blade to fluid flow velocity [14].

$$\mathrm{TSR} = \frac{\omega \mathrm{D}}{2\mathrm{U}} \tag{3}$$

$$\omega = \frac{2\pi n}{60} \tag{4}$$

Coefficient Torsi (C_t) and Coefficient Power (C_p):

$$C_p = \frac{P_{Rotor}}{P_{Available}} \tag{5}$$

$$C_t = \frac{C_p}{TSR} \tag{6}$$

where, A is the area of the rotor, ρ for water density, ω represents the angular velocity, C_t for the torque coefficient, T is Torque, and C_p for the Power coefficient, n is the rotating velocity, and V is the free flow velocity [14].

3 Analysis of Preparation

This simulation method is carried out using ANSYS Software and CFX Solver. The research used modification shape of straight blade angle placement. The study was conducted using the rotor with the horizontal axis and five blades [15]. The position of straight blades is 75°, 80°, 85°, and 90°. The first step is to create a 3D geometry of the rotor to be used, as shown in Fig. 1. The study was conducted using an aspect ratio (H/D) of 0.5, endplate parameter (D_O/D) of 1.1, and an overlapping ratio of 0 [7]. This study uses a blade of 5 without straight blades variation and with variations for each modification of the blade shape as shown in Fig. 2. The design dimensions in Fig. 1 are shown in Table 1.

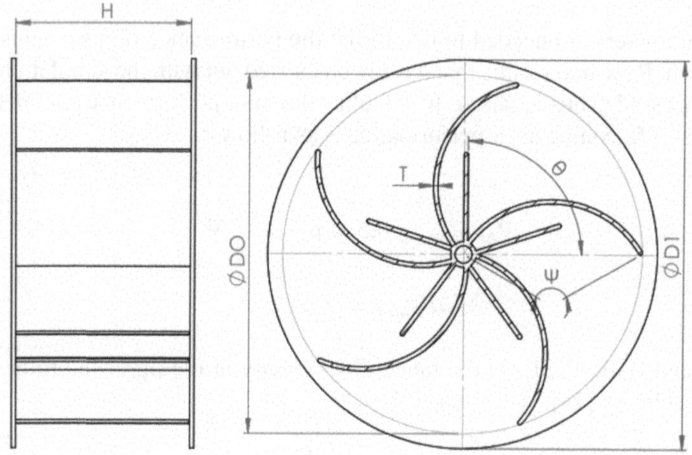

Fig. 1 Schematic of the design to be used in 2D

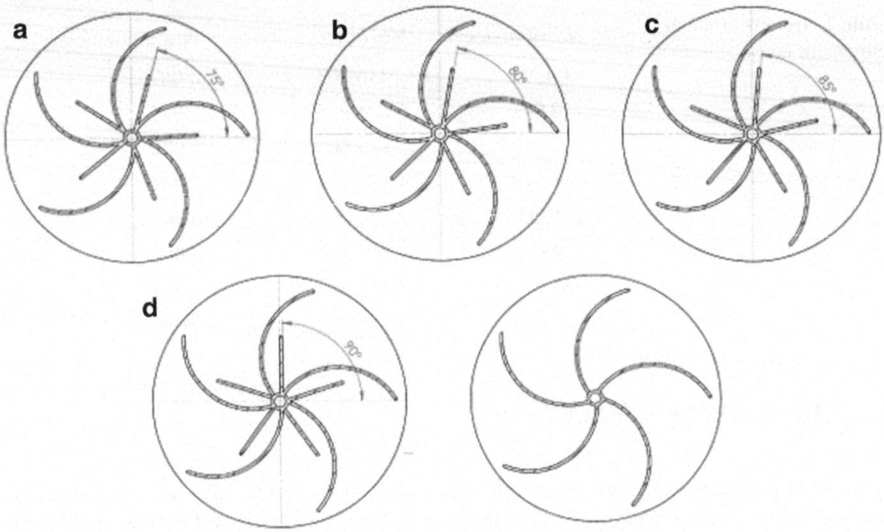

Fig. 2 Variations in an angular placement to be tested in an experiment. **a** 75°; **b** 80°; **c** 85°; **d** 90° and without Straight Blade

Table 1 Dimension of rotor

Parameter	Dimension	Information
H (mm)	82	Rotor height
DO (mm)	164	Rotor diameter
D_I (mm)	180,4	End plate diameter
T (mm)	2	Thickness
θ (°)	75°, 80°, 85°, 95°	Placement angle
Ψ (°)	120°	Curvature angle

4 Method and Numerical Model

4.1 Mesh Independence Study

Research has been carried out using Ansys CFX Solver software. Three-dimensional simulations have been used in this study. Before the simulation process, the meshing process is done first to get the best simulation results. Therefore, mesh reviews in the simulation process are needed to obtain a sufficient number of elements and nodes by increasing the number of elements [16]. The results of the mesh study that have been carried out can be seen in Table 2. Table 2 shows the relation between a number of elements and nodes used with the power coefficient. The results of the mesh study, the number of valid elements, and nodes are 1,359,190 and 250,636.

Table 2 Details of mesh refinement levels

No. of nodes	No. of elements	Power coefficient
61,470	316,303	0.067
122,041	641,975	0.074
162,110	865,605	0.078
218,349	1,176,031	0.081
236,000	1,276,081	0.085
250,636	1,359,190	0.086
265,800	1,441,947	0.078
340,265	1,851,288	0.076

4.2 *Mesh Independence Study*

In the first stage, the connection made on Ansys workbench. The meshing method that has been used is tetrahedral. The meshing solving method used in this simulation is CFX. Based on statistical values, the number of elements and nodes used based on independence meshing is 1,359,190 and 250,636. Meshing results that have been done can be seen in Fig. 3 for the static domain and rotary domain. Meshing results for each variation are shown in Fig. 4.

The meshing domain consists 2 domains, static domain and rotary domain. The stationary domain consists of Inlet, Outlet, Wall, and Interface. The Rotary Domain section is shown in Fig. 5. Besides showing the stationary domain, Fig. 5 also shows the rotary domain. The rotary domain consists of the rear interface, front interface, interface, and rotor. The combination of a domain and boundary conditions is shown in Fig. 6.

Meshing generated in the rotary and static domains were exported to the Ansys CFX solver setting then set boundary conditions that will be needed in the simulation.

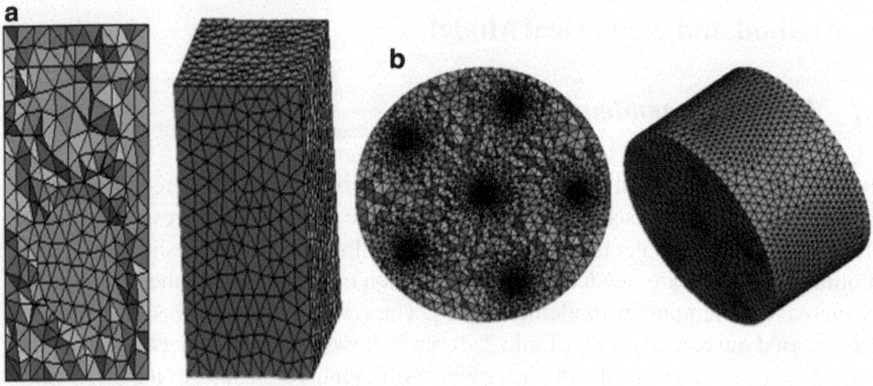

Fig. 3 Mesh at (**a**) rotary domain, (**b**) stationary domain

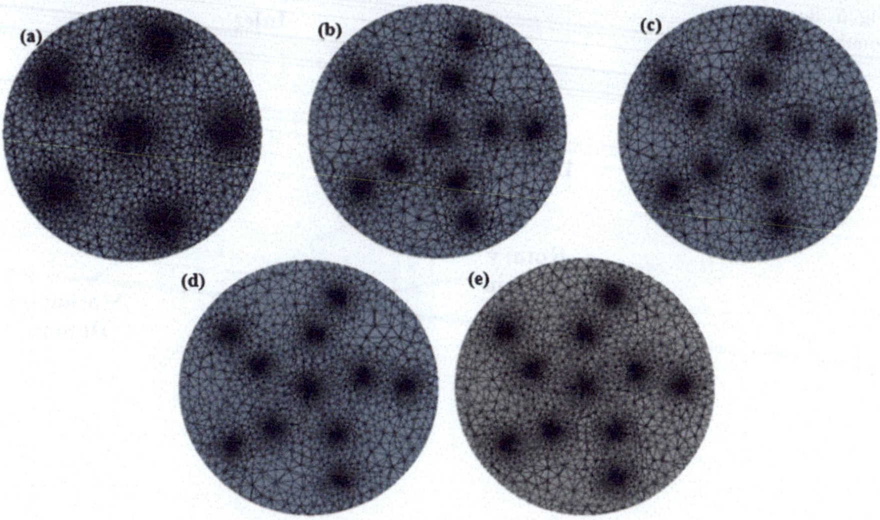

Fig. 4 Mesh of Rotor conventional (**a**) without Straight Blade, and with straight blade at angle of (**b**) 75°, (**c**) 80°, (**d**) 85°, and (**e**) 90°

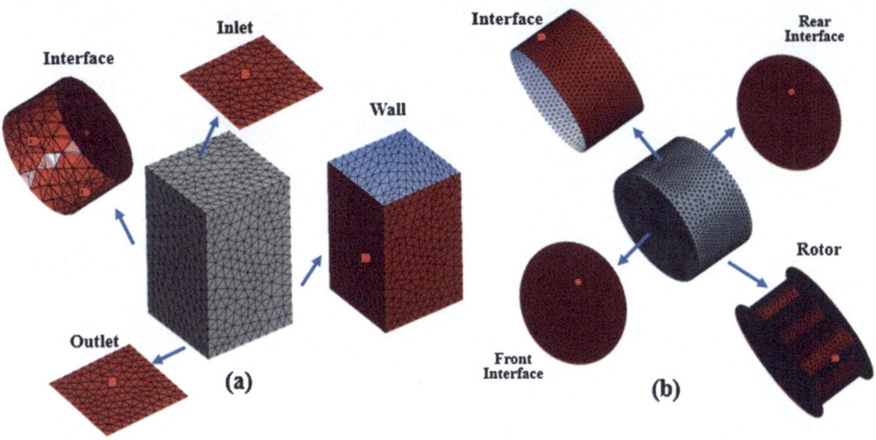

Fig. 5 (**a**) Stationary domain; (**b**) Rotary domain

Fluid properties and operating parameters for the boundary conditions are shown in Table 3.

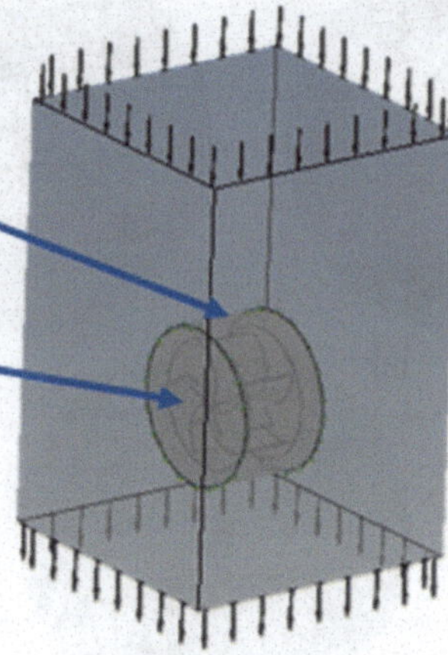

Fig. 6 Schematic of simulation

Table 3 Boundary condition of simulation

No.	Parameter	Value
1	Domain motion	Rotating
2	Boundary condition inlet flow	Subsonic
3	Fluid type	Water
4	Density (kg/m^3)	1000
5	Water velocity (m/s)	1.5
6	Outlet pressure (atm)	1
7	Tekanan Referensi (atm)	1
8	Gravity (m/s^2)	9.81
9	Turbulence model	K epsilon

5 Result and Discussion

Simulations from Savonius' three-dimensional rotor research have been carried out on these five rotors. The results of this study will be displayed in the form of torque value, pressure distribution, velocity vector, and velocity distribution. Torque values are converted into power coefficient and torque coefficient. The results of the C_P (power coefficient) and C_T (torque coefficient) will be presented in graphical form, and the results of the pressure distribution, velocity distribution, and velocity vector will be presented in the form of two-dimensional images.

Power coefficient (C_P) and torque coefficient (C_T) values produced from 5 blade rotors without a straight blade are shown in Fig. 7. The C_{Pmax} is 0.069 at TSR 0.7. C_{Tmax} value is 0.134 at TSR 0.5. C_{Pmax} and C_{Tmax} values are not on the same TSR value. The results of the power coefficient (C_P) and the torque coefficient (C_T) that

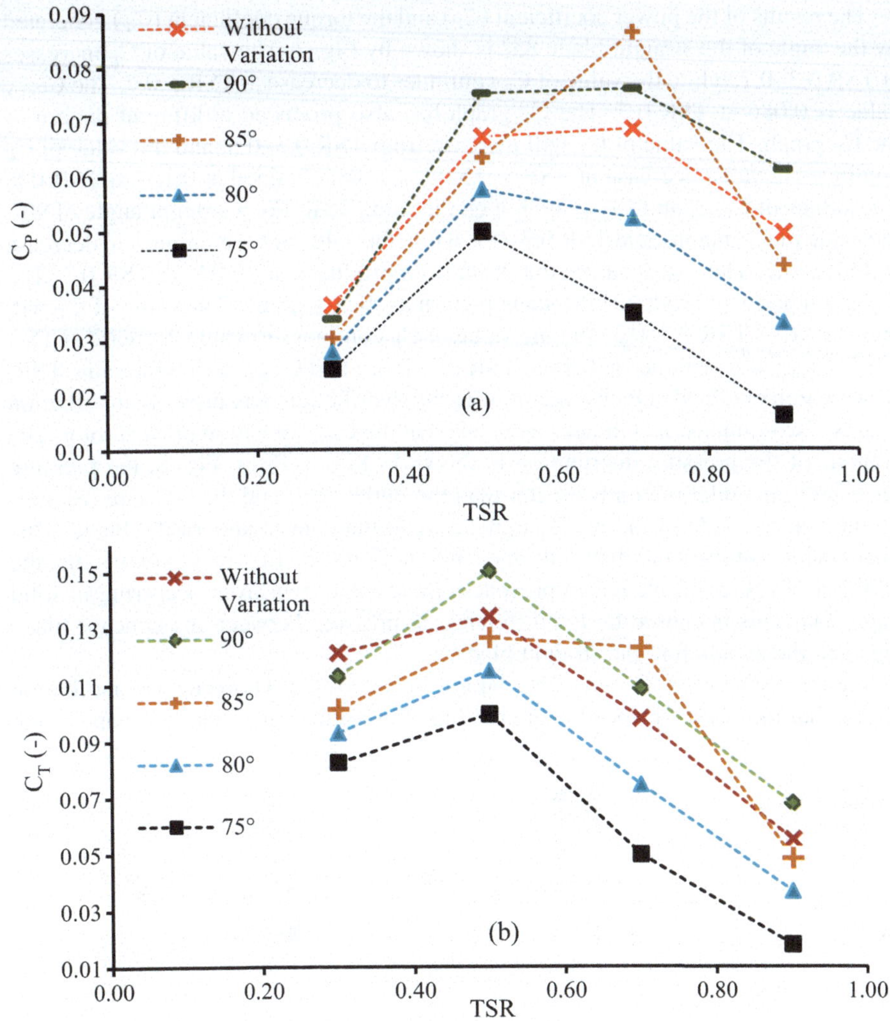

Fig. 7 Graph of (**a**) C_P with TSR and (**b**) C_T with TSR

has been generated by the angle of the straight blade 75°. The value of C_p increases at TSR 0.3–0.5, while the value of C_P continues to decrease at TSR > 0.5. The C_{Pmax} value is 0.05 at TSR 0.5. The C_T graph has also produced the same pattern as the C_P graph. The value of C_T will increase from TSR 0.3–0.5, and the value of C_P continues to value decrease at TSR > 0.5. C_{Tmax} was 0.099 at TSR 0.5. C_{Pmax} and C_{Tmax} are on the same TSR. For a straight angle of 80°, the value of C_p increases at TSR 0.3–0.5 while the value of C_p continues to decrease at TSR > 0.5. The C_{Pmax} value from Fig. 9 is 0.058 at TSR 0.5. The C_T graph has also produced the same pattern as the C_P graph. C_{Tmax} was obtained at 0.115 at TSR 0.5.

The results of the power coefficient (C_P) and the torque coefficient (C_T) generated by the angle of the straight blade 85° is shown by Fig. 7. The value of C_p increases at TSR 0.3–0.7, while the value of C_p continues to decrease at TSR > 0.7. The C_{Pmax} value is 0.086 at TSR 0.7. The C_T graph has also produced a different pattern as the C_P graph. The value of C_T will increase from TSR 0.3–0.5, and the value of C_P continues to value decrease at TSR > 0.5. C_{Tmax} was obtained at 0.127 at TSR 0.5. The values of C_{Pmax} and C_{Tmax} are not on the same TSR. For a straight angle of 90°, the value of C_P increases at TSR 0.3–0.7, while the value of C_p continues to decrease at TSR > 0.7. The C_{Pmax} values that have been produced are 0.076 at TSR 0.7. The C_T graph has also produced the same pattern as the C_P graph. The value of C_T will increase from TSR 0.3–0.5, and the value of C_P continues to value decrease at TSR > 0.5. C_{Tmax} was obtained at 0.15 at TSR 0.5. C_{Pmax} and C_{Tmax} are on the same TSR.

Research obtained velocity vector, velocity distribution, and pressure distribution results. Two-dimensional images were taken at the C_{Pmax} position of each rotor. The contour of the pressure distribution is shown in Fig. 8. From Fig. 8, the pressure received in the inlet region is greater than the outlet area, and the pressure received on the concave side of the blade is greater than the convex side of the blade. This phenomenon of pressure distribution occurs in all rotors. Figure 8 explains that the addition of a straight blade can prevent compressible flow from occurring in solid gaps [11]. This is caused by the difference in pressure between the concave blade area and the area before the straight blade.

Figure 9 shows the velocity distribution in each rotor. The velocity distribution shows that the blade tip speed is greater than the blade center, and the shape of the

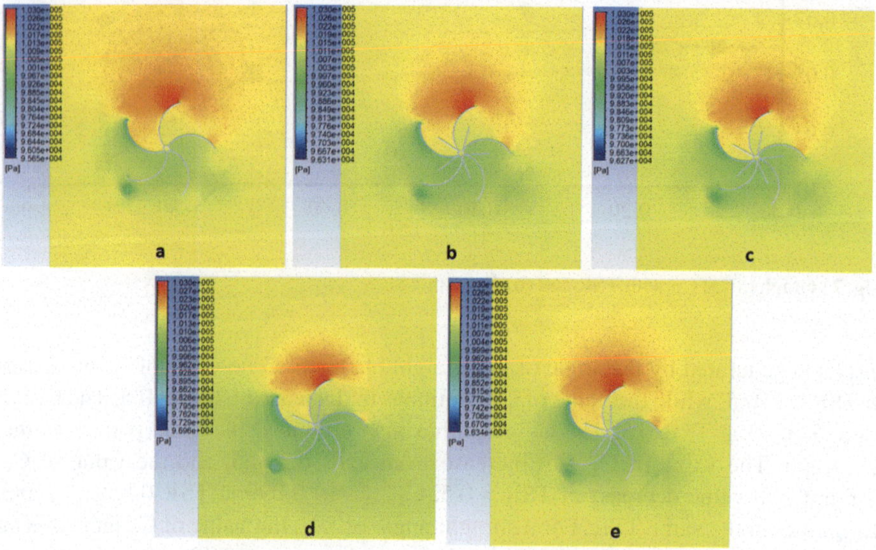

Fig. 8 Distribution of pressure (**a**) without the straight blade, with the straight blade at angle of placement of (**b**) 75°, (**c**) 80°, (**d**) 85°, and (**e**) 90°

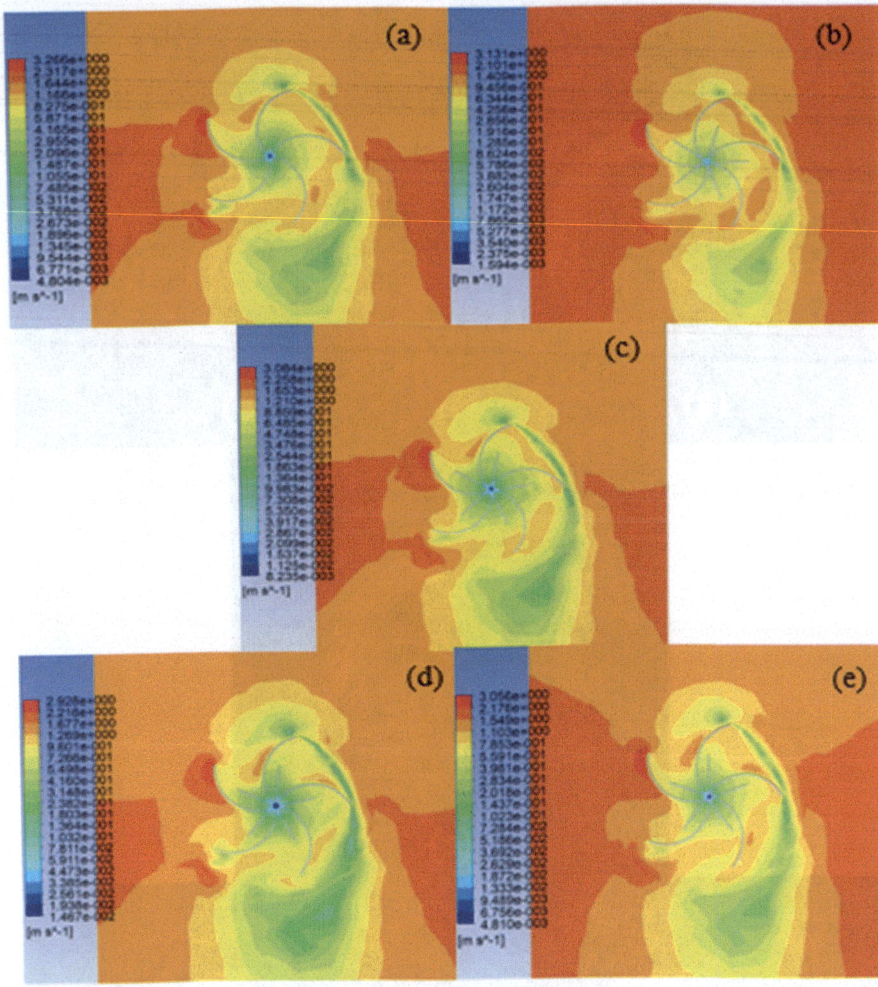

Fig. 9 Distribution of velocity (**a**) without the straight blade, and with straight blade at angle of placement of (**b**) 75°, (**c**) 80°, (**d**) 85°, and (**e**) 90°

velocity contour has the same contour pattern on the five rotors. The direction of the velocity vector in each rotor is shown in Fig. 10. The direction of the velocity vector in Fig. 10 has the same color gradation with Fig. 9. The velocity vector at the tip of the blade has a color gradation that is more dominant than the other blade side. This indicated that the largest fluid velocity flow is at the tip of the blade. The addition of a straight blade can increase the power coefficient. This is similar to the study of Altan et al. on the addition of a straight blade. However, Altan et al. [11] conducted a study of three blades and wind turbine, while this research was carried out on five blades and water fluids. This study shows a greater percentage increase in the power

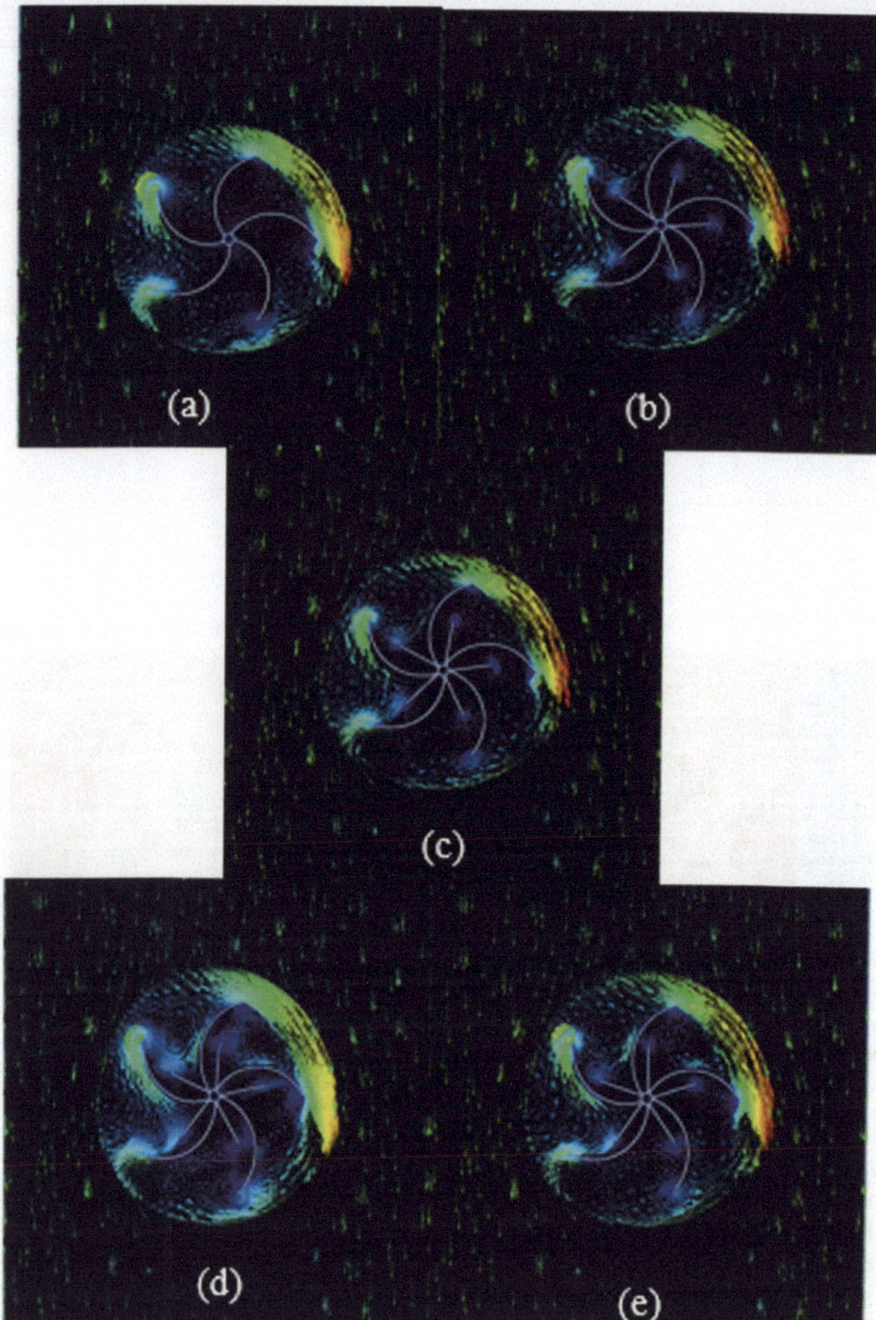

Fig. 10 Vector of velocity (**a**) without the straight blade, and with straight blade at angle of placement of (**b**) 75°, (**c**) 80°, (**d**) 85°, and (**e**) 90°

coefficient produced by 26% than the study of Altan et al. which produce 20% of turbines without a straight blade [11].

6 Conclusion

The performance of Savonius water turbines with the addition of straight blades with angles of placement 75°, 80°, 85°, 90°, and Savonius water turbines without straight blades has been evaluated using the Ansys CFX solver software. The conclusion is that the rotor with a straight blade with angle of placement of 85° has the highest C_{Pmax} value compared to other C_{Pmax} rotor values. The addition of a straight blade can increase the power coefficient by 26% of the turbine rotor comparing to rotor without the addition of a straight blade.

References

1. Ministry of energy and mineral resources of Indonesia. In: Handbook of energy & economic statistics of Indonesia 2018. Jakarta
2. Nguyen HT, Skitmore M, Gray M, Zhang X, Olanipekun AO (2017) Will green building development take off? An exploratory study of barriers to green building in Vietnam. Resour Conserv Recycle 127(August):8–20
3. Statistik BP (2017) Jumlah Curah Hujan dan Jumlah Hari Hujan di Stasiun Pengamatan BMKG, 2011–2015 Tahun Provinsi. Badan Meteorologi, Klimatologi, dan Geofisika, 2017. Available: https://www.bps.go.id/statictable/2017/02/08/1959/jumlah-curah-hujandan-jumlah-hari-hujan-di-stasiun-pengamatan-bmkg-2011-2015.html. Accessed 05 Nov 2019
4. Kumar A, Saini RP (2016) Performance parameters of Savonius type hydrokinetic turbine—a review. Renew Sustain Energy Rev 64:289–310
5. Sarma NK, Biswas A, Misra RD (2014) Experimental and computational evaluation of Savonius hydrokinetic turbine for low-velocity condition with comparison to Savonius wind turbine at the same input power. Energy Convers Manag 83:88–98
6. Talukdar PK, Sardar A, Kulkarni V, Saha UK (2017) Parametric analysis of model Savonius hydrokinetic turbines through experimental and computational investigations. Energy Convers Manag 158:36–49
7. Wahyudi B, Soeparman S, Hoeijmakers HWM (2015) Optimization design of Savonius diffuser blade with moving deflector for hydrokinetic cross flow turbine rotor. Energy Procedia 68:244–253
8. Taufan Apha Sanditya SH, Prasetyo A, Kristiawan B (2018) Effect of blade curvature angle of savonius horizontal axis water turbine to the power generation. IOP Conf Ser: J Phys: Conf Ser 979:012044
9. Hamzah I, Prasetyo A, Tjahjana DDDP, Hadi S (2018) Effect of blades number to performance of Savonius water turbine in water pipe. In: AIP conference proceedings, vol 1931, pp 4–8
10. Kamoji MA, Kedare SB, Prabhu SV (2009) Experimental investigations on single stage modified Savonius rotor. Appl Energy 86(7–8):1064–1073
11. Deda Altan B, Altan G, Kovan V (2016) Investigation of 3D printed Savonius rotor performance. Renew Energy 99:584–591
12. Prabowoputra DM, Hadi S, Prabowo AR Sohn JM (2020) Performance investigation of the savonius horizontal water turbine accounting for stage rotor design. Int J Mech Eng Robot Res 9(2):184–189

13. Hadi S, Khuluqi H, Prabowoputra DM, Prasetyo A, Tjahjana DDDP, Farkhan A (2019) Performance of savonius horizontal axis water turbine in free flow vertical pipe as effect of blade overlap. J Adv Res Fluid Mech Therm Sci 58(2):219–223
14. Patel V, Eldho TI, Prabhu SV (2018) Theoretical study on the prediction of the hydrodynamic performance of a Savonius turbine based on stagnation pressure and impulse momentum principle. Energy Convers Manag 168:545–563
15. Hindasageri V, Ramesh H Gaurav A (2012), Effect of variation of wave height and ocean depth on the performance of savonius rotors utilizing the orbital motion of ocean waves in shallow waters, vol 3, pp 53–57
16. Thakur N, Biswas A, Kumar Y, Basumatary M (2019) CFD analysis of performance improvement of the Savonius water turbine by using an impinging jet duct design. Chin J Chem Eng 27(4):794–801

Uniform Dispersion of Carbonyl Iron Particles in Bulk Magnetorheological Flexible Foam

Rizuan Norhaniza, Nur Azmah Nordin, Saiful Amri Mazlan, Ubaidillah, and Siti Aishah Abdul Aziz

Abstract Magnetorheological (MR) foams are smart materials which their rheological properties can be manipulated by stimulation of the external magnetic field. However, these materials suffer from low homogeneity of magnetic particles in MR foams. This paper seeks to remedy this issue by using flexible polyurethane foam. Two types of foams were fabricated, one without magnetic particles which refer as polyurethane (PU) foam and the other with 55 wt% magnetic particles of carbonyl iron (CIPs), known as MR foam. Magnetic properties of both foams were characterised using vibrating sample magnetometer (VSM). Concurrently, the MR foam was characterised at three different positions; top, middle and bottom in order to validate the distribution of CIPs. VSM analysis shows that the PU foam exhibited diamagnetic material while MR foam exhibit superparamagnetic material. The most important relevant finding is that the magnetic saturation of MR foam are 68.512, 70.367 and 68.879 Am^2/kg at top, middle and bottom position respectively. Moreover, the standard deviation for all positions was only 1.990 Am^2/kg indicating that the distributions of CIPs in the MR foam are uniform. Similar findings are also observed for the coercivity, remanence and permeability of MR foam with standard deviation of 0.029×10^{-3} kA/m, 0.009 and 0.041×10^{-2} Am^2/kg, respectively for all positions. The result is believed to be significant in contributing high performance of MR foam with uniform magnetic properties for various products and applications.

Keywords Magnetic saturation · Coercivity · Superparamagnetic · Diamagnetic · Dispersion of CIPs · Foam

R. Norhaniza · N. A. Nordin (✉) · S. A. Mazlan · S. A. A. Aziz
Engineering Materials and Structures (eMast) iKohza, Malaysia-Japan International Institute of Technology (MJIIT), Universiti Teknologi Malaysia, Jalan Sultan Yahya Petra, 54100 Kuala Lumpur, Malaysia
e-mail: nurazmah.nordin@utm.my

S. A. Mazlan
Advanced Vehicle System (AVS) Research Group, Universiti Teknologi Malaysia, Jalan Sultan Yahya Petra, 54100 Kuala Lumpur, Malaysia

Ubaidillah
Faculty of Engineering, Universitas Sebelas Maret, Jalan Ir. Sutami 36A, Kentingan, Surakarta 57126, Central Java, Indonesia

U. Sabino et al. (eds.), *Proceedings of the 6th International Conference and Exhibition on Sustainable Energy and Advanced Materials*, Lecture Notes in Mechanical Engineering,
https://doi.org/10.1007/978-981-15-4481-1_25

1 Introduction

MR foam is a new type of lightweight controllable materials that can be stimulated by the magnetic field. The stimulus will activate the magnetic particles of carbonyl iron particles (CIPs) that are embedded in the porous polymer matrix. The exposure of MR foam to the magnetic field would make the CIPs rearrange their magnetic moment to align with the external magnetic fields, simultaneously will change the arrangement of the porous matrix properties such as stiffness, storage modulus, young modulus and strain. But up until now, it is still challenging to produce functioning bulk MR foam. The MR foam that has been cured usually suffers from non-uniform dispersion of CIPs which is due to collapse in their bulk structure become a major drawback of the bulk MR foam.

Uniformity of CIP's distribution is a main factor that could enhance the performance of MR foam and led to a stronger MR effect. Several studies have been done to fabricate MR foam with different materials and concentration of magnetic particles. At such, Schumann et al. [1] studied the concentration of CIPs in MR foams, from 0 to 25 wt%. The results revealed that the possibility of reproducible MR foam with optimum properties is at 10 wt% CIP as less or more concentration tend to cause collapsing in the MR foam cylinder dimensions. Besides, they also highlighted the uneven distribution of MR foam which due to the foam that sticks to the mould during expansion process. In other hand, the effect of different types of magnetic particles which are iron micron-particles and barium ferrite in MR foam were also studied by other researchers [2]. The authors proved that the combination of polymer materials and magnetic particles played an important role in producing bulk MR foam. In the meantime, the results depicted that the iron micron-particles with concentration up to 23 wt% is an appropriate amount to produce MR foam as compared to barium ferrites which only up to 20 wt%.

In other study, Nikje et al. [3] synthesized organic nano size of magnetic particles iron oxide in which the iron oxide surface had been coated in order to avoid agglomeration issue during the fabrication process and capable to produce uniform distribution of particles in the MR foam. The SEM result showed the improvement of magnetic particles dispersion in MR foam with coated iron oxide. However, the MR foam produced had a low magnetic saturation, about 1.38 Am^2/kg, make it difficult to be the stimulus by external magnetic field. Meanwhile, study by Volpe et al. [4] identified that the type of polymer matrix also played an important role in producing homogenous MR foam. Two polymer matrixes namely Engage and Eva had been chosen to prepare various MR foams using foam injection moulding method. The preparation involved the control of pressure under nitrogen gas ambient. It was observed that morphology of Engage MR foam is not uniform from bottom to the top. Core-skin morphology was observed at the top of bulk MR foam. Differences in the density were also observed at different positions of both bulk MR foams. These non-uniform densities resulted into different performance of MR foams.

Overall, previous studies have highlighted that the uniformity of the MR foam is affected by the types of polymer, magnetic particles, mould, and parameter of preparation method including temperature and pressure [1–4]. Therefore, in this work, the non-sticky mould was used in the MR foam fabrication. In the meantime, the flexible polyurethane (PU) foam is chosen as polymer matrix due to high durability and stable reaction with time. The magnetic properties of bulk MR foam were studied at three different positions, particularly top, middle and bottom in order to validate the uniformity of the MR foam. The findings are important since the results are assumed to contribute to the advanced material field, particularly in the applications of controllable actuator.

2 Experimental

In this experiment, two types of samples were prepared without and with CIPs, namely PU foam and MR foam. Both foams were prepared using the same polyurethane (PU) matrix from combination of polyurethane polyol and 4-4 MDI diisocyanate. The average sizes of CIPs used in MR foam are from 3 to 5 μm.

Firstly, the polyol was stirred for 20 s using a mechanical stirrer with a speed of 500 rpm. Then the diisocyanate was poured into the polyol and mixed again for another 20 s. The mixture was then poured into the cylindrical smooth mould for foaming process.

In order to prepare the MR foam, 55 wt% of CIPs were mixed in polyol then stirred for 20 s also at the speed of 500 rpm. Diisocyanate was then added to the mixture and stirred again at same speed and period. This final mixture was poured into a cylindrical smooth mould for foaming process. The PU and MR foams were left for one day before the characterization process.

The bulk MR foam of height 10 cm had been cut at three different positions; top, middle and bottom. Then, eight-point of a circle samples from each MR foam position with a diameter of 0.6 cm and thickness of 1 mm were prepared in order to investigate the magnetic properties using Vibrating Sample Magnetometer (VSM). Magnetisation characterisation on PU foam and CIPs powder are also observed to understand the nature of the MR foam.

3 Results and Discussion

Images of bulk PU foam and three positions of MR foam are shown in Fig. 1. The PU foam is in white matrix and when the same liquid PU matrix is added with CIPs to produce MR foam, the foam matrix has turned to dark grey. From the naked eyes, all the MR foam samples show similarity in their tone colour over the whole area. This similarity might be related to the uniform dispersion of CIPs in the MR foam matrix from bottom to the top of MR foam matrix. This MR foam images contradicted with the MR foam prepared by Volpe et al. [4]. Their MR foam showed different colour tone at different positions due to the difference in the density which had been mentioned in their paper.

Fig. 1 Top view images of pure and MR foams

For magnetic property analyses, hysteresis loop of PU foams is shown in Fig. 2a. The loop shows a low magnetization of PU foam because it is not reacted to the external magnetic field supplied from the VSM. This denotes that the PU foam used in the study is a diamagnetic material [5, 6]. In contrast, Fig. 2b shows a hysteresis loop of CIPs. The loop displays the magnetic saturation, M_s reached 195 Am^2/kg, showing the magnetic property in CIPs. It is also observed that the loop displays small area which represent the superparamagnetic behaviour of the CIPs [7, 8]. Details of magnetic properties of PU foam and CIPs were tabulated in Table 1 and are discussed with the MR foam results.

Furthermore, Fig. 3 shows the hysteresis loop of MR foam at three different positions; top, middle and bottom. The M_s of the MR foam are 68.512, 70.367 and 68.879 Am^2/kg for top, middle and bottom respectively. This similar trend showed

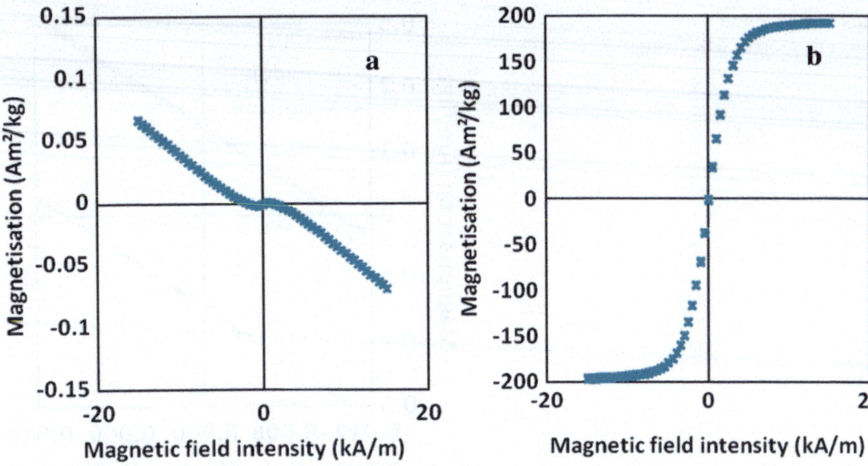

Fig. 2 Hysteresis loops of **a** PU foam and **b** CIPs

Table 1 Details on magnetic properties of CIPs, PU foam and at different positions of MR foams

Samples		M_s (Am²/kg)	$H_c \times 10^{-3}$ (kA/m)	M_r (Am²/kg)	Permeability × 10^{-2} (Am²/kg)
CIPs		195.000	9.993	0.708	7.500
PU foam		0.045	9.501	0.001	3.500
MR foam	Top	68.512	9.940	0.141	1.455
	Middle	70.367	9.938	0.145	1.463
	Bottom	68.879	9.947	0.147	1.461
Standard deviation		1.990	0.029	0.009	0.041

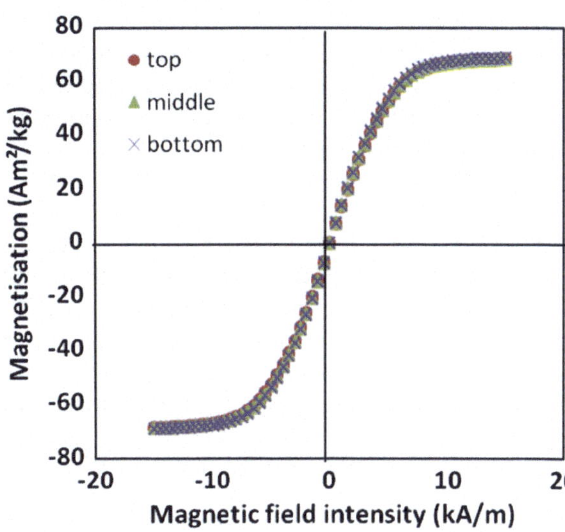

Fig. 3 Hysteresis loop of MR foam at three different positions; top, middle and bottom

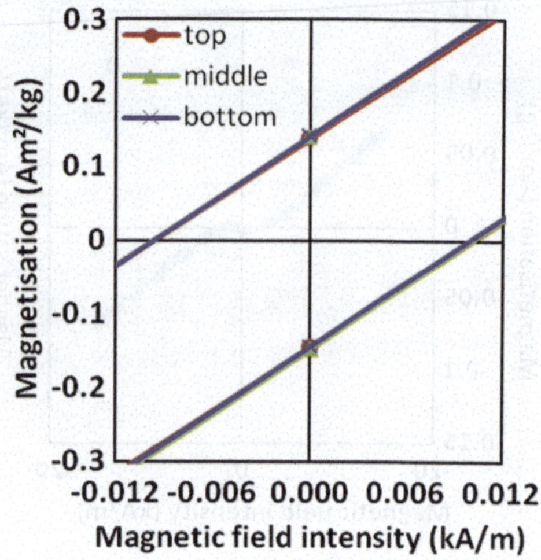

Fig. 4 Intercept of hysteresis loop of MR foam

a strong effect of CIPs on the PU matrix. However, the lower value of M_s of MR foams compared to CIP might be due to the layer of PU matrix acts as a shield to the CIPs in the PU matrix [9, 10] and reduces the strength of magnetic field stimulus to reach the CIPs. As compared with the CIPs loop, a small area of the MR foam's loop shows that the MR foam still maintains the superparamagnetic behaviour of the CIPs. Figure 4 is zoomed up from the hysteresis loop of Fig. 3 to show the intercepts of the *x*-axis that represent the coercivity value, H_c and intercept of *y*-axis which represent the remanence value, M_R of MR foams. The data are evaluated and presented in Table 1.

Table 1 lists the magnetic saturation, M_s, coercivity, H_c, remanence, M_r and permeability of CIPs, PU foam and MR foam. From the table, the standard deviation of M_s between all three positions is only 1.990 Am^2/kg showing similar property among them. This might be due to the uniformity of CIPs dispersion in the MR foam. The uniform distribution of M_s in the prepared bulk MR foam is very important in order to make the MR foam reacts at the same capabilities at any positions when stimulated with the magnetic field.

Meanwhile, the coercivity value of CIPs is 9.993×10^{-3} kA/m while the average coercivity for MR foam is 9.941×10^{-3} kA/m with drops by 0.05×10^{-3} kA/m. Coercivity, H_c is the intensity of magnetic field necessary to demagnetize the magnetized materials. As tabulated in Table 1, the coercivity of all different positions showed insignificant changes which might be due to less contact between CIPs [11]. Since the CIPs are dispersed through the whole bulk of MR foam matrix, making only a small volume of CIPs embedded in the spot area of the bulk MR foam [12]. Once the CIPs were stimulated, the magnetic field is easier to dissipate into surrounding thus resulted in decrement of the resistance to demagnetization and the H_c. Similar values of H_c are observed between those three positions with a standard deviation of only

0.029×10^{-3} kA/m. The results showed similarity in H_c values among three positions showing the whole prepared bulk MR foam is having a constant demagnetizing process.

In this study, the M_r value of CIPs is 0.708 Am^2/kg while for MR foam the average M_r is 0.144 Am^2/kg with a difference of 0.614 Am^2/kg. The low value in M_r could be due to low amount in CIPs in the test sample of the MR foam compared to CIPs sample during characterisation. As previously discussed, the CIPs are less contact to each other making it easier to self-demagnetize thus reduces the M_r and the H_c. The M_r values are similar for all three positions of MR foam with standard deviation among the positions is only 0.009 Am^2/kg indicating the uniform dispersion of M_r property in bulk MR foam. Similar drop is detected for an MR foam permeability where the average value is 1.460×10^{-2} Am^2/kg, lower than CIPs permeability 7.5 $\times 10^{-2}$ Am^2/kg. Meanwhile, the standard deviation between each position is 0.041 $\times 10^{-2}$ Am^2/kg represents the similarity of MR foam permeability throughout the prepared sample. Low permeability of MR foam compared to CIPs is due to the presence of foam matrix. When the CIPs are embedded in the matrix, the CIPs face difficulty to move once reacted by magnetic field thus limited their movement inside the matrix [13].

4 Conclusion

In this work, the MR foam has been successfully prepared in bulk size without sticking to the mould and without collapsing in the structure. The results show the diamagnetic PU foam can be turned into superparamagnetic properties with the addition of CIPs. The results also depicted that the MR foam shows an excellent uniform value of magnetic saturation, coercivity, remanence and permeability owing to the uniformity of CIPs dispersion from top to bottom side of the prepared samples. Thus, it is believed that this kind of bulk MR foam which exhibits uniform magnetic properties is beneficial for applications in sensor and actuator field.

Acknowledgements This study was financially supported by the Universiti Teknologi Malaysia under Trans-disciplinary Research Grant (Vot No: 07G13) and Fundamental Research Grant Scheme (Vot No: 5F001).

References

1. Schümann M, Seelig N, Odenbach S (2014) The effect of external magnetic fields on the pore structure of polyurethane foams loaded with magnetic microparticles Smart Mater Struct 24:105028
2. D'Auria M, Davino D, Pantani R, Sorrentino L (2016) Polymeric foam-ferromagnet composites as smart lightweight materials. Smart Mater Struct 25:055014

3. Nikje MMA, Kalishomi RG, Akbar R (2015) Preparation of polyurethane flexible foam nanocomposites by incorporation of Fe_3O_4 nanoparticles modified by reaction product of GPTS and APTS. Cell Polym 34:249–264
4. Volpe V, D'Auria M, Sorrentino L, Davino D, Pantani R (2018) Magneto-mechanical behavior of elastomeric carbonyl iron particles composite foams produced by foam injection molding. J Magn Magn Mater 466:44–54
5. Mao X, Xu Y, Xue Q, Wang W, Gao D (2013) Ferromagnetism in exfoliated tungsten disulfide nanosheets. Nanoscale Res Lett 8:430
6. Murali A, Gurusamy-Thangavelu SA, Jaisankar SN, Mandal AB (2014) Augmentation of properties on sparingly loaded nanocomposites via functionalized single-walled carbon nanotubes using a covalent approach. RSC Adv 4:62947–62950
7. Li ZT, Lin B, Jiang LW, Lin EC, Chen J, Zhang SJ, Tang YW, He FA, Li DH (2018) Effective preparation of magnetic superhydrophobic Fe_3O_4/PU sponge for oil-water separation. Appl Surf Sci 427:56–64
8. Sun M, Dai B, Liu K, Yao K, Zhao J, Lyu Z, Wang P, Ding Y, Yang L, Han J, Zhu J (2018) Enhancement in thermal conductivity of polymer composites using aligned diamonds coated with superparamagnetic magnetite. Compos Sci Technol 164:129–135
9. Wu L, Li L, Li B, Zhang J, Wang A (2015) Magnetic, durable, and superhydrophobic polyurethane@ Fe_3O_4@ SiO_2@ fluoropolymer sponges for selective oil absorption and oil/water separation. ACS Appl Mater Interfaces 7:4936–4946
10. Zhuang Y-T, Gao W, Yu Y-L, Wang J-H (2016) A three-dimensional magnetic carbon framework derived from Prussian blue and amylopectin impregnated polyurethane sponge for lead removal. Carbon NY 108:190–198
11. Stabik J, Dybowska A (2010) Magnetic induction of polymer composites filled with ferrite powders. Arch Mater Sci Eng 41:13–20
12. Sung Lee J, Myung Cha J, Young Yoon H, Lee J-K, Keun Kim Y (2015) Magnetic multigranule nanoclusters: a model system that exhibits universal size effect of magnetic coercivity. Sci Rep Nat 5:12135
13. De Vicente J, Bossis G, Lacis S, Guyot M (2002) Permeability measurements in cobalt ferrite and carbonyl iron powders and suspensions. J Magn Magn Mater 251:100–108

Effect of Barium on the Structure and Characteristics of Mg_2Si Reinforced Particles Al–Mg_2Si–Cu in Situ Composite

Nur Azmah Nordin, Saeed Farahany, Tuty Asma Abu Bakar, Ali Ourdjini, Saiful Amri Mazlan, Siti Aishah Abdul Aziz, and Hafizal Yahaya

Abstract Addition of barium (Ba) in various concentrations is susceptible to cause changes on Mg_2Si reinforced particles in Al–Mg_2Si–Cu in situ composite. In this study, six samples of the composite with different concentrations of Ba (0.1–0.8 wt%) were prepared. The alteration of Mg_2Si structure, phase reaction characteristics and cooling curves behaviour of the composite were investigated via optical microscope, scanning electron microscope (SEM), and computer aided cooling curve thermal analysis (CACCTA). The results depicted that 0.2 wt% exhibit the appropriate concentration of Ba added in order to modify and refine the Mg_2Si particles. The skeleton and dendrite shape of Mg_2Si particles have been transformed into fine polygonal shape accompanied with decreased in average size from 1178.5 μm of the unmodified particles to 289.1 μm. In fact, the refinement of Mg_2Si particles is associated with the increased of nucleation temperature, T_{N} of the respective phase together with the least undercooling, ΔU correspond to the easiness of the particles to be formed prior to its growth. Meanwhile, the decrement of T_N respective to other concentrations of Ba indicates the opposite refinement effect of the particles as it became coarser. Besides, the refinement of Mg_2Si has induced more nucleation of the particles resulting the increment of the density of particles and better distribution over the composite area. Therefore, the corresponding mechanical and tribological properties of the composite are believed to be improved accordingly.

N. A. Nordin (✉) · S. A. Mazlan · S. A. A. Aziz · H. Yahaya
Engineering Materials and Structures (eMast) iKohza, Malaysia-Japan International Institute of Technology (MJIIT), Universiti Teknologi Malaysia, Jalan Sultan Yahya Petra, 54100 Kuala Lumpur, Malaysia
e-mail: nurazmah.nordin@utm.my

S. Farahany
Faculty of Materials and Mechanical Engineering, Buein Zahra Technical University, Buein Zahra 3451745346, Iran

T. A. Abu Bakar
Faculty of Engineering, School of Mechanical Engineering, Universiti Teknologi Malaysia, 81310 Johor Bahru, Malaysia

A. Ourdjini
Department of Mechanical Engineering, Faculty of Engineering, University of Ottawa, Ottawa, ON, Canada

U. Sabino et al. (eds.), *Proceedings of the 6th International Conference and Exhibition on Sustainable Energy and Advanced Materials*, Lecture Notes in Mechanical Engineering,
https://doi.org/10.1007/978-981-15-4481-1_26

Keywords Mg_2Si particles · Barium · In situ composite · Modification · Refinement · Thermal analysis

1 Introduction

Hyper-eutectic Al–Si alloy with high Mg content is belong to Al metal matrix composite (MMCs) containing hard particles of Mg_2Si. Al–Mg_2Si in recent years has emerged to be used for automotive components which involve with tribological aspect such as brake disc material, cylinder head, cylinder frames and piston [1, 2], also being a good potential for aerospace applications [3]. The intermetallic compound (IMC) of Mg_2Si become increasingly attractive as reinforcing phase in Al- or Mg-based composites owing to its high melting temperature (1085 °C), low density (1.99×10^3 kg m^{-3}), high hardness (4.5×10 Nm^{-2}), low thermal expansion coefficient (7.5×10^{-6} K^{-1}) and reasonably high elastic modulus (120 GPa) [2, 4, 5], result in advantages of the Al–Mg_2Si composite. Among numerous methods of fabricating the MMCs such as hot extrusion, superheating melt treatment and vibration [6–8], in situ process of common gravity casting is known to be the appropriate and economical method due to even distribution of reinforcing particles, good particle's wetting, stability of matrix and reinforcement interfaces, less steps of processing for industrial utilization as well as low processing cost [9, 10]. However, primary Mg_2Si phase in normal casting route exhibit in coarse and rough structure while some particles exist in dendritic form which subsequently deteriorated the properties of the Al composite [5, 10, 11]. Therefore, modification and refinement of the coarse structure become a significant important in order to improve the morphology and acquire enhanced mechanical properties of the composite. It is believed that high mechanical strength and ductility of the composite could recuperate the quality of casting products especially when facing further application demands [1].

It is known that the refinement of the coarse silicon in hypo-eutectic Al–Si alloy was made through the addition of certain elements such as boron, B [12], scandium, Sc [13] and strontium, Sr [14] while cerium + phosphorus, Ce + P [15] and lanthanum, La [16] that were added to hypereutectic Al–Si alloy also succeeded to modify and refine the primary structure. Due to great similarity system between Al–Si and Al–Mg–Si alloys [17], this melt treatment approach is therefore being applied in Al–Mg_2Si composite in order to achieve refined Mg_2Si particle with beneficial shapes and distributions. Li et al. [18] and Qin et al. [19] have reported that the addition of Al–P master alloy and P, respectively on the Al–Mg–Si composite possessed favourable refining and modifying effect since both elements have changed the dendritic structure of Mg_2Si to polygonal shape as well as refining the size. Other several works also indicated that the addition of bismuth, Bi [1, 20], anthimony, Sb [10, 21] and Sr [21] can modify the Mg_2Si structure to a polygon shape with softening the sharp angle of the particle's shape. Rare earth elements such as Gd [2], Ce [9] and yttrium, Y [22] also have been reported to be capable of modifying the Mg_2Si phase in the Al-based composites. These refined particles have successively improved the strength, ductility as well as hardness of the composite. However, research efforts regarding the modification and refinement effect of Ba in Al–Mg–Si alloy is rather limited, especially on the solidification behaviour and thermal characteristics of the Al composite's melt. Therefore, the current study has been carried out in order to evaluate the effect of Ba in the Al–Mg_2Si–Cu in situ composite. The aim of this study is to investigate the effect of different Ba concentrations on the modification and refinement effect of Mg_2Si particles as well as the correlation towards the results of thermal analysis.

2 Experimental Procedure

A 2 kg ingot of commercial Al–Mg_2Si–Cu in situ composite was melted in a silicon carbide crucible using a muffle furnace with a melt temperature of 750 ± 5 °C. Once the ingot was completely melt, Ba additions with few concentrations of 0.1, 0.2, 0.4, 0.6 and 0.8 wt% were introduced, separately. The melt was held for around 20 min to allow for dissolution and homogenization, then it was stirred and the surface was skimmed. In the meantime, the thermal analysis was set up by embedding the K-type thermocouples in the centre and wall, 15 mm above the bottom part of a cylindrical ceramic mould (outer diameter 40 mm, height 40 mm and wall thickness of 7 mm), which would be preheated at 750 °C for 15 min, prior to the thermal analysis set up. The thermocouple was calibrated before each test and bended around a thin rod of stainless steel to position it precisely in places. The diameter of wire was ~0.35 mm. In a short time, the molten alloy was taken out of the furnace and was pouring into the preheated ceramic mould and the thermal analysis connected would record the temperature-time data respective to the cooling condition of Al composite using a high-speed data acquisition system (EPAD-TH8-K), that linked to a computer with a DEWESoft 7.5 software. Meanwhile, another software of FlexPro9.0.31 data analysis was used for smoothing the curves and plotting the cooling curve respective to the first and second derivative curves for extracting the characteristic temperatures of any required phase. Figure 1 shows the experimental casting method of the Al–Mg_2Si–Cu composite, along with the thermal analysis set up.

Samples for metallography were sectioned horizontally and were prepared by standard grinding procedures. The ground specimens were then subjected to a final polishing with OPS colloidal silica suspension and obtaining mirror like surfaces. Microstructures were analysed using an optical microscope (Nikon-ECLIPSE-LV150 Digital Microscope).

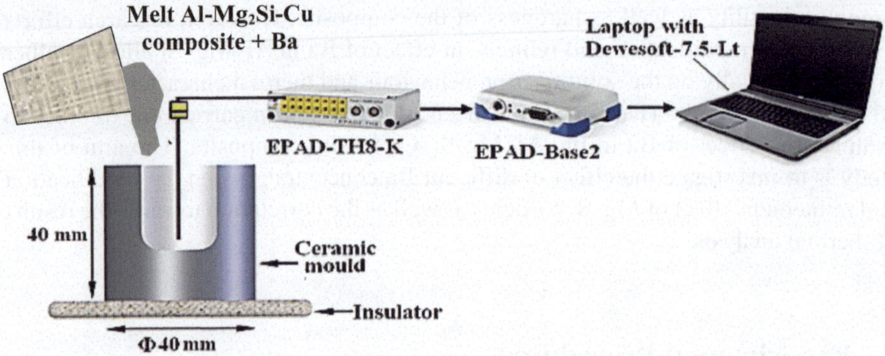

Fig. 1 Schematic diagram of casting process set up together with the thermal analysis

3 Results and Discussion

3.1 *Effect of Barium on the Mg_2Si Particles*

Figure 2 shows the characteristics of Mg_2Si particles, including the normalized particle's area, grain size (perimeter), aspect ratio and number of particles per unit area (mm^2) as a function of Ba addition. The figure also equipped with the morphology effects that correspond to different concentrations of Ba. It can be clearly observed that 0.2 wt% Ba caused the appropriate modification effect on the reinforced Mg_2Si particles. In fact, the average size of the Mg_2Si particles has been reduced by about 76%, from 1178.51 μm (unmodified) to 289.1 μm. This is followed by the normalized average area that decreased from 1 (considering 1 for the unmodified structure of Mg_2Si particle) to 0.67. Similarly, the aspect ratio also showed the improvement since it decreased from 1.22 in the 0 Ba-based Al composite to 1.12 with 0.2 wt% Ba. However, with the increased of Ba content from 0.4 to 0.8 wt%, the normalized average area increased even more than the unmodified Mg_2Si structure due to particles getting larger in size as shown in the Fig. 2. Correspondingly, with further increase in Ba content exceeding 0.2 wt% (0.4–0.8 wt%), the aspect ratio of the Mg_2Si particles re-increased indicating that the width and height of the structure become non-equivalent. Furthermore, 0.2 wt% Ba caused the highest number of Mg_2Si_P particles formed per unit area (mm^2), with 41 compared to 9 in the untreated composite, which also reflected to the morphology shown in Fig. 2.

As shown in the Fig. 2 as well, the nature form of Mg_2Si_P particles (unmodified) existed in skeleton and coarse structure (hollow), as this outcome also was reported in previous research [9–11]. This inappropriate shape of particles will lead to internal crack formation via its defect and sharp edges and the formed crack would continue to propagate throughout the eutectic phase therefore weakened the strength of the in situ composite. Introducing Ba into the composite melt are shown to cause a significant change to the Mg_2Si structure as the particles are observed to transform from dendrite

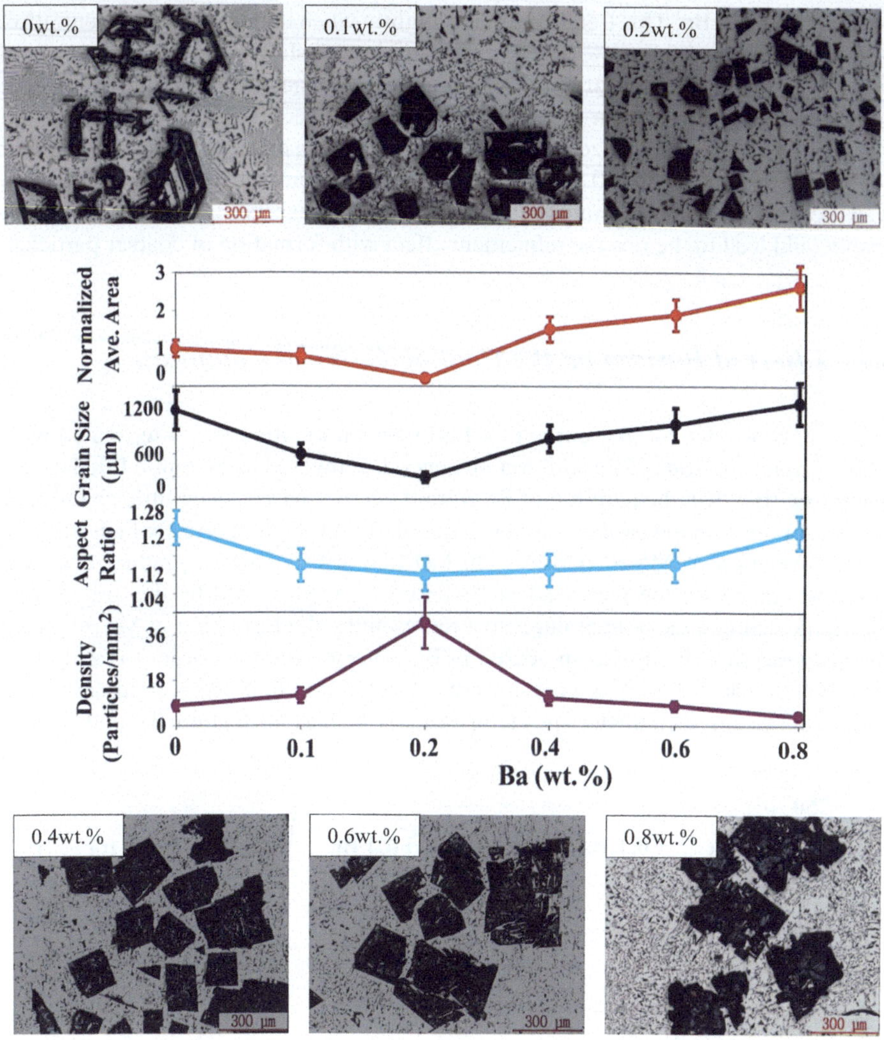

Fig. 2 Characteristics particles of Mg_2Si phase, together with the optical micrographs as a result of various Ba concentrations

to polygonal shape accompanied with refining its average size. As observed in Fig. 2, the sharp ends of the Mg_2Si_P particles started to fade and corresponding skeleton begin transformed to polyhedral shape with 0.1 wt% Ba. Meanwhile, the structure has been further refined with 0.2 wt% Ba with some particles appeared in triangle and rectangular form. The average size also noted to be decreased in both concentrations. However, further increased the Ba up to 0.4 wt%, the refinement effect seems to disappear as the particles regrowth in coarser structure and re-increased in size. Some particles retained in polygonal shape but the sizes are extremely large compared to the effect of 0.2 wt% Ba. This trend continued slightly when increased the concentration of Ba up to 0.6 and 0.8 wt%. The Mg_2Si particles are shown to further increase in size and having a coarser structure with damage at edges of the particles. Indeed, no significant difference on the Mg_2Si particles between 0.4–0.8 wt% Ba as the particles size have similar size and even bigger than those observed in the unmodified composite. It appears that 0.2 wt% Ba is a turning point for Mg_2Si refinement as they are shown to have the most refined structure and exceeding this concentration level would lead to the reverse refinement effect with formation of coarser particles.

3.2 *Effect of Barium on the Cooling Curve al Composite*

Figure 3 shows the cooling curve of Al–Mg_2Si–Cu in situ composite treated with various concentrations of Ba addition. In general, compared to the unmodified in situ composite (0 wt%), the addition of Ba from 0.1 to 0.8 wt% caused little changes in the arrest points on the cooling curves, particularly for the first arrest which belongs to the formation of Mg_2Si particles. In fact, the arrests show a gradual decrease in temperature with the increased of Ba addition, as presented by the grey arrow. This indicated that the temperatures that representing the formation of Mg_2Si phase showed gradual reduction as increased in Ba concentrations. In order to verify this information, each of cooling curves were projected into first and second derivative curves and required characteristics temperatures such as nucleation, T_N and growth

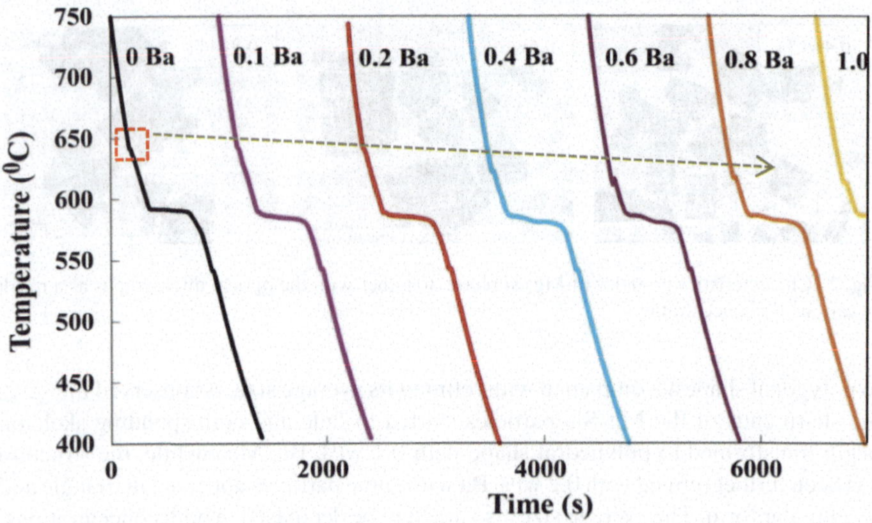

Fig. 3 Cooling curves of Al–Mg_2Si–Cu composite as a result of Ba in different concentrations: 0.1, 0.2, 0.4, 0.6, 0.8 wt%

temperatures, T_G corresponding to nucleation of Mg_2Si particles could be extracted precisely [9, 23].

The change in nucleation temperature, T_N for Mg_2Si particles has been further interpreted in Fig. 4a. In particular, Fig. 4a is a magnified area of the first peaks correspond to formation of Mg_2Si particles, for all concentrations of Ba. It can be observed that the addition of Ba in different amounts caused $T_N^{Mg_2Si}$ to increase from 647.3 °C for the untreated composite to 650.3 and 664.8 °C with additions of 0.1 and 0.2 wt% Ba, respectively. Similar to the effect of Ce addition [9], the solidification

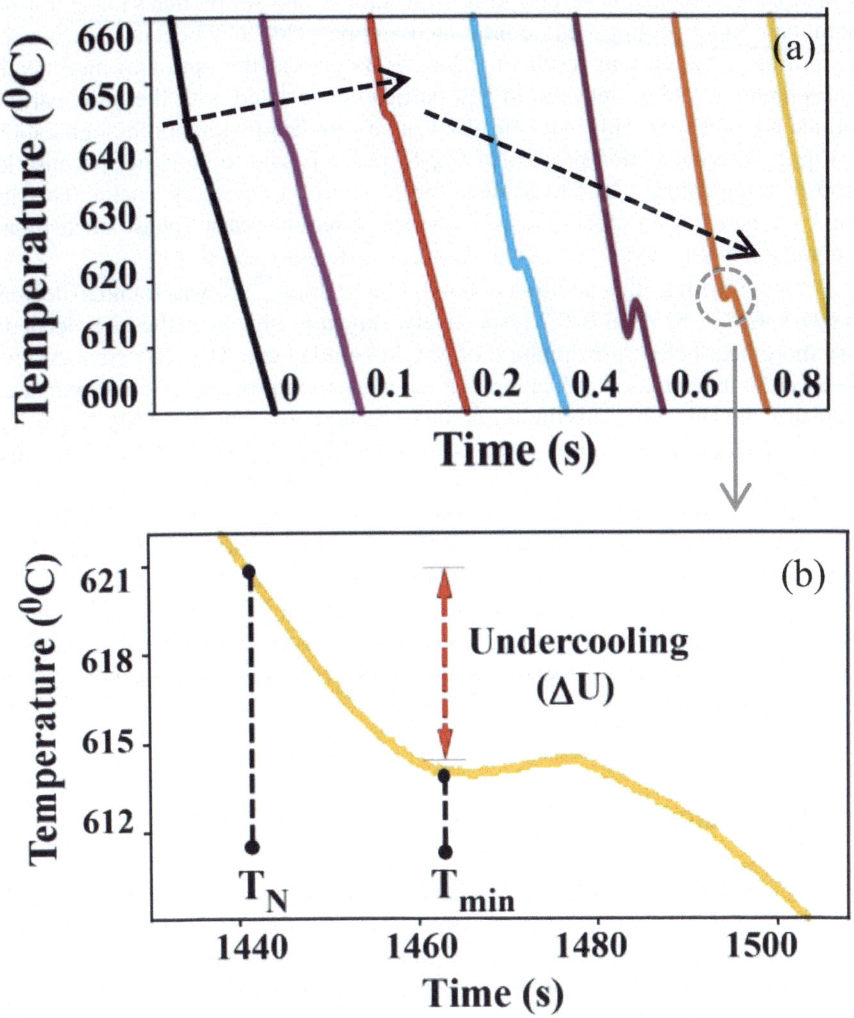

Fig. 4 **a** Change of T_N peaks respective to formation of Mg_2Si_P phase, **b** example magnified peak of Mg_2Si_P formation indicating undercooling (ΔU) point of interest

process that started at higher temperature indicates the readiness of Mg_2Si particles to nuclcate and grow. Along with the change of $T_N^{Mg_2Si}$, the effect also equipped with a change in undercooling (U) of Mg_2Si nucleation as can be seen in Fig. 4a while, example of undercooling existence for the phase formation is illustrated in Fig. 4b, for 0.8 wt% Ba. In general, with the addition of 0.1 and 0.2 wt% Ba, the undercooling difference, ΔU has decreased or appeared no undercooling as compared to the untreated composite. It was reported that if higher number of nucleation sites for Mg_2Si phase exists, little undercooling is required as the particles are easy to nucleate and become a stable nuclei. Thus, more nucleation of particles with refined structure are obtained. In reverse, if few favourable sites for particles nuclei formed, significant undercooling will be presented and particles may become coarser [23]. Accordingly, the current result of 0.2 wt% Ba caused the optimum modification and refinement effect onto the Mg_2Si particles associated with the most particles formed per unit area. This outcome also can be correlated with morphologies shown in Fig. 2. The nucleation process of Mg_2Si phase is said to be easier because less drop of temperature (ΔU) to achieve the formation of primary nuclei. This in a good agreement with other researches which stated that easier phase nucleation is accompanied with decreased in undercooling difference, ΔU [24].

However, with further addition of 0.4–0.8 wt% Ba, $T_N^{Mg_2Si_P}$ was found to decrease to 631.8, 622.3, 623 and 620.7 respectively, shown in Fig. 4a as the melt needed to lose more heat before precipitation of Mg_2Si could occur. The respective ΔU also observed to be increased indicating that the nucleation process for primary phase is more difficult (Fig. 4b). This finding could be related to the change of Mg_2Si particles to coarser shape as the mount of Ba is increased from 0.4 up to 0.8 wt% Ba, as in Fig. 2.

As the addition of Ba could alter the structure in the composite, it is believed that the modification and refinement of the coarse Mg_2Si particles is susceptible to improve the mechanical properties of the in situ composite by reducing stress concentration site on the respective finer polygonal structure. Therefore, further investigation need to be carried out in order to clarify the effect of Ba additions on the in situ composite and towards the mechanical properties enhancement, especially with the 0.2 wt% addition.

4 Conclusion

As a conclusion, addition of Ba in various concentrations could modify the course Mg_2Si particulate reinforced Al–Mg_2Si–Cu in situ composite. However, the adequate refinement effect was achieved with 0.2 wt% Ba while exceeding this value caused the reverse refinement and modification effect. The skeleton and dendrite shape of Mg_2Si particles have been modified to a polygonal form accompanied with decreased in average grain size by 75%, followed by 8% reduction in aspect ratio and increase in particles density increased by about 78%. The modification also led to better distribution of particles over the composite area. It is believed that the modification of the structure has a significant influence towards the mechanical properties enhancement of the in situ composite.

Acknowledgements This study was financially supported by the Universiti Teknologi Malaysia under Trans-disciplinary Research Grant (Vot No: 07G13) and Fundamental Research Grant Scheme (Vot No: 5F001).

References

1. Razavykia A, Farahany S, Yusof NM (2015) Evaluation of cutting force and surface roughness in the dry turning of Al–Mg_2Si in-situ metal matrix composite inoculated with bismuth using DOE approach. Measurement 76:170–182
2. Ghandvar H, Idris MH, Ahmad N, Emamy M (2018) Effect of gadolinium addition on microstructural evolution and solidification characteristics of Al-15%Mg_2Si in-situ composite. Mater Charact 135:57–70
3. Fadavi A, Boostani S, Tahamtan ZY Jiang D, Wei S, Yazdani R, Khosroshahi A, Taherzadeh Mousavian R, Xu J, Zhang X, Gong D (2015) Enhanced tensile properties of aluminium matrix composites reinforced with graphene encapsulated SiC nanoparticles. Comp Part A: App Sci Manuf 68:155–163
4. Tebib M, Samuel AM, Ajersch F, Chen XG (2014) Effect of P and Sr additions on the microstructure of hypereutectic Al–15Si–14Mg–4Cu alloy. Mater Charact 89:112–123
5. Emamy M, Nodooshan HRJ, Malekan A (2011) The Microstructure, hardness and tensile properties of Al–15%Mg_2Si in-situ composite with yttrium addition. Mater Des 32:4559–4566
6. Bahrami A, Razaghian A, Emamy M, Khorshidi R (2012) The effect of Zr on the microstructure and tensile properties of hot-extruded Al–Mg_2Si composite. Mater Des 36:323–330
7. Abu Bakar NANTA, Hamzah E, Farahany S, Ourdjini A (2017) Effect of superheating melt treatment on Mg_2Si particulate reinforced in Al–Mg_2Si–Cu in situ composite Procedia Eng. 184 595–603
8. Li M, Tamura T, Miwa K (2007) Controlling microstructures of AZ31 magnesium alloys by an electromagnetic vibration technique during solidification: From experimental observation to theoretical understanding. Acta Mater 55:4635–4643
9. Nordin NA, Farahany S, Abu Bakar TA, Hamzah E, Ourdjini A (2015) Microstructure development, phase reaction characteristics and mechanical properties of a commercial Al-20%Mg_2Si-xCe in situ composite solidified at a slow cooling rate. J Alloys Compd 650:821–834
10. Farahany S, Ghandvar H, Nordin NA, Ourdjini A (2016) Microstructure characterization, mechanical, and tribological properties of slow-cooled Sb-treated Al–20Mg_2Si–Cu in situ composites. J Mater Eng Perf 26:1685–1700
11. Hong-Chen Y, Wang H-Y, Chen L, Zha M, Wang C, Li C, Jiang Q-C (2016) Influence of Li_2Sb additions on microstructure and mechanical properties of Al–20Mg_2Si alloy. Materials 9:243–251
12. Chen Z, Kang H, Fan G, Li J, Yiping L, Jie J, Zhang Y, Li T, Jian X, Wang T (2016) Grain refinement of hypoeutectic Al–Si alloys with B. Acta Mater 120:168–178
13. Prukkanon W, Srisukhumbowornchai N, Limmaneevichitr C (2009) Modification of hypoeutectic Al–Si alloys with scandium. J Alloy Comp 477:454–460
14. Farahany S, Ourdjini A, Idris MH, Shabestari SG (2013) Evaluation of the effect of Bi, Sb, Sr and cooling condition on eutectic phases in an Al–Si ADC12 alloy by in-situ thermal analysis. Thermochim Acta 559:56–86

15. Wang A, Zhang L, Xie J (2013) Effects of cerium and phosphorus on microstructures and properties of hypereutectic Al–21%Si alloy. Mater Perf Charact 31(5):522–525
16. Elgallad EM, Doty HW, Alkahtani SA, Samuel FH (2016) Effects of La and Ce addition on the modification of Al–Si based alloys. Adv Mater Sci Eng 1–13
17. Zhang J, Fan Z, Wang YQ, Zhou BL (2000) Microstructural development of Al–15wt%Mg_2Si in situ composite with mischmetal addition. Mater Sci Eng A 281:102–144
18. Li Z, Li C, Gao Z, Liu Y, Liu X, Guo Q, Yu L, Li H (2015) Corrosion behaviour of Al-Mg_2Si alloys with/without addition of Al-P master alloy. Mater Charact 110:170–174 (2015)
19. Nasiri N, Emamy M, Malekan A, Norouzi MH (2012) Microstructure and tensile properties of Cast Al–15%Mg_2Si composite: effects of phosphorous addition and heat treatment. Mat Sci Eng A 556:446–453
20. Barzani MM, Farahany S, Songmen V (2017) Machinability Characteristics, thermal and mechanical properties of Al–Mg_2 in-situ composite with bismuth. Measurement 110:263–274 (2017)
21. Nordin NA, Farahany S, Ourdjini A, Bakar TA, Hamzah E (2013) Refinement of Mg_2Si reinforcement in a commercial Al–20%Mg_2Si in-situ composite with bismuth, antimony and strontium. Mater Charact 86:97–107
22. Jafari Nodooshan HR, Liu W, Wu G, Bahrami A, Pech-Canul MI, Emamy M (2014) Mechanical and tribological characterization of Al–Mg_2Si composites after yttrium addition and heat treatment. J Mater Eng Perf 23(4):1146–1156
23. Malekan M, Shabestari SG (2011) Computer-aided cooling curve thermal analysis used to predict the quality of aluminum alloys. J Therm Anal Calorim 103:453–458
24. Donald RA, Pradeep PF, Wendelin JW (2010) The science and engineering of materials. 6th edn. Christopher M. Shortt, US

Extreme Learning Machine Based-Shear Stress Model of Magnetorheological Fluid for a Valve Design

Irfan Bahiuddin, Abdul Yasser Abd Fatah, Saiful Amri Mazlan, Fitrian Imaduddin, Mohd Hatta Mohammed Ariff, Dewi Utami, and Nurhazimah Nazmi

Abstract Magnetorheological fluid (MRF) models are important tools in the design of the material based-valve in a damper. Although Bingham Plastic and polynomial models are quite widely employed, these data-driven models have disadvantages in term of the accuracy and narrow applicable operating conditions, including magnetic field and shear rate. Therefore, this paper aims to utilize an extreme learning machine (ELM) based-shear stress model to design an magnetorheological (MR) valve. Firstly, an MRF model of MRF 132DG is built using rotational rheometer test results. Secondly, the model is employed to model a meandering MR valve drop pressure utilizing the known design parameters and a finite element magnetic method (FEMM) results. The comparison of the steady state pressure between the simulation and experimental results (in literature) has shown a good agreement in term of the pattern and accuracy with error of less than 3%. In summary, ELM has shown its potential to model MRF behavior while employing it to an MR device.

Keywords Magnetorheological fluids · Machine learning · Extreme learning machine · Meandering magnetorheological valve

---

I. Bahiuddin · S. A. Mazlan · M. H. M. Ariff · D. Utami · N. Nazmi
Malaysia-Japan International Institute of Technology, Universiti Teknologi Malaysia, Jalan Sultan Yahya Petra, 54100 Kuala Lumpur, Wilayah Persekutuan Kuala Lumpur, Malaysia

I. Bahiuddin
Department of Mechanical Engineering, Vocational College, Universitas Gadjah Mada, Jl. Yacaranda Sekip Unit IV, Yogyakarta, Daerah Istimewa Yogyakarta 55281, Indonesia

A. Y. A. Fatah (✉)
Razak Faculty of Technology and Informatics, Universiti Teknologi Malaysia, Kuala Lumpur, Malaysia
e-mail: yasser.kl@utm.my

F. Imaduddin
Mechanical Engineering Program, Faculty of Engineering, Universitas Sebelas Maret, Jl. Ir. Sutami 36 A, Kentingan, Surakarta 57126, Central Java, Indonesia

U. Sabino et al. (eds.), *Proceedings of the 6th International Conference and Exhibition on Sustainable Energy and Advanced Materials*, Lecture Notes in Mechanical Engineering,
https://doi.org/10.1007/978-981-15-4481-1_27

1 Introduction

Magnetorheological fluids (MRF) can be classified as a type of smart materials with its controllable properties with the presence of magnetic field. The viscosity can be harder with the higher applied value of magnetic field as a result of the aligning magnetic particle across the fluid. Another property of the fluid that can be affected is yield stress or a critical value to represent the minimum pressure to make the fluid flow [1, 2]. Another important parameter of MRF is shear stress as a direct result of the rheometer measurement. Together with shear rate, shear stress can form a flow curve that can be derived as apparent viscosity, plastic viscosity, and yield stress. The mentioned parameters need to be taken into account when designing a device utilizing MRF, such as magnetorheological (MR) damper, MR brake, MR valve. To accommodate those parameters, models were developed at various devices [3, 4].

One of the widely known models in the designing process of MR devices is steady state model. This kind of model is simple and easy to understand, but quite a powerful tool to know the predicted performance of a device in general. For example, MR valve in [5] was designed first supported by employing steady state models to calculate the pressure drop prediction that can be derived to force or torque of the device. The steady state model is a combination of some equations based on theoretical, empirical observation on the similar devices, and the rheological properties of MRF. The utilized rheological properties are plastic viscosity and yield stress [5–7]. Both parameters are derived based on Bingham plastic model fitted at a flow curve. In the previous works, the steady state models have shown its capability to reproduce the similar pattern with the investigated MR devices [5, 8]. The error of the comparison with the experimental results may be acceptable while discussing the general behavior and the pattern at different conditions of magnetic field or frequency or other operating condition variables. Nevertheless, it will be beneficial if the error can be reduced. Furthermore, the existing models only cover two parameters, yield stress and plastic viscosity. Meanwhile, the other rheological variables such as shear rate and other ex-tended operating variables were neglected. To overcome the limitations, the rheological behavior of MRF can be represented by neural network model. Recent works have gain high accuracy while neural networks models compared with experimental data at basic case [9] or extended cases [10, 11]. The algorithm can be selected from artificial neural networks [12] or Extreme Learning Machine [9, 13].

Therefore, this paper aims to provide an alternative steady state model by considering the shear rate and accommodating a more comprehensive rheological behavior of MRF using machine learning based-rheological models. The study case for the pro-posed concept is the MR valve with the concept of meandering path [5]. The paper first discusses the general method of the proposed method and its positions in the existing steady state model. Then, the methodology is discussed in detail, including the MR valve design as the study case, the general description of the model building. Finally, the results are discussed and compared with the experimental results and the existing models.

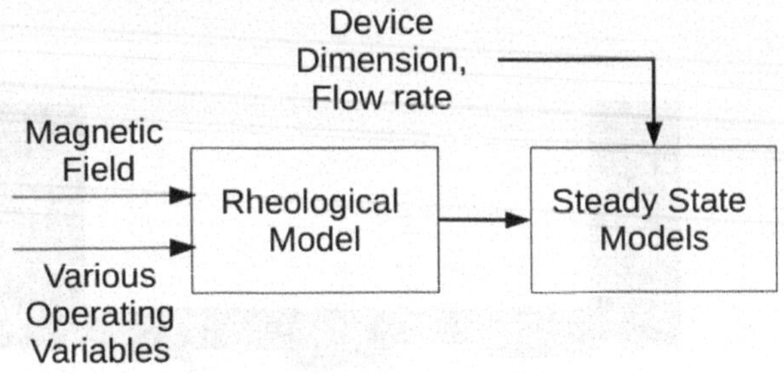

Fig. 1 The general concept of the modeling device

2 The Proposed Concept and Methodology

2.1 *The General Concept of the MR Device Modeling*

Figure 1 shows the general concept to model a device for predicting the model performance at various magnetic fields and operating variables. The first part is the rheological model to represent the MRF behavior. The existing models try to capture the material behavior using various methods, such as Bingham Plastic, Herschel-Bulkley, Papanastasiou [4], polynomial [5], and power-law equation [14]. The disadvantages of the existing models are the accuracy is acceptable within a narrow range, the inflexibility of the inputs or additional variables, such as temperature and normal force. Therefore, a machine learning based-rheological model has been proposed [9, 12]. In this paper, ELM based-machine learning is employed to capture the material behavior instead of the conventional model.

Meanwhile, the steady state models consist of the linear models to capture the behavior when the device receives a certain pattern of inputs, such as ramp, sine, and step. The steady state models of each device can be different depending on the dimension, geometry, and operating condition. Although steady state models contain simplified equation and various assumptions neglecting most of the nonlinear behavior of the device, the model has successfully captured general pattern at various operating condition as long as the pre-determined condition or assumption is achieved.

2.2 *The MR Valve Modeling*

The meandering MR valve has the purpose of optimizing the effective area of the MR valve device and has been proven to produce higher pressure drop compared to the previous MR valve types [5]. This device type consists of 3 types of paths, which are annular, radial, and orifice path, as shown in Fig. 2. Using FEMM (finite element method magnetic) simulation as discussed in the previous work [5], the magnetic fields (B) at various applied current for each gap zone are described in Table 1. For

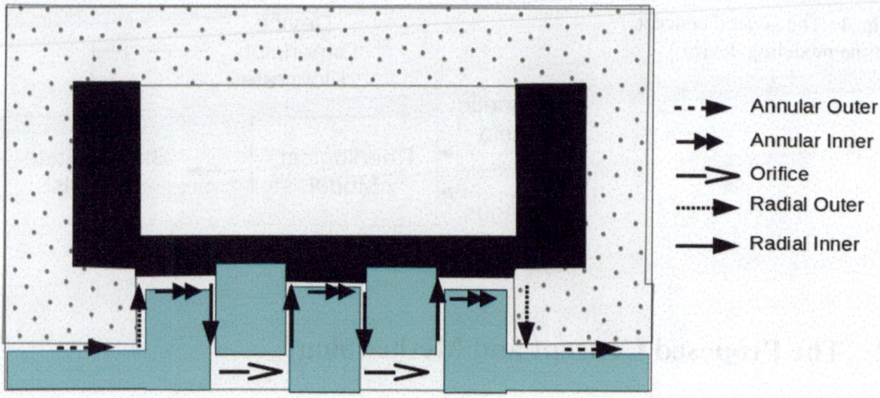

Fig. 2 Gap zone in meandering type of MR valve

Table 1 Magnetic value for applied current at each gap zone

Current (A)	Magnetic field (Tesla)		
	Outer annular	Outer radial	Inner radial
0	0.00	0.00	0.00
0.2	0.11	0.13	0.25
0.4	0.17	0.25	0.43
0.6	0.19	0.38	0.57

the inner annular and orifice zone, the magnetic fields are not listed because the value is negligible or equivalent to zero. The description and the values employed in the simulation are described in Table 3. The models of the pressure drop (ΔP) calculation for each gap zone type is listed in Table 2 with c is expressed by:

$$c = 2.07 + 12\eta Q/12\eta Q + 0.8\pi Rd^2\tau_y(B) \quad (1)$$

The total pressure drop calculation of meandering valve is in Eq. (2) [5].

Table 2 Equations to calculate pressure drops [11]

Passage type	Pressure drop viscous	Pressure drop yield
Annular	$\Delta P_v = \frac{6\eta Q L_a}{\pi d^3 R_a}$ (4)	$\Delta P_y = \frac{c\tau(B) L_a}{d}$ (5)
Radial	$\Delta P_v = \frac{6\eta Q}{\pi d^3}\ln\left(\frac{R}{R_0}\right)$ (6)	$\Delta P_y = \frac{c\tau(B)}{d}(R - R_0)$ (7)
Orifice	$\Delta P_v = \frac{8\eta Q L_o}{\pi R_o^4}$ (8)	–

$$\Delta P_{total} = 2\Delta P_{outer_{annular}} + 4\Delta P_{inner_{radial}} + 2\Delta P_{orifice} + 3\Delta P_{inner_{annular}} + 2\Delta P_{outer\_radial} \tag{2}$$

$$\begin{aligned}\Delta P_{total} = {} & 2\left(\frac{6\eta Q L_{ao}}{\pi d^3 R_{0Outer}} + \frac{c_{ao}\tau_{ao}(B)L_{ao}}{d}\right) \\ & + 4\left(\frac{6\eta Q}{\pi d^3}\ln\left(\frac{R_1}{R_{0inner}}\right) + \frac{c_{ri}\tau_{ri}(B)}{d}(R_1 - R_{0inner})\right) \\ & + 2\left(\frac{8\eta Q L_o}{\pi R_{0inner}^4}\right) + 3\left(\frac{6\eta Q L_{ao}}{\pi d^3 R_{0Outer}}\right) \\ & + 2\left(\frac{6\eta Q}{\pi d^3}\ln\left(\frac{R_1}{R_{0outer}}\right) + \frac{c_{ro}\tau_{ro}(B)}{d}(R_1 - R_{0outer})\right)\end{aligned} \tag{3}$$

2.3 *Machine Learning Based-Rheological Model*

The usual equation for modeling the rheological properties is polynomial model fitted to the so called yield stress graph over magnetic field. The yield stress variables can be obtained from the fitting process of the Bingham Plastic equation to the flow curve as a result of rheometer measurement. For MRF 132DG, the yield stress equation is described in the Eq. (9) [5].

$$\tau_y(B) = \begin{cases} -58.92B^3 + 7466B^2 + 35.74B - 3.387 & for\ \tau_y(B) > 0 \\ 0 & for\ \tau_y(B) \le 0 \end{cases} \tag{9}$$

The proposed platform attempted to replace the polynomial equation. In general, the proposed platform has the process as described in Fig. 3 to replace the first box of Fig. 1 named "rheological model." The platform has two parts, which are the shear stress model and parameter estimator. The first part is built using a machine learning model. Therefore, the platform user can give magnetic field inputs and shear rate to determine the prediction zone. Then, plastic viscosity and yield stress are obtained from the platform output.

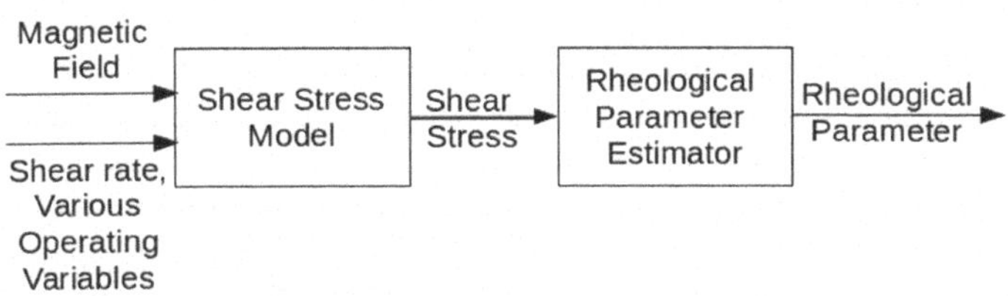

Fig. 3 The machine learning based-shear stress model and rheological parameter estimation to replace the rheological model part in Fig. 1

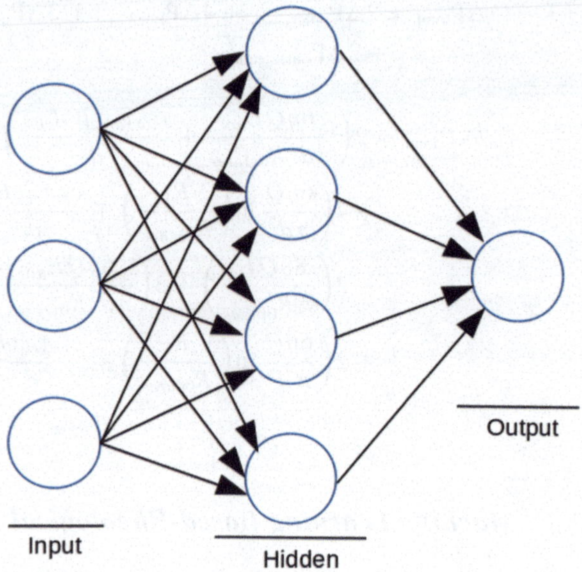

Fig. 4 Illustration of neural networks model with single hidden layer

For the first part, the shear stress model is a multilayer perceptron with 3 layers or single hidden layer feedforward neural networks (SLFNs), as illustrated in Fig. 4. In general, the model is a neural network as a shear stress predictor as a function of shear rate $\dot{\gamma}$ and magnetic field B as shown in Eq. (10).

$$\tau = f(B, \dot{\gamma}) \quad (10)$$

The neural network model was trained using Extreme Learning Machine (ELM) method. ELM has been proven to gain better generalization and faster training time than traditional ANN [15]. Meanwhile, the second part of the model consists of parameter estimator inspired by Papanastasiou equation and optimization algorithm, as discussed in [16].

2.4 Simulation Setup and Scenario

The simulation description can be divided into two parts, which are the steady state model and the rheological model. For the steady state model, the parameters are described in Table 3. Meanwhile, the rheological model is in the form of a sinusoid activation function, 10,000 hidden, and one hidden layer. The defined inputs are shear rate and magnetic field standardized using logarithm normalization and linear normalization, respectively. Meanwhile, the output is shear stress. The shear stress and shear rate are inputted to the parameter estimator. The optimization and objective function algorithm parameters are adopted from [16]. While the plastic viscosity is

Table 3 Definition and value of the MR valve parameters

Symbol	Remarks	Value
Q	Flow rate	37, 55, 73 ml/s
I	Current	0, 0.2, 0.4, 0.6 A
$d = d_{ao} = d_r = d_{rO} = d_{ri}$	Tunnel gap	0.5 mm
d_{ai}	Tunnel gap of inner annular	1 mm
L_{ai}	Tunnel length of inner annular	$(6 - 2d_r)$ mm
L_{ao}	Tunnel length of outer annular	10 mm
L_o	Tunnel length of orifice	5 mm
R_1	Radius of the inner annular path	6.5 mm
R_{0outer}	Radius of the outer annular path	3 mm
R_{0inner}	Radius of orifice path	2.5 mm
η	Plastic viscosity	0.092 Pa s
τ/τ_y	Yield stress	Calculated based on the model and MRF behavior
B	Magnetic field	Referring to Table 1

predetermined as 0.092 Pa s, the same as off state properties provided in the technical documentation from the manufacturer, the calculated yield stress is used to calculate the pressure drop.

The shear stress model was built based on material from Lord Corporation called MRF 132 DG as in [16]. The material was tested using a twin parallel plate rotational rheometer MCR 302 manufactured by Anton Paar, Physical, GmbH, Austria. The magnetic field and shear rate were set at 0–500 mT and 0.01–2000 s^{-1}, respectively.

The predicted yield stress is divided into two scenarios. While the first scenario is based on a low shear rate range called ELM 1, the second one is predicted based on a high shear rate range denoted as ELM 2. As a benchmark, the mathematical model in Eq. (9) is employed called as 'poly.' Meanwhile, the experimental data from [Fitrian paper] is utilized as the reference called 'exp.'

The model performance is evaluated using root mean square error (RMSE) with the following formulation:

$$RMSE = \sqrt{\frac{\sum_{h=1}^{K}\left(\Delta P_{\mathrm{e}} - \Delta P_{\mathrm{p}}\right)^2}{K}} \tag{11}$$

where K, ΔP_e, and ΔP_p are the number of the data, measured pressure drop data, and predicted pressure drop data, respectively.

3 Results and Discussion

The RMSE in Table 4 show that the proposed methods (ELM 1 and ELM 2) have shown a good agreement with the experimental results at various flow rates and better than Poly scheme. In general, ELM 2 has the least error compared to other schemes at various flow rates. ELM 1 RMSEs are slightly higher than ELM 2's because it utilizes high shear rate region to predict the yield stress. The visual observation of the comparison among the schemes can be checked in Fig. 5. ELM 1 and ELM 2 follow closely the experimental data. Meanwhile, the prediction by 'Poly' or the conventional model shows the relatively high deviation at high magnetic field or applied current. The higher shear rate region can be related to the measured peak pressure drop while the sine displacement signal is given to the MR valve installed in a damper. Peak pressure drop means that the flow rate is at the highest value in the given sine cycle that has higher possibility to be related to high shear rate region than low shear rate region.

Table 4 The model error at various flow rates

Scheme	Flowrate (ml/s)			
	37	55	73	All data
ELM 1	0.29	0.25	0.32	0.29
ELM 2	0.22	0.22	0.23	0.22
Poly	0.35	0.31	0.4	0.36

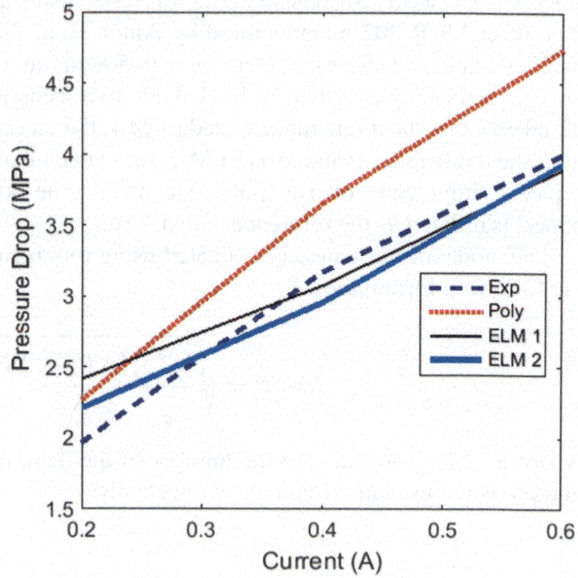

Fig. 5 Comparison of the predicted pressure drop at various schemes at 73 ml/s

Some additional finding can also be discussed based on the obtained results. Firstly, a polynomial model is a powerful tool to describe the general behavior of material and a simple curve of yield stress as a function of magnetic field. However, the possibility to include more complex behavior of MRF, including shear rate and temperature, can be difficult. The benefit is the simplicity of the model. Meanwhile, machine learning-based model can be a tool to achieve more accurate prediction and open the possibility to extend the model to other inputs as long the experimental data is available. The disadvantage of the machine learning based-rheological model is the coverage is only limited to the available measurement or empirical data, which is also the disadvantage of the polynomial model. Furthermore, the mathematical model of the neural networks is also complex. If the other users want to use the machine learning based-model, the model file needs to be copied instead of copying the mathematical equation by handwriting. Alternatively, the model can be uploaded in a certain website; therefore, various researchers can use the model. The current investigation is only limited to the narrow operating range of magnetic field and flow rate. The model also is developed based on a relatively lower range magnetic field (up to 0.5 T) than the possible occurred magnetic field in the valve (up to 1 T) if considering the possible applied current range on the coil.

4 Conclusion

In this work, a machine learning platform is integrated with a traditional steady state model to demonstrate the model capability to predict MR devices, which in this case is MR valve. The platform consists of a shear stress model and steady state model. The shear-stress model is a neural network model with one hidden layer trained by Extreme Learning Machine (ELM), while the parameter estimator contained an optimization algorithm. The results have shown a good agreement with the measured data with RMSE less than of the conventional model. The platform has shown its potential to overcome the conventional model limitation and also opens the possibility to add more inputs that is easier than when using the conventional model. In the future, the proposed model performance needs to be investigated at a wider operating range condition of magnetic field and flow rate.,

Acknowledgements The authors gratefully acknowledge the financial support of the Ministry of Education in Malaysia and Universiti Teknologi Malaysia under grant vote no: PDRU (04E02).

References

1. de Vicente J, Klingenberg DJ, Hidalgo-Alvarez R (2011) Magnetorheological fluids: a review. Soft Matter 7:3701. https://doi.org/10.1039/c0sm01221a

2. El Wahed A, Balkhoyor L (2017) Characteristics of magnetorheological fluids under single and mixed modes. Proc Inst Mech Eng Part C J Mech Eng Sci 231:3798–3809. https://doi.org/10.1177/0954406216653621
3. Imaduddin F, Mazlan SA, Zamzuri H (2013) A design and modelling review of rotary magnetorheological damper. Mater Des 51:575–591. https://doi.org/10.1016/j.matdes.2013.04.042
4. Ghaffari A, Hashemabadi SH, Ashtiani M (2015) A review on the simulation and modeling of magnetorheological fluids. J Intell Mater Syst Struct 26:881–904. https://doi.org/10.1177/1045389X14546650
5. Imaduddin F, Amri Mazlan S, Azizi Abdul Rahman M, Zamzuri H, Ubaidillah, Ichwan B (2014) A high performance magnetorheological valve with a meandering flow path. Smart Mater Struct 23:65017. https://doi.org/10.1088/0964-1726/23/6/065017
6. Bahiuddin I, Mazlan SA, Imaduddin F, Ubaidillah, Ichwan B (2016) Magnetorheological valve based actuator for improvement of passively controlled turbocharger system. In: AIP conference proceedings. https://doi.org/10.1063/1.4943431
7. Wang DH, Ai HX, Liao WH (2009) A magnetorheological valve with both annular and radial fluid flow resistance gaps. Smart Mater Struct 18:115001. https://doi.org/10.1088/0964-1726/18/11/115001
8. Li WH, Wang XY, Zhang XZ, Zhou Y (2009) Development and analysis of a variable stiffness damper using an MR bladder. Smart Mater Struct 18:74007. https://doi.org/10.1088/0964-1726/18/7/074007
9. Bahiuddin I, Mazlan SA, Shapiai MI, Choi S-B, Imaduddin F, Rahman MAA, Ariff MHM (2018) A new constitutive model of a magneto-rheological fluid actuator using an extreme learning machine method. Sens Actuat A Phys 281:209–221. https://doi.org/10.1016/j.sna.2018.09.010
10. Bahiuddin I, Wahab NAA, Shapiai MI, Mazlan SA, Mohamad N, Imaduddin F, Ubaidillah (2019) Prediction of field-dependent rheological properties of magnetorheological grease using extreme learning machine method. J Intell Mater Syst Struct (in press). https://doi.org/10.1177/1045389X19844007
11. Bahiuddin I, Mazlan SA, Shapiai I, Imaduddin F, Ubaidillah, Choi S-B (2018) Constitutive models of magnetorheological fluids having temperature-dependent prediction parameter. Smart Mater Struct 27:95001. https://doi.org/10.1088/1361-665X/aac237
12. Rabbani Y, Shirvani M, Hashemabadi SH, Keshavarz M (2017) Application of artificial neural networks and support vector regression modeling in prediction of magnetorheological fluid rheometery. Colloids Surf A Physicochem Eng Asp 520:268–278. https://doi.org/10.1016/j.colsurfa.2017.01.081
13. Bahiuddin I, Mazlan SA, Shapiai MI, Imaduddin F, Ubaidillah (2017) Study of extreme learning machine activation functions for magnetorheological fluid modelling in medical devices application. In: 2017 International conference on robotics, automation and sciences (ICORAS). IEEE, pp 1–5. https://doi.org/10.1109/ICORAS.2017.8308053
14. Jung ID, Kim M, Park SJ (2016) A comprehensive viscosity model for micro magnetic particle dispersed in silicone oil. J Magn Magn Mater 404:40–44. https://doi.org/10.1016/j.jmmm.2015.12.024
15. Huang G-B, Zhu Q-Y, Siew C-K (2006) Extreme learning machine: Theory and applications. Neurocomputing. 70:489–501. https://doi.org/10.1016/j.neucom.2005.12.126
16. Bahiuddin I, Mazlan SA, Shapiai MI, Imaduddin F, Ubaidillah, Choi S-B (2019) A new platform for the prediction of field-dependent yield stress and plastic viscosity of magnetorheological fluids using particle swarm optimization. Appl Soft Comput 76:615–628. https://doi.org/10.1016/j.asoc.2018.12.038

Enhancement of Isotropic Magnetorheological Elastomer Properties by Silicone Oil

M. H. A. Khairi, Siti Aishah Abdul Aziz, N. M. Hapipi, Saiful Amri Mazlan, Nur Azmah Nordin, Ubaidillah, and N. I. N. Ismail

Abstract This article focused on the influence of additional silicone oil as an plasticizer additive in a magnetorheological elastomer (MRE) upon its rheological properties. Here, a silicone rubber is used as a base material of the MRE with CIPs act as the filler. Silicone oil (SO) as a plasticizer is added to the base ingredients to improve the viscosity and dispersion of magnetic particles and enhance the MRE properties. Various tests comprised of magnetic, morphology, and rheology tests were conducted for MRE characterization purpose. The results showed that the addition of SO on the MRE had increased 19% of magnetic properties compared to non-SO based MRE. Dispersion of magnetic particles is improved by the addition of SO as observed through Low Vacuum Scanning Electron Microscope (LVSEM). In rheology test, both absolute and relative MR effects were increased by 0.3 MPa and 343%, respectively, with the incorporation of 15 wt% SO. The introduction of SO has proven to resolve the agglomeration issues in isotropic MRE which degrade performances of MRE application devices and systems.

Keywords Magnetorheological elastomer · Silicone oil · Plasticizer

1 Introduction

Magnetorheological elastomer (MRE) is smart materials whose rheological properties can be controlled by the applied external magnetic field. Under a stronger applied

M. H. A. Khairi · S. A. A. Aziz · N. M. Hapipi · S. A. Mazlan (✉) · N. A. Nordin
Engineering Materials and Structures (eMast), Research Group, Malaysian-Japan International Institute of Technology, Universiti Teknologi Malaysia, Jalan Sultan Yahya Petra, 54100 Kuala Lumpur, Malaysia
e-mail: amri.kl@utm.my

Ubaidillah
Mechanical Engineering Department, Faculty of Engineering, Universitas Sebelas Maret, Jl. Ir. Sutami 36A Kentingan Jebres, Surakarta 57126, Indonesia

N. I. N. Ismail
Technology and Engineering Division, Advanced Rubber Technology Unit, Rubber Research Institute Malaysia (RRIM), 47000 Sungai Buloh, Selangor Darul Ehsan, Malaysia

U. Sabino et al. (eds.), *Proceedings of the 6th International Conference and Exhibition on Sustainable Energy and Advanced Materials*, Lecture Notes in Mechanical Engineering,
https://doi.org/10.1007/978-981-15-4481-1_28

magnetic field, the MRE perform a higher stiffness. Because of their unique properties, MRE have recently been used in a broad range of possible applications such as vibration absorbers [1], sensing devices [2] and morphing structure [3]. Advantage of MREs is no need container or seals to store or prevent leakage as compared to magnetorheological fluid (MRF) [4, 5]. The MR effect, which calculated from the storage modulus is frequently adopted to evaluate the MRE performance [6]. The main components of MRE was magnetically polarized particles, non-magnetic matrix, and additives. Common matrix materials for MREs include silicone rubber [7], natural rubber [8] and polyurethane (PU) rubber [9]. MREs are often classified into two different type depending on the curing process which are isotropic and anisotropic [10]. Isotropic MREs can be characterized by the randomly distribution of magnetic particle in the matrix. Meanwhile, anisotropic MREs are curing under a strong magnetic field resulted of chain-like arrangement of magnetic particles in a matrix.

In term of particles type, carbonyl iron particles (CIPs) were used as the magnetic particles for the MRE due to their high permeability, low remnant magnetization and high saturation magnetization. Lokander and Stenberg [11] have shown that the maximum MR effect on MRE has been reported to be achieved when the iron particle fill rate at about 70 wt%. It is known that higher particle loading in MRE would lead to higher storage modulus and enhanced both absolute and relative MR effect. However, too many particle loadings will increase the viscosity of polymer composite, significantly, would result in higher initial storage modulus. This would reduce the corresponding MR effect of the MRE. Therefore, to reduce the respective initial storage modulus, plasticizers are introduced to the MRE. Gong et al. [12] and Tian and Nakano [13] agreed that the movement of the magnetic particles is governed by the elastomer matrix. In this way, softer matrices obtained by adding plasticizers cause an increase in the MR effect by providing lower zero-field moduli and improving the interaction between CIPs.

Therefore, in this study, presence of SO in the MRE is believed to improve the interaction between particles due to less restriction from the plasticized-matrix. Consequently, the main technical contribution of this work is to solidly investigate the influence of SO on the magnetic properties and morphology of the CIPs in the matrix. The shear storage modulus and MR effect were further investigated using a rheometer.

2 Fabrication of MRE

The MRE samples used in this work were prepared by compounding CIPs having an average diameter of 6 μm (BASF, type OM) and room temperature vulcanization (RTV) silicone rubber. The RTV silicone rubber in the form of liquid was added with CIPs content of 70 wt% and silicone oil (SO) content of 15 wt% and mixed homogeneously using a mechanical stirrer for 10 min. To prepare the isotropic MRE samples, the compounding was molded into a customized mold for two hours at room temperature with 1.1 mm thickness (Fig. 1). MRE sample without SO also prepared as a benchmark to the MRE with 15 wt% SO. The magnetic characteristics of the fabricated MREs were measured using the vibrating sample magnetometer (VSM). Meanwhile, the microstructure observation was performed using low vacuum scanning electron micrography (LV-SEM) at an accelerating voltage of 15 kV with 1000 times magnification. The influence of the ramped magnetic flux on the storage modulus of MRE samples was measured by the rheometer (Anton Paar, Physica MCR 302, Germany). The samples were evaluated under constant strain amplitudes of 0.01% and at excitation frequency of 1 Hz.

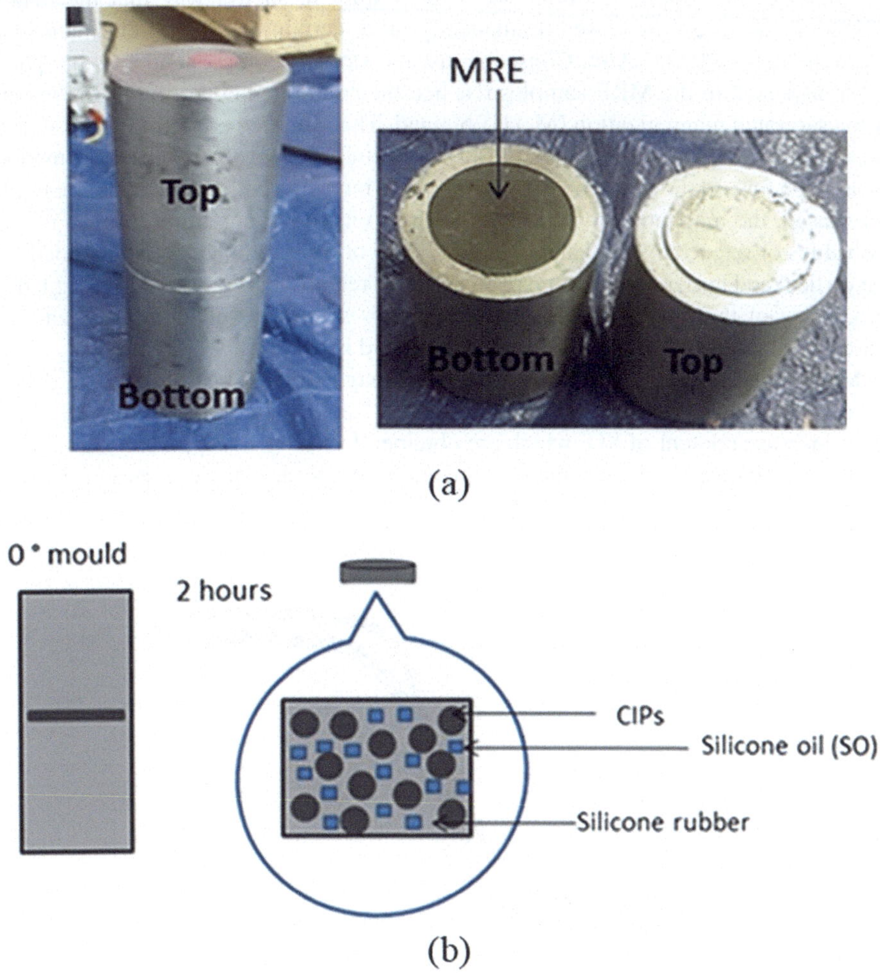

Fig. 1 Fabrication of MRE. **a** Mould and MRE sample, **b** illustration of isotropic MRE

3 Results and Discussions

3.1 Magnetic Properties of MRE

Figure 2 illustrates the magnetization curves of isotropic MRE samples with and without SO content. It is apparent that the magnetic properties of isotropic MREs with and without SO are induced when subjected to the application of magnetic field. The magnetization curves reveal narrow magnetic hysteresis loops for both samples. This fact indicates that both MREs exhibit a soft magnetic characteristic. The trend of the graph is similar for both samples in such a way that the magnetization curves are increased dramatically up to certain magnetic field intensity, approximately 500,000 A/m. Consequently, the slope becomes slender as the particles dispersed in the MRE samples reached their saturated condition as the value of the saturated magnetization (M_s) is obtained. The saturated condition refers to no increment of magnetization when all the magnetic moments in magnetic domains are aligned toward the direction of magnetic field intensity. Unless the magnetic field is released, the magnetic moments may return to its initial condition. From Fig. 2, the values of retentivity (M_r) and coercivity (H_c) of each sample can be obtained at the vertical and horizontal axes, respectively. The retentivity is defined as a magnetic remanence of the magnetic material that implies a measure of the magnetization which is remained after the applied magnetic field is released.

Meanwhile, the coercivity is a measure of magnetizing force required to drive the reverse magnetization after being saturated. Table 1 presents the M_S, M_r, and H_c at increase content of SO, which are obtained from Fig. 1. Apparently, addition of SO contributes to enhancement of magnetic properties of the isotropic MRE.

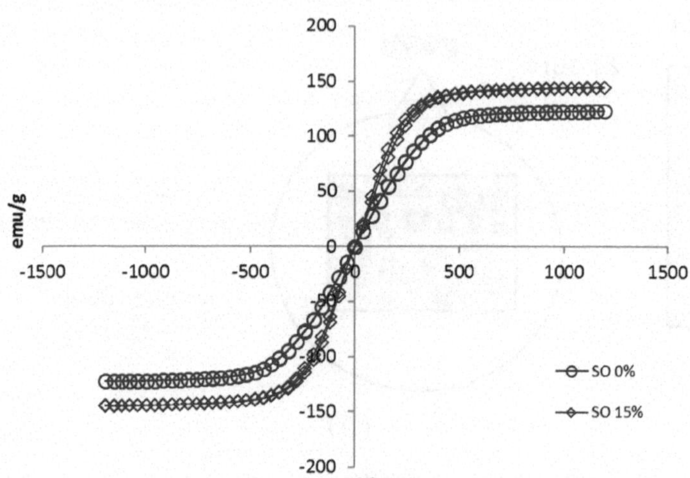

Fig. 2 Magnetization curve of isotropic MRE with and without SO

Table 1 Magnetic properties of MRE with and without SO

Sample wt%	M_s (emu/g)	H_c (mT)	M_r (emu/g)
0	121	7.80	0.22
15	144	7.32	0.28

This is proven by increasing of Ms with SO content as referred to the Table 1. The Ms values presented in Table 1 illustrate that the Ms is increased from 121 to 144 emu/g, for the zero and highest SO content, respectively. Besides, the M_r is also proportionally increased with M_S. However, the values are relatively small because the isotropic MRE with SO exhibits soft magnetic material characteristic. The H_c values for both samples are very close to each other but not identical. This is due the characteristic of soft magnetic materials. The results demonstrate that incorporation of SO in the isotropic MREs with SO can contribute to a potential of altering the magnetic properties by adjusting the content of SO and the applied magnetic field.

3.2 *Morphology of MRE*

The cross sectioned area surface morphology of isotropic MRE with and without SO are presented in Fig. 3. It can be observed that for the MRE either with or without SO, the CIPs are dispersed randomly in the matrix as shown in Fig. 3a, b, respectively. Apparently, there are significant differences in the distribution of the CIPs in the matrix when SO was added. In the presence of 15 wt% of SO, the distribution of CIPs has improved as no big aggregates can be seen in the fractured surface of MRE as shown in Fig. 3b. Furthermore, from fractography observation, the SO can enhance the compatibility between the CIPs and the silicone rubber matrix as there are no such cavities seen as compared to matrix without SO, indicating that CIPs have been well embedded in the matrix. This is indicating that the introduction of SO has enhanced the distribution and texture of the isotropic MRE.

3.3 *Rheological of MRE*

The influence of the ramped magnetic flux on the shear storage modulus of MREs was evaluated under constant strain amplitudes of 0.01%, excitation frequency of 1 Hz at ambient temperature. The ramp magnetic flux was realized by driving an electric current from 0 to 5 A in a linear fashion. The flux densities were measured and provided more than 0.8 T for both samples. Figure 4 shows the shear storage modulus under the increasing strengths of the magnetic field for MRE with and without SO. In this work, keeping the CIPs weight percentage of 70% constant, both sample's shear modulus was increased linearly as the magnetic field increased. This proves that both MREs samples exhibited noticeable increment of MR effects. Even though both

Fig. 3 Fractured surface morphology of isotropic MREs. **a** Without SO and **b** with 15 wt% SO

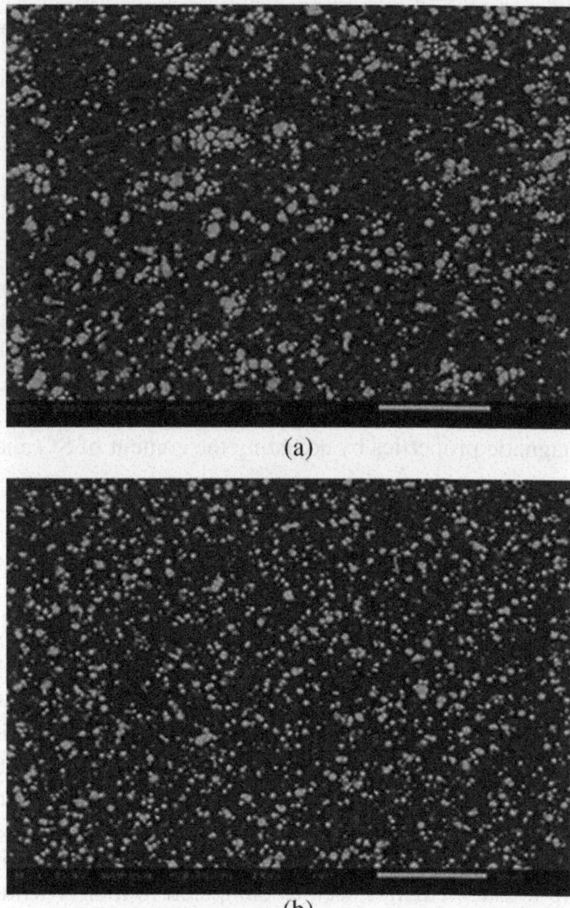

MREs samples showed an inclined trend as the magnetic field increased, however MRE with 15 wt% SO showed greatly improvement in the MR effect as shown in Fig. 4b. The MR effect of MRE with SO was 343% (0.3 MPa) under the magnetic field of 0.8 T, almost six times higher as compared to MRE without SO, which only 55%. This is due to the uniformly distribution between particles that greatly influence the magneto-induced modulus which resulted in strong interaction between particles, thus lead to stiffer matrix provided higher MR effect. Meanwhile, the absolute modulus of MRE with and without SO was 0.12 and 0.29 MPa, respectively.

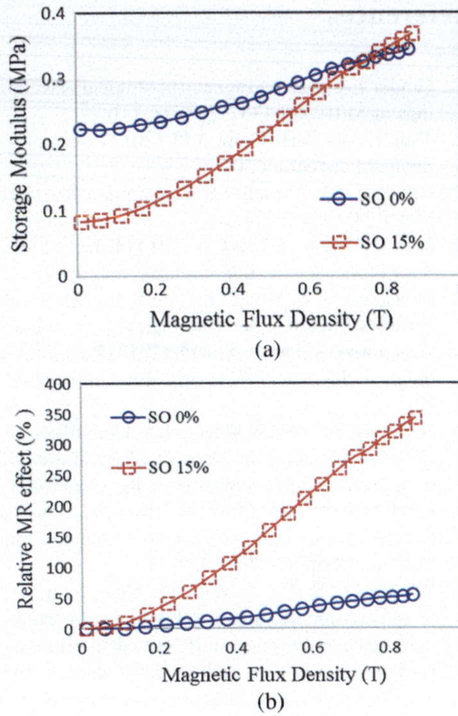

Fig. 4 MRE with and without SO. **a** Storage modulus, **b** relative MR effect as a function of sweep magnetic field

4 Conclusion

The effect of SO as a plasticizer additive on the performance of MREs based on silicone rubber has been experimentally investigated. The addition of the SO helps CIPs to disperse well with only a few agglomerations structures. The experimental results have demonstrated that the addition of SO increase the magnetization by 19%. In terms of rheology, the MRE with SO had the lower zero-field modulus and the MR effect of MRE with 15 wt% SO had increased from 55 to 343% when SO was added to the rubber matrix with absolute modulus of 0.3 MPa.

Acknowledgements This study was financially supported by the Universiti Teknologi Malaysia under Transdisciplinary Research Grant (Vot No. 06G77) and Professional Development Research University Grant (Vot No. 04E02).

References

1. Ginder JM (2001) Magnetorheological elastomers in tunable vibration absorbers. In: Proceedings of SPIE, vol 4331, pp 103–110
2. Tian TF, Li WH, Deng YM (2011) Sensing capabilities of graphite based MR elastomers. Smart Mater Struct 20
3. Nguyen VQ, Ahmed AS, Ramanujan RV (2012) Morphing soft magnetic composites. Adv Mater 24:4041–4054
4. Lian C, Lee K-H, Lee CH (2017) Effect of temperature and relative humidity on friction and wear properties of silicone-based magnetorheological elastomer. Tribol Trans 0:1–9
5. Shi G, Wang W, Wang G, Yang F, Rui X (2019) Dynamic mechanical properties of FeSi alloy particles-filled magnetorheological elastomers elastomers. Polym Technol Mater 00:1–13
6. Ge L, Gong X, Fan Y, Xuan S (2013) Preparation and mechanical properties of the magnetorheological elastomer based on natural rubber/rosin glycerin hybrid matrix. Smart Mater Struct 22
7. Hapipi N, Aziz SAA Mazlan SA, Ubaidillah, Choi SB, Mohamad N, Khairi MHA, Fatah AYA (2019) The field-dependent rheological properties of plate-like carbonyl iron particle-based magnetorheological elastomers. Results Phys 12:2146–2154
8. Lee CJ, Kwon SH, Choi HJ, Chung KH, Jung JH (2018) Enhanced magnetorheological performance of carbonyl iron/natural rubber composite elastomer with gamma-ferrite additive. Colloid Polym Sci 296:1609–1613
9. Yu M, Qi S, Fu J, Zhu M, Chen D (2017) Understanding the reinforcing behaviors of polyaniline-modified carbonyl iron particles in magnetorheological elastomer based on polyurethane/epoxy resin IPNs matrix. Compos Sci Technol 139:36–46
10. Kallio M, Lindroos T, Aalto S, Järvinen E, Kärnä T, Meinander T (2007) Dynamic compression testing of a tunable spring element consisting of a magnetorheological elastomer. Smart Mater Struct 16:506–514
11. Lokander M, Stenberg B (2003) Performance of isotropic magnetorheological rubber materials. Polym Test 22:245–251
12. Wu JK, Gong XL, Fan YC, Xia HS (2012) Improving the magnetorheological properties of polyurethane magnetorheological elastomer through plasticization. J Appl Polym Sci 123:2476–2484
13. Tian T, Nakano M (2018) Fabrication and characterisation of anisotropic magnetorheological elastomer with 45° iron particle alignment at various silicone oil concentrations. J Intell Mater Syst Struct 29:151–159

Frequency-Dependent on the Magnetorheological Effect of Magnetorheological Plastomer

N. M. Hapipi, Saiful Amri Mazlan, Siti Aishah Abdul Aziz, M. H. A. Khairi, Ubaidillah, Mohd Hatta Mohammed Ariff, and Abdul Yasser Abd Fatah

Abstract A low cross-linked magnetic polymer matrix also known as magnetorheological plastomer (MR plastomer) containing micron-sized carbonyl iron particles (CIPs) is a new kind of MR materials. MR plastomer can be prepared by two main methods which are physically and chemically crosslinking. However, the study on the dynamic properties of involving chemically crosslinked MR plastomer particularly the viscoelastic properties especially frequency-dependent are not systematically investigated. Therefore, in this study, the effect of the frequency on the MR effect and damping performance of chemically crosslinking MR plastomer under oscillatory modes condition were analysed. The magnetic particles namely CIPs were inserted into a plasticine-like polymer matrix causing the materials to exhibit an MR effect in response to an external magnetic field. Polymer base matrix was prepared using poly-vinyl alcohol (PVA), and boric acid (BA) was used as a cross-linking agent for chemically crosslinked MR plastomer. The MR plastomer samples were prepared using 70 wt% of CIPs as magnetic particles. The samples were tested using a rheometer with different test frequencies, which are 1, 5, and 10 Hz at the on-state condition. The experimental results revealed that the frequency has a significant correlation with the MR effect of samples where the MR effect of the sample decreased with the increment of test frequency. The MR effect for each sample at 1, 5, and 10 Hz are 6793, 5049, and 3131% respectively. In contrast, for the frequency sweep test, the storage modulus of the sample showed an increasing trend with the increment of test current, while the loss factor revealed an opposite result. The results proved that this kind of MR materials has the potential to be used in various of applications like soft actuator, vibration absorber, and force sensor.

N. M. Hapipi · S. A. Mazlan · S. A. A. Aziz · M. H. A. Khairi · M. H. M. Ariff (✉)
AVS Research Laboratory, Malaysia-Japan International Institute of Technology (MJIIT), Universiti Teknologi Malaysia, Jalan Sultan Yahya Petra, 54100 Kuala Lumpur, Malaysia
e-mail: mohdhatta.kl@utm.my

Ubaidillah
Mechanical Engineering Department, Faculty of Engineering, Universitas Sebelas Maret, Jl. Ir. Sutami 36A Kentingan, Surakarta 57126, Central Java, Indonesia

A. Y. A. Fatah
Department of Engineering, Razak School of Engineering and Advanced Technology, Universiti Teknologi Malaysia, Jalan Sultan Yahya Petra, 54100 Kuala Lumpur, Malaysia

U. Sabino et al. (eds.), *Proceedings of the 6th International Conference and Exhibition on Sustainable Energy and Advanced Materials*, Lecture Notes in Mechanical Engineering,
https://doi.org/10.1007/978-981-15-4481-1_29

Keywords Plastomer · Plasticine-like · Rheology · Magnetorheological · Frequency

1 Introduction

Magnetorheological (MR) material is a material that can altered its rheological properties by the application of an external magnetic field. Recently, MR materials have gained increasing attentions due to their unique properties such as stiffness and damping that can be controlled rapidly, continuously and reversibly. MR materials are divided into several groups with a variety of matrices which are MR fluid, MR elastomer, MR gel, MR foam, and MR grease. Even though they are made with different kind of matrices, they all exhibited same properties where the viscosity, stiffness, and damping properties can be controlled by an external magnetic field. Currently, a new kind of MR materials has been discovered by researchers that are known as MR plastomer which exhibit plastic properties at room temperature [1–4]. Basically, MR plastomer consisted of magnetizable particles embedded in a low cross-linking polymer as a carrying matrix. The carrying matrix can be polyurethane (PU), polyvinyl alcohol (PVA), paraffin, carrageenan and guar gum. Due to the low or weak cross-linking matrix polymer, the particles inside the matrix are highly mobile and causing the MR plastomer to exhibit high MR effect. Moreover, in comparison to the MR fluid, MR plastomer shows no tendency of particle sedimentation problems as the matrix more viscous. To this end, MR plastomer has becoming a promising candidate for practical applications like vibration energy absorbers and isolators [5].

To date, a variety of MR plastomers have been extensively studied by researchers. Shiga et al. [6] reported the first discoveries of MR plastomer with the name of plastic MR materials using silicone gel as matrix. Then, the studies on the MR plastomer starting to appear in the current relevant research field. According to Xuan et al. [1], the mechanical properties of plastic MR materials can be enhanced by using more viscous matrices like polymer and PU matrices. It will cause the MR plastomer to exhibit higher stability than the MR fluid. Commonly, PU matrix prepared by using toluene diisocyanate (TDI) and polypropylene glycol (PPG-1000) as a basic material is highly used to prepare PU based MR plastomer [7–9]. For example, Xu et al. [8], studied the influence of iron particles concentration and magnetic field on the creep and recovery behaviors of MR plastomer under various conditions by using PU based MR plastomer such as TDI and PPG-1000 as the carrying matrix. The results showed that the existence of magnetic field made the MR plastomer stiffer and decreases the creep strain. Moreover, MR plastomer acts as viscose liquid without magnetic field.

Recently, a new kind of MR plastomer fabricated from hydrogel has been reported. The fabrication of hydrogel MR plastomer is more economical and convenient compared to PU MR plastomer. Basically, hydrogel MR plastomer is synthesized by using water or an organic swollen solvent like PVA. PVA hydrogels can be prepared by using two methods; physically crosslinked or chemically crosslinked. Physically crosslinked PVA hydrogel has been studied extensively by few researchers [10–12].

Wu et al. [10] investigated the physically crosslinked hydrogel MR plastomer using PVA as carrying matrix. The result demonstrated that the MR effect achieved by hydrogel MR plastomer can reaches up to ~230%, and the tensile strength is close to the value exhibited by natural rubber. These results verified that the hydrogel MR plastomer also can be a good and reliable matrix for MR materials fabrication. However, there have been little discussions about the dynamic behavior for chemically crosslinked PVA based hydrogel MR plastomer. In order to ensure that this kind of hydrogel MR plastomer is a competence material to be used in engineering application, the study on its dynamic behavior is necessarily important. The dynamic property like oscillatory shear test is the most commonly used measurement to study the rheological properties of the MR materials. Therefore, the focus of this paper is to experimentally investigate the frequency-dependent on the MR effect of chemically crosslinked PVA hydrogel MR plastomer. Both magnetic-induced modulus and the loss factor are presented in this paper.

2 Experimental Procedure

2.1 *Materials*

Poly(vinyl) alcohol (PVA) was bought from Merck Company, Germany (Mw 60,000 $\geq$ 98.0%). Sodium tetraborate decahydrate (boric) was purchased from a drug store and dimethyl sulfoxide (DMSO) brand ChemAR was supplied by Systerm Chemicals. Carbonyl iron particles (CIPs) from BASF were used as magnetic particles. Deionized (DI) water was used to prepare the aqueous solution.

2.2 *Sample Preparations*

A polymer solution of 7.5% (w/v) PVA plastomer was prepared by dissolving the PVA beads in DMSO at 80°C using a hotplate for 2 h. The mixture was then cooled down to room temperature after it completely dissolved. Then, the CIPs were added into the PVA solution with concentrations of 70 wt%. Next, the boric acid solution of 3% (w/v) was prepared by mixing the crystal with DI water. The PVA solution that contained CIPs and borax solution were mixed with constant stirring using mechanical stirrer for 10 min with 25 rpm. The mixture was kept overnight in order to obtain a uniform distribution of MR plastomer sample as shown in Fig. 1. Next, the dynamic properties of the MR plastomer samples were then examined using a parallel-plate rheometer (Physica MCR 302, Anton Paar Co., Germany) equipped by magneto-controllable accessory.

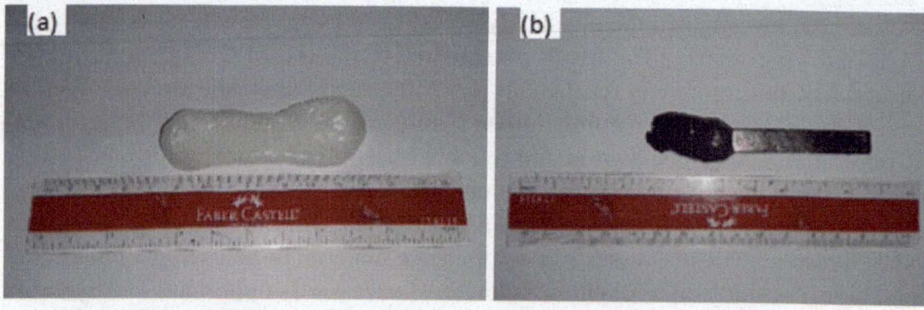

Fig. 1 **a** The PVA hydrogel MR plastomer without CIPs and **b** PVA hydrogel MR plastomer with CIPs in the present of permanent magnet

3 Result and Discussions

Rheological Properties The frequency dependence of the storage modulus and loss factor for PVA hydrogel MR plastomer were investigated. Generally, the storage modulus represents the elasticity of the materials related to the capability of the materials to store energy during deformation. Meanwhile, the loss factor described the damping properties of the materials which are the ability of the materials to dissipate energy. Thus, the graph of the storage modulus-frequency of the MR plastomer sample contain 70 wt% CIPs with different magnetic flux density is presented in Fig. 2.

As shown in Fig. 2, at off-state condition (0 mT) the storage modulus is almost unchanged with the increment of frequency. Whilst, at on-state condition (200 and 400 mT) the storage modulus of the sample increases with the increment of frequency. Meanwhile, at low frequency below than <5 Hz, the storage modulus dramatically increases with the increasing of shear frequency for sample tested at on-state conditions. However, as noticed in Fig. 2, at higher frequency above >5 Hz, the storage modulus is not depending on the frequency. This phenomenon caused by the changes of the internal polymer chains of the PVA-based MR plastomer. The polymer chain of PVA-based matrix started to intertwine and breaking the hydrogen bonds between PVA chains and didiol group in BA [13].

Moreover, the storage modulus increases with the increment of the external magnetic field. The higher the magnetic field supplied toward the sample during the test, the larger the storage modulus exhibited by the samples. The reason is due to the formation of chain-like structures triggered by the application of external magnetic field. The curve of loss factor as a function of frequency is illustrated in Fig. 3. The loss factor can determine the dissipation energy or known as damping capacity of materials.

From the graph in Fig. 3, at off-state conditions (0 mT), the loss factor sharply increased with the increment of frequency at beyond 40 Hz. However, the trend insignificantly changed at on-state conditions (200 and 400 mT) as the loss factor almost unchanged with the increasing of frequency. Meanwhile, at low frequency

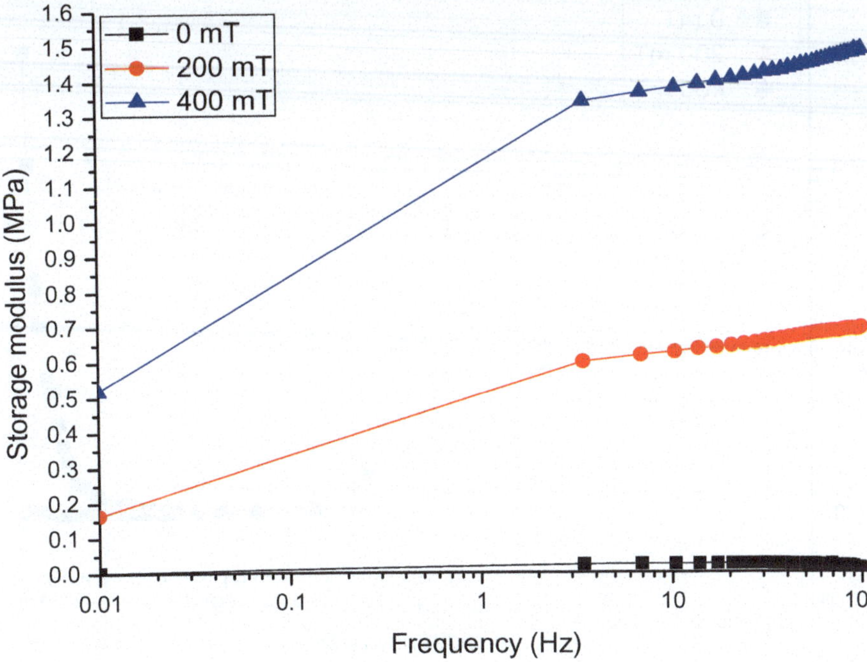

Fig. 2 Storage modulus as a function of frequency

(<5 Hz), the loss factor of the samples decreases with the increasing of magnetic flux density. At higher frequency above 5 Hz, the value of loss factor does not depend on the frequency for on-state condition. The increment of loss factor at off-state conditions started to increase again above ~5 Hz which means that the dissipation energy for the MR plastomer sample is higher compared to the MR plastomer sample at on-state conditions. In this situation, generally at 0 mT, CIPs are randomly distributed and the particles tend to cluster, consequently increased the energy dissipation [14]. Oppositely, the result revealed that in respond of magnetic field, the damping properties of the sample is lowered due to the enhancement of particles chain-like structure. When the sample is exposed to the magnetic field, the CIPs started to form chain-like structure thus reduced the distance between them. Therefore, less friction occurs between the particles, resulted in the increment of energy dissipation decreased. Besides that, samples tested without magnetic field showed drastic increment of loss factor at higher frequency because the polymer chain cannot keep up with the force and the intermolecular chains of the polymer started to break.

The effect of the test frequency on the magneto-induced modulus of the MR plastomer with 70 wt% CIPs at 1, 5 and 10 Hz are shown in Fig. 4. The zero-field modulus (G_0), the absolute MR effect ($\Delta G' = G_{max} - G_0$), and the relative MR effect of the samples are tabulated in Table 1. The relative MR effect is the key

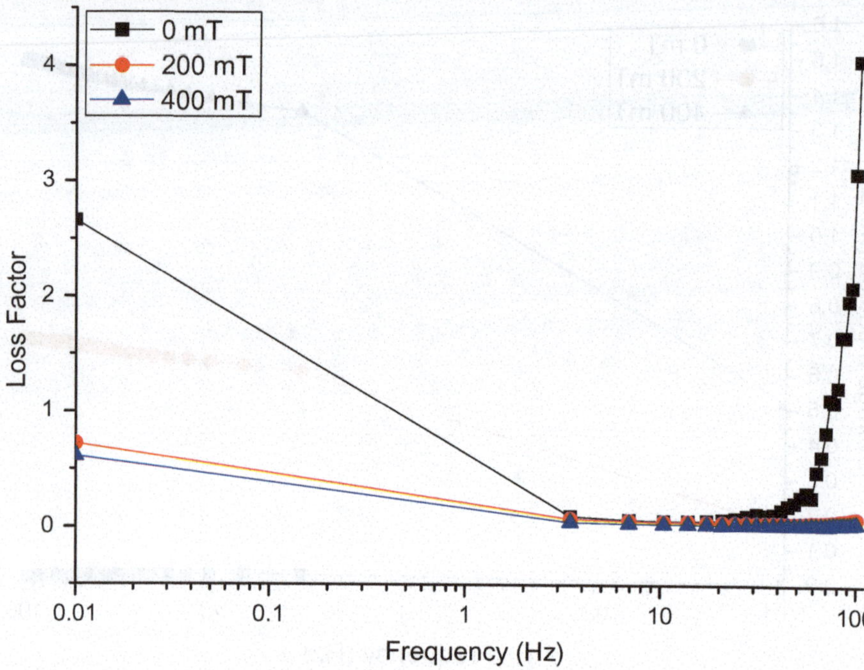

Fig. 3 Loss factor as a function of frequency

parameter to determine the performance of the MR materials and can be calculated using Eq. 1.

$$G_{MR\,effect} = \left[\frac{G'_{max} - G_0}{G_0}\right] \times 100\% \quad (1)$$

A shown in Fig. 4, it is shown that the storage modulus showed similar trends for all the tested frequencies (1, 5, 10 Hz) where they are increased along the increment of magnetic flux density. Furthermore, it is noted that the zero-field modulus, G_0 increases with increasing of the test frequency. The increment of G_0 is due to the frequency-dependent of the polymer matrix because by increasing the frequency, the motion of PVA molecular chains cannot be sustained with the external stimuli. Then, the matrix tested with higher test frequency tend to be stiffer and more rigid resulting in an increment of G_0. Additionally, with the increase of test frequency from 1 to 10 Hz, the absolute MR effect decreases from 1.765 to 1.407 MPa. The MR effect of the MR plastomer sample decreases with the increasing of test frequency as listed in Table 1.

From Table 1, the MR effect achieved by the sample with a different test frequency of 1, 5, and 10 Hz is 6793, 5049, and 3131% respectively. It shows that by increasing the test frequency from 1 to 10 Hz, the MR effect of the sample does not increase

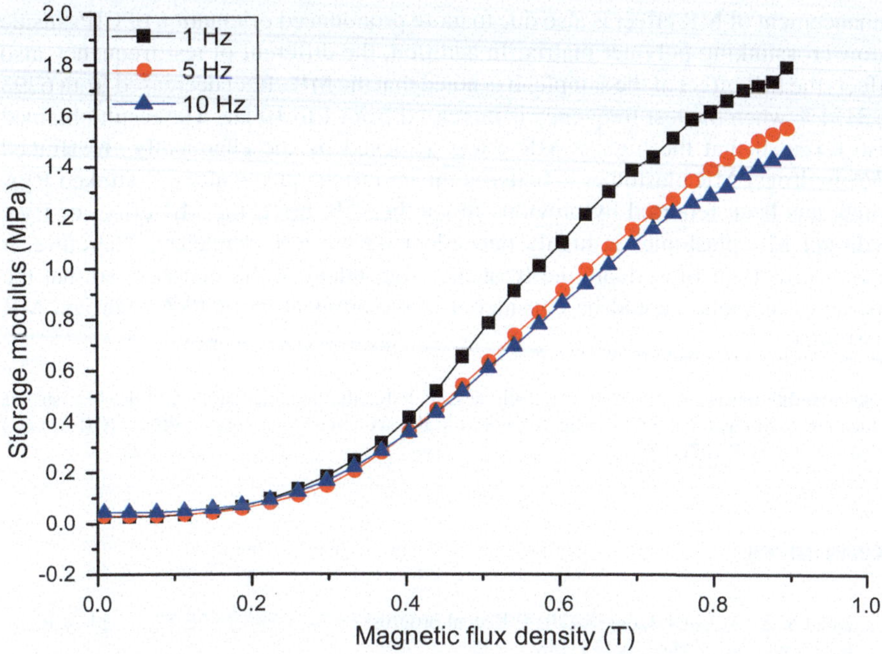

Fig. 4 The effect of test frequency on the storage modulus of MR plastomer

Table 1 The zero-field modulus and MR effect of PVA hydrogel MR plastomer

Test frequency (Hz)	Zero-field modulus (G_0/MPa)	Absolute MR effect (MPa)	Relative MR effect (%)
1	0.026	1.765	6793
5	0.030	1.524	5049
10	0.044	1.407	3131

but decreases. This is due to the breakage of polymer chains inside the matrix as it cannot keep up with the force.

4 Conclusions

The chemically cross-linked PVA hydrogel MR plastomer with 70 wt% of CIPs were prepared through a reaction between BA and PVA. From the results, it has been proven that this kind of matrix can become a good alternative as a carrier matrix for MR materials. Moreover, the chain-like structures of the CIPs inside the matrix are moveable and resulted in enhancement of MR effect. Furthermore, the enhancement of MR effect is also due to more pronounced orientation of CIPs inside a low crosslinking polymer matrix. In addition, the different of test frequency also affects the MR effect of the sample. It is noted that the MR effect decreased from 6793 to 3131% when the test frequency is increased from 1 to 10 Hz. The result obtained also revealed that the highest MR effect achieved by the chemically crosslinked PVA hydrogel MR plastomer is better compared to the physically crosslinked PVA which has been reported by previous researchers. In brief, the MR effect of PVA hydrogel MR plastomer is highly dependent on the test frequency. Therefore, it is very important to evaluate the frequency-dependent of the materials so that the operating frequency could be considered in the application of PVA hydrogel MR plastomer.

Acknowledgements The authors gratefully acknowledge the financial support of the Ministry of Education in Malaysia and Universiti Teknologi Malaysia under grant vote no PDRU (04E02) and grant vote no TDR (07G19)

References

1. Xuan S, Xu Y, Liu T, Gong X (2015) Recent progress on the magnetorheological plastomers. Int J Smart Nano Mater 6:135–148
2. Zhao W, Pang H, Gong X (2017) Novel magnetorheological plastomer filled with NdFeB particles: preparation, characterization, and magnetic-mechanic coupling properties. Ind Eng Chem Res 56:8857–8863
3. Liu T, Xu Y, Gong X, Pang H, Xuan S (2013) Magneto-induced normal stress of magnetorheological plastomer. AIP Adv 3
4. Xu Y, Gong X, Xuan S, Zhang W, Fan Y (2011) A high-performance magnetorheological material: preparation, characterization and magnetic-mechanic coupling properties. Soft Matter 7:5246–5254
5. Pang H, Xuan S, Liu T, Gong X (2015) Magnetic field dependent electro-conductivity of the graphite doped magnetorheological plastomers. Soft Matter 11:6893–6902
6. Shiga T, Kurauchi T, Okada A (1993) Electroviscoelastic effect of polymer blends consisting. Macromolecules 26:6958–6963
7. Xu J, Wang P, Pang H, Wang Y, Wu J, Xuan S et al (2018) The dynamic mechanical properties of magnetorheological plastomers under high strain rate. Compos Sci Technol 159:50–58
8. Xu Y, Gong X, Xuan S, Li X, Qin L, Jiang W (2012) Creep and recovery behaviors of magnetorheological plastomer and its magnetic-dependent properties. Soft Matter 8:8483
9. Xu J, Xuan S, Pang H, Gong X (2017) The strengthening effect of 1D carbon materials on magnetorheological plastomers: mechanical properties and conductivity. Smart Mater Struct 26:035044
10. Wu J, Gong X, Fan Y, Xia H (2011) Physically crosslinked poly(vinyl alcohol) hydrogels with magnetic field controlled modulus. Soft Matter 7:6205
11. Liu TY, Hu SH, Liu TY, Liu DM, Chen SY (2006) Magnetic-sensitive behavior of intelligent ferrogels for controlled release of drug. Langmuir 22:5974–5978
12. Baqiya MA, Taufiq A, Sunaryono M, Sari DP, Dwihapsari Y et al (2018) Development of PVA/Fe_3O_4 as smart magnetic hydrogels for biomedical applications. Hydrogels
13. Wang H, Shyr T, Hu M (1999) The elastic property of polyvinyl alcohol gel with boric acid as a crosslinking agent. J Appl Polym Sci 74:3046–3052
14. Xu Y, Gong X, Peng C, Sun Y, Jiang W, Zhang Z (2010) Shear thickening fluids based on additives with different concentrations and molecular chain lengths. Chine J Chem Phys 23:342–346

Effect of TiO_2/Ag Nanocomposite Loading on the Optical Properties of Chitosan Film

Melda Taspika, Resetiana Dwi Desiati, and Eni Sugiarti

Abstract Effect of TiO_2/Ag loading on the optical properties of chitosan film will be studied in this report. Herein, TiO_2/Ag-chitosan composite films were prepared by solution casting method in wet phase separation using potassium hydroxide (KOH) to produce free-standing films. Field emission scanning electron microscopy (FE-SEM) images show the encapsulated TiO_2/Ag particles and the agglomeration of TiO_2/Ag particles in chitosan which contributes to the absorption intensity and the band-gap energy of composite films. The increase of TiO_2/Ag loading in chitosan leads to the increase of the absorbance intensity and the decrease of optical band-gap energy from 5.2 to 4.6 eV. Additionally, the transparency of composite films decreases with the increase of dispersed TiO_2/Ag particles.

Keywords TiO_2/Ag composite · Chitosan film · Wet phase separation · Optical properties

1 Introduction

Titanium dioxide (TiO_2) has a great deal of attention due to its outstanding properties such as antibacterial activity, tunable band-gap, and optical properties [1–3]. The addition of silver (Ag) reduced the band-gap of TiO_2 which increased its photocatalytic activity [4]. Based on our previous report, calcination temperature influenced antibacterial activity of TiO_2/Ag nanocomposites depending on their structure and morphology [5]. TiO_2/Ag nanocomposite calcinated at temperature 500 °C is the best on inhibiting the growth of *S. aureus* bacterium which potentially contaminates raw meat products [6]. However, the as-produced TiO_2/Ag nanocomposite was in powder form that confines its application as photocatalyst and packaging. Therefore, a polymer as a matrix is necessary to produce a solid-state film.

Chitosan is a pseudonatural cationic polymer which is largely used in many applications because of its abundance, high elasticity, biodegradability, and antibacterial effectiveness [7–9]. There are several researchers who have reported about

M. Taspika (✉) · R. D. Desiati · E. Sugiarti
Research Center for Physics, Indonesian Institute of Sciences, South Tangerang, Indonesia
e-mail: melda1992taspika@gmail.com

U. Sabino et al. (eds.), *Proceedings of the 6th International Conference and Exhibition on Sustainable Energy and Advanced Materials*, Lecture Notes in Mechanical Engineering,
https://doi.org/10.1007/978-981-15-4481-1_30

TiO_2/Ag-chitosan composite films. Apjok et al. synthesized cellulose-chitosan-Ag/TiO_2 nanocomposite for coating with food grade cellulose paper sheet as a substrate [10, 11]. Antibacterial activity of the sheet decreases after storage for six months because of TiO_2/Ag-chitosan corrosion effect. Jbeli et al. coated a glass substrate with a thin film of chitosan following with dipped in silver and titanium solution respectively to forming a rigid chitosan-Ag–TiO_2 solid state film [12]. Natarajan et al. prepared TiO_2/Ag chitosan nanocomposite film using solution casting method on a glass plate and peeling it off in dry phase separation which leads the film being sturdy [13–15]. In this research, the influence of TiO_2/Ag loading on the optical properties of chitosan film will be reported. A various concentration of TiO_2/Ag powder nanocomposite will be dispersed in chitosan solution gel. The composite films will be prepared by solution casting method following with wet phase separation from the substrate to produce free-standing films. The properties of the as-prepared films will be studied by using FE-SEM and UV-vis.

2 Experimental

2.1 Materials

Chitosan with 80 mesh particle size was purchased from home industry in Indramayu, Indonesia. TiO_2/Ag nanocomposite powder was synthesized by sol-gel method based on our previous report [5]. Glacial acetic acid (CH_3COOH) 96% and potassium hydroxide (KOH) were analytical grade and purchased from Merck. All materials were used without further purification. Glacial acetic acid will be diluted in distilled water with 1%v/v concentration.

2.2 Preparation of TiO_2/Ag-Chitosan Composite Films

TiO_2/Ag-chitosan composite films will be prepared by solution casting method. For detail, 0.5 g chitosan was added into 50 mL acetic acid (1% v/v) while vigorously stirred for 24 h to form homogeneous solution. After that, TiO_2/Ag powder with various concentration (0, 4, and 10 wt%) was dispersed into gel solution. The mixed solution was stirred continuously for 90 h. Then, the solution would be spread on circular petri dish and dried at 50 °C for 24 h. The as-produced films would be separated from the substrate in wet phase. The dried films would be immersed in 1 M potassium hydroxide (KOH) solution for 24 h to produce free-standing films. The solidified films were dried at room temperature.

2.3 *Characterization of TiO_2/Ag-Chitosan Composite Films*

The morphological analysis was carried out using field emission scanning electron microscope (FE-SEM) (JEOL JIB 4610f) to observe the effect of TiO_2/Ag loading on the surface structure of composite films. All samples were prepared under liquid nitrogen to observe TiO_2/Ag particles in chitosan polymer. The samples were coated with gold to avoid charging under electron beam. The UV-visible absorption spectrum was recorded by a HITACHI UH5300 spectrometer to study the effect of TiO_2/Ag loading on the optical properties.

3 Results and Discussions

FE-SEM observation of the morphology of pure chitosan and various TiO_2/Ag-content in chitosan films is provided in Fig. 1 which depicts the agglomeration of TiO_2/Ag particle in chitosan polymer. The more concentration of dispersed TiO_2/Ag in chitosan film leads the more agglomeration in the composite films. Dispersed TiO_2/Ag particles were enclosed by chitosan polymer so that liquid nitrogen was used to peel the layer off.

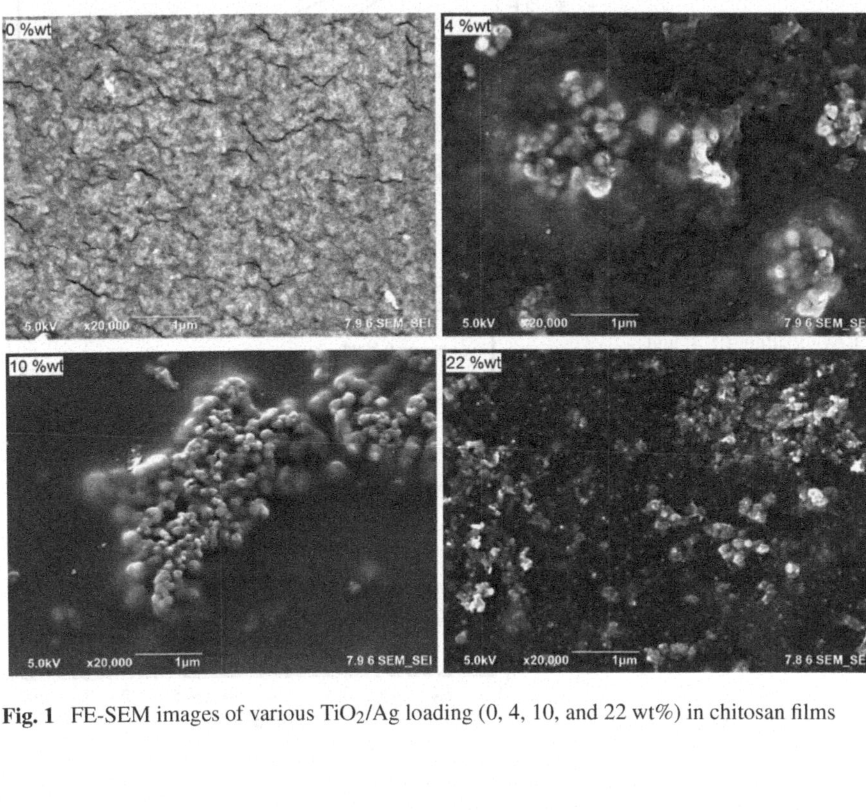

Fig. 1 FE-SEM images of various TiO_2/Ag loading (0, 4, 10, and 22 wt%) in chitosan films

The absorbance spectra of the as-prepared TiO_2/Ag-chitosan films were depicted in Fig. 2. The increase of dispersed TiO_2/Ag in chitosan polymer enhanced the absorbance intensity of the as-produced films. In the other hand, the transparency of the films decreases with the increase of dispersed TiO_2/Ag concentration. The increase of TiO_2/Ag concentration enlarged the absorbance peak of TiO_2/Ag-chitosan films. Besides, the increase of TiO_2/Ag concentration shifted the absorbance peak to the larger wavelength which is contributed to the agglomeration of TiO_2/Ag particles in the polymer [16].

The optical band-gap energy of TiO_2/Ag-chitosan films have been calculated by using Tauc relation from the absorption spectra [17]. The absorption coefficient vs photon energy of the as-produced films was shown in Fig. 3. The absorption coefficient increases as dispersed TiO_2/Ag was added to the chitosan polymer. The

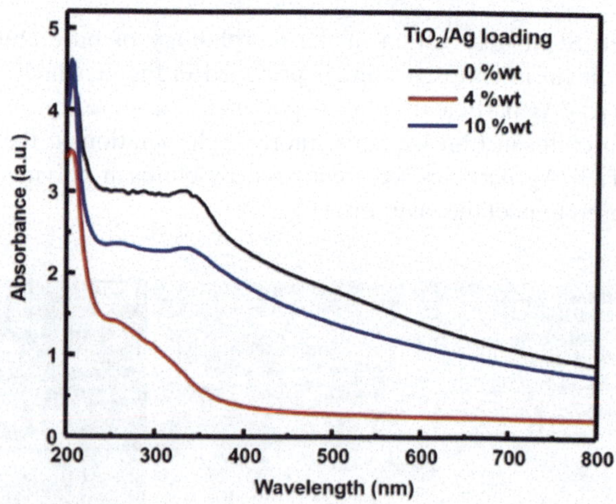

Fig. 2 Absorbance spectra of various TiO_2/Ag loading (0, 4, and 10 wt%) in chitosan films

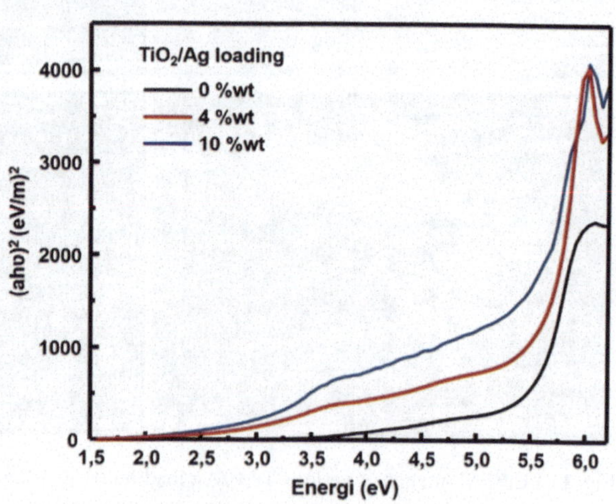

Fig. 3 Band-gap energy using Tauc' plot of various TiO_2/Ag loading (0, 4, and 10 wt%) in chitosan films

band-gap energy of pure chitosan film was found to be 5.2 eV. When TiO_2/Ag powder was added to the chitosan, the band-gap energy was found to be shifted to the lower energy value because of the interaction between TiO_2/Ag particles and chitosan. The decrease of optical band-gap was attributed to the electron-hole interaction in valence and conduction band [18]. TiO_2/Ag particles possibly modify the electronic structure of chitosan which leads the alteration of chitosan energy states [19].

4 Conclusions

Herein, the effect of TiO_2/Ag loading on the properties of chitosan film has been initially studied from FE-SEM and UV-vis results. The films were prepared by solution casting method in wet phase separation using potassium hydroxide to produce free-standing films. Field emission scanning electron microscopy (FE-SEM) images show the encapsulated TiO_2/Ag particles and the agglomeration of TiO_2/Ag particles in chitosan which contributes to the absorbance intensity and the band-gap energy of composite films. The transparency of composite films decreases with the increase of dispersed TiO_2/Ag particles. The more TiO_2/Ag concentration in the polymer increases the absorbance intensity of the as-produced composite films. The increase of TiO_2/Ag loading in chitosan leads to the decrease of optical band-gap energy from 5.2 to 4.6 eV.

References

1. Dette C et al (2014) TiO_2 anatase with a bandgap in the visible region. Nano Lett 14(11):6533–6538
2. Zimbone M et al (2015) Photocatalytical and antibacterial activity of TiO_2 nanoparticles obtained by laser ablation in water. Appl Catal B 165:487–494
3. Evtushenko YM et al (2015) Optical properties of TiO_2 thin films. Phys Proc 73:100–107
4. Harikishore M et al (2014) Effect of Ag doping on antibacterial and photocatalytic activity of nanocrystalline TiO_2. Proc Mater Sci 6:557–566
5. Desiati RD, Taspika M, Sugiarti E (2019) Effect of calcination temperature on the antibacterial activity of TiO_2/Ag nanocomposite. Mater Res Express
6. Wu S et al (2018) *Staphylococcus aureus* isolated from retail meat and meat products in China: incidence, antibiotic resistance and genetic diversity. Front Microbiol 9:2767
7. Rinaudo M (2006) Chitin and chitosan: properties and applications. Prog Polym Sci 31(7):603–632
8. Pang Y et al (2017) Biodegradable and biocompatible high elastic chitosan scaffold is cell-friendly both in vitro and in vivo. Oncotarget 8(22):35583
9. Goy RC, Morais STB, Assis OBG (2016) Evaluation of the antimicrobial activity of chitosan and its quaternized derivative on *E. coli* and *S. aureus* growth. Revista Brasileira de Farmacognosia 26(1):122–127
10. Apjok R et al (2019) Active packaging based on cellulose-chitosan-Ag/TiO_2 nanocomposite for storage of clarified butter. Cellulose 26(3):1923–1946
11. Cozmuta AM et al (2018) Active papers coated with chitosan and containing TiO_2 and Ag/TiO_2 nanoparticles for increasing the shelf-life of walnut kernels. Cellulose 25(9):5205–5225

12. Jbeli A et al (2018) Chitosan-Ag–TiO_2 films: an effective photocatalyst under visible light. Carbohyd Polym 199:31–40
13. Natarajan S et al (2016) Antibacterial and antifouling activities of chitosan/TiO_2/Ag NPs nanocomposite films against packaged drinking water bacterial isolates. Environ Sci Pollut Res 23(19):19529–19540
14. Natarajan S et al (2017) Antifouling activities of pristine and nanocomposite chitosan/TiO_2/Ag films against freshwater algae. RSC Adv 7(44):27645–27655
15. Natarajan S et al (2018) Antifouling and anti-algal effects of chitosan nanocomposite (TiO_2/Ag) and pristine (TiO_2 and Ag) films on marine microalgae *Dunaliella salina*. J Environ Chem Eng 6(6):6870–6880
16. Vaseeharan B, Sivakamavalli J, Thaya R (2015) Synthesis and characterization of chitosan-ZnO composite and its antibiofilm activity against aquatic bacteria. J Compos Mater 49(2):177–184
17. Saleviter S et al (2018) Optical and structural characterization of immobilized 4-(2-pyridylazo) resorcinol in chitosan-graphene oxide composite thin film and its potential for Co_2+ sensing using surface plasmon resonance technique. Results Phys 11:118–122
18. Pushparekha et al (2019) Design, fabrication and studies on optical properties of new hybrid chitosan films doped with 1,3,4-oxadiazole derivatives for down conversion and photoluminescence applications. Opt Mater 89:80–91
19. Abdullah OG et al (2015) Reducing the optical band gap of polyvinyl alcohol (PVA) based nanocomposite. J Mater Sci: Mater Electron 26(7):5303–5309

Effect of Sea Sand Content on Hardness of Novel Aluminium Metal Matrix Composite AA6061/Sea Sand

Hammar Ilham Akbar, Eko Surojo, and Dody Ariawan

Abstract Sea sand containing ceramics compounds can be used as raw material for lightweight metal matrix composite material. A new composite AA6061/Sea sand with stir casting manufactured has been developed yet. Process parameters as particle treatment, stirring speed, stirring time, and the fraction of particles should be studied to high hardness of material. The effect of the fraction of sea sand particles on homogeneity and hardness is studied in this paper. The fraction 2–6 %wt of sea sand particles are used for manufactured the composite. The result showed that hardness is affected by density, porosity and dispersed particles higher dispersed particles resulted in higher hardness of the material.

Keywords Sea sand · Metal matrix composite · Hardness

1 Introduction

Demand for aluminium composites in various industries has increased in recent years [1]. This is due to the widespread AMC commercialization, such as the automotive, aerospace, marine and defence [2]. In the automotive industry, aluminium composites are used as a component of the braking system. AMC is very suitable for use in the automotive field because of its excellent weight and strength ratio [3], lightweight material that can support the fuel saving program [4].

AMC manufacturing generally uses ceramic particles such as Al_2O_3, SiO_2, SiC, TiO_2, graphite as an reinforcement in AMC [5]. The use of ceramic particles results in higher mechanical properties, but has expensive production costs [6]. The researchers developed AMC composites with reinforcement from waste and local natural resources [3] such as aloe vera powder [7], bamboo leaf ash [8], quarry dust

H. I. Akbar (✉) · E. Surojo (✉) · D. Ariawan
Department of Mechanical Engineering, Faculty of Engineering, Universitas Sebelas Maret, Surakarta, Indonesia
e-mail: hammarilham@gmail.com

E. Surojo
e-mail: esurojo@ft.uns.ac.id

U. Sabino et al. (eds.), *Proceedings of the 6th International Conference and Exhibition on Sustainable Energy and Advanced Materials*, Lecture Notes in Mechanical Engineering,
https://doi.org/10.1007/978-981-15-4481-1_31

[2], this proposed to reduce production costs while maintaining its mechanical properties. Indonesia as an archipelagic country has outstanding natural resources in the marine field, one of which is sea sand. Oxide minerals such as SiO_2, Al_2O_3, Fe_2O_3, TiO_2 which are mostly contained in sea sand [5, 9]. It allows sea sand to be used as high-tech materials such as reinforcing materials on metal matrix composites. The use of sea sand particles as a reinforcing material on metal matrix composites has never been studied before.

Metal matrix composite production processes are generally carried out in several methods including: solid phase processing, liquid state processing and infiltration technique [10]. The stir casting method is the simplest method and has a large production capacity compared to other methods [11]. The stir casting process is carried out by melting the matrix to melt and then stirring, this is done to form a vortex flow so that the reinforcing particles are evenly mixed throughout the liquid metal [12]. Behind the simplicity of the stir casting method, this method also has a major problem which is homogeneity.

The homogeneity of the composite material is affected by several factors, such as the reinforcing particle fraction, the stirring speed, the particle treatment [5]. The number of fractions of the reinforcement affects the mechanical properties of the metal matrix composites. Previous studies have shown that the influence of mineral oxide reinforcing fractions such as Al_2O_3, SiC, and TiC influences the mechanical properties of AMCs [13–16].

This research focuses on the influence of the fraction of sea sand on the hardness of aluminium matrix composites. The application of sea sand particles is a way to substitute conventional reinforcement such as Al_2O_3 and SiC with natural resource materials to reduce production cost.

2 Material and Method

This research Aluminium 6061 and sea sand particles were used as matrix and reinforcement, respectively. The sea sand is taken from Samas in Special Region of Yogyakarta, Indonesia. The sea sand has ball milled to obtain 200 mesh sizes as shown in Fig. 1. The average dimension of sea sand particles is shown in Fig. 2.

The stir casting equipment is shown by Fig. 3. Stirring liquid metal was carried out with 4 impeller blades 45° coated by TiO_2 as shown in Fig. 4. Stirring temperature was carried out at 720–750 °C for 5 min with stirring speed 400 rpm. Mg powder with 1% wt was added to liquid metal as the wettability agent. After stirring, the molten metal was poured into a permanent mould which has been heated at 500 °C. Hardness test was carried out using Brinell Hardness Testing with 2.5 mm steel ball indenter and 62.5 kg load based on ASTM E-10. Density and porosity was conducted by the Archimedes method.

Fig. 1 Sea sand after ball mill

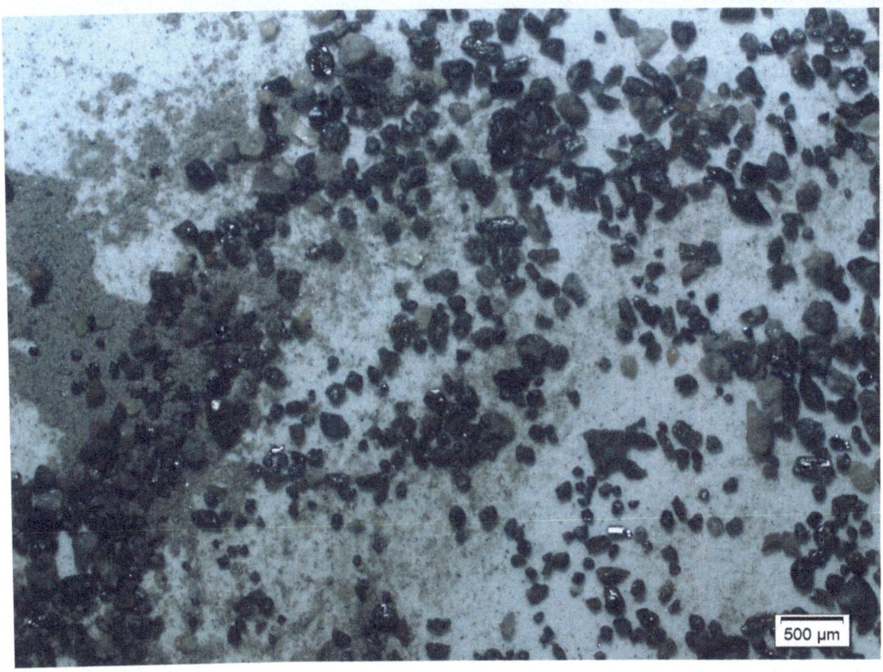

Fig. 2 Sea sand particles size

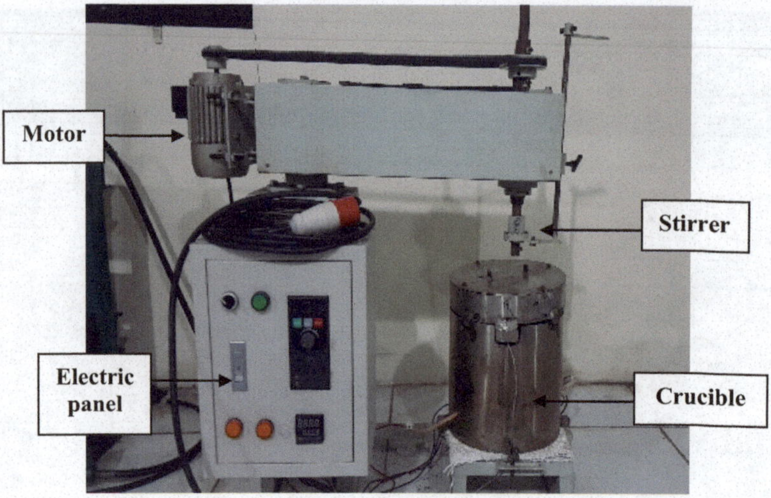

Fig. 3 Stir casting equipment

3 Result and Discussion

XRD results showed the sea sand contains Magnesioferrite ($MgFe_2O_4$) compound as shown in Fig. 5. Magnesioferrite is a compound mainly formed by MgO and Fe_2O_3 [17]. The hardness of Magnesioferrite compound is 800–900 kg/mm^2 [17], therefore sea sand can be used as a reinforcement at AMC. Sea sand reinforcement can reduce the cost of AMC production rather than using conventional reinforcement such as Al_2O_3 and SiC. The physical properties of the reinforcements are shown in Table 1.

The stir casting process showed that the sea sand particles can be mixed with molten aluminium. The dispersed and undispersed particle by molten aluminium is shown in Fig. 6. The higher content of particles resulted higher undispersed particles. The particles that undispersed tend to be constant at 12–14%. Higher undispersed particle is caused by size the particles. Particles with size more than 200 μm tends to sink in the bottom of crucible [20].

Figure 7 shows the density of composite. The addition of 2 %wt sea sand particles produced higher density at 2.69 g/cm^3. Meanwhile, addition of 4 and 6 %wt particle resulted stable density at 2.67 g/cm^3. Composite density decreased with the increasing weight of sea sand particles, this is due to lower sea sand density which is 2.5 g/cm^3 [19].

Figure 8 presents the effect of addition sea sand particles on porosity. It shows the porosity rise with increasing the sea sand particles fraction. It occur caused larger particle size tend not to be dispersed if added in higher content [20] and higher content of particles tends to increasing opportunity the porosity [21]. The lowest porosity of the composite was obtained in 2 %wt of sea sand. Higher porosity is shown in 4 and 6 %wt of sea sand. The size and number of particles affected the composite density.

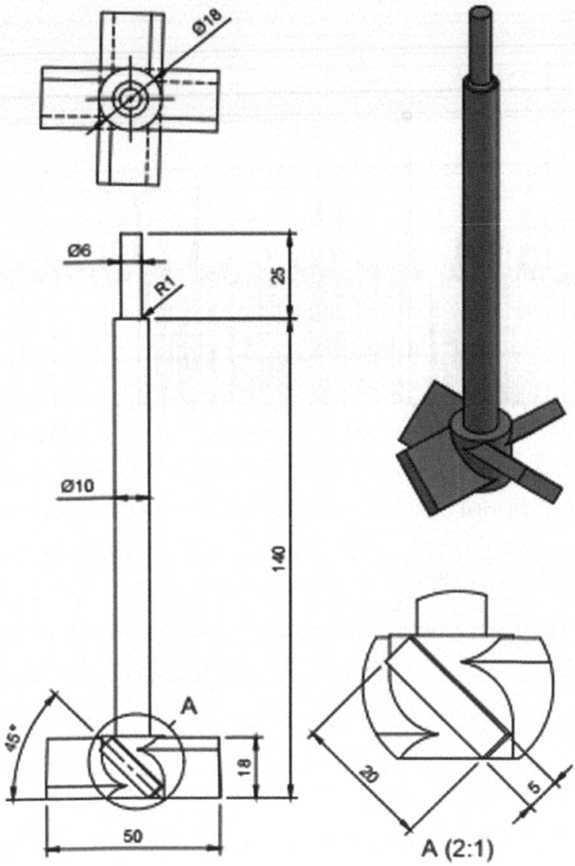

Fig. 4 Four blade impeller (mm)

Higher content particle increased the possibility of agglomeration, which results in a decrease in density.

The hardness of composite is shown in Fig. 9. The highest hardness was obtained in addition of 6 %wt of sea sand particles. While the lower hardness was found in the 2 %wt sea sand addition. The composite hardness increased with increasing weight of the sea sand particles. That phenomenon was due to the fact that sea sand contained ceramics particle such as Fe_2O_3, and MgO which has higher hardness than aluminium [11]. Hardness improve with increasing particles content caused the higher particles that dispersed in aluminium matrix, in the other hand the levels of porosity also rising, due to more surface contact between molten aluminium and particles [21].

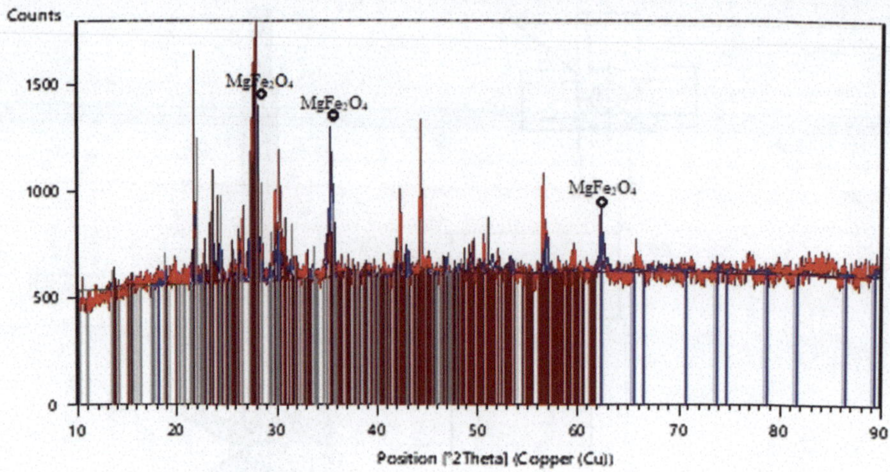

Fig. 5 XRD result of sea sand

Table 1 Properties of reinforcement [17–19]

Properties	Sea sand	SiC	Al_2O_3
Density (g/cm^3)	2.5	3.21	3.87
Young modulus (GPa)	124.28	200–480	380
Hardness (Mohs)	6–6.5	9.7	9.0

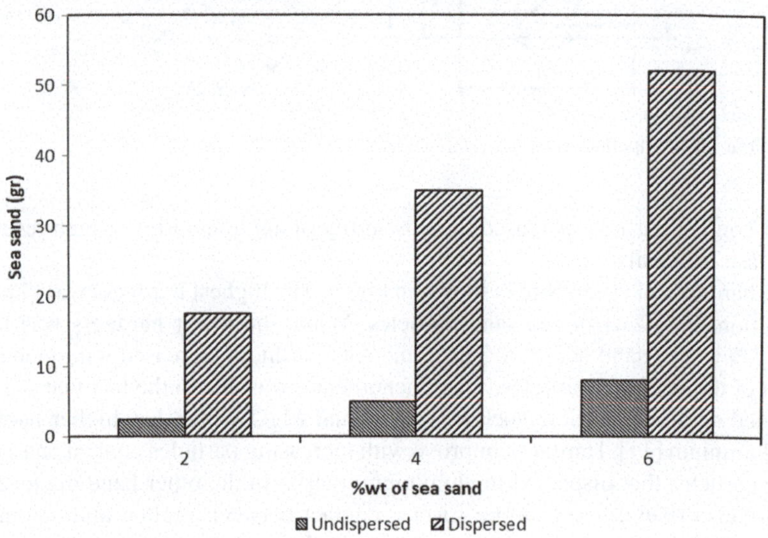

Fig. 6 Dispersion of sea sand particles

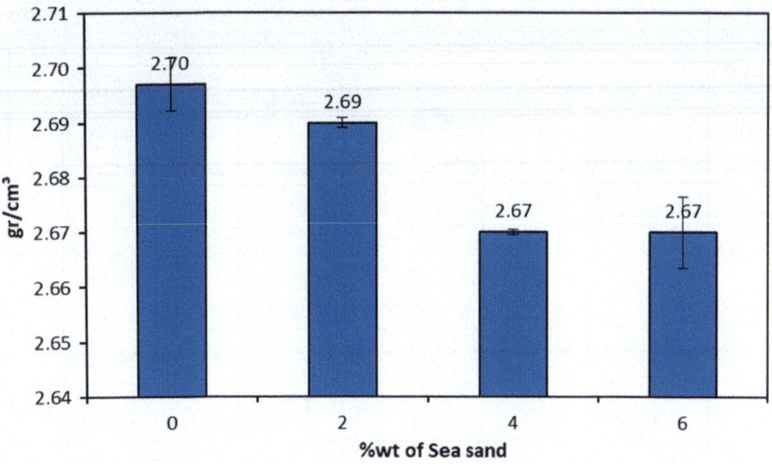

Fig. 7 Density of composite

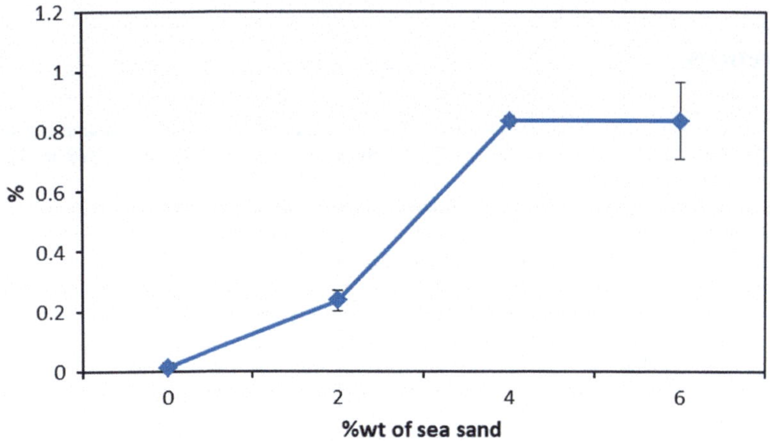

Fig. 8 Porosity of composite

4 Conclusion

From the experiment showed sea sand particles can be dispersed into aluminum liquid. The sea sand particles content affected the mechanical and physical properties of composite. Composite density decreased, porosity increased and increased hardness with increasing sea sand fraction.

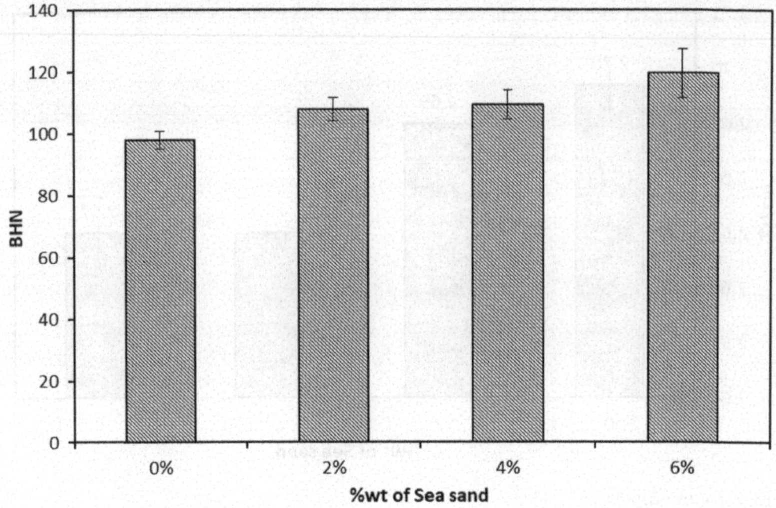

Fig. 9 Hardness of composite

References

1. Akbar HI, Surojo E, Ariawan D (2018) Effect of quenching agent on dimension stability of Al 6061-Al_2O_3 composite. In: MATEC web conference, vol 159. https://doi.org/10.1051/matecconf/201815902047
2. Alaneme KK, Bamike BJ (2018) Characterization of mechanical and wear properties of aluminium based composites reinforced with quarry dust and silicon carbide. Ain Shams Eng J 9:2815–2821. https://doi.org/10.1016/j.asej.2017.10.009
3. Thirumoorthy A, Arjunan TV, Senthil Kumar KL (2018) Latest research development in aluminum matrix with particulate reinforcement composites—A review. Mater Today Proc 5:1657–1665. https://doi.org/10.1016/j.matpr.2017.11.260
4. Chandra B, Singh H (2017) ScienceDirect fabrication and characterisation of Al_2O_3/aluminium alloy 6061 composites fabricated by Stir casting. Mater Today Proc 4:2783–2792. https://doi.org/10.1016/j.matpr.2017.02.157
5. Bisane MMP, Sable YS, Sonawane D (2015) Recent development and challenges in processing of ceramic reinforced Al matrix composite through stir casting: a review. Int J Eng Res Appl Sci 21:11–16
6. Ravindran S, Mani N, Balaji S, Abhijith M, Surendaran K (2019) Mechanical behaviour of aluminium hybrid metal matrix composites—A review. Mater Today Proc 16:1020–1033. https://doi.org/10.1016/j.matpr.2019.05.191
7. Hima Gireesh C, Durga Prasad KG, Ramji K, Vinay PV (2018) Mechanical characterization of aluminium metal matrix composite reinforced with Aloe vera powder. Mater Today Proc 5:3289–3297. https://doi.org/10.1016/j.matpr.2017.11.571
8. Kumar BP, Birru AK (2017) Microstructure and mechanical properties of aluminium metal matrix composites with addition of bamboo leaf ash by stir casting method. Trans Nonferrous Met Soc China (English Ed) 27:2555–2572. https://doi.org/10.1016/s1003-6326(17)60284-x
9. Alimi M, Putri SE (2016) Analisis Kandungan Mineral Pasir Pantai Losari Kota Makassar Menggunakan Xrf Dan Xrd. J Chem 17:19–23 in Indonesian
10. Singh J, Chauhan A (2017) Fabrication characteristics and tensile strength of novel Al_2024/SiC/red mud composites processed via stir casting route. Trans Nonferrous Met Soc China (English Ed) 27:2573–2586. https://doi.org/10.1016/s1003-6326(17)60285-1

11. Zhou W, Xu ZM (1997) Casting of SiC reinforced metal matrix composites. J Mater Process Technol 63:358–363. https://doi.org/10.1016/S0924-0136(96)02647-7
12. Annigeri UK, Veeresh Kumar GB (2017) Method of stir casting of aluminum metal matrix composites: a review. Mater Today Proc 4:1140–1146. https://doi.org/10.1016/j.matpr.2017.01.130
13. Afkham Y, Khosroshahi RA, Rahimpour S, Aavani C, Brabazon D, Mousavian RT (2018) Enhanced mechanical properties of in situ aluminium matrix composites reinforced by alumina nanoparticles. Arch Civ Mech Eng 18:215–226. https://doi.org/10.1016/j.acme.2017.06.011
14. Saenpong P, Talangkun S, Laonapakul T, Boonma A (2018) Microstructures and hardness of A356-SiC composites produced by the mechanical stir casting. Mater Today Proc 5:9489–9496. https://doi.org/10.1016/j.matpr.2017.10.129
15. Gopalakrishnan S, Murugan N (2012) Production and wear characterisation of AA 6061 matrix titanium carbide particulate reinforced composite by enhanced stir casting method. Compos Part B Eng 43:302–308. https://doi.org/10.1016/j.compositesb.2011.08.049
16. Dhaneswara D, Syahrial AZ, Ayman MT (2017) Mechanical properties of nano SiC-reinforced aluminum A356 with Sr modifier fabricated by stir casting method. Procedia Eng 216:43–50. https://doi.org/10.1016/j.proeng.2018.02.087
17. Zhao YM, Zhang YN, Bi CS, Guo LH (1998) The discovery of magnesioferrite from Au (Fe, Cu) magnesian Skarn deposits and study of the magnesioferrite- magnesiomagnetite series. Acta Geol Sin Ed 72:382–391
18. Kainer KU, Basics of metal matrix composites. https://doi.org/10.1002/3527608117.ch1
19. Hilman PM, Supraprto SJ, Sunuhadi DN, Tampubolon A, Wahyuningsih R, Widhyatna D, Pardiarto B, Gunradi R, Franklin F, Yudawinata K, Sutisna DT, Dinarsih D, Sukaesih S, Yuningsih ET, Candra C, Oktaviani P, Rahmawati R, Ulfa RM, Sukmayana I, Ostman I: Pasir Besi di Indonesia, Geologi, Eksplorasi dan Pemanfaatannya. Pusat Sumber Daya Geologi, Bandung (2014) in Indonesian
20. Bihari B, Singh AK (2017) An overview on different processing parameters in particulate reinforced metal matrix composite fabricated by stir casting process. Int J Eng Res Appl 7:42–48. https://doi.org/10.9790/9622-0701034248
21. Garg P, Jamwal A, Kumar D, Sadasivuni KK, Hussain CM, Gupta P (2019) Advance research progresses in aluminium matrix composites: manufacturing & applications. J Mater Res Technol 8:4924–4939. https://doi.org/10.1016/j.jmrt.2019.06.028

Energy Saving Investigation on Undesignated Campus Mosques

Bangun I. R. Harsritanto, Satrio Nugroho, Gentina Pratama Putra, and Aditya Rio Prabowo

Abstract Mosque is a unique building by the function as Moslem prayer room. Indonesia as the world biggest moslem country has mandated each buildings to provide prayer room especially mosque by issuance of Ministry of Public Work-Housing Settlement decree no. 14/2017. Academic building like campus also mandated to have the campus mosque to facilitate the Moslem students or lecturers in performing their prayer. However the campus buildings commonly were not designed with mosque facility in purpose. Thus campus designer performed remodeling their building to meet this demand. The importance of thermal comfort in mosque to facilitate worshippers perform their prayer in profane room with minimum energy requirement, make energy saving investigation is critical for room performance, This study purpose is to assess and compare the energy saving over time on several type of undesignated campus mosques as a step towards comprehensive study about mosque and energy saving strategies. Literature study, Mosque "energy audit" evaluation on site observation, and edge building simulation were performed to investigate the energy saving. The study presents the analysis of several undesignated campus mosques type in tropical region's performance toward energy saving effort.

Keywords Energy saving · Building physics · Undesignated campus mosque · Edge buildings

1 Introduction

Mosques are unique building which having various frequency rates at different days and hours for Moslems gathering, resting and especially praying. The design and

---

B. I. R. Harsritanto (✉) · S. Nugroho
Universitas Diponegoro, Semarang, Indonesia
e-mail: Bangunirh@arsitektur.undip.ac.id

G. P. Putra
Universitas Mercu Buana, Jakarta, Indonesia

A. R. Prabowo
Universitas Sebelas Maret, Surakarta, Indonesia

U. Sabino et al. (eds.), *Proceedings of the 6th International Conference and Exhibition on Sustainable Energy and Advanced Materials*, Lecture Notes in Mechanical Engineering,
https://doi.org/10.1007/978-981-15-4481-1_32

features of the mosques were varied by contextual site, cultural influence and geographical factors [1]. Mosques were simply designed to have prayer space scheme where the long edge is directing to qibla (the bait of Allah on Mecca) and ablution room to support the prayer [2]. Nowadays mosques has equipped with acoustic comfort by audio systems, visual comfort via lighting-visual arts and thermal comfort by air conditioning systems. Recently, many researches tried to improve thermal comfort and energy saving of various building types such as: office, mall, and not least mosque [3–7]. The energy should efficiently use to maintain the proper comfort area for the mosque users. In the literature there are limited number of studies about mosque and rarely also about the campus mosque which not designated to be mosque.

Almost a quarter of the world's people are Moslem. The Moslem community has built many mosques throughout the earth. Mosque is integral part of all Moslem community in the world and act as identity in Moslem majority countries as well as any other country with Moslem population [8]. Indonesia as the world biggest moslem country has mandated each buildings to provide prayer room especially mosque by issuance of Ministry of Public Work-Housing Settlement decree no. 14/2017. The decree also strictly stated that mushola must installs: clear direction to Qibla, distinctive separation between profane and common room, and proper thermal and lighting comfort [9].

Campus in Indonesia usually has central mosque. However the demand to pray right on prayer time brought every buildings to provide a space to perform prayer (undesignated mosque). The condition of undesignated mosque on campus usually fully equipped with ablution area and prayer area. Thus campus designer performed remodeling their building to meet this demand. The importance of thermal comfort in mosque to facilitate worshippers perform their prayer in profane room with minimum energy requirement, make energy saving investigation is critical for room performance.

Some Studies shown comparison of buildings of the same area and mosques have a higher energy usage and also higher energy expenditure for cooling purposes [10]. Moreover, there is fact that wall and roof thermal treatment and preventing air leakage, a maximum 25% of energy savings can be achieved [11]. There also methods for cooler (refrigerant) is not necessary risen the thermal comfort of the mosque users and thus the consumed energy proven to waste [12]. Through a comparison of several studies, occupancy zoning or building insulation is provided, more acceptable thermal comfort level will be achieved. The same time, the expenses reduced unto half of other mosques through less energy usage. Some studies also conclude that the existing scales for measuring thermal comfort levels were varied for the building typology of mosques [13].

This study purpose is to assess and compare the energy saving over time on several types of undesignated campus mosques as a step towards comprehensive study about mosque and energy saving strategies. Literature study, Mosque "energy audit" evaluation on site observation, and edge building simulation were performed to investigate the energy saving.

Fig. 1 Condition of Mosque in Pukyong National University, Busan, South Korea

2 Materials and Methods

2.1 *South Korea Case*

Pukyong National University is one of the campus in South Korea which having a mosque inside the education area. The condition of big number of Moslem students brought a mosque in 2014 and moved it into bigger room of 30 worshippers capacities in 2017 (see Fig. 1).

This mosque was underground bunker before being renovated to be class and now mosque, made this mosque is sample of undesignated mosque.

2.2 *Indonesia Case*

Architecture campus of Universitas Diponegoro is having three undesignated mosque in the building A, C and D as the result of majority academia are moslem that need to pray at least dzuhr and asr at campus and maybe maghrib, isya and shubh on campus off schedules. The room dimensions were varied according to the undesignated mosque conditions. Later on the conditions would be explained on discussion sections.

The condition of both countries is undesignated mosque on campus with ablution area and prayer area. Thus campus designer performed remodeling their building to meet these minimum requirements of mosque.

2.3 Literature Study

This method was included on introduction section as background of this scientific writing to set the research basis. Some research of refrigerant, heater which closely related to energy consumption to support thermal comfort inside mosque has been studied. Furthermore the energy consumption calculated by edge building simulation to predict the energy saving score.

2.4 Site Observation

This method was performed to gain data's from the mosques in Pukyong National University and Architecture campus of Undip, especially: dimension and energy consuming features. The measurements and documentation were taken using digital devices to make high precision.

2.5 Energy Evaluation Using Edge Building Simulation

Edge building simulation is an online building simulation by IFC sponsored by World Bank. The physical data obtained on site observation in this phase being filled on edgebuilding.com application to be calculated the energy saving prediction. The score of energy saving and water saving later on being reported as this study results.

3 Result and Discussion

3.1 South Korea Case

The condition of Al Pukyong in Daeyeon campus can be seen in Figs. 1 and 2. The capacities of 30 prayers taken by the direct counting while preparing communal jumat prayer. During the counting, more than 50 worshippers were came and performed their prayer until the ally between the mosque and ablution area. The dimension of 12 m * 5 m is not enough to fit the number above 30 users.

The features of centralized Air Conditioner with COP 3.5, undertile heater (ondol) and LED bulb were using the electricity as main power. Underground location of the mosque brought thermal comfort problems on the passive design analysis directly. However the energy saving simulation may show the other calculations.

The edge building simulation being input the data of mosque dimension with AC central (COP 3.5) bring score in energy saving: 5.2%, and undertile heater shown

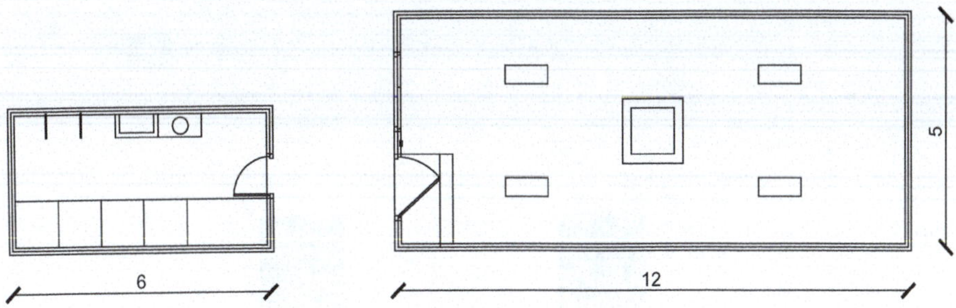

Fig. 2 Mosque-ablution plan in Computer Science Bld, Pukyong National University

0.02% of energy saving while application of LED bulb brought 0.21% saving. The total energy saving of this undesignated mosque is 5.44% (see Fig. 3).

The next phase of edge building simulation shown that water consumption on mosque Al Pukyong is 1.24% compared to the edge standard by using the low flow faucet 6L. This water devices consumed around 246 kL/unit/year according to the simulation result (see Fig. 4).

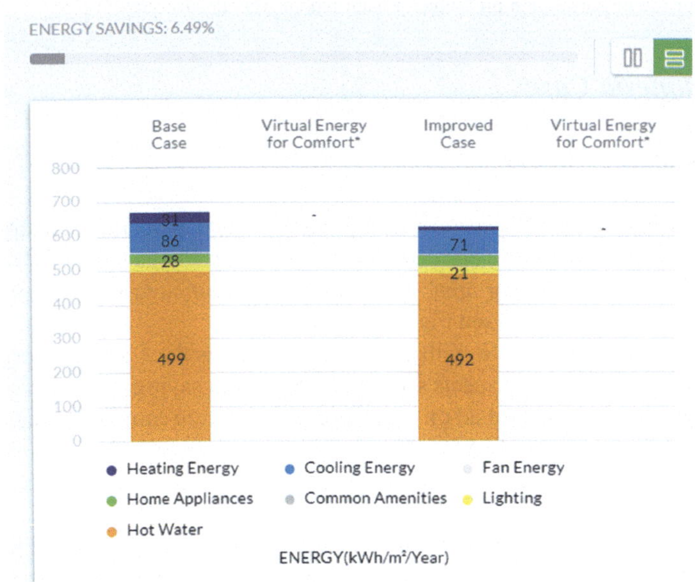

Fig. 3 Energy saving simulation on Pukyong National University Mosque

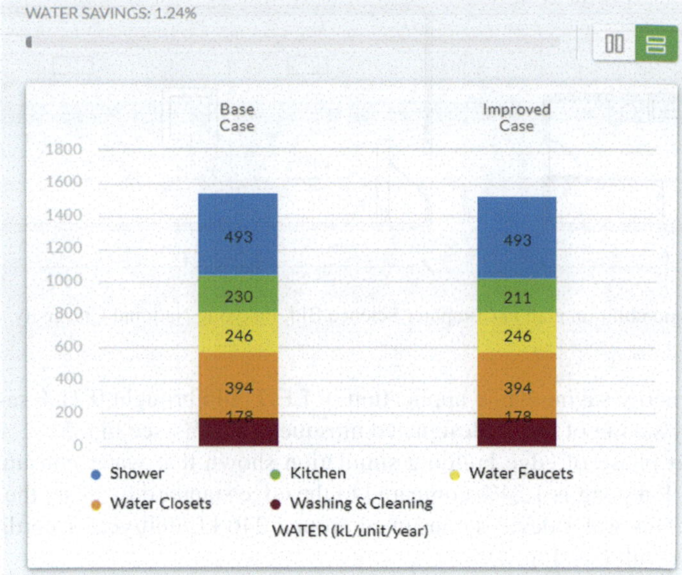

Fig. 4 Water saving simulation on Pukyong National University Mosque

3.2 *Indonesia Case*

3.2.1 Building a Mosque

Mosque in Building A Architecture Undip is functioned as lecturer mosque so that only this mosque equipped by air conditioner type split package with 0.5 PK installed on this 7.2 m * 2.1 m prayer area. However the ablution area is only 1.9 m * 1 m with three low flow faucets (see Fig. 5).

The installment of AC split with COP 3.5 brought result of 3.59% energy saving (see Fig. 6), while low flow faucets scored 11.85% water saving (see Fig. 7). Later on the low flow faucets also applied on other three architecture campus so that the water saving for each places were 11.85%. The situation of mosque that located inside room with no outside windows and ventilation worsen the energy consumption compared to the earlier Pukyong mosque.

This mosque was a lecturers/professors room that modified to meet the prayer demand amongst lecturers. The architecture lecturers itself reach 38 persons so that couldn't fit to perform prayer (salat) at the same time in this undesignated mosque. The maximum capacity is 9 worshippers with composition of 3 rows and 3 columns include imam (leader) who will take about 10 cm more forward than the follower (makmum).

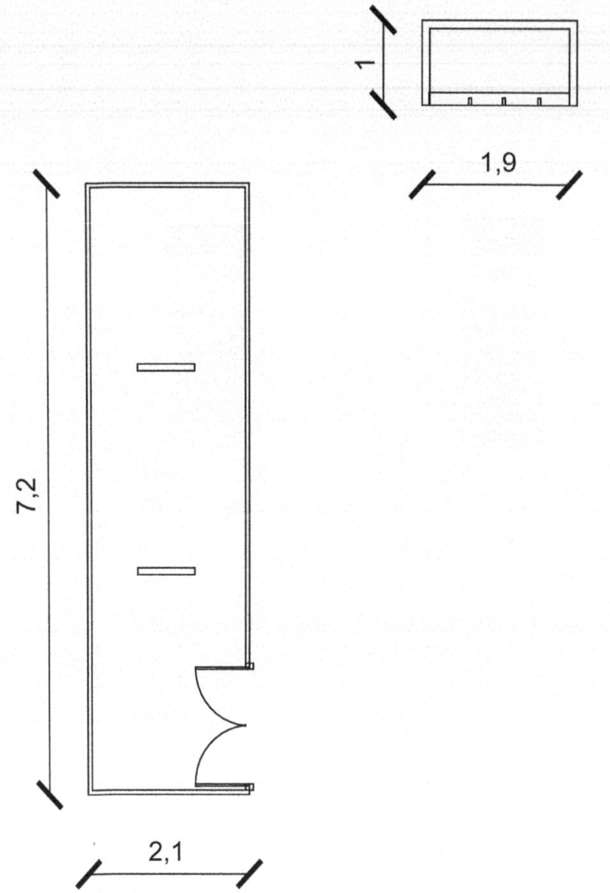

Fig. 5 Mosque-ablution plan in Building A Architecture Undip

3.2.2 Building C Mosque

Mosque in Building C Architecture Undip is located in 2nd floor of the building. This mosque is equipped by stand-wall fans and window panel installed on this 6 m * 3 m prayer area. However the ablution area is only 1 m * 1 m with three low flow faucets inside 2 m * 3.9 m lavatory (see Fig. 8). This mosque only fit for 12 caused by the 3 m of width which allow more person in a row rather than mosque in building A (Fig. 9).

The energy saving simulation result of ceiling fans installment is 7.59% and natural ventilation (windows) is 5.79%. So that the total energy saving was 9.21% with energy consumption of fans 14 kWH/m^2/year and lighting 8 kWH/m^2/year. Same with previous building A, the water saving with three low flow faucet resulted 11. 85% score (see Fig. 7).

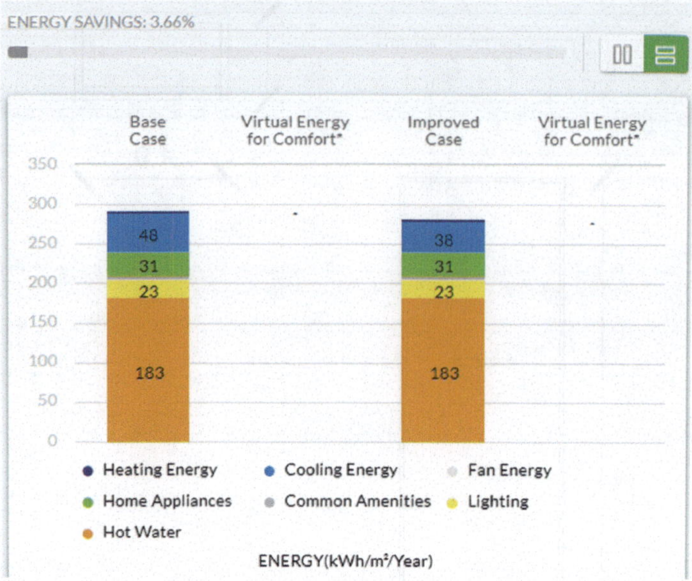

Fig. 6 Energy saving simulation on Building A Architecture Undip

Fig. 7 Water saving simulation on Building A, C and D Architecture Undip

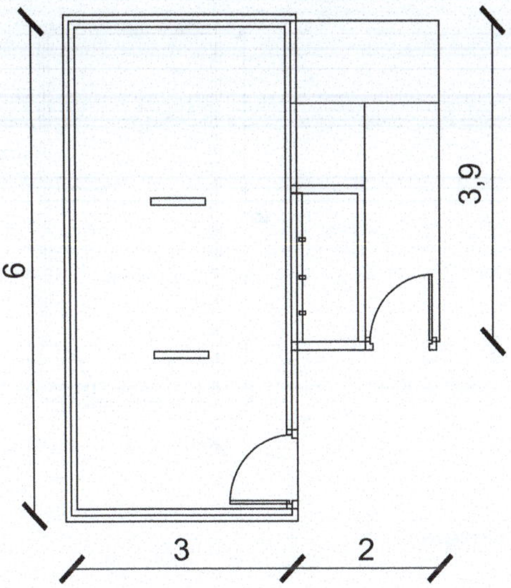

Fig. 8 Mosque-ablution plan in Building C Architecture Undip

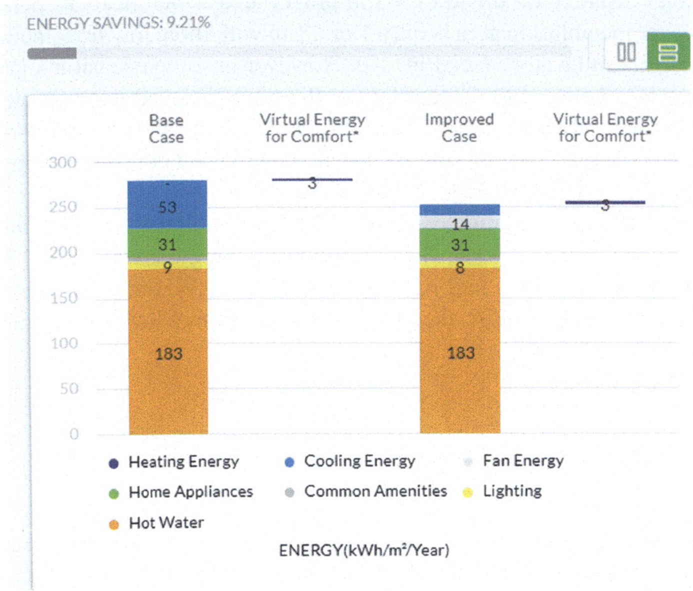

Fig. 9 Energy saving simulation in Building C Architecture Undip

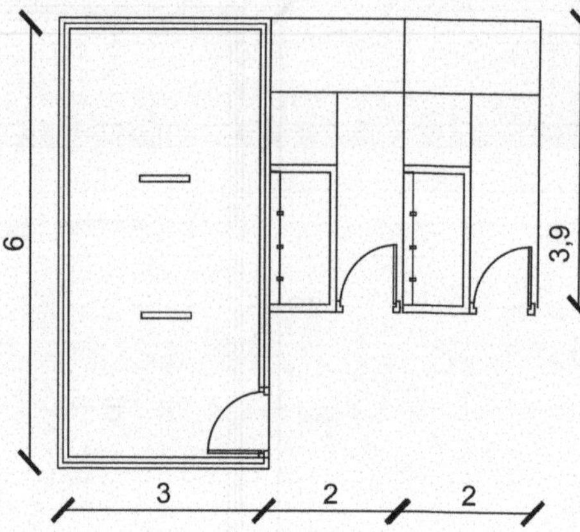

Fig. 10 Mosque-ablution plan in Building D Architecture Undip

3.2.3 Building D Mosque

The final mosque in Building D Architecture Undip is located in 1st floor of the post graduate building (plus vocational school). This mosque is installed a stand fan and window panel installed on this 6 m * 3 m prayer area which fit to 12 persons on 3 rows. However the ablution area is only 1 m * 1 m with three low flow faucets inside two of 2 m * 3.9 m lavatory (see Fig. 10). However in some occasion a lavatory on this building was being closed according to the plumbing and door problems. This mosque using natural ventilation as natural lighting source and air circulators so that can reduce the energy consumption for thermal and visual comforts during campus working hours of 7AM-5PM from Monday to Friday.

The energy saving simulation result of stand fan installment is 7.79% and natural ventilation (windows) is 7.73%. So that the total energy saving was 9.38% with energy consumption of fans and lighting of 6 and 37 kWH/m^2/year (see Fig. 11). Same with previous building A, the water saving with three low flow faucet resulted 11. 85% score (see Fig. 7).

3.3 The Comparison Studies

The earlier discussion of each mosque cases on Pukyong National University and Architecture campus of Universitas Diponegoro can be resumed and compared into Table 1. The AC was attached on PKNU and building A, but wall/stand fan was installed non-permanent on building C and D. The windows exist on building C and D but not in building A (as result of space inside a room) and Al Pukyong (as result of underground space). The lowest energy saving is building A caused by the contextual location and the highest is building D with application of fan on the small room rather than AC and position which having windows on a side of it. However

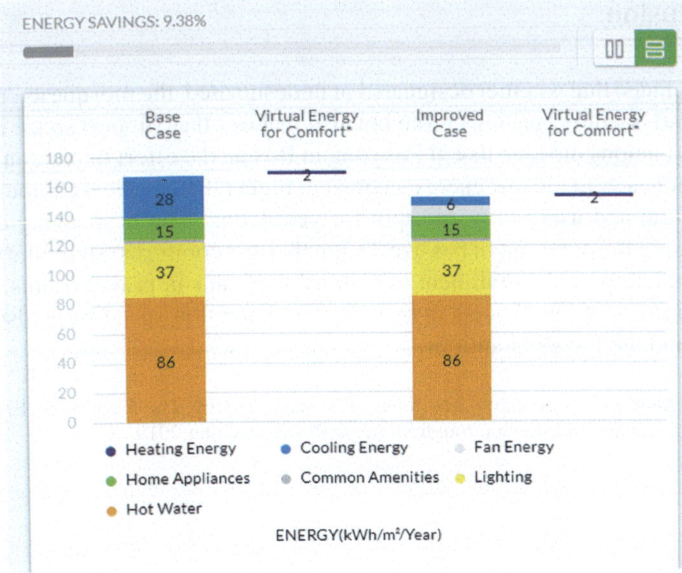

Fig. 11 Energy saving simulation in Building D Architecture Undip

Table 1 Comparison of the studied mosques

Location	Features	Energy saving (%)	Water saving (%)
PKNU	AC central + sub floor heater	5.44	6.49
A bld Undip	AC split-windows	3.59	11.85
C bld Undip	Windows + wall fan	9.21	11.85
D bld Undip	Windows + wall fan	9.38	11.85

the water saving of Al Pukyong is lowered than other since the area of ablution is bigger.

The passive design strategy had shown its effect on this study by evaluation of edge building simulations. The comfort standard on each place is incomparable [13] since Korea is subtropical area (demand AC and heater) while Indonesia is tropical area. However the mosque main space of prayer room and ablution [2] must be provided on this undesignated mosque.

Furthermore the thermal conditioner of AC and heater shall be more calculated since it cost highest energy consumption (see Figs. 3, 6, 9 and 11) and highly related to the area of the prayer space [10].

4 Conclusion

We can concluded that whether designated or undesignated, the mosque as communal space demand not only spacious room but also features that support space comforts. In subtropic campus mosque like al Pukyong in Busan, the effort in Air conditioning and sub floor heater made low energy saving but higher than tropical campus mosque with no natural and using only AC split for ventilating. The best saving energy in tropical is using the potential of passive design thermal comfort system like windows (natural ventilation) and installment of ceiling/stand fans in prayer rooms. In water saving, the efforts of low faucets in bathroom and ablution space were also depend to the area and the faucet quantities.

Acknowledgements This research was financially supported by The Faculty of Engineering, Diponegoro University, Indonesia through Strategic Research Grant 2019.

References

1. Atmaca A, Zorer Gedik G (2019) Evaluation of mosques in terms of thermal comfort and energy consumption in a temperate-humid climate. Energy Build 195:195–204
2. Harsritanto BIR (2018) Pengenalan Arsitektur Masjid. Suara Merdeka
3. Aghniaey S, Lawrence T (2018) The impact of increased cooling setpoint temperature during demand response events on occupants thermal comfort in commercial buildings: a review. Energy Build 173:19–27
4. Cardoso V, Ramos N, Almedia R, Barreira E, Martins J, Simoes M et al (2018) A discussion about thermal comfort evaluation in a bus terminal. Energy Build 168:86–96
5. Ioannou A, Itard L, Agarwal T (2018) In-situ real time measurements of thermal comfort and comparison with the adaptive comfort theory in Dutch residential dwellings. Energy Build 170:229–241
6. Li H, Lee W, Jia J (2016) Applying a novel extra low temperature dedicated a out- door air system in office buildings for energy efficiency and thermal comfort. Energy Convers Manage 121:162–173
7. Zomorodian Z, Tahsildoost M, Hafezi M (2016) Thermal comfort in educational buildings: a review article. Renew Sustain Energy Rev 59:895–906
8. Abdul-Matin I (2010) Green Deen: what Islam teaches about protecting the planet. Berrett-Koehler Publishers, San Francisco
9. Settlement, Ministry (2017) Persyaratan Kemudahan Bangunan Gedung Permen PUPR no.14/2017. Kementrian Pekerjaan Umum dan Perumahan Rakyat, Jakarta
10. Al-Homoud M, Abdou A, Budaiwi I (2005) Mosque energy performance, Part II: monitoring of energy end use in a hot-humid climate. Eng Sci 16(1)
11. Budaiwi I (2011) Envelope thermal design for energy savings in mosques in hot-humid climate. J Build Perform Simul 4(1):49–61
12. Al-Homoud M, Abdou A, Budaiwi I (2009) Assessment of monitored energy use and thermal comfort conditions in mosques in hot-humid climates. Energy Build 41:607–614
13. Al-ajmi F (2010) Thermal comfort in air-conditioned mosques in the dry desert climate. Build Environ 45(11):2407–2413

University Student's Knowledge Toward Energy Conservation and the Implementation on Their Design Project

Bangun I. R. Harsritanto, Hana F. S. Rusyda, Gentina Pratama Putra, and Aditya Rio Prabowo

Abstract Energy conservation performed important role on responding the climate changes on personal and organizational position. Personal responses are known as crucial part toward lower carbon emission, energy consumption, water usage, etc. Nowadays every architecture students in the world has responsibility to design more green building or environment friendly city during their study at campus so they will become agent of change on their future world. However there is less evidence of the on those global movement such endeavors on the student's behavior and actions. This study objective is to analyze the knowledge of energy conservation from architecture students and their implementation on design project. Literature studies, questionnaires, and statistical analysis were performed to pursue the objective. In the result, we can summarize that student energy conservation awareness and design project were related in unique ways.

Keywords Energy conservation · Perception · University student · Design project

1 Introduction

Energy issues play the role on daily life topic of social, technical, economic, environment as the limited fossil fuel resources and worse world conditions. The worlds were faced with defining new ways to respect the energy consumption, energy resources and alteration toward energy freedom. This new ways will be determined by every citizen who participated in societies. Through the energy choices that are integral to decisions of daily life. A well information and energy literate people will be a good start to make more responsible and meaningful energy-related decisions and actions.

B. I. R. Harsritanto (✉) · H. F. S. Rusyda
Universitas Diponegoro, Semarang, Indonesia
e-mail: Bangunirh@arsitektur.undip.ac.id

G. P. Putra
Universitas Mercu Buana, Jakarta, Indonesia

A. R. Prabowo
Universitas Sebelas Maret, Surakarta, Indonesia

U. Sabino et al. (eds.), *Proceedings of the 6th International Conference and Exhibition on Sustainable Energy and Advanced Materials*, Lecture Notes in Mechanical Engineering,
https://doi.org/10.1007/978-981-15-4481-1_33

Energy literacy becomes an important life skill to empower this generation students as well as the common public [1]. Similar to technological literacy [2] and environment literacy [3–5], the component of energy literacy defined into three domains of cognitive (knowledge), affective (attitude, values) and behavior [6].

Combination of social environment and education problems include the public's inadequate access to proper information, people apathy and misunderstanding of external costs of fossil fuel resources, and psychological factors such as resistance on behavioral change are slowing the energy conservation efforts [7]. Effective energy education programs that improve energy literacy play important role on improving this situation, because childhood education may bring strong formula about energy awareness and how to respond the current issues [8–10].

This study purpose is to investigate the knowledge of energy conservation from architecture students at several university in Indonesia and how they implementation the energy literacy on design project. After a brief of literature, the questionnaires have been spread to obtain the student cognitive-affective-behavior on their daily life and during designing, and statistical analysis were performed to pursue the objective. In the result, we can summarize that student energy conservation awareness and design project were related in unique ways.

2 Materials and Methods

2.1 *Survey Instruments*

The instrument for measuring energy conservation literacy is a written questionnaire which designed for architecture students in Indonesia. The instruments were developed as an important part of this study. The instruments contained three parameters of cognitive-affective-behavior on two subscales of student habit (4 items) and their implementation on project design (5 items). The response scales were using modification of three part likert's which could be coded as -1, 0, 1 to evaluate their habit and implementation on design. The energy conservation literacy questionnaire was managed to established psychometric principles and methodologies in the sociology and architecture education [11, 12].

In this research, I coded the likert -1 into 1, 0 into 2, and 1 into 3. The score 1 is indicated no effort of energy saving, 2 is shown low effort and 3 is more effort and updated the knowledge of energy saving.

The instrument's content objectives were guided by a framework that established comprehensive criteria for energy conservation literacy based on a literature review, educational standards [6] and input from a diverse panel of experts in architecture energy and buildings.

Fig. 1 Overall survey results

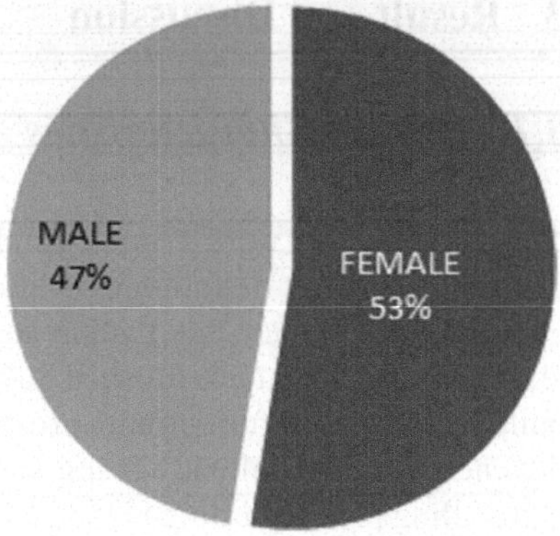

2.2 Samples

The samples were taken from the architecture student from several universities in Indonesia. The number of the participants is 100 with random distribution. The composition of 50:50 on gender couldn't be reached since the participants were not limited by gender basis. The participants were 52% female students and 48% male students from four Indonesia architecture campuses (see Fig. 1). The participating professors were encouraged their student to fill the questionnaire after finished the energy conservation based design studio projects.

2.3 Data Collection and Analysis

The students recorded their questionnaire answers through online goggle form, which automatically transposed into excel tabulations for analysis. The responded items were converted into code of numerical points to represent cognitive-affective-behavior on their habit and design. The modificated likerts of three scale (−1, 0, 1) were designed to represent implementation of energy conservation on their habit and design, such as: wasting energy (−1), neutral (0), saving energy (1). However in the excel charts presentation the alteration code of −1 into 1, 0 into 2 and 1 into 3 were performed to show the rank hierarchy. The statically analysis were performed and visualized using Undip's campus facilities of Microsoft (MS) Excel and Statistical Package for Social Sciences (SPSS).

3 Result and Discussion

3.1 Subject Characteristics

The respondences were 47 males and 53% females from four Indonesia architecture campuses (Institut Teknologi Bandung, Universitas Diponegoro, Universitas Mercubuana, and Universitas Parahiyangan). All of them were final year students (at least 3rd year) who have finished 80% of the course subjects and already learn about building physics or furthermore green building courses. Those conditions may reflect the energy literacy already being studied by the respondences during the questionnaire filling [1]. Regarding to the gender (male and females) proportion didn't reflect of any condition that may result energy conservation value, since this study was not to determine the gender criteria but only the student habit and their implementation in the project design related to energy conservation criteria's [12, 13].

The domain and parameters of energy conservation defined on questionnaire through selected questions. The purposive samples and closed questions were applied to focus the energy conservation studies on academic people [14]. Nine questions were representing the habit and design implementation of the architecture students (see Table 1). Later on, the definitions would be describing this study analysis.

Table 1 Domain, parameter and definition of the energy conservation questions

Domain	Parameter	Definition
Habit	Transportation preference	Identify carbon footprint moda
	Preparation leaving home	Identify potential unused energy
	Energy saving advocacy	motivate energy saving campaign
	Energy saving effort	Identify energy saving acts
Design implementation	Score energy saving	calculate energy saving
	AC system	understand AC consumption
	Artificial lighting system	Understand lighting energy consume
	Glassing type	Understand the energy saver glass

3.2 Energy Saving Habit

Summary of student transportation preference shown that 77% were using personal moda (1), 6% were hitchhiking their friends (2), and 17% were using mass transportation (3) or just walking (see Fig. 2). The usage of personal moda indicated the preference of students to choose high carbon footprint transportation and also the students didn't understand about the concept of energy conservation in transportations. Regarding to the knowledge they have, the result was also response of the pressure of unproper mass transportation on the Bandung and Semarang. Social effect of norms pressured student to conform to desired habits [15] as a result of feelings of a social problem adaptation [16].

Summary of student before leaving home habits described that 55% were locking the door (1), 45% were also checking the unused gas/bulbs (3), and 0% were got their laundries (2) before left (see Fig. 3). This situations shown that energy saving was not become common habit before leaving the room/home. In the university course the campaign to save energy by turning off the light/gas on unused condition is a must do actions. However these results showed the other way. There is no significant influences of social messages on energy conservation [17] seems realistic in this situations. The majority of participants were choosing to act a common safety

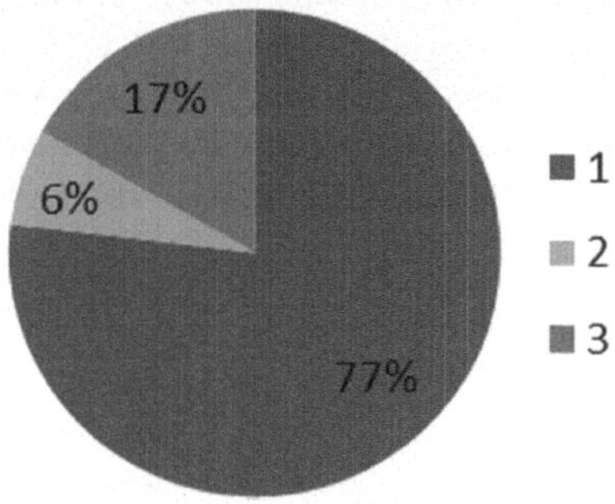

Fig. 2 Student's transportation preference habit

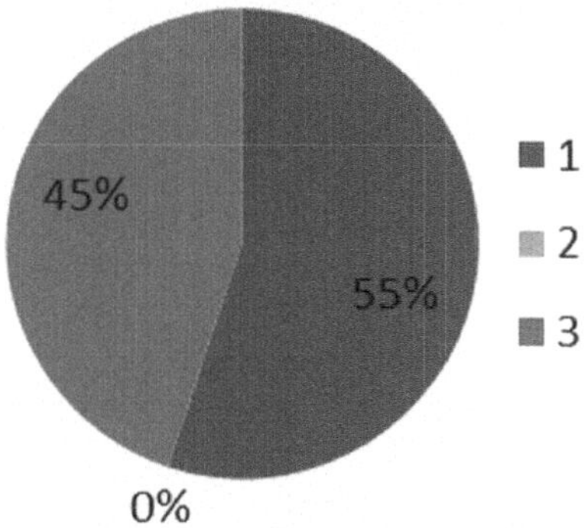

Fig. 3 Students leaving home habit

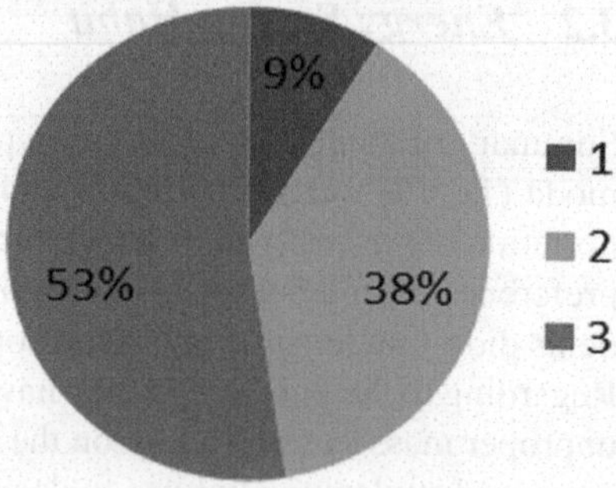

Fig. 4 Student habit for advocating energy conservation

procedure as locking the door and slightly lower participants were ready to save the unused electricity and gas.

Student tendency to provoke their friends and relative to save energy reflected on Fig. 4. The 53% students were rarely campaigned (2), followed by 38% of them who often persuade their inner circle to understand energy conservation strategies (3), and 9% never explain (1) about it (see Fig. 4). Since the unclear connection between energy conservation policy and information amongst people, the motivation to support this campaign might be influenced [14].

The limit of information about energy conservation in Indonesia was still on the reduce consumption shown on Fig. 5. The above charts shown 83% students try to reduce consumption (2) and 17% already understand that using renewable energy (3) is the next phase of energy saving effort, while none of student who didn't understand the terminology of energy conservation concept of reduce and produce from the course they had [14].

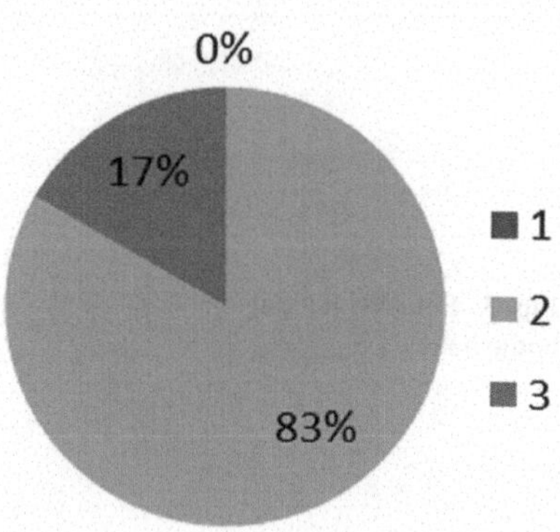

Fig. 5 Student's energy saving efforts

3.3 Energy Saving Implementation Design

The next variable of energy saving is the action in implementing the concept on their design projects, since von Goethe said "Knowing is not enough, we must apply. Willing is not enough, we must do" [14]. The Fig. 6 described that 50% students were applied the energy conservation concept (3) until they got "plus indicators on energy saving" which means they make more efficient energy consumption rather than the standards. However the 38% of them were didn't use the concepts (1) and 12% try to calculate the energy consumption (2) then found that they wasted more energy rather than the proper building model. This energy saving behavior demanded more advocacy and supporting environment to widen the impact on the academic people [13].

The site analysis about solar-chart on building physics class affect massive basic knowledge of passive design. The application of the building orientation theory in architecture students (see Fig. 7) showed that 74% of them were designed the building orientation is North–South (3) which lowered the building exposure from the solar heats. 21% of students forced their building to consume more energy by designed the orientation toward sun directly on the west-east (1), while 5% didn't apply the orientation analysis (2). The memorable education may bring rigid response toward current issues [9]. Curriculum and continuously task to analyze the site and building toward solar chart bring the architecture student's passive design is energy conservation based.

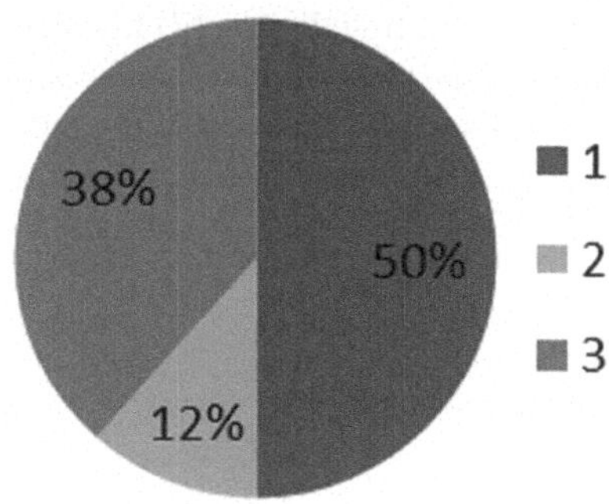

Fig. 6 Student energy saving on last project

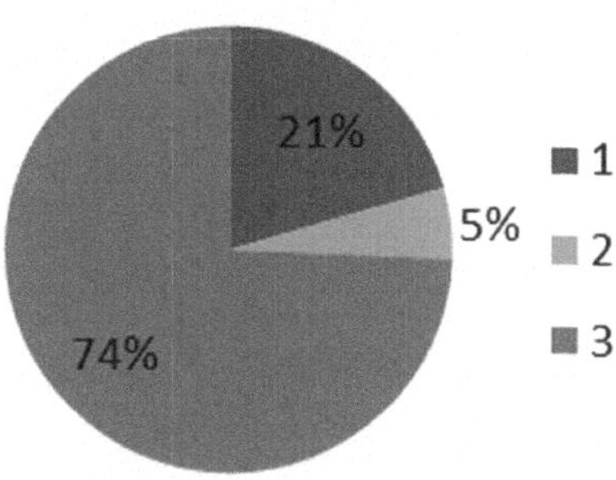

Fig. 7 Students building orientation on last project

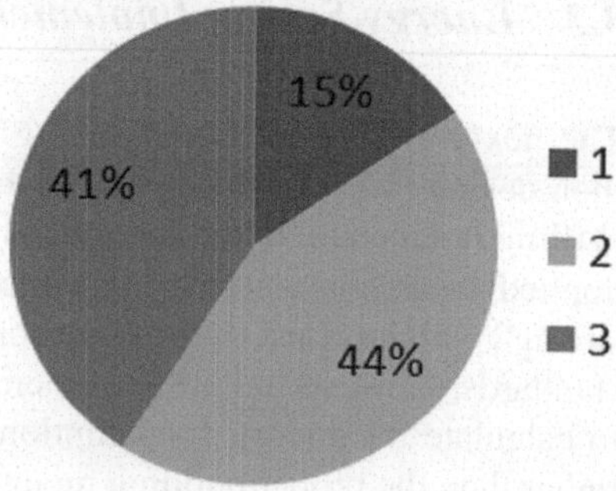

Fig. 8 Student AC system on last project

The usage of AC system of centralized and split showed that even in tropical area, the building user demanded a stable temperature. However the AC central system was more energy efficient rather than split one is a technology improvement which not be followed by students [2]. The 44% of student choose centralized (3), 41% use AC split (2) and 15% didn't designed the air conditioner system (1) for their buildings (see Fig. 8). Technology is a thing that academic people must update every time as their agent of change role and sometimes the improvements is very exponential which make people could adapt quickly.

The glass and lighting application were related closely since the proper (transparent and reflective) glass appliance may reduce the artificial lighting installment. From Fig. 9, 80% students use LED (3), 14% neon lighting (2) and 6% didn't designed it (1). While Fig. 10 shown 50% use low e glass (3) which saving more energy rather 32% chosen tinted glass (2) that bring anti-glare effects and 18% didn't understand about glass technology (1). The situations showed that LED is better campaigned rather than low-e glass system amongst students. In another side the result of 6 and 18% students who didn't understand about energy saving technology brought big challenges for lecturers to attract student to understand it.

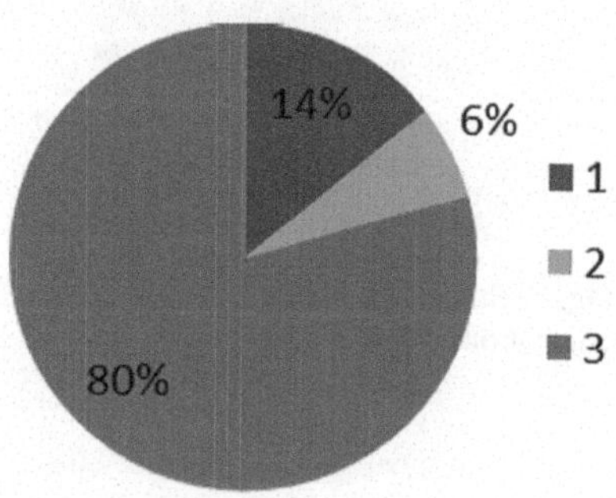

Fig. 9 Student artificial lighting on last project

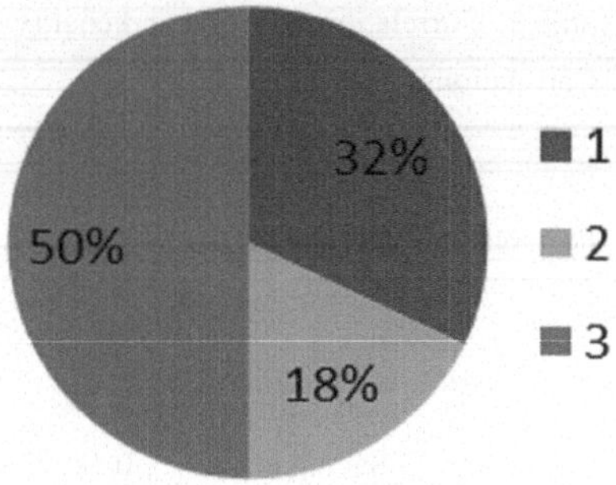

Fig. 10 Students glass type on last project

3.4 Relationship Between Habit and Implementation Design

In the method sections, the conversion codes from likert scales −1, 0, 1 were transformed into 1, 2, 3 code to explain about the degree of energy saving in habit and design preferences. In SPSS this conversion being used to make the calculation formula works in proper.

From the parameters student's habit result, we make statistic calculation using Pearson correlation on SPSS software to analyze the habit relations to the building design implementation of energy conservations. The four habits of transportation preferences, leaving home habits, energy saving advocating and the efforts correlation to the energy saving design is mentioned on Table 2 shown that advocating efforts (0.360), saving energy efforts (0.189), transportation preference (0.141) and before leaving room habit (0.028) were close related to the project design respectfully.

The advocating in energy saving is the nearest relation to the energy conservation design (0.36) and the farthest is before leaving the room/house habits (0.028). Those conditions warned the motivators, lectures or persuaders to give more campaign in "before leaving home habit". Furthermore we could innovate the technology that automatically turnoff the energy device after detecting that no one are in the house/room under the term of smart building.

The statistic also reflect that marginally significant formula of ($0.05 < p < 0.1$) may indicated that the habit of before leaving home is not significant while the others were significant especially advocating efforts. The transportation preferences and energy saving efforts shown the middle score that affect the energy conservation design based appliance.

4 Conclusion

In this study, we investigated the knowledge of energy conservation from architecture students and their implementation on design project. We specifically found that simple term of energy conservation habits like reduce the energy consumptions, promote energy conservation were easy to understand and applied in student's life rather than

Table 2 Correlation of habits and energy conservation design

Correlations

		Energy saving	Transportation	Habit	Advocating	Saving effort
Pearson correlation	Energy saving	1.000	0.141	0.028	0.360	0.189
	Transportation	0.141	1.000	−0.085	0.012	0.036
	Habit	0.028	−0.085	1.000	0.204	−0.051
	Advocating	0.360	0.012	0.204	1.000	0.169
	Saving effort	0.189	0.036	−0.051	0.169	1.000
Sig. (1-tailed)	Energy saving	.	0.141	0.417	0.002	0.074
	Transportation	0.141	.	0.259	0.463	0.394
	Habit	0.417	0.259	.	0.059	0.349
	Advocating	0.002	0.463	0.059	.	0.099
	Saving effort	0.074	0.394	0.349	0.099	.
N	Energy saving	60	60	60	60	60
	Transportation	60	60	60	60	60
	Habit	60	60	60	60	60
	Advocating	60	60	60	60	60
	Saving effort	60	60	60	60	60

next phase of usage of renewable energy and turnoff the energy devices if we don't use it. While environment pressure to use own car/motorbike might the hardest.

Problems to be solved since it related to mass transport supplies and policies. We concluded also the knowledge of energy saving need to be applied and informed quickly after delivered on architecture course. We found that most of students didn't update the information on air conditioning and glass technologies. And lastly, we suggest that the energy saving knowledge were urgently applied in our habit and design requirement if we want to make this architect generation become more agent of energy conservation.

Acknowledgements This research was financially supported by The Faculty of Engineering, Diponegoro University, Indonesia through Strategic Research Grant 2019.

References

1. Powers SE, De Waters JE (2011) Energy literacy of secondary students in New York State (USA): a measure of knowledge, affect, and behavior. Energy Policy 39:1699–1710
2. Pearson G, Young A (2002) Technically speaking: why all Americans need to know more about technology. National Academy Press, Washington, DC
3. Disinger J, Roth C (1992) Environmental literacy. ERIC/CSMEE Digest 1–7

4. Roth C (1992) Environmental literacy: its roots, evolution, and directions in the 1990s. ERIC/SMEAC Information Reference Center, Ohio
5. Wilke R (1995) Environmental education literacy/needs assessment project: assessing environmental literacy of students and environmental education needs for teachers. The University of Wisconsin, Wisconsin
6. Powers SE, De Waters JE (2007) Developing an energy literacy scale. In: The 114th annual ASEE conference & exposition, pp 23–28
7. Sovacool B (2009) The cultural barriers to renewable energy and energy efficiency in the United States. Technol Soc 31:365–373
8. Stern P (1992) What psychology knows about energy conservation. Am Psychol 47:12–24
9. Zografakis N, Menegaki A, Tsagarakis K (2008) Effective education for energy efficiency. Energy Policy 36:3226–3232
10. Benson J, Clark F (1982) A guide for instrument development and validation. Am J Occup Therapy 36:789–800
11. Abdel-Gaid S, Trueblood C, Shrigley R (1986) A systematic procedure for constructing a valid micro computer attitude scale. Journal of Research in Science Teaching 23:823–839
12. Xua X, Maki A, Chen C-F, Dong B, Day JK (2017) Investigating willingness to save energy and communication about energy use in the American workplace with the attitude-behavior-context model. Energy Res Soc Sci 32:13–22
13. Kaplowitz MD, Thorp L, Coleman K, Yeboah FK (2012) Energy conservation attitudes, knowledge, and behaviors in science laboratories. Energy Policy 50:581–591
14. Lee L-S, Lee Y-F, Altschuld JW, Pan Y-J (2015) Energy literacy: evaluating knowledge, affect, and behavior of students in Taiwan. Energy Policy 76:98–106
15. Cialdini R (2003) Crafting normative messages to protect the environment. Curr Dir Psychol Sci 12(4):105–109
16. Cooter R, Feldman M, Feldman Y (2008) The misperception of norms: the psychology of bias and the economics of equilibrium. Rev Law Econ 4(3):889–911
17. Khashe S, Heydarian A, Becerik-Gerber B, Wood W (2016) Exploring the effectiveness of social messages on promoting energy conservation behaviour in the buildings. Build Environ 102:83–94

The Change of Behavior of Magnetorheological Damper with a Single-Stage Meandering Valve After Long-Term Operation

Dewi Utami, Ubaidillah, Saiful Amri Mazlan, H. D. R. Tamrin, Irfan Bahiuddin, Nur Azmah Nordin, and Siti Aishah Abdul Aziz

Abstract A magnetorheological (MR) damper with meandering valve type is known for its capability to produce higher force than other MR valve types while maintaining the compactness. It is essential to investigate the properties of MR damper in long-term operation. Therefore, this paper aims to investigate the MR damper with single-stage meandering valve performance after long-term operation. The experiment will be executed by applying the usage of MR fluid in a bypass double-ended damper. Firstly, the damper was characterized to know the initial damper performance. Secondly, the damper performance will be observed in long-term operation

---

D. Utami · S. A. Mazlan · H. D. R. Tamrin · I. Bahiuddin · N. A. Nordin · S. A. A. Aziz
Engineering Materials and Structures (eMast) iKohza, Malaysia-Japan International Institute of Technology (MJIIT), Universiti Teknologi Malaysia, Jalan Sultan Yahya Petra, 54100 Kuala Lumpur, Malaysia
e-mail: dewiutami11@gmail.com

S. A. Mazlan
e-mail: amri.kl@utm.my

H. D. R. Tamrin
e-mail: nhusnidaim@gmail.com

I. Bahiuddin
e-mail: irfan.bahiuddin@ugm.ac.id

N. A. Nordin
e-mail: nurazmah.nordin@utm.my

S. A. A. Aziz
e-mail: aishah118@gmail.com

Ubaidillah (✉)
Mechanical Engineering Department, Faculty of Engineering, Universitas Sebelas Maret (UNS), Jl. Ir. Sutami 36A, Kentingan, Surakarta 57126, Central Java, Indonesia
e-mail: ubaidillah_ft@staff.uns.ac.id

S. A. Mazlan
Advanced Vehicle System (AVS) Research Group, Universiti Teknologi Malaysia, Jalan Sultan Yahya Petra, 54100 Kuala Lumpur, Malaysia

I. Bahiuddin
Department of Mechanical Engineering, Vocational College, Universitas Gadjah Mada, Jl. Yacaranda Sekip Unit IV, 55281 Yogyakarta, Daerah Istimewa Yogyakarta, Indonesia

U. Sabino et al. (eds.), *Proceedings of the 6th International Conference and Exhibition on Sustainable Energy and Advanced Materials*, Lecture Notes in Mechanical Engineering,
https://doi.org/10.1007/978-981-15-4481-1_34

up to 100,000 cycles. Based on the force versus displacement graph, the peak to peak force has increased by about 33.3% after 28,000 cycles. In other words, the damper becomes stiffer after a long-term operation test will affect the performance of MR damper.

Keywords Durability · Long term operation · Magnetorheological damper · Magnetorheological fluid · Meandering valve

1 Introduction

Magnetorheological (MR) fluid is a viscoelastic smart material which means that it will exhibit the characteristics of viscosity and elasticity when it experiences deformation. Sonawane [1] stated that the first MR fluid was developed by Jacob Rabinow in 1948 which then has caused the fluid to get huge interest due to its unique rheological properties and its high potency in various applications. MR fluid can function as a carrier fluid that has the ability to change its original liquid state to a solid-like state instantaneously, and additionally, the process is conveniently reversible [2–4]. Commonly, carbonyl iron particles (CIP) is frequently utilized as magnetic particles in carrier fluids. Vicente et al. [2] have made a discovery which then proves that the major reason for the usage of CIP particles is due to their high magnetization potency.

The high-speed reaction towards magnetic field had causes this high-performance smart materials to be used widely in vehicles' suspension, biomedical engineering mechanisms, and fan clutches for engines in the form of MR damper, MR brake, and MR clutches [5–7]. The most popular MR device is MR dampers and has been applied in various commercial cars [8]. The most common challenge of the MR dampers is the balance between the compactness of the design and the operation range of the produced peak force. One of the recent studies about the damper shape is a damper based on Meandering MR valve that has been proven to be more compact and higher peak force or pressure drop than annular MR valve, radial MR valve, and Annular radial MR valve [9, 10]. Because of the complexity of the shapes, Ichwan et al. [11] proposed a modular version of the meandering valve, therefore, can be easily scalable. Nevertheless, the most troubling issue is revealed when it comes to maintaining the performance of the MR fluid inside those mechanisms when the damper is subjected to repetitive and long-term operations which involve in continuous force applications.

Long-term and continuous usage may instigate a major problem for MR fluid in a damper that is known as In-Use-Thickening (IUT) which is a phenomenon where the MR damper loses its effectiveness due to the damage that was received from high shear rate and high shear stress over time [12]. After experiencing the IUT phenomenon, the MR fluid within the damper will eventually lose the ability to maintain its liquid state even if the fluid is not subjected to a magnetic field anymore which will then disqualifies it from being a suitable carrier fluid in any devices [12, 13]. Since that the MR fluid is a smart material that is in high demand, studies have been focusing on the problems regarding its performance in continuous operation and the way to manage the performance of MR devices [12]. For the past years, durability research regarding the usage of MR fluid with shear mode has been developed and as the results, MR fluid has been extensively utilized in truck seat damper [12], MR clutches [14], a prosthetic knee joint [15], and fan clutches [16]. However, the study about the effect of long-term operation for MR damper by employing the use of a single-stage meandering valve has not been reported as the authors' best knowledge. Therefore, the objective of this research is to execute long term operation on MR damper using a single-stage meandering MR valve. During the operation, the MR damper was subjected to a continuous load for 28,800 cycles since it was observed the changed of MR damper behaviour already occured. The observation is done before and after the repetitive, long-term operation.

2 Methodology

2.1 *Preparation of Device*

In this research, the damper being employed for the experimental work was a double-ended damper with a bypass system. The bypass damper is a damper with the valve being applied outside of the piston housing. Damper with double ends has a piston with identical diameter compressing and depressing the carrier fluid from both ends of the damper housing.

The valve being used was a meandering valve which is a type of valve that was designed by combining multiple annular and radial flow channels [11]. The combination of both annular and radial flow channels has enabled the maximum utilization of the effective area in the valve. In this paper, the meandering valve being used as a single stage. Figure 1 below shows a visual representation of a single-stage meandering valve. The dimensions were designed as in the previous work [11], which are 0.5 mm for annular gap size, 0.5 mm for annular radial size, 5 mm for radial gap length, 7.5 mm for orifice gap length, 9.5 mm for outer radius radial, 2.5 mm for inner/orifice radial.

2.2 *Material Preparation*

As we know, an MR damper operates just like any other common damper. The factor that differentiates it from other damper was the usage of MR fluid inside the damper instead of a typical carrier fluid. In this study, the MR fluid being utilized was MR fluid LORD 140CG. Table 1shows the principal properties of MR fluid LORD 140CG.

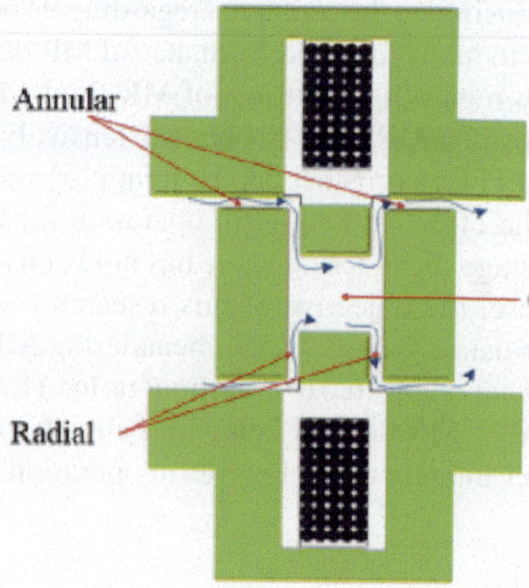

Fig. 1 Schematic configuration of a single-stage meandering valve

Table 1 Principle properties of MR fluid LORD 140CG

Variables	Value
Solid weight percentage	85.44%
Density	3.54–3.74 g/cm^3
Operating temperature	−40 to +130 °C
Viscosity	0.280 ± 0.070 Pa s at 40°C

2.3 Testing Procedures

The long-term operation method that was executed in this research was fatigue dynamic test. The experimenting works of this research employed the usage of a machine named Shimadzu Servopulser Machine. This test can evaluate the performance of the MR fluid in the damper. Figure 2 shows a visual representation of an MR damper being attached to the Shimadzu Servopulser machine. The valve being

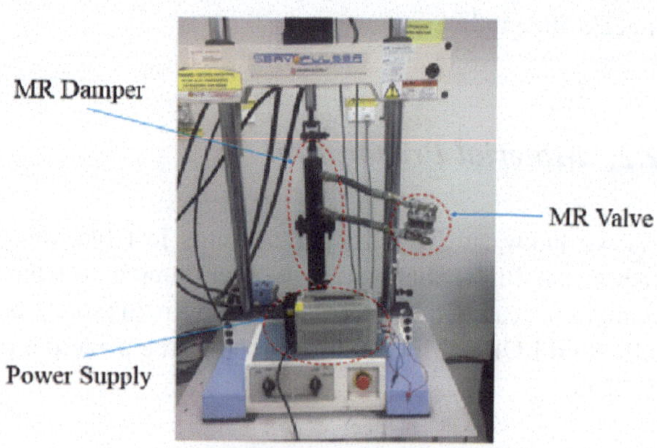

Fig. 2 Fatigue dynamic test setup

used in a bypass damper system was connected to the power supply which has been supplying the current to the damper during the fatigue dynamic test.

Various frequencies and currents were tested before the long-term operation test. The initial test parameters were set as followings respectively; frequency of 0.6 Hz with current at 0.2 and 0.5 A, frequency of 0.8 Hz with current at 0.2 and 0.5 A, frequency of 1.0 Hz with current at 0.1, 0.5, 0.7, 1.0 and 1.2 A. There were 4 sessions of fatigue dynamic test being conducted with the maximum value of 10 h for the first three session. A total cycle of 28,800 was reached for each of those three sessions. As for the final session, the test was stopped after the cycle of 13,600 had been reached. This was because of the total cycles of 100,000 cycles had been achieved. The total 100,000 cycles used as a parameter since it was expected the MR fluid behaviour already change, in fact, it could show the change of MR fluid behaviour since the first 28,800 cycles in the first session. The control parameter of this experiment was current (0.5 A), frequency (0.8 Hz), and length of stroke (5 mm).

3 Results and Discussion

During the whole duration of the fatigue dynamic test, the data that was obtained was the damping force over the number of cycles. All those data has been recorded and plotted as shown in Figs. 3 and 4. Referring to Fig. 3, which shows the graph of force over the displacement, it was observed that the damping force of the damper increases as the number of cycles increases. The highest cycle which is 28,800th cycle has generated forces with a peak value of 0.458 kN while the lowest cycle which is 200 cycles has produced a force with a peak value of 0.406 kN. The data is plotted in Fig. 3 generated graphs with hysteresis curves. From Fig. 4, the visual representation of cycles to the value of peak-to-peak value force (F_p) has been provided. During the 5000th cycle, the F_p generated was 0.646 kN. The value of F_p increases at cycle 10,000th with the value of 0.676 kN and at the 25000th cycle the F_p value reached

Fig. 3 Damping Force over the length of stroke

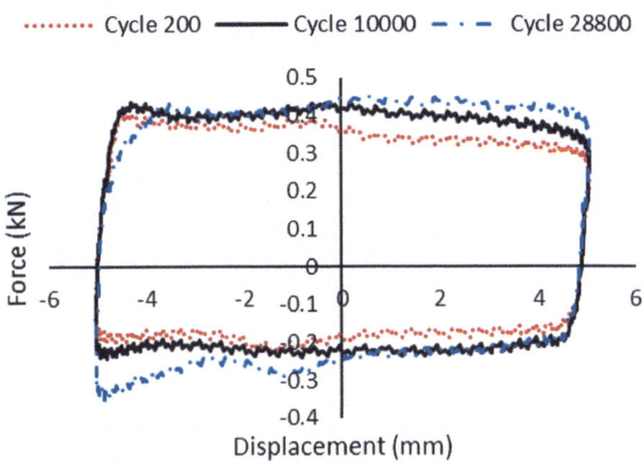

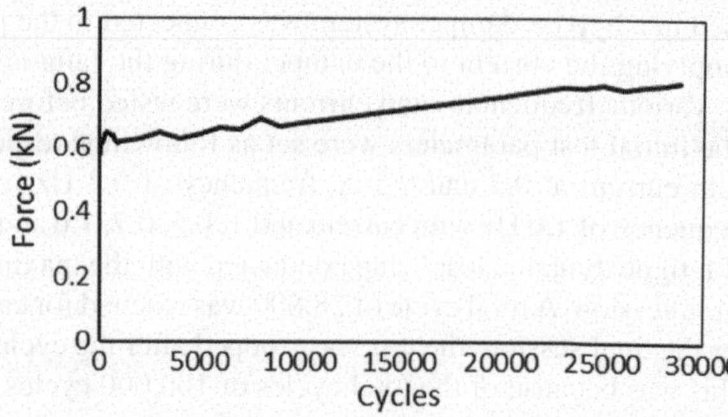

Fig. 4 Increment of damping force with respect of cycles

0.798 kN. The F_p value being generated during those particular cycles indicated that the value of F_p increases as the cycle increases.

The observation from Figs. 3 and 4 showed the increase of the viscosity (η) and yield stress (τ_y) of the MR fluid after has been subjected to a long term operation. The conducted fatigue dynamic test at long duration may have instigated the IUT phenomenon caused by the MR fluid change of behaviour as described by Carlson [12]. The study had proven that when a certain MR fluid was used in a longer time period, there were also increment occurred in terms of viscosity of the fluid even without the presence of a magnetic field. Consequently, this may be a contributing factor in the value of force (F) increasing overtime during the usage of any MR devices as shown in Eq. (1) where $F_{viscous}$ is produced force as a function of flow rate (Q) and plastic viscosity (η); and F_{yield} is produced as a function of the magnetic field (B) and yield stress (τ_y).

$$F = F_{viscous}(Q, \eta) + F_{yield}(B, \tau_y) \tag{1}$$

The alternative scenario of the increment of the damping force is the magnetic particle that has been trapped in the MR valve. During fatigue dynamic test, the magnetic field had been applied on the MR valve through the power supply with a current of 0.5 A. Because of the long term duration of the fatigue dynamic test, the current that was applied on the valve was ensured to be flowing during the whole duration of the operation until the experiment was completely done. During the off-state condition, the MR fluid will be able to freely flow through and forth the MR valve. Meanwhile, during the on-state condition, the current constantly being supplied and the magnetic field continuously being applied. This has caused the tendency of the magnetic particles being trapped within the flow passage of the MR fluid becomes higher. This situation can be related to the IUT phenomenon. With the IUT occurring at the MR valve, it was predicted that the particles agglomerated and caused the MR fluid to clump and eventually lead to clotting. Those chain reactions eventually lead to interruption of the flow of MR fluid for the fatigue dynamic test. Over time, the interruption caused the damping force that was produced to become higher as the operation proceeded. These odds can be considered highly possible by analyzing the data that was obtained from this research. Furthermore, it is believed that the damping force will continue to increase if the fatigue dynamic test is to be continued even after the 100,000th cycle because there will be more particles being clumped at the MR valve as the experiment continues. The similar works that have this kind of scenario are as reported by Utami et al. [13, 17].

The agglomeration of the magnetic particle or the trapped magnetic particle in the MR valve may cause a higher pressure drop than in the normal operation. The MR valve is basically a device used for controlling the passage flow of the MR fluid. In the damper system of this study, the valve primarily created pressure drop when the MR fluid went through this particular device. Just like in any piping system with a flowing matter, when the matter flows through the orifice there will be a pressure difference between the matter before entering and after entering the orifice. In a damper, the pressure drop was utilized conveniently in order to produce some kind of obstacle. The manipulation of pressure drop somehow also enables the manipulation of the damping force. With the presence of this particular obstacle inside the damper system, it will be more difficult for the piston to undergo compression. To put it simply, the damping force is generated will be higher when a more potent pressure drop is produced. Moreover, in this research, the orifice was blocked by a clump of MR fluid that experience IUT. With the IUT occurring at the valve, it is possible that the pressure drop became higher because of the clump of MR fluid blocking the orifice.

Bypass damper configuration may also contribute to the increase of the peak force of the MR damper. The main purpose of utilizing a damper in car suspension systems is to decrease the impact of driving on bumpy roads. The slight difference between a bypass damper and a conventional damper is the attached valve outside of the housing which then enables the addition of extra resistance to minimize the impact. However, by employing the application of a bypass configuration, the viscosity of the MR fluid tends to increase. This is due to the components of the whole MR valve device being in a static motion. With the absence of any moving parts inside the MR valve the probability of the MR fluid in the valve to become more viscous will multiply because there is no device that scrambles the trapped magnetic particles within the MR valve device. Therefore, the CIP has more tendencies to be extracted from the MR fluid during the operation. Consequently, the potency of the damping force being generated for the whole damping system will increase as a result of the increase of viscosity and elasticity of the MR fluid.

4 Conclusion

In this paper, the fatigue dynamic test on an MR fluid in an MR damper with a single-stage meandering valve had been conducted. The value of the damping force over cycles has been recorded and analysed. The damping forces increment confirmed through the damping force over displacement graph. It showed that that being subjected to a long-term operation may affect the performance of any MR devices since all of those devices employ the utilization of MR fluid.

Acknowledgements This study was financially supported by the Universiti Teknologi Malaysia under Transdisciplinary Research Grant (Vot No: 06G77) and Professional Development Research University grant (Vot No: 04E02).

References

1. Sonawane AV (2016) A study of properties, preparation and testing of magneto-rheological (MR) fluid. Int J Innov Res Sci Technol 2(09):82–86
2. de Vicente J, Klingenberg DJ, Hidalgo-Alvarez R (2011) Magnetorheological fluids: a review. Soft Matter 7(8):3701. Available from http://xlink.rsc.org/?DOI=c0sm01221a
3. Ashtiani M, Hashemabadi SH, Ghaffari A (2015) A review on the magnetorheological fluid preparation and stabilization. J Magn Magn Mater 374:711–715. Available from http://dx.doi.org/10.1016/j.jmmm.2014.09.020
4. Bahiuddin I, Mazlan SA, Shapiai MI, Choi SB, Imaduddin F, Rahman MAA et al (2018) A new constitutive model of a magneto-rheological fluid actuator using an extreme learning machine method. Sens Actuat A: Phys 281:209–221 Available from https://doi.org/10.1016/j.sna.2018.09.010
5. Baranwal D, Deshmukh TS (2012) MR-fluid technology and its application—a review. Int J Emerg Technol Adv Eng 2(12):563–569
6. Abd Fatah AY, Mazlan SA, Koga T, Zamzuri H, Zeinali M, Imaduddin F (2015) A review of design and modeling of magnetorheological valve. Int J Mod Phys B 29(04):1530004
7. Imaduddin F, Mazlan SA, Zamzuri H (2013) A design and modelling review of rotary magnetorheological damper. Mater Des 51:575–591. Available from http://dx.doi.org/10.1016/j.matdes.2013.04.042
8. Rossi A, Orsini F, Scorza A, Botta F, Belfiore N, Sciuto S (2018) A review on parametric dynamic models of magnetorheological dampers and their characterization methods. Actuators 7(2):16
9. Imaduddin F, Amri Mazlan S, Azizi Abdul Rahman M, Zamzuri H, Ichwan B (2014) A high performance magnetorheological valve with a meandering flow path. Smart Mater Struct 23(6):065017. Available from http://stacks.iop.org/0964-1726/23/i=6/a=065017?key=crossref.6dc89ea13ff0faecff7d2548de85f6fd. Accessed on cited 17 Jan 2017
10. Imaduddin F, Mazlan SA, Ubaidillah, Zamzuri H, Fatah AYA (2016) Testing and parametric modeling of magnetorheological valve with meandering flow path. Nonlinear Dyn 85(1):287–302
11. Ichwan B, Mazlan SA, Imaduddin F, Ubaidillah, Koga T, Idris MH (2016) Development of a modular MR valve using meandering flow path structure. Smart Materi Struct 25(3):037001. Available from http://stacks.iop.org/0964-1726/25/i=3/a=037001?key=crossref.e05bed0668764c2ffc31a0f32eb618ab
12. Carlson JD (2003) Critical factors for MR fluids in vehicle systems. Int J Veh Des 33(1/2/3):207–217
13. Utami D, Ubaidillah, Mazlan S, Imaduddin F, Nordin N, Bahiuddin I et al (2018) Material characterization of a magnetorheological fluid subjected to long-term operation in damper. Materials 11(11):2195
14. Zhu X, Jing X, Cheng L (2012) Magnetorheological fluid dampers: a review on structure design and analysis. J Intell Mat Syst Struct 23(8):839–873. Available from http://journals.sagepub.com/doi/10.1177/1045389X12436735
15. Ubaidillah, Sutrisno J, Purwanto A, Mazlan SA, Ubaidillah B, Sutrisno J et al (2015) Recent progress on magnetorheological solids: materials, fabrication, testing, and applications. Adv Eng Mater 17(5):563–597

16. Ulicny JC, Hayden CA, Hanley PM, Eckel DF (2007) Magnetorheological fluid durability test-organics analysis. Mater Sci Eng A 464(1–2):269–273
17. Utami D, Mazlan SA, Ubaidillah, Bahiuddin I, Imaduddin F, Nordin NA (2018) The changed of behaviour of MR fluid in MR damper after a long-term operation. Key Eng Mater 775:171–176

Performance Assessment of Water Turbine Subjected to Geometrical Alteration of Savonius Rotor

Dandun Mahesa Prabowoputra, Syamsul Hadi, Aditya Rio Prabowo, and Jung Min Sohn

Abstract The availability of fossil fuels as an energy source has decreased every year due to the increment of public and industry consumptions, as well as its characteristic as a non-renewable resource. Thus, various efforts are concentrated to develop another source to decrease dependency on fossil fuel. Hydropower is one of the best renewable energy sources because it is environmentally friendly and its management is relatively cheaper than other alternatives. The abundant availability of water flow in Indonesia makes hydropower an alternative new energy source that is easily developed, one of which is research on water turbines. The study is aimed to conduct a performance assessment on the savonius water turbine accounting for the effect of modifying the rotor geometry. The research subject will use a 110 mm of diameters single-stage rotor that will be modified into number and shape of the blade. Variations in number of blade, were applied at the number of 2 and 3 with two types of the blade (type letter S and letter L). Calculation of the research was carried out using computational fluid dynamic technique of the Ansys code in which the interacting fluid flow is explicitly idealized. The results of simulation and analysis indicated that maximum C_p was achieved in Savonius with modified Type L rotors. This configuration is suitable to be applied to the river and sewage water pipes, to utilize the flow of wastewater as a new energy source and reduce the level of emissions using fossil resources.

Keywords Renewable energy · Savonius rotor · Multi-stage rotor · Hydropower · Water turbine · Coefficient of power · CFD techniques

D. M. Prabowoputra
Graduate School of Mechanical Engineering, Universitas Sebelas Maret, Jl. Ir. Sutami 36A, Surakarta 57126, Indonesia

S. Hadi (✉) · A. R. Prabowo
Department of Mechanical Engineering, Universitas Sebelas Maret, Jl. Ir. Sutami 36A, Surakarta 57126, Indonesia
e-mail: syamsulhadi@ft.uns.ac.id

J. M. Sohn
Department of Naval Architecture and Marine Systems Engineering, Pukyong National University, Busan 48513, South Korea

U. Sabino et al. (eds.), *Proceedings of the 6th International Conference and Exhibition on Sustainable Energy and Advanced Materials*, Lecture Notes in Mechanical Engineering,
https://doi.org/10.1007/978-981-15-4481-1_35

1 Introduction

The demand for electricity grows rapidly every year. Growth is expected to increase by 14% every two years for the period 2016–2024. In 2016, the capacity of national and new and renewable energy-based generators was 6.9 GW or around 12% of the total national power plant [1]. The data shows that new and renewable energy-based power plants still need to be improved, especially with the fact that the availability of fossil fuels is always decreasing.

Geothermal, wind energy, hydro energy, and mini-micro hydro are new and renewable energy sources in Indonesia. The mini-micro hydro has considerable potential, which is 19.38 MW. But the installation that has been done is only 2.6 MW or around 13% [1]. The average production capacity growth in 2012–2014 was 18%. However, in 2015–2017, there was no increase in production capacity. This shows that mini-micro hydro has great potential to be developed.

Research on micro-mini hydro plants has been developed, one of the developments in the design of water turbines. Water turbines have many types, generally divided into two, namely the horizontal axis and vertical axis. For the horizontal axis that has been used in this study is the Savonius type. In 1920, Savonius published his rotor design, which was tested with water and wind fluids. The Savonius Rotor Design is an adaptation of the Rotor System at the Flettner Rotor Principle.

Research on the effect of geometric modification on rotor savonius performance was carried out. This modification includes the design in the savonius rotor without the shaft between the endplate. The results of the study, it was found that removing the shaft on the rotor can increase the coefficient of Power (C_p) in the Savonius wind turbine [2]. Besides this modification, there is a modification of the helix rotor without endplate in the savonius turbine [3–5]. Research on modification of the rotor geometry of Savonius has been carried out in this decade, both experimentally and simulations. In another study regarding the overlap ratio that has been carried out on horizontal axis water turbines, it shows that $C_{p\max}$ is obtained in 0.3 overlaps. The $C_{p\max}$ produced in the study was 0.19 at *TSR* 0.79 [6]. Modifications made are in the form of rotor [2–4], variation of helix angle, number of blades, and Stage [7, 8], overlap ratio and other modifications [6, 9]. Several studies of Savonius rotors have been carried out simulations [10]. In these studies, geometric changes were associated with parameters that affected the performance of savonius turbines. The Performance Parameters used are Power Coefficient (C_p).

Savonius wind turbine research is more developed compared to the savonius water turbine. Therefore that research on the savonius water turbine still has many opportunities to develop. Analysis has been carried out using a numerical method to compare the performance of the savonius wind turbine with the savonius water turbine. At the same boundary conditions, the performance of the Savonius Water Turbine is better than the savonius wind turbine. Where the dimensions of the rotor, Power output, and C_p are the same, wind turbines, require wind velocity nine times greater than the Velocity of water flow [11].

Savonius turbine research is still growing today. One of the developments that have been made is about the modification of geometry. A large amount of research on geometry changes in the Savonius turbine has shown researchers' interest in this topic. Therefore, it is a challenge to improve the performance of the water turbine savonius. Several studies have been conducted and presented in the form of literature/scientific journals. Therefore based on the literature that has been studied, it is necessary to develop modifications shape of blade and number of blade in the Savonius turbine rotor to improve turbine performance.

2 Savonius Rotor

Savonius Rotor has a simple structure and easy to modify geometric designs. Also, the Savonius Rotor is also easy to use on water turbines and wind turbines. The shape of the Rotor in Savonius has the ability to receive flow from all directions. Therefore, research on the rotor Savonius in turbines continues to develop, both experimental and simulation studies.

2.1 Principle of Savonius Rotor

Rotor Savonius is an adaptation of the Rotor System at the Flettner Rotor Principle; therefore, that the savonius rotor working principle is similar to the Flettner Rotor Principle. The work of the savonius rotor is to use a different resistance coefficient between two blades in the turbine. The torque produced by the concave Blade is higher than the advancing blade therefore, that rotation occurs. At the same time, some fluid flows arrive at a convex underwater surface through a fluid tunnel that produces torque.

2.2 Rotor Performance Parameter Equation

The results of the research that have been carried out through simulation, then processed using equations to calculate the performance parameters. These equations are:
Power:

$$\mathrm{P_{Available}} = 0.5 \times \rho \times \mathrm{A} \times \mathrm{V}^3 \tag{1}$$

$$P_{Rotor} = T \times \omega \tag{2}$$

TSR (Tip Speed Ratio):

$$\mathrm{TSR} = \frac{\omega D}{2U} \tag{3}$$

Coefficient Torsi (C_t) and Coefficient Power (C_p):

$$C_p = \frac{P_{\mathrm{Rotor}}}{P_{\mathrm{Available}}} \tag{4}$$

$$C_t = \frac{C_p}{\mathrm{TSR}} \tag{5}$$

where V is the free flow velocity, ρ for water density, A is the Area of the Rotor, T is Torque, ω represents the angular velocity, C_t for the torque coefficient, and C_p for the Power coefficient [12].

3 Analysis of Preparation

The present simulation is done using ANSYS Software and CFX Solver. This study was carried out using the rotor with a horizontal axis. The variation that has been done with modification shape of blade and number of blade. The dimensions for each rotor are shown in Table 1. The number of rotors tested is four rotors. The design of the rotor is shown in Fig. 1. Research conducted using the aspect ratio (*D/H*) of 1 and the overlap ratio of 0. Table 1 shows that D_1 is the endplate diameter, d is diameter of blade, d_0 is diameter variation of blade, H is the rotor height, L is the length of variation blade, and T is the thickness. The research used number of blades two and three for each modification shape of blade.

Table 1 Dimension of rotor

Parameter	Jenis rotor			
	Conventional rotor		Modification rotor	
	Two blade	Three-blade	Two blade	Three-blade
D_1 (mm)	110	110	110	110
d (mm)	50	50	50	50
H (mm)	110	110	110	110
d_0 (mm)	–	–	23	23
T (mm)	2	2	2	2
L (mm)	–	–	15	15

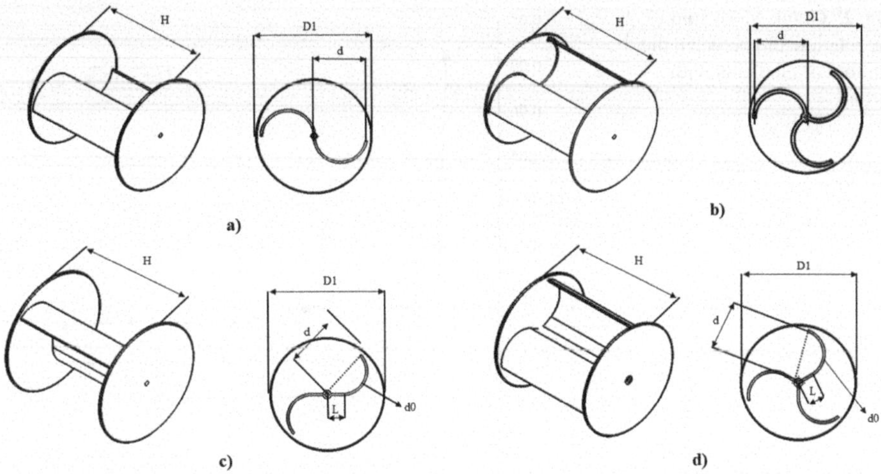

Fig. 1 The geometry of rotor conventional **a** two-blade, **b** three-blade, rotor modification **c** two-blade, and **d** three-blade

4 Methodology

In this study, CFX software has been used for computing. Using this software, all rotor designs have been analyzed. Equation of momentum, turbulent kinetic energy has been solved numerically using the software. Where the governing equation is shown in Eqs. 6–9:
Continuity Equation:

$$\frac{\partial \rho}{\partial t}+\frac{\partial(\rho u)}{\partial x}+\frac{\partial(\rho v)}{\partial y}+\frac{\partial(\rho w)}{\partial z}=0 \tag{6}$$

X-Momentum:

$$\frac{\partial \rho u}{\partial t}+\frac{\partial\left(\rho u^2\right)}{\partial x}+\frac{\partial(\rho u v)}{\partial y}+\frac{\partial(\rho u w)}{\partial z}=-\frac{\partial \rho}{\partial x}+\frac{1}{Re}\left(\frac{\partial \tau_{xx}}{\partial x}+\frac{\partial \tau_{xy}}{\partial y}+\frac{\partial \tau_{xz}}{\partial z}\right) \tag{7}$$

Y-Momentum:

$$\frac{\partial \rho u}{\partial t}+\frac{\partial(\rho u v)}{\partial x}+\frac{\partial\left(\rho v^2\right)}{\partial y}+\frac{\partial(\rho v w)}{\partial z}=-\frac{\partial \rho}{\partial y}+\frac{1}{Re}\left(\frac{\partial \tau_{xy}}{\partial x}+\frac{\partial \tau_{yy}}{\partial y}+\frac{\partial \tau_{yz}}{\partial z}\right) \tag{8}$$

Z-Momentum:

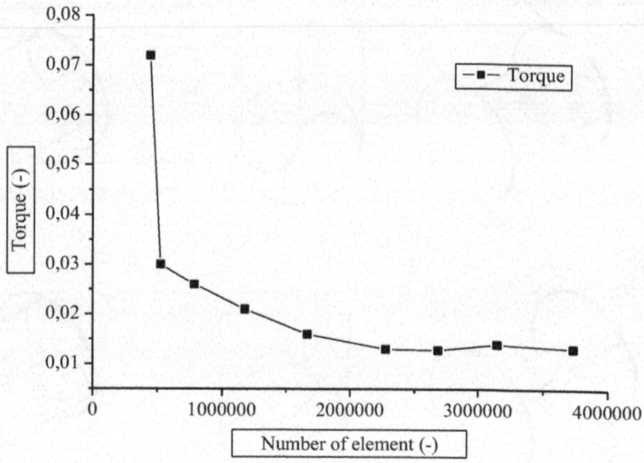

Fig. 2 Graph Mesh studies the relationship between the number of elements with torque

$$\frac{\partial \rho u}{\partial t}+\frac{\partial(\rho u w)}{\partial x}+\frac{\partial(\rho v w)}{\partial y}+\frac{\partial\left(\rho v^2\right)}{\partial z}=-\frac{\partial \rho}{\partial y}+\frac{1}{Re}\left(\frac{\partial \tau_{xz}}{\partial x}+\frac{\partial \tau_{yz}}{\partial y}+\frac{\partial \tau_{zz}}{\partial z}\right) \tag{9}$$

where ρ is density, t is time, u, v, and w are velocity, and x, y, and z are the coordinates.

4.1 Mesh Independence Study and Benchmarking

Research has been carried out in three-dimensional simulation. Water Turbine Savonius simulations have been carried out using the software Ansys CFX solver. Before being implemented, simulation, has made the process of mesh study. Mesh study is done by increasing the number of elements to get an effective number of elements [13]. The result of the mesh study that has been done is shown in Fig. 2. The figure shows the relationship between the number of elements and the torque value. From the results of the mesh study, the number of effective elements is 2,700,000.

Validation was carried out on Roy S. et al. research [14]. Validation was carried out at the position of *TSR* 0.74 where $C_{p\max}$ was reached. C_p at 0.65 *TSR* obtained from the simulation was 0.252, wherein the Roy S. et al. study it was 0.52. The data shows a 3.14% error rate. Benchmark has shown in Fig. 3.

4.2 Numerical Model

Meshing has been done on a workbench. The meshing method that has been used is tetrahedral. The number of elements used is based on the results of the Mesh independence study, which is 2,700,000. Each domain has meshed separately. Figures 4 and 5 show the results of meshing for each domain. The rotor uses inflation.

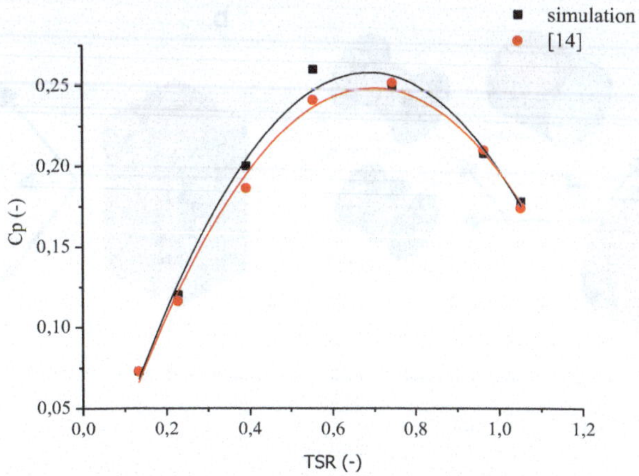

Fig. 3 Graph of benchmarking simulation with Roy's experiment

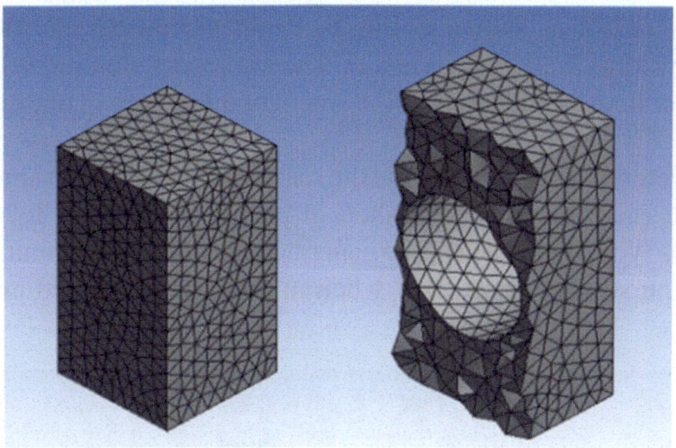

Fig. 4 Mesh at stationary domain

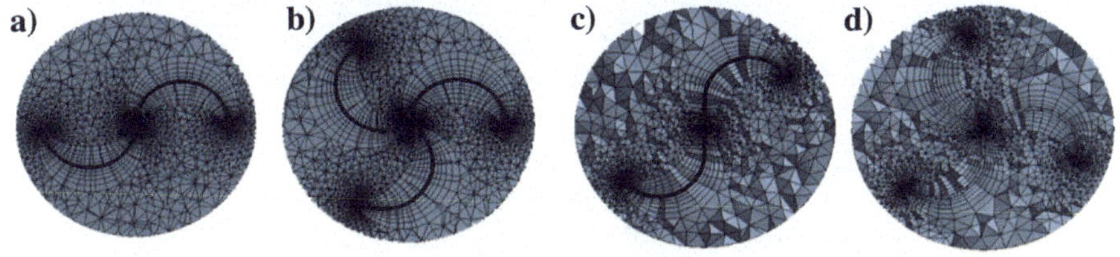

Fig. 5 Mesh of rotor conventional **a** two-blade, **b** three-blade, rotor modification **c** two-blade, and **d** three-blade

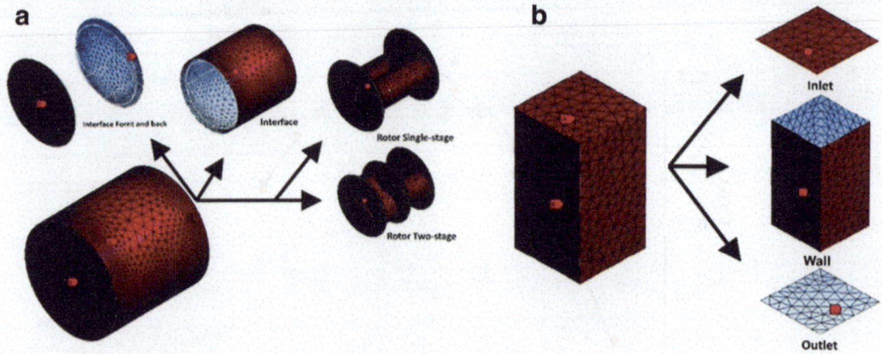

Fig. 6 **a** Rotary domain, **b** stationary domain

Domain Simulation consists of 2 parts, Rotary Domain, and Stationery Domain. A rotary domain consists of front interface, interface, rear interface, and rotor. The Rotary Domain section is shown in Fig. 6. Besides showing the rotating domain, Fig. 6 also shows the Stationary domain. The stationary domain consists of Inlet, Outlet, Wall, and Interface. The overall domain and boundary conditions are shown in Fig. 7.

A complete reference should provide the reader with enough information to locate the article concerned, whether published in print or electronic form and should, depending on the type of reference, The generated mesh on the computational domain is exported to the solver Ansys CFX wherein the case is defined and solved with

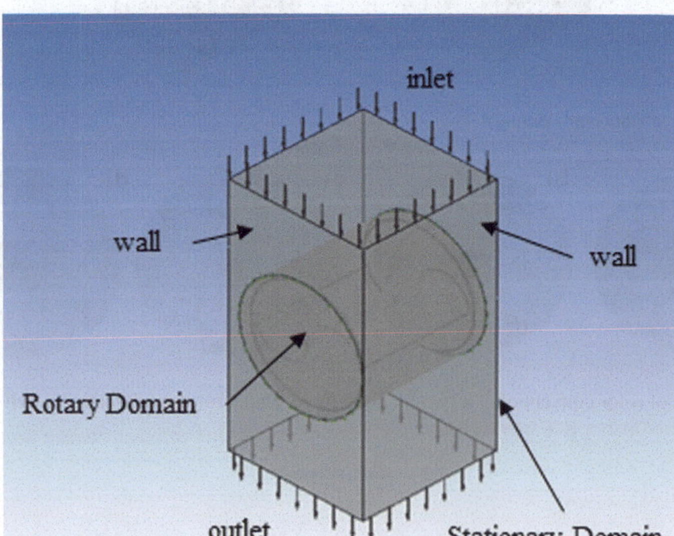

Fig. 7 Schematic of simulation

Table 2 Boundary condition of simulation

No.	Parameter	Value
1.	Fluid type	Water
2.	Density (kg/m^3)	1000
3.	Domain motion	Rotating
4.	Boundary condition inlet flow	Subsonic
5.	Water velocity (m/s)	0.8
6.	Outlet pressure (atm)	1
7	Turbulence model	K epsilon
8	Gravity (m/s^2)	9.81
9	Pressure (atm)	1

necessary boundary conditions. Boundary conditions, operating parameters and fluid properties used in simulation are shown in Table 2.

5 Results and Discussion

Three-dimensional simulation research has been carried out on these four rotors. The results taken from the simulation are the torque value (converted to the coefficient of power and coefficient of torque), pressure distribution, velocity distribution, and velocity vector. The results of C_P and C_T will be presented in the form of graphs, while the distribution of pressure, velocity and velocity vector are in the form of two-dimensional images.

Simulation results are shown in Figs. 8 and 9. Where Fig. 8 shows the relationship between C_p and TSR, and Fig. 9 shows C_t with TSR. The information in the picture shown that rotor A is a conventional rotor with two blades, rotor B is a variation

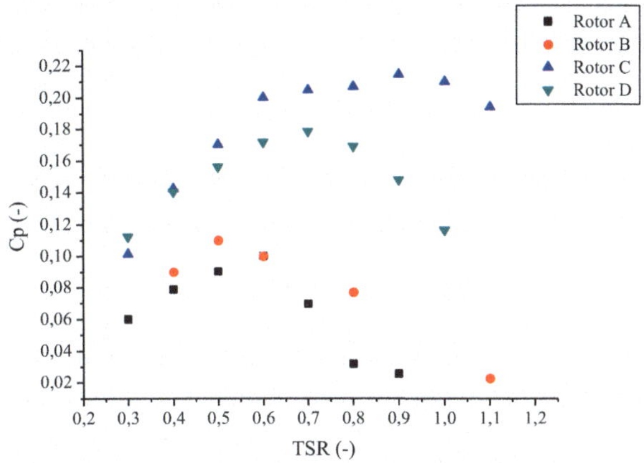

Fig. 8 Graph of relationship between C_P with TSR

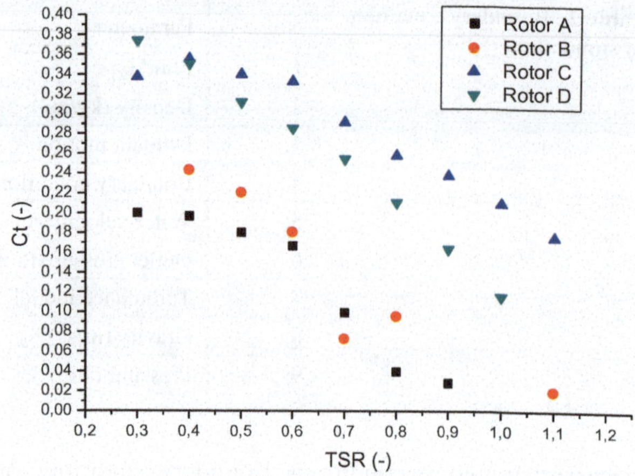

Fig. 9 Graph of the relationship between C_T with TSR

rotor with two blades, rotor C is a traditional rotor with three blades, and rotor D is a variation rotor with three blades.

The C_P and C_T produced by Conventional 2 blade rotors are shown in Figs. 8 and 9. The graph shown that the $C_{P\max}$ is 0.10 at *TSR* 0.6. C_{Tmax} is reached at 0.196 at *TSR* 0.4. $C_{P\max}$ and $C_{T\max}$ are not on the same *TSR*. C_P generated by the Modified Rotor 2 blades continues to increase at *TSR* < 0.5 and decrease at *TSR* > 0.5. The resulting $C_{P\max}$ was 0.11 at *TSR* 0.5. $C_{T\max}$ was obtained at 0.24 at *TSR* 0.4. C_P produced by the Conventional Rotor 3 blade continues to increase at *TSR* < 0.7 and decrease at *TSR* > 0.7. The resulting $C_{P\max}$ is 0.20 at TSR 0.7. $C_{T\max}$ was obtained at 0.35 at TSR 0.4. The graph shows produced by the Modified Rotor 3 blades $C_{p\max}$ generated by 0.18 at TSR 0.7, and $C_{T\max}$ is obtained at 0.37 at TSR 0.3.

From the simulation research that has been done, the results of pressure distribution, velocity distribution contour, and velocity vector are obtained. Where the advantage of this research is to get the contours of the pressure distribution, velocity distribution, and vector velocity. The picture is taken at the $C_{P\max}$ position for each rotor. The contour of the pressure distribution of each rotor is shown in Fig. 10. The figure shows that overall, the pressure at the inlet flow is higher than the outlet flow. This phenomenon occurs in all rotor variations. Figure 11 shows the distribution of velocity at each rotor. The contours of the velocity show the same pattern between conventional two-rotor rotors and rotor modifications, as well as three blades. The vector direction of velocity is shown in Fig. 12. The picture shows the vector flow direction of the velocity of each rotor. The value of velocity (color gradation) is the same as shown in Fig. 11.

Factorial design analysis has been carried out to determine the effect of each factor [15]. There are two factors analyzed, namely geometric shape and TSR. The results of the factorial design analysis are shown in Table 3. The results are compared with Table 4 to determine the level of influence of these factors. Where is the comparison between F_0 and F in Table 4? Obtained that F_0 rotor geometry > F from Table 4, F_0 TSR < F from Table 4, and F_0 interaction < F from Table 4.

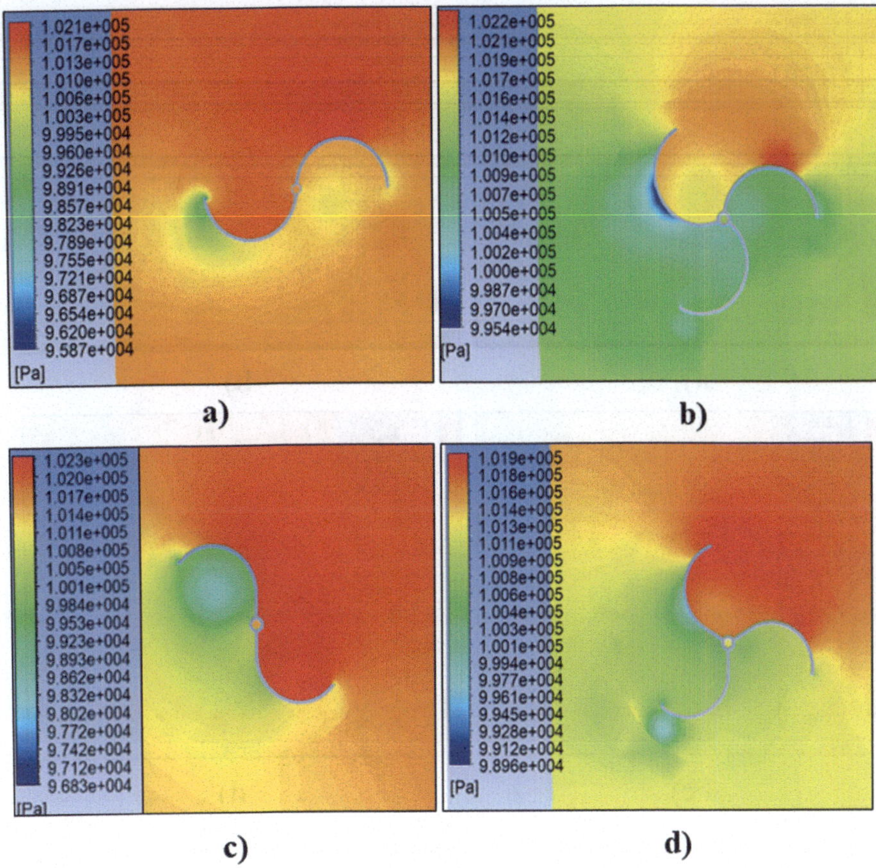

Fig. 10 Distribution of pressure on rotor conventional **a** two-blade, **b** three-blade, rotor modification, **c** two-blade, and **d** three-blade

6 Conclusion

A simulation of four different types of savonius rotors has been carried out, conventional rotor two blades, three blades, and Modification rotor two blades, three blades. The results of this study conclude that the modification Rotor with three-blade has the highest coefficient power maximum (C_{Pmax}) value compared to another rotor. Therefore the modification rotor design with a three-blade is a design that can improve the performance of water turbine savonius.

From the factorial design analysis, it was concluded that the shape of the rotor geometry has a great influence on the value of C_P. This is indicated by the value of $F_0 > F$ table while the TSR and geometric shapes are factors that do not affect each other (no interaction), where this is indicated by $F_0 < F$ table.

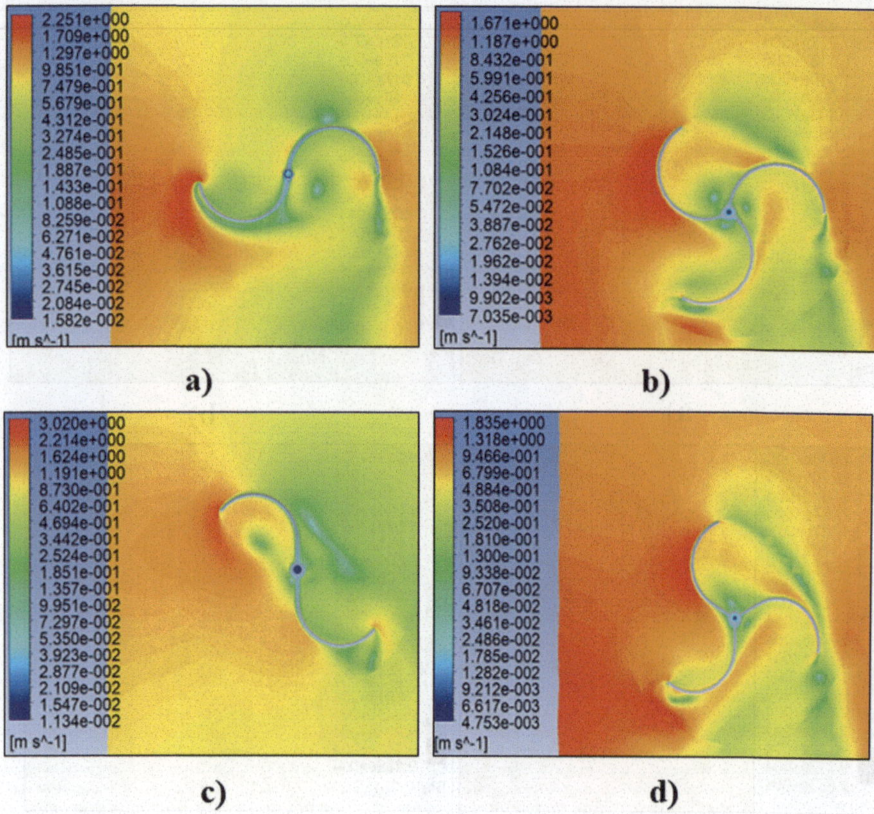

Fig. 11 Distribution of velocity on rotor conventional **a** two-blade, **b** three-blade, rotor modification, **c** two-blade, and **d** three-blade

Where the advantage of this research is to get the contours of the pressure distribution, velocity distribution, and vector velocity, through this contour, we can find out the position of the highest pressure, lowest pressure, highest velocity, lowest velocity therefore that we can make even better development of the next design. It cannot be obtained through experimental research.

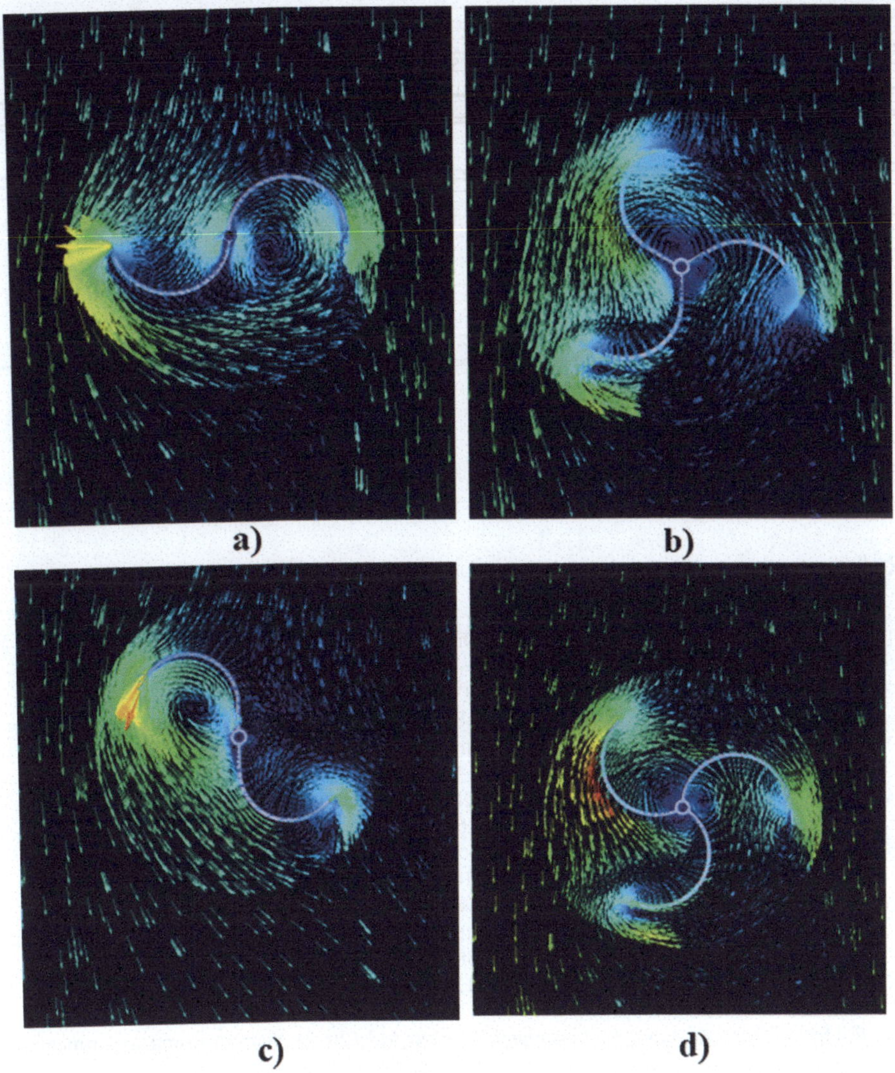

Fig. 12 Vector of velocity on rotor conventional **a** two-blade, **b** three-blade, rotor modification, **c** two-blade, and **d** three-blade

Table 3 Analysis of variance for coefficient power

Source of variation	Sum of squares	Degrees of freedom	Mean square	F_0	*P*-value
TSR	0.006	5	0.00	1.78	0.1809
Rotor geometry	0.043	3	0.01	22.05	0.0000
Nonaddivity	0.002	1	0.0023	3.57	0.0797
Error	0.009	14	0.0007		
Total	**0.06**	**23**			

Table 4 Percentage points of the F distribution [15]

	df		*F*
$F_{0.05}$	5	17	2.7
$F_{0.05}$	3	17	3.7
$F_{0.05}$	1	19	4.38

References

1. Yudiartono A, Rahardjo I, Fitriana I (2018) Outlook energy Indonesia 2018 : sustainable energy for land transportation, vol 134, no 4
2. Kamoji MA, Kedare SB, Prabhu SV (2009) Experimental investigations on single-stage modified savonius rotor. Appl Energy 86(7–8):1064–1073
3. Damak A, Driss Z, Abid MS (2013) Experimental investigation of helical savonius rotor with a twist of 180. Renew Energy 52:136–142
4. Soo K, Ik J, Pan J, Ryu K (2015) Effects of end plates with various shapes and sizes on helical savonius wind turbines. Renew Energy 79:167–176
5. Deb B, Gupta R, Misra RD (2014) Experimental analysis of a 20° twist helical savonius rotor at different overlap conditions, vol 594, pp 1060–1064
6. Hadi S, Khuluqi H, Prabowoputra DM, Prasetyo A, Tjahjana DDDP, Farkhan A (2019) Performance of savonius horizontal axis water turbine in free flow vertical pipe as effect of blade overlap. J Adv Res Fluid Mech Therm Sci 58(2):219–223
7. Kamoji MA, Kedare SB, Prabhu SV (2011) Experimental investigations on two and three stage modified savonius rotor, pp. 483–509
8. Frikha S, Driss Z, Ayadi E, Masmoudi Z, Abid MS (2016) Numerical and experimental characterization of multi-stage savonius rotors. Energy 114:382–404
9. Ramar SK, Seralathan S, Hariram V (2018) Numerical analysis of different blade shapes of a savonius style vertical axis wind turbine, vol 8, no 3
10. Prabowoputra DM, Hadi S, Prabowo AR, Sohn JM (2020) Performance investigation of the savonius horizontal water turbine accounting for stage rotor design. International Journal of Mechanical Engineering and Robotics Research 9(2):184–189
11. Abulnaga BE (1988) Water power without waterfalls. Water Resources Journal 57–60
12. Patel V, Eldho TI, Prabhu SV (2019) Velocity and performance correction methodology for hydrokinetic turbines experimented with different geometry of the channel. Renew Energy 131:1300–1317
13. Thakur N, Biswas A, Kumar Y, Basumatary M (2019) CFD analysis of performance improvement of the savonius water turbine by using an impinging jet duct design. Chin J Chem Eng 27(4):794–801

14. Roy S, Saha UK (2015) Wind tunnel experiments of a newly developed two-bladed savonius-style wind turbine. Appl Energy 137:117–125
15. Montgomery DC (2012) Design and analysis of experiments, 8th edn., vol 2

Numerical Study of the Wingtip Fence on the Wing Airfoil E562 with Fence Height Variations

S. P. Setyo Hariyadi, Sutardi, Wawan Aries Widodo, and Bambang Juni Pitoyo

Abstract Winglets are equipment that must be used by aircraft and unmanned aerial vehicles (UAV) in the modern era. The function of the winglet is to block the fluid flow jump from the lower side to the upper side on an aircraft wing. With the reduction of fluid flow jumps, it is expected that the wing's aerodynamic performance will increase. The researchers mostly made geometric variations of the winglets to obtain maximum results, namely the increase of lift and decreasing drag. This research was conducted using Ansys Fluent 19.0 with a turbulent model k-ω SST. The freestream flow rate to be used is 10 m/s (Re = 2.34×10^4) with the angle of attack (α) = 0°, 2°, 4°, 6°, 8°, 10°,12°, 15°, 16°, 17°, 19° and 20°. The model of the specimen is an Eppler 562 airfoil with forward and rearward wingtip fence. From this study, it was found that a winglet with fence height equals the chord line length resulting in better aerodynamic performance than winglet with fence height equal to half the chord line length. It this shows that the winglet with fence height equals the chord line length which can reduce the fluid flow jump from the lower side to the upper side. It was also found that the use of the wingtip fence reduced the strength of vorticity magnitude on the z-axis compared to plain wings. This is due to the vorticity effect of the plain wing and wingtip fence on the flow of fluid passing through it.

Keywords Winglet · Wingtip fence · Lift · Drag · E562

1 Introduction

The world of aviation requires continuous innovation. The intended innovation is not only in the use of fuel but also in aspects related to fuel savings. These fuel savings are closely related to the aerodynamic performance of the aircraft. Geometry modification is the thing most often used by researchers to improve the aerodynamic

S. P. Setyo Hariyadi (✉) · B. J. Pitoyo
Aviation Polytechnic of Surabaya, Surabaya, Indonesia
e-mail: hudzaifahsetyo@gmail.com

Sutardi · W. A. Widodo
Mechanical Engineering Department, Industrial Technology Faculty, Sepuluh Nopember Institute of Technology, Surabaya, Indonesia

U. Sabino et al. (eds.), *Proceedings of the 6th International Conference and Exhibition on Sustainable Energy and Advanced Materials*, Lecture Notes in Mechanical Engineering,
https://doi.org/10.1007/978-981-15-4481-1_36

performance of an object. This geometry modification has the main goal of increasing lift force as high as possible and reducing drag as low as possible.

Geometry modifications in aircraft are mostly carried out on wings. Airplane wings are one of the most important parts of designing an airplane. This is because the aircraft's wings play an important role in producing lift force on the aircraft. The main force components in an aircraft consist of weight, lift force, thrust, and drag. Lifting force works against the direction of the load so that the aircraft can fly.

Sohn and Chang [8] conducted experimental research to see the effect of adding Whitcomb's winglet to vortex wingtip on NACA 631-212 airfoil. The results obtained are wingtip vortices on the wings with Whitcomb's winglet having a weaker and more spreading configuration than simple wing fairing. This causes an increase in the Whitcomb's winglet C_L/C_D.

Yen and Fei [10] also conducted an experimental study to see the effect of the addition of dihedral winglet to vortex wingtip on NACA 0012 airfoil with variations in cant angle $\delta = -30°$, $\delta = 15°$, $\delta = 30°$, $\delta = 45°$, $\delta = 90°$, $\delta = 135°$. The results obtained are at $\delta = 90°$ the value of the lift-to-drag ratio under stall conditions and the maximum value of the lift-to-drag ratio is 32 and 17% greater than without winglet.

The research of Myilsamy et al. [5] produced a comparison chart of C_L, C_D and C_L/C_D for each type of winglet against the angle of attack. In this study, it was shown that the C_L ratio for each type of winglet against the angle of attack, the C_L graph tends to rise to the angle of attack 14°. In the C_D comparison for each type of winglet against the angle of attack, it tends to rise to the angle of attack 14°. In the ratio of lift to drag ratio for each type of wing-let to the angle of attack, C_L/C_D tends to rise to the angle of attack 8° and then decreases to the angle of attack 14°.

In this simulation study comparing the height of the wingtip fence so that it can be seen its effect on aerodynamic performance and other properties behind the UAV wing.

2 Research Methodology

2.1 Computational Fluids Dynamic

This research was carried out using Ansys 19.0 with turbulent k-ω SST models. The flow velocity used is 10 m/s ($Re = 2.34 \times 10^4$) with the angle of attack (α) = 0°, 2°, 4°, 6°, 8°, 10°,12°, 15°, 16°, 17°, 19° and 20°. The test specimen is in the form of plain wing airfoil E562 with variations in winglet geometry. The winglet used is the wingtip fence in the form of forward and rearward. Reynolds Number is determined by chord length. In this case, the length of the chord used is 20 cm. Figure 1 is a simulation domain and boundary conditions used in the simulation Mulvany et al. [4]. The specimen model is a plain wing airfoil type Eppler 562 with a winglet attached to the tip of a wingtip fence as shown in Fig. 2.

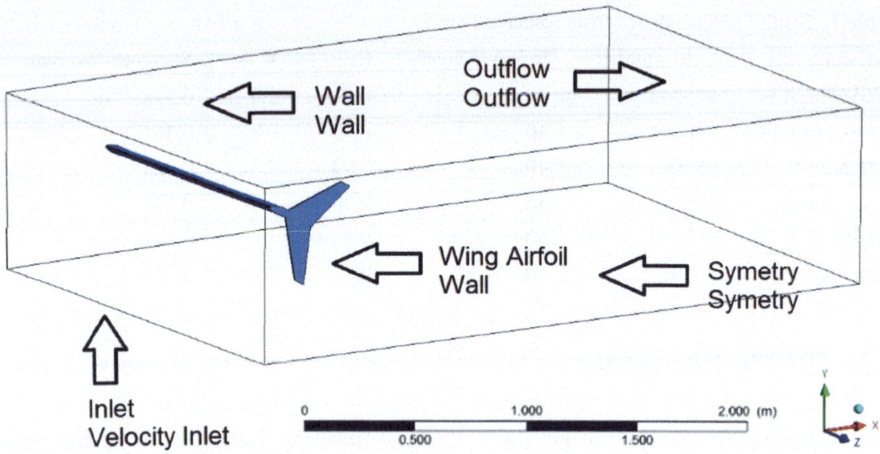

Fig. 1 Simulation domain and boundary conditions

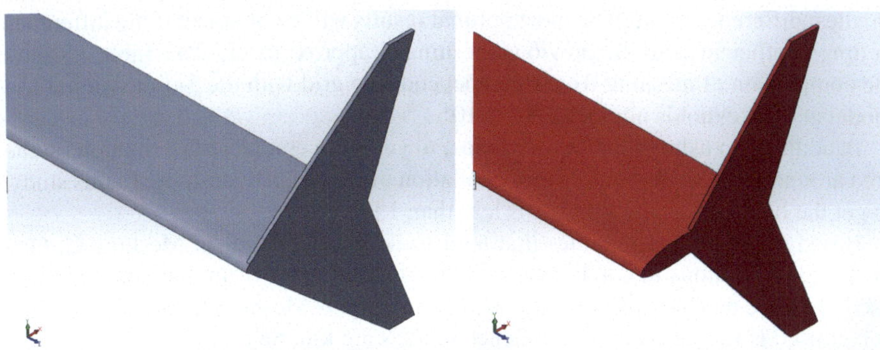

(a) Forward Wingtip Fence (b) Rearward Wingtip Fence

Fig. 2 Specimens model

In numerical simulations, geometry, meshing, and boundary types are made for the test objects. The next step is to do the meshing process and determine the boundary type of the specimens.

The conditions of the environment are entered at a temperature of 30 °C and a pressure of 1 atm. The working fluid conditions included include viscosity (μ) = 1.86×10^{-5} Ns/m^2, (ρ) = 1.17 kg/m^3, length scale at the inlet side of 0.024 m, and turbulence intensity in numerical modeling this is 0.8%.

Table 1 Grid of independence from research [7]

Meshing type	Cells number	Inflation layer	C_D	y^+	Skewness average
Meshing A	875962	40	0.91	1.4	0.348
Meshing B	768003	40	0.90	0.8	0.351
Meshing C	685063	40	0.92	2.1	0.334
Meshing D	569233	40	0.90	0.8	0.343
Meshing E	469620	40	0.88	1.4	0.347
Meshing F	367075	40	0.86	2.1	0.346

2.2 Grid Independence

Grid independence is done to get the amount of meshing that tends to be constant so that the modeling results are close. In this independence grid, the amount of meshing is divided into several types of meshing. The grid independence criteria use the resulting C_D results. The numerical C_D difference is chosen based on the smallest difference value. The most optimal results will be obtained if the difference in drag coefficient with the previous meshing is approximately 2%. Table 1 shows the comparison of meshing from the independence grid with the 3-dimensional test model at the Reynolds number 2.34×10^4.

In addition to using the disc difference, to get more complete information on the area around the wall, it requires the calculation of y+ on each meshing. In this study, to get the best results, the y+ used is less than 1 [3].

Based on Table 1, C_D values that tend to be smaller occur in Meshing C. One of the considerations in conducting numerical simulations is the time and memory used, then the meshing used for the next simulation is Meshing C. In this study the solver used is for steady and Turbulence models are k-ω SST.

3 Result and Discussion

3.1 Drag Coefficient Analysis

In Fig. 3, it is shown that the addition of winglets will increase drag compared to the plain wing. In this study, it can be seen that the height of the fence h = 0.5c results in a higher drag compared to the fence h = c. This applies when compared to the forward wingtip fence cant angle 90° [2] and rearward wingtip fence cant angle 90° [6]. At fence h = 0.5c also produces a higher drag than Turanoguz's research results [9]. From Fig. 3 it is found that the aspect that influences the height of the drag from the fence h = 0.5c is not just the pressure and viscous drag produced, but it must be seen other aspects that are a result of the geometry used. Other aspects include vorticity, which is one of the effects of induced drag.

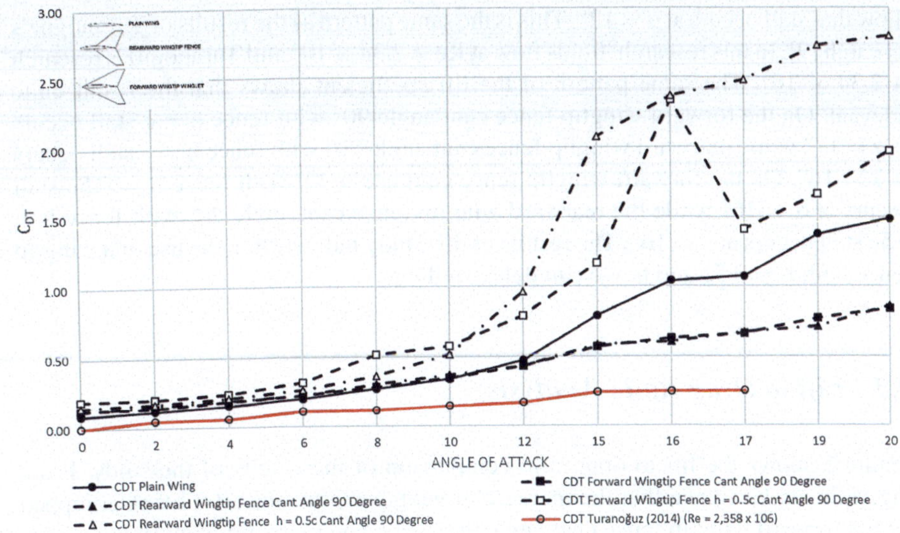

Fig. 3 C_D comparison of research results

3.2 *Lift Coefficient Analysis*

Figure 4 shows the C_L results of the study. Figure 4 also shows that the lift coefficient increases with increasing angle of attack. In the plain wing, the results of the study

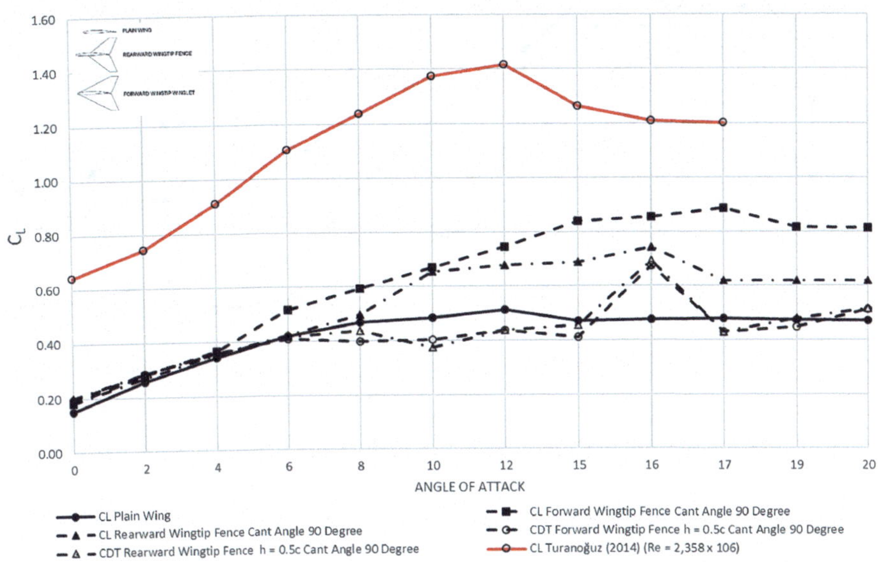

Fig. 4 C_L comparison of research results

show that stall occurs at $\alpha = 12°$. This is the same pattern as the results of Turanoguz's research [9]. In this research, fluids flow at $Re = 2.34 \times 10^4$ and Turanoguz's research at 2.34×10^6. The same pattern of the lift coefficient shows that the result quite relevant. On the forward wingtip fence cant angle 90° with fence h = c stall occurs at $\alpha = 17°$ while rearward wingtip fence cant angle 90° with fence h = c stall occurs at $\alpha = 16°$. On the forward wingtip fence cant angle 90° with fence h = 0.5c stall occurs at $\alpha = 16°$ while the rearward wingtip fence cant angle 90° with fence h = 0.5c stall occurs at $\alpha = 16°$. The results of this study indicate that the use of a wingtip fence both h = 0.5c and h = c can delay stalling.

3.3 *Lift to Drag Ratio Analysis*

Figure 5 shows the lift to drag ratio comparison of the results of the study. From Fig. 5, it can be shown that the increase in wing performance after being equipped with a forward wingtip fence cant angle 90° h = c and rearward wingtip fence cant angle 90° h = c. However, this did not occur in the forward wingtip fence cant angle 90° h = 0.5c and the rearward wingtip fence cant angle 90° h = 0.5c when compared to the plain wing. On the forward wingtip fence cant angle 90° results in higher performance than plain wing at $\alpha = 6°$ while the rearward wingtip fence cant angle

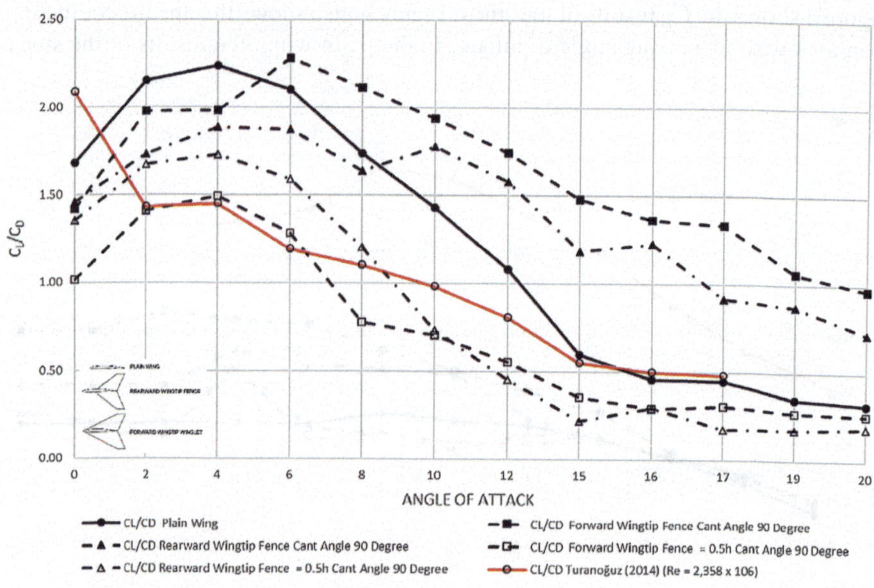

Fig. 5 C_L/C_D comparison of research results

90° produces a better performance at $\alpha = 10°$. The use of forward wingtip fence cant angle 90° h = 0.5c, and rearward wingtip fence cant angle 90° h = 0.5c have decreased aerodynamic performance at all angles of attack compared to the plain wing.

3.4 Vorticity Magnitude Analysis

The tendency of flow to leak around the wingtips has an important effect on the aerodynamics of the wings. This flow forms a rotating motion that leads to downstream in the trailing vortex. This vortex tip induces a small component of the velocity of air around the wing downward. This bottom component is called downwash.

The higher the angle of attack the wider the vortex formed. Vortex formation starts from the end of the trailing edge [1]. This results in a much greater vortex if the vortex has reached the leading edge. Therefore, the function of the winglet is to reduce the size of the vortex will be more perfect if the formation of the vortex is blocked from the trailing edge.

The width and concentration of the vortex is one indication of the great fluid jump from the lower surface to the upper surface of the wing. Figure 6 shows the evolution and comparison of the area of vorticity magnitude of plain wing, forward wingtip fence cant angle 90° h = c, rearward wingtip fence cant angle 90° h = c, forward wingtip fence cant angle 90° h = 0.5c, and rearward wingtip fence cant angle 90° h = 0.5c at a distance of 1 C behind the trailing edge (x = 2c). At the $\alpha = 0°$, the effect of the presence of winglets is less visible, because the vortex tip phenomenon is not very well-formed, so the results of the visualization cannot be so distinguished. Different at $\alpha = 12°$ and $\alpha = 17°$, phenomena from vortex tip can be seen more so that the difference in the form of vortex tip on the plain wing and airfoil with winglet fence can be distinguished.

From $\alpha = 12°$ and $\alpha = 17°$, it can be seen that the tip vortex on the plain wing is more concentrated in the center of the vortex. On Eppler 562 airfoil with forward wingtip fence cant angle 90° h = c, and rearward wingtip fence cant angle 90° h = c, the vortex tip that occurs is split into 2 parts.

On Eppler 562 airfoil with forward wingtip fence cant angle 90° h = 0.5c, and rearward wingtip fence cant angle 90° h = 0.5c, tend to have the shape of a plain wing but have greater value and area. It is possible that the forward and the rearward wingtip fence cant angle 90° h = 0.5c have a greater drag value.

From this visualization phenomenon it can be concluded that the use of forward wingtip fence cant angle 90° h = c and rearward wingtip fence cant angle 90° h = c is more effective to be used than the forward wingtip fence cant angle 90° h = 0.5c and the rearward wingtip fence cant angle 90° h = 0.5c because the characteristic area of vorticity is smaller than plain wing.

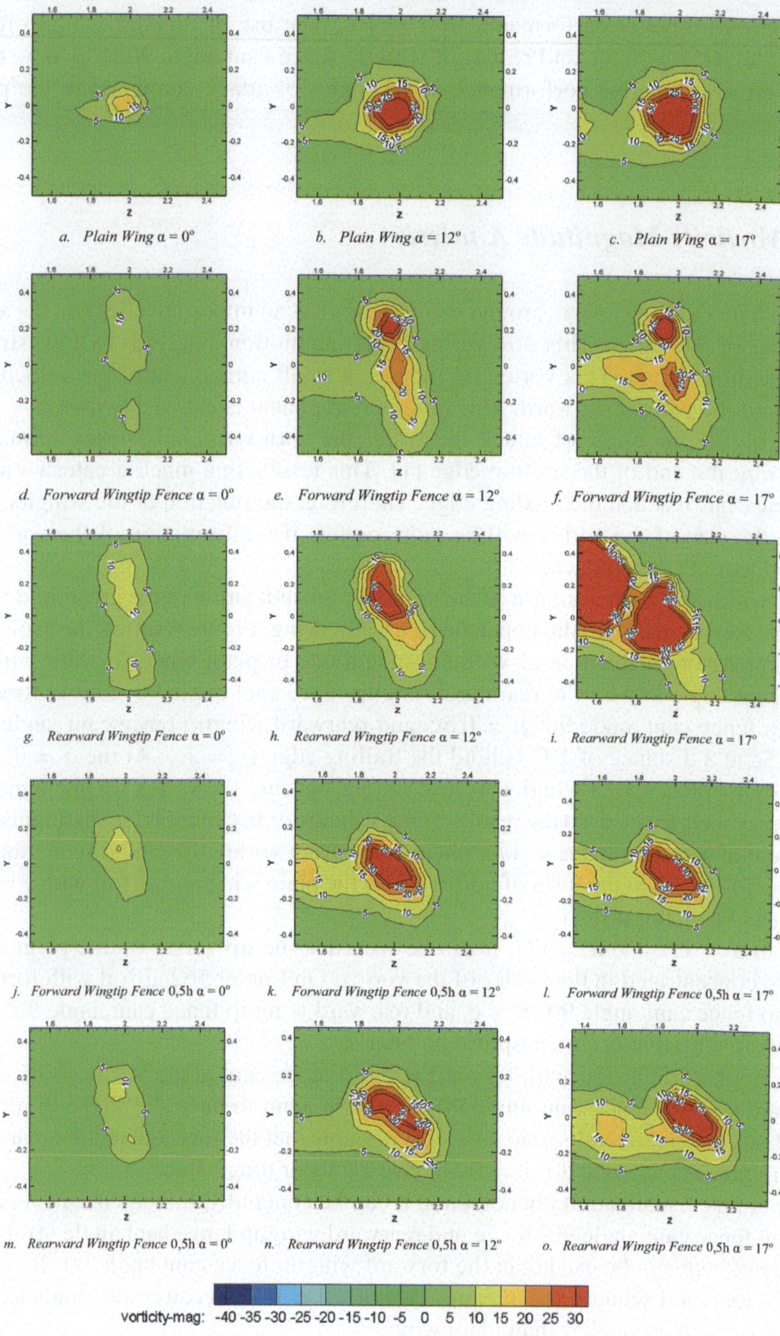

Fig. 6 Comparison of magnitude vorticity area as far as 1 C behind the wing in s^{-1}

4 Conclusion

This research has shown the effect of using winglet shape variations significantly on aerodynamic performance and vorticity. From the numerical simulation, several conclusions were shown, including:

1. Wingtip fence h = 0.5c produces higher drag compared to wingtip fence h = c. This is very much influenced by the induce drag formed by the endplate geometry.
2. The use of wingtip fence delays stalling. On the forward wingtip fence cant angle 90° with fence h = c, a stall occurs at $\alpha = 17°$. Rearward wingtip fence cant angle 90° with fence h = c, stall occurs at $\alpha = 16°$. In the forward wingtip fence cant angle 90° with fence h = 0.5c, stall occurs at $\alpha = 16°$ while rearward wingtip fence cant angle 90° with fence h = 0.5c, stall occurs at $\alpha = 16°$.
3. Forward wingtip fence cant angle 90° h = c and rearward wingtip fence cant angle 90° h = c produce an increase in aerodynamic performance. Whereas the forward wingtip fence cant angle 90° h = 0.5c and the rearward wingtip fence cant angle 90° h = 0.5c result in a decrease in the lift to drag ratio when compared to the plain wing.
4. On the forward wingtip fence cant angle 90° h = c and rearward wingtip fence cant angle 90° h = c, resulting in vortices which are divided into 2 parts. Whereas the forward wingtip fence cant angle 90° h = 0.5c and the rearward wingtip fence cant angle 90° h = 0.5c, produce vortices such as plain wing but have greater value and area.

References

1. Genç MS, Özkan G, Açikel HH et al (2016) Effect of tip vortices on flow over NACA4412 aerofoil with different aspect ratios. EPJ Web Conf 114:2–5. https://doi.org/10.1051/epjconf/201611402027
2. Hariyadi SSP, Sutardi, Widodo WA (2018) Drag reduction analysis of wing airfoil E562 with forward wingtip fence at cant angle variations of 75° and 90°. In: AIP conference proceedings
3. Kontogiannis SG, Mazarakos DE, Kostopoulos V (2016) ATLAS IV wing aerodynamic design: from conceptual approach to detailed optimization. Aerosp Sci Technol 56:135–147. https://doi.org/10.1016/j.ast.2016.07.002
4. Mulvany N, Chen L, Tu J, Anderson B (2004) Steady-state evaluation of two-equation RANS (Reynolds-Averaged Navier-Stokes) turbulence models for high-reynolds number hydrodynamic flow simulations. Dep Defence, Aust Gov 1–54
5. Myilsamy D, Thirumalai, Yokesh PP (2015) Performance investigation of an aircraft wing at various cant angles of winglets using CFD simulation. Altair Technol Conf 23
6. Setyo Hariyadi SP, Sutardi, Widodo WA (2018a) Numerical study of flow characteristics around wing airfoil Eppler 562 with variations of rearward wingtip fence. In: AIP Conference Proceedings 1983. 10.1063/1.5046207
7. Setyo Hariyadi SP, Sutardi, Widodo WA, Mustaghfirin MA (2018b) Aerodynamics analysis of the wingtip fence effect on UAV wing. Int Rev Mech Eng 12:837–846. https://doi.org/10.15866/ireme.v12ihttps://doi.org/10.15517

8. Sohn MH, Chang JW (2012) Visualization and PIV study of wing-tip vortices for three different tip configurations. Aerosp Sci Technol 16:40–46. https://doi.org/10.1016/j.ast.2011.02.005
9. Turanoguz E, Alemdaroglu N (2015) Design of a medium range tactical UAV and improvement of its performance by using winglets. In: International Conference on Unmanned Aircraft Systems, ICUAS 2015, pp 1074–1083. https://doi.org/10.1109/icuas.2015.7152399
10. Yen SC, Fei YF (2011) Winglet dihedral effect on flow behavior and aerodynamic performance of NACA0012 wings. J Fluids Eng Trans ASME 133. 10.1115/1.4004420

Simulation of DC Motor Speed Control System Uses PSO to Determine Controller Parameters

R. Lullus Lambang G. Hidayat, Budi Santoso, Wibowo, and Iwan Istanto

Abstract This paper presents a new method for determining gains in a DC motor speed control system using PSO algorithm. The DC motor with a speed control system contains 2 feedback loops i.e. back emf loop and speed loop for the controller input. The controller used is the integral (I) controller and proportional-integral (PI) controller. The PSO algorithm is used to obtain the best gains for I controller and PI controller. In this paper, the PSO algorithms are PSO standard, PSO inertia and PSO constriction algorithms. The objective function is based on integral time absolute error (ITAE) criteria. The time response of DC motor with the speed controller obtained by each of PSO algorithm and the speed controller obtained by Ziegler-Nichols (ZN) method are compared. The simulation results show the transient response of the system with I and PI controllers tuned by PSO is better than the system with controller tuned by ZN method. The performance of each PSO algorithm is then qualitatively determined with a boxplot representation and hypothesis testing is performed to determine the performance of 3 pairs of PSO algorithms. The results show that the PSO-co algorithm is a better method to determine parameters of a DC motor speed controller than other PSO algorithms.

Keywords DC motor · PSO constriction · PSO inertia · PSO standard · Speed control

1 Introduction

Permanent magnet DC motor is widely used as a servomotor. The structure of servomotor are permanent magnet DC motor and a control system equipped with speed and/or position feedback. In general, DC motor has back electromotive force which

R. L. L. G. Hidayat (✉) · B. Santoso · Wibowo
Mechanical Engineering Department, Faculty of Engineering, Universitas Sebelas Maret, Surakarta, Indonesia
e-mail: lulus_l@staff.uns.ac.id; lulus_l@yahoo.com

I. Istanto
BPPT-Balai MEPPO, Klaster Teknologi Produksi, Maritim, dan Transportasi, Puspiptek Serpong, Tangerang Selatan, Indonesia

U. Sabino et al. (eds.), *Proceedings of the 6th International Conference and Exhibition on Sustainable Energy and Advanced Materials*, Lecture Notes in Mechanical Engineering,
https://doi.org/10.1007/978-981-15-4481-1_37

depend son the speed of motor. The control system of DC motor performs a measurement of actual motor speed and fed to the controller. Type of proportional-integral controller (PI) [1] and proportional-integral-derivative (PID) controller [2, 3] are commonly applied in industry.

Controller performance is affected by controller parameters. Case of PID controller, the problems of a controller design is to determine proper value of proportional, K_p, integral, K_i and derivative, K_d parameters so that the output system requirement is achieved. They are overshoot percentage, rise time, settling time, and steady state error which should be minimum for servomotor applications. Currently, PID controller tuning methods using optimization techniques has been developed. Particle Swarm Optimization (PSO) algorithm is an optimization technique that are preferably applied to solve PID controller design problem. Alruim Alhasan [4] used standard PSO, time varying inertia weight PSO and adaptive PSO to determine PID controller parameter of DC motor system. Performance of adaptive PSO was a superior compare to other algorithms in term of overshoot and settling time, also it provided fast convergence rate. Wu [5] proposed interactive evolution PSO, linear weight decrease PSO (LWDPSO) and Stochastic PSO (SPSO) to properly determine PID controller parameters. The result of this advanced research showed that IEPSO provide better performance than LWDPSO and SPSO. IEPSO constructed control effect that is better than ZN method, LWDPSO and SPSO. Qian et al. [6] proposed PSO-PID, that is PID controller-PSO algorithm, used to control an evaporation process in organic Rankine Cycle, whereas the optimum control of lights in office rooms had been used by applying PSO-PID as in [7].

Genetic algorithm is an artificial intelligent technique that is proven to produce a better transient response in rise time and settling time, compared to the ZN method [8]. In [9] the performance of the Particle Swarm Optimization (PSO) algorithm with the conventional ZN method and the Cohen Con method have compared and it was proved that the PSO algorithm is better for tuning PID controller. In addition, research of comparing the optimization algorithm based on GA, PSO and simulated annealing has been conducted [10]. The results showed that simulated annealing is a better PID controller optimization method compared to other artificial intelligent methods.

This research investigate the I and PI controller applied to DC motor control system. PI controller for servo motor and industrial application has extensively discussed in [11], however the method to determine parameter of PI controller is still an open problem. It starts with investigation of inner loop and outer loop applied to DC motor and then PID controller parameter is obtained by using PSO algorithm. PSO standard, PSO inertia and PSO constriction are applied and their performances are investigated. This research determines the best of the algorithms by using step input and statistical analytics of boxplot representation and hypothesis testing [12]. The result shows qualitative explanations and conclusions on the performance of the PSO standard, PSO inertia and PSO constriction to obtain parameter of PI controller.

This paper is organized as follows. Section 2 discusses the DC motor speed control model, controller I and PI strategies, and DC motor parameters. Section 3 briefly discusses optimization techniques with PSO that are applied to determine I/PI controller parameters. Section 4 discusses the research method that is controller performance simulation and statistical analysis to determine the performance of PSO std, PSO-in and PSO-co. Section 5 explains the results of the controller system simulation and the performance of each PSO algorithm. Furthermore conclusions are provided in Sect. 6.

2 DC Servomotor Speed Control Model

2.1 *Servomotor with I and PI Controller*

The equation of DC motor voltage regardless of the inductance is as follows [11]:

$$V = RI + C_m\omega_m, \tag{1}$$

C_m is the back emf constant, ω_m is the rotor rotational speed. The torque produced by the motor is proportional to the current: $T = K_t.\ I$, where K_t is the torque constant. The torque rotates the motor inertia and C is the viscous friction coefficient, the torque is:

$$T = Js\omega_m + C\omega_m \tag{2.a}$$

Substitution Eqs. (1) and (2) to the torque equation: $T = K_t.\ I$,

$$K_t(V - C_m\omega_m) = R(Js\omega_m + C\omega_m). \tag{2.b}$$

After some algebraic operation to Eq. (2.b), a first-order equation is obtained as follows:

$$\frac{\omega_m}{V} = \frac{K_t}{RJs + RC + K_tC_m}. \tag{3}$$

Figure 1 shows a DC servomotor block diagram with speed feedback and integral controller, *K/s*. The output speed transfer function of ω_m to set value ω_i is:

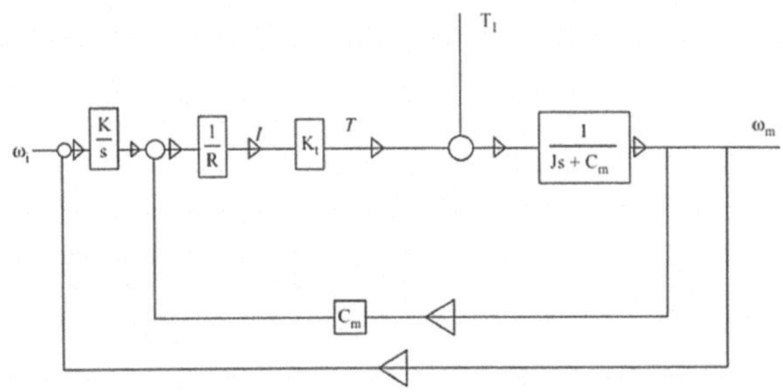

Fig. 1 DC servomotor block diagram with speed feedback [11]

Table 1 DC servomotor parameters [11]

Parameters	Symbol	Value	Unit
Resistance	R	7	Ω
Inductance	L	0.008436	H
Torque constant	K_t	0.094	N m/A
Back emf constant	C_m	0.094	V/rad/s
Viscous friction coefficient	C	1. e-5	N m/rad/s
Motor inertia	J	2.2097.e-4	N m/rad/s

$$\frac{\omega_m}{\omega_i} = \frac{KK_t}{RJs^2 + (RC + C_m K_t)s + KK_t}. \tag{4}$$

Equation (4) shows that in a steady state ($s = 0$), the output speed is ω_m, is equal to the input speed, ω_i.

In the integral controller, a proportional controller is added to obtain a fast rise time. Proportional and integral gain must be determined so that with a proper damping, a stable system is produced. The closed-loop system transfer function with a PI is:

$$\frac{\omega_m}{\omega_i} = \frac{K_t(K_p s + K_i)}{RJs^2 + (K_p K_t + CR + C_m K_t)s + K_i K_t}. \tag{5}$$

2.2 DC Servomotor Parameters

The servomotor parameters used in this study are shown in Table 1.

3 PSO Review

3.1 PSO Standard

PSO is a population-based optimization method developed by Kennedy and Eberhart in 1995 [13]. PSO is easy to do and the calculations do not employ gradients. PSO performs a solution to the optimization problem using particle populations (swarm). Each particle is a candidate for the problem solution. When an iteration is conducted, a particle changes its speed and position in the search space until a solution is found. The concept of particle velocity and position update in PSO is shown in Fig. 2.

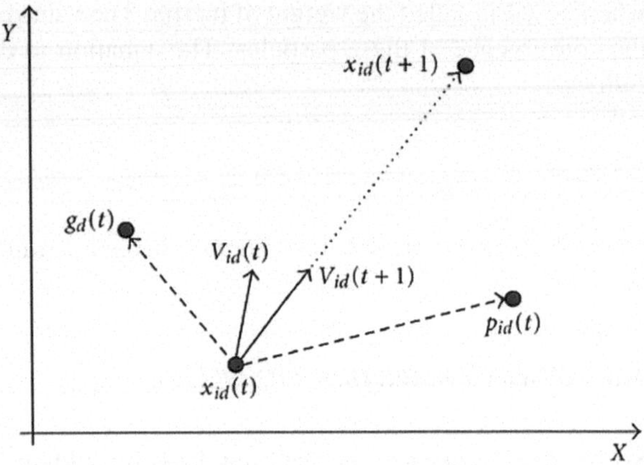

Fig. 2 Position and speed update of particle *i*

Swarm consist of M particles, particle *i* will have 3 vectors:

1. current position vector $x_{i,d} = \left[x_{i,d}^{1}, x_{i,d}^{2}, x_{i,d}^{3}, \ldots, x_{i,d}^{N}\right]$
2. current velocity vector: $v_{i,d} = \left[v_{i,d}^{1}, v_{i,d}^{2}, v_{i,d}^{3}, \ldots, v_{i,d}^{N}\right]$
3. best position vector: $P_{i,d} = \left[P_{i,d}^{1}, P_{i,d}^{2}, P_{i,d}^{3}, \ldots, P_{i,d}^{N}\right]$

N is a vector dimension. At the iteration of $d = d + 1$, the particle velocity is updated with the equation:

$$v_{i,d+1} = v_{i,d} + c_1.R1_{i,d}\left(P_{i,d} - x_{i,d}\right) + c_2.R2_{i,d}\left(G_d - x_{i,d}\right), \quad (6)$$

where c_1 and c_2 are acceleration coefficients, that is determined as $c_1 = c_2 = 2$. The parameters $R1_{i,d}$ and $R2_{i,d}$ are arrays of random numbers whose values are uniformly distributed between 0 and 1. G_d is the best global position. The position of *i*-particle is then updated with the equation:

$$x_{i,d+1} = x_{i,d} + v_{i,d}. \quad (7)$$

Equations (6) and (7) constitute the PSO standard algorithm that was first developed. Furthermore, in this research, two modifications to the PSO algorithm are discussed, i.e. PSO inertia and PSO constriction.

3.2 PSO Inertia (PSO-in)

The first modification of the PSO standard algorithm is a PSO with an inertia weighted or PSO-inertia algorithm. In this algorithm, the particle speed is updated using the formula:

$$v_{i,d+1} = w.v_{i,d} + c_1.R1_{i,d}\left(P_{i,d} - x_{i,d}\right) + c_2.R2_{i,d}\left(G_d - x_{i,d}\right), \quad (8)$$

where $w < 1$ is called the weight of inertia. The value of w is less than 1 to accelerate the convergence of these particles. The equation to determine the value of w is as follows:

$$w = w_{max} - (w_{max} - w_{min}) \times \frac{d}{d_{max}},$$

where $w_{max} = 0.9$, $w_{min} = 0.4$ and d_{max} is the maximum iteration.

3.3 PSO Constriction (PSO-Co)

Second modification of the PSO standard algorithm is PSO uses a constriction factor χ to control speed. In this algorithm, the speed is determined by:

$$v_{i,d+1} = \left[v_{i,d} + c_1.R1_{i,d}\left(P_{i,d} - x_{i,d}\right) + c_2.R2_{i,d}\left(G_d - x_{i,d}\right)\right]. \tag{9}$$

χ is calculated by

$$\chi = \frac{2}{\left|2 - \varphi - \sqrt{\varphi^2 - 4\varphi}\right|} \tag{10}$$

so that $\varphi = c_1 + c_2$, and c_1 and c_2 are determined so that $\varphi > 4$. This version of PSO algorithm is known as PSO constriction (PSO-co). PSO-in and PSO-co algorithms can accelerate the convergence rate of a swarm during iteration process effectively.

3.4 Objective Functions

DC servomotor speed controller is intended to obtain transient response of Eqs. (4) and (5) each using an integral controller and a proportional-integral controller. In the PID controller design, generally the performance criteria of Integral Square Error (ISE) and Integral Time absolute Error (ITAE) are used. The disadvantage of minimizing ISE criteria is the response with a small overshoot but long settling time because the same weight applied to the error does not depend on time. ITAE criteria can overcome the disadvantages of ISE criteria, but the process of deriving analytic formulas is complicated and time-consuming. The ISE and ITAE performance criteria are as follows:

$$ISE = \int_0^\infty e^2(t)dt \tag{11}$$

$$ITAE = \int_0^{\infty} t|e(t)|dt \tag{12}$$

In this research the minimization of ITAE is used as an objective function for PSO algorithm. The controller parameters obtained by minimizing ITAE will result in a minimal overshoot, rise time, settling time and steady-state error.

3.5 PSO Algorithm Procedure

The PSO algorithm is shown as the following pseudocode [13]:

Begin

Initialize the current position $X_{i,0}$, velocity $V_{i,0}$ and personal best position $P_{i,0}$ (setting $P_{i,0}$ = $X_{i,0}$) of each particle, evaluate their fitness values, and find the global best position G_0.

While (termination condition = false)

Do

Set d = d + 1;

Choose a suitable value of w; (for PSO-In)

for (i =1 to M)

  for j=1 to N

  $v^j_{i,d+1} = v^j_{i,d} + c_1 \cdot R1^j_{i,d}(P^j_{i,d} - x^j_{i,d}) + c_2 \cdot R2^j_{i,d}(G^j_{i,d} - x^j_{i,d})$ (for PSO orig);

  (or $v^j_{i,d+1} = w \cdot v^j_{i,d} + c_1 \cdot R1^j_{i,d}(P^j_{i,d} - x^j_{i,d}) + c_2 \cdot R2^j_{i,d}(G^j_{i,d} - x^j_{i,d})$ for PSO-In));

  (or $v^j_{i,d+1} = X[v^j_{i,d} + c_1 \cdot R1^j_{i,d}(P^j_{i,d} - x^j_{i,d}) + c_2 R2^j_{i,d}(G^j_{i,d} - x^j_{i,d})]$ (for PSO-Co));

  If $v^j_{i,d+1} > V_{max}$; $v^j_{i,d+1} = V_{max}$; end if

  If $v^j_{i,d+1} < -V_{max}$; $v^j_{i,d+1} = -V_{max}$; end if

$$x^j_{i,d+1} = x^j_{i,d} + v^j_{i,d+1}$$

  end for

  Evaluate the fitness value of $x_{i,d+1}$, that is, the objective function value $f(x_{i,d+1})$;

  Update $P_{i,d}$ and G_0

  end for

end do

end

N is the number of elements in vectors $x_{i,d}$ and $v_{i,d}$. At the initialization, the positions $x_{i,d}$ and velocity $v_{i,d}$ are in the range [$-xmax$, $xmax$] and [$-vmax$, $vmax$], respectively.

4 Research Method

4.1 ZN Method

The ZN tuning method is used to determine the integral controller parameter K in Eq. (4) and proportional-integral controller parameters, K_p and K_i on the Eq. (5). The ZN tuning method is based on the open-loop response to the transfer function as follows:

$$G(s) = \frac{K_t}{RJs + RC_m + K_t C_m} \tag{13}$$

Proportional-integral controller is determined using the Matlab code `[Gc, Kp, Ti] = ziegler (2, [K, L, T, 10])`. The output is an integral proportional controller $G_c = K_p + K_i/s$. Furthermore, the application of ZN tuning method to determine the integral controller parameters, K_i is performed on the transfer function:

$$G(s) = \frac{K_p s + K_i}{s} \frac{1}{K_p s + K_i} \frac{K_t}{RJs + RC_m + K_t C_m} \tag{14}$$

The integral controller is determined using the Matlab code `[Gc, Kp, Ti] = ziegler (2, [K, L, T, 10])`, based on the open-loop response of the transfer function *G(s)* in Eq. (14). The closed-loop system response with integral controller and proportional-integral controller with the ZN tuning method is then compared with the closed-loop system response with the controller parameters determined by PSO algorithms.

The Matlab script of ziegler function is as follows:

```
function [Gc, Kp, Ti, Td, H] = ziegler(key, vars)
Ti = []; Td = []; H = 1;
if length(vars) == 4;
        K = vars(1); L = vars(2); T = vars(3); N = vars(4); a = K*L/T;
if key == 1, Kp = 1/a;
        elseif key == 2, Kp = 0.9/a; Ti = 3.33*L;
        elseif key == 3 || key == 4, Kp = 1.2/a; Ti = 2*L; Td = L/2;
        end
elseif length(vars) == 3,
        K = vars(1); Tc = vars(2); N = vars(3);
if key == 1, Kp = 0.5*K;
        elseif key == 2, Kp = 0.4*K; Ti = 0.8*Tc;
        elseif key == 3||key == 4, Kp = 0.6*K; Ti=0.5*Tc;Td = 0.12*Tc;
end
elseif length(vars) == 5,
        K = vars(1); Tc = vars(2); rb = vars(3); N = vars(5);
        pb = pi*vars(4)/180; Kp = K*rb*cos(pb);
if key == 2, Ti = Tc/(2*pi*tan(pb));
elseif key == 3||key == 4,Ti=Tc*(1+sin(pb))/(pi*cos(pb));Td = Ti/4;
    end
end
[Gc,H] = writepid(Kp, Ti, Td, N, key, H);
```

Table 2 PSO parameters [13]

Parameters		Symbol	Value
Population size		M	50
Maximum iteration		d_{max}	50
Limit of position x		$[x_{min}, x_{max}]$	[−10, 10]
Limit of velocity v		$[v_{min}, v_{max}]$	[−5, 5]
PSO-std	Acceleration coefficients	C_1, C_2	2, 2
PSO-in	Inertia weight	$[w_{max}, w_{min}]$	[0.9, 0.4]
PSO-co	Acceleration coefficients	C_1, C_2	2.05, 2.05
	Constriction factor	χ	Equation (10)

4.2 *PSO Implementation*

The PSO algorithm consists of 2 stages, the first stage is the initialization of position, speed and *pbest* and the second stage is the implementation of PSO for each particle. PSO algorithm parameters as shown in Table 2.

4.3 *Statistical Representation and Hypothesis Test*

At each iteration of PSO algorithm, a fitness value is obtained. The fitness value, X is the objective function value of each particle. The sample data consists of 50 fitness values is acquired and then the mean best fitness value, $\overline{X}$ and its standard deviation, s are calculated.

The performance of the PSO-std, PSO-in and PSO-co algorithms is represented as boxplots. Boxplots is representations of value data samples that show the lower quartile, median, and upper quartile. While outliers are shown as data points outside the boxplot.

In this research, the hypothesis test is performed to make a decision regarding to a PSO algorithm pair, whether the mean best fitness value of the 2 algorithms is the same (Null hypothesis: H_0) and whether the best mean fitness value of the first algorithm is greater than the mean best fitness second algorithm (Alternative hypothesis: H_a). The hypothesis test consists of the following steps:

1. Determine the null hypothesis and alternative hypothesis. In this research, the null hypothesis and the alternative hypothesis are H_0: $\overline{X}_1 \geq \overline{X}_0$ and H_a: $\overline{X}_1 < \overline{X}_0$.
2. Determine the significance level. The significance level is the probability that the null hypothesis is rejected while it is actually true. In this study, the hypothesis test is one-tailed test and the significance level is chosen equal to 0.05.

3. Determine the type of statistical test. This study uses the assumption of a normal distribution because of the large number of sample data ($n > 30$), so the statistical test uses the standard z normal distribution, i.e.

$$z = \frac{\overline{X}_1 - \overline{X}_2}{\sqrt{\left(s_1^2/n_1\right) + \left(s_2^2/n_2\right)}}, \quad (15)$$

where $\overline{X}_1$ and $\overline{X}_2$ are the mean best fitness of the first and second PSO algorithms, s_l and s_2 are the standard deviations of sample of fitness values of the first and second PSO algorithm, respectively and $n_1 = n_2 = n$ is the PSO sample size. The probability value, p is determined from the z value.
4. Formulate decision making rules. Decision rules are formulated based on H_0 and H_a, level of significance and statistical tests.
5. Accept or reject the null hypothesis, H_0 and interpret the results based on the probability value on the level of significance. If p is greater than significance level, the mean best fitness value of the first PSO algorithm occurs by accident whereas if p is smaller than the significance level, the value of the mean best fitness of the first PSO algorithm is greater than the value of the second PSO algorithm at 0.05 level of significance (determined in step 2).

5 Results and Discussion

5.1 *Response System with Integral Controller*

Figure 3 shows responses of the system with the integral controller obtained by ZN, PSO-std, PSO-in and PSO-co tuning methods. Table 3 shows the transient response characteristics of the system to step function. Therefore the selected characteristics of DC servomotor rotation speed control is fast rise time and settling time, according to the application in industries.

The system responses with the PSO-I controller shows almost the same transient characteristics of percent overshoot, rise time, settling time and the K value. The PSO algorithm produces a larger gain K than the ZN method, causing a larger percent overshoot. Integral gain, K is more like a proportional gain for second order system. Furthermore, a proportional-integral controller is applied so that a response with small percent overshoot and faster settling time than the system response with an integral controller only, is obtained.

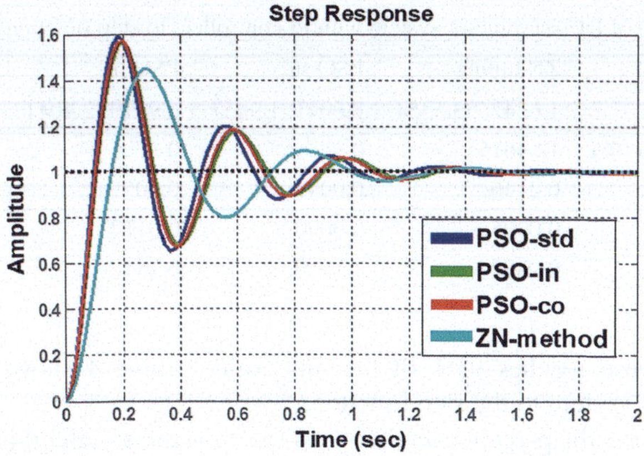

Fig. 3 The response of the stairs of the system with an integral controller

Table 3 Response characteristics of DC servomotor system ladders with integral controller

Characteristics	ZN tuning	PSO std	PSO in	PSO co
K	2.2224[a]	4.4146	4.4276	4.2726
Percent overshoot (%)	44.6870	56.8314	56.8533	56.4886
Rise time (s)	0.1088	0.0733	0.0732	0.0747
Settling time (s)	1.2158	1.2306	1.2290	1.2485

[a]ZN tuning, K is obtained by trial and error: i.e. $1e5 \times K$

5.2 *Response System with Proportional-Integral Controller*

Figure 4 shows the responses of the proportional-integral controller system obtained by ZN, PSO-std, PSO-in and PSO-co methods. Figure 4b is enlargement of Fig. 4a

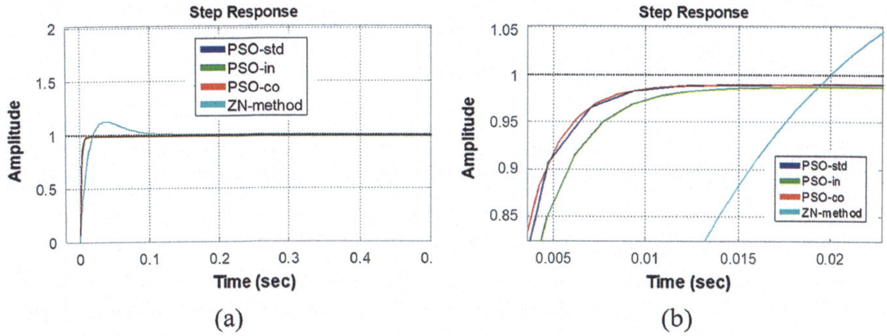

Fig. 4 Step response of a system with proportional-integral controller

Table 4 Response of DC servomotor system with PI controllers to step function

Characteristics	ZN tuning	PSO std	PSO in	PSO co
K_p; K_i	1.6242; 53.5354	8.6041; 1.9657	6.9019; 2.4041	8.6149; 4.3937
Percent overshoot (%)	12.4615	0	0	0
Rise time (s)	0.0146	0.0043	0.0055	0.0043
Settling time (s)	0.0905	0.0090	0.0117	0.0087
Peak time (s)	0.0382	1.1929	0.7842	0.5341

when the response reaches 95% of the final value. Table 4 shows the transient responses of the system to the step function.

Figure 4a shows the percent overshoot and rise time are smaller than the response with an integral controller only (Fig. 3). Figure 4b shows the rise time and settling time of each PSO-PI controller. PI controller with PSO-in algorithm provides the slowest rise time compared to other PSO-PI algorithms, because its K_p value is smaller than the K_p value of other PSO-PI controllers.

Table 4 shows that the PSO-PI controller can eliminate percent overshoot and improve rise time, and settling time. Peak time is determined when system response reaches 95% of the final value when PSO-PI controller is applied. Table 4 shows the PI controller with the PSO-co algorithm produces the fastest peak time. This controller should be applied for the DC servomotor speed controller because the motor reaches the full speed fastest.

From the results of systems with integral controllers and systems with proportional-integral controllers, it can be seen that the PSO-PI controller provides better system response characteristics compared to controllers with parameters determined by the ZN method. The PSO-PI controller is superior, especially in peak overshoot because it can determine the K_p parameter values that increases damping of DC servomotor control systems.

Furthermore, this research examines the performance of each PSO-PI algorithm using bloxplot representation and hypothesis testing.

5.3 PSO-PI Algorithm Performance

Boxplot graph The PSO-PI program is run and fitness value data is obtained. The fitness value is the best global value of ITAE minimization for each iteration. Each PSO-PI algorithm runs 50 iterations. Figure 5 shows samples of fitness value data from PSO-PI controller algorithm as boxplots.

The performance of PSO-std, PSO-in and PSO-co to determine the proportional-integral controller parameters are shown in Fig. 5a–c, respectively. The sequence of boxplots shows that a solution obtained is increasingly converging after 20 iterations. The data distribution is indicated by the size of boxplot that contains the lower quartile, median and upper quartile, while the outliers are shown as data points

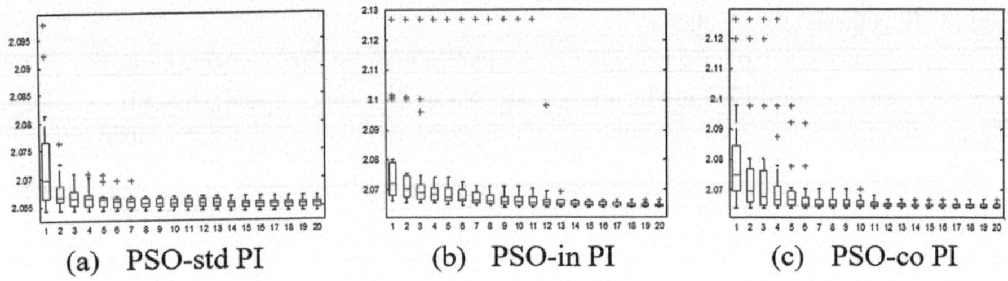

Fig. 5 Boxplot of sample data for PSO-PI objective values (data for 20 iterations)

Table 5 Mean best fitness and standard deviation of fitness value samples

Controller	Mean best fitness, $\overline{X}$	Deviasi standar, s
PSO-std PI	1.8886e−005	8.4454e−005
PSO-in PI	9.8491e−009	4.4045e−009
PSO-co PI	3.5376e−009	1.2173e−008

(+sign). PSO-co PI shows a convergent solution with decreasing outlier value by the growing iteration. PSO-in PI shows that there are still large outliers until the 10th iteration. Whereas the PSO-std PI has a distribution of fitness values larger than distribution of fitness values of PSO-in PI and PSO-co PI.

Hypothesis Test. The PSO algorithm uses 50 particles and 50 iterations. For each iteration, the average value of fitness X is obtained and the mean best fitness value, $\overline{X}$ and standard deviation, s is calculated. Table 5 shows the mean best fitness values and standard deviations for each PSO-PI algorithm.

The smallest mean best fitness value is generated by the PSO-co PI algorithm, while the smallest standard deviation value is generated by the PSO-in PI algorithm. This result is in accordance with the boxplot graph representation.

Furthermore the hypothesis test is conducted to make a decision whether the mean best fitness value of a PSO-PI algorithm pair is the same or the best mean fitness value of the first PSO-PI algorithm is greater than the mean best fitness algorithm of the 2nd algorithm. Hypothesis testing also determines if the mean best fitness value is different, if it happens by chance. Hypothesis testing follows 5 steps as in Sect. 4.3.

Statistical values to perform hypothesis test are shown in Table 6.

Table 6 shows that significance occurred in the pair of PSO-in and PSO-co PI algorithms. In this pair, it was concluded to reject the null hypothesis and then accept an alternative hypothesis with a level of confidence of 95%.

6 Conclusion

This research develops the method to determine the I and PI controllers parameters for DC servomotor speed control. Compared to the I controller, PSO-PI controller

Table 6 Hypothesis test calculation

Compared Statistics	PSO-std versus PSO-in PI	PSO-std versus PSO-co PI	PSO-in versus PSO-co PI
mean difference: $\overline{X}_0 - \overline{X}_1$	1.8876e−005	1.8882e−005	6.3115e−009
Standard error: $s = \sqrt{(s_0^2/n_1) + (s_1^2/n_2)}$	1.1944e−005	1.1944e−005	1.8307e−009
95% confidence interval: $\overline{X}_0 - \overline{X}_1 \pm 1.96s$	[−4.5333e−006, 4.2286e−005]	[−4.5270e−006, 4.2292e−005]	[2.7232e−009, 9.8998e−009]
z value	1.5804	1.5810	3.4475
p value	0.057008	0.056939	0.000283
Significance ($p <$ 0.05)	Not significant	Not significant	Significant

has a smaller percent overshoot and faster rise time. PSO-PI controller parameters obtained using the PSO-co algorithm produce the best response compared to other PSO algorithms (PSO-std and PSO-in). The data representation of the PSO-PID co-calculation process with the boxplot shows a convergent solution. Hypothesis testing shows that the PSO-co PI algorithm provides a significantly smaller mean best fitness compared to the PSO-in PI algorithm.

Acknowledgements This work was supported by DP2M DIKTI (Directorate of Research and Public Service of Directorate General of Higher Education) Ministry of Research, Technology and Higher Education Indonesia with contract number 718/UN27.21/2019.

References

1. Romasevych Y, Loveikin V, Usenko S (2019) PI-controller tuning optimization via PSO-based technique. National University of Life and Environmental Sciences of Ukraine. https://doi.org/10.15199/48/2019.07.08
2. Ziegler JG, Nichols NB, Rochester NY (1942) Optimum setting of automatic controllers. Trans ASME
3. Ogata K (2010) Modern control engineering, 5th edn. Prentice Hall
4. Alruim Alhasan H, Gunes M (2017) A new adaptive particle swarm optimization based on self-tuning of PID controller for DC motor system. Cukrova Univ J Fac Eng Archit 32(3):243–249
5. Wu S (2017) A PID controller parameter tuning method based on improved PSO. Int J Adv Comput Res 8(34). ISSN (Print): 2249-7277. ISSN (Online): 2277-7970. https://doi.org/10.19101/ijacr.2017.733026
6. Qian W, Wang Y, Chen L, Fu W (2017) PSO-PID control applied in evaporator of organic Rankine cycle steam. GRC Trans 41

7. Copot C, Mac Thi T, Ionescu C (2018) PID based particle swarm optimization in offices light control. In: 3rd IFAC conference on advances in proportional integral derivative control, Ghent, Belgium, 9–11 May 2018
8. Elsrogy WM, Fkirin MA, Moustafa Hassan MA (2013) Speed control of DC motor using PID controller based on artificial intelligence techniques. In: Conference: proceeding of 1st IEEE international conference on control, decision and information technology (CoDIT 2013), Tunisia
9. Mishra AK, Narain A (2013) Speed control of DC motor using particle swarm optimization. Int J Res Eng Res Technol (IJERT) 2(6). ISSN: 2278-0181
10. Singh SK, Katal N, Modani SG (2014) Optimization of PID controller for brushless DC motor by using bio-inspired algorithms. Res J Appl Sci Eng Technol 7(7):1302–1308
11. Firoozian R (2014) Servo motors and industrial control theory, 2nd edn. Springer, Berlin. ISSN 2192-063X (electronic), ISBN 978-3-0727503 (eBook)
12. Walpole RE, Myers RH, Myers SL, Ye K (2007) Probability & statistics for engineers & scientist, 8th edn. Pearson Prentice Hall, Upper Saddle River, N.I 07458 (Box plot)
13. Sun J, Lai C-H, Wu X-J (2012) Particle Swarm Optimisation-classical and quantum perspectives. Chapman & Hall/CRC Numerical Analysis and Scientific Computing (Statistical Comparison)

Polytetrafluoroethylene-Packaged Singlemode-Multimode-Singlemode Fiber Structure for Temperature Sensor

Rima Fitria Adiati and Agus Muhamad Hatta

Abstract We investigated a packaging scheme for a singlemode-multimode-singlemode (SMS) optical fiber structure for a temperature sensor. The packaging material functions to protect the small and fragile sensor against the harsh environment as well as to increase the sensor sensitivity. *Polytetrafluoroethylene* (PTFE) material was chosen due to its excellent dielectric properties; hence the sensor applies to areas with electromagnetic interference. The effect of PTFE packaging is observed numerically and experimentally on a 43-mm multimode fiber-length SMS fiber sensor at the temperature range of 25–100°C. The temperature change causes thermal expansion of optical fiber and the packaging, therefore, induces changes in the optical power output of the sensor. The numerical calculation is based on modal propagation analysis of the light in the SMS fiber structure, while taken into consideration the coefficient of thermal expansion (CTE) of packaging material. The experiment includes the packaging in a rectangular shape, with 1, 2, 3, 5, 6, and 10-mm PTFE thickness, applied in single-sided design. Intensity-based interrogation method shows that the sensitivity of the sensor increases when PTFE packaging is applied, compared to the bare SMS fiber structure sensor.

Keywords Optical fiber sensor · Temperature sensor · SMS fiber structure · Sensor packaging · PTFE

1 Introduction

Optical sensing has been used in several applications due to its unique characteristics and advantages compared to electronic sensors that have been widely known. The optical sensor provides a certain depth of information, with longer lifetime and small size [1] enables specific application in which conventional sensor is unable to provide. The ability of fiber optic sensors has been enhanced to substitute traditional sensors for acoustics, pressure, electric and magnetic field measurement,

R. F. Adiati (✉) · A. M. Hatta
Photonics Engineering Laboratory, Department of Engineering Physics, Faculty of Industrial Technology, Institut Teknologi Sepuluh Nopember, Surabaya 60111, Indonesia
e-mail: rima.adiati@gmail.com

U. Sabino et al. (eds.), *Proceedings of the 6th International Conference and Exhibition on Sustainable Energy and Advanced Materials*, Lecture Notes in Mechanical Engineering,
https://doi.org/10.1007/978-981-15-4481-1_38

vibration, acceleration, rotation, temperature, linear and angular position, strain, viscosity, humidity, chemical measurements and some others. They can be used in high voltage, high temperature, or corrosive environments due to its dielectric property; Besides, these sensors are compatible with communications systems and can carry out remote sensing [2]. These advantages are the cause for the application of an optical fiber sensor for monitoring system, one includes a temperature sensor.

One of the possible fiber structures for temperature sensing is singlemode-multimode-singlemode (SMS). Using the multimode interference phenomena, the structure provides sensitive measurement while maintaining easy fabrication, low-cost, and stable configuration [3]. Aside from the temperature, the SMS fiber structure has been applied as various sensors, e.g. refractive index [4], strain [5], vibration [6], or even simultaneous measurements [7, 8].

Real-world application of SMS fiber structure requires other materials as packaging, given the fragile nature of the optical fiber. Previous researches show the utilization of metals as sensor packaging to enhance the sensitivity of the sensor, for example, nickel, steel [9], and aluminum [10]. Polypropylene, epoxy, quartz glue [11], PMMA [12], and Polyurethane-acrylate [13] are some examples in which polymers applied as the packaging of the optical fiber sensors.

This paper observes the design of sensor packaging as a protection as well as to increase the sensor performance, i.e., its sensitivity. The material chosen for the packaging is Polythetaflouroethylene (PTFE), which has the commercial name of Teflon. PTFE is a polymer with excellent chemical resistance and is immune to electromagnetic interference [14]. The effect of packaging is observed both numerically and experimentally on a 43-mm MMF-length SMS fiber sensor at the temperature range of 25-100°C. This research stands as a preliminary step of optical fiber sensor application in areas with the electromagnetic field; hence the use of polymer as packaging. The effect of packaging was observed in the power output analysis.

2 Method

This paper explores the steps in both numerical analysis and experiments, such as literature studies, formulation of packaging effects to the sensor output, selection of SMS fiber parameters, fabrication of SMS sensor, application of packaging, and characterization of sensors.

2.1 Numerical Analysis

The numerical analysis carried out in this paper is based on Modal Propagation Analysis (MPA). This method explains the propagation of light inside the SMS fiber structure. The schematics of SMS fiber is shown in Fig. 1. In the multimode (MMF) part of SMS fiber structure, the light of one mode interferes with one another and

Fig. 1 Schematics of singlemode-multimode-singlemode fiber structure [15]

creates a periodical constructive-destructive fringe. This particular phenomenon is called the multimode interference (MMI), while constructive interference is called self-imaging points. The self-imaging points located periodically along the light propagation in optical fiber, resulting in different power outputs for different MMF lengths. Therefore, the MMF length becomes one important parameter in designing an SMS fiber structure as a sensor. Using the following calculation, one can calculate the response of the SMS sensor for each variation of MMF length, in which the best variation will be used in the experiment stage.

In the input singlemode fiber (SMF), there is only one fundamental mode with field profile $E_{(r,0)}$. This mode will divide itself into several linearly polarized $\left(LP_{n,m}\right)$ mode with field profile $F_{m(r)}$. The mathematical formula for field distribution is [16]:

$$\int_0^\infty |E(r,0)|^2 r dr = \int_0^\infty |F_m(r)|^2 r dr \tag{1}$$

$$E(r,0) = \sum_{m=1}^{M} c_m F_m(r) \tag{2}$$

where c_m is the excitation coefficient for each m modes estimated by:

$$c_m = \frac{\int_0^\infty E(r,0) F_m(r) r dr}{\int_0^\infty F_m(r) F_m(r) r dr} \tag{3}$$

The electric field distribution varies during the propagation of light, affected by the distance z,

$$E(r,z) = \sum_{m=1}^{M} c_m F_m(r) \exp(i\beta_m z) \tag{4}$$

where

r	Core radius of MMF
n_c	Refractive index of the core
n_{cl}	Refractive index of the cladding
λ	Wavelength
c_m	Excitation coefficient of mode number m
$F_m(r)$	Field profile of mode number m

β_m Propagation constant

Equation (4) shows that the propagation length affects the field distribution. The coupling efficiency, then the SMS structure output is influenced by the MMF length. Under the assumption that lead-in and lead-out singlemode fibers are identical, the coupling loss Ls of SMS optical fiber structure stated by:

$$L_s(z) = 10\log_{10}\left(\left|\sum_{m=1}^{M} c_m^2 \exp(i\beta_m z)\right|^2\right) \tag{5}$$

The output power P_{out}(dB) at the connection between MMF end and SMF lead-out can be calculated with [16]:

$$P_{out} = 10\log_{10}\frac{\left|\int_0^\infty E(r,z)E_0(r)rdr\right|^2}{\int_0^\infty |E(r,z)|^2 rdr \int_0^\infty |E_o(r,z)|^2 rdr} \tag{6}$$

MMI phenomena happening in the MMF can be affected by external stimuli such as temperature changes. The temperature inflicts changes in SMS fiber structure parameters; hence the changes in output power. The numerical analysis method used in this paper to show the effect of temperature changes toward the SMS fiber sensor is intensity-based interrogation system [17].

For a temperature sensor application, a coefficient of thermal expansion (CTE) for packaging material α_1, and silica fiber optics α_2, play important roles. When the temperature varies, the difference CTE values between packaging material and silica fiber induces an axial strain, as follows [18]

$$\varepsilon = (\alpha_2 - \alpha_1)\Delta T \tag{7}$$

where ΔT is a temperature variation in °C, $\alpha_{1,2}$ is CTE in 1/°C and ε is an axial strain in $\mu\varepsilon$. The values of CTE for PTFE and other materials used in simulation are shown in Table 1.

The temperature variation, as well as the axial strain induced by temperature, affect the MMF length, fiber core radius, and its refractive indices given by [22]:

$$\Delta L = L\varepsilon + \alpha_1 L\Delta T \tag{8}$$

Table 1 CTE values of selected materials

Material	CTE (/°C)	Reference
Silica (optical fiber)	5×10^{-7}	[19]
Steel	1.1×10^{-5}	[18]
Aluminium	2.36×10^{-5}	[20]
PTFE	1.24×10^{-4}	[21]

$$\Delta r = -\sigma a_{SMF,MMF}\varepsilon + \sigma r_{SMF,MMF}\Delta T \quad (9)$$

$$\Delta n_{SMF,MMF} = -p_e\varepsilon + \beta n_{SMF,MMF}\Delta T \quad (10)$$

where ΔL is an MMF length variation, Δr is core radius variation, σ is the Poisson ratio, $\Delta n_{SMF,MMF}$ is the refractive index variation of core and cladding, p_e is the effective strain-optic coefficient, and β is a thermo-optic coefficient.

Substituting eq. (7) into eqs. (8–10), it can be obtained:

$$\Delta L = L(\alpha_2 - \alpha_1)\Delta T + \alpha_1 L\Delta T + \alpha_2 L\Delta T \quad (11)$$

$$\Delta a = -\sigma r_{SMF,MMF}(\alpha_2 - \alpha_1)\Delta T + \sigma r_{SMF,MMF}\Delta T \quad (12)$$

$$\Delta n_{SMF,MMF} = -p_e(\alpha_2 - \alpha_1)\Delta T + \beta n_{SMF,MMF}\Delta T \quad (13)$$

The presence of packaging material affects the parameters of the SMS fiber structure. The numerical analysis shows how different packaging material causes changes in eq. (11) to (13) and therefore different power output as shown in eq. (6).

2.2 *Experiment*

Fabrication of the SMS fiber structure is the first step of the experiment stage. Fusion splicer is used to splice the 43 mm-length MMF to SMF pigtail at each end. The decision of using 43 mm-length of MMF is based on the previous calculation, where the response has the most linearity compared to other lengths between 40–45 mm length. The packaging design is shown at Fig. 2a. The dimension of the PTFE material is 75 × 20 mm, with 1, 2, 3, 5, 6, and 10 mm thickness variation. Epoxy resin is used for permanent adhesive.

The characterization of the SMS sensor firstly carried out using an intensity-based interrogation method. The output power is measured at the temperature range of 25 °C–100 °C, with 5 °C interval. Set up for the method is shown at Fig. 2b. The light source used has operational wavelength of 1550 nm and output power −6 dBm. The power versus temperature graph then compared for bare sensor and sensor with packaging.

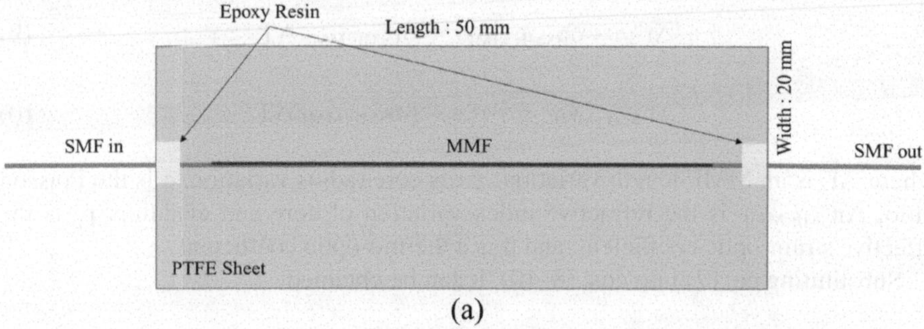

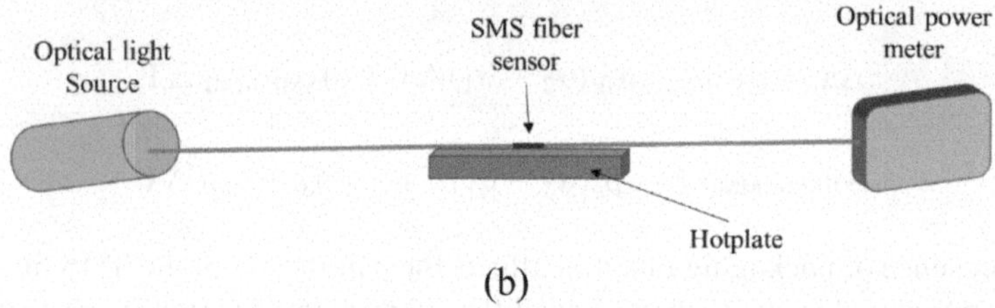

Fig. 2 Experimental set up of proposed SMS fiber temperature sensor, **a** top view of packaging design, **b** intensity measurement set up

3 Results and Discussions

3.1 Numerical Analysis

Figure 3 shows the calculated output power at $\lambda = 1550$ nm with temperature range 0 °C–100 °C. There is consistence of increasing or decreasing response for all sensors, with or without packaging. The sensitivity of the sensor is the slope of the power versus temperature line. It is shown that for the bare sensor, the gradient has a positive value, while sensor with packaging has negative value and higher sensitivity. A negative slope indicates that the CTE of the packaging material is high enough, and the SMS fiber structure-based devices were overcompensated. The sensitivity of the PTFE-packaged sensor is the highest among other packaging materials. It is caused by PTFE characteristic as thermoplastic with high-temperature conductivity. It is also an alternative that PTFE becomes a solution when metal packaging cannot be used in such situations as in free magnetic interference areas.

In these numerical calculations, the sensitivity of the bare sensor is 0.011 dB/°C, while the PTFE-packaged sensor reaches −0.105 dB/°C. This calculation shows that the packaging application can enhance the sensitivity of SMS fiber structure up to 9.55 times higher than the bare sensor. The sensitivity is in line with CTE values shown in Table 1, where the higher CTE value of the material, the sensitivity of the packaged sensor is also higher. However, the numerical calculation cannot show the effect of packaging thickness and shape to the sensitivity.

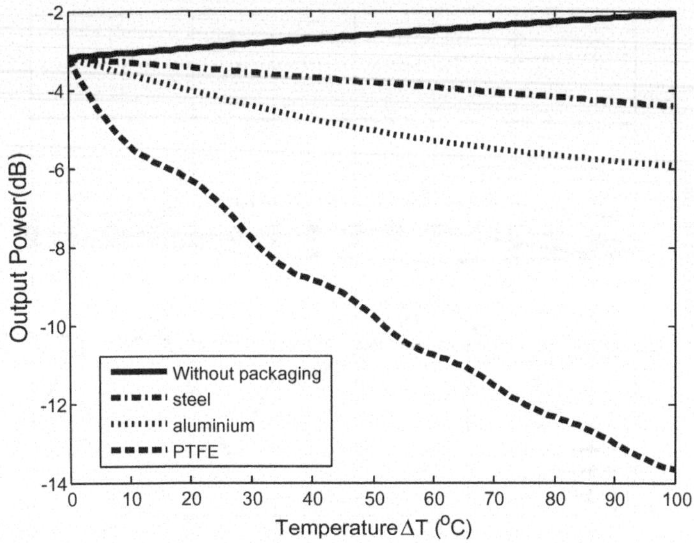

Fig. 3 Simulation result: the output response of SMS fiber structure temperature sensor with MMF length = 43 mm at various packaging material

3.2 Experiment Results

Figure 4 shows the SMS fiber sensor with PTFE packaging. The thickness variation is 1, 2, 3, 5, 6, and 10 mm, and the shape is a one-sided bar. As shown in the figure, epoxy resin is used in each tip of the fiber. A very thin v-groove was cut into the PTFE material to increase the contact area between PTFE and optical fiber.

The response of the SMS fiber as temperature sensor is shown in Fig. 5. Similar to Fig. 3, the sensitivity of the sensor can be calculated as the gradient of the power versus temperature lines. It can be examined for the bare sensor that the gradient is positive, similar to the numerical calculation with a sensitivity of 0.0102 dB/°C. The response for a PTFE-packaged sensor has different values of sensitivity depends of the thickness.

The measured output power in the sensor with 2 mm packaging has the most fluctuation among others. For thickness variation of 5, 6, and 10 mm, linear results only obtained in the certain temperature range. The sensitivity at all measurements

Fig. 4 SMS fiber sensor with PTFE packaging

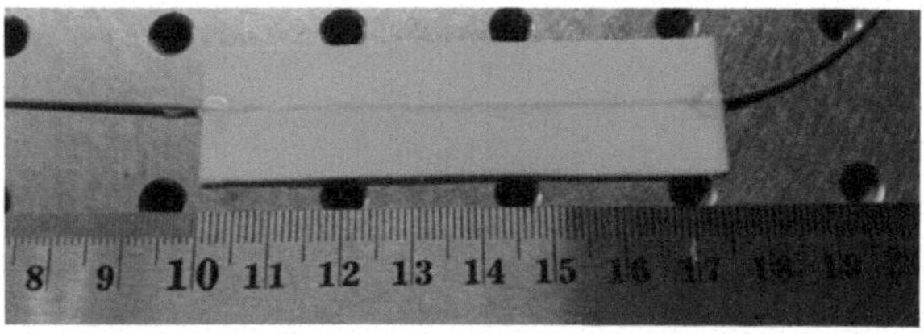

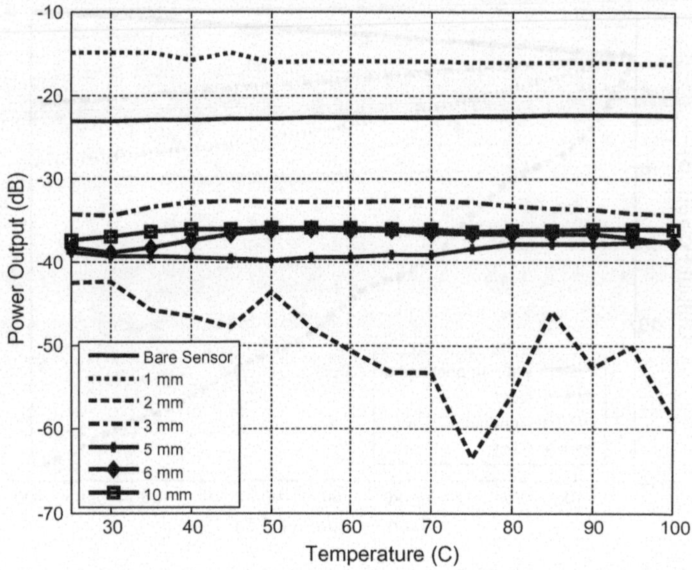

Fig. 5 Experiment results: the output response of SMS fiber structure temperature sensor at different packaging thickness, epoxy resin as glue

are also lower than simulated values. These unexpected measurements are the result of imperfect thermal expansion of PTFE. Direct observation indicates that the high difference of CTE between optical fiber (silica material) and PTFE, combined with epoxy resin locking the material expansion, causing the bending of PTFE material. The heat transfer process that happens from one direction (from below the sensor) also increases the probability of fluctuated results. In Table 2, it is presented the sensitivity comparison between numerical calculation and experimental.

Table 2 shows that the simulation cannot yet determine the effect of packaging thickness because it was assumed that the thermal expansion happens only axially.

Table 2 The sensitivity of PTFE-packaged SMS fiber temperature sensor

Type of PTFE packaging	Thickness	Sensitivity (dB/°C)		
		Simulation	Experiment	remarks
Bare sensor	–	0.0110	0.0102	
One-sided packaging	1 mm	−0.1050	−0.0179	
	2 mm	–	−0.2155	
	3 mm	–	−0.0471	for 65 °C–100 °C
	5 mm	–	−0.0437	for 25 °C–50 °C
	6 mm	–	−0.0352	for 55 °C–100 °C
	10 mm	–	0.0176	

However, from the experiment, the 2 mm packaging offers the highest sensitivity among others. But, the results also show fluctuations and inconsistency of trendline. It comes as a consideration while designing the packaging dimension because packaging thickness is related to heat distribution, hence the time response of the sensor. A very thin PTFE packaging material creates a fast-responding sensor, but less steady response. If the packaging is too thick, the sensor cannot provide an accurate measurement.

The results in Table 2 also have a different measurement range, caused by the usage of different SMS sensor. To assign one sensor for various PTFE thickness is not possible since epoxy resin is permanent glue. The proposed solution is using impermanent glue such as scotch tape or masking tape. The result of this experiment is shown in Fig. 6, while Table 3 points out the sensitivity of the sensor.

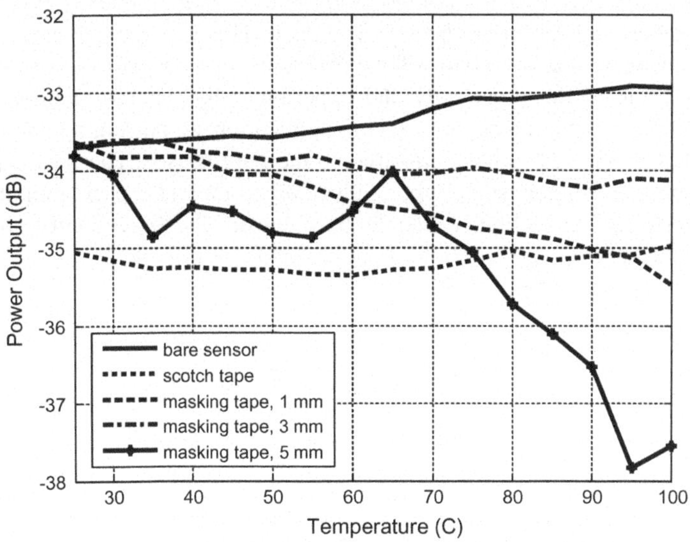

Fig. 6 Experiment results: the output response of the SMS fiber structure temperature sensor at different glues and packaging thickness

Table 3 The sensitivity of PTFE-packaged SMS fiber temperature sensor for various glue material

Type of glue	Thickness	Sensitivity (dB/°C)	Remarks
Bare sensor	–	0.0103	
Scotch tape	1 mm	0.0095	for 60 °C–100 °C
Masking tape	1 mm	−0.0244	
	3 mm	−0.0060	
	5 mm	−0.0498	

As shown in Fig. 6, the sensor using 1 mm packaging and masking tape yields the best result in terms of sensitivity and linearity. For 1 mm packaging thickness, the sensitivity is −0.02443 dB/°C, with 87.61% linearity. 3 mm packaging using masking tape also has similarly linear result but lower sensitivity, while 5 mm packaging has fluctuations in the effect. The usage of scotch tape as glue cannot provide linear result outside the range of 60 °C–100 °C.

Possible future research is to find the best combination of packaging material, thickness, glue material, and possibly, the shape of the packaging.

4 Conclusions

PTFE packaging scheme for a singlemode-multimode-singlemode (SMS) fiber structure as a temperature sensor has been investigated. The packaging material has been proved to increase the sensitivity of the SMS fiber sensor, with a 43-mm length of MMF section, both numerically and experimentally. The simulation shows in intensity-based measurement, how PTFE application as packaging increases the sensitivity of about 9.55 times compared to the bare sensor. The best experimental result, namely using masking tape as glue and 1 mm PTFE packaging thickness, has sensitivity 2.37 times higher than the bare sensor. The experiment also presents the different responses for variation of packaging thickness.

References

1. Rao YJ, Huang S (2002) Applications of fiber optic sensors. In: Yu FT, Yin S (eds) Fiber optic sensors. Marcel Dekker Inc, New York, pp 449–490
2. Sabri N, Aljunid SA, Salim MS, Ahmad RB, Kamaruddin R (2013) Toward Optical Sensors: Review and Applications. J Phys: Conf Ser. no. 012064, p 423
3. Wang P, Zhao H, Wang X, Farrell G, Brambilla G (2018) A review of multimode interference in tapered optical fibers and related applications. Sensors 18(858):1–29
4. Li Y, Liu Z, Jian S (2014) Multimode interference refractive index sensor based on coreless fiber. Photonics Sens 4(1):21–27
5. Sari D, Hatta AM, Pratama DY (2016) Packaging aluminium impacting on optical fiber with SMS fiber structure
6. Waluyo TB, Bayuwati D (2016) An SMS (single mode—multi mode—single mode) fiber structure for vibration sensing. J Phys: Conf Ser 817(012035):1–6
7. Sun Y, Liu D, Lu P, Sun Q, Yang W, Wang S, Liu L, Zhang J (2017) Dual-Parameters optical fiber sensor with enhanced resolution using twisted MMF based on SMS structure. IEEE Sens J 17(10):3045–3051
8. Wu Q, Hatta AM, Wang P, Semenova Y, Farrell G (2011) Use of a bent single SMS fiber structure for simultaneous measurement of displacement and temperature sensing. IEEE Photonics Technol Lett 23(2):130–132
9. Lin YB, Chang KC, Chern JC, Wang LA (2005) Packaging methods of fiber-bragg grating sensors in civil structure applications. IEEE Sens J 5(3):419–423
10. Chotimah AK, Hatta AM, Pratama DY (2016) The effect of packaging material on optical fiber temperature sensor with singlemode multimode singlemode (SMS) structure

11. Zhang Y, Tian X, Xue L, Zhang Q, Yang L, Zhu B (2013) Super-high sensitivity of fiber temperature sensor based on leaky-mode bent SMS structure (2013). IEEE Photonics Technol Lett 25(6):560–563
12. Dreyer U, de Sousa KM, Somenzi J, de Lourenço I Jr, de Oliveira V, Kalinowski HJ (2013) A technique to package fiber bragg grating sensors for strain and temperature measurements. J Microwaves, Optoelectron Electromagnetic Appl 12(2):638–646
13. Zhang Y, Xue L, Wang T, Yang L, Zhu B, Zhang Q (2014) High performance temperature sensing of single mode-multimode-single mode fiber with thermo-optic polymer as cladding of multimode fiber segment. IEEE Sens J 14(4):1143–1147
14. Polyfluor Plastics BV PTFE (polytetrafluoroethylene). [Online]. Accessed 1 Oct 2018
15. Wang X, Lewis E, Wang P (2017) Investigation of the self-imaging position of a singlemode-multimode-singlemode optical fiber structure. Microwave Opt Technol Lett 59:1645–1651
16. Wang Q, Farrell G, Yan W (2008) Investigation on single-mode–multimode–single-mode fiber structure. J Lightwave Technol 26(5):512–519
17. Hatta AM, Indriawati K, Bestariyan T, Humada T, Sekartedjo (2013) SMS fiber structure for temperature measurement using an OTDR. Photonic Sens 3(3):262–266
18. Li E (2007) Temperature compensation of multimode interference-based fiber devices. Opt Lett 32(14):2064–2066
19. Li E, Peng G-D (2008) Wavelength-encoded fiber-optic temperature sensor with ultra-high sensitivity. Opt Commun 281:5768–5770
20. Wang W-H, Feng Y-J, Shi W-Q, Xiong Z-Y, Li S-D, Wu W, Lin J-X (2009) Analysis of packaging material impacting on FBG temperature. In: Proceedings of SPIE. Shanghai, China
21. Kirby RK (1956) Thermal expansion of polytetrafluoroethylene (Teflon) from −190 to +300 C. J Res Nat Bur Stan 57(2)
22. Tripathi SM, Kumar A, Varshney RK, Kumar YBP, Marin E, Meunier J-P (2009) Strain and temperature sensing characteristics of single-mode–multimode–single-mode structures. J Lightwave Technol 27(13):2348–2356

Speed Control of Permanent Magnet Synchronous Motor Using Universal Bridge and PID Controller

Rifdian Indrianto Sudjoko, Hartono, and Prasetyo Iswahyudi

Abstract The permanent magnet synchronous motor (PMSM) are commonly used in industrial application due to their high steady state torque, high power, high efficiency, low inertia, and simple control to their drives compared to the other motor drives. This paper suggest the use of the proportional integral derivatives (PID) controller to control the speed of PMSM. The closed loop speed control for this type of motors with voltage sources inverter and universal bridge are designed. The whole drive system is simulated in Matlab/Simulink based on the mathematical model of the system device including PMSM and inverter. The aim of the drive system is to have speed control over wide speed range. Simulation results show that the speed controller has a good dynamic response and can control the PMSM successfully with a better performance.

Keywords Permanent magnet synchronous motor · PID controller · Voltage source inverter · Universal bridge

1 Introduction

1.1 Permanent Magnet Synchronous Motor (PMSM)

Synchronous motor and brushless DC motor included as the type of AC permanent magnet motor. The construction of this motor consist of wound stator and wound rotor. The armature winding of this motor has a single phase winding or three phase winding. The armature winding also called as a stator winding. The rotor fabricated as permanent magnet.

AC permanent magnet motor can perform both as a synchronous and brushless dc motor. In each case, the motor consist of a wound stator and rotor. The stator may have a single phase or multiphase winding which is sometimes called the armature winding.

R. I. Sudjoko (✉) · Hartono · P. Iswahyudi
Aviation Polytechnic of Surabaya, Surabaya, Indonesia
e-mail: rifdian.anto@gmail.com

U. Sabino et al. (eds.), *Proceedings of the 6th International Conference and Exhibition on Sustainable Energy and Advanced Materials*, Lecture Notes in Mechanical Engineering,
https://doi.org/10.1007/978-981-15-4481-1_39

PMSM has two type of rotor, salient pole rotor (Fig. 1) and cylindrical rotor (Fig. 2). Salient pole rotor usually used for low speed machines and cylindrical rotor used for high speed machines.

In both case, PMSM can be attached to the rotor surface (Fig. 1) or it can be buried (Fig. 2).

PMSM motor has the same rotating speed between rotor speed and magnetic field speed that is generated by stator windings. Rotor in PMSM supplied through the brushes or without brushes (brushless). The operation PMSM in using brushes or brushless had different electromechanical characteristic. This characteristics also

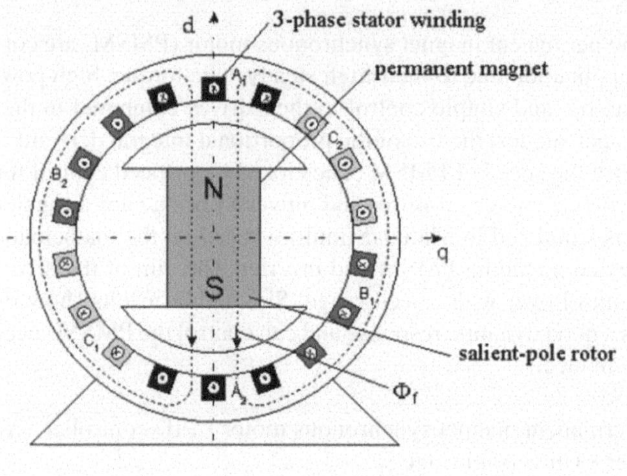

Fig. 1 Construction of permanent magnet synchronous motor with salient pole rotor

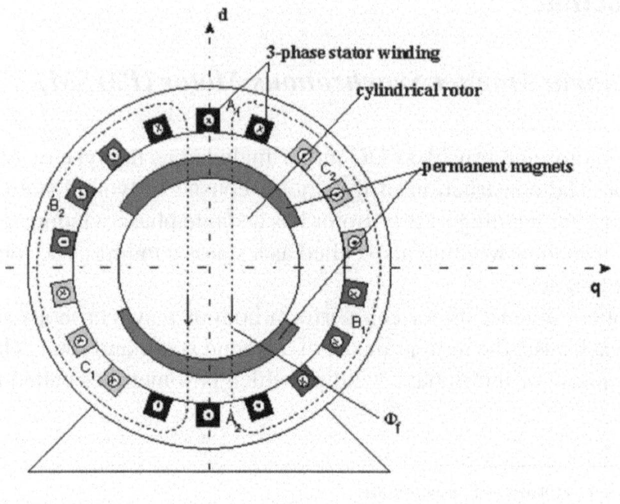

Fig. 2 Construction of permanent magnet synchronous motor with cylindrical rotor

different when PMSM supplied by single phase and three phase source. PMSM machine also called as BLDC (brushless dc) when the rotor supplied without brushes [1–3].

1.2 Permanent Magnet Synchronous Motor Operation

Synchronous motor with brushes and brushless included as the type of AC permanent magnet motor. The operation mode of brushes and brushless depend on the supply and control circuit. Figure 3 represent the construction of the motor with stator conventional windings placed in slot and distributed symmetrically in circumference.

Armature winding with three phase of PMSM is connected to 3-phase AC supply. Rotating magnetic flux is produced by three phase stator current. This condition also happened in three phase permanent magnet motor. The magnetic flux of PMSM is more steady with relationship to the rotor. In the PMSM, rotor must had the same speed as the rotating flux that generated by stator [1–3].

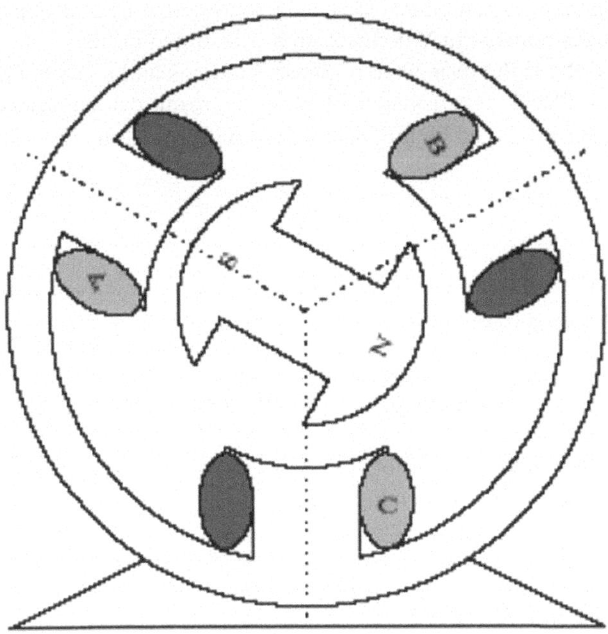

Fig. 3 Construction of permanent magnet synchronous motor with three phase windings

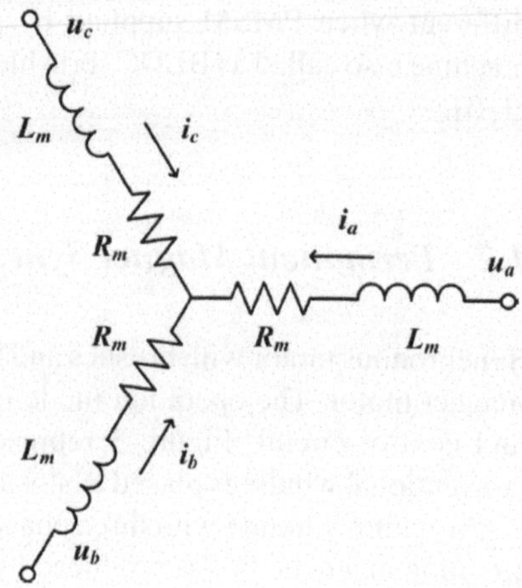

Fig. 4 Modelling of PMSM representation

1.3 Modelling of Permanent Magnet Synchronous Motor

Electrical component system of PMSM can be represent as three group of resistances and inductances connected in Y-connection as illustrated in Fig. 4.

From Fig. 4, R_m is the resistance of phase windings and L_m is the inductance of phase windings. PMSM electromechanical torque result can be represent from the mechanical power produced by the machine through equation

$$\tau_e = \frac{P_m}{\omega_m} \tag{1}$$

PMSM modelling from mechanical system can be described using Newton's second law through the equation

$$J\frac{d\omega_m}{dt} = \tau_e - B_{fric}\omega_m \tag{2}$$

From Eq. 2, the field working of PMSM usually ignored the friction coefficient because the coefficient is small enough [4].

1.4 Three Phase Inverter

In high power application, Three phase dc to ac converter (inverter) are widely used for adjustable frequency drive applications. A basic principle of three phase dc to ac converter containing of three single phase switches which connected to one of the three load terminal ports.

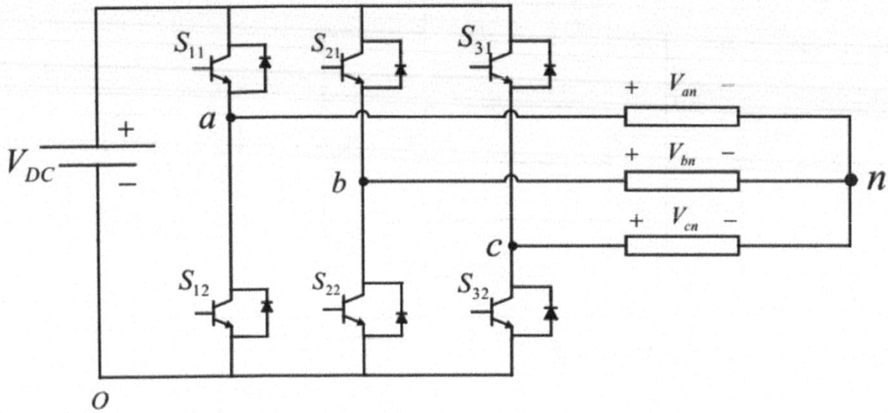

Fig. 5 Three phase dc to ac converter circuit

Inverter is a circuit that is used to convert a DC voltage source into an AC voltage source. The power semiconductor components used can be in the form of SCRs, transistors, and MOSFETs that operate as switches and converters. Three phase inverter can be shown in Fig. 5.

Judging from the conversion process, inverters can be divided into three types, namely inverters: series, parallel, and bridge. Bridge inverters can be divided into halfwave bridge inverters and full wave bridges. The resulting output voltage can be in the form of one phase or three phases.

1.5 Proportional Integral Derivative Controller

PID control systems are the most widely used in control systems industry. The success of the PID controller depends on its accuracy in determining PID constant (reinforcement). Practically the determination process PID constants are based on human expertise based on rules called rules of thumb. If the right result has been obtained, then this PID constant used for further control. This of course has weaknesses because this constant is the same for every error value that occurs and requires tuning reset if there are changes in plant parameters in the PID constant. For To overcome this, a method is needed to determine the PID constant exactly according to the plant. It is hoped that the performance of the PID control can be improved. In Fig. 6 shows the principle of PID controller.

Common characteristics used in the regulation of an intermediate system others include stability, accuracy, response speed and sensitivity. In action proportional control, the output of the control system will be proportional to the input. The output signal is a reinforcement of the error signal with certain factors [5]. This reinforcement factor is a proportional constant of the system, which is expressed by K_p, where K_p has a high/fast response. In integral control, the output always changes during

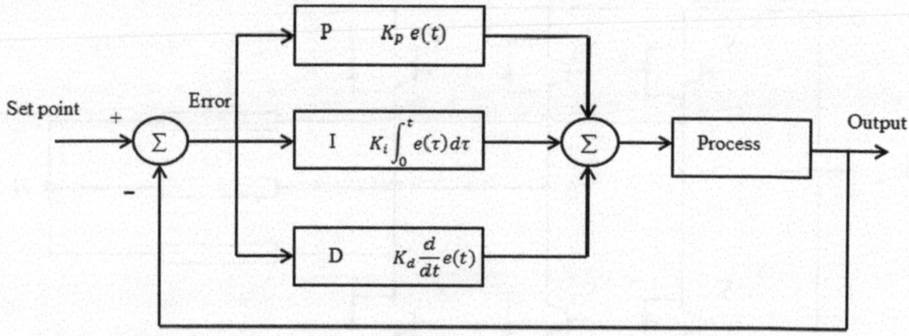

Fig. 6 Diagram of PID control circuit

deviations, and the speed at which the output changes is proportional to the deviation, a constant expressed as K_i, where K_i has a sensitivity which is high, i.e. by reducing the error generated from the feedback signal. The greater the value of K_i, the higher the sensitivity, but time needed to achieve stability more quickly, likewise the opposite. Where as derivative control works by rate deviation changes, so this type of control is always used together with proportional and integral controllers. Constants expressed in K_d, where K_d is affecting the stability of the system, because of this control action able to reduce errors. By combining these PID control actions then expected to get a response that has a level of stability tall one [6].

2 Methodology

There are tools which will be used for the effective implementation of this project, they include Matlab tool, Simulink tool and Simpower system tool.

2.1 System Overview

The proposed circuit of the system had been designed and simulated using Matlab/Simpowersystem. Figures 7 and 8 shows the simulink circuit of the proposed model.

2.2 Open Loop System

The open loop system consist of four main component. They are DC voltage supply, universal bridge, PWM generator and PMSM machine block. In open loop system, there are no feed back to control the speed. So the speed depend on the value of dc voltage supply. The open loop control system is traditionally control that used in several permanent magnet synchronous motor. This control is simple because contain a few component to implemented (Fig. 7).

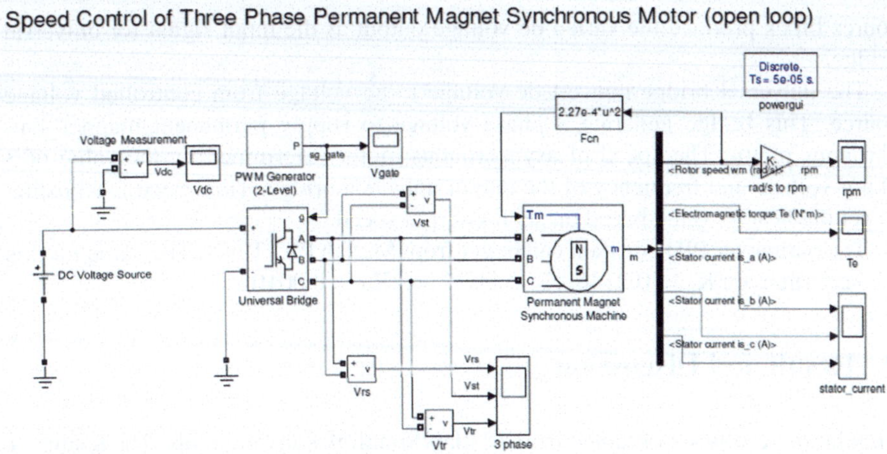

Fig. 7 Simulink model of open loop system

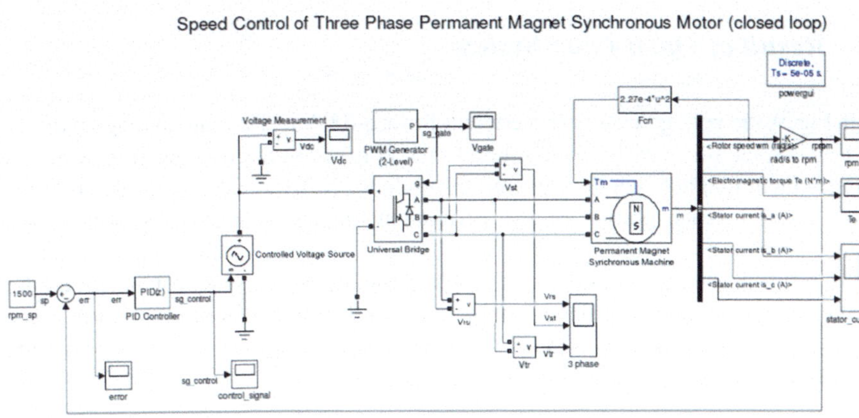

Fig. 8 Simulink model of closed loop system

2.3 *Closed Loop System*

The improvement of open loop is closed loop system. The closed loop system consist of five main component. The component are controlled voltage source, PID controller block, universal bridge block, PWM generator block and permanent magnet synchronous machine block. The circuit diagram of closed loop system shows in Fig. 8.

In closed loop system, speed sensor used as feed back signal to set point reference. The error is the differences between actual speed and rpm set point. The PID block process this error as signal control to controlled voltage sources. Controlled voltage source block produce the varied dc voltage output as the input signal for universal bridge.

The universal bridge convert dc voltage to ac voltage from controlled voltage source. This bridge generate 3 phase voltage to supply permanent magnet synchronous motor. The speed of asynchronous motor determined by the value of 3 phase voltage and frequency of the output universal bridge. This speed is adjusting to get the nearest value based on set point rpm speed.

The parameter PID constant can be set from Matlab PID Tuner. The value for this PID constant are $K_p = 0.2146$, $K_i = 117.7$ and $K_d = 0.00102$.

3 Result and Discussion

Simulation results are obtained from Matlab/Simulink software tools. The results of line voltage, phase voltage, current in the line are listed along with rpm speed and torque results of permanent magnet synchronous motor in the following figures.

3.1 Result of Open Loop System

From Fig. 9, the voltage value of inverter output are 400 Vac. Permanent magnet motor use this voltage for operation according to the nominal voltage rating (Fig. 10).

Permanent magnet synchronous motor produce 35 A for stator current. This current is equal to each phase because permanent magnet synchronous motor are a balanced load.

From Fig. 11, the steady state speed of permanent magnet synchronous motor are 1500 rpm. The response of open loop system can be shown from this figure. Open loop system respons are rise time (t_r) = 0.02 s, delay time (t_d) = 0.01 s, peak time (t_p) = 0.04 s and steady state time (t_s) = 0.2 s.

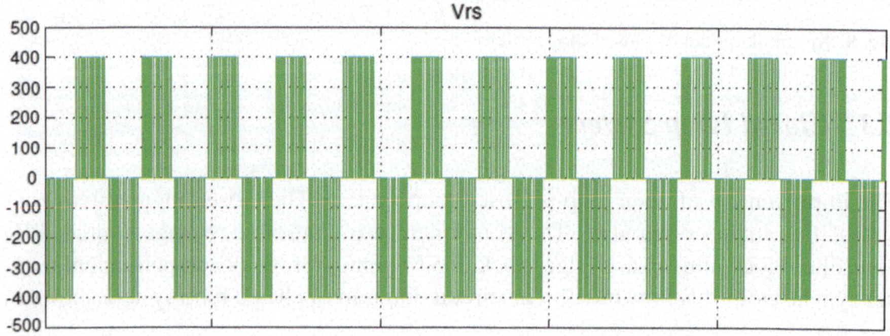

Fig. 9 Line voltage output of universal bridge

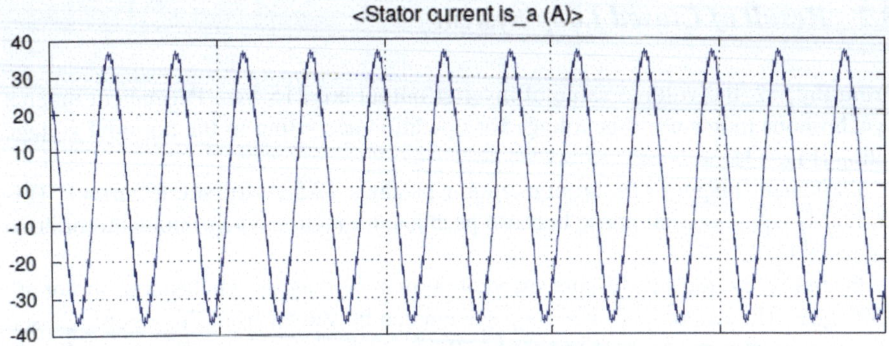

Fig. 10 Stator current of permanent magnet synchronous motor

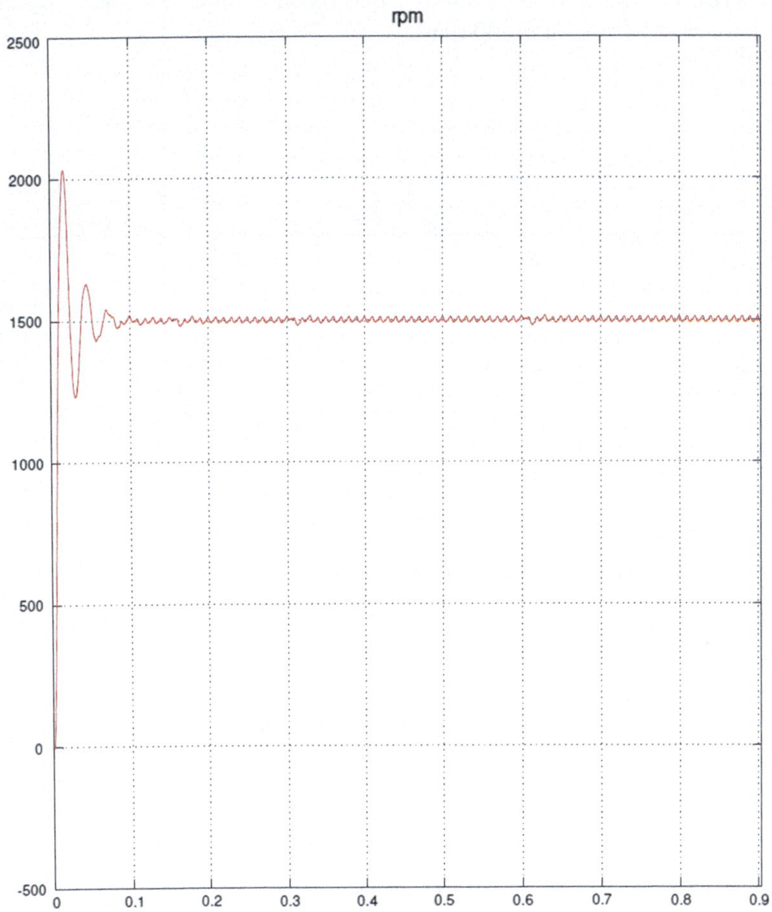

Fig. 11 Rpm output of permanent magnet synchronous motor

3.2 Result of Closed Loop System

From Fig. 12, the voltage value of inverter output are 215 Vac. Permanent magnet synchronous motor use this voltage for operation according to the nominal voltage rating (Fig. 13).

Permanent magnet synchronous motor produce 12.5 A for stator current. This current is equal to each phase because permanent magnet synchronous motor are a balanced load.

From Fig. 14, the steady state speed of permanent magnet synchronous motor are 1500 rpm. The respons of open loop system can be shown from this figure. Closed loop system respons are rise time (t_r) = 0.02 s, delay time (t_d) = 0.01 s, peak time (t_p) = 0.04 s and steady state time (t_s) = 0.07 s.

The performance from Fig. 14 shows that the system have a good response to reach motor speed set point (1500 rpm).

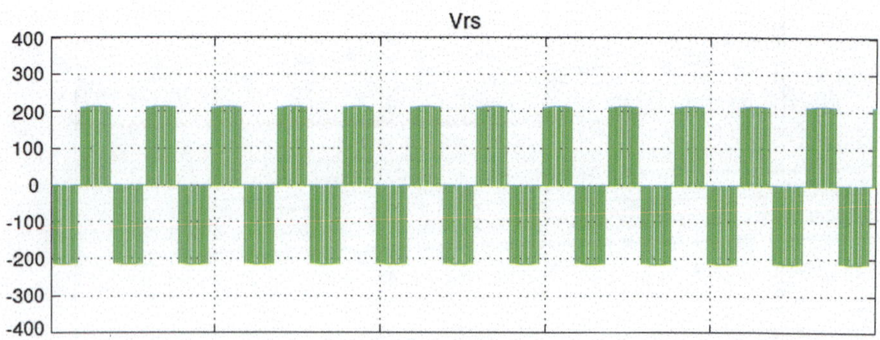

Fig. 12 Line voltage output of universal bridge

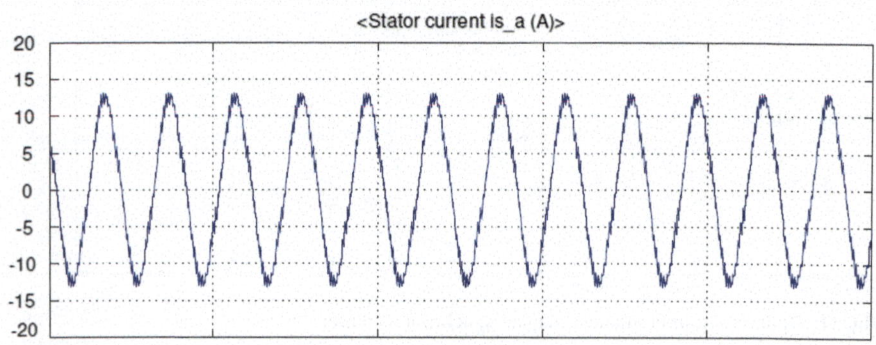

Fig. 13 Stator current of permanent magnet synchronous motor

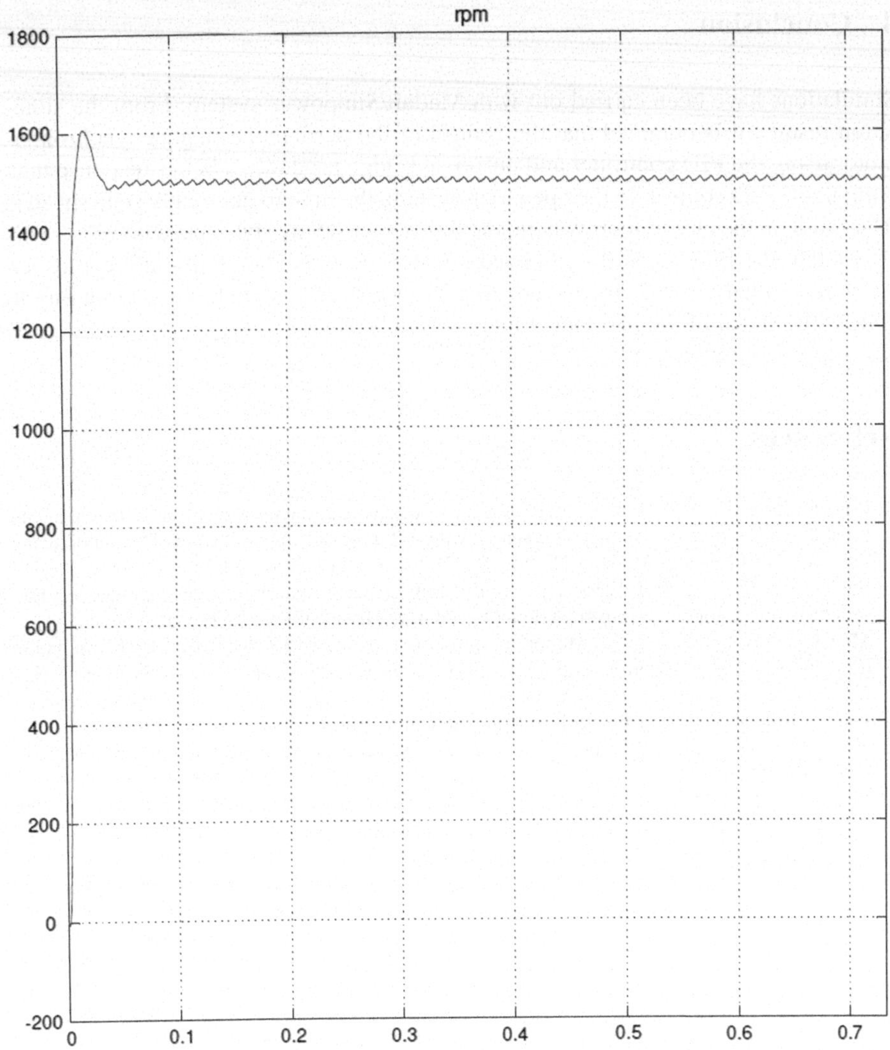

Fig. 14 Stator current of permanent magnet synchronous motor

4 Conclusion

Simulations have been carried out with Matlab/Simpower system. From the simulation results, it is obtained that the control of the permanent magnet synchronous motor using the PID controller and universal bridge produces a better response than without the PID control. In the open loop system, the time to get steady state speed is 0.2 s while in the closed loop system, the time to get steady state speed is 0.07 s. This shows that the PID controller obtained a faster response than without a controller. After all simulation we can conclude that the proposed system has a good ability to control the speed of the permanent magnet synchronous motor.

References

1. Qazeem Jabir F (2017) Dynamic performance of permanent magnet synchronous machine and brushless DC motor fed single phase source with different starting circuits in closed loop using simulink. Int J Eng Res Manag (IJERM) 04(02). ISSN: 2349–2058 (2017)
2. Sophie Sekalala F (2006) Performance of three phase permanent magnet motor operating as a synchronous motor and a brushless dc motor. Master Theses of Louisiana State University
3. Miss Avanti B, Tayade F (2014) Modelling and simulation of a BLDC by using matlab/simulation tool. IOSR J Electr Electron Eng 55–62. The national conference on electrical engineering research and advancement (EERA-2014)
4. Anders Roland P (2013) Control and parameter identification of a permanent magnet synchronous motor with a LC filter. Master thesis, Department of Energy Technology, Aalborg University, Denmark
5. Yusoff SBM (2014) PID control technique for three phase induction motor using matlab simulink and arduino. Master thesis of electrical engineering, Faculty of Electrical and Electronic Engineering, University Tun Hussein Onn, Malaysia
6. Dawood SM (2015) PID controller based multiple (master/slaves) permanent magnet synchronous motors speed control. Iraq J Electr Electron Eng 11(22)

Development of Cr Coated AISI 304 Material for Artificial Hip Joint

Joko Triyono, Giffari Muhammad Ghiats, Eko Surojo, Eko Pujiyanto, and Suyatmi

Abstract The aim of this study is to determine the mechanical strength of the Cr coated Stainless Steel (AISI) 304 material which will be applied to the artificial hip joint. The coating method used by electroplating thin dense chrome. The variables were used in this research are coating time, Cr acid concentration, and voltage. The Tests carried out were hardness test and compressive strength. The highest hardness test was reached at variations in coating time of 15 min, concentration of Cr acid 200 mol, and voltage 6 V (287 HVN). This value is higher than the hardness of human bone (50 HVN). The highest compressive strength was obtained in coating time of 15 min, the concentration of Cr acid 200 mol, and the voltage of 6 V (248 MPa). This value is also higher than the compressive strength of human bones (230 MPa). These shows that the Cr coated AISI 304 material has a good mechanical strength to be applied as an artificial hip joint material because it has a higher hardness and compressive strength than the human bone.

Keywords Cr coated AISI 304 · Material · Artificial hip joint

1 Introduction

Hip joint is the joint that connecting the pelvic bone and femur and is one of the largest joints in the body. The joint is supporting about 3/8 of total body weight [1]. The joints are susceptible to damage due to accidents, wear, tearing of the cartilage, osteoarthritis, infection, and bone degeneration [1]. Figure 1 shows the hip joint.

The damaged hip joints can be overcome by hip replacement surgery which can be done by removing cartilage and damaged bone from the hip joint by using an

J. Triyono (✉) · G. M. Ghiats · E. Surojo
Mechanical Engineering Department, Sebelas Maret University, Surakarta, Indonesia
e-mail: jokotriyono@staff.uns.ac.id

E. Pujiyanto
Industrial Engineering Department, Sebelas Maret University, Surakarta, Indonesia

Suyatmi
Medical Faculty, Sebelas Maret University, Surakarta, Indonesia

U. Sabino et al. (eds.), *Proceedings of the 6th International Conference and Exhibition on Sustainable Energy and Advanced Materials*, Lecture Notes in Mechanical Engineering,
https://doi.org/10.1007/978-981-15-4481-1_40

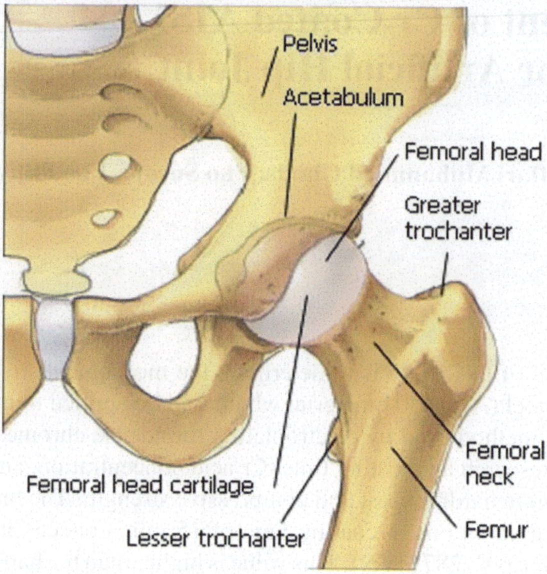

Fig. 1 Hip joint [2]

artificial hip joint made of biomaterial [2]. Figure 2 shows the normal, damaged and artificial hip joint.

Biomaterials used as implant materials must have several properties, i.e. good mechanical properties, good chemical properties and good biological properties [4].

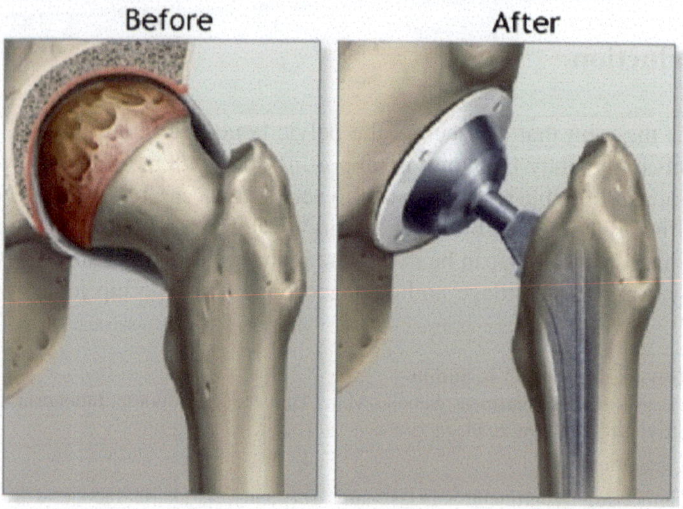

Fig. 2 The damaged hip joint and artificial hip joint [3]

The aim of this study is to determine the physical and mechanical properties of Cr coated AISI 304 material. The mechanical properties studied are compressive strength and hardness material. The obtained of the mechanical properties are compared with the mechanical properties of commercial product and human bones. The results of this study are expected to be useful in the field of orthopedics.

2 Methods

Specimens of hardness and compressive strength were made by a lathe based on ASTM E92-82 and E9-89A standards. The specimens can be seen in Figs. 3 and 4. Figure 3 is specimens of hardness and compressive strength of stainless steel 304. Figure 4 is specimens of hardness and compressive strength of Cr coated stainless steel 304.

In this research, two types of materials were used, namely AISI 304 and chrome coated AISI 304. The production of Cr coated AISI 304 was carried out by hard-chrome plating type electroplating method. One method of plating was thin dense chrome plating. Thin dense chrome plating was carried out by dipping a working specimen in the cathode and lead in the anode. The solution used was a Cr acid solution. The advantages of this coating are increasing fatigue strength, smooth surface and high corrosion resistance. It has a very thin layer (0.005–0.015 mm), dense, and has no porosity [5].

The testing conducted in this study were test of chemical composition, hardness and compressive strength. The chemical composition testing using the spectrometer that It can be seen in the Fig. 5.

Hardness testing was carried out by the Hardness Vickers Number testing machine (see Fig. 6), 200 kgf load cell and indentor suppression time of 10 s. Hardness testing was carried out by according to ASTM E 92-82 standards.

Fig. 3 Specimens of hardness and compressive strength of AISI 304

Fig. 4 Specimens of hardness and compressive strength of Cr coated AISI 304

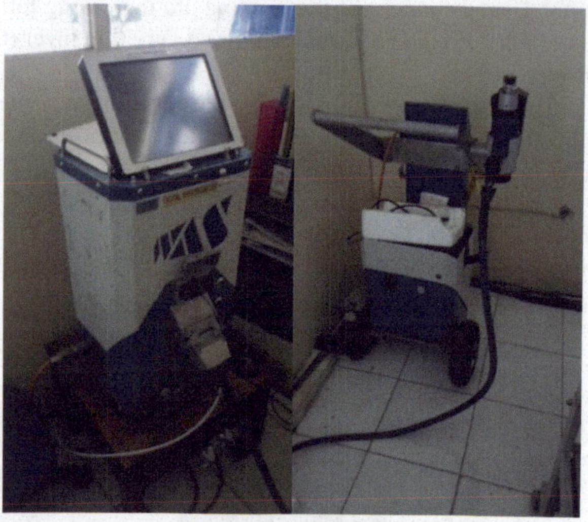

Fig. 5 Spectrometer

Compressive strength testing was performed by Universal Testing Machine (see Fig. 7) according to ASTM E 9-89A standard.

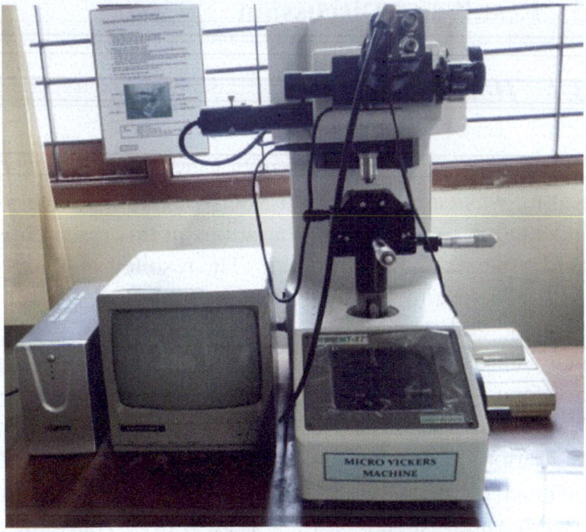

Fig. 6 Hardness vickers number (HVN)

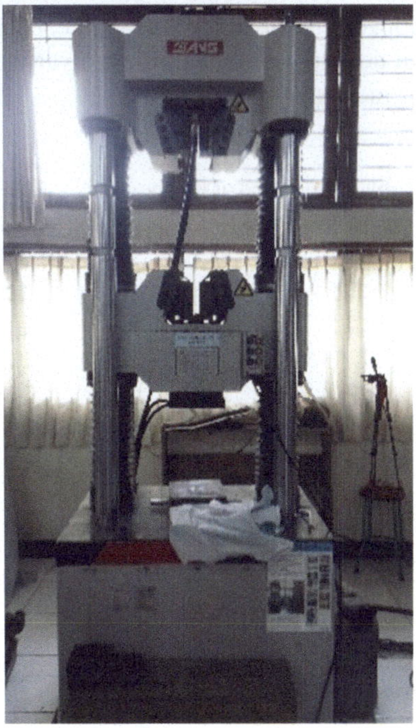

Fig. 7 Universal testing machine

3 Result and Discussion

3.1 Hardness Testing

The specimens used in this study were AISI 304 and Cr coated AISI 304. Hardness testing was carried out by a Microvickers machine with a loading of 200 kgf and indentation time for 10 s. Variations in this research were concentration of chrome acid, voltage, and coating time. The resulted of hardness of AISI 304 and Cr coated AISI 304 specimens can be seen in Table 1.

Table 2 shows the resulted of hardness test of Cr coated AISI 304 Vickers thin dense chrome plating with independent variables of coating time, Cr acid concentration, and voltage. A Taguchi method was used to analysis data. The highest hardness was obtained in variation 9 with a duration of 15 min of coating time, 200 mol concentration of Cr acid, and a voltage of 6 V (287 HVN). The lowest hardness was obtained in variation 1 with a coating duration of 5 min, 100 mol of Cr acid concentration, and a voltage of 4.5 V (258 HVN). The value is higher than the hardness of human bones (50 MPa) commercial products of AISI 316L material (234 MPa). The longer duration of coating time, concentration and voltage can increase the amount of deposition of the chrome layer on the specimen surface and causing the hardness is increasing.

Table 1 Resulted of hardness test of Cr coated AISI 304

Variation	Coating time (min)	Concentration of Cr (mol)	Voltage (V)	Hardness (HVN)
1	5	100	4.5	258
2	5	150	6	273
3	5	200	7.5	273
4	10	100	6	280
5	10	150	7.5	273
6	10	200	4.5	270
7	15	100	7.5	272
8	15	150	4.5	278

Table 2 Response of hardness

Response table for means			
Level	A	B	C
1	268.0	270.0	268.7
2	274.3	274.7	280.0
3	279.0	276.7	272.7
Delta	11.0	6.7	11.3
Rank	2	3	1

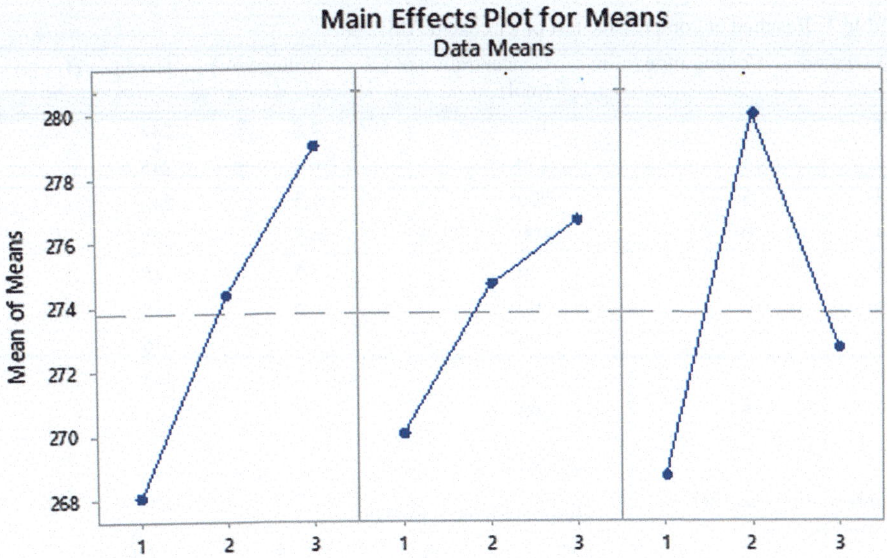

Fig. 8 Main factor of hardness test

Table 2 shows the average response of the sequence of factors that have an influence on the hardness of Cr-coated AISI 304. The largest value to the smallest are voltage, coating time, and Cr acid concentration. Table 2 can be displayed graphically as shown in Fig. 8.

Figure 8 shows the average ratio of the main factors of hardness test. The resulted of thin dense chrome plating shows that the process can increase the hardness of AISI 304. The highest hardness was achieved in variation 9 (287 HVN). This value is higher than the hardness of the material before it is coated (258 HVN), human bones (50 HVN) and commercial products (182 HVN).

Adding the concentration of Cr acid can increase the amount of deposition of the chrome layer on the surface of the specimen during the coating process and causing increased material hardness [6]. The increasing in hardness due to thin dense chrome plating coating increases linearly in proportion to the increase in coating time, the increased concentration of Cr acid solution and the voltage.

3.2 *Compressive Testing*

Compressive strength testing was carried out by Universal Testing Machine, pressing speed of 10 mm/min. The resulted of compressive can be seen in Table 3. Table 3 shows the resulted of compressive test of thin dense chrome-plating of Cr coated AISI 304. The highest hardness was found in variation 9 with a duration of 15 min of coating, concentration of 200 mol of acid, and a voltage of 6 V (248 MPa). The

Table 3 Resulted of compressive test of Cr coated AISI 304

Variation	Coating time (min)	Concentration of Cr (mol)	Voltage (V)	Hardness (HVN)
1	5	100	4.5	244
2	5	150	6	245
3	5	200	7.5	245
4	10	100	6	245
5	10	150	7.5	245
6	10	200	4.5	246
7	15	100	7.5	246
8	15	150	4.5	247
9	15	200	6	248

Table 4 Response of compressive

Response table for means			
Level	A	B	C
1	244.7	245.0	245.7
2	245.3	245.7	246.0
3	247.0	246.3	245.3
Delta	2.3	1.3	0.7
Rank	1	2	3

lowest hardness was found in variation 1 with a coating duration of 5 min, 100 mol of Cr acid concentration, and a voltage of 4.5 V (244 MPa).

Table 4 shows the average response of the sequence factors that have an influence on the compressive of Cr coated AISI 304. The largest value to the smallest are coating time, concentration Cr acid and voltage. Table 4 can be displayed graphically as shown in Fig. 9.

Figure 9 shows the average ratio of the main factors of compressive test. The resulted of thin dense chrome plating shows that the process can increase compressive of AISI 304. The highest compressive was achieved in variation 9 (248 MPa). The value is higher than the compressive of the material before it is coated (239 MPa), human bones (230 MPa) but lower than commercial products (320 MPa).

Adding the concentration of Cr acid can increase the amount of deposition of the chrome layer on the surface of the specimen during the coating process and causing increased material compressive [6]. The increasing in compressive due to thin dense chrome-plating coating increases linearly in proportion to the increase in coating time, the increased concentration of Cr acid solution and the voltage.

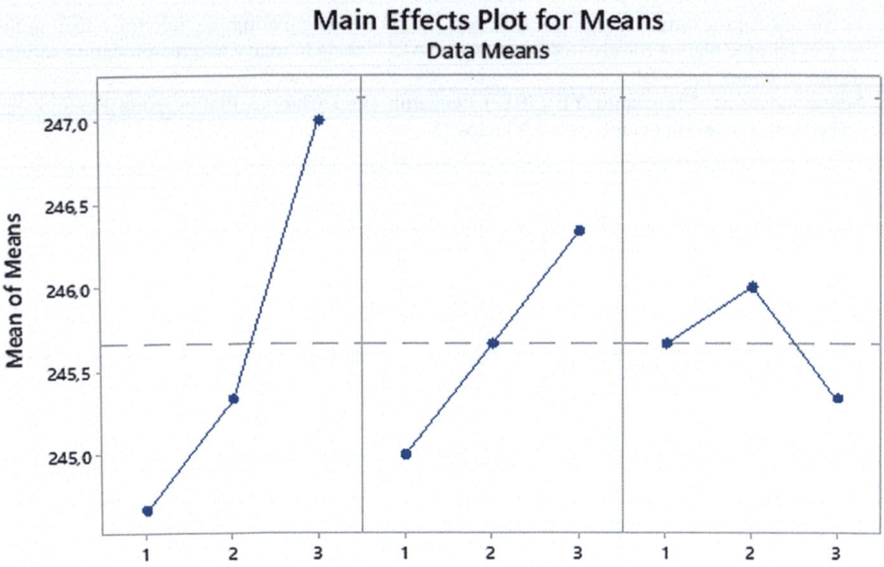

Fig. 9 Main factor of compressive

4 Conclusions

Some of the conclusions of this study are as follows:

1. The highest hardness and compressive strength of Cr AISI 304 material was obtained in variations 9 (287 HVN and 248 MPa). These values are higher than the hardness and compressive strength of materials without coating (258 HVN and 239 MPa).
2. The highest hardness and compressive strength of Cr AISI 304 material (287 HVN and 248 MPa) is higher than the hardness and compressive strength of human bones (50 HVN and 230 MPa).

References

1. Amirouche F, Solitro GF (2011) Challenges in modeling total knee arthroplasty and total hip replacement. Procedia IUTAM 2:1825. https://doi.org/10.1016/j.piutam.2011.04.003
2. Shapi A (2011) Digital preoperative planning for total hip replacement using two dimensional X-ray imaging. Int J Comput Appl 17(2):20–27
3. Axer A (1972) Total hip replacement. Harefuah 82(10):460–464. Available at: http://www.ncbi.nlm.nih.gov/entrez/query.fcgi?cmd=Retrieve&db=PubMed&dopt=Citation&list_uids=4636269
4. Hendian (2013) Perancangan Dan Simulasi Pembentukan Hip Joint Prosthetic, Mar 2006, pp 1–3

5. Service, Reciprocating, and Technical Report (2015) Chrome plating: a guide for selecting the type of chrome plating for use in contact with BAL^{TM} seals in rotary and reciprocating service technical report, pp 1–10
6. Setyahandana B, Christianto YE (2017) Pengaruh Hard Chrome Plating pada Peningkatan Kekerasan Baja Komponen Kincir 12(1):26–35

Scrutinizing the Prospect of *Cerbera manghas* Seed and Its De-oiled Cake for a Fuel: Physicochemical Properties and Thermal Behavior

M. Muzayyin, S. Sukarni, and R. Wulandari

Abstract Biomass is a renewable fuel derived from biological substances such as crops, waste, animals, and residues, which has been ranked as the third-largest source of energy in the world. One of the biological materials that potential as biomass fuel from the perspective of the abundance is *Cerbera manghas,* which is a mangrove plant, poisonous, and therefore it does not compete with the needs of the foods. This research aims to study the physicochemical properties and thermal characteristics of *Cerbera manghas* seed and its de-oiled cake. The seed of *Cerbera manghas* samples is dried in an oven at 80 °C for 4 h. The de-oiled cake is obtained by the pressing of the seed at 33.81 N/m^2 to release its oil content. The higher heating values (HHVs) of the seed and de-oiled cake are examined by using an adiabatic bomb calorimeter. Physical properties are evaluated by proximate analysis test to comprehend the respective moisture (M), volatile matter (VM), fixed carbon (FC), and ash (A) content. The chemical properties, especially C, H, O, N, and S, are analyzed by the ultimate analysis test. The thermal behavior of the seed and de-oiled cake are studied by a thermal analyzer. It can be reported that the seed and de-oiled cake are dominated by volatile matters, which are 91.41 and 84.93%, respectively, and they have HHVs of 32.46 MJ/kg for seed and 27.95 MJ/kg for de-oiled cake. Their chemical properties are dominated by 78.01% carbon for the seed and 67.34% carbon for the de-oiled cake. From the thermal test under the inert atmosphere, it was known that they have the temperature of active pyrolysis zone at around 150 until 500 °C, and resulted in relatively low residuals at the end of the pyrolysis process those are 8.22% for the seed and 15.43% for the de-oiled cake. These overall results

M. Muzayyin
Master Program of Mechanical Engineering, Graduate School, Universitas Negeri Malang, Jl. Semarang no. 5, Malang 65145, Indonesia

S. Sukarni (✉) · R. Wulandari
Department of Mechanical Engineering, Center for Renewable and Sustainable Energy Engineering (CRSEE), Universitas Negeri Malang, Jl. Semarang no 5, Malang 65145, Indonesia
e-mail: sukarni.ft@um.ac.id

S. Sukarni
Centre of Advanced Materials for Renewable Energy (CAMRY), Universitas Negeri Malang, Jl. Semarang no 5, Malang 65145, Indonesia

U. Sabino et al. (eds.), *Proceedings of the 6th International Conference and Exhibition on Sustainable Energy and Advanced Materials*, Lecture Notes in Mechanical Engineering,
https://doi.org/10.1007/978-981-15-4481-1_41

strengthened the potential of *Cerbera manghas* seed and its de-oiled cake being the fuel feedstock.

Keywords *Cerbera manghas* seed · De-oiled cake · Thermal behavior · Physicochemical properties

1 Introduction

The energy requirement in Indonesia is estimated continuously rising from 795 million barrels of oil equivalent (BOE) in 2016 to 4569 million BOE in 2050 [1]. In contrast, the fossil energy reserve is continuously depleting. Therefore, exploration the renewable energy sources to replace or substitute fossil energy is a crucial issue recently. Biomass is the potential candidate for renewable and sustainable energy sources to replace fossil-based energy, primarily because of its abundance and its ability to reduce greenhouse gas emissions [2].

Biomass is a renewable fuel derived from biological substances such as crops, waste, animals, and residues [3]. A variety of plants, including pine, miscanthus, canary grass, tall fescue, and brief rotation coppice willow, were used as biomass to generate energy [4]. Biomass is the third-largest energy source after coal and fuel oil [5], but until now, the utilization of biomass has not been able to meet the increasing energy needs. The essential parameter of the biomass being fuel's feedstock is its abundance so that availability is guaranteed [6]. One of the tropical plants that are very easy to cultivate and a non-food crop is *Cerbera manghas*; therefore, it to be potential biomass to support energy needs in the future [2]. However, thoroughly evaluation related to the potential of *Cerbera manghas* to be fuel is rarely found in the literature.

Cerbera manghas have a seeded fruit with lignocellulose fibers that resemble coconuts. They are mostly known as city shading plants, and their economic value has never been explored [2]. *Cerbera manghas* plants can grow in soil with less nutritious, easy to cultivate, and spread throughout Indonesia. They contain primary elements of cellulose, hemicellulose, lignin, starch, and pectin [5]. The content of lignin and cellulose in *Cerbera manghas* is potential as an energy source. *Cerbera manghas* fruit consists of 8% seeds and 92% fruit flesh. The seeds themselves are divided into 14% shells and 86% seed meat. Their seeds contain oil between 35 and 50% [7].

The deep understanding of biomass physicochemical properties can be performed by the proximate and ultimate method [8, 9]. These properties are essential parameters to determine their feasibility as the fuel feedstock. Therefore, scrutinizing the *Cerbera manghas* physicochemical properties is required to reliably estimate its potential as a fuel.

The essential parameters resulted from the proximate analyzer are the values of moisture (M), volatile matter (VM), fixed carbon (FC), and ash [10]. Moisture content is critical content to be considered in solid fuels because it significantly impacts the decomposition behavior during the pyrolysis [11]. Volatile matter is the substances that are decomposed if the biomass is heated at a certain temperature and time and released as the gas during the pyrolysis, and they determine the combustion reactivity during the oxidation process. Fixed carbon (FC) is the remaining content after the VM is fully released, and ash (A) is a residue that is left after the decomposition process.

The main parameters resulted from the ultimate analysis are the content of carbon (C), hydrogen (H), nitrogen (N), oxygen (O), and sulfur (S) of the sample [12]. Carbon and hydrogen content is an essential element in fuel [13], in which both elements contribute significantly to the amount of energy on biomass. The nitrogen content associated with the contribution to the formation of nitric oxide and air pollution and does not significantly influence the heating value [10]. The oxygen content contributes to the pyrolysis decomposition process of the sample, where high values of oxygen and volatile matter will affect the formation concentration of methane (CH_4) produced during the devolatilization process [14]. The oxygen released during the thermal process of organic components contributes to the reactivity in the pyrolysis process [4].

Pyrolysis is considered to be the most efficient process among the thermochemical conversions technique because the ratio of feed to fuel is high compared to the combustion and gasification process [15]. This research is aimed to find the physicochemical properties and thermal characteristics of *Cerbera manghas* seed and its de-oiled cake to evaluate their potential as a fuel feedstock.

2 Materials and Methods

2.1 Materials

The *Cerbera manghas* fruit was originated from Malang City, East Java, Indonesia. The selected sample was ripe fruit that has fallen itself from the tree and characterized by brown color. The samples were cleaned and then were separated their seeds from the flesh. Furthermore, the fruit seeds were dried in an oven for 4 h at 80 °C. The dried samples then were mashed using a mortar being the porridge-like material and the de-oiled cake was obtained from the pressing process of porridge-like material at a 33.81 N/m^2 to separate the oil content. From this process, the sample of *Cerbera manghas* seed and its de-oiled cake were achieved and subsequently being analyzed their physicochemical properties and thermal decomposition characteristics.

2.2 *Physicochemical Analyses*

The physical properties of both samples are evaluated by proximate analysis according to ASTM D 3173-17, ASTM D 3175-17, ASTM D 3172-13, and ASTM D 3174-12 to understand the respective moisture, volatile matter, fixed carbon, and ash content. The calorific value was examined using an adiabatic bomb calorimeter in line with the ASTM D 5865-13 standard method. The chemical elements of C, H, O, N, and S were analyzed by ultimate analysis based on the standard of ASTM D 5373-14, ASTM D 5373-14, ASTM D 3176-15, ASTM D 5373-14, and ASTM D 4239-14e in that order [8, 9].

2.3 *Thermal Analyses Methods*

The thermogravimetric analyzer (Mettler Toledo TG-DSC1) was used for analyzing the thermal behavior of both samples. Each sample of weighing 20 mg was inserted into the crucible made from Al_2O_3 for every test at a heating program of 15 °C/min from 25 to 900 °C. A 50 ml/min of nitrogen has flowed continuously during the process. The thermogravimetric (TG) and derivative thermogravimetric (DTG) curves were recorded automatically during the experiment [16].

3 Results and Discussions

3.1 *Biomass Samples*

The samples of *Cerbera manghas* were classified as the seed (a) and de-oiled cake (b) as viewed in Fig. 1.

Fig. 1 The seed (a) and de-oiled cake (b) of *Cerbera manghas*

3.2 Physicochemical Properties

From the proximate and ultimate analyses, the respective physical and chemical properties of both samples were shown in Table 1.

Table 1 showed the value of volatile matter (VM) content in the sample. It was observed that VM present as the most dominant content in both samples, where the VM content of the seed was 91.41%, and the de-oiled cake was 84.93%. The high volatile matter content in both samples made them more easily decomposed during the pyrolysis process. The value of volatile matter was influenced by the content of chemical elements in the sample including carbon (C), oxygen (O), and hydrogen (H).

The second-largest content was fixed carbon (FC), which was the material that remains after the VM was fully released, excluded ash. The FC of the respective seed and its de-oiled cake were 4.41 and 8.46%. Table 1 indicated that the carbon content (C) in the samples would be released along with the volatile matter during the pyrolysis process, considering that a less amount of carbon formed the FC. Yin [10] explained that the formation of VM and FC contents were influenced by carbon and hydrogen elements, where both C and H determined a significant contribution to the energy content of biomass samples.

The third highest ingredient in the samples was the ash content, which was a residual substance after the decomposition process entirely occurred. The proximate analysis indicated that a relatively small ash value of 3.12% in the seed and of 4.22% in the de-oiled cake. Sheng et al. [17] explained that the increasing value of ash content in biomass led to a decrease in the heating value. Table 1 indicated that the seed sample had a higher calorific value of 32.46 MJ/kg with 3.12% ash content, whereas the de-oiled cake sample had a higher calorific value of 27.95 MJ/kg with 4.22% ash content.

The smallest result of the proximate analysis was the moisture content, which was an essential substance in biomass to be analyzed because this content greatly influenced the behavior of the biomass during the thermal decomposition process [11]. The value of moisture content in the sample was 1.06% for the seed and 2.39% for the de-oiled cake. The smaller the water content in the sample, the less energy was used for the evaporation process.

The chemical properties of the seed and de-oiled cake were depicted in Table 1, as well. The results showed that the elements contained in the sample are dominated by carbon (C), which was 78.01% for the seed and 67.34% for de-oiled cake and oxygen (O), which was 13.04% for the seed and 19.45% for de-oiled cake. Carbon plays a significant role in determining the heat value of the fuel feedstock, whereas most of the oxygen would be released due to the thermal attack and provided the oxygen requirement during the combustion [18].

Table 1 Physicochemical properties of the seed and de-oiled cake of *Cerbera manghas*

Sample	Proximate analysis (wt%)				Ultimate analysis (wt%)					Colorific value (MJ/kg)
	M	VM	FC	A	C	H	O	N	S	
Seed	1.06	91.41	4.41	3.12	78.01	2.01	13.04	2.54	0.21	32.46
De-oiled cake	2.39	84.93	8.46	4.22	67.34	3.01	19.45	3.3	0.29	27.95

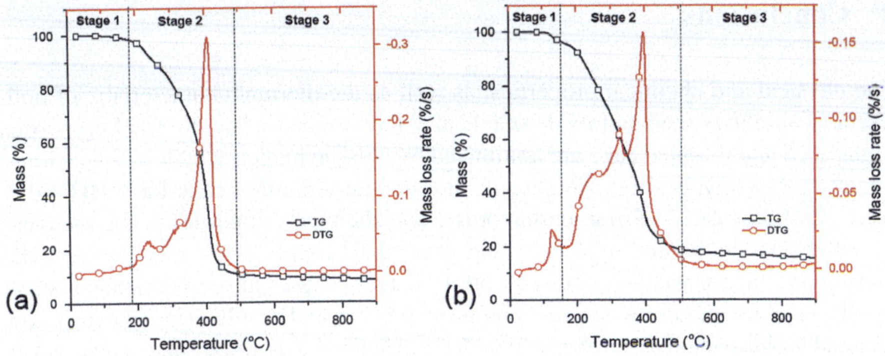

Fig. 2 The TG-DTG curves of the seed (a) and its de-oiled cake (b) of *Cerbera manghas*

3.3 Thermogravimetric Analysis

The thermal degradation profile of the seed and de-oiled cake of *Cerbera manghas* could be seen in Fig. 2. It was observed that both samples degraded into three stages during the thermal processes, and the active pyrolysis occurred at temperatures between around 150–500 °C.

The first stage decomposition process of the seed started at a temperature of 25 °C and finished at 185 °C with a mass loss of 1.79%, and the de-oiled cake started at 25 °C and ended at 153 °C with a mass loss of 3.43%. This stage was associated with the release of moisture and light volatile elements in the material [19]. The proximate results indicated that moisture content of the respective seed and its de-oiled cake were 1.06 and 2.39%, and therefore the loss of the mass in this stage was strongly thought due to the moisture element in both samples as well as the materials that were easily decomposed at low temperatures and liberated as the light volatile.

The second stage of the seed decomposition process was taken place at a temperature range of 185–488 °C that was accompanied by a mass loss of 87.53% and its de-oiled cake at a temperature range of 153–508 °C with mass loss of 77.92%. This stage was attributed to the release of volatile matter (VM), and it was an active pyrolysis stage considering that a huge amount of the mass was lost in this stage.

The third stage was the final stage, where for the seed, this was stretched at a temperature range of 488–900 °C with a mass loss of 2.46% and de-oiled cake, this was expanded at a temperature range of 508–900 °C with a mass loss of 3.22%. This stage was related to the decomposition of char to ashes and called as the passive pyrolysis stage [20]. The residual mass at the end of the pyrolysis process for the seed and de-oiled cake were 8.22% and 15.43%, in that order.

4 Conclusions

The physical and chemical properties, as well as the thermal characteristic of both *Cerbera manghas* seed and its de-oiled cake, have studied. The physical properties of the seed and de-oiled cake are dominated by volatile matter, which was 91.41 and 84.93%, respectively. Both samples of the seed and de-oiled cake have HHV that were 32.46 and 27.95 MJ/kg, in that order. The chemical properties of the seed and de-oiled cake dominated by carbon that were 78.01 and 67.34%, respectively. The thermogravimetric analysis showed both biomass samples have experienced active pyrolysis at temperatures around 150 until 500 °C and resulted in relatively low residual yields of 8.22% for the seed and 15.43% for the de-oiled cake. These entire results strengthened the potential of *Cerbera manghas* seed and its de-oiled cake for the future fuel feedstock.

References

1. KESDM (2016) Data Inventory Emisi GRK Sektor Energi. Kementerian Energi dan Sumber Daya Mineral, Jakarta (2016)
2. Iman G, Handoko T (2011) Pengolahan Buah Bintaro sebagai Sumber Bioetanol dan Karbon Aktif 1–5
3. Wang T, Zhai Y, Zhu Y, Li C, Zeng G (2018) A review of the hydrothermal carbonization of biomass waste for hydrochar formation: process conditions, fundamentals, and physicochemical properties. Renew Sustain Energy Rev 90:223–247. https://doi.org/10.1016/j.rser.2018.03.071
4. Gillespie GD, Everard CD, Fagan CC, McDonnell KP (2013) Prediction of quality parameters of biomass pellets from proximate and ultimate analysis. Fuel 111:771–777. https://doi.org/10.1016/j.fuel.2013.05.002
5. Chen C, Ma X, He Y (2012) Co-pyrolysis characteristics of microalgae Chlorella vulgaris and coal through TGA. Bioresour Technol 117:264–273. https://doi.org/10.1016/j.biortech.2012.04.077
6. Sukarni, Sudjito, Hamidi N, Yanuhar U, Wardana ING (2014) Potential and properties of marine microalgae *Nannochloropsis oculata* as biomass fuel feedstock. Int J Energy Environ Eng 5:279–290. https://doi.org/10.1007/s40095-014-0138-9
7. Suryani I, Yusuf Permana MU, Hatta Dahlan M (2012) Pembuatan Briket Arang Dari Campuran Buah Bintaro Dan Tempurung Kelapa Menggunakan Perekat Amilum 18:24–29
8. Sukarni S, Zakaria Y, Sumarli S, Wulandari R, Ayu Permanasari A, Suhermanto M (2019) Physical and chemical properties of water hyacinth (*Eichhornia crassipes*) as a sustainable biofuel feedstock. In: IOP conference series: materials science and engineering. IOP Publishing, pp 012070. https://doi.org/10.1088/1757-899X/515/1/012070
9. Qian H, Zhu W, Fan S, Liu C, Lu X, Wang Z, Huang D, Chen W (2017) Prediction models for chemical exergy of biomass on dry basis from ultimate analysis using available electron concepts. Energy 131:251–258. https://doi.org/10.1016/j.energy.2017.05.037
10. Yin C-Y (2011) Prediction of higher heating values of biomass from proximate and ultimate analyses. Fuel 90:1128–1132. https://doi.org/10.1016/j.fuel.2010.11.031
11. Sukarni S (2016) Exploring the potential of municipal solid waste (MSW) as solid fuel for energy generation: case study in the Malang City, Indonesia. In: AIP conference proceedings. p 020003. https://doi.org/10.1063/1.4965733

12. Marcilla A, Catalá L, García-Quesada JC, Valdés FJ, Hernández MR (2013) A review of thermochemical conversion of microalgae. Renew Sustain Energy Rev 27:11–19. https://doi.org/10.1016/j.rser.2013.06.032
13. Ozyuguran A, Akturk A, Yaman S (2018) Optimal use of condensed parameters of ultimate analysis to predict the calorific value of biomass. Fuel 214:640–646. https://doi.org/10.1016/j.fuel.2017.10.082
14. Gil MV, García R, Rubiera F, Pevida C (2019) Assessing the influence of biomass properties on the gasification process using multivariate data analysis. Energy Convers Manag 184:649–660. https://doi.org/10.1016/j.enconman.2019.01.093
15. Chee A, Lim R, Lai B, Chin F, Abbas Z, Ling K (2016) Kinetic analysis of rice husk pyrolysis using Kissinger-Akahira-Sunose (KAS) method. Procedia Eng 148:1247–1251. https://doi.org/10.1016/j.proeng.2016.06.486
16. Sukarni, Sudjito, Hamidi, N., Yanuhar, U., Wardana, I.N.G.: Thermogravimetric kinetic analysis of Nannochloropsis oculata combustion in air atmosphere. Front Energy 9:125–133. https://doi.org/10.1007/s11708-015-0346-x
17. Sheng C, Azevedo JLTÃ (2005) Estimating the higher heating value of biomass fuels from basic analysis data. Biomass Bioenergy 28:499–507. https://doi.org/10.1016/j.biombioe.2004.11.008
18. Sukarni S, Sumarli S, Nauri IM, Purnami P, Al Mufid A, Yanuhar U (2018) Exploring the prospect of marine microalgae *Isochrysis galbana* as sustainable solid biofuel feedstock. J Appl Res Technol 16:53–66
19. Singh YD, Mahanta P, Bora U (2017) Comprehensive characterization of lignocellulosic biomass through proximate, ultimate and compositional analysis for bioenergy production. Renew Energy 103:490–500. https://doi.org/10.1016/j.renene.2016.11.039
20. Açikalin K (2012) Pyrolytic characteristics and kinetics of pistachio shell by thermogravimetric analysis. J Therm Anal Calorim 109:227–235. https://doi.org/10.1007/s10973-011-1714-3

Improving the Performance of Photovoltaic Panels by Using Aluminum Heat Sink

Ian Guardian, Bayu Sutanto, Rendy Adhi Rachmanto, Syamsul Hadi, and Zainal Arifin

Abstract Solar photovoltaic (PV) system has drawn the tremendous attention of researchers and governments in the past recent years. However, operating temperature of PV has major role to its efficiency and lifetime. This paper reported an experiment of aluminum *heat sink* as a cooling device to reduce the operating temperature and to improve the performance of PV panels. The performance of PV panels with and without aluminum *heat sink* were compared. Result showed that solid heat sink increased the average power output by 3.14 W and efficiency by 1.61%. Modification by perforating the fins increased the average output power by 1.79 W and efficiency by 0.60% compared to the solid heat sink. This cooling method was an effective and potential to be applied on commercial PV panels.

Keywords Photovoltaic system · Cooling system · Experiment · Aluminum heat sink

1 Introduction

A limited amount of fossil based energy demands an increase in the use of renewable alternative energy. Solar energy is one of the abundant alternative energies. Indonesia has a daily solar energy potential of 4.8 kWh/m^2 [1].

Solar energy consists of light and heat that is emitted to the earth within a specified period. Light energy and solar heat can be exploited using an energy converter. Solar heat can be used by solar collectors as a fluid heater. Sunlight can be converted by PV panels into electrical energy.

First-generation photovoltaic (PV) cells are still the most widely used in daily life. These PV cells have the advantage of the highest efficiency among other PV cells generations. Currently, silicon-based PV cells on the market have efficiencies ranging from 14 to 19% in which the excess energies will waste as heat [2, 3].

The operational temperature of PV panels could reach 78–88 °C at 1000 W/m^2 solar radiation intensity. The optimal operating temperature for PV panels is generally

---

I. Guardian · B. Sutanto · R. A. Rachmanto · S. Hadi · Z. Arifin (✉)
Universitas Sebelas Maret, Surakarta, Indonesia
e-mail: Zainal_arifin@staff.uns.ac.id

U. Sabino et al. (eds.), *Proceedings of the 6th International Conference and Exhibition on Sustainable Energy and Advanced Materials*, Lecture Notes in Mechanical Engineering,
https://doi.org/10.1007/978-981-15-4481-1_42

25 °C. An increase in temperature of 1 °C could cause a reduction in the output power of PV panels by 0.4–0.65% of the total power produced [4, 5]. Related to the high operating temperature, PV panels will directly reduce the performance of PV panels. Therefore, one effective way to increase the efficiency of PV panels is to increase the temperature of PV panels while working [6].

Based on existing research, an active cooling solution could reduce temperature of PV panels at 7.5 °C greater than passive cooling solution [5]. However, the active cooling solution needs an external power input, so using an active solar cell cooler will reduce the overall efficiency of the PV panels system.

One form of passive cooling solution is heat sinks. Adding heat sinks could reduce operating temperatures by 4.2%, increase efficiency by 3.26% and increase the output power by 5.5% [7]. Other studies mention that heat sinks could reduce temperatures by 6 °C, increase efficiency by 1.77% and increase power by 1.87 W [8].

Variations in the shape of the heat sink geometry affect the performance in removing heat. Heat sinks with perforated fin geometry, have a higher heat transfer coefficient than solid fins [9, 10]. Therefore, this research used variations of the heat sink's fin attached to the PV panels module. This passive cooling method is expected to increase the performances of PV panels on small scale.

2 Theory

PV panels convert energy from sunlight into photons into electrical energy. The amount of energy converted by PV panels is divided into two main factors. The first factor is the capacity of these PV panels, and the second factor is environmental conditions such as temperature, humidity and wind speed.

2.1 Principle of PV Panel

The structure of a solar cell consists of a diodes (junction) connection between two thin layers made of semiconductor materials, p-type (positive) and n-type (negative) semiconductor. The connection between p-type and n-type (p-n junction) semiconductors will form a depletion region. When photons with a specific energy enter the layers of PV panels, n-type silicon will excite a large numbers of electron. Then the p-type silicon will produce many holes.

Negative charge will move up to front contact which is a metal conductor. At the same time, positive charge will move downward to back contact. The process of electron flow causes a difference in voltage and generates a direct current between the front and rear contacts. The electric current can be channeled to the outside circuit.

2.2 PV Panel Performance Parameters

PV panel performance parameters consist of

1. Open circuit voltage (V_{OC}) is the maximum voltage capacity that can be reached when no current flows in the circuit.
2. Short circuit current (I_{SC}) is the maximum electric current when there is no resistance in the circuit.
3. Maximum power point (P_{MPP}) is the point on the I-V curve producing the maximum multiplication of current and voltage.
4. Fill factor (FF) is the division of maximum power (P_{MPP}) on the V_{oc} and I_{sc} products. The equation of FF is shown below:

$$\mathrm{FF} = \frac{P_{MPP}}{I_{SC} \times V_{OC}} = \frac{I_{MPP} \times V_{MPP}}{I_{SC} \times V_{OC}} \tag{1}$$

5. Efficiency (η) is the ratio between the maximum power (P_{MPP}) to the power of solar radiation received by PV panels (P_{light}). The equation used to find Efficiency is shown below:

$$\eta = \frac{P_{MPP}}{I_{light}} = \frac{P_{MPP}}{I_{rad} \times A} = \frac{I_{SC} \times V_{OC} \times FF}{I_{rad} \times A} \tag{2}$$

2.3 Heat Sink

The heat sink is used to expand the heat transfer area. The wider the heat transfer area, the greater the heat that can be released from the system. In general, heat sinks use a conductive metal material. The ability of the heat sink is measured by its effectiveness (ε_{fin}). which can be determined by the following equation [11].

$$\varepsilon_{fin} = \frac{Q_{total\ fin}}{Q_{total\ nofin}} = \frac{h\left(A_{unfin} + \eta_{fin} A_{fin}\right)(T_b - T_\infty)}{h A_{no\ fin}(T_b - T_\infty)} \tag{3}$$

3 Experimental Setup

This study aimed to determine the effect of the addition of a heat sink cooler to the PV panels performance. Comparative experiments were carried out on the variation of the heat sink used. The variations used in this experiment were PV panels without cooling, solid heat sink and perforated heat sink.

Experiments were carried out using 50 Wp PV panels. PV panels were connected with variations in load. Voltmeter and ammeters were used to get voltage and current data. Pyranometer was used to get radiation intensity data. K-type thermocouples were used to obtain the temperature of the PV panels. Thermocouples were placed on the top surface of the PV panels. Fans were used to regulate wind speed.

The experiments were carried out in July and on the rooftop building of the Faculty of Engineering, Sebelas Maret University (Latitude −7.56234 °S, Longitude 110.8538 °E), Surakarta, Indonesia. The experiments started at 07:30 AM and ended at 11:30 AM. Presentation of experimental data is limited to radiation intensity of 1000 W/m^2 as shown in Fig. 1.

The heat sink used was made of series 6 aluminum plate. Figure 2 is the geometries and dimensions of the heat sink used on the experiments. The base of the heat sink used a plate with a thickness of 3 mm and the heat sink fins used a plate with a thickness of 2 mm. The number of fins used was ten. Heat sinks had total effectiveness number of 2.63 based on Eq. 3. The heat sink was attached to the bottom of the PV panels with a thermal grease coated on the contact side.

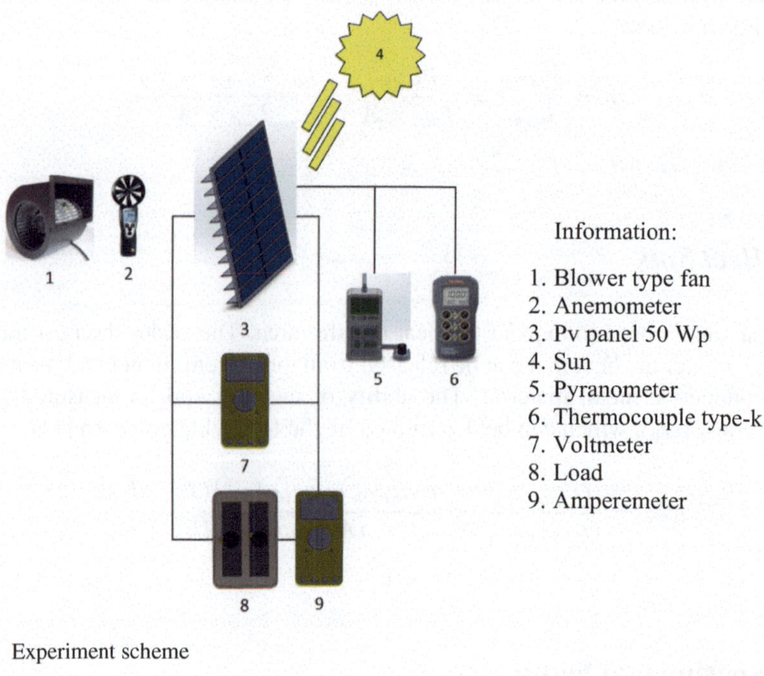

Fig. 1 Experiment scheme

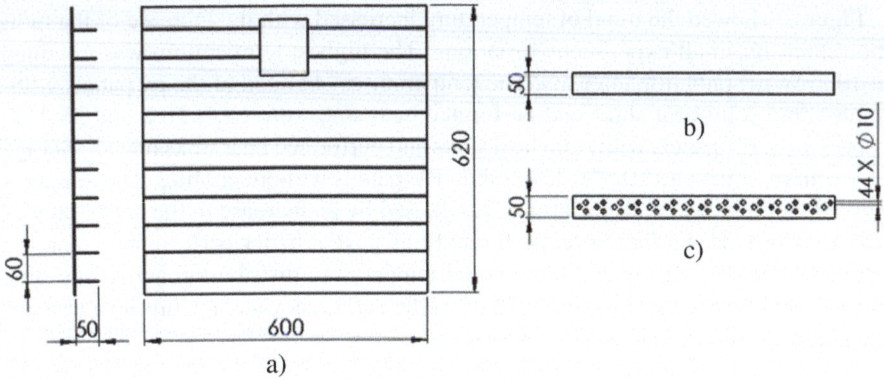

Fig. 2 Dimensions in mm. **a** Heat sink, **b** solid fin, **c** perforated fin

4 Result and Discussion

Quantitative data obtained in this study were temperature (T), currents and voltages. Furthermore, the data was processed to obtain PV panels performance parameters consisting of open-circuit voltage (V_{oc}), short circuit current (I_{sc}), maximum power (P_{MPP}) and efficiency (η).

4.1 Temperature

The high operating temperature could affect the performance of the silicon PV panels [4, 5, 12]. The addition of a heat sink would increase the heat transfer rate of the system. Figure 3 showed experiments temperature data.

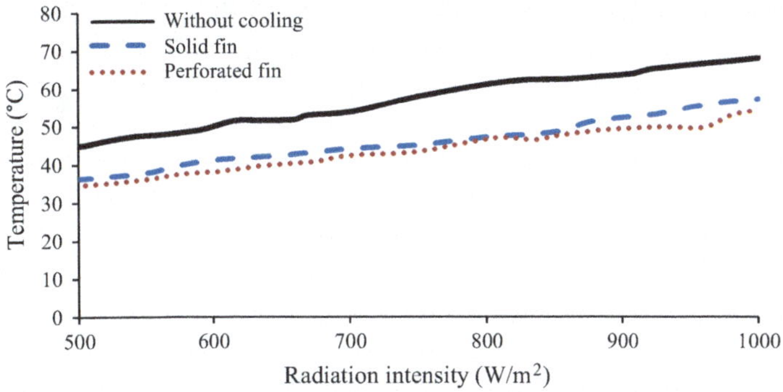

Fig. 3 Graph of the relationship between solar radiation intensity and PV panels temperature

Figure 3 showed the trend of temperature increased with the increase of the radiation intensity in all experiment variations. The highest temperature was obtained on 1000 W/m^2 radiation intensity. The temperatures obtained of the pv panels without cooling, solid heat sinks and perforated heat sink were 68.1, 58.2 and 55.4 °C respectively. PV panels with solid heat sinks and perforated heat sinks had an average temperature of 9.3 and 10.5 °C lower than PV panels without cooling. The temperature reduction in heat sink variations was caused by an increase in the active area of heat transfer [11]. Furthermore, perforated fins gave a better performance than the solid one because it provided more equal temperature distribution across the fin's area [12–14] which led to the increase of the fins efficiency and the fins heat transfer rate [11, 12].

4.2 *Open Circuit Voltage (V_{oc})*

The voltage on a PV panels was influenced by the magnitude of solar radiation intensity and operational temperature [5]. Open circuit voltage (V_{oc}) is the voltage value obtained when there is no current flowing in the open circuit.

Figure 4 showed that the V_{oc} value decreased with the increase of temperature in all variations of solar radiation intensity. At the solar radiation intensity of 1000 W/m^2, the Voc value decreased from 21 to 19.2 V followed by the temperature increase from 55.4 to 68.1 °C. The decrease in the amount of V_{oc} when the temperature rising was due to the effect of a reverse saturation current that occurred at p-n junction [15]. With the more significant impact of reverse saturation current, the voltage difference

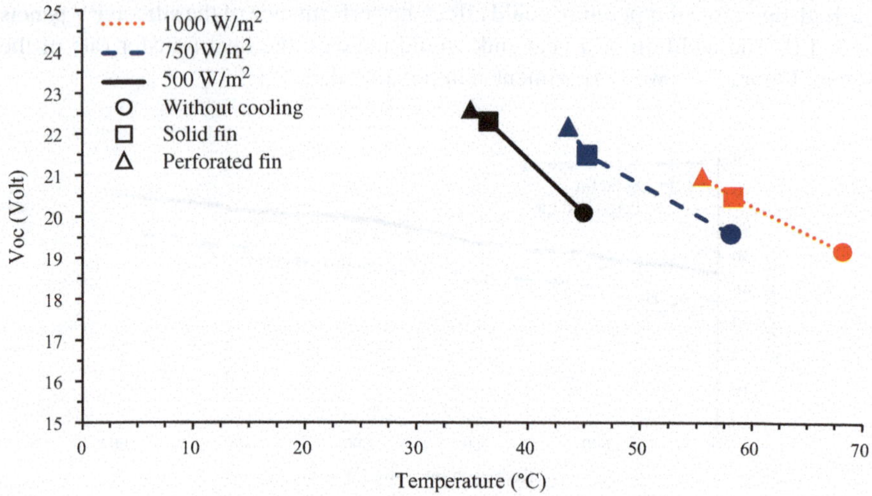

Fig. 4 Graph of the relationship between pv panels temperature and V_{oc}

of type-p and type-n silicon would decrease. The effect of reverse saturation current was strongly influenced by the magnitude of temperature [16].

4.3 Short Circuit Current (I_{sc})

Short circuit current (I_{sc}) was strongly influenced by the magnitude of the intensity of solar radiation [5]. I_{sc} is measured in a very small resistance state so that the maximum current flows in the circuit.

Figure 5 showed that the value of I_{sc} showed a modest increase with the rise in temperature in all variations of solar radiation intensity. At the intensity of solar radiation of 1000 W/m^2, the I_{sc} value increased from 2.52 to 2.55 A followed with the increase of temperature from 55.4 to 68.1 °C. An increase in the value of I_{sc} when the temperature rising was caused by a decrease in the value of the band-gap energy

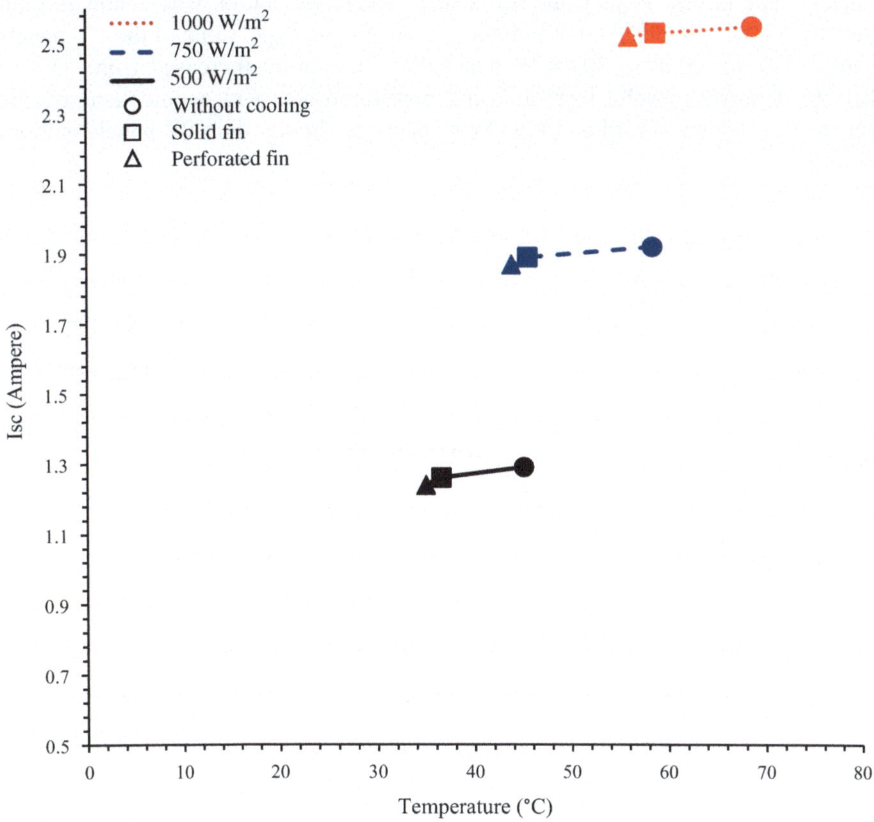

Fig. 5 Graph of the relationship between PV panels temperature and I_{sc}

needed to excite the electron. It resulted an increase of the current flows. However, the increasing value of the current was minimal [16].

4.4 Maximum Power (P_{MPP})

Power is the amount of energy per unit of time. The maximum output power of silicon PV panels is obtained from the product of the voltage and current multiplication.

Figure 6 showed that the value of P_{MPP} increased with the increase in radiation intensity. The use of perforated heat sink gave the largest average of P_{MPP} value than other variations. This showed that the decrease in temperature was greatly affected by the amount of P_{MPP} PV panels. V_{oc} was the main parameter that increased when the temperature decreased. Whereas, changes in I_{sc} values did not significantly affect P_{MPP}.

Figure 7 showed the effect of temperature on P_{MPP} on three variations of radiation intensity. The largest P_{MPP} value for each PV panels variation was obtained at an intensity of 1000 W/m^2. At that radiation intensity, the P_{MPP} value of the PV panels decreased from 35.98 to 30.44 W when the temperature increased from 55.4 to 68.1 °C. The use of solid heat sink and perforated heat sink would increase the average P_{MPP} value of 3.14 and 4.95 W respectively higher than PV panels without cooling.

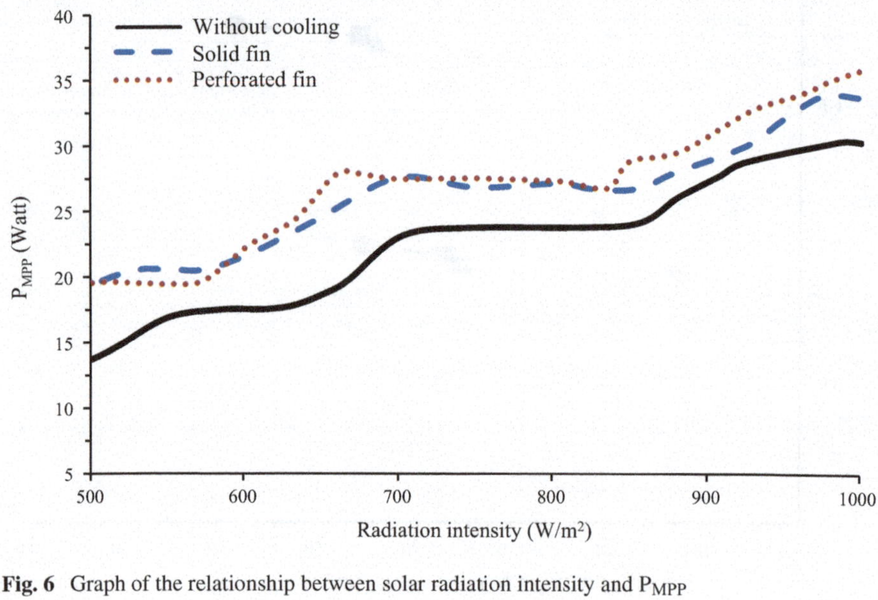

Fig. 6 Graph of the relationship between solar radiation intensity and P_{MPP}

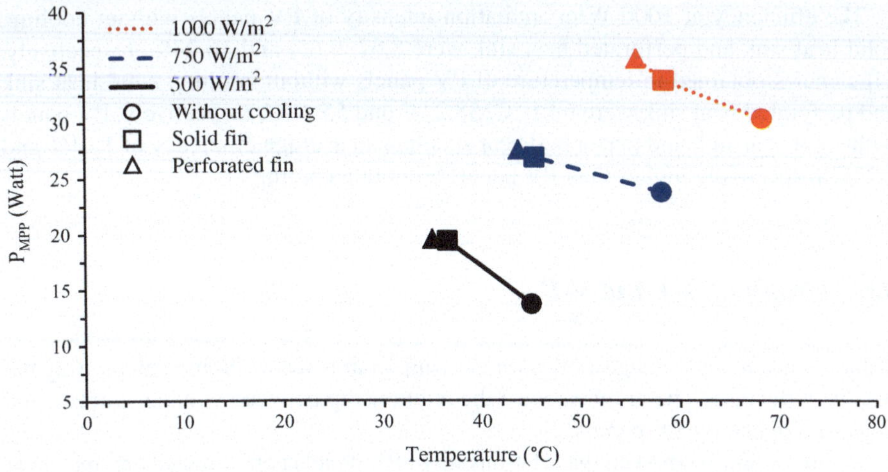

Fig. 7 Graph of the relationship between PV panels temperature and P_{MPP}

4.5 *Efficiency*

The efficiency of silicon PV panels is obtained from calculations using Eq. 2.

Figure 8 showed the effect of temperature on the efficiency of PV panels. From the graph, it was found that an increase in temperature would reduce the value of the efficiency of the PV panels. It happened in all variations of radiation intensity. PV panels using perforated heat sinks had the highest efficiency in each variation of radiation intensity. At a radiation intensity of 1000 W/m^2, the efficiency of the PV panels decreased from 10.27 to 8.67% with the increase in temperature from 55.4 to 68.1 °C.

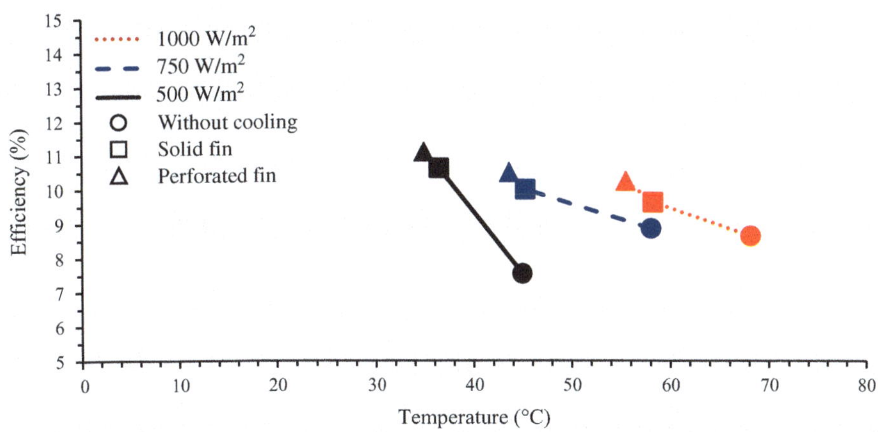

Fig. 8 Graph of the relationship between PV panels temperature and efficiency (η)

The efficiency at 1000 W/m^2 radiation intensity of PV panels without cooling, solid heat sink and perforated heat sink were 8.67, 9.65, and 10.27%, respectively. Efficiencies obtained in temperature of PV panels without cooling, solid heat sink and perforated heat sink were 68.1 °C, 58.2 °C and 55.4 °C, respectively. PV panels with solid heat sink and perforated heat sink had an average efficiency of 1.61% and 2.21% respectively higher than PV panels without a cooling.

4.6 Graph of V-I and V-P

Voltage-current and voltage-power relationship graphs were obtained by plotting the results of the data. The graph showed the relationship between voltage, current, and power produced by PV panels.

Figures 9 and 10 showed the relationship of PV panels performance parameters at 1000 W/m^2 radiation intensity. The temperature obtained of the PV panels without cooling, solid heat sink and perforated heat sink were 68.1 °C, 58.2 °C and 55.4 °C, respectively. Figure 9 showed that the lower the temperature, the higher the voltage generated. V_{oc} value obtained by PV panels without a cooling, solid heat sink and perforated heat sink were 19.2 V, 20.5 V, and 21 V, respectively. However, the value of the current changed a minimum. Figure 10 showed the higher the voltage, the

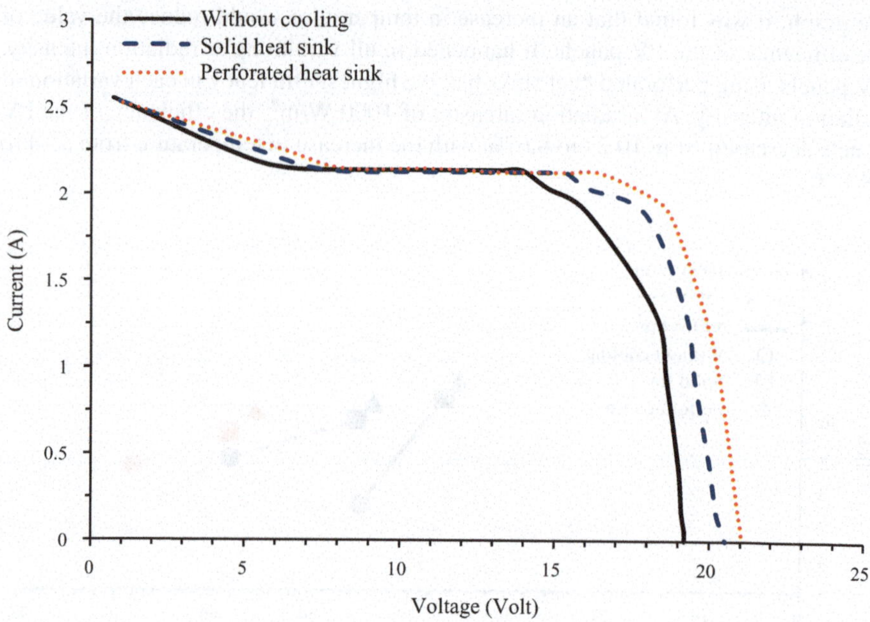

Fig. 9 Graph of relationship between voltage and pv panels currents at radiation intensity of 1000 W/m^2

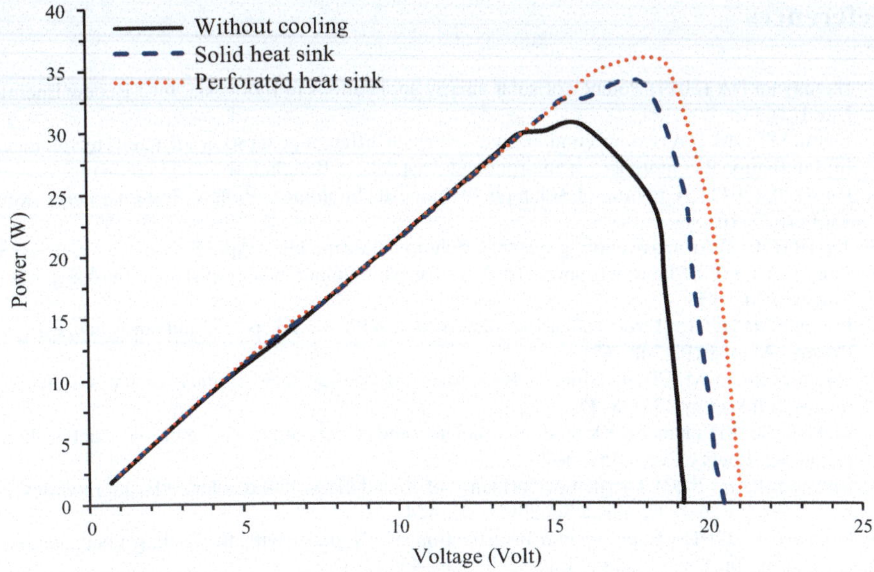

Fig. 10 Graph of relationship between voltage and PV panels power at radiation intensity of 1000 W/m^2

higher P_{MPP}. P_{MPP}s obtained of the pv panels without a cooling, solid heat sink and perforated heat sink were 30.40 W, 33.83 W, and 35.98 W, respectively.

5 Conclusion

Experimental studies were carried out using PV panels with variations of no cooling, solid heat sink, and perforated heat sink. The conclusion of this study is that the use of solid heat sink could reduce the average temperature by 9.3 °C, increasing the average output power by 3.14 W and efficiency by 1.61% compared to PV panels without cooling. Modification by perforating the fins could reduce the average temperature of 1.26 °C, increasing the average output power by 1.81 W and efficiency by 0.60% compared to the solid heat sink.

Acknowledgements This work partially supported by the grant of PDUPT from the Ministry of Research, Technology, and Higher Education, the Republic of Indonesia with contract number 719/UN.27.21/PN/2019 for FY 2019.

References

1. Handayani NA (2019) Potency of solar energy applications in Indonesia. Int J Renew Energy Dev 1(2):33
2. Kibria MT (2014) A review: comparative studies on different generation solar cells technology. In: International conference on environmental aspects of Bangladesh 51–53 (2014)
3. Parida B (2011) A review of solar photovoltaic technologies. Renew Sustain Energy Rev 15(3):1625–1636
4. Teo HG (2012) An active cooling system for photovoltaic modules. Appl Energy 90(1):309–315
5. Jiang J-A (2005) Maximum power tracking for photovoltaic power systems. Tamkang J Sci Eng 8(2):147–153
6. Rahman MM (2015) Effects of various parameters on PV-module power and efficiency. Energy Convers Manag 103:348–358
7. Hasanuzzaman M (2016) Global advancement of cooling technologies for PV systems: a review. Sol Energy 137:25–45
8. Idoko L (2018) Enhancing PV modules efficiency and power output using multi-concept cooling technique. Energy Rep 4:357–369
9. Radziemska E (2003) Thermal performance of Si and GaAs based solar cells and modules: a review. Prog Energy Combust Sci 29(5):407–424
10. Gotmare JA (2015) Experimental investigation of PV panel with fin cooling under natural convection. Int J Adv Technol Eng Sci 03:447–454 (2015)
11. Cengel Y (2014) Heat and mass transfer: fundamentals and applications. McGraw-Hill Higher Education
12. Doori WHARA (2011) Enhancement of natural convection heat transfer from the rectangular fins by circular perforations. Int J Autom Mech Eng 4:428–436
13. Mahmod AM (2008) Determination of the temperature distribution the perforated fins under natural convection. Tikrit J Eng Sci 15(2):63–78
14. Jassem RR (2013) Effect the form of perforation on the heat transfer in the perforated fins. Acad Res Int 4(3)
15. Modul Surya Terapung Dengan Metode Thermosiphon (2018) Institut Teknologi Bandung
16. Tobnaghi DM (2013) Investigation of light intensity and temperature dependency of solar cells electric parameters. Electr Power Eng Contol Syst 90–93

The Effect of Fins Number Variation on Aluminum Heat Sink to the Photovoltaic Performance

Musthofa Jamaluddin, Rendy Adhi Rachmanto, Syamsul Hadi, Chico Hermanu Brillianto Apribowo, Trismawati, and Zainal Arifin

Abstract Optimization of energy becomes a popular topic nowadays due to the rise in global energy demand, the decrease in fossil fuel, and environmental problems. One of the solutions to cope with those problems is solar energy. Solar energy has a problem, which is the increased working temperature of the panel. Working temperature increase in panel photovoltaic (PV) panels will cause a decrease in performance and damage to PV panels. One way to reduce the working temperature in PV panel is a cooling method using a heat sink. The variations used were no fin, 5 fins, and 10 fins. This study to obtain the I-V curve obtained by measuring the voltage and current values. Panel without heat sink (ground) was obtained efficiency of 8.72%, maximum power of 34.18 W, and working temperature of 72.6 °C with the same intensity 1100 W/m^2. Heat sink panel no fins obtained an efficiency of 9.13%, maximum power of 39.65 W, and working temperature of 69 °C. Heat sink panel with 5 fins, the efficiency is 9.58%, the maximum power is 39.65 W, and the working temperature is 66.5 °C. Heat sink panel with 10 fins, the efficiency is 10.21%, the maximum power is 40.17 W, and the working temperature is 63.4 °C. The results showed that the number of fins the efficiency increased from 0.56 to 1.8%, increased the maximum power by 2.56–5.55 W, and the working temperature decreased by 3.6 to 9.1 °C compared to the panel without heat sink.

Keywords Solar cells · Cooling · Heat sink · Fins

1 Introduction

Energy is one of the global challenges in the world, especially with the increase in energy demands every year. Solar energy is a renewable energy that can be a solution to solve the problem of energy shortages in many countries, one of which is in Indonesia. Indonesia has a tropical climate because it is located on the equator,

---

M. Jamaluddin · R. A. Rachmanto · S. Hadi · C. H. B. Apribowo · Z. Arifin (✉)
Sebelas Maret University, Surakarta, Central Java 57126, Indonesia
e-mail: Zainal_arifin@staff.uns.ac.id

Trismawati
Panca Marga University, Probolinggo, East Java 67271, Indonesia

U. Sabino et al. (eds.), *Proceedings of the 6th International Conference and Exhibition on Sustainable Energy and Advanced Materials*, Lecture Notes in Mechanical Engineering,
https://doi.org/10.1007/978-981-15-4481-1_43

causing a very large amount of acceptable sunlight. Solar energy is an energy in the form of light and heat from the sun, which can be used as a renewable energy source. The tool for converting solar energy into electrical energy is solar panels/solar cells. Solar panels convert solar energy into electrical energy by utilizing the photovoltaic process. The photovoltaic process utilizes photons contained in sunlight at certain wavelengths to excite the electrons in the semiconductor material so that electron flow occurs [1]. The flow of electrons is converted into electrical energy by photovoltaic.

Photovoltaic panels are increasingly being used throughout the world because of the ability of panels to work on diffusion radiation. So, it is necessary to know that photovoltaics respond to conditions in different climates. The ability of the panel to work will affect the resulting conversion efficiency. The conversion efficiency of photovoltaic is relatively low, which means some of the heat energy absorbed is not used. This heat energy will cause an extraordinary increase in temperature on the panel. In practice, only 15–20% of solar irradiation can be converted into electrical energy. The rest will be wasted into heat energy. Heat energy affects the efficiency of the panel to decrease by about 0.40–0.65% at each panel temperature rise by one degree [2].

The efficiency of the panel in each solar cell varies depending on the type of panel. There are three generations of solar cells that have been developed to date. Solar cells have been developed from generation to generation with different components and different characteristics. The first generation of solar cells is based on silicon material. Meanwhile, the second generation in the form of thin layers and the third generation in the form of dye-based solar cells called Dye-Sensitized Solar Cell (DSSC). The first generation successfully developed by researchers is a technology that uses a single crystal silicon material and polycrystals. This technology can produce solar cells with high efficiency [3].

One of the causes of decreased photovoltaic efficiency is the working temperature. The increase in working temperature affects the output produced by photovoltaics. The increase in working temperature in photovoltaic can reach 80 °C in the blazing region [4]. A high increase in working temperature will cause damage to the panel (when wasted heat is not dissipated efficiently), which not only reduces the efficiency of the conversion of light into electricity but also damages cells and other module materials [5]. The increase in temperature also causes the open-circuit voltage and load parameters to decrease. Furthermore, by extracting thermal energy and cooling the photovoltaic, the photovoltaic efficiency can be increased [6].

The reduction in efficiency in photovoltaics caused by rising temperatures can be prevented by several methods: passive and active methods. Passive methods do not require additional energy to operate. In this case, the passive method utilizes natural convection to decrease the temperature on the panel. Various photovoltaic cooling processes not only increase panel efficiency but also lead to an increase in panel lifetime due to reduced thermal pressure. Besides, it makes high durability and relatively low cost. It is possible to utilize heat transfer for other important purposes [7, 8]. Passive cooling consists of cooling the heat sink, PCM (Phase Change Material) cooling, and reflector cooling. Some cooling processes that will be used can be planned so that the temperature drops steadily to increase efficiency. The active method for the process requires or consumes energy from outside, such as pumps, fans, electricity, etc. [3]. The active cooling consists of water cooling, wind cooling, pipe cooling and nanofluid cooling [9]. According to several existing studies, cooling panels in the form of both passive and active which use a heat sink are possible for further research.

Popovici [10] used ANSYS simulation, which adjusted the angle of the fins with an angle variation of 45°, 90°, and 135°. The results of the Popovici study [10] shown the best temperature reduction was at the angle of 45° from the initial temperature of 56–41.87 °C, at the angle of 90°, also decreased to 42.35 °C with a difference of approximately 1 °C compared to a 45° angle. The provision of heat sinks with aluminum material and variations in fins according to previous studies by regulating the number, distance, width or height of the fins will affect the performance of the heat sink in transferring the heat [11, 10]. Therefore, this research uses a passive cooling method in the form of regulating the number of heat sink fins on the solar cell module. This cooling method is expected to improve the efficiency of small-scale solar cell modules effectively.

2 Research Methodology

The model in this research referred to the one made by Popovici [10] and Chen [11]. This study aimed to determine the effect of the number of fins on the heat sink to the performance of solar cells. This experiment was carried out using variations in the number of fins, while the panel used was a 50 Wp solar panel.

The solar panels were arranged according to Fig. 1 test scheme. In this study, data collection will be conducted outdoors. Then, data retrieval was done using a resistor variable as the load used. The load used is 17 points of load variation.

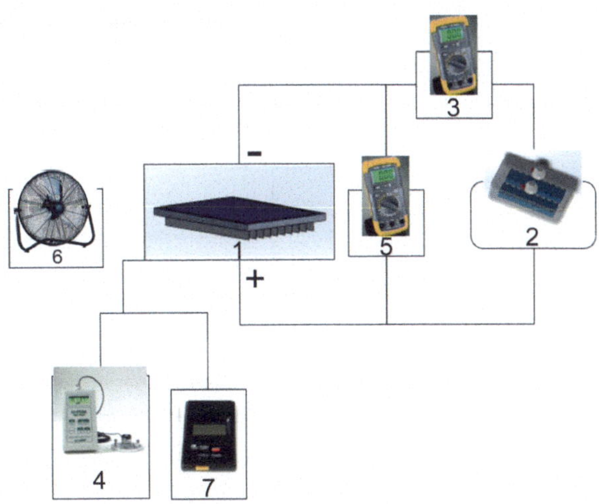

Fig. 1 Test circuit scheme. Information: (1) solar panel 50 Wp, (2) resistor variable, (3) ammeters, (4) pyranometer, (5) voltmeter, (6) blower fan, (7) K-type thermocouple

Radiation intensity data (W/m^2) was obtained from the pyranometer, while a k-type thermocouple was used to get the temperature. The placement of the thermocouple is shown in Fig. 2.

The heat sink used in this study was made of an aluminum plate. At the bottom of the heat sink (heat sink base) was a 3 mm thick plate, and at the fin section of the heat sink was a plate with a thickness of 2 mm and a height of 5 mm, as shown in Fig. 5. Figures 3 and 4 show the variation used in this study was in the form of the number of fins. The placement of the heat sink was on the bottom of the solar panel, which has been coated with thermal grace on the contact side.

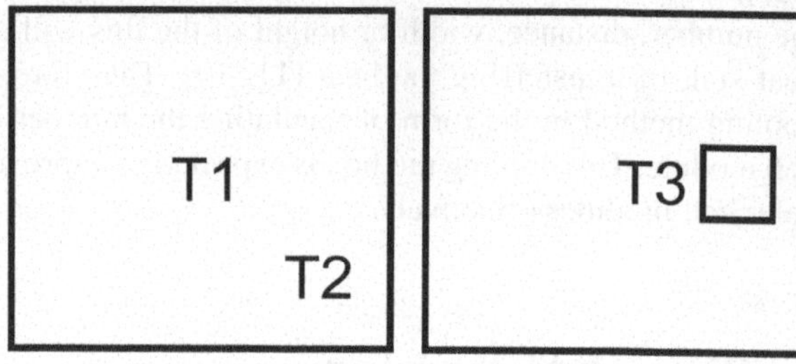

Fig. 2 Thermocouple placement

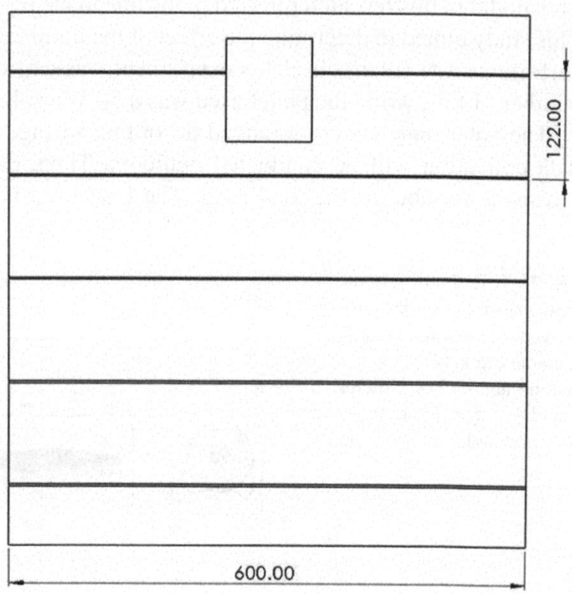

Fig. 3 Heat sink design variation 1

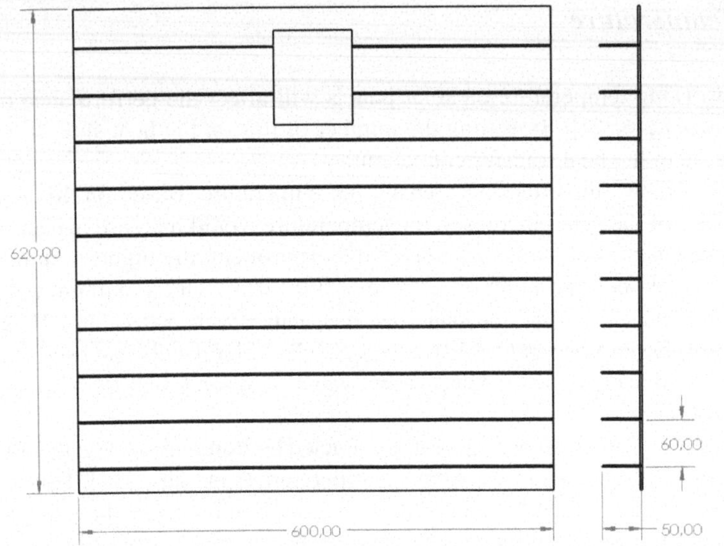

Fig. 4 Heat sink design variation 2

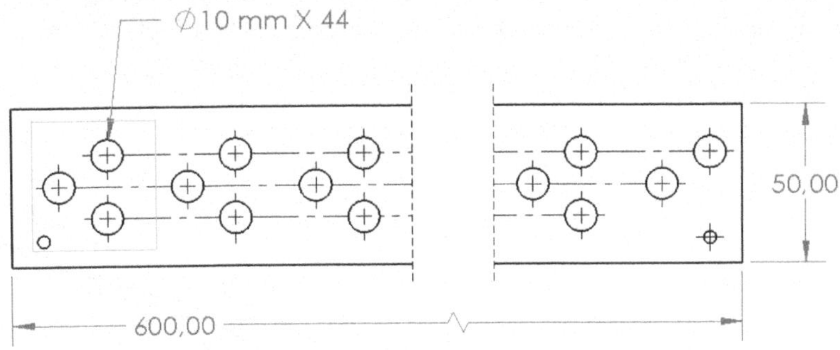

Fig. 5 Fins dimension

3 Result and Discussion

Data analysis was performed based on the heat sink cooling variations used in this study. The data collection method in this study used the quantitative method. Quantitative data obtained in this study were temperature (T), short circuit current (I_{sc}), open-circuit voltage (V_{oc}), maximum power (P_{MPP}), and efficiency (η).

3.1 Temperature

A high working temperature on solar panels will affect the performance of silicon solar panels [2, 6, 12]. By setting the number of fins on the heat sink, it will affect the increase in the heat transfer coefficient.

Figure 6 shows the effect of intensity on temperature. Based on this graphic, as the intensity of the panel increases, the temperature would increase in each variation of the experiment. For each variation of the experiment, the highest temperature in the solar panel occurred at an intensity of 1100 W/m^2. The maximum temperature obtained on the ground panel, panel no fins, panel with 5 fins, and 10 fins were 72.6 °C, 69 °C, 66.5 °C, and 63.4 °C, respectively. Solar panels that used the cooling system had an average lower temperature value of around 3.6–9.2 °C compared to panels ground.

The results of previous experiments conducted by Rahman showed that increasing the intensity of the sun caused the panel temperature to increase [13]. This is because the higher the sun's intensity received, the more photon energy the panel received. The effect of this large photon energy would make the process of electron excitation increase, which increased temperature on the panel [3].

The effect of using fins on the heat sink was an increase in the heat transfer process on the panel. This heat transfer process was due to the increased transfer area in the panel and heat dissipation through fins [14]. Therefore, solar panels with heat sinks and a higher number of fins had lower temperatures than grounds.

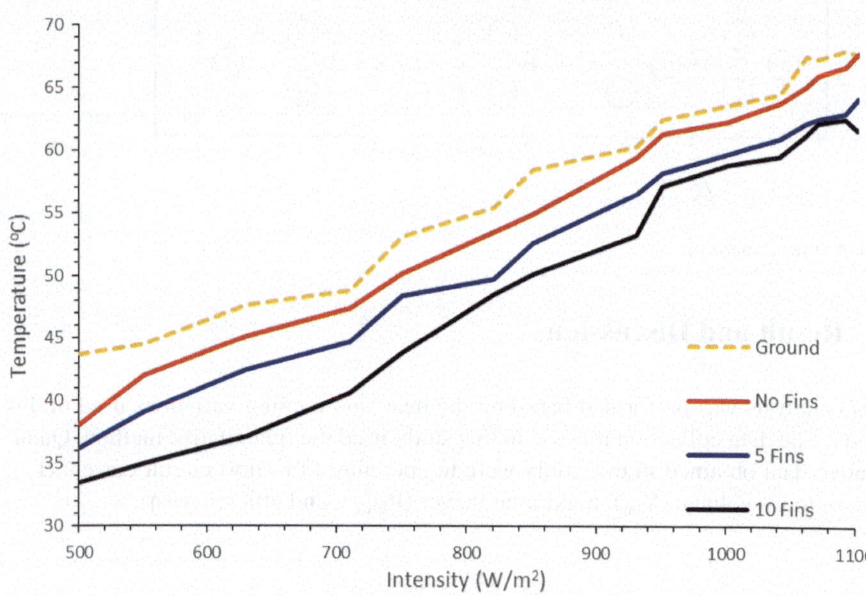

Fig. 6 Relationship between intensity and temperature

The heat sink used was the one with fins that speeded up the heat transfer process so that it was evenly distributed. Meanwhile, the provision of perforated fins was to accelerate the process of heat transfer in the heat sink. The effect of giving fins caused the power output to increase and temperature to decrease [2]. Therefore, adjusting the number of fins and the distance in the fins used would cause an increase in performance and temperature drops.

3.2 *Short Circuit Current (I_{SC})*

Short circuit current (I_{SC}) is a condition at the time of maximum current without resistance to the panel. The amount of received solar intensity will influence the short circuit current (I_{SC}) [13].

Figure 7 shows that I_{sc} increased with the increase in radiation intensity. At a maximum intensity of 1100 W/m^2, the maximum I_{sc} was obtained for all variations. The maximum I_{sc} values obtained on the ground solar panel, no fins, 5 fins, and 10 fins in a row were 2.68 A, 2.74 A, 2.76 A, and 2.78 A, respectively. It was obtained at the temperature of the ground solar panel, with no fin panel, panel with 5 fins, and 10 fins were 72.6 °C, 69 °C, 66.5 °C, and 63.4 °C, respectively.

The increase in radiation intensity caused an increase in I_{sc} because the greater the intensity of the sun would make the photon energy, which was received by the panel greater. Photon energy influences the electron excitation process. The greater the photon energy received, the more electron excitation process made. Electrons can be excited when the photon that hits them has energy with a level above the bandgap energy (1.11 eV). Therefore, the more electrons excited, the more electric current flowed [15].

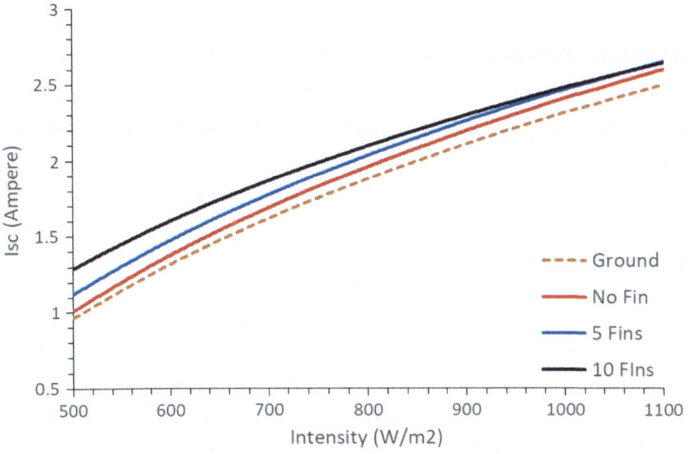

Fig. 7 Relationship between intensity and I_{sc}

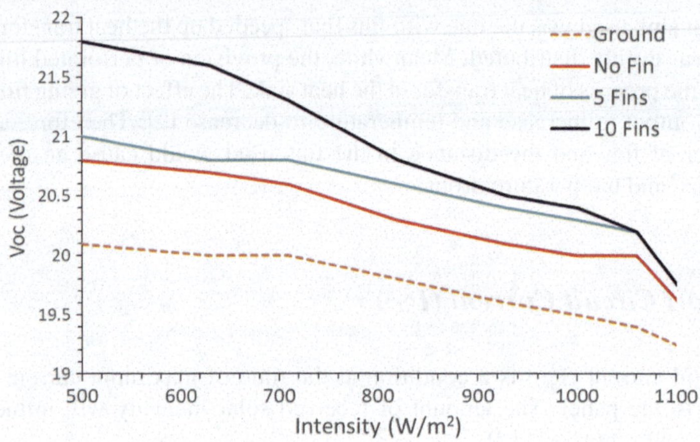

Fig. 8 Relationship between intensity and V_{OC}

3.3 Open Circuit Voltage (V_{OC})

Open circuit voltage (V_{OC}) refers to the maximum voltage capacity obtained in a condition when there is no current in the open circuit. V_{OC} is influenced by the intensity of the radiation received and the temperature on the panel. When the temperature and intensity of the sun increases, the Open circuit voltage decreases [13].

Figure 8 shows the influence of intensity on the open-circuit voltage (V_{OC}). At a maximum intensity of 1100 W/m^2, V_{OC} was obtained on the ground solar panel, with no fins, 5 fins, and 10 fins at 19.2 A, 19.5 A, 19.6 A, and 19.6 A respectively. The V_{OC} value decreased when the intensity and temperature on the panel increased.

3.4 Maximum Power (P_{MPP})

The maximum power (P_{MPP}) is the amount of energy per unit time. Maximum power is obtained from the I-V curve. I-V curve for maximum power is obtained from the maximum multiplication between current and voltage in the circuit, as in Eq. 1. The research data was obtained with the help of the resistors variable to get the results of the I-V curve, as shown in Fig. 9.

$$P_{MPP} = I_{MPP} \times V_{MPP} \tag{1}$$

Figure 9 showed the results of the I-V curve were obtained at an intensity of 1100 W/m^2 in each variation. Data retrieval is done by using a variable resistor as the load used. The load used is 17 points of load variation from 0; 2.5; 3.5; 4.7; 5.4; 5.8; 6; 6.4; 6.6; 6.9; 7.4; 8.5; 13.8; 19.6; 42.5; 111 and 330 Ω. Figure 10 shows the effect

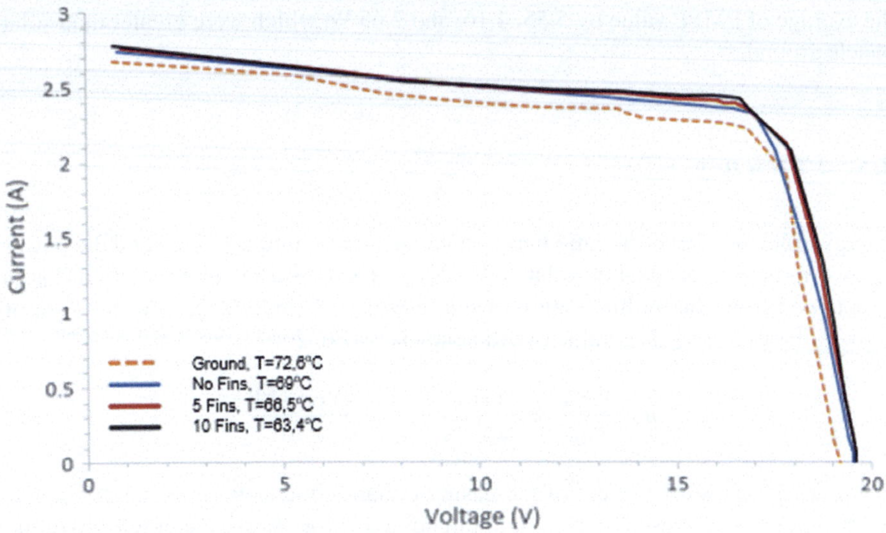

Fig. 9 I-V curve at 1100 W/m^2 in each variation

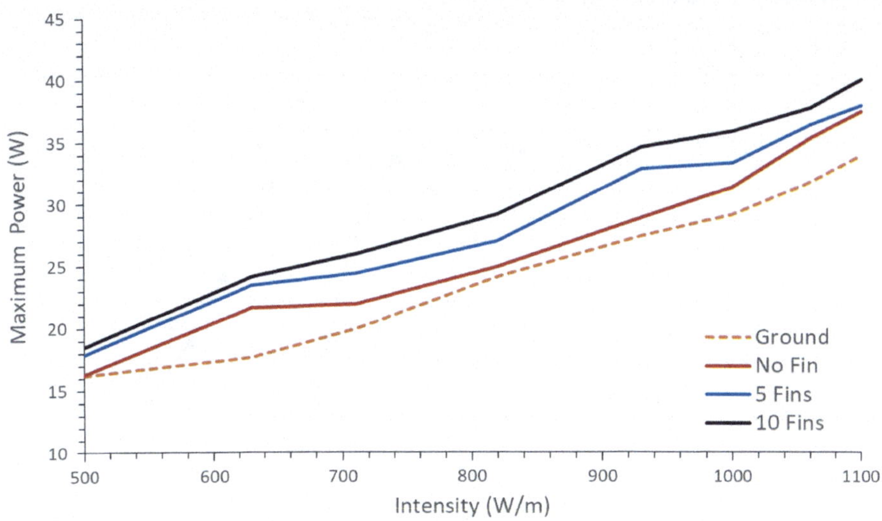

Fig. 10 Relationship between intensity and maximum power

of intensity on maximum power. Each increase in radiation intensity would cause maximum power to increase. The largest P_{MPP} value for each increase in variation was obtained at an intensity of 1100 W/m^2. At this intensity, the value of the P_{MPP} solar panel decreased from 40.17 W to 34.18 W when the temperature increased from 63.4 to 72.6 °C. The use of heat sinks with 10 fins, 5 fins, and no fins increased the average of PMPP value by 5.55, 4.16, and 2.65 W, which were greater than solar panels ground.

3.5 *Efficiency*

Energy efficiency (η) is the ratio between the maximum power (P_{MPP}) and the power of solar radiation received by solar cells (P_{light}). The power of solar radiation (P_{light}) is obtained from the multiplication of the intensity of sunlight (I_{rad}) to the area of active solar cells (A). The value of efficiency is by Eq. 2.

$$\eta = \frac{P_{MPP}}{I_{light}} = \frac{P_{MPP}}{I_{rad} \times A} = \frac{I_{SC} \times V_{OC} \times FF}{I_{rad} \times A} \tag{2}$$

Figure 11 shows the effect of increasing radiation intensity on panel efficiency. As the intensity increased to the maximum point of 1100 W/m^2, the efficiency of the panel also increased. A significant increase was obtained in 10 fins by 1.8% from the initial efficiency between 8.72 and 10.21%. The increase also occurred in 5 fins and no fins at 1.1 and 0.54%, respectively.

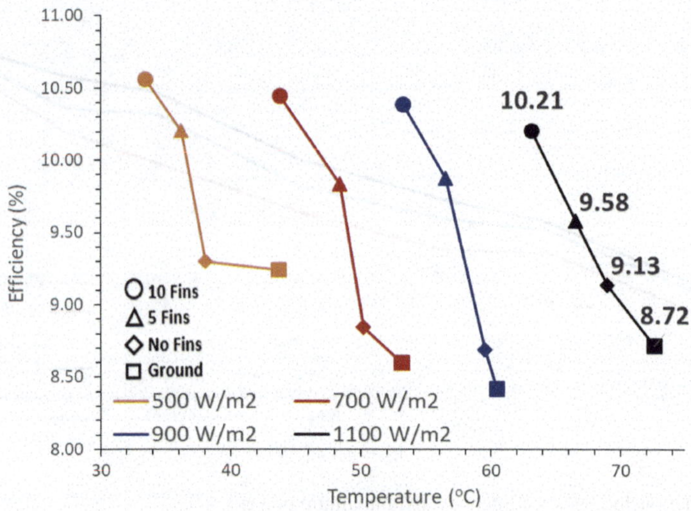

Fig. 11 Relationship between intensity and efficiency

4 Conclusion

In every test, it has been done using variations of solar panels ground, no fins, 5 fins, and 10 fins. The conclusion drawn from this study was that the use of heat sinks with 10 fins, 5 fins, and no fins could reduce the average temperature from 3.6 to 9.1 °C, increase the average output power by 2.65–5.55 W and increase efficiency by 0.56–1.8% of solar panels without a heat sink.

Acknowledgements This work partially supported by the grant of PDUPT from the Ministry of Research, Technology, and Higher Education, the Republic of Indonesia with contract number 719/UN.27.21/PN/2019 for FY 2019.

References

1. Bilgili M (2015) An overview of renewable electric power capacity and progress in new technologies in the world. Renew Sustain Energy Rev 49:323–334
2. Teo HG (2012) An active cooling system for photovoltaic modules. Appl Energy 90(1):309–315
3. Sutanto B (2018) Simulasi dan Pengujian Sistem Pendingin untuk Modul Surya Terapung dengan Metode Thermosiphon 23116301
4. Reddy SR (2015) A review of PV-T systems: thermal management and efficiency with single phase cooling. Int J Heat Mass Transf 91:861–871
5. Xu H (2017) International Journal of Heat and Mass Transfer Coupled natural convection and radiation heat transfer of hybrid solar energy conversion system. Int J Heat Mass Transf 107:468–483
6. Nadda R (2018) Efficiency improvement of solar photovoltaic/ solar air collectors by using impingement jets: a review. Renew Sustain Energy Rev 93:331–353
7. Siecker J (2017) A review of solar photovoltaic systems cooling technologies. Renew Sustain Energy Rev 79(May):192–203
8. Hasanuzzaman M (2016) Global advancement of cooling technologies for PV systems: a review. Sol Energy 137:25–45
9. Nižetić S (2018) Comprehensive analysis and general economic-environmental evaluation of cooling techniques for photovoltaic panels, Part II: active cooling techniques. Energy Convers Manag 155:301–323
10. Popovici CG (2016) Efficiency improvement of photovoltaic panels by using air cooled heat sinks. Energy Procedia 85:425–432
11. Chen H (2015) Comparative study on the performance improvement of photovoltaic panel with passive cooling under natural ventilation. Int J Smart Grid Clean Energy 374–379
12. Reddy SR (2015) International journal of heat and mass transfer a review of PV–T systems: thermal management and efficiency with single phase cooling. Heat Mass Transf 91:861–871
13. Rahman MM (2015) Effects of various parameters on PV-module power and efficiency. Energy Convers Manag 103:348–358
14. Incropera FP, Lavine AS, Bergman TL et al (2007) Fundamentals of heat and mass transfer. Wiley, New York
15. Liang P (2015) A review of concentrator silicon solar cells. Renew Sustain Energy Rev 51:1697–1708

Gain Scheduling Model Predictive Path Tracking Controller for Autonomous Vehicle on Highway Scenario

Zulkarnain Ali Leman, Mohd Hatta Mohammed Ariff, Hairi Zamzuri, Mohd Azizi Abdul Rahman, and Fitri Yakub

Abstract The design of the controller tracking path is one of the important factors in the development of autonomous vehicles. One problem for autonomous vehicle operating on highway road must be able to do a satisfactory path tracking so any accidents do not occur. This paper will discuss designing tracking path controller using combination a model predictive controller (MPC), feed forward (FF) and particle swarm optimization (PSO) based on scenario road courses on the highway with several variations of the vehicle speed. The PSO algorithm used to determine optimal weighting gains on the cost function of the MPC and the FF used to reduce the lateral error of the vehicle to the desired trajectory. The approach solves a single adaptive FF-MPC problem for tracking road trajectories. The vehicle model was developed based on *3 DOF* non-linear vehicle model. This controller model was developed based on *X*, *Y* global position and yaw rate to get input in the form of front steering to the vehicle dynamic system. For path tracking strategy, comparisons with the Stanley controller are done to analyse MPC reliability as non-linear controller in low and middle speed scenario. Simulation results have found that the FF-gain scheduling MPC controller has the significant performance on tracking trajectory at mid and high of the vehicle speeds. In addition, with the using of feed forward and optimal gain weighting on MPC controller made the actuator lifetime is longer than Stanley controller due to reduce the actuator aggressiveness.

Keywords Gain scheduling MPC · Feed forward · Weighting gain · Autonomous vehicle · PSO

Z. A. Leman · M. H. M. Ariff (✉) · H. Zamzuri · M. A. A. Rahman
Advanced Vehicle System, Malaysia-Japan International Institute of Technology, Universiti Teknologi Malaysia, Kuala Lumpur, Malaysia
e-mail: mohdhatta.kl@utm.my

Z. A. Leman
Jurusan Teknik Mesin, Fakultas Teknik, Universitas Sriwijaya, Palembang, Indonesia

F. Yakub
Malaysia-Japan International Institute of Technology, Universiti Teknologi Malaysia, Kuala Lumpur, Malaysia

U. Sabino et al. (eds.), *Proceedings of the 6th International Conference and Exhibition on Sustainable Energy and Advanced Materials*, Lecture Notes in Mechanical Engineering,
https://doi.org/10.1007/978-981-15-4481-1_44

1 Introduction

Path tracking and collision avoidance is currently a main issue of discussion for automotive vehicle safety. The path tracking particularly using active front steering (AFS) technology is consider as one of a feasible solution due to its potential to reduce traffic accidents as well as improving handling performance of the vehicle [1–6]. Most studies on path tracking design use dynamic or kinematic models to represent the behaviour of vehicle models [7–11]. The kinematic model ignores the dynamic effects. On the other hand, dynamic model considers the forces acting on the vehicle when moving.

The MPC controller have been applied in the automotive industry and applications for decades [2, 3]. The MPC able to plan tasks and carry out tasks is a major consideration in relation to planning and control systems. MPC uses a mathematical dynamic process model of the system to predict future values and optimize control process performance [10, 11]. This method also performs optimization to get the optimal value for input to the plant.

In Falcone [10] showed that model predictive control was implemented to predict an optimal steering input for obstacle avoidance task using both nonlinear and linear-time-varying (LTV) MPC. However, the implementation of this approach will require high computational resources especially in solving the optimisation problem in real time. Reference [12–14] implemented MPC in autonomous vehicles for orchard environment, while Tomatsu et al. [15] implemented MPC for path tracking on an excavator in digging operation with slow speed. More applications MPC controller for slow speed can be seen in previous studies [16–18]. Beal [19] applies predictive control models using custom C-Code tested on autonomous vehicles that can solved optimization. Also, in several studies that discussed solving MPC optimization problems using metaheuristic algorithms. This can be seen in two studies [20, 21] using the metaheuristic optimization method to implement the real-time optimization process. Merabti et al. [20] discusses three types of metaheuristic optimization algorithms to complete optimization of nonlinear MPC for control of tracking the mobile robot path. Falcone et al. [10] developed MPC combined with path planning based on bicycle vehicle model. Yakub and Mori [22] developed MPC based on Falcone et al. [10] concept combined with feed forward controller. However, all these studies assume the vertical force acting on the tire is constant and the actuator is still operating quite aggressively because larger prediction and control horizons are required in order to stabilize the vehicle along the path. In addition, the determination of the weighting gain value in standard MPC is done by trial-error based on the linearization process at a certain speed.

Therefore, the standard MPC controller developed with this method requires the optimal weighting gain tuning to minimize the cost function. In addition, the FF controller is used to reduce the aggressiveness of the actuator during manoeuvring. The main methodology used in this study is to use the gain scheduling MPC controller as a basic control law to increase its trajectory tracking performance to variations in speed parameters. Then the GS-MPC controller is combined with the FF controller.

The weighting gain of the lateral error, yaw error and wheel steering knowledge database are built by optimizing controller parameters for each combination of speed using particles swarm optimization (PSO). PSO control shows the potential in the application of the research field lane change manoeuvre and path tracking [2, 7]. So that it can be combined with the MPC controller to get the optimal tracking path in various trajectory scenarios. The PSO algorithms has used to determine the optimal weighting gain for MPC controllers, so vehicles can perform trajectory tracking with good performance at some speeds.

To develop a path tracking controller for an autonomous vehicle with input in the form of steering vehicles to 3 DOF non-linear vehicle model for various trajectory courses. The proposed controller was tested by simulating using MATLAB/Simulink to get vehicle response characteristics when doing manoeuvre on various trajectory. Weighting gain of the MPC values are calculated based on vehicle longitudinal speed parameters. The weighting gain parameter were optimized for the adaptive path tracking controller during the optimization process. Controller parameters will automatically be tuned based on vehicle speed and various trajectory courses. This was translated into three sets of MPC weighting gain which correspond to seven different vehicle speeds and three different sets of trajectory courses. This proposed controller called is Gain-Scheduling MPC (GS-MPC). Then the GS-MPC controller will be compared to the adaptive Stanley controller and standard MPC controller as used on reference [2, 23].

The paper is organized as follows: in Sect. 2 will present the non-linear model of the vehicle. The paper will discuss the standard MPC controller restrictions as a requirement for the vehicle to remain in a trajectory without collisions when operating in a rural space where the driver is deemed capable of maneuvering the vehicle. In Sects. 3 and 4, the methodology and the proposed GS-MPC controller were presented and in Sect. 5 we present the results of the validation of the proposed method. Finally, in Sect. 6 will provide some conclusions and describe future work.

2 Vehicle Dynamic Modelling

2.1 Non-linear Vehicle Model

The main source of forces acting on the vehicle are tire forces. The lateral longitudinal and vertical tire forces are the result of contact between the tire and the road [10, 11]. For path tracking, the control object is the non-linear dynamics of the vehicle. The vehicle on a horizontal road without considering the influence of the road gradient, establish two coordinate systems (Fig. 1), one coordinate system is fixed on the inertial space (X-Y), the other coordinate system is fixed on the vehicle body (x-y). Define the center of rotation of the car when steering is CG.

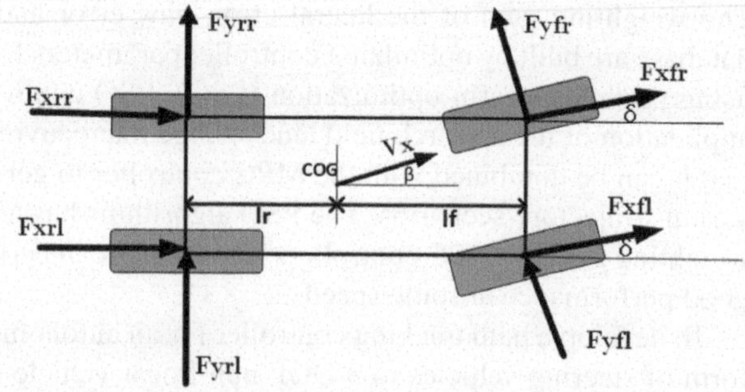

Fig. 1 The non-linear vehicle dynamic model

$$\begin{cases} m_b(\dot{v}_x - v_y\dot{\psi}) = (F_{xfl} + F_{xfr})\cos\delta - (F_{yfl} + F_{yfr})\sin\delta + F_{xrl} + F_{xrr} \\ m_b(\dot{v}_y + v_x\dot{\psi}) = (F_{yfl} + F_{yfr})\cos\delta + (F_{xfl} + F_{xfr})\sin\delta + F_{yrl} + F_{yrr} \\ I_z\ddot{\psi} = l_f((F_{yfl} + F_{yfr})\cos\delta + (F_{xfl} + F_{xfr})\sin\delta) - l_r(F_{yrl} + F_{yrr}) \\ \quad + \frac{w}{2}((-F_{xfl} + F_{xfr})\cos\delta + (F_{yfl} - F_{yfr})\sin\delta - F_{xrl} + F_{xrr}) \\ \dot{X} = \dot{x}\sin\psi + \dot{y}\cos\psi \\ \dot{Y} = \dot{x}\cos\psi - \dot{y}\sin\psi \end{cases} \quad (1)$$

The vehicle dynamics used in Eq. 1 can be compactly written as:

$$\dot{\xi} = f(\xi(t), u(t)) \quad (2)$$

where $\xi(t) \in R^n$ is the state of the system; $u(t) \in R^m$ is the input; $n = 4$ is the number of states; $m = 1$ is the number of inputs. The four states are the lateral and longitudinal velocities in the vehicle body, the yaw angle, yaw rate and the lateral and longitudinal vehicle coordinates in the inertial frame, respectively. These factors are denoted as '$\varsigma = [X, Y, \dot{\psi}, \psi]$'. The inputs are denoted as δ is the front steering angle.

Where m_b is the mass of the car, ψ is the yaw angle of the car (direction angle); v_x is the component of the vehicle speed in the x-axis direction; F_{yf}, F_{yr} are the lateral force of the front and the rear wheel; l_f and l_r are the distances of the front and rear axles from the center of gravity, I_z, the moment of inertia of the car.

2.2 Tire Model

This paper use non-linier tire properties based on the work of Nagai et al. [23]. In the dynamic nature of the wheel, the effect of load transfer is an important factor to be considered. Load transfer is a characteristic that is affected by longitudinal and lateral acceleration during running and disturbances. The static forces of the wheel are calculated by the following Eq. 3. It is assumed that the two coaxial wheels maintain the same angle of rotation. When the side angle is small, the following relationship exists between the tire lateral force and the side angle. In performing a maneuver where the tire makes a large slip angle, the tire forces will be in the nonlinear region. In this case, vehicles experience instability, so the use of linear tire models will get a performance that is not as expected. Therefore, nonlinear tire model should be used in controller design.

$$\begin{aligned} F_{z0f} &= \frac{mgl_r}{2L} \\ F_{z0r} &= \frac{mgl_f}{2L} \end{aligned} \tag{3}$$

$$\begin{aligned} \Delta F_{zfr} &= \frac{ma_x h}{2L} + \frac{ma_y hl_r}{wL} \\ \Delta F_{zfl} &= -\frac{ma_x h}{2L} + \frac{ma_y hl_r}{wL} \\ \Delta F_{zrr} &= -\frac{ma_x h}{2L} + \frac{ma_y hl_f}{wL} \\ \Delta F_{zrl} &= -\frac{ma_x h}{2L} - \frac{ma_y hl_f}{wL} \end{aligned} \tag{4}$$

where h denotes the height of center of gravity. The dynamic model of the vehicle 3 degrees of freedom can be seen in the Fig. 1 and Eq. 1. In general, vehicles operate by braking and traction during driving on the highway. Based on the concept of friction circle, the addition and reduction of longitudinal and lateral forces will cause a reduction in the tire's cornering force. The cornering force which is influenced by the nonlinear characteristics of the tire and the braking/traction action can be calculated using the equation.

$$\begin{aligned} F_{yi} &= K_{xi}\left[\frac{2}{\pi}(F_{z0i} + \Delta F_{zi})\right]\tan^{-1}\frac{\pi}{2\pi(F_{z0i} + \Delta F_{zi})}C_i\beta_i \\ K_{xi} &= \sqrt{1 - \left[\frac{F_{xi}}{\mu(F_{z0i} + \Delta F_{zi})}\right]^2} \end{aligned} \tag{5}$$

While the longitudinal forces, F_x, of the wheel are calculated based on the concept of friction force circles.

$$\begin{aligned} &\sqrt{F_y^2 + F_x^2} \le \mu F_z \\ &F_x \le \sqrt{(\mu F_z)^2 - F_y^2} \end{aligned} \tag{6}$$

where, C_{af} and C_{ar} are the cornering stiffnesses of the front and rear wheels respectively. Equations (3)-(6) fully considers the force characteristics of the car, the lateral motion law and the lateral deflection characteristics of the tire. Even if the vehicle is in complex road conditions such as high-speed cornering, the model can still better

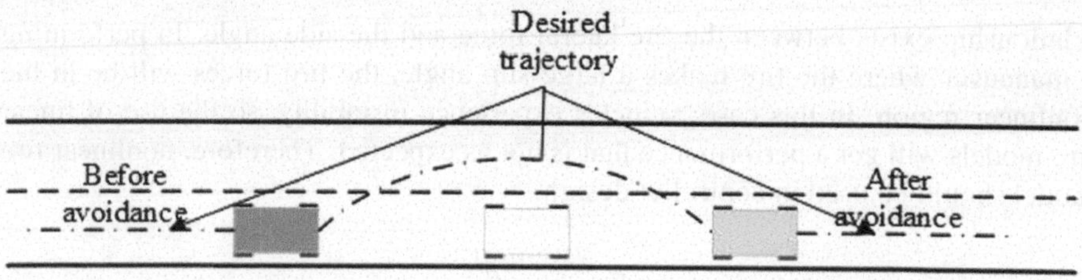

Fig. 2 double lane change scenario

reflect the lateral movement of the car, providing a basis for MPC-based steering control strategy design.

2.3 *Trajectory Generation*

During the steering of the car, the desired yaw rate ψ_{ref} should be determined by the road radius of curvature R_{ref} and the speed of the vehicle:

$$\dot{\psi}_{ref} = \frac{V_x}{R_{ref}} \tag{7}$$

The typical road shown in Fig. 2, the vehicle starts on a straight road at the initial moment, then enters the curve and then drives again. Out of the corner into the straight road.

Vehicle collisions are a major cause of road accidents. The design of vehicle control algorithms to avoid accidents is the goal of any collision avoidance system. There are two options that allow for collision avoidance maneuvers [22], namely longitudinal control (emergency braking only) and lateral control (active steering only). The reference trajectory yaw (ψ_{ref}), yaw rate (ψ_{ref}), lateral position (Y_{ref}), and longitudinal position X_{ref} fed to the controller.

3 Automatic Path Tracking Control Algorithm Based on Gain Scheduling MPC

Because MPC [23, 24] adopts the control ideas of multi-step prediction, feedback correction and rolling optimization, it has the advantages of good control effect and strong robustness, plus it can better balance control objectives and system constraints and maintain the system. The stability and performance optimization, so this paper designs the automatic path tracking control algorithm in the framework of MPC. Through analysis, the purpose of path tracking is to control the wheel steering to make the vehicle travel along the centreline of the lane. That is, during the driving process, the vehicle centre of mass and the centre displacement of the lane are zero, and the vehicle body orientation is consistent with the lane direction,

$$\begin{aligned} \xi(k+1) &= f(\xi(k), g(\Delta u(k))) \\ g(\Delta u(k)) &= u(k) - u(k-1) \\ y(k) &= C\xi(k) \end{aligned} \tag{8}$$

$$\begin{aligned} x(k+1) &= f(x(k), g(\Delta u(k))) \\ x(k+2) &= f(x(k+1), g(\Delta u(k+1))) \\ &= f(f(x(k), g(\Delta u(k))), g(\Delta u(k+1))) \\ &\vdots \\ x(k+Hp|k) &= f(\ldots f(x(k), g(\Delta u(k))), g(\Delta u(k+Hc-1))) \end{aligned} \tag{9}$$

where $u(k)$ and $\Delta u(k)$ are the control input and the increment of control input of step k, respectively; ξ(k) is the state variable of step k; $y(k)$ is the prediction output of step k. The specific expressions are as follows:

$$\begin{aligned} J(\xi(t), u(t), \Delta u(t)) = &\sum_{i=1}^{Hp} \|y(k+i|k) - r(k+i|k)\|_Q^2 + \\ &\sum_{i=1}^{Hc} (\|\Delta u(k+i-1)\|_R^2 + \|u(k+i-1)\|_S^2) \end{aligned} \tag{10}$$

Considering the limitation of the self-control ability of the car during the driving process, the front wheel steering angle has the following constraints:

$$\delta_{\text{min}} \le \delta \le \delta_{\text{max}} \tag{11}$$

where δ_{min} and δ_{max} are the minimum and maximum steering angles of the front wheels, respectively. In order to smooth the dynamic response of the system during the steering process and improve the ride comfort of the vehicle, when optimizing the performance index, the expected reference trajectory in the form of exponential decay is introduced, so that the variable to be optimized approaches the optimal trajectory along the smooth value. Under the MPC framework, the purpose of automatic lane keeping control can be expressed as the following quadratic performance indicators, where Q, R and S are the weight coefficients in the cost function, Hp and Hc are the prediction time domain and control time domain of MPC respectively.

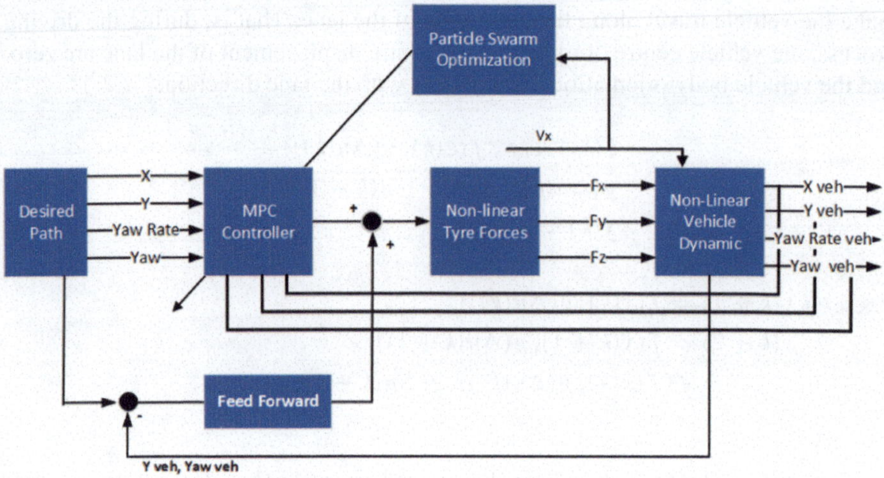

Fig. 3 Model predictive control for non-linear vehicle dynamic model

4 Control Architecture

In this study, a path tracking control system based on the FF-MPC is proposed. As is shown in Fig. 3, the control system consists of an online FF-MPC calculation module, a PSO module and a non-linear vehicle model. The reference trajectory [ψ_{ref}, ψ_{ref}, Y_{ref}, X_{ref}] is fed to the control algorithm.

The optimal inputs optimized by the MPC controller are sent to non-linear vehicle model. The state information of vehicle model is fed to the MPC controller again for the next receding horizon optimization. The algorithm for determining the optimal value of the gain cost function based on the PSO concept for speed variations is adopted based on Amer et al. [2]. While the FF controller design is based on the concepts discussed in Yakub and Mori [23].

5 Simulation Results

In order to verify the designed steering control algorithm, this paper carries out a simulation test under the MATLAB platform. The vehicle parameters are shown in Table 1.

$T_s = 0.1$ s, $H_p = 10$, $H_c = 2$, $\delta_{min} = -40°$, $\delta_{max} = 40°$. In the simulation test, the vehicle is simulated at a low speed of 10 m/s and a high speed of 25 m/s, respectively (Table 2).

Root mean square error (RMSE) lateral position and yaw error for MPC controller and Stanley controller summarized in Table 2. It has reported the maximum RMSE lateral and yaw deviation of each proposed controller with constants longitudinal

Table 1 Vehicle parameters

Parameters	Value	Units
Mass (m)	1573	kg
Front to COG	1.1	m
Rear to COG	1.58	m
Inertia mass moment	2873	kg m^2
Front cornering stiffness	80,000	N/rad
Rear cornering stiffness	40,600	N/rad
Lane width	6	m
Lane distance	250	m

Table 2 Root mean square yaw rate and lateral position for double lane change trajectory

	GS-MPC		MPC		Adaptive Stanley	
Vehicle speed (m/s)	Yaw	Lateral position	Yaw	Lateral position	Yaw	Lateral position
10	0.0073	0.0471	0.0351	0.2604	0.0086	0.0141
12.5	0.0082	0.0544	0.0275	0.2605	0.0179	0.1611
15	0.0103	0.0713	0.0263	0.2591	0.0137	0.1906
17.5	0.0169	0.0717	0.0552	0.2645	0.0135	0.2575
20	0.0171	0.0493	0.1731	0.3454	0.0149	0.3367
22.5	0.0216	0.0565	0.038	0.3804	0.0191	0.4459
25	0.0306	0.1078	0.0415	0.4201	0.0226	0.5254

vehicle speed for double lane change scenario. The GS-MPC controller shows good performance for trajectory tracking. The GS-MPC controller shows good performance compared to adaptive Stanley and standard MPC in trajectory tracking. This can be seen in the RMSE lateral error which is smaller than the RMSE lateral error using conventional MPC and adaptive Stanley. The RMSE lateral and yaw using the GS-MPC is smaller in all vehicle speed ranges that showed in Table 2.

Figure 4 shows a comparison of adaptive Stanley, standard MPC and GS-MPC controller for path tracking performance for vehicle speed of 20 m/s. The GS-MPC controller can do tracking path well when compared to Adaptive Stanley controller and MPC. This happens because GS-MPC with the optimal weighting gain value, so that the GS-MPC controller can minimize lateral errors when doing path tracking and avoidance obstacle maneuvers. However, the opposite applies with adaptive Stanley controllers, the control law of the controller is based on geometric calculations of vehicles and roads. So, the controller cannot anticipate the possibility that will happen in the future. Table 2 reported the maximum RMSE lateral and yaw deviation of each proposed controller with constants longitudinal vehicle speed for double lane change scenario. The GS-MPC controller shows good performance for double lane change, lane change and curvature trajectory tracking in vehicle speed of 10–25 m/s.

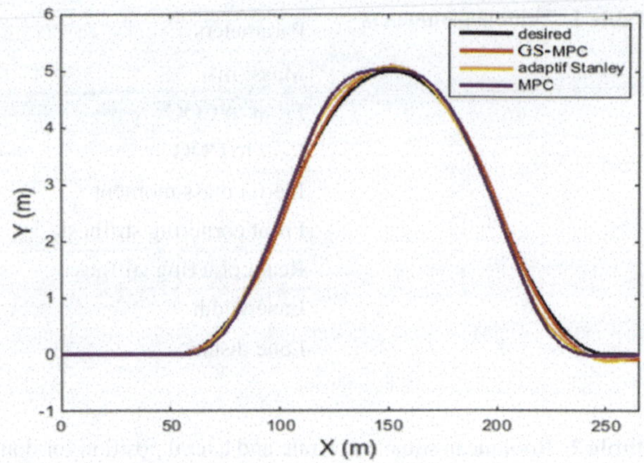

Fig. 4 Simulation results: comparison path tracking performance for double lane change trajectory at vehicle speed of 20 m/s

Figure 5 showed trajectory tracking performance of GS-MPC controller on the double lane change to test handling maneuver. Figure 5a shows the results of trajectory tracking the vehicle obtained using control law using GS-MPC. The autonomous vehicle can approach the desired yaw dan lateral position (Fig. 5a). The GS-MPC controller managed to guide the autonomous vehicle successfully along the trajectory as shown in Fig. 5 for the vehicle speed 20 m/s. This simulation has been conducted with road surface coefficient of 0.8. The simulation results using the GS-MPC controller better than adaptive Stanley controller and standard MPC to follow the trajectory properly in vehicle speed of 10–25 m/s. As can been seen lateral and yaw error on Fig. 5 and Table 1 which shows that the GS-MPC controller managed to produce lower lateral and yaw error than adaptive Stanley controller and standard MPC. Figure 5b shows the wheel steering as input control values of the MPC controller. Input control consists wheel steering with constraints ±0.5 rad.

6 Conclusion

Through simulation experiments, it is proved that the control strategy can quickly eliminate the lateral displacement deviation and yaw angle deviation, ensure the vehicle travels along the road centerline, and effectively smooth the dynamic response of the system, whether at low speed or high speed, showing better adaptability and robustness.

This study proposed a path tracking controller based on Gain Scheduling MPC. The controller can track the given reference path by regulating the front steering angle. The GS-MPC system uses single track vehicle model and non-linear tire model to optimize the control inputs. The deviation of yaw angle and lateral displacement relative to the reference are added to the cost function to reflect the tracking performance. At the same time, the control inputs and control input increments constraints are applied to prevent actuator saturation as well. Simulation experiments verified

Fig. 5 Simulation results: **a** lateral and yaw error for 20 m/s vehicle speed, **b** wheel steering input at 20 m/s vehicle speed in double lane change manoeuvre

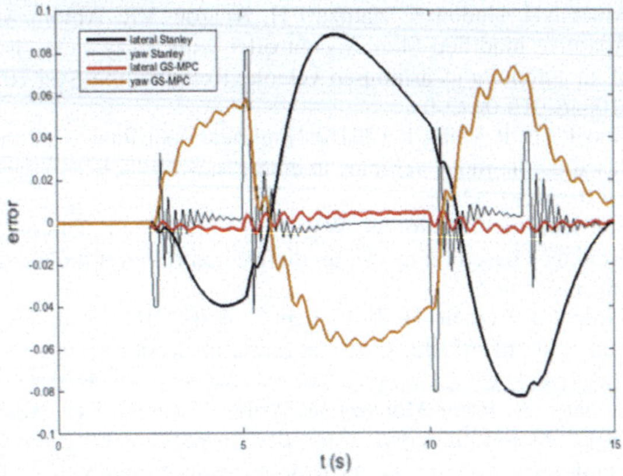

(a)

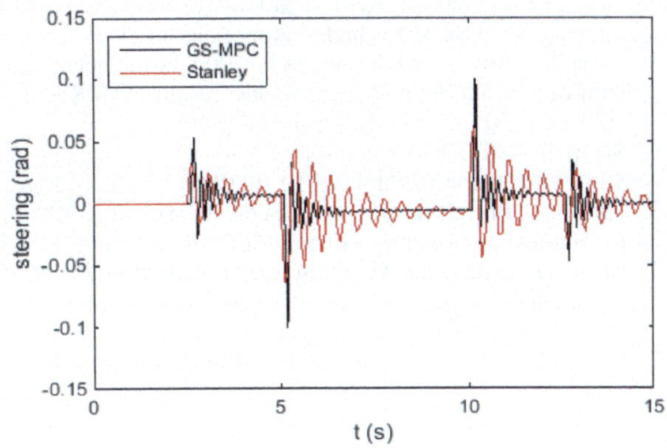

(b)

the effectiveness and accuracy of the path tracking in comparison with linear MPC controller under the speed of 80 km/h. The feasibility and stability of the GS-MPC are also confirmed in solving path tracking under speed of 80 km/h. In future work, we will focus on a linearized tire model that can represent the changing trend of the nonlinear tire force in the predictive horizon to reduce the computational burden.

Acknowledgements The work presented in this study is funded by Ministry of Higher Education Malaysia through Fundamental Research Grant Scheme, Universiti Teknologi Malaysia, FGRS/1/2019/TK08/UTM/02/10, VOT Number: 5F204.

References

1. Lee J, Chang H (2018) Analysis of explicit model predictive control for path-following control. Plos One 1–19. https://doi.org/10.1371/journal.pone.0194110j

2. Amer NH, Hudha K, Zamzuri H, Aparow VR, Abidin AFZ, Kadir ZA, Murrad M (2018) Adaptive modified Stanley controller with fuzzy supervisory system for trajectory tracking of an autonomous armoured vehicle. Robot Auton Syst 105:94–111. https://doi.org/10.1007/s10846-016-0442-0
3. Guo J, Hu P, Wang R (2016) Nonlinear coordinated steering and braking control of vision-based autonomous vehicles in emergency obstacle avoidance. IEEE Trans Intell Transp Syst 17(10):3230–3240
4. Ji J, Khajepour A, Melek WW et al (2017) Path planning and tracking for vehicle collision avoidance based on model predictive control with multiconstraints. IEEE Trans Veh Technol 66(2):952–964
5. Saruchi S, Zamzuri H, Zulkarnain N, Ariff MHM, Wahid N (2017) Composite nonlinear feedback with disturbance observer for active front steering. Indonesian J Electr Eng Comput Sci 7(2):434–441
6. Leman ZA, Hatta Mohammad Ariff M, Zamzuri H, Abdul Rahman MA, Amri Mazlan S (2019) Model predictive controller for path tracking and obstacle avoidance manoeuvre on autonomous vehicle. In: 12th Asian control conference (ASCC), Kitakyushu-shi, Japan, pp 1271–1276
7. Tan Q, Dai P, Zhang Z, Katupitiya J (2018) MPC and PSO Based control methodology for path tracking of 4WS4WD vehicles. Appl Sci 8:1–24
8. Sharp R, Casanova D, Symonds P (2000) A mathematical model for driver steering control, with design, tuning and performance results. Veh Syst Dyn 33:289–326. https://doi.org/10.1076/0042-3114(200005)33:5;1-Q;FT289
9. Sharp R (2012) Rider control of a motorcycle near to its cornering limits. Veh Syst Dyn 50:1193–1208. https://doi.org/10.1080/00423114.2011.607899
10. Falcone P, Borrelli F, Asgari J, Tseng HE, Hrovat D (2007) Predictive active steering control for autonomous vehicle systems. IEEE Trans Control Syst Technol 15:566–580
11. Bayar G, Bergerman M, Koku AB, Konukseven EI (2015) Localization and control of an autonomous orchard vehicle. Comput Electron Agric 115:118–128. https://doi.org/10.1109/TCST.2007.894653
12. Pérez TL (1979) An algorithm for planning collision-free paths among polyhedral obstacles. Commun ACM 22(10):560–570
13. Perez TL (1983) Spatial planning: a configuration space approach. IEEE Trans Comput 100(2):108–120
14. Bayar G, Bergerman M, Koku AB (2016) Improving the trajectory tracking performance of autonomous orchard vehicles using wheel slip compensation. Biosyst Eng (in press). https://doi.org/10.1016/j.biosystemseng.2015.12.019
15. Tomatsu T, Nonaka K, Sekiguchi K, Suzuki K (2015) Model predictive trajectory tracking control for hydraulic excavator on digging operation. In IEEE conference on control applications (CCA), pp 1136–1141. https://doi.org/10.1109/cca.2015.7320765
16. Yamashita AS, Alexandre PM, Zanin AC, Odloak D (2016) Reference trajectory tuning of model predictive control. Control Eng Pract 50:1–11. https://doi.org/10.1016/j.conengprac.2016.02.003
17. Prodan I, Olaru S, Fontes FACC, Pereira FL, Borges de Sousa J, Maniu JS (2015) Predictive control for path-following. From trajectory generation to the parametrization of the discrete tracking sequences. In: Developments in model-based optimization and control: distributed control and industrial applications. Springer International Publishing, Cham, pp 161–181. https://doi.org/10.1007/978-3-319-26687-9_8
18. Raffo GV, Gomes GK, Rico JEN, Kelber CR, Becker LB (2009) A predictive controller for autonomous vehicle path tracking. IEEE Trans Intell Transp Syst 10:92–102. https://doi.org/10.1109/TITS.2008.2011697
19. Beal CE (2011) Applications of MPC to vehicle dynamics for active safety and stability. PhD, Department of Mechanical Engineering, Stanford University
20. Merabti H, Belarbi K, Bouchemal B (2016) Nonlinear predictive control of a mobile robot: a solution using metaheuristcs. J Chinese Inst Eng 39:282–290. https://doi.org/10.1080/02533839.2015.1091276

21. Xue T, Li R, Tokgo M, Ri J, Han G (2015) Trajectory planning for autonomous mobile robot using a hybrid improved QPSO algorithm. Soft Comput 1–17. https://doi.org/10.1007/s00500-015-1956-2
22. Yakub F, Mori Y (2015) Comparative study of autonomous path-following vehicle control via model predictive control and linear quadratic control. J Automob Eng 229(12):1695–1713. https://doi.org/10.1177/0954407014566031
23. Nagai M, Shino M, Gao F (2012) Study on integrated control of active front steer angle and direct yaw moment. JSAE Rev 233:309–315. https://doi.org/10.1016/S0389-4304(02)00189-3
24. Hoffmann GM, Tomlin CJ, Montemerlo D, Thrun S (2007) Autonomous automobile trajectory tracking for off-road driving: controller design, experimental validation and racing. In: American control conference, 2007, ACC'07, pp 2296–2301. https://doi.org/10.1109/acc.2007.4282788

Effect of Glass Powder on Frictional Properties of Composite Friction Brake

Martinus Heru Palmiyanto, Eko Surojo, and Dody Ariawan

Abstract Organic friction composites have been developed as a lightweight vehicle brake material. Organic composites consist of phenolic resins that act as a binder with reinforcing fibres and add component additives based on the functional requirements. One such functional filler, for providing friction properties in the brake material, is abrasives. Abrasives, which clean the film layer on the surface of the friction contact, are added to the composite material for increasing and maintaining its coefficient of friction. Silicon oxide (SiO_2) is a commonly used abrasive material. SiO_2, apart from quartz sand, can also be obtained from soda lime glass waste, in which SiO_2 is the main content. In this study, SiO_2 particles were obtained from glass waste through the milling process of up to 200 mesh. Glass powder (GP) added to the composite composition is regulated by varying the volume fraction of 0% volume to GP 2%, GP 4% and GP 6%. The purpose of this study was to determine the effect of GP as an abrasive modifier on the coefficient of friction of non-asbestos organic composites. We varied the sliding speed and contact pressure in the pin-on-disc tribology testing process to determine the behaviour of the coefficient of friction. The results showed that the addition of GP to friction composites increases the coefficient of friction. The coefficient of friction in all variations of the composition decreases as sliding speed and contact pressure increase.

Keywords Glass powder · Abrasive · Brake friction · Organic composite

M. H. Palmiyanto (✉) · E. Surojo · D. Ariawan
Mechanical Engineering Department, Faculty of Engineering, Universitas Sebelas Maret, Surakarta 57126, Indonesia
e-mail: martinus76palmiyanto@gmail.com

E. Surojo
e-mail: esurojo@ft.uns.ac.id

M. H. Palmiyanto
Mechanical Engineering Department, Akademi Teknologi Warga Surakarta, Sukoharjo 57552, Indonesia

U. Sabino et al. (eds.), *Proceedings of the 6th International Conference and Exhibition on Sustainable Energy and Advanced Materials*, Lecture Notes in Mechanical Engineering,
https://doi.org/10.1007/978-981-15-4481-1_45

1 Introduction

The primary function of brake elements is to reduce vehicle speed and stop the vehicle as desired. Brakes also ensure safety and control while driving, both for humans and the environment. Composite brake material has been widely used in land vehicles, especially light vehicles. Friction brakes consisting of non-asbestos organic composite materials offer a number of advantages. They are soft, quiet, lightweight and environment friendly. The main ingredients of non-asbestos organic composite brakes are matrices as binders, fibres for reinforcement, fillers and additional friction materials [1].

The principle of braking is based on the occurrence of friction from two contact surfaces, namely brake lining and discs. The friction process converts the kinetic energy of the disc rotation, in accordance with the speed of the vehicle, into heat energy. Subsequently, the heat induced from the contact surface is distributed to the surrounding environment. The repeated pressure and friction will lead to the erosion of the brake shoe material and the dust that accumulates between the two contact surfaces causes wear. Dust accumulation on the contact surface under repetitive pressure leads to the formation of a thin layer called glaze. As the thickness of the glaze increases, the coefficient of friction on the brake lining decreases, affecting the brake performance.

One of the additional materials that affect friction characteristics is the abrasive material. It increases the coefficient of friction. It also maintains the coefficient of friction during braking at high temperatures. Friction materials that have been developed by researchers include quartz (SiO_2) and silicon carbide (SiC). Several researchers have studied friction materials for the development of non-asbestos organic friction composite brake materials. Kim et al. [2] examined the effect of various abrasive particle contents such as SiO_2 and SiC on the friction and vibration properties of composite materials in automotive vehicle brakes. Bjwe et al. [3] compared the performance of nano-SiC, micro-SiC; nano-SiO_2 and micro-SiO_2 with that of micro alumina and macro alumina on the friction properties and composite wear friction of brake canvas materials. Sugozu [4] examined the effect of SiO_2 content in nano-scale particles and micro particles on the friction and wear properties of composites. The researchers are searching for new abrasive materials that can be added to non-asbestos organic composite materials. Cai et al. [5] reported that mullite contains alumina and silica and has the potential to be developed as an added abrasive material. Zhengjia et al. [6] found that the presence of silica impurities and $NaAl_3\,(SO_4)_2OH_6$ in mullite increases the stability of friction in brake composites.

The abrasive material with silica content is a potential candidate to be developed as filler in non-asbestos organic friction composites. Silica is processed from quartz obtained from natural mineral mining. Sustained mining will lead to the depletion of quartz, as it is not naturally replenished. Mining causes environmental damage too. In this backdrop, we propose the utilisation of soda lime glass waste as the abrasive material. According to Galvao et al. [7], sodalime glass has SiO_2 as the main content and Na_2O and CaO as givers. The presence of SiO_2 content makes it suitable to be developed as a non-asbestos organic friction material. The purpose of this study was to determine the effect of adding glass powder (GP) as an abrasive modifier on the coefficient of friction of non-asbestos organic composites. We varied the sliding speed and contact pressure in the pin-on-disc tribology testing process to determine the corresponding changes in behaviour of the coefficient of friction.

2 Materials and Methods

2.1 Material Preparation

Sodalime glass used for the manufacture of GP was obtained from unused glass waste. Soda lime glass was crushed to obtain fine powder over two stages as shown in Fig. 1. In the first stage, the glass was pounded initially with mortar and pestle and then by using a ball mill to obtain a finer powder which can pass the screening of 200 mesh. In the second stage, the powder was heated in an oven at a temperature of 800 °C for 3 h, for obtaining GP with the required purity.

The non-asbestos organic composites are shown in Table 1. The basic mixture consists of novalac phenolic resin (25%), nitrile butadiene rubber (5%), MoS_2 (5%), graphite (10%), and cashew dust (10%). Glass fibre (5%) and kenaf fibre (6%) are chosen as reinforcements. Variations in GP were adjusted to compensate for the amount of residue from the $CaCO_3$ filler.

Fig. 1 Glass powder (GP) refining process. **a** Collision with mortar and pestle, **b** fineness by milling method [7], **c** the result of fine glass powder

Table 1 Composition of specimen

Composition	% volume		
	Basic mixture	Glass powder	$CaCO_3$
1	67	0	33
2	67	2	31
3	67	4	29
4	67	6	27

The specimens were made as per the following method. First, the material is weighed according to Table 1. The mixing process begins with the insertion of glass fibre and kenaf fibre for 5 min. Then, other material powders are added to make composite specimens for the next 3 min. After all the ingredients are mixed, they are put into a pin-shaped mould to proceed with the pre-forming process. It is subjected to a pressure of 40 MPa for 10 min on the pre-forming process. It is then heated at a temperature of 150–160 °C. At this stage, the gas is released six times in 1 min. The heating process is continued for 9 min.

After the specimen is formed, the post-curing process begins. Post-curing procedures are carried out to increase the density of crosslinking in phenolic resins [8]. The post-curing process starts with the heating of the specimen in an oven where the temperature is increased from room temperature to 140 °C within 1 h. The heating continues for a further 6 h when the temperature increased from 140 to 180 °C. The next step is the cooling process in which the temperature is reduced to room temperature over 30 h. In this research, we used three specimens with varying compositions.

2.2 *Friction Test*

Tribology testing using pin-on-disc tribometer is shown in Fig. 2. A rotor disk, with 150 mm diameter and 15 mm thickness, is made of wear-resistant alloy steel hardened with the DIN X 153 CrMov series 12. The rotor disk is connected to a drive motor having 5 HP power. The panel box contains a series of electrical and monitor displays for machine operation.

We varied the sliding speed by controlling the alternating current frequency that flows to the electric motor using a three-phase inverter. We varied the contact pressure by controlling the air pressure that moves the pneumatic pistons. Pneumatic pistons were used as movers that exert a normal force pushing the pin specimen to the surface of a rotating disk. A pair of hinged arms, equipped with ballast as a counterweight and load cell, was needed to support the specimen holder. When the specimen rubs against the rotating disk rotor, the arm will swing down pressing the load cell. The response given by the load cell was recorded using data acquisition, which was then measured as the frictional force acting on the specimen. The tests were carried out to obtain approximately 500 data records. The coefficient of friction was calculated as the ratio between the frictional force and the normal force. Friction tests were performed using parameters as shown in Table 2. Contact pressure was calculated by dividing the normal force by the area of cross section of the pin specimen.

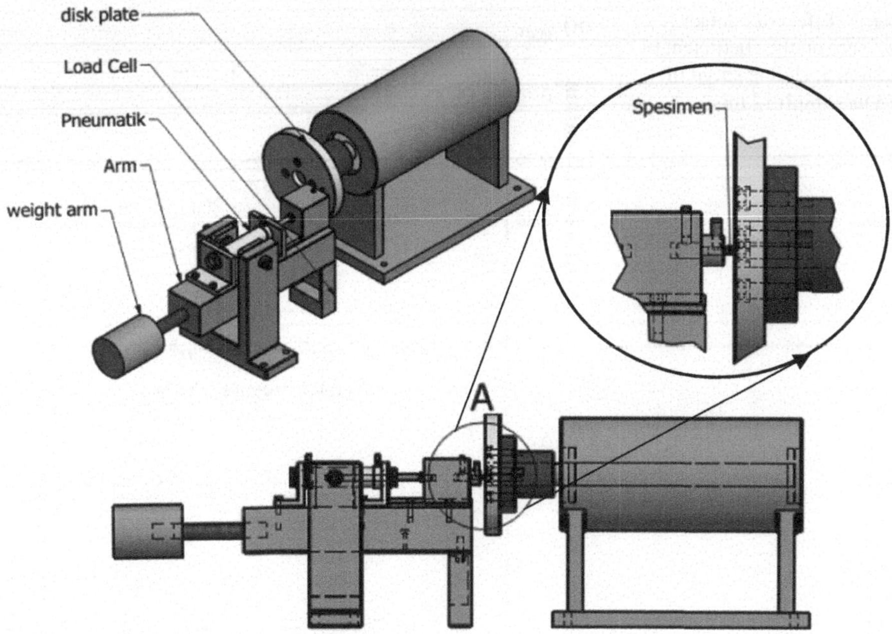

Fig. 2 Schematic pin-on-disc tribometer testing

Table 2 Parameter of pin-on-disc tribometer testing

Variation of parameter	Testing parameter
Contact pressure (MPa)	0.25; 0.5; 0.75; 1 Sliding speed at 3 and 12 m/s
Sliding speed (m/s)	3; 6; 9; 12 Contact pressure at 0.25 and 1 MPa

3 Results and Discussion

3.1 Effect of Glass Powder on the Coefficient of Friction

The behaviour of GP composites on the coefficient of friction was evaluated from pin-on-disc testing. The sliding speed and contact pressure were varied to evaluate the performance of the GP added to the composite composition. The measurement results are shown in Figs. 3 and 4. The coefficient of friction increases with increasing GP content in the specimen. The highest coefficient of friction of 0.408 is shown at 6% GP at a sliding speed of 3 m/s and at a contact pressure of 0.25 MPa, while the GP specimen of 0% has the lowest friction coefficient of 0.235 at a sliding speed of 12 m/s and at a contact pressure of 1 MPa.

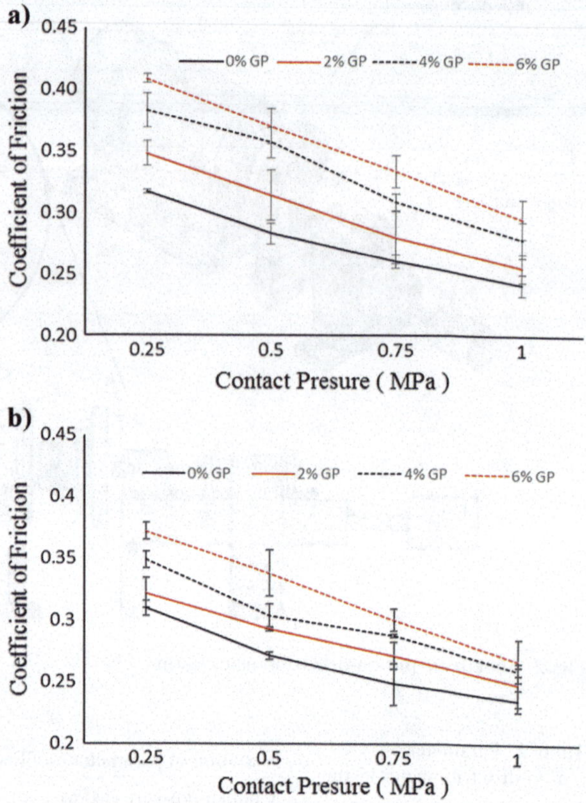

Fig. 3 Effect of contact pressure on the coefficient of friction at sliding speed of **a** 3 m/s and **b** 12 m/s

The coefficient of friction of the composites added with GP in sequence varied in the following sequence: 6% GP > 4% GP > 2% GP > 0% GP. This phenomenon occurs in all variations of contact pressure and sliding speed. This behaviour was related to the fact that the friction material with GP contains a large amount of abrasive material such as silica. As explained before, GP has a dominant SiO_2 content [7]. The results show that the presence of GP in composite specimens has a positive effect on increasing the friction properties of the specimens. GP, which has a hard Mohs value, can be an abrasive alternative for reducing the use of silica from natural minerals.

3.2 *Effect of Contact Pressure on the Coefficient of Friction*

The coefficient of friction as a function of contact pressure at a constant sliding speed of 3 m/s is presented in Fig. 3a. It shows that specimens of 6% GP, 4% GP and 2% GP produce the coefficients of friction within a range of 0.295–0.408, 0.278–0.38 and 0.255–0.377, respectively. Specimens without GP (0% GP) result in a coefficient

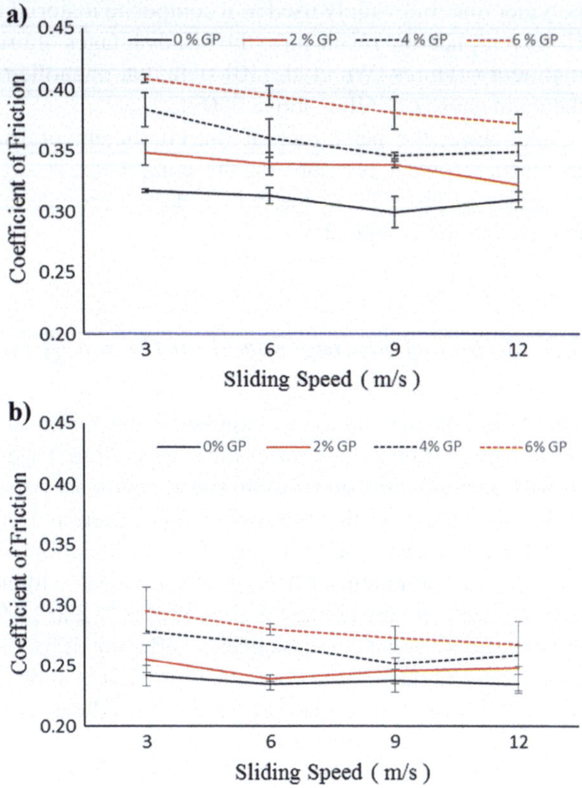

Fig. 4 Effect of sliding speed on the coefficient of friction at contact pressure of **a** 0.25 MPa and **b** 1 MPa

of friction range of 0.241–0.316. The coefficient of friction at the sliding speed of 12 m/s is shown in Fig. 3b, which expresses the same trend as shown in Fig. 3a, where the specimen with 6% GP had the highest coefficient of friction, and the specimen without GP has the lowest coefficient of friction.

Figure 3a, b show a significant decreasing coefficient of friction increasing contact pressure. It is the same trend for both the 0% GP friction composite and the composite glass friction composite (2% GP, 4% GP and 6% GP). The coefficient of friction decreases with increasing contact pressure.

Previous studies also reported the same behaviour: decreasing coefficient of friction with increasing contact pressure [2, 9]. Bijwe et al. [3] state that Amonton's first law (coefficient of friction proportional to the applied load) does not apply in the case of polymer matrix composite friction materials. The coefficient of friction varies due to various pressures and speeds.

The coefficient of friction of the decreased friction test specimen is closely related to the presence of heat. Test specimens undergo repetitive friction processes, following the heat generated by the pressure on the contact surface (specimen and disk). The heat generated through the transition glass will reduce the performance of the matrix as a binding material for the friction composite. Phenolic resin is a thermoset type polymer material widely used as a composite material for brake friction composites. However, phenolic resins have the disadvantages of brittleness and low resistance to high temperatures. Wu et al. [10] state that phenolic resins are often considered as the main cause of fading above 350 °C.

Therefore, the behaviour of the coefficient of friction, which decreases with increasing contact pressure on the composite, is strongly influenced by the performance of phenolic resins, which do not provide sufficient bond strength to the particles in the composite.

3.3 Effect of Sliding Speed on the Coefficient of Friction

The sliding speed variation was applied to the pin-on-disc tribology test to find out the composite friction behaviour of the composition. Figure 4a, b show the relationship of friction coefficient and sliding speed at contact pressures of 0.25 and 1 MPa.

Figure 4a, b show that the coefficient of friction decreases with increasing sliding speed. But the decrease in the coefficient of friction was less real at various sliding speeds. The coefficient of friction of specimens without GP content (0% GP) tends to be constant at both contact pressures of 0.25 and 1 MPa, while specimens with GP content show a slight decrease in the coefficient of friction. This is different compared to the trend of decreasing coefficient of friction shown in Fig. 3a, b, which is more due to the increased contact pressure. This indicates that composite specimens have resistance to friction at various sliding speeds. Variations in increased friction have also been reported by Cai et al. [5], who found that an increase in sliding speed from 1 to 1.5 m/s decreased the coefficient of friction. Furthermore, an increase in sliding speed from 1.5 to 2.0 m/s increased the coefficient of friction again, which became stable at 2.5 m/s.

4 Conclusion

The effect of adding GP as an abrasive modifier on the non-asbestos organic composite friction coefficient was evaluated by pin-on-disc tribology testing. We varied the sliding speed and contact pressure in the process of pin-on-disc tribology testing. The following are the conclusions of testing and discussion:

1. GP on the friction composite can increase the coefficient of friction.
2. GP can be an alternative abrasive substitute for silica obtained from natural minerals.
3. Increased contact pressure indicates a significant decreasing trend in the coefficient of friction in all friction composites.
4. The coefficient of friction decreases less with increasing sliding speed in all friction composites.

Acknowledgements The authors wish to thank the Ministry of Technology Research and Higher Education for supporting and funding research. Through the Fundamental Research Grant with Contract No. 719/UNS27.21/PN/2019.

References

1. Blau PJ (2001) Compositions, functions, and testing of friction brake materials and their additives. Oak Ridge National Laboratory ORNL/TM-2001/64 (2001)
2. Kim SS, Hwang HJ, Shin MW, Jang H (2011) Friction and vibration of automotive brake pads containing different abrasive particles. Wear 271:1194–1202
3. Bijwe J, Aranganathan N, Sharma S, Dureja N, Kumar R (2012) Nano-abrasives in friction materials—influence on tribological properties. Wear 296(1/2):693–701
4. Sugozu KB (2015) Friction and wear properties of friction materials containing nano/micro-sized SiO2 particles. Ind Lubric Tribol 68(2):259–266
5. Cai P, Wang Y, Wang T, Wang Q (2015) Effect of resins on thermal, mechanical and tribological properties of friction materials. Tribol Int 87:1–10
6. Ji Z, Luo WY, Zhou K, Hou S, Zhang Q, Li J, Jin H (2018) Effects of the shapes and dimensions of mullite whisker on the friction and wear behaviors of resin-based friction materials. Wear. https://doi.org/10.1016/j.wear.2018.03.018
7. Galvão ACP, Farias ACM, Mendes JUL (2015) Characterization of waste of soda-lime glass generated from lapping process to reuse as filler in composite materials as thermal insulation. Cerâmica 61:367–373
8. Surojo E, Jamasri, Malau V, Ilman MN (2014) Effects of hot molding pressure and post curing time on flexural strength of brake shoe composite. In: Seminar Nasional Metalurgi dan Material VII, pp. 149–153. Institut Teknologi Bandung, Bandung
9. Arsada R, Surojo E, Ariawan D, Muhayat N, Raharjo WW (2018) Effect of NBR (Nitrile Butadiene Rubber) on flexural strength of composite friction brake. In: 2nd Nommensen International Conference on Technology and Engineering, IOP Conf. Series: Materials Science and Engineering 420, 012057
10. Wu Y, Zeng M, Xu Q, Hou S, Jin H, Fan L (2012) Effects of glass-to-rubber transition of thermosetting resin matrix on the friction and wear properties of friction materials. Tribol Int 54:51–57

Feasibility of Electric Generation from Municipal Solid Wastes by Incineration and Gasification

Suyitno, Evi Gravitiani, Zainal Arifin, Mohamad Muqoffa, and Syamsul Hadi

Abstract Municipal solids wastes have become a major problem in several cities in developing countries. Waste management by sorting, recycling, reusing does not necessarily reduce the volume of waste significantly. Therefore, some technologies such as incineration and gasification are very interesting to study along with their potential to produce electricity. The study was conducted by simulation where the waste data was obtained experimentally. Financial analysis was done to calculate the initial cost, maintenance and operational cost, total revenue, internal rate of return, payback period, net presence value, and benefit/cost ratio. The results of the study show that gasification technology was more feasible compared to incineration technology. In a 5 MW electricity capacity, there were several scenarios that need to be implemented so that the electric power plant from municipal solid wastes becomes feasible.

Keywords Feasibility · Electric generation · Municipal solid wastes · Incineration · Gasification

1 Introduction

Municipal solids wastes have become a major problem in several cities worldwide because of increasing population and industrial activities. It is estimated 1.3 billion tons per year of wastes generated and increases to 2.2 billion tons per year by 2025 [1]. Waste management by sorting, recycling, and reusing does not necessarily reduce the volume of waste significantly. Therefore, some technologies such as incineration and gasification are very interesting to study along with their potential to produce electricity.

An incineration process is well known and proven technology for combusting wastes using near stoichiometric air in the combustion chamber at temperatures of about 1000 °C or above. For electric generation, the hot flue gases are used to heat the boilers and produce superheated steam. The superheated steam enters the steam

Suyitno (✉) · E. Gravitiani · Z. Arifin · M. Muqoffa · S. Hadi
Universitas Sebelas Maret, Jl. Ir. Sutami 36A, Surakarta, Indonesia
e-mail: suyitno@uns.ac.id

U. Sabino et al. (eds.), *Proceedings of the 6th International Conference and Exhibition on Sustainable Energy and Advanced Materials*, Lecture Notes in Mechanical Engineering,
https://doi.org/10.1007/978-981-15-4481-1_46

turbine and moves the generator to produce electric energy. The different between incineration and gasification is on the ratio between air to MSWs used. An incinerator uses a higher amount of air than that used by a gasifier. Gasifiers work in a partial oxidation at elevated temperature (600–1700 °C) [2] and produce intermediate synthetic gases (CO, H_2, CH_4) which can be further combusted, meanwhile incinerators generate hot flue gases (CO_2, H_2O).

In Indonesia, the MSWs are collected in temporary dumpsite and landfills. Wastes are collected in landfills approximately varying from 200 to 5228 tons/day for each landfills depending on the cities [3]. The amount of MSWs can produce electricity at a capacity of about 1 MW or more. Moreover, a hypothetic gasification plant is capable to generate electricity from 794 to 1065 kW/ton MSWs in Brazil [2]. However, the capability of gasification to produce electricity is strongly depend on the type of gasifier used. Gasifiers [4] with air produce a syngas with heating value ranging from 4.0 to 6.0 MJ/Nm^3, while when using oxygen, they produce a syngas with a medium heating value ranging from 10.0 to 20.0 MJ/Nm^3.

To implementing the gasifier and incinerator, the aspect is not only considered from technology but at least from financial also. The project of electric generation from MSWs from financial study must calculate net present value (NPV), internal rate of return (IRR), and benefit to cost ratio (B/C). Therefore, the aim of this study is to investigate the feasibility of technology for processing municipal solid wastes of 833 kg/h into electricity between gasification technology and incineration technology.

2 Methods

The study was conducted by simulation where the waste data was obtained experimentally. The technology analysis was done in ASPEN HYSYS. The MSWs was treated as nonconventional solids where the ultimate and proximate analysis are shown in Table 1. During the pyrolysis process, the MSWs were thermally reacted without air and produced gas and char. The chars were treated as solid. The properties of conventional substances were solved by Redlich Kwong-Soave (RK-Soave).

The capacity of MSWs which processed was 833 kg/h. The ratio of air to MSWs used for gasification and incineration was varied from 0.05 to 0.5 and 2.74 to 3.4, respectively. For steam turbine, the mass flow rate of steam was 20,000 kg/h.

The gasifier and incinerator concept to produce electricity from MSWs are shown in Fig. 1. In gasification power plant, the MSWs were first separated and sorted. The MSWs were then dried and pyrolyzed at the temperature of 1000 °C to produce char,

Table 1 Ultimate and proximate analysis of the MSW (%wt)

C	H	N	S	Cl	O	Ash	Moisture	VM	FC
29.6	3.94	0.9	0.16	0.74	25.2	40.5	60.2	24.4	35.1

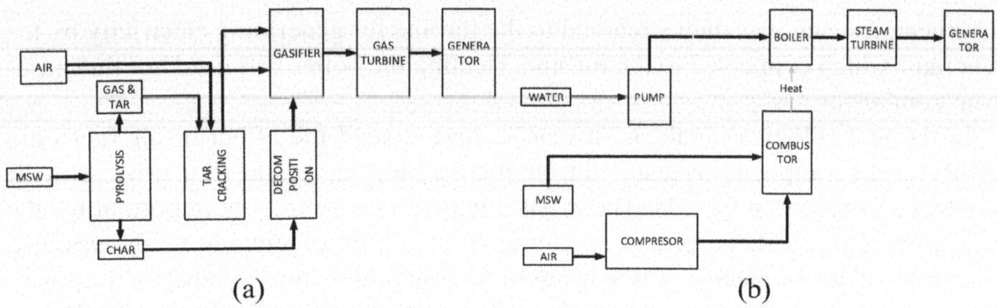

Fig. 1 Scheme of electric generation from **a** gasification, **b** incineration

gas, and tar. Tars were further treatment and cracked thermally. The gas produced from tar cracking was entered into the gasifier. Moreover, the chars were flowed into gasifiers to produce syngas and ash. The syngas was used in gas turbine for producing mechanical energy for rotating the generator and produced electric energy. During the gasification process, chars were treated as carbon and the possible reactions that takes placed in the reactor are as following:

$$C + O_2 \rightarrow CO_2 \tag{1}$$

$$C + \frac{1}{2}O_2 \rightarrow CO \tag{2}$$

$$C + H_2O \rightarrow CO + H_2 \tag{3}$$

$$C + CO_2 \rightarrow 2CO \tag{4}$$

$$C + 2H_2 \rightarrow CH_4 \tag{5}$$

$$H_2 + \frac{1}{2}O_2 \rightarrow H_2O \tag{6}$$

$$CO + \frac{1}{2}O_2 \rightarrow CO_2 \tag{7}$$

$$CH_4 + 2O_2 \rightarrow CO_2 + 2H_2O \tag{8}$$

$$CO + H_2O \rightarrow CO_2 + H_2 \tag{9}$$

$$CH_4 + H_2O \rightarrow CO + 3H_2 \tag{10}$$

In the incineration power plant, the sorted MSWs were dried and combusted in the incinerator. The hot flue gases were then used to heat the water in boiler. The superheated steam was then streamed to the turbine for generating electricity by the generator which connected to the turbine. Heating the boiler was used hot flue gases from combustor.

In the technological analysis, it can be investigated the effect of air fuel ratio (AFR) on gas compositions resulted from incineration and gasification process. AFR is calculated based on Eq. (1). These gas compositions were very important for the next process of gasification and incineration. The technological analysis was also used to calculate the electric power generation. In addition, the efficiency of the power plants of the two technologies was also calculated using Eq. (2). In the simulation, pumps, compressors, and turbines have a 72% isentropic efficiency.

$$\text{AFR} = \frac{\text{mass flow of air}}{\text{mass flow of MSWs}} \tag{11}$$

$$\eta = \frac{W_{nett}}{\dot{m}_{MSWs}.HV_{MSWs}} \tag{12}$$

Meanwhile, the financial analysis was done to calculate the initial cost, operational and maintenance (O&M) cost, internal rate of return (IRR), payback period (PP), net present value (NPV), benefit to cost ratio (B/C), and energy production cost. The expected NPV and B/C ratio is positive and higher than 1, respectively for a financially attractive project. Meanwhile IRR is that internal rate of return percentage and if greater than the IRR required by investors, the investment is financially attractive [5].

3 Results and Discussion

The simulation of the power generation process from MSWs using gasification and incineration process technology has been carried out. The composition of gas resulted from gasification (syngas) and incineration (flue gas) can be seen in Fig. 2.

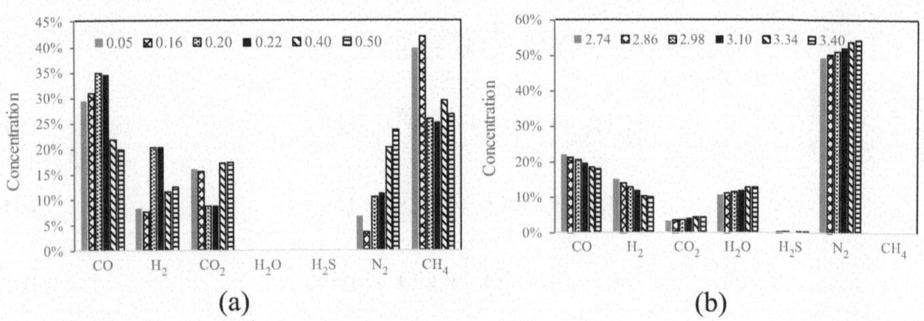

Fig. 2 Gas composition from **a** gasifier, **b** incinerator

The syngas was consisted of CO, CH_4, H_2, CO_2, and N_2. However, H_2O and H_2S were not found in the syngas. The highest concentration of CH_4 was achieved at AFR of 0.16. At low AFR, the large amount of CH_2 was formed, but as the AFR increased, chars reacting with O_2 were increasing thereby increasing the content of CO. Meanwhile, as can be seen in Fig. 2, in the flue gas resulted from the incineration, CO and H_2 contents decreased with increasing amount of air. CO and H_2 reacted with O_2 and produced CO_2 and H_2O so that the amount of CO_2 and H_2O increased. Therefore, the content of gas in syngas and flue gas is strongly affected by AFR or equivalence ratio [6].

The concentration of CO in syngas and flue gas ranged between 20–35% and 16–22%, respectively. The CH_4 content in syngas was above 25% and higher than the H_2 content in syngas. This showed that the quality of syngas was influenced by CH_4, CO, and H_2. The content of the three types of gas in syngas is very important, because all three gases are combustible [7]. Meanwhile the content of CO and H_2 in the flue gas were still quite high. It indicating that the burning quality in incineration was still imperfect.

AFR also affected the electrical energy produced. The higher amount of air entered into gasification and incineration caused the electric energy generated to increase. This is caused by increasing concentrations of combustible gas from gasification and increasing flue gas temperatures so that the energy entering the turbine also increased.

At the same flow rate of MSWs, the electric energy generated from incineration was almost double compared to the electrical energy from gasification. However, flowing air into the incinerator required the high compressor power so that the net power from gasification was still higher than the net power from incineration. The net power from gasification and incineration varied from 3 to 4 MW and from 1.7 to 3.4 MW, respectively as shown in Table 2. In addition, the electricity efficiency of gasification and incineration ranged from 43.3 to 56.3% and from 24.2 to

Table 2 Work and electric generated by gasification and incineration technology

Technology	AFR	Electric generation (kW)	Work Nett (kW)	Efficiency
Gasification	0.05	3188	3090	0.433
	0.16	3556	3300	0.447
	0.20	3698	3385	0.475
	0.22	3767	3426	0.504
	0.40	4404	3806	0.533
	0.50	4758	4017	0.563
Incineration	2.74	5530	1727	0.242
	2.86	6010	2035	0.285
	2.98	6501	2184	0.306
	3.10	6961	2644	0.370
	3.34	7893	3233	0.453
	3.40	8116	3370	0.472

47.2%, respectively. It is consistent with previous study stating that the equivalence ratio increased from 0.18 to 0.22, the gasification efficiency increased from 61.43 to 68.15% and then decreased to 59.56% as the equivalence ratio was further increased to 0.28 [8].

Furthermore, the financial analysis was carried out at a similar output power, which were at AFR of 0.20 and 3.40 for gasification and incineration, respectively. In general, the initial cost of incineration technology was much greater than gasification technology because it needed more equipment used for incineration, especially boilers and superheaters. The turbines used in gasification technology were gas turbines, meanwhile in incineration technology used steam turbines. The operational and maintenance costs (O&M) of incineration technology were also still higher than the O&M costs of gasification technology.

The scenario was carried out on three energy production costs, namely 0.158, 0.207, and 0.229 USD/kWh. In general, the energy production cost of gasification technology was higher than the energy production cost of incineration technology. In gasification technology, the financially feasible value of the plant at the selling price of electricity was 0.229 USD/kWh while the incineration technology can be achieved at a lower selling price of 0.207 USD/kWh. In addition, the incineration technology also has NPV and B/C that was much higher than that of gasification technology. The payback period from incineration technology could be achieved at 5.6 years while the gasification technology was 8.5 years as shown in Table 3.

If seen from the ability to produce electrical energy and electric power efficiency, the choice of the gasification technology is more attractive compared to the incineration technology. Meanwhile, when compared with financial analysis, incineration technology was still more prospective compared to the gasification technology. However, the initial cost and O&M costs for incineration are still higher than those of gasification technology. Moreover, considering the circular economy, the gasification technology provides various amount of combustible gas which can be purified or separated for another process. Besides, the gasification technology results also

Table 3 Financial analysis of gasification and incineration technology for producing electricity from MSWs

Technology	Initial cost (USD)	O&M Cost (USD)	IRR (%)	NPV	PP	B/C	Energy production cost (USD/kWh)
Gasification technology	29,739,576	4,025,140	−2	−6,513,205	>20	0.78	0.158
	29,739,576	4,025,140	8	35,238,545	10.6	2.2	0.207
	29,739,576	4,025,140	11.2	53,984,228	8.5	2.8	0.229
Incineration technology	46,707,445	6,646,092	9.5	68,386,588	9.6	2.5	0.158
	46,707,445	6,646,092	18.6	161,125,139	5.6	4.4	0.207
	46,707,445	6,646,092	22.2	202,762,856	4.8	5.3	0.229

the tars which can be used further for biofuel production, instead of to be cleaned. Therefore, the choice of gasification technology will ultimately be more attractive.

4 Conclusions

The processing of MSWs as much as 833 kg/h for producing electricity has been analyzed using gasification technology and incineration technology. The electric energy generated from incineration technology is still higher, but the net power is low because to deliver pressurized air at large quantities requires high electrical energy. The net power and efficiency of gasification technology are ranging from 3 to 4 MW and from 43.3 to 56.3%, respectively. Meanwhile, the net power and efficiency of incineration technology each ranges from 1.7 to 3.4 MW and from 24.2 to 47.2%. The initial and maintenance costs of the incineration technology are high. The payback period from incineration technology can be achieved at 5.6 years while the gasification technology is 8.5 years. Therefore, the choice of gasification technology will ultimately be more attractive.

Acknowledgements The authors thank to the Rector of Universitas Sebelas Maret for supporting the research through the Mandiri Research Grant in FY 2019/2020. The authors thank also to Dr. Sunu H. Pranolo from Universitas Sebelas Maret who has enabled the use of ASPEN HYSYS software.

References

1. Moya D, Aldás C, López G, Kaparaju PJEP (2017) Municipal solid waste as a valuable renewable energy resource: a worldwide opportunity of energy recovery by using waste-to-energy technologies. Energy Procedia 134:286–295
2. Luz FC, Rocha MH, Lora EES, Venturini OJ, Andrade RV, Leme MMV, del Olmo OAJEC (2015) Management: techno-economic analysis of municipal solid waste gasification for electricity generation in Brazil 103:321–337
3. Chaerul M, Tanaka M, Shekdar AVJ (2007) Municipal solid waste management in Indonesia: status and the strategic actions 12:41–49
4. Lettner F, Haselbacher P, Timmerer HL, Leitner P, Suyitno S, Rasch B (2007) Latest results of CLEANSTGAS-staged biomass gasification CHP. In: 15th European biomass conference and exhibition for research to market deployment. ETA
5. Kierulff HJBH (2008) MIRR: a better measure 51:321–329
6. Arena UJWm (2012) Process and technological aspects of municipal solid waste gasification. A Rev 32:625–639
7. Haselbacher P, Lettner F, Timmerer HL, Suyitno S, Rasch B (2005) Experimental gas quality results from staged gasification, paper and poster. In: 14th European conference & exhibition: biomass for energy, industry and climate protection
8. Watson J, Zhang Y, Si B, Chen W-T, de Souza RJR, Reviews SE (2018) Gasification of biowaste: a critical review and outlooks. Renew Sustain Energy Rev 83:1–17

Investigation of the Angle Variations of the Guide Vane's Bottom Guide Plate Againsts the Inflow of Banki Turbine Blades

Sirojuddin, Lukman K. Wardhana, Obit Rizky, Regina Ibnawati, and Junior R. Syahri

Abstract Banki turbine is a crossflow turbine, usually for micro-hydro power plant applications. The output power is the main important thing. To get the maximum output power, researchers had done various solutions, one of them was to regulate the velocity of the water entering the runner blade from the first stage to the second stage with a good flow trajectory and didn't strike the shaft. This research aims to investigate the angle variations of the guide vane's bottom guide plate against the inflow of banki turbine blades, so that a uniform flow and velocity were obtained, did not strike the shaft and maximize power. To find the best variant, 7 lower plate angle variants and 2 runner casing variants were made. Each geometry of variant was created in the 2D and 3D Software. The flow test simulation and analysis were carried out in the CFD Simulation. The discharge of water was 2 m^3/min and head 5.5 m. After the CFD test simulation, the following results were obtained that the greater the angle the better the flow trajectory and not struck the shaft. In the KB-7 variant, it flows through blade 1, did not struck the runner shaft, then fully developed flow trajectory jumped to the blade 2 so that could stabilize the rotation and increased the power of the turbine. It was found that KB-7 was the best variant.

Keywords Investigation · Angle variation · Guide plate · Guide vane · Flow trajectory

1 Introduction

Banki turbine usually used in micro hydroelectric power plants. Microscale is below 1 MW [1]. The selection of the turbine is based on the cheap price, manufacture and easy transportation, simple assembly and maintenance, it suitable for remote areas who do not have electricity lines.

The important point in the Banki turbine is how to obtain the maximum output power. Kaniecki [2] investigated the pattern of the outflow of the turbine by adding a draft tube to eliminate vortices. The vortex could be removed but the flow of water

Sirojuddin (✉) · L. K. Wardhana · O. Rizky · R. Ibnawati · J. R. Syahri
State University of Jakarta, Jakarta, Indonesia
e-mail: sirojuddin@unj.ac.id

U. Sabino et al. (eds.), *Proceedings of the 6th International Conference and Exhibition on Sustainable Energy and Advanced Materials*, Lecture Notes in Mechanical Engineering,
https://doi.org/10.1007/978-981-15-4481-1_47

still struck the shaft. Using CFD, [3] examined the effect of angle and length of inlet guide vane in the cross-flow fan to increase efficiency using wind fluid. It was found that the best fan performance when using a guide vane angle 50° and length of 300 mm (L/D1 = 1.5) length. [4] investigated the influence of the turbine nozzle shape on the performance and internal characteristics of the turbine using CFD analysis. The results were affecting the performance of the turbine nozzle shape but seen the flow struck the turbine shaft. De Andrade et al. [5] analyzed water flow and pressure in several rounds of the turbine runner. Inflows made in tapered form without guide vane and the results have shown that the flow trajectory not inline and still struck the shaft runner. Kokubu et al. [6] investigated the flow of water into four types of the casing with an opening width of the guide vane 100%. It was found that the highest efficiency was 4th modeled but the flow for all types still struck the shaft runner. Sammartano et al. [7] researched turbine efficiency based on the number of blades and water angle of attack. The result showed that the 35 blades and 22° water angle of attack got the efficiency of 86% but the inflow discharge regulated only by the width of the nozzle (not used a guide vane) and the flow still struck the shaft. Kaunda Chiyembekezo et al. [8] conducted a numerical investigation on the inlet and outlet water of the blade runner. Inflows of water regulated by the nozzle. The flow was slightly struck the shaft and the cavitation still appear. Girma and Dribssa [9] analyzed the flow of banki turbines with horizontal inflow. The result showed that the water fully struck the shaft. The banki turbine made without guide vane's bottom guide plate. Adhikari et al. [10] studied cavitation that occurs in banki turbines with horizontal inflow. Results of the study shown that there was cavitation on the blade and the flow trajectory struck the shaft.

Kaunda Chiyembekezo et al. [8] In their conclusion that incidence losses were greatly reduced when the jet did not strike the shaft in the runner.

The research of [9] without the guide vane's bottom guide plate and the flow still struck the shaft. In this research was tried to investigate the variation of the angle of the guide vane's bottom guide plate against the inflow of banki turbine blades, in order to obtain a good flow trajectory and did not strike the shaft as research conducted by Kaunda Chiyembekezo et al. [8], so that the output power of turbine was expected to increase.

2 Method

This research was conducted at the Design Laboratory of Mechanical Engineering, State University of Jakarta, using CAD and CFD Software. The Flow diagram can be seen in Fig. 1.

Fig. 1 Flow chart

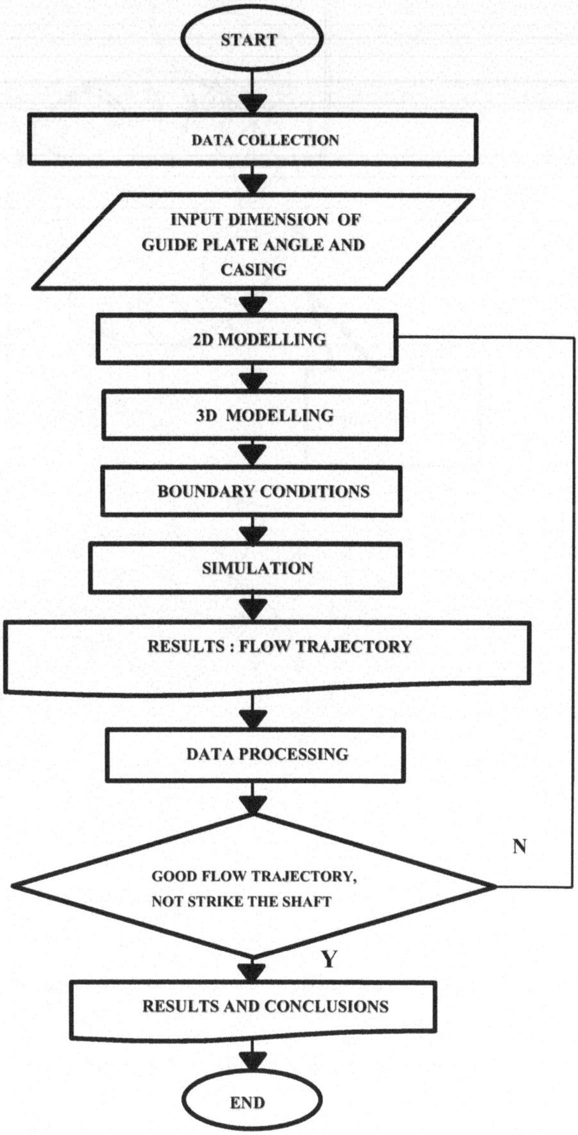

2.1 Plate Angle and Runner Case Variant

7 lower plate angle variant and 2 runner casing variant. Side view of Rounded–Rounded Case as shown in Fig. 2a and side view of Rounded-Square casing as shown in Fig. 2b.

Each variant created in 2D CAD software and then created in 3D software and flow trajectory using CFD software.

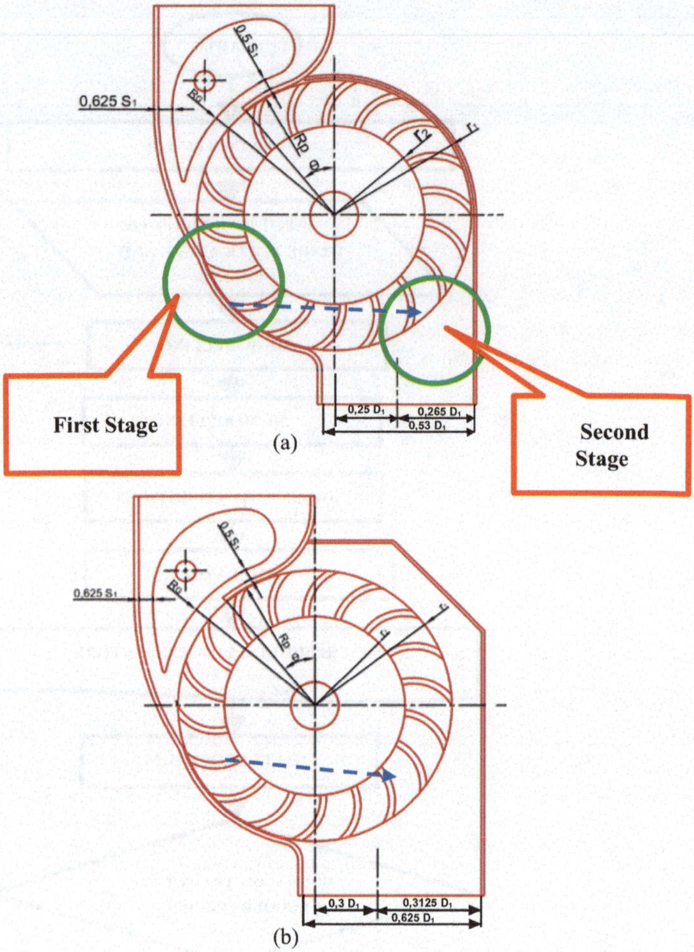

Fig. 2 Side view of rounded–rounded casing (**a**), side view of square-rounded casing (**b**)

2.2 *Parameters and Boundary Conditions*

The thickness of the water inlet to S1 was 16 mm. Width of water inflow through the guide vane was divided into two parts, left and right. The thickness of the left side is 5/8 S1 and ½ S1 on the right side. The radius of the bottom guide vane was ½ D + 12 mm and the radius of the bottom plate is ½ D + 0.8 mm. The focus of this study was the modification of theta (Θ). Parameter turbine can be seen in Tables 1, 2, 3 and 4:

Table 1 Parameters and data of turbine design

Design parameter	Data calculation	Design dimension	Description
α_1	16°	6°	Angle of entrance velocity
β_1	30°	30°	Angle of blade
D_1	200 mm	200 mm	Outer diameter of runner
D_2	131.97 mm	132 mm	Inner diameter runner
n	18 pieces	18 pieces	Number of blades
V_1	10.18 m/s	10.18 m/s	Absolute velocity of water
L	200 mm	200 mm	Width of wheel/runner
H	5.5 m	5.5 m	Head of water
N	468.84 rpm	470 rpm	Runner speed
Q	2 m^3/min	2 m^3/min	Water discharge
Pth	1798.5 W	–	Theoretical power
Ds	30 mm	30 mm	Shaft diameter
Df	52 mm	52 mm	Hub diameter
S1	16 mm	16 mm	Thickness of jet

2.3 *Governing Equation*

Flow Simulation solves the Navier-Stokes equations, which are formulations of mass, momentum and energy conservation laws for fluid flows. The equations are supplemented by fluid state equations defining the nature of the fluid, and by empirical dependencies of fluid density, viscosity and thermal conductivity on temperature. Inelastic non-Newtonian fluids are considered by introducing a dependency of their dynamic viscosity on flow shear rate and temperature, and compressible liquids are considered by introducing a dependency of their density on pressure. A particular

Table 2 Boundary conditions

Item	Value
Turbulence model	K-epsilon
Rotation	Non-rotation
Material	Water
Density	998.2 kg/m^3
Discharge	0.0333 m^3/s
Reference axis	Y
Gravitation	9.81 m/s^2
Calculation type	Steady state
Environment pressure	101.325 Pa
Temperature	293.20 K
Turbulence parameters	Boundary layer parameters

Table 3 Size of computational domain

X min	0.004 m
X max	0.218 m
Y min	−0.002 m
Y max	0.539 m
Z min	0.116 m
Z max	0.349 m

Table 4 Basic mesh dimensions

Number of cells in X	14
Number of cells in Y	36
Number of cells in Z	16

problem is finally specified by the definition of its geometry, boundary and initial conditions.

Flow Simulation is capable of predicting both laminar and turbulent flows. Laminar flows occur at low values of the Reynolds number, which is defined as the product of representative scales of velocity and length divided by the kinematic viscosity. When the Reynolds number exceeds a certain critical value, the flow becomes turbulent, i.e. flow parameters start to fluctuate randomly.

Most of the fluid flows encountered in engineering practice are turbulent, so Flow Simulation was mainly developed to simulate and study turbulent flows. To predict turbulent flows, the Favre-averaged Navier-Stokes equations are used, where time-averaged effects of the flow turbulence as the Reynolds stresses appear in the equations for which additional information on the flow parameters are considered, whereas the other, i.e. large-scale, time-dependent phenomena are taken into account directly. Through this procedure, extra terms known must be provided. To close this system of equations, Flow Simulation employs transport equations for the turbulent kinetic energy and its dissipation rate, the so-called k-ε model [11].

The k-ε model can be derived from the incompressible Navier-Stokes equations [12].

$$\rho\left(\frac{\partial u_i}{\partial t}+\sum_j uj\frac{\partial u_i}{\partial x_j}\right)=-\frac{\partial p}{\partial x_i}+\eta\nabla^2 u_i \quad (1)$$

where u(x, t) represents the velocity vector field, p(x, t) is the pressure field, p(x, t) is the pressure field, ρ is the density constant, η is the dynamic viscosity, and $u = \frac{\eta}{\rho}$ is the kinematic viscosity.

Noting Navier-Stokes are derived from the equations for conservation of mass, momentum, and energy, we have that

$$\frac{\partial\rho}{\partial \mathrm{t}}+\sum_{\mathrm{j}} u_j\frac{\partial_\rho}{\partial \mathrm{x_j}}=\rho\sum_j\frac{\partial u_j}{dx_i}=0 \quad (2)$$

Applying statistical averaging to (N S) produces the Reynolds equations:

$$\rho\frac{\partial\overline{u_i}}{\partial t}+\sum_j\left(\rho\overline{u_j}\frac{\partial\overline{u_i}}{\partial x_j}+\rho\overline{\frac{\partial u_i'}{\partial x_j}u_j'}\right)=-\frac{\partial\bar{p}}{\partial x_i}+\sum_j\frac{\partial\overline{\tau_{ij}}}{\partial x_j} \quad (3)$$

with $u = \bar{u} + u'$ written in the mean plus fluctuation decomposition,

$$\overline{\tau_{ij}}=\eta\left(\frac{\partial\overline{u_i}}{\partial x_j}+\frac{\partial\overline{u_j}}{\partial x_i}\right),\ \eta\nabla^2 u_i=\sum_j\frac{\partial\tau_{ij}}{\partial x_j} \quad (4)$$

3 Results and Discussions

From CFD simulation test results of several variants runner casing are shown in Fig. 3:

3.1 Velocity Contour

- **Rounded-Rounded Case (BB)**
  See Fig. 3.
- **Square-Rounded Case (KB)**

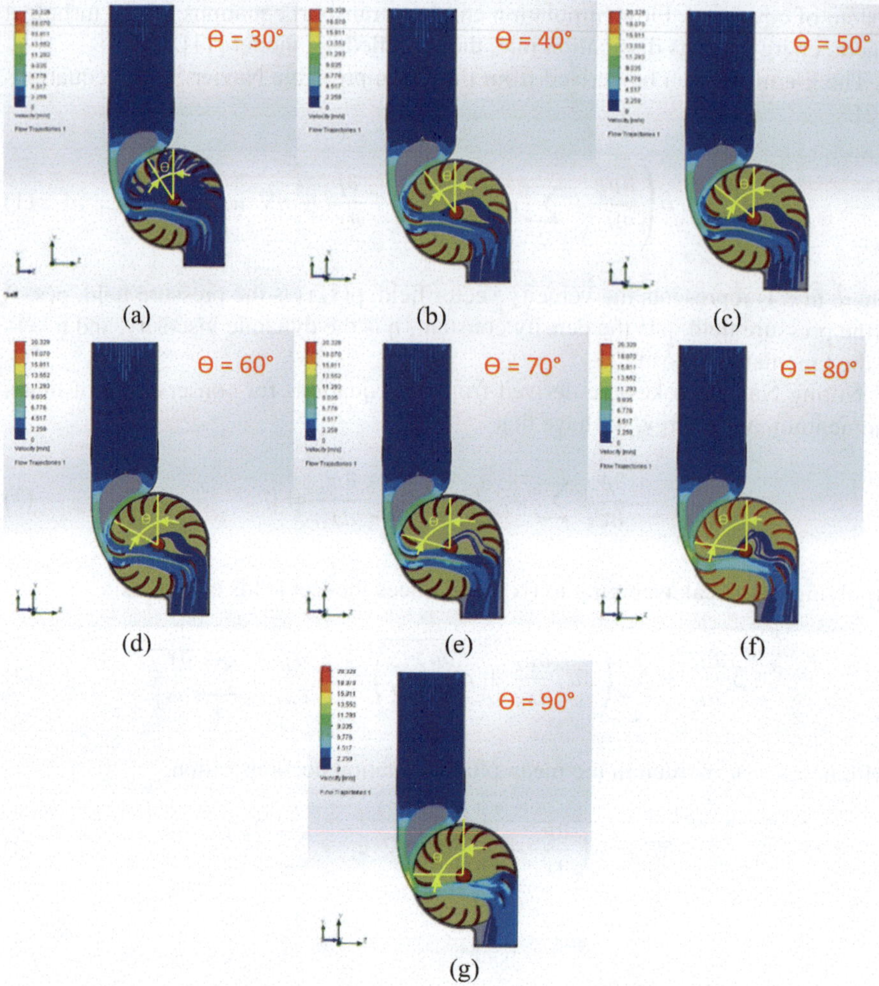

Fig. 3 Velocity contour, BB-1 (**a**), BB-2 (**b**), BB-3 (**c**), BB-4 (**d**), BB-5 (**e**), BB-6 (**f**), BB-7 (**g**)

See Fig. 4.

3.2 *Pressure Contour*

- **Rounded-Rounded Case (BB)**
  See Fig. 5.
- **Square-Rounded Case (KB)**
  See Fig. 6.

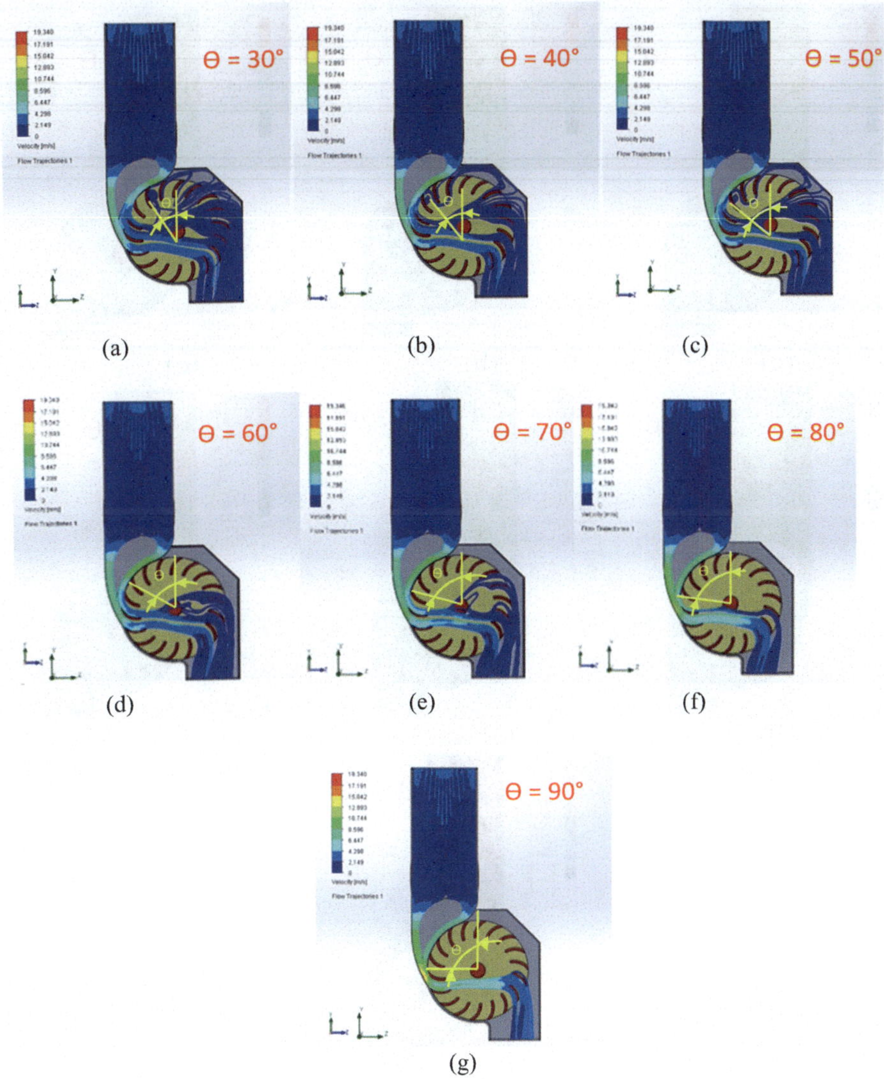

Fig. 4 Velocity contour, KB-1 (**a**), KB-2 (**b**), KB-3 (**c**), KB-4 (**d**), KB-5 (**e**), KB-6 (**f**), KB-7 (**g**)

3.3 Discussion

From Fig. 3 velocity contour, it was found that for rounded-rounded case design variant BB-1 to BB-6, when the angle of theta (θ) at 30°, 40°, 50°, 60°, 70°, 80°, the flow trajectory separated in the first stage then struck the shaft, as shown in Fig. 3a–f. While the angle of θ at 90° the flow trajectory was very good and nearly not struck the shaft as shown in Fig. 3g.

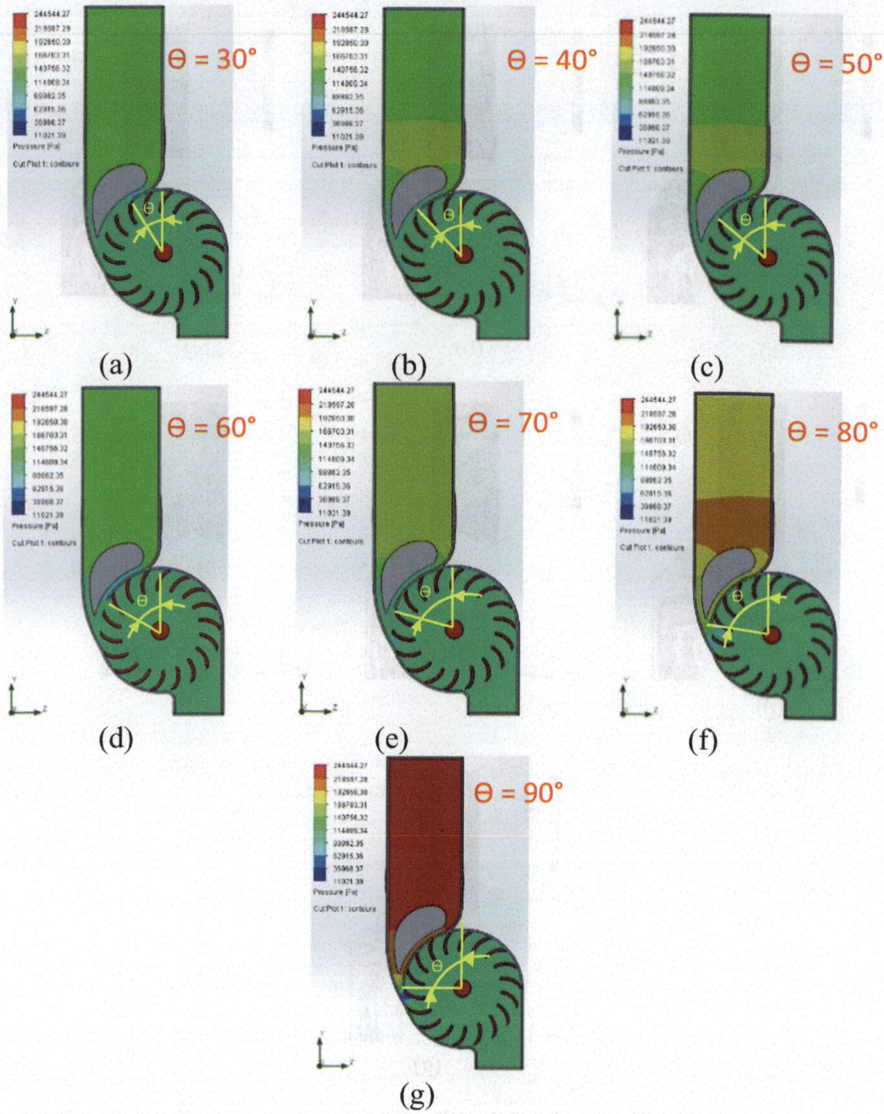

Fig. 5 Pressure Contour, BB-1 (**a**), BB-2 (**b**), BB-3 (**c**), BB-4 (**d**), BB-5 (**e**), BB-6 (f), BB-7 (**g**)

From Fig. 4 velocity contour, it was found that for square-rounded case design variant KB-1 to KB-6, when the angle of θ at 30°, 40°, 50°, 60°, 70°, 80°, the flow trajectory still separated in the first stage then struck the shaft as shown in Fig. 4a–f. When the angle of θ at 90° for square-rounded case the flow trajectory shown very good in the first stage and second stage and also not struck the shaft as shown in Fig. 4g.

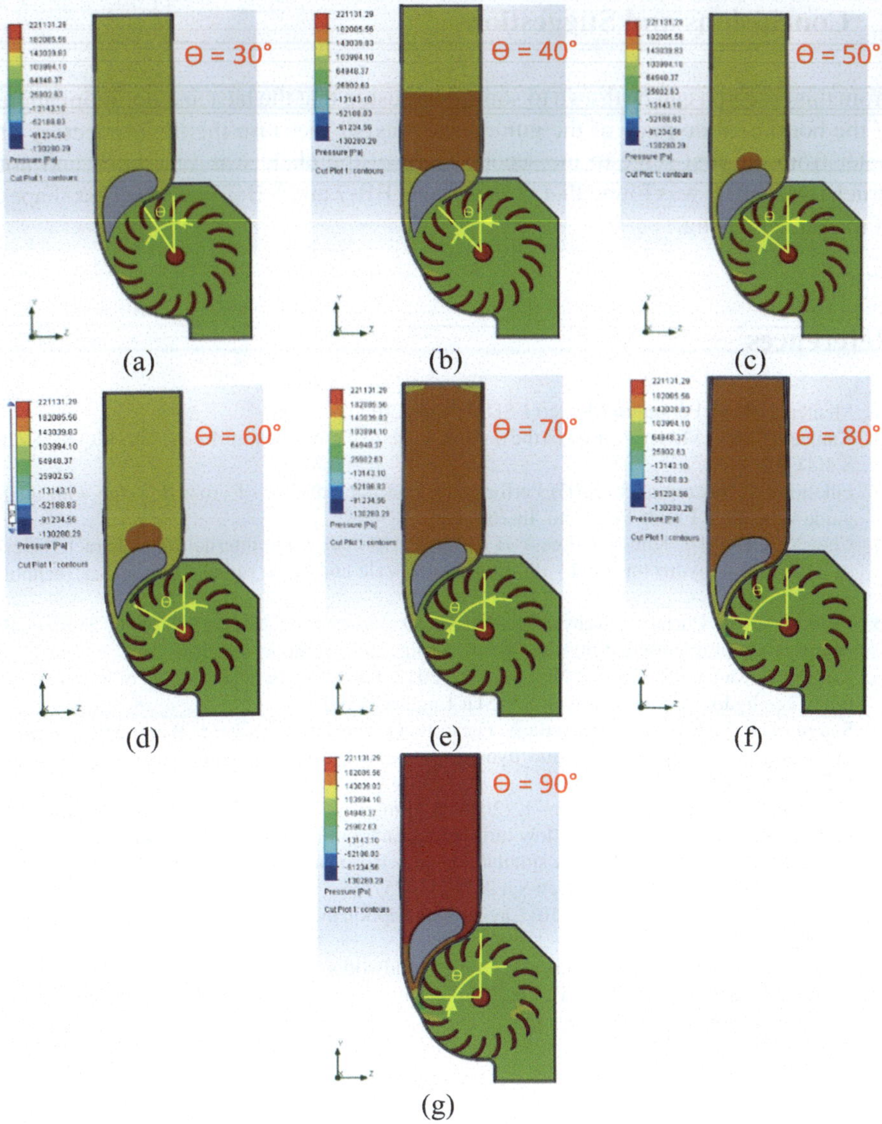

Fig. 6 Pressure Contour, KB-1 (**a**), KB-2 (**b**), KB-3 (**c**), KB-4 (**d**), KB-5 (**e**), KB-6 (**f**), KB-7 (**g**)

From Figs. 5 and 6 pressure contour, it was found that on the diffuser of variant BB-7 and KB-7 the pressure was higher than other variants, these because of the angle of guide vane's bottom guide plate is bigger than other variant and also reduce the inflow area of water through the blade of the runner. The angle of 90° was the optimum angle.

4 Conclusions and Suggestions

From this research can be drawn to some conclusion that the best and optimum angle of the bottom guide plate of the guide vane was 90° because the flow trajectory of water from the first stage to the second stage of the blade was very good and not struck the shaft. It was found that both variants BB-7 and KB-7 were the best shapes of the runner casing.

References

1. Menteri ESDM: Peraturan Menteri ESDM No.8 (2011)
2. Kaniecki (2014) Modernization of the outflow system of cross-flow turbines. TASK Q Sci Bull S 4(4):601–608
3. Fukutomi J, Nakamura R (2005) Performance and internal flow of cross-flow fan with inlet guide vane. Proc Fluids Eng Conf 48:763–769
4. Choi Y-D, Lim J-I, Kim Y-T, Lee Y-H (2008) Performance and internal flow characteristics of a cross-flow hydro turbine by the shapes of nozzle and blade runner. J Fluid Sci Technol 3(3):398–409
5. De Andrade J, Curiel C, Kenyery F, Aguilln O, Vásquez A, Asuaje M (2011) Numerical investigation of the internal flow in a Banki turbine. Int J Rotating Mach, 1–12
6. Kokubu K, Kanemoto T, Son S-W, Choi Y-D (2012) Performance improvement of a cross-flow micro eco hydro turbine. J Korean Soc Mar Eng 36(7):902–909
7. Sammartano V, Arico C, Carravetta A, Fecarotta O, Tucciarelli T (2013) Banki-Michell optimal design by computational fluid dynamics testing and hydrodynamic analysis. Energies 6(5):2362–2385
8. Kaunda CS, Kimambo CZ Nielsen K (2014) A numerical investigation of the flow profile and performance of a low cost Crossflow turbine. Int Energy Environ 5(3):275–296
9. Girma M, Dribssa E (2014) Flow simulation and performance prediction of cross flow turbine using CFD tool. Int J Eng Res Gen Sci 2(6):747–757
10. Adhikari RC, Vaz J, Wood D (2016) Cavitation inception in crossflow hydro turbines. Energies 9:1–12
11. Systemes Dassault (2016) Technical reference solidworks flow simulation
12. Scott-Pomerantz CD (2004) The K-Epsilon model in the theory of turbulance. Ph.D. Desertation. University of Pittsburgh

Modification of Blade Profile the Banki Water Turbine to Increase Power

Sirojuddin, Lukman K. Wardhana, Obit Rizky, Regina Ibnawati, and Junior R. Syahri

Abstract Banki Turbine is very popular, especially for micro hydroelectric power plant. Increasing power and efficiency still the main topic for researchers. One of the point to increase the power by modification of the blade profile. The purpose of this research was to find out the best blade profile to increase the power of banki water turbine. The observation focus on output power and flow contour inside the runner of the turbine. In this research, the blade profile would be made in 5 variants. Variant P1 as a standard and other 4 variants P2, P3, P4, and P5 were the modification profile. The experiment would be conducted by CFD software simulation. The head of water was 5.5 m and discharge 2 m^3/min. Based on the simulation test result showed that variant P 4 was the highest with power increase more than 10% compare to the variant standard P1 and the good flow trajectory contour compare to others.

Keywords Blade profile · Banki water turbine · Power increase · Flow trajectory

1 Introduction

For hilly regions and remote areas, cross-flow turbine is a suitable type for producing electricity. Banki turbine is a cross-flow type turbine usually for micro-hydro power. In Indonesia, scale limitation according to the ministry of ESDM RI No. 8 Year 2011 is less than 1 MW [1].

The blade profile is one of the most important part of turbine banki, because it will affect the output power. The design of the blade runner profile must be analyzed and researched seriously to get a good flow trajectory so that increasing power can be obtained. Mockmore and Merryfield [2] issued Bulletin No. 25 about the translation of Donat Banki's turbine paper to show the results of a series of tests on a laboratory turbine built according to the specification of Banki. According to the paper maximum efficiency would be 87.8% but experimentally reached only 68%. Zar Ni Tin et al. [3] Investigated the performance of the constructed cross-flow turbine and verify that all parts and systems installed worked fine and performing their functions

Sirojuddin (✉) · L. K. Wardhana · O. Rizky · R. Ibnawati · J. R. Syahri
State University of Jakarta, Jakarta, Indonesia
e-mail: sirojuddin@unj.ac.id

U. Sabino et al. (eds.), *Proceedings of the 6th International Conference and Exhibition on Sustainable Energy and Advanced Materials*, Lecture Notes in Mechanical Engineering,
https://doi.org/10.1007/978-981-15-4481-1_48

correctly. The designed cross-flow turbine was capable to produce up to 300 W AC power at the head of 6 m and flow rate of 0.0091 m^3/s. The blade profile that used as variant P-1 in this research. Maximum efficiency was obtained 86% compared to 300 W design power, but if compared to theoretical power the efficiency was 48% only. Chattha et al. [4] Conducted research to get the optimum profile of the leading edges of the blade. Four different profiles (Flat tip blade profile, round tip blade profile, pointed-tip blade profile and oval tip blade profile) were modeled and simulated with ANSYS CFX and then compared their performance in terms of their efficiencies. The complete turbine was analyzed for the determination of the leading edge of the blade in the first stage and the trailing edge in the second stage. The design and simulation conditions were based on the installed CFT (Cross-Flow Turbine). The results showed that efficiency of the round tip blade profile was better than the other three profiles of the blade. Jasa et al. [5] Investigated for 3 variants of blade radius that were 0.334 r1, 0.326 r1, 0.323 r1 and 3 variants attack of angle that were 15°,16°, 17°. Kaunda Chiyembekezo et al. [6] conducted a numerical investigation on the inlet and outlet water of the blade runner. Inflows of water regulated by the nozzle. The flow was slightly struck the shaft and the cavitation still appear. Choi et al. [7] Investigated the influence of the turbine nozzle shape on the performance and internal characteristics of the turbine using CFD analysis. The results were affecting the performance of the turbine nozzle shape but seen the flow still struck the turbine shaft. De Andrade et al. [8] analyzed water flow and pressure in several rounds of the turbine runner. Inflows made in tapered form without guide vane and the results showed that the flow trajectory not inline and still struck the shaft runner.

Sammartano et al. [9] researched turbine efficiency based on the number of blades and water angles of attack. The result showed that the 35 blades and 22° water angle of attack got the efficiency of 86% but the inflow discharge regulated only by the width of the nozzle (not used a guide vane) and the flow still struck the shaft. Girma and Dribssa [10] analyzed the flow of banki turbines with horizontal inflow. The result showed that the water fully struck the shaft.

Rajput [11] explain in various engineering fields that encountered with the problems which involve the flow of fluid around submerged bodies/objects. In such problems either a fluid maybe flowing around a stationary submerged body or a body maybe flowing through a large mass of stationary fluid or both the body and the fluid may be in motion. There are two components when a body wholly immersed in a real fluid may be subjected to following kind of forces due to relative motion between the body and the fluid: (a) Drag Force, The components of force in the direction of flow (free stream) on a submerged body is called the drag force, F_d. (b) Lift Force. The components of forces at right angles to the direction of flow is called the lift force, F_l. The application for immerged bodies that having drag and/or lift force: (a) A tall chimney exposed to wind, (b) Flow of water past a bridge pier, (c) Flow of fluids past blades, in fans, blowers, compressors, turbines, etc. (d) Motion of aeroplanes, submarines, torpedoes, etc.

This research would be a focus on modification of blade profile the banki water turbine to increase power to obtain maximum drag force in the first stage of the blade

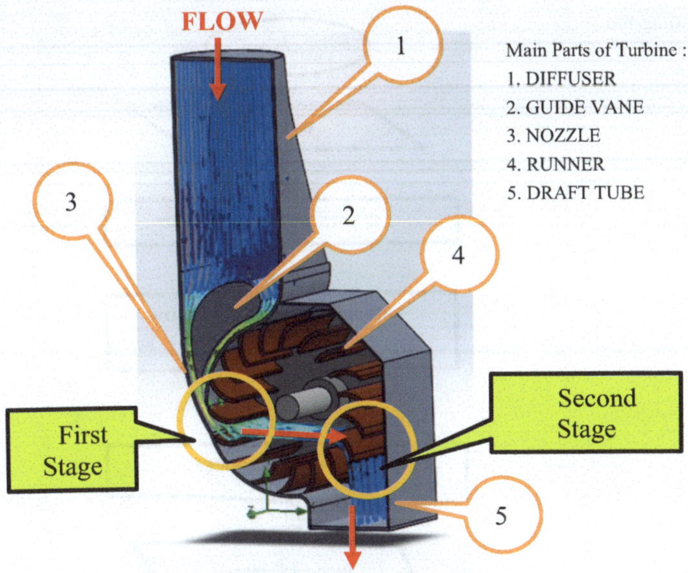

Fig. 1 Main parts of Banki turbine

using flow around submerged bodies theory-drag and lift [11], a good flow trajectory so that the power generated was expected to be maximized.

Main parts of Banki turbine can be seen ini Fig. 1:

2 Method

This research was conducted at the Design Laboratory of Mechanical Engineering, State University of Jakarta, using CAD and CFD Software. The Flow diagram can be seen in Fig. 2:

2.1 Blade Profile and Variants

See Figs. 3, 4 and 5.

2.1.1 Parameters and Boundary Conditions

Parameter and boundary conditions of turbine can be seen in Tables 1 and 2:

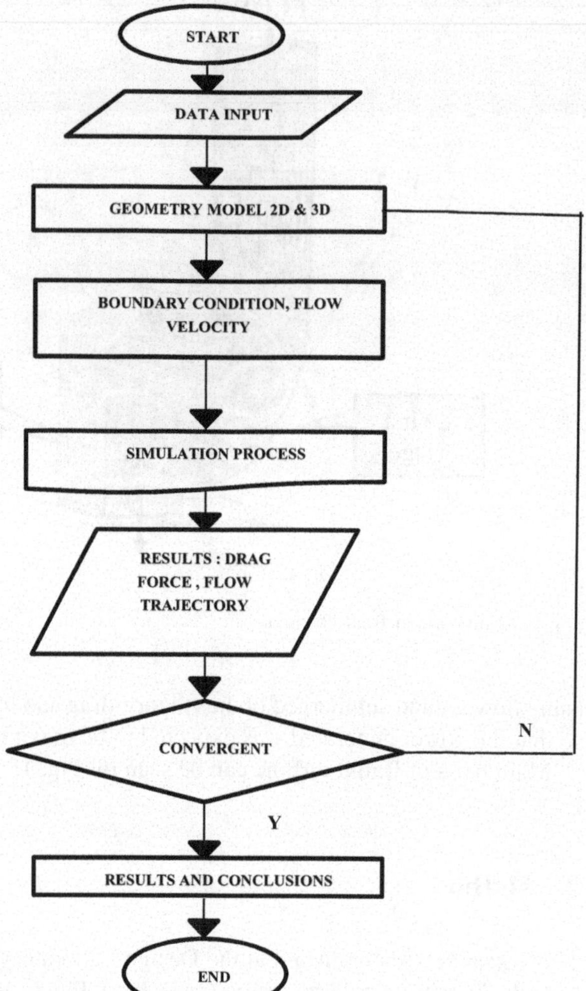

Fig. 2 Flow diagram

2.2 *Governing Equation*

Flow Simulation solves the Navier-Stokes equations, which are formulations of mass, momentum and energy conservation laws for fluid flows. The equations are supplemented by fluid state equations defining the nature of the fluid, and by empirical dependencies of fluid density, viscosity and thermal conductivity on temperature. Inelastic non-Newtonian fluids are considered by introducing a dependency of their dynamic viscosity on flow shear rate and temperature, and compressible liquids are considered by introducing a dependency of their density on pressure. A particular problem is finally specified by the definition of its geometry, boundary and initial conditions.

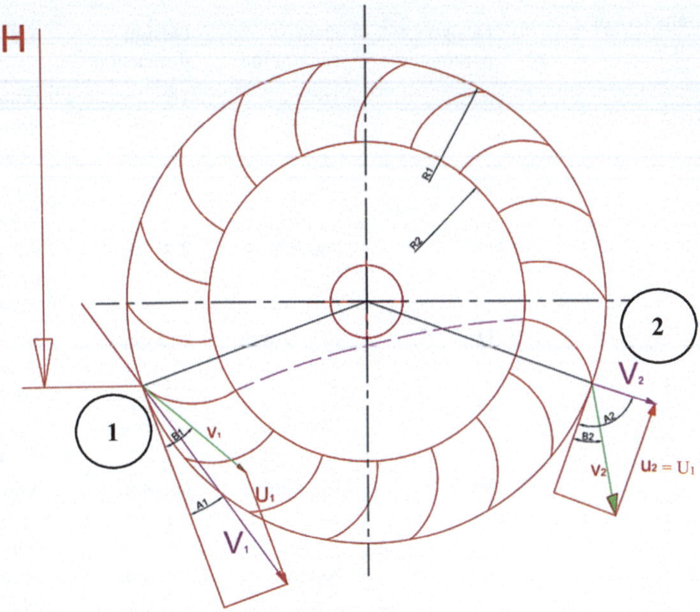

Fig. 3 Velocity profile inside the blade runner

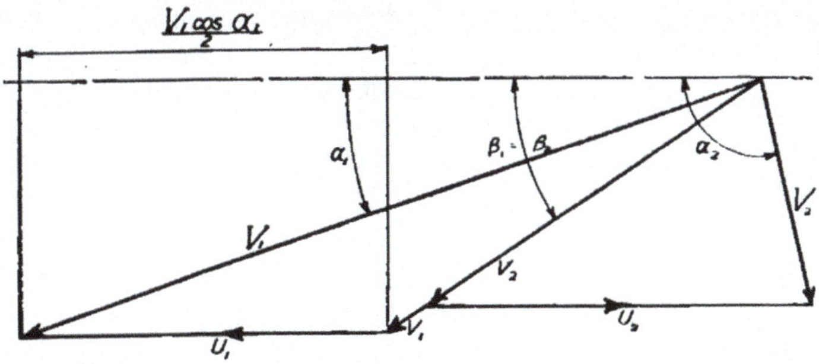

Fig. 4 Velocity diagram

Fig. 5 Variants of blade profile

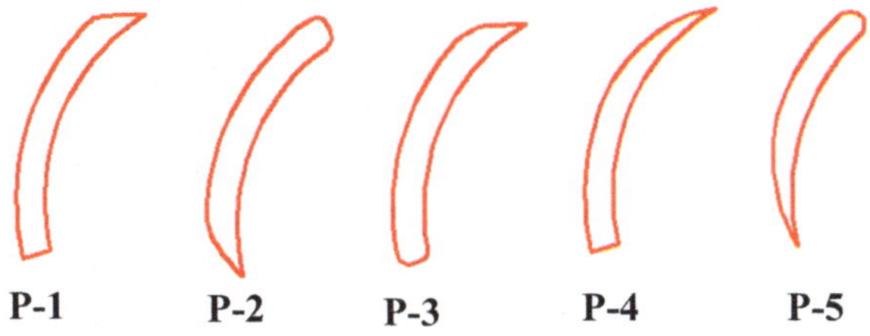

Table 1 Parameters of turbine design

Design parameter	Data calculation	Design dimension	Description
α_1	16°	16°	Angle of entrance velocity
β_1	30°	30°	Angle of blade
D_1	200 mm	200 mm	Outer diameter of runner
D_2	131.97 mm	132 mm	Inner diameter runner
n	18 pieces	18 pieces	Number of blades
V_1	10.18 m/s	10.18 m/s	Absolute velocity of water
L	200 mm	200 mm	Width of wheel/runner
H	5.5 m	5.5 m	Head of water
N	468.84 rpm	470 rpm	Runner speed
Q	2 m^3/min	2 m^3/min	Water discharge
Ds	30 mm	30 mm	Shaft diameter
Df	52 mm	52 mm	Hub diameter
S1	16 mm	16 mm	Thickness of jet

Flow Simulation is capable of predicting both laminar and turbulent flows. Laminar flows occur at low values of the Reynolds number, which is defined as the product of representative scales of velocity and length divided by the kinematic viscosity. When the Reynolds number exceeds a certain critical value, the flow becomes turbulent, i.e. flow parameters start to fluctuate randomly.

Most of the fluid flows encountered in engineering practice are turbulent, so Flow Simulation was mainly developed to simulate and study turbulent flows. To predict turbulent flows, the Favre-averaged Navier-Stokes equations are used, where time-averaged effects of the flow turbulence as the Reynolds stresses appear in the equations for which additional information on the flow parameters are considered, whereas the other, i.e. large-scale, time-dependent phenomena are taken into account directly. Through this procedure, extra terms known must be provided. To close this

Table 2 Boundary conditions

Item	Value
Turbulence model	K-epsilon
Rotation	Non-rotation
Material	Water
Density	998.2 kg/m^3
Discharge	0.0333 m$^{3/}$s
Reference axis	Y
Gravitation	9.81 m/s^2
Calculation type	Steady + state
Environment pressure	101.325 Pa
Temperature	293.20 K
Turbulence parameters	Boundary layer parameters

system of equations, Flow Simulation employs transport equations for the turbulent kinetic energy and its dissipation rate, the so-called k-ε model [12].

The k-ε model can be derived from the incompressible Navier-Stokes equations [13].

$$\rho\left(\frac{\partial u_i}{\partial t} + \sum_j uj\frac{\partial u_i}{\partial x_j}\right) = -\frac{\partial p}{\partial x_i} + \eta\nabla^2 u_i \tag{1}$$

where u(x, t) represents the velocity vector field, p(x, t) is the pressure field, p(x, t) is the pressure field, ρ is the density constant, η is the dynamic viscosity, and $\upsilon = \frac{\eta}{\rho}$ is the kinematic viscosity.

Noting Navier-Stokes are derived from the equations for conservation of mass, momentum, and energy, we have that

$$\frac{\partial \rho}{\partial \mathrm{t}} + \sum_{\mathrm{j}} u_j \frac{\partial_\rho}{\partial \mathrm{x_j}} = \rho\sum_j \frac{\partial u_j}{dx_i} = 0 \tag{2}$$

Applying statistical averaging to (N S) produces the Reynolds equations:

$$\rho\frac{\partial \overline{u_i}}{\partial t} + \sum_j\left(\rho\overline{u_j}\frac{\partial \overline{u_i}}{\partial x_j} + \overline{\rho\frac{\partial u_i'}{\partial x_j}u_j'}\right) = -\frac{\partial \overline{p}}{\partial x_i} + \sum_j \frac{\partial \overline{\tau_{ij}}}{\partial x_j} \tag{3}$$

with u = $\bar{u} + u'$ written in the mean plus fluctuation decomposition,

$$\overline{\tau_{ij}} = \eta\left(\frac{\partial \overline{u_i}}{\partial x_j} + \frac{\partial \overline{u_{ja}}}{\partial x_i}\right), \eta\nabla^2 u_i = \sum_j \frac{\partial \tau_{ij}}{\partial x_j} \tag{4}$$

From [2] theoretical power equation as follows:

$$\mathrm{Pth} = \rho.\mathrm{g.Q.H} = \frac{\rho.g.Q.V_1^2}{C^2.2.g} \tag{5}$$

Pth is theoretical power (Watt), Q is discharged (m^3/s), ρ is the density (kg/m^3), H is head (m), V_1 (m/s) is absolute velocity of water = $C(2gH)^{0.5}$ and C = 0.98.

Brake Power [2] calculated as follows:

$$\mathrm{BP} = \{(\rho.\mathrm{g.Q})(\mathrm{V_1} \cos \alpha_1)\mathrm{u_1}\} + \{(\rho.\mathrm{g.Q})(\mathrm{V_2} \cos \alpha_2)\mathrm{u_2}\} \tag{6}$$

where: $u_1 = u_2$

So that:

$$\mathrm{BP} = (\rho.\mathrm{g.Q})(\mathrm{V_1} \cos \alpha_1 + \mathrm{V_2} \cos \alpha_2)\mathrm{u_1} \tag{7}$$

$$V_2 \cos \alpha_2 = \upsilon_2 \cos \beta_2 - u_1 \tag{8}$$

$$\upsilon_2 = \varphi\, \upsilon_1$$

where φ is an empirical coefficient = 0.98. From the velocity diagram,

$$\upsilon_1 = (V_1 \cos \alpha_1 - u_1)/(\cos \beta_1) \tag{9}$$

$$U_1 = \frac{V_{1 \cos \alpha_1}}{2} \tag{10}$$

$$\beta_1 = \beta_2 \tag{11}$$

$$\mathrm{u_1} = \frac{\pi.D_1.N}{60} \tag{12}$$

The calculation approach with drag force is carried out in this research because the velocity triangle in Fig. 4 and the equations above cannot predict the value of forces due to the shape of the blade and the situation in the runner, therefore through this CFD software, it can be done by clicking the surface of the blade which the water pass from the first stage to the second stage of blade in the runner.

From flow around submerged bodies-theory [11] we can get the drag force (Fd) of the flow of fluids past blades in the turbine (Fig. 6):

$$F_d = \int_A p.dA.\sin \theta c + \int_A \tau_0.dA \cos \theta$$

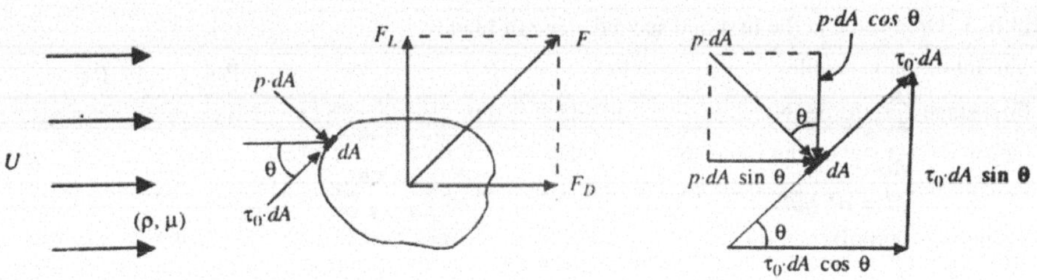

Fig. 6 Pressure and frictional forces on an elementary surface of an immersed body

- The term $\int_A$ p.dA sin θ is called pressure drag
- The term $\int_A \tau_0.dA \sin\theta$ is called friction drag or skin drag or shear drag.

On the element of area dA on the surface of the body, let p and τ represent the static pressure and shear stress, and let θ be the inclination of the tangent to the element with the direction of flow. The component of the force, due to p and τ, along the direction of motion is known as Drag Force, F_d.

$$F_d = \frac{1}{2}\rho.C_d.A.U^2$$

where C_d is Coefficient of drag, ρ is density of fluid, U is relative velocity of fluid about the body and A is some characteristic area usually the area projected on a plane perpendicular to the relative motion of the fluid.

In this case drag power (P_d) calculated as follows:

$$P_d = F_d \, Cos\alpha_1.u_1 \tag{13}$$

F_d is total drag forces at the first and second stage of blades (N), P_d is the drag power (Watt), α_1 is angle of entrance velocity and u_1 is tangential velocity of the runner = u_2.

Efficiency of turbine:

$$\eta = \frac{Pd}{Pth} \times 100 \tag{14}$$

Table 3 Drag force at the first and second stage of blades

Variant of blade profile	P-1	P-2	P-3	P-4	P-5
First stage drag force (N)	302.48	319.26	277.15	353.39	318.33
Upper second stage drag force (N)	3.991	4.693	17.414	4.901	23.402
Lower second stage drag force (N)	52.482	47.679	25.436	55.659	14.816
Total drag force (N)	56.823	52.717	43.142	60.952	38.569

Table 4 Drag power at the first stage and the second stage of blades

Variant of blade profile	Drag power (W)
P-1	1422.65
P-2	1501.57
P-3	1303.51
P-4	1662.09
P-5	1497.19

3 Results and Discussions

3.1 Results

3.1.1 Drag Force and Drag Power

From CFD simulation obtained the value of drag force for the first stage and second stage as shown in Tables 3 and 4:

3.1.2 Efficiency

Efficiency of turbine can be seen in Tables 5:

Table 5 Efficiency of turbine

Variant of blade profile	Theoretical power (W)	Efficiency (%)
P-1	1798.5	79.1
P-2		83.5
P-3		72.5
P-4		92.4
P-5		83.2

3.1.3 Velocity Contour

See Fig. 7.

3.1.4 Pressure Contour

See Fig. 8.

3.2 *Discussions*

From Table 3 it was found that the drag force of variant P-1 = 358.9 N, P-2 = 371.6 N, P-3 = 320 N, P-4 = 413.9 N, and P-5 = 356.5 N. The highest force was variant P-4. While the drag power of variants P-1, P-2, P-3, P-4, P-5, P-6 were 1422.6, 1501.6, 1303.5, 1662.1, and 1497.2 W. It was found that the efficiency of variants P-1, P-2, P-3, P-4, P-5, P-6 were 79.1, 83.5, 72.5, 92.4, and 83.2%.

From Fig. 7 velocity contour, the flow trajectory of variants P-2, P-4 and P-5 seem to be a good trajectory whether in the first stage and the second stage. While in variant P-2 and P-5 the flow trajectory only good in the first stage but not good in the second stage because the flow struck the shaft. From Fig. 8 pressure contour, the pressure at P-1 and 1 bar, at P-2 and P-5 approx 1.1 bar but the flow trajectory to the second stage of blade struck the shaft. P-3 less than 1 bar, P-4 close to the first stage blade the pressure approx 1.1 bar but surrounding its approx less than 1 bar. The best variant was P-4 with efficiency 92.4%. Drag force occurs on the surface of the first stage and second stage of the blade and calculated by CFD SolidWorks Flow Simulation.

4 Conclusions

From simulation results and calculation above could be concluded as follows:

1. Blade profile affected the value of drag force in the first and second stage of the blade.
2. Efficiency variant P-4 = 92.4%, was the highest efficiency compared to others especially to variant P-1.
3. Good flow trajectory were variant P-2, P-4, and P-5.

Acknowledgements This research funded by Dana BLU POK Engineering Faculty of State University of Jakarta No: 461.a/SP/2018 dated 23rd May 2018.

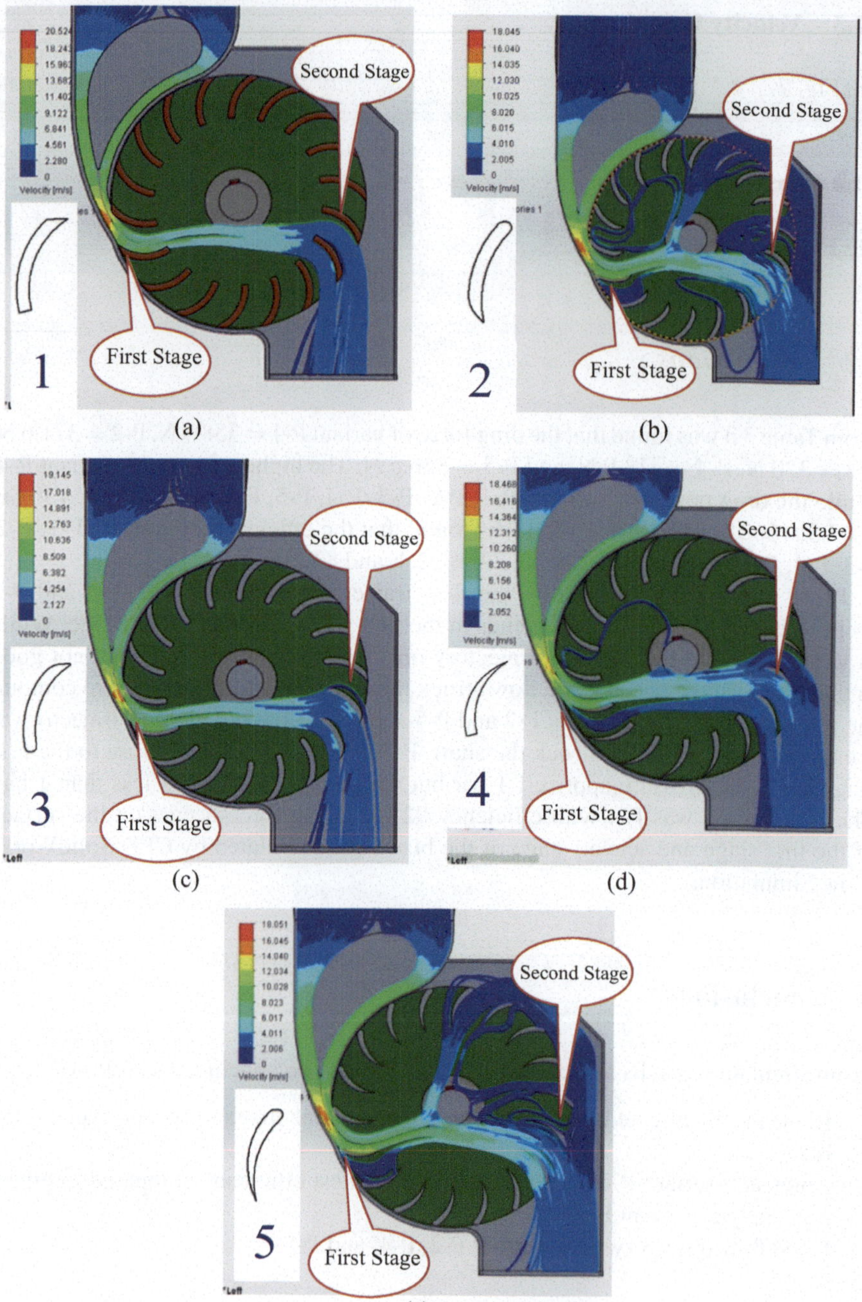

Fig. 7 Velocity contour. **a** Variant P-1. **b** Variant P-2. **c** Variant P-3. **d** Variant P-4. **e** Variant P-5

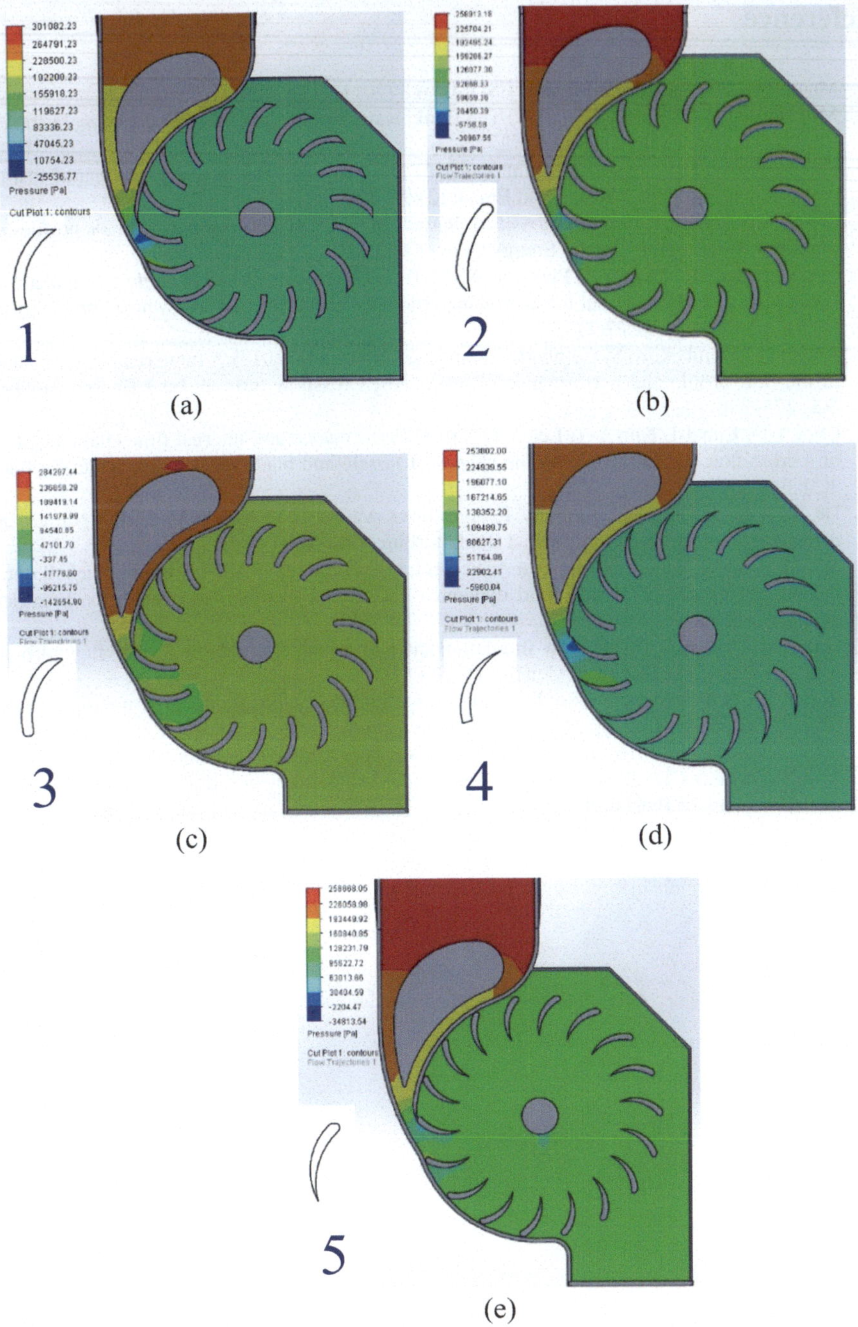

Fig. 8 Pressure contour. **a** Variant P-1. **b** Variant P-2. **c** Variant P-3. **d** Variant P-4. **e** Variant P-5

References

1. Menteri ESDM: Peraturan Menteri ESDM No.8 (2011)
2. Mockmore CA, Merryfield F (1949) The Banki water turbine, engineering experiment station (of Oregon State College). Bull Ser 2
3. Zar Ni Tin W, Htay Htay W, Myint T (2016) Design, construction and performance test of cross-flow turbine. Int J Mech Prod Eng 4(12):95–100
4. Chattha JA, Zaffar A, Ibrahim B, Asif M, Sarwar MA (2018) Optimisation of blade profiles of cross flow turbine. Int J Power Energy Conv 9(4):311
5. Jasa L, Ardana IP, Priyadi A, Purnomo MH (2017) Investigate curvature angle of the blade of Banki's water turbine model for improving efficiency by means particle swarm optimization. Int J Renew Energy Res 7(1)
6. Kaunda Chiyembekezo S, Cuthbert Kimambo Z, Nielsen K (2014) A numerical investigation of the flow profile and performance of a low cost Crossflow turbine. Int J Energy Environ 5(3):275–296
7. Choi Y-D, Lim J-I, Kim Y-T, Lee Y-H (2008) Performance and internal flow characteristics of a cross-flow hydro turbine by the shapes of nozzle and blade runner. JFluid Sci Technol 3(3):398–409
8. De Andrade J, Curiel C, Kenyery F, Aguilln O, Vásquez A, Asuaje M (2011) Numerical investigation of the internal flow in a Banki turbine. Int J Rotat Mach, 1–12
9. Sammartano V, Arico C, Carravetta A, Fecarotta O, Tucciarelli T (2013) Banki-Michell optimal design by computational fluid dynamics testing and hydrodynamic analysis. Energies 6(5):2362–2385
10. Girma M, Dribssa E (2014) Flow simulation and performance prediction of cross flow turbine using CFD tool. Int J Eng Res Gen Sci 2(6):747–757
11. Rajput RK (1998) A textbook of hydraulics for engineering student of degree and diploma. S. Chand & Company Ltd., pp 312–316
12. Systemes Dassault (2016) Technical reference solidworks flow simulation
13. Scott-Pomerantz CD (2004) The K-Epsilon model in the theory of turbulance. Ph.D Dissertation. University of Pittsburgh

Stress Analysis of Thick-Walled Cylinder for Rocket Motor Case Under Pressure

Lasinta Ari Nendra Wibawa, Kuncoro Diharjo, Wijang Wisnu Raharjo, and Bagus H. Jihad

Abstract This paper investigated the stresses developed in a thick-walled cylinder for rocket motor case under internal pressure. Stress analysis used the finite element method with ANSYS software for rocket motor case selection. The simulation used thick-walled cylinders with wall thickness 6, 7, 8, 9, and 10 mm having 300 mm long and outer diameter of 122 mm. It used an internal pressure of 2, 4, 6, 8, and 10 MPa. The material variation in this research used Aluminium 6061, CFRP, and GFRP. Comparing the hoop and longitudinal stress values between analytics and simulations were for the validation process. The simulation results show that the thicker the cylinder wall, the von Mises stress decreases. Aluminium 6061 and CFRP have a safety factor greater than 1 for all wall thickness and internal pressure variations. GFRP has a safety factor greater than 1 for all internal pressure variations when the wall thickness of 10 mm.

Keywords Finite element method · Pressure vessel · Rocket motor case · Stress analysis · Thick-walled cylinder

1 Introduction

The rocket motor is an essential part of a rocket that contains fuels used to boost a missile [1]. It consists of a case, nozzle, cap, insulator, igniter, etc. It works using principles such as pressure vessels because it stores propellant fuel. Generally, rocket motor works in environments with high pressure and temperature [2].

The design of a rocket motor case depends on the internal pressure and material used. The rocket motor case that works at high internal pressure generally uses a thick-walled cylinder, while at low internal pressure uses a thin-walled. Materials

L. A. N. Wibawa · K. Diharjo (✉) · W. W. Raharjo
Department of Mechanical Engineering, Faculty of Engineering, Sebelas Maret University, Surakarta, Indonesia
e-mail: kuncorodiharjo@ft.uns.ac.id

B. H. Jihad
National Institute of Aeronautics and Space (LAPAN), Pulo Gadung, Indonesia

U. Sabino et al. (eds.), *Proceedings of the 6th International Conference and Exhibition on Sustainable Energy and Advanced Materials*, Lecture Notes in Mechanical Engineering,
https://doi.org/10.1007/978-981-15-4481-1_49

have high yield strength typically use a thin-walled cylinder, while materials have low yield strength use a thick-walled.

The design of the rocket motor case uses a cylindrical theory to calculate the effect of the thickness of the case on the von Mises stress under internal pressure. Stress analysis used the finite element method with ANSYS software for rocket motor case selection. The simulation used thick-walled cylinders with wall thickness 6, 7, 8, 9, and 10 mm having 300 mm long and outer diameter of 122 mm. It used an internal pressure of 2, 4, 6, 8, and 10 MPa.

The finite element analysis is numerical mathematical techniques for calculating the strength and structural behavior of engineering components with dividing objects into mesh shapes. This simulation technique allows every design and product to be analyzed in great detail. It is widely used to measure stress and stress concentration in a thick-walled and thin-walled cylinder [3].

The goals of this investigation are to predict hoop and longitudinal stresses in thick-walled cylinders, von Mises stress distribution over the thickness of the thick-walled cylinder with closed ends, and select of materials which suite to these cases.

2 Stress in Thick-Walled Cylinder

The thin-walled and thick-walled are classifications of pressure vessels based on their dimensions. The thin-walled pressure vessel in which the wall thickness (t) is less than 1/20 of the internal diameter (d). On the other hand, the thick-walled in which the wall thickness (t) is more significant than 1/20 of it (d).

The rocket motor case in this study is assumed to be a thick-walled cylinder. In the case of a thick-walled, as shown in Fig. 1a, stress on parts of the wall cannot be assumed to be evenly distributed [4]. The tangential and radial stresses develop by considering values that depend on the radius of the element. Figure 1b, c show the stress distributions on thick-walled cylinders.

Theory of Elasticity is the primary theory for the thick-walled cylinder, which produces a stress condition as a continuous function of the radius above the wall of

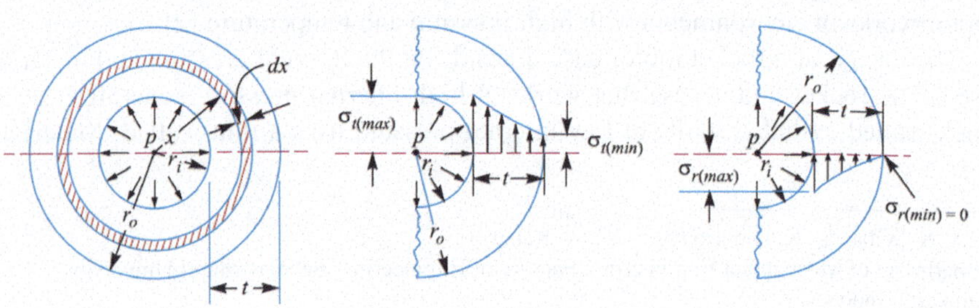

a. Thick-walled cylinder b.Tangential stress distribution c. Radial stress distribution

Fig. 1 Stress distribution in thick-walled cylinder subjected to internal pressure [4]

the pressure vessel [5]. The stress in a cylindrical pressure vessel depends on the ratio of the inner radius to the outer radius (r_o/r_i) rather than cylinder size.

The thick-walled cylinder pressure will occur in three directions, namely hoop (circumferential or tangential), longitudinal (axial), and radial. The maximum stress equations in thick-walled cylinders in the hoop, longitudinal, and radial directions are:

$$(\sigma_h)_{max} = p_i\left(\frac{r_o^2 + r_i^2}{r_o^2 - r_i^2}\right) \quad (1)$$

$$(\sigma_l)_{max} = p_i\left(\frac{r_i^2}{r_o^2 - r_i^2}\right) \quad (2)$$

$$(\sigma_r)_{max} = -p_i \quad (3)$$

where:

p_i = internal pressure
r_o = outer radius of cylinder
r_i = inner radius of cylinder

3 Finite Element Analysis

Figure 2 shows the model geometry of the cylinder. Since the geometry exhibits asymmetry in the axial (Y), then it can model only the top half of it [6]. The finite element analysis used a 90-degree segment of the solid model of the cylinder. The symmetric nature of the geometry and loading means that displacements are zero in

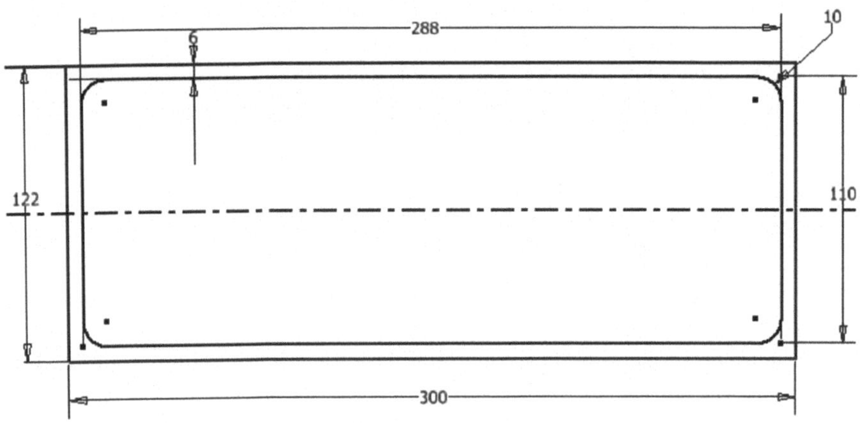

Fig. 2 Design of thick-walled cylinder with a wall thickness of 6 mm

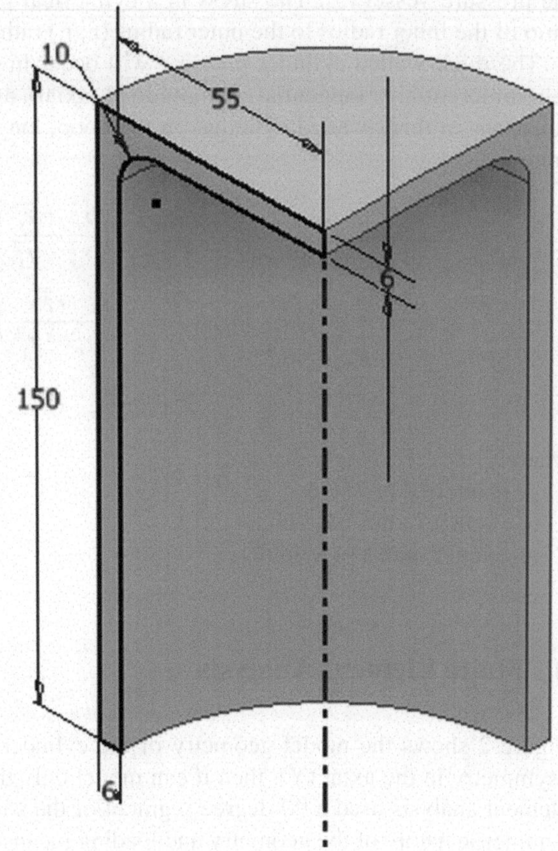

Fig. 3 The one-eight segment of the cylinder with a wall thickness of 6 mm

directions normal to the face exposed by the vertical and horizontal cuts employed create this one-eight segment of the cylinder (Fig. 3) [7].

The following assumptions are made theoretically on thick-walled cylinders:

- Aluminium 6061 is homogeneous and isotropic. CFRP and GFRP are linearly elastic, initially free of stress and defect. The fiber and matrix are homogenous, linearly elastic, and isotropic.
- The cylinder section remains flat even after applying internal pressure.
- All cylinder surface will expand or contract independently.

The boundary conditions of the finite element method using ANSYS software are shown in Fig. 4 and input parameters in Table 1:

The criteria for rocket motor case material are lightweight and high strength. Aluminium 6061, Carbon Fiber Reinforced Polymer (CFRP), and Glass Fiber Reinforced Polymer (GFRP) are chosen for it [8–11]. The material properties of Aluminium 6061, CFRP, and GFRP are shown in Table 2. The data of mechanical properties of materials using Autodesk Material Library [12].

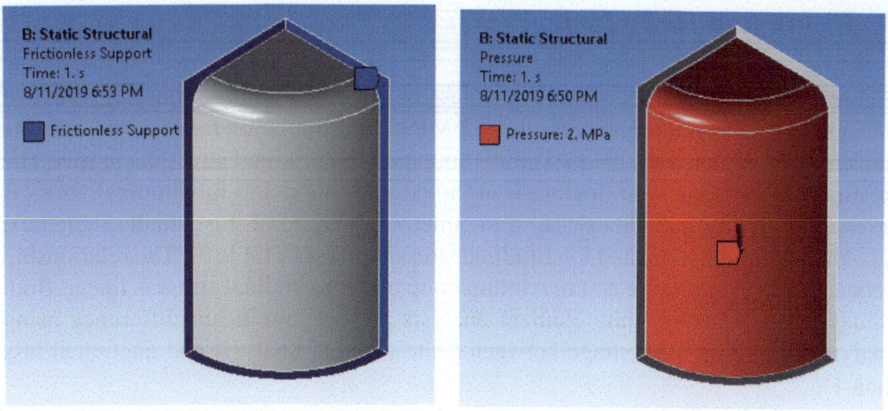

Fig. 4 The boundary conditions of the finite element method using ANSYS software

Table 1 Input parameters

Parameters	Symbol	Value
Length of cylinder	L	300 mm
Outer diameter of cylinder	D_o	122 mm
Wall thickness	t	6, 7, 8, 9, and 10 mm
Inner radius of cylinder	r_i	55, 54, 53, 52, and 51 mm
Internal pressure	P_i	2, 4, 6, 8, and 10 MPa
Element size	–	3 mm
Number of nodes	–	25,194, 23,872, 27,096, 26,341, 26,926
Number of elements	–	13,867, 13,041, 15,321, 14,952, 15,343

Table 2 Mechanical properties of materials

Material	Density (gr/cm^3)	Young modulus (GPa)	Tensile strength (MPa)
Aluminium 6061	2.70	68.9	400
CFRP	1.43	133	577
GFRP	1.75	13.9	194

4 Result and Discussion

Figure 5 shows the effect of wall thickness on the hoop and longitudinal stress of the cylinder with an internal pressure of 2 MPa. The simulation results show that the thicker the cylinder wall, the maximum hoop and longitudinal stress decreases. The relationship between wall thickness and maximum hoop and longitudinal stress is inversely proportional. Otherwise, if the internal pressure of the cylinder increases, then the maximum hoop and longitudinal stress increases (Table 3). The relationship between internal pressure and maximum hoop and longitudinal stress is linear. Both values of stress using finite element analysis are not significant difference using analytic—the error percentage between finite element analysis and analytical less than 1%.

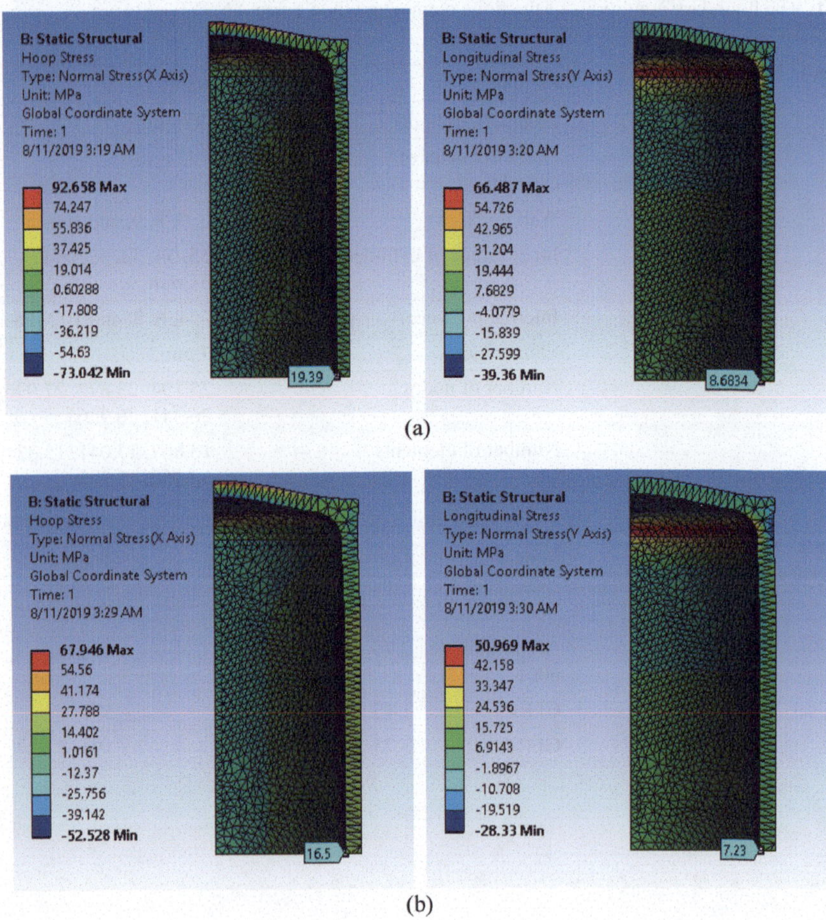

Fig. 5 Hoop and longitudinal stress of cylinder with wall thickness difference (**a**) 6 mm, (**b**) 7 mm, (**c**) 8 mm, (**d**) 9 mm, and (**e**) 10 mm with internal pressure of 2 MPa

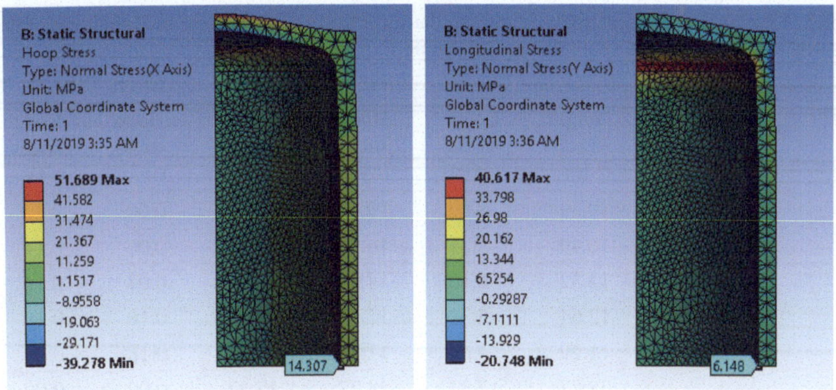

(c)

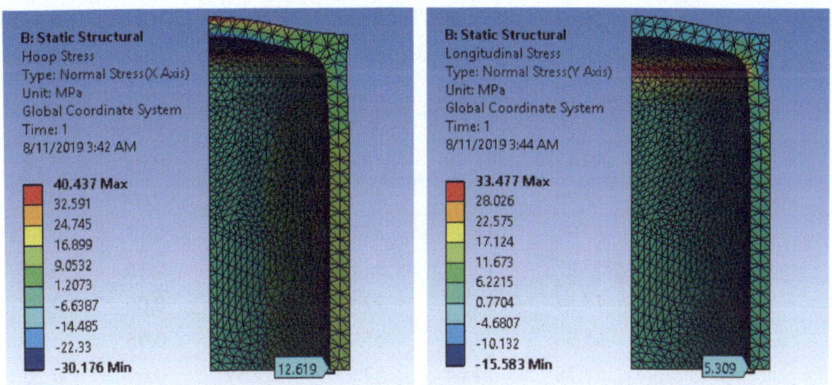

(d)

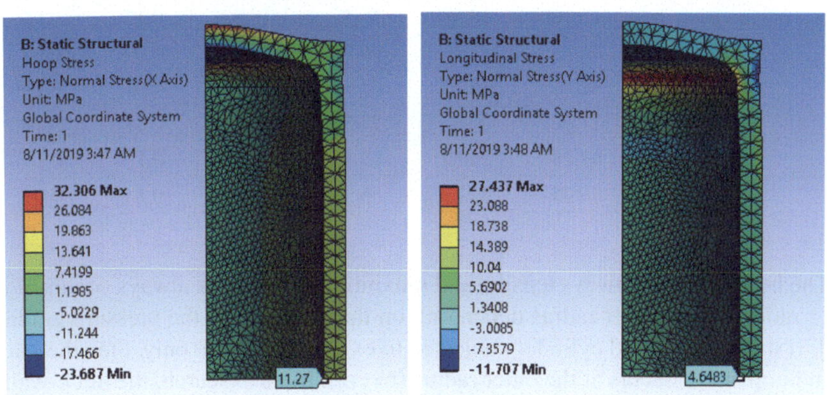

(e)

Fig. 5 (continued)

Table 3 Comparison between analytic and simulation of the maximum hoop and longitudinal stress with wall thickness and internal pressure variations

Thickness (mm)	Pressure (MPa)	Analytic		FEA		Error	
		Hoop stress (MPa)	Long. stress (MPa)	Hoop stress (MPa)	Long. stress (MPa)	Hoop stress (%)	Long. stress (%)
6	2	19.39	8.69	19.39	8.68	0.00	0.12
7	2	16.49	7.24	16.5	7.23	0.06	0.14
8	2	14.32	6.16	14.31	6.15	0.07	0.16
9	2	12.64	5.32	12.62	5.31	0.16	0.19
10	2	11.29	4.64	11.27	4.65	0.18	0.22
6	4	38.77	17.37	38.77	17.37	0.00	0.00
7	4	32.98	14.49	32.99	14.46	0.03	0.21
8	4	28.64	12.32	28.62	12.3	0.07	0.16
9	4	25.27	10.64	25.24	10.62	0.12	0.19
10	4	22.58	9.29	22.55	9.30	0.13	0.11
6	6	58.16	26.08	58.16	26.05	0.00	0.12
7	6	49.47	21.73	49.49	21.69	0.04	0.18
8	6	42.96	18.48	42.92	18.44	0.09	0.22
9	6	37.91	15.95	37.86	15.93	0.13	0.13
10	6	33.87	13.93	33.82	13.95	0.15	0.14
6	8	77.54	34.77	77.54	34.73	0.00	0.12
7	8	65.96	28.98	65.99	28.92	0.05	0.21
8	8	57.28	24.64	57.23	24.59	0.09	0.20
9	8	50.54	21.27	50.48	21.24	0.12	0.14
10	8	45.16	18.58	45.1	18.59	0.13	0.05
6	10	96.93	43.46	96.93	43.42	0.00	0.09
7	10	82.45	36.22	82.49	36.15	0.05	0.19
8	10	71.6	30.8	71.54	30.74	0.08	0.19
9	10	63.18	26.59	63.1	26.55	0.13	0.15
10	10	56.45	23.22	56.37	23.24	0.14	0.09

The hoop stress is always tensile, and maximum hoop stress always occurs at the inner radius or the outer radius depending on the direction of the pressure gradient [13]. If the thick-walled cylinder subjected to external pressure only, then the maximum hoop stress occurs at the outer radius ($r = r_o$). In the research, the thick-walled cylinder subjected to internal pressure only, then the maximum hoop stress occurs at the inner radius ($r = r_i$).

When the cylinder has closed ends the internal pressure acts on them to develop a force along the axis of the cylinder. It is called the longitudinal stress and is usually less than the hoop stress. In the research, the hoop stress higher than longitudinal stress. The longitudinal stress values are about 4144% than the hoop stress.

Aluminium 6061, Carbon Fiber Reinforced Polymer (CFRP), and Glass Fiber Reinforced Polymer (GFRP) can withstand maximum hoop and longitudinal stress. The three materials have a tensile strength exceeding the two maximum stresses (hoop and longitudinal). It means the materials can be used as rocket material case without end caps.

One of the theories of failure is based on maximum distortion energy, which is known as the von Mises yield criterion. The von Mises stress is also known as the equivalent stress. The von Mises stress counts what combination of stress at a certain point will cause failure. In other words, it says that the material will fail when the equivalent stress exceeds the yield strength of the material [14, 15].

Figure 6 shows the maximum von Mises stress of cylinder for wall thickness of 6 mm with an internal pressure difference. The simulation results show that the thicker the cylinder wall, the von Mises stress decreases (Table 4). The relationship between wall thickness and von Mises stress is inversely proportional. Otherwise, if the internal pressure of the cylinder increases, then the maximum von Mises stress increases (Table 4). The relationship between internal pressure and maximum von Mises stress is linear.

The maximum von Mises stress always occurs at the outer radius of the end cap. It means the outer radius of the end cap is a critical area. The maximum von Mises stress can be decreased with increasing the end cap thickness, although without increasing the wall thickness. In other words, the method is more efficient because the material is lighter than increase overall cylinder thickness. It can also be done by combining different materials. For example, the tube material is made of GFRP while the end cap material is made of Aluminum 6061.

The safety factor is used to perform the evaluation process so that the components or structures design is guaranteed safe even though they used the minimum dimensions [16]. It is one indicator of the strength of an element. Parts are said to be safe when the maximum von Mises stress is lower than the strength of the material. In other words, the component is safe when the safety factor value is more than 1.

The value of the safety factor for each material shown in Fig. 7. Aluminium 6061 and CFRP have a safety factor greater than 1 for all wall thickness and internal pressure variations. GFRP has a safety factor greater than 1 when the wall thickness of 8, 9, and 10 mm at an internal pressure of 6 MPa. At the internal pressure of 8 MPa, GFRP has a safety factor greater than 1 when the wall thickness of 9 and 10 mm. At the internal pressure of 10 MPa, GFRP has a safety factor greater than 1 when the wall thickness of 10 mm. In other words, GFRP has a safety factor greater than 1 for all internal pressure variations when the wall thickness of 10 mm.

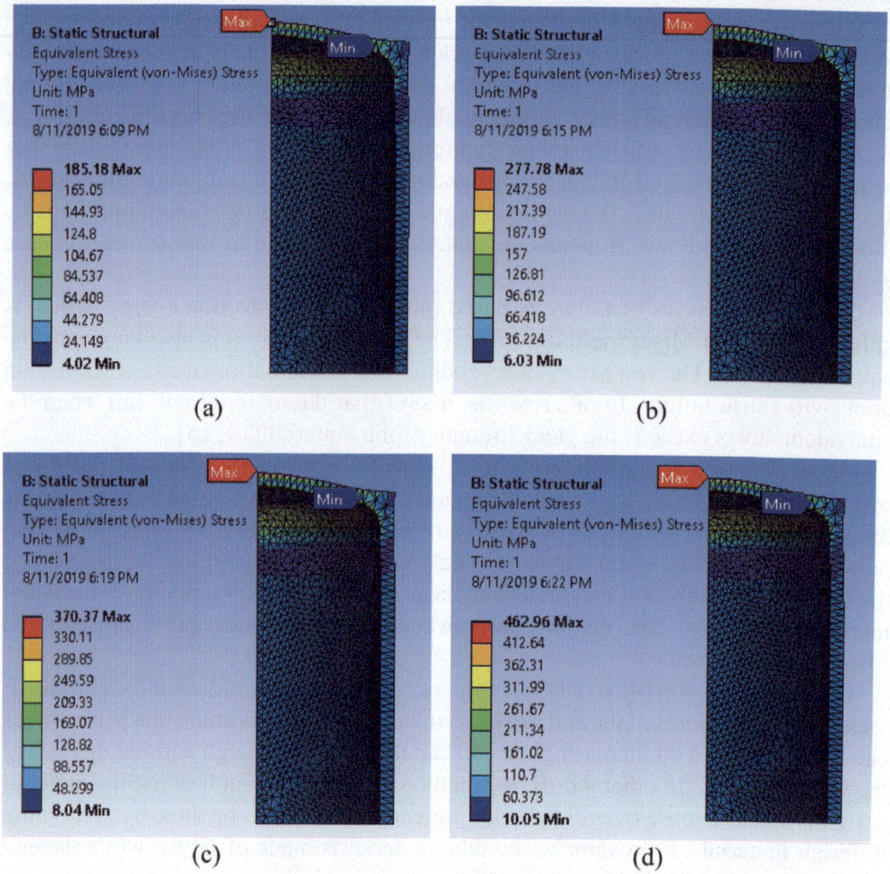

Fig. 6 The maximum von Mises stress of cylinder for wall thickness of 6 mm with an internal pressure of (**a**) 4 MPa, (**b**) 6 MPa, (**c**) 8 MPa, and (**d**) 10 MPa

5 Conclusion

The simulation results show that the thicker the cylinder wall, the maximum hoop and longitudinal stress decreases. The validation of analytical results is done with finite element analysis for the maximum hoop and longitudinal stress with fixed boundary condition. The results showed the error percentage of less than 1 percent.

Von Mises is chosen as a failure criterion. The simulation results show that the thicker the cylinder wall, the von Mises stress decreases. Aluminium 6061 and CFRP have a safety factor greater than 1 for all wall thickness and internal pressure variations. GFRP has a safety factor greater than 1 for all internal pressure variations when the wall thickness of 10 mm.

Table 4 The value of maximum von Mises stress

Thickness (mm)	Pressure (MPa)	Von Mises stress (MPa)
6	2	92.59
7	2	67.93
8	2	51.72
9	2	40.47
10	2	32.33
6	4	185.18
7	4	135.86
8	4	103.45
9	4	80.93
10	4	64.66
6	6	277.78
7	6	203.79
8	6	155.17
9	6	121.40
10	6	96.99
6	8	370.37
7	8	271.72
8	8	206.90
9	8	161.86
10	8	129.32
6	10	462.96
7	10	339.65
8	10	258.62
9	10	202.33
10	10	161.65

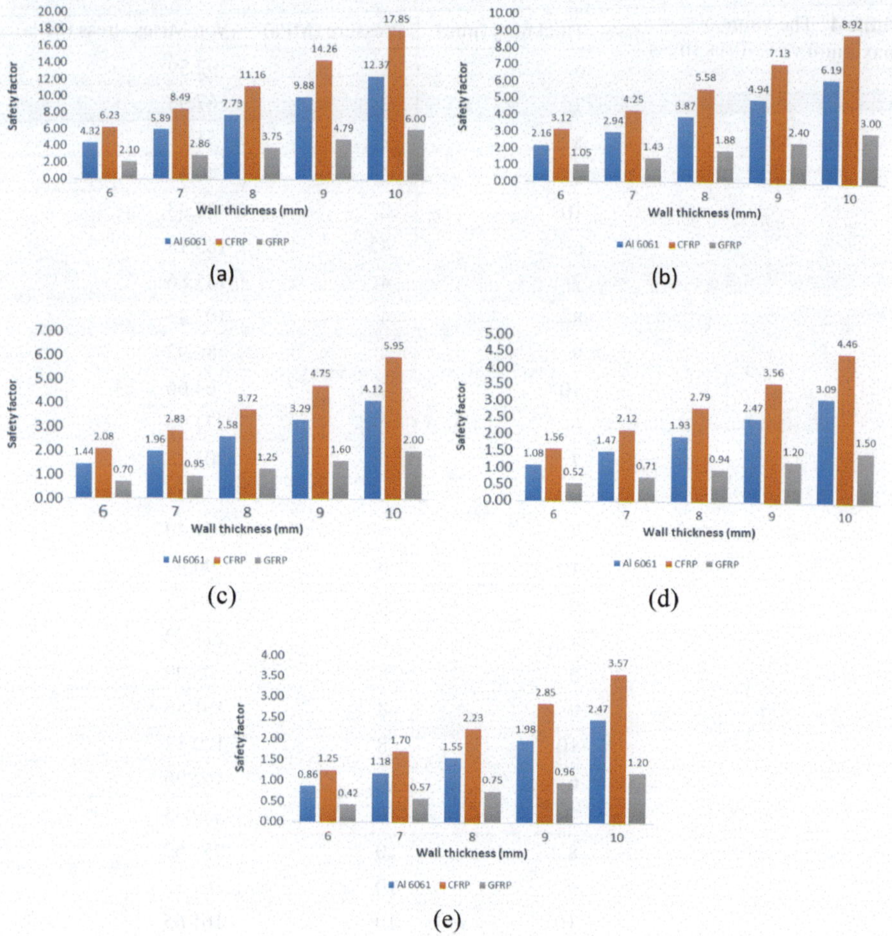

Fig. 7 Effect of the wall thickness on safety factor using different materials with internal pressure of (**a**) 2 MPa, (**b**) 4 MPa, (**c**) 6 MPa, (**d**) 8 MPa, and (**e**) 10 MPa

Acknowledgements The author would thank to the Indonesian National Institute of Aeronautics and Space (LAPAN), especially the Bureau of Human Resources, Organization, and Law for financial research support by LAPAN Scholarship.

References

1. Wibawa LAN (2018) Merancang Komponen Roket 3D dengan Autodesk Inventor Professional 2017. Buku Katta
2. Emrich W (2016) Rocket engine fundamentals In: Principles of nuclear rocket propulsion, pp 11–20

3. Mohamed AF (2018) Finite element analysis for stresses in thin-walled pressurized steel cylinders. Int J Sci Eng Res 9(3):201–204
4. Khurmi RS, Gupta JK (2005) A textbook of machine design, 14th ed. (no. I). Eurasia Publishing House, New Delhi
5. Masikh QS, Tariq M, Sinha PK (2014) Analysis of a thin and thick walled pressure vessel for different materials. Int J Mech Eng Technol 5(10):9–19
6. Dadkhah F, Zecher J (2008) ANSYS workbench software tutorial with multimedia CD release 11. Schroff Development Corporation
7. Lawrence KL (2012) Ansys workbench tutorial release 14. SDC Publications
8. Diharjo K et al (2013) Study of flexural strength on hybrid composite of glass/carbon-bisphenol for developing car body of electrical vehicle. In: Proceedings of the 2013 joint international conference on rural information and communication technology and electric-vehicle technology, rICT and ICEV-T 2013, pp 6–9
9. Zainudin M, Diharjo K, Kaavessina M, Setyanto D (2018) The properties degradation of exposed GFRP roof. In: AIP conference proceedings, vol 1931, pp 1–8
10. Diharjo K, Susilo DD, Sudargo PH, Kaleg S (2018) Vibration-damping factor of glass/kenaf/polyester hybrid composite. Key Eng Mater, 772 KEM:38–42
11. Kaleg S, Ariawan D, Diharjo K (2018) The flexural strength of glass fiber reinforced polyester filled with aluminum tri-hydroxide and montmorillonite. Key Eng Mater 772 KEM: 28–32
12. What is the autodesk material library? 2019. (Online). Available: http://help.autodesk.com/cloudhelp/2014/ENU/MoldflowAdvisor/files/GUID-20607781-D7C1-460E-A025-0F2543B3A4FE.htm. Accessed 13 Nov 2019
13. Skinner L (2018) Snubbing Theory and Calculations. In: Hydraulic rig technology and operations. Gulf Professional Publishing, pp 189–275
14. Raju G, Babu KH, Siva Nagaraju N, Kiran K (2015) Design and analysis of stress on thick walled cylinder with and without holes. J Eng Res Appl 5(1):75–83
15. Pasha SM, Satish JTV, Ramakrishna G (2019) Design and analysis of thick walled cylinder with holes. Int J Sci Res Rev 07(03):1166–1172
16. Wibawa LAN (2018) Simulasi Kekuatan Komponen Sarana Pengujian Roket Menggunakan autodesk inventor professional 2017. Buku Katta

Ankle Foot Orthotic (AFO) for Deformity Patients: The Design and Manufacturing of Shoes Orthotics

P. W. Anggoro, B. Bawono, T. Yuniarto, J. Jamari, and A. P. Bayuseno

Abstract A foot is a part of the human body that is crucial and functions when doing an activity of standing, walking, jumping, or running. When the body is on the move, a good footwear that is required to support weight is shoes. The design and manufacturing of shoes has currently experienced a very rapid development, to meet the needs of users. However, some people have an abnormal foot deformity so it tends to always get pain or uncomfortable while wearing shoes. These people obviously need special shoes with ankle foot orthosis (AFO). Research employing AFO reverse innovative design (RID) of the phase leg scanning, design, and manufacturing of shoes that fit the contour of the patient's foot is still rare in Indonesia. This process takes approximately 34–45 working days in the making of AFO with the traditional concept by simply following specific instructions from the physician or orthopedic, handmade, not precision, and comfortable use. The application of the Computer-Aided Reverse Engineering System (CARE System) is used to assist researchers to design and manufacture AFO which fits with the contour of the patient's foot. Curve base surface modeling (CBS-modeling) in RID methods was applied to obtain the optimal design of the insole and shoe last. The research output of shoe orthotics (insole, shoe lasts, and shoes) have been used and enjoyed by patients with diabetes very well. This technology is significantly more efficient than the manual process. The manufacturing process was shortened up to 64%.

Keywords AFO · CARE system · Base curve surface modeling · Foot deformities · RID

P. W. Anggoro (✉) · B. Bawono · T. Yuniarto
Department of Industrial Engineering, Faculty of Industrial Technology, University of Atma Jaya Yogyakarta, Jl. Babarsari 44, Yogyakarta 55281, Indonesia
e-mail: pauluswisnuanggoro@ymail.com

J. Jamari · A. P. Bayuseno
Department of Mechanical Engineering, University of Diponegoro, Jl. Prof. Soedarto, SH. Tembalang, Semarang 50275, Indonesia

U. Sabino et al. (eds.), *Proceedings of the 6th International Conference and Exhibition on Sustainable Energy and Advanced Materials*, Lecture Notes in Mechanical Engineering,
https://doi.org/10.1007/978-981-15-4481-1_50

1 Introduction

A foot is part of the human body that serves to do a lot of activities, such as standing, walking, jumping, and running. When the body is on the move, a good footwear that is required to support the weight shoes. The design and manufacturing of footwear have developed very fast and rapidly. However, some people have a foot deformity that is not normal (deformity foot) as reported by Refs. [1–3]. Foot deformity can be classified as "significant or not" according to the condition of the patient's leg. Prior research (…)has suggested that significant deformity is closely related to the alignment of the foot mechanics related to aspects of normal shoes, namely heel, ball, and toe. If the conditions of the foot are not aligned or aligned with all of these aspects, the shoe design parameters will potentially cause friction, shear forces. Furthermore, due to such pressure on the foot, the user will feel pain or discomfort. This is what causes the foot to have a significant deformity. Foot deformities, especially in diabetic patients, can be shown in Fig. 1.

People with such disorders (see Fig. 1) tend to experience pain when using the shoes as footwear when moving. Shoes commonly do not last long because the friction between the foot with the shoe upper part tears the shoe material. Problems to find the right shape and size of the shoe that fits the dimensions of the patient's legs are also a major factor. In Indonesia, it is not easy to find costumized shoes for people with a foot deformity due to the lengthy duration of the shoe production. (AFO). Therefore, there is an urgent need to produce orthotic shoe, known as the

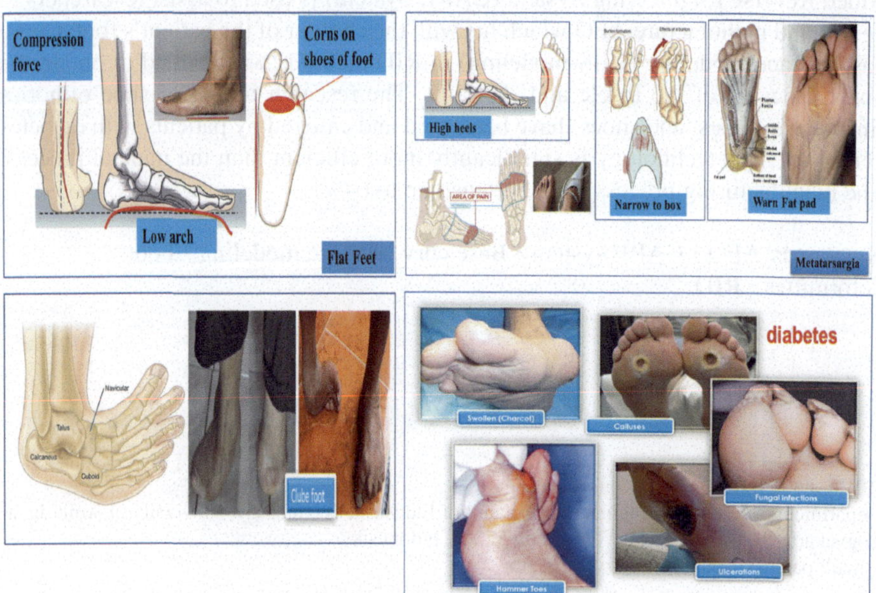

Fig. 1 Foot deformity in patients with foot deformity

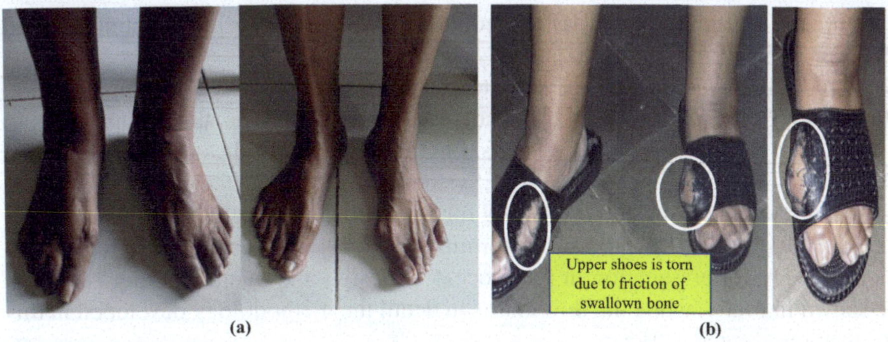

Fig. 2 Foot diabetic patients **a** Feet diabetic patients with high risk scale. **b** Shoe upper tear due to friction with the swollen bone

ankle-foot orthosis (AFO). This type of shoe is designed specifically for people who have foot deformities, including patients with diabetes, club foot, flat foot, and high heel injury.

This research was conducted to address some complaints on prolonged pain, difficulty in choosing comfortable shoes, and the appearance of the second leg swollen bone of patients in Fig. 2a. This led to a short lifespan of shoes because the shoes are easily torn at upper section Fig. 2b. The number of AFO shoe-last makers and orthotic footwear industry in Ganjuran Bantul, Yogyakarta is limited. Most of the industries are still based on manual procedures, experiences, and follow the specific instructions given by a doctor or orthopedic. Consequently, the products may not be precise and inconvenient for the users. Moreover, this process requires a considerable amount of time. Normally, it takes around 35–45 working days to complete the process; from the diagnosis stage through the fabrication of AFO shoes.

AFO shoe is an equipment for footwear designed specifically to support the ankle and foot section during daily activities and help the healing process or the reduction of pain in patients with deformity during the rehabilitation process, as has been reported by Refs. [4]. Some researchers such as [4–6] introduced the methods of design and production of customized shoe insole orthotics utilizing digital technology surface capture, CAD/CAM, CNC, and adaptive manufacturing, but all these studies are still too broad and not integrated from the design phase through the manufacturing process and testing the product in patients. Some good research about orthotic shoes by Refs. [3, 7–13] in research explains that the manufacturing process AFO generally always use the technology of reverse engineering (RE), both conventional and non-conventional phases of the methodology. The orthotic shoes in this paper, according to [10], are included in the category of organic shapes, which refer to a complex foot relief with different contours for each foot. Therefore, both feet should be scanned in detail and separately.

That is why Reverse Innovative Design (RID) is the most appropriate RE procedure conducted in this research. The RID application can be done very well in an industrial or laboratory system when the system has a set of technologies that support the process. Automation of RID according to [14] is often called Computer Aided Reverse Engineering System (CARESystem).

In Indonesia, research that elaborated an integrated process of producing AFO shoes with CARESystem, which include design, manufacture, and product tests to users, is not much, as reported by Refs. [4, 15]. Previous studies demonstrated that the use of modern technology CARESystem significantly accelerated the design time and manufacturing of AFO shoes for patients with diabetes mellitus orthotics [3, 7–9, 11–13, 16–18] and other deformities. The detailed stages of CARESystem methodology for application in the process of design and manufacturing AFO can be shown in Fig. 4, while the flow charts in detail the methodology developed in this study are presented in Fig. 5.

In Indonesia, the use of commercial software to make shoe lasts is still uncommon. Mostly, shoe lasts and the prototypes for academic purposes are made with conventional technology or hand made. This study successfully demonstrated the product development of AFO (insole, outsole, shoe last and orthotic shoes) that are based on modern technology integrating 3D HandySCANN 700TM, PowerSHAPE 2016 CAD, CAM PowerMILL 2016 and 1020EV20A YCM CNC 3 axis. All of the existing infrastructures in this research are fully integrated with other tools during the study (Fig. 4a). This process has enabled to shift away from the manual production process of AFO shoes to the innovative production process that produces more precise and accurate shoes for the patient's feet. The stages used in this study to design orthotic shoes are described in detail are presented in Fig. 3.

RID-based AFO development [10] in this research was used to simplify the stages (see Fig. 3). and followed for the development of research in the years 2019–2022 in patients with clap foot, flat feet and diabetes due to amputation.

2 Context Problems and Methodology

This research was conducted to respond to the need for the development stages of design, manufacturing and fabrication AFO shoes on foot deformity patients, especially in patients with diabetes mellitus using modern technology in the field orthotic CARESystem. This technology is very necessary to lessen the design and manufacturing time of AFO orthotic shoes. AFO footwear product consists of four parts, namely: insole, outsole, shoe last and shoes upper. Those parts are shown in Fig. 4. The assembly of thesecomponents will form customized AFO shoes that can solve the concerns of patients and related industries.

There were two diabetic patients, aged between 55–74 years old and weight between 60–70 kgs were selected as research subjects. These two patients had diabetes for more than 10 years and according to previous research [1, 2] both were included in the "high risk" scale category.

The purpose of this study is to produce customized orthotic shoe lasts for patients with high-risk diabetes mellitus using CARESystem (Fig. 4a). This modern technology, which included a set of 3D scanners, CAD, CAE, CAM, CNC, and RP was employed to design and manufacture the orthotic shoes.

There are two types of AFO shoes that were produced in this study, namely: AFO shoes type 1 and type 2. Type 1 is orthotic shoes with hand-made shoe lasts (Figs. 3b and 4b). The second type is shoes with AFO shoe last made with a CNC machine (Figs. 3c and 4c).

This project began with scanning the patient's foot using the Handyscan 3D image 700 to obtain the insole and outsole model in files with IGES format (Fig. 5a). Then, the scanning results were processed usingthe curve based surface modeling, as has been done successfully by Refs. [3, 7, 8, 10, 16, 17] (Figure 5b). The process of making this prototype conducted by researchers to verify the results of the size of the foot with the insole and shoe last that have been machined on CNC. The accuracy of each geometry has been done [7] with the result of an error on each foot measured dimension is less than 0.05 mm (Fig. 5c).

The process of manufacturing orthotic shoes insole on a CNC machine with EVA FOAM material as illustrated in Fig. 5d. Manufacturing output in the form insole with

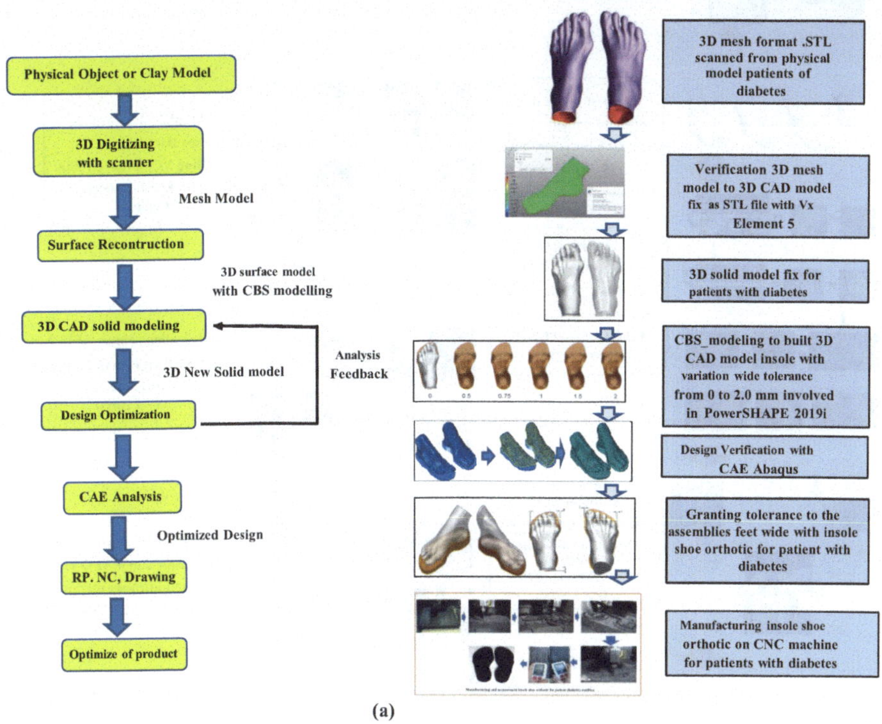

Fig. 3 Flowchart method CARESystem AFO RID technology in patients with diabetes mellitus **a** RID AFO. **b** Conventional methods AFO. **c** Methods of shoes CARESystem AFO

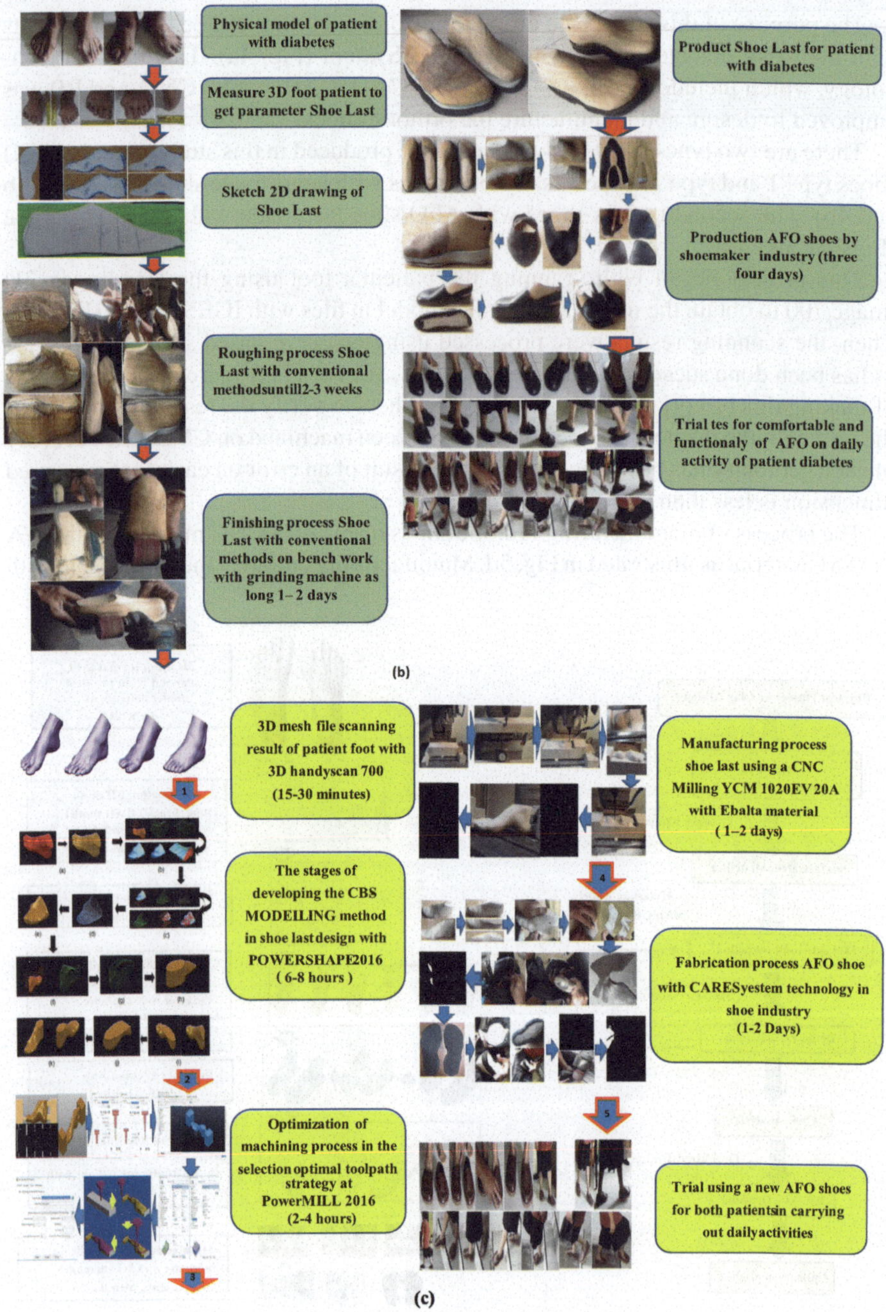

Fig. 3 (continued)

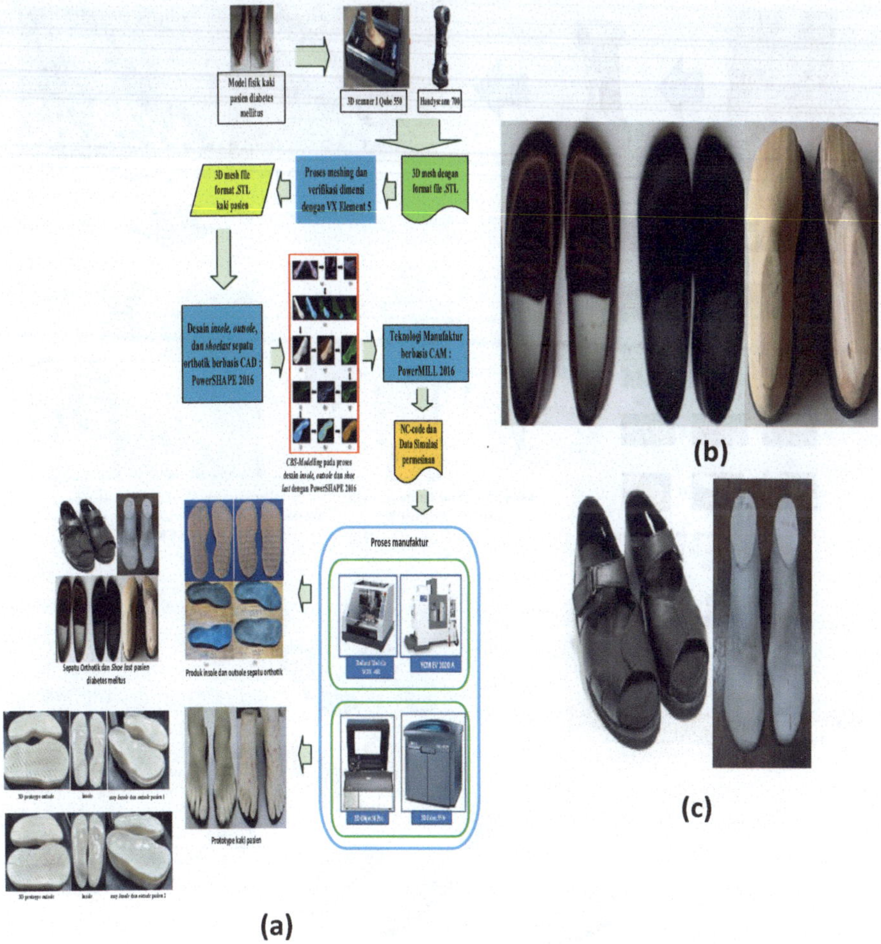

Fig. 4 **a** Architectural design and manufacturing shoes CARESystem based AFO. **b** Footwear AFO type 1. **c** Footwear AFO type 2

surface roughness up to the maximum set a maximum price of fewer than 10 μm. For the shoe last that was hand-made, it is a necessary outsole product made from Polyurethan (PU). This outsole was processed using CNC machine YCM 1020EV20a (Fig. 4a and Fig. 5e). After the shoe last footwear was completed, the manufacturing process was carried out on a custom AFO shoe industry in the area Ganjuran Bantul, Yogyakarta, Indonesia (Fig. 5f). The research procedures is described in detail in Fig. 5.

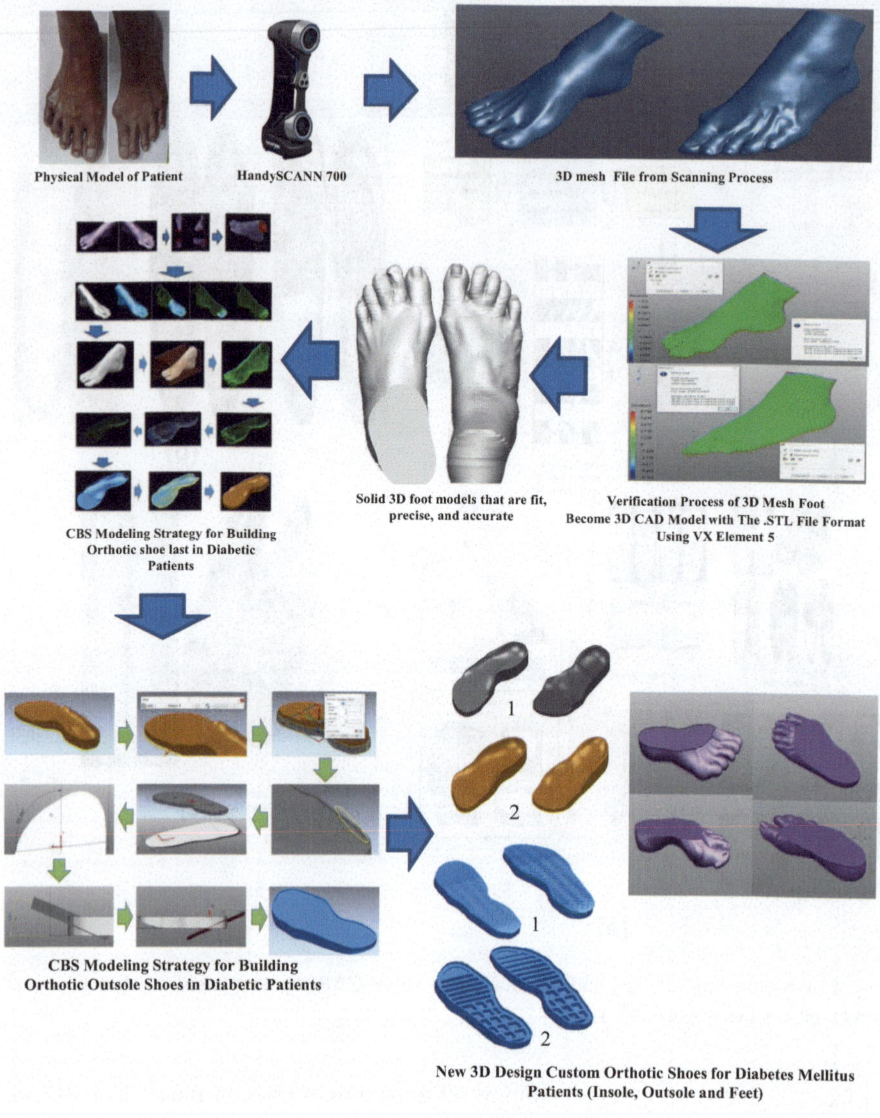

(a)

Fig. 5 Stages of methodology research **a** CBS-modeling to get a 3D CAD model of the insole, outsole, and shoe last with PowerSHAPE CAD. **b** Manufacturing prototype insole, outsole patient's legs and shoe orthotics using EDEN 350. **c** Verification of 3D CAD feet, insole, outsole and foot prototype 3D printer engine results. **d** Manufacturing orthotic shoe insole in CNC machine Rolland models MDX 40R. **e** Manufacturing and shoe last shoe outsole orthotic in CNC machine. **f** Fabrication AFO in shoe industry, Ganjuran, Bantul, Yogyakarta

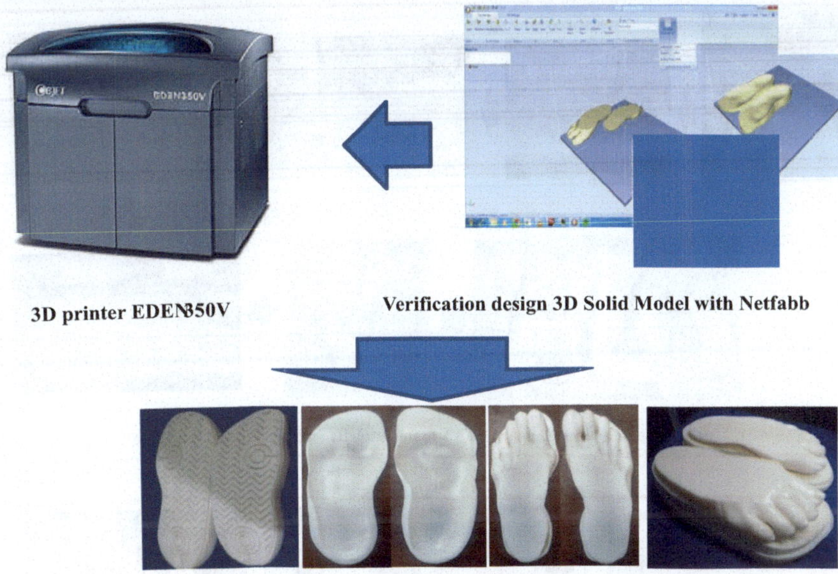

Prototype product of Orthotic Shoes with the assembly part

(b)

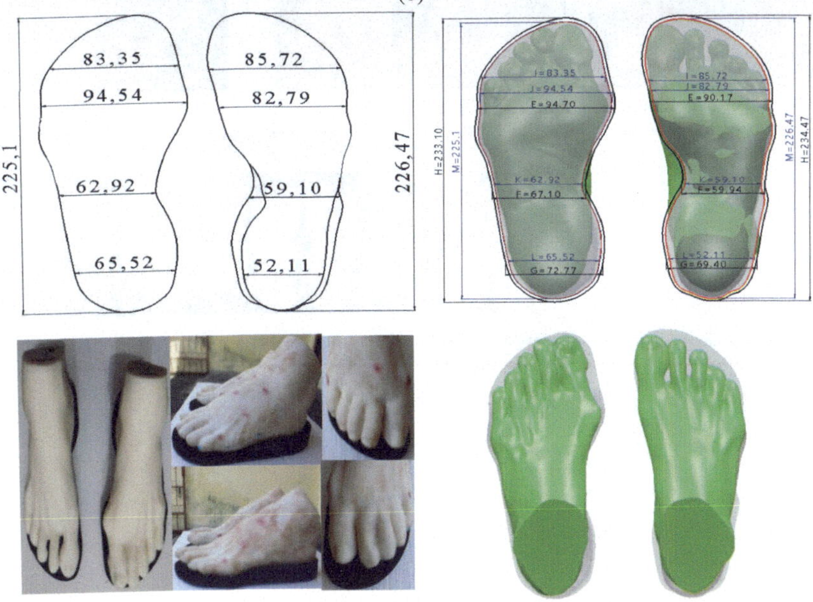

Prototype and 3D CAD Feet Models from 3D printer

(c)

Fig. 5 (continued)

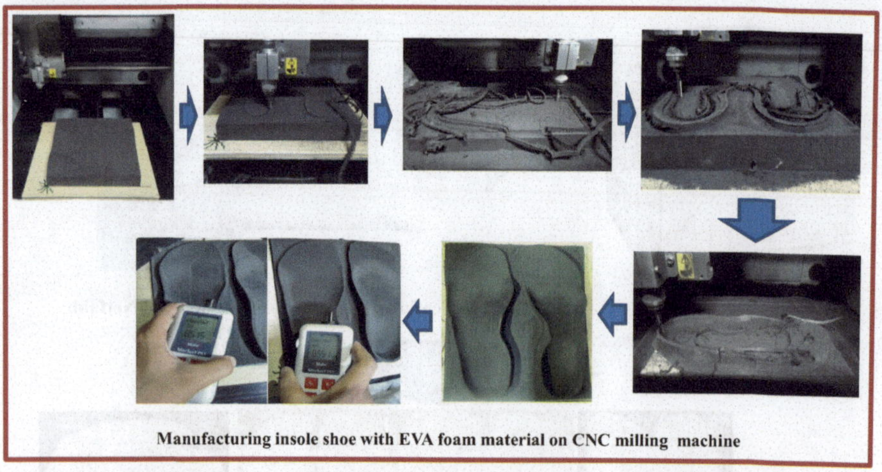

(d)

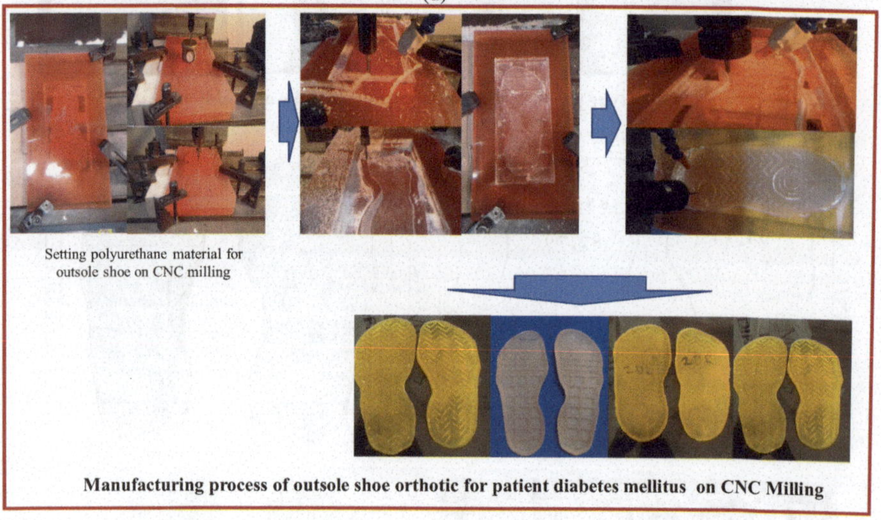

(e)

Fig. 5 (continued)

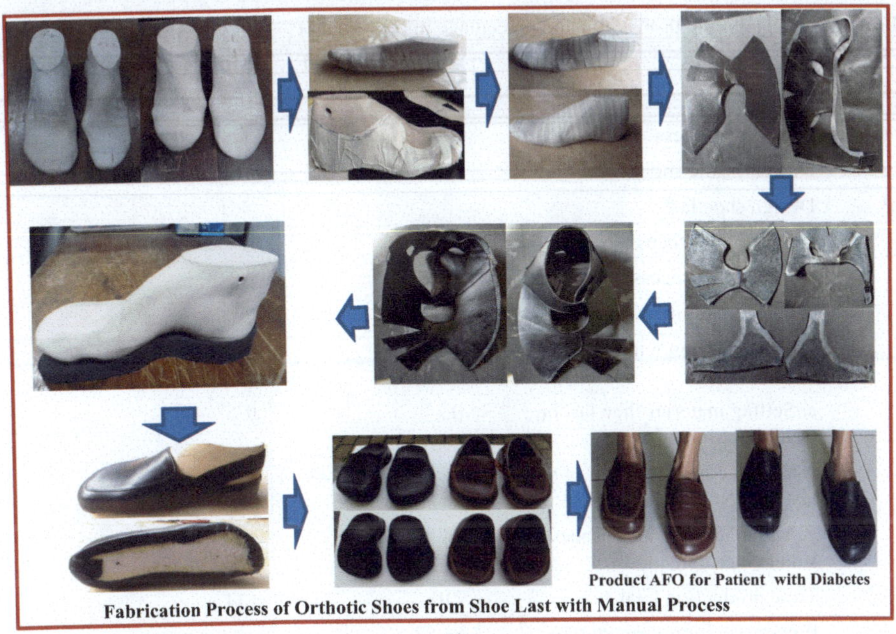

(f)

Fig. 5 (continued)

3 Result and Discussion

This paper describes the design, manufacturing and fabrication technology CARESystem of orthotic shoes (See Fig. 4). These shoes were specially made for two high-risk diabetic patients [1, 2] with foot deformities (see Fig. 1). Both patients were women. The condition of the patients' feet with their usual footwear is presented in Fig. 2. It can be seen that the footwear is rapidly destroyed by the growth of the patients' bones in both swollen feet. Consequently, the upper parts of the shoes can be easily torn and the shoes will not last long. For patients from the middle and lower economic class, this situation is very disturbing.

The reverse innovative design (RID) of shoe insoles and shoe lasts were performed successfully with the stages described in Fig. 3. The sequence has also been successfully undertaken by Refs. [7, 10, 19, 20]. The RID results demonstrated that the shoe insoles designed and manufactured on CNC machines were really precise, accurate and comfortable to be used by both patients. The error measurement results between 3D CAD insole, prototype and product insole with the patient's feet are 0.2 mm [7, 8, 21].

Figure 3b shows the total time to make manual AFO shoe last is 210 h, while the AFO shoe processing timeusing the CARESystem technology is for 134 h. The detailed explanation of the work of the two types of shoes is presented in Table 1.

Tabel 1 Compare time production and percentage improvement (%) of AFO shoes

No	Description	Shoe last manual (hours)	Shoe last CARESystem (hours)
1	Scanning process	8	1
2	Design Insole shoes	2	18
3	Design shoe last		
	a. Roughing process shoe last	144	0
	b. Finishing process shoe last	32	0
	c. CBS modeling of shoe last	0	24
	d. Optimazation with CAM software	0	8
	e. Setting material shoe last on CNC machine	0	9
	f. Manufacturing process of shoe last on CNC machine	0	50
	g. Manufacturing AFO shoes on small enterprice industry	24	24
5	Time production total	210	134
6	Procentase Improvement (%)	63.57	

Table 1 indicates that there is a time reduction in the shoe manufacturing process using the AFO CARESystem by 63.57% or about 64% faster compared to manual methods. In addition to the lack of precision in terms of size, the hand-made production of larger shoe lasts, approximately 5–8 mm in size, from the physical model of the patient's feet took longer timenearly 3–5 weeks. This time will multiply if the deformity forms are more complicated, such as for patients with club foot, flat foot, or high heel injury. This current finding was consistent with previous research [5, 6, 8, 9].

The present AFO products have significantly cut down the production time. With the technology, the orthotic shoes can be completed 64% faster than the hand-made production. [9]. Both patients admitted that the geometry quality and comfort had no significant difference between the traditional and innovative processes. However, in terms of the model and shape of shoes, the two patients agreed that the innovative products with the modern technology produced orthotic shoes that met the patients' expectations. Therefore, CARE System that use of CAD technology (PowerSHAPE2016) to design development and manufacture orthotic shoes is definitely feasible and can be recommended for further processing in the shoe industry.

The application of CARE System technology as shown in Figs. 4 and 5 has succeeded in proving that the technology is fast, precise, and accurate in the stages of design, manufacturing and fabrication of AFO shoes for diabetic patients, particularly patients with foot deformities. This study [9, 21] also has successfully demonstrated that the technology made the shoe production time 64% faster than the hand-made orthotic shoes as presented in Fig. 4 (c) and Tabel 1. This finding is important for

AFO shoe designers to produce AFO shoes that are convenient, accurate, precise, and faster. These results show an increase of 24% compared with other research [6]

Our research output is a pair of custom orthotic (insole, shoe last and upper shoe) shoes for diabetic patients designed using existing CAD software that have been tested and used every day by two patients and have been shown to be tested in an orthotic laboratory (PT. Pratama Sentra Rehabilitasi, Jakarta, Indonesia). Both the development process of this technology (traditional and CARESystem) have also been carried out to obtain a significant influence on the concerns experienced by both patients with diabetes mellitus during this time.

Figure 2 illustrated the two types of shoes AFO with the CARE System. Both types of AFO shoes have been used by the patients with diabetes for less than eight months. They feel comfortable in the shoes. The AFO shoes type 1 in Fig. 4b is not an artistic form [7], giving rise to the complaint in two patients but successfully repaired with the making shoe last (Fig. 4c) with CNC machines according to the shape of shoes in general, as already reported by Refs. [8, 9] and is able to provide comfort for the patient. Regarding the geometry quality and shape, the shoes have already met both DM patients [3].

To examine the quality of the AFO shoes, both patients were asked to wear and perform daily activities such as sitting, standing about 5 min, walking slowly, walking fast and walking on the stairs. Running was not performed considering the age of the patients. The patients wore the shoes about 2–5 min. The test was conducted slowly, smoothly and both patients said that they were very comfortable in using these AFO shoes (Fig. 5). The present finding supports other researchers [11].

Meanwhile, the use of optimum cutting parameters in the CNC milling machine including spindle speed, feed rate, step over, depth of cut, toolpath strategy to determine the type of design and the right materials, EVA foam, have also successfully been produced by previous researchers [8, 11, 18, 22]. The machining parameters can be used as a reference for the shoe industry or laboratory to use CNC machines in the process of shoe last and the insole made of EVA Foam.

The current findings can be developed to solve the case design, manufacture and fabrication of orthotic shoes for patients with other deformities (club foot, flat feet, and other types of diabetes feet problem). Future research can produce innovative findings to speed up and improve the quality of the AFO shoe design process for DM patients.

4 Conclusion

The applications of CARESystem technology based on RID successfully improved the design and fabrication stages of AFO shoes specifically for DM patients with high-risk scale. AFO shoe insoles offer an exact surface contour, dimensional accuracy, and precision appropriate for patient's feet. and a good comfort level to reduce the pain. In the future, CARESystem can be used to design and fabrication of various AFO shoes types of patients such as club foot, flat feet, high heel syndrome, metatarsalgia and disability amputation due to diabetes.

The design ofinsole, outsole and shoe orthotics with existing CAD software has been tested in patients and has been tested in the orthotic laboratory. The process of the technology (traditional and CARESystem) has been carried out to address the concerns experienced by both patients with diabetes mellitus. The patients participating in this study felt very satisfied and comfortable with the products after wearing for about six months.

AFO new product trials in both patients have shown savings of working time on average, for the entire production process AFO approximately 64%. Both patients admitted geometry quality and comfort no significant difference with traditional and innovative processes. But in terms of the model and the shape of shoes, two patients agreed that innovative processes with CARESystem (see Fig. 3c and Fig. 4c) AFO was able to produce shoes that fit with the expectations of the patient so that the use of CAD technology (PowerSHAPE2016) in design development and manufacturing of shoe orthotics a decent design process and can be recommended for further processing in the shoe industry.

Collaborative research has been conducted by researchers together with a team from SIBAD UNDIP Group Research. This collaborationwill lead to the establishment of MoU and MoA between Diponegoro University with other universities and the private sector which has been a partner in the implementation of research inbiomedical engineering,

Acknowledgements We would like to gratefully thank you for PUTP Polytechnic ATMI Surakarta, Tribology Laboratory of the Department of Mechanical Engineering, University of Diponegoro in Semarang and CV EMA Pacific Surakarta that already provide full support in the form of infrastructure support CAM, CAD, and RE during the design, developed process as well as the writing of this paper.

Funding This research was funded by Directorate of Research and Community Service, Directorate General of Research and Development Strengthening, Ministry of Research, Technology and Higher Education Fiscal Year 2019, grant Number: 257-93/UN7.P4.3/PP/2019 and Number: 257-73/UN7.P4.3/PP/2019.

References

1. Munro W (2005) Orthotic prescription process for the diabetic foot. Diabet Foot 8(2):72–82
2. Uccioli L, Giacomozzi C (2012) The role of footwear in the prevention of diabetic foot problems. The diabetic foot: medical and surgical management, contemporary diabetes, https://doi.org/10.1007/978-1-61779-791-0_26,@springer Science+Business Media, LLC 2012
3. Anggoro PW (2018) Application of computer aided reverse engineering system (CARESystem) in the design and manufacturing stages of orthotic shoes. Dissertation. Mechanical engineering doctoral program. Department of mechanical engineering, Faculty of engineering. Diponegoro University, Semarang

4. Walbran K, Turner, McDaid AJ (2016) Customized 3D printed ankle-foot orthosis with adaptable carbon fiber composite spring joint. Cogent Eng 3(1):1–11 https://doi.org/10.1080/23311916.2016.1227022
5. Miguel D, Michele G, Marco M, Maura M, Enrique M, Roberto R (2011) Shoe customization design tools for the "Diabetic Foot". Comput Aided Des Appl 8(5):693–711
6. Mandolini M, Vitali MC, Macchione A, Raffaeli R, Germani M (2015) A CAD tool to design bespoke insoles for severe orthopaedic treatments. Comput Aided Des Appl 12(6):700–709
7. Anggoro PW, Tauviqirrahman M, Jamari J, Bayuseno AP, Bawono B, Avellina MM (2018) Computer-aided reverse engineering system in the design and production of orthotic insole shoes for patients with diabetes. Journal Cogent Engineering 5(1):1–20
8. Anggoro PW, Bawono B, Tauviqirrahman M, Jamari J, Bayuseno AP, Wicaksono A (2019) Reverse innovative design of insole shoe orthotic for diabetic patient. J Eng Appl Sci 14(1):106–113
9. Anggoro PW, Tauviqirrahman M, Jamari J, Bayuseno AP, Wibowo J, Saputro YD (2019) The optimal design and manufacturing shoe last product for an ankle-foot orthotic for the patient with diabetes. Int J Manufacturing, Material, and Mechanical Engineering 9(2):62–80
10. Ye X, Liu H, Chen L, Chen Z, Pan X, Zhang S (2008) Reverse innovative design—an integrated product design methodology. Comput Aided Des 40(7):812–827
11. Bawono B, Anggoro PW, Tauviqirrahman M, Jamari J, Bayuseno AP (2019) Milling strategy optimized for orthotics insole to enhance surface roughness and machining time by Taguchi and response surface methodology. J Ind Prod Eng 36(4):1–12
12. Bawono B, Anggoro PW, Tauviqirrahman M, Jamari J, Bayuseno AP (2019) The evaluation of the use of AFO (Ankle Foot Orthotics) with the MOXFQ (Manchester-Oxford Foot Questionnaire Method). Atlantis Highlights Eng (AHE) 1:657–662
13. Bawono B, Anggoro PW, Tauviqirrahman M, Jamari J, Bayuseno AP, Anthony AA (2019) Optimization parameters tooling design to increase the surface quality of orthotic insole shoes using the Taguchi approach and surface response methods, The 6th international conference on advanced material science and technology published in materials today: proceedings vol 13. Elseveir, Semarang, pp 47–52
14. Oancea G, Ivan NV, Pescaru R (2013) Computer aided reverse engineering system used for customized products. Acad J Manuf Eng 11(4):1–20
15. Chapman JD, Preece S, Braunstein B, Höhne A, Nester CJ, Brueggemann P, dan Hutchins s (2013) Effect rocker shoe design features on forefoot plantar pressure in people with and without diabetes. Crit Biomech 28(6):679–685
16. Anggoro PW, Bawono B, Wibowo J, Jamari J, Bayuseno AP (2017) Optimization of manufacturing process parameters for the product of ISO-Diabetes patients with high-risk classes. J Adv Sci Lett 23(12):11910–11917
17. Anggoro PW, Saputra E, Tauviqirrahman M, Jamari J, Bayuseno AP (2017) A 3-Dimensional finite element analysis of the insole shoe orthotic for foot deformities. Int J Appl Eng Res 12(15):5254–5260
18. Anggoro PW, Bawono B, Tauviqirrahman M, Jamari J, Bayuseno AP (2019) Design and manufacturing insole shoes orthotic for optimal surface roughness using CNC milling machine. J Eng Sci Technol 14(4):1799–1819
19. Anggoro PW, Bawono B, Sujatmiko I (2015) Reverse engineering technology in redesign process ceramics: application for CNN plate, Procedia Manufacturing vol 4. Elsevier, pp 521–527
20. Anggoro PW, Bawono B, Wijayanto A, Jamari J, Bayuseno AP (2016) Parameter optimation of strategies at CNC milling machines Rolland modela MDX 40R CAM against surface roughness made insole shoe orthotic EVA rubber foam. Int J Mechatron Mech Eng 06(4):96–104
21. http://lifestyle.bisnis.com/read/20181105/106/856489/-sepatu-orthotik-untuk-penderita-diabetes, last accessed 2018/12/30
22. Anggoro PW, Anthony AA, Tauviqirrahman M, Jamari J, Bayuseno AP, Nugroho A (2019) CNC milling of eva foam with varying haradness for custom orthotic shoe insoles and process parameter optimization. J Mechanical Eng Sci 13(3):5347–5370

Puzzle Islamic Floral Patterns Product Tiles for Wall and Ceiling to Decorate of Al Huda Mosque Indonesia—Design, Manufacturing, and Fabrication

P. W. Anggoro, A. T. Yuniarto, M. Tauviqirrahman, J. Jamari, A. P. Bayuseno, K. B. Purwanto, and O. K. W. Widyanarka

Abstract Ceramic wall ornaments that are commonly used as interior building and encountered in hotels, restaurants, residential elite, museums and places of worship. Ceramic is a decoration signifying strong characters or special characteristics of the building. This character is usually in the form of ornaments that adorn the walls and are designed according to the history of the development of the times. The basic characteristic of Islamic Ornament generally shaped floral such as leaves, stems, buds, flowers, and palms. Wall tiles with special needs with unique ornaments and precision are very unlikely produced by some industrial ceramics in Indonesia at this time which still depends on hand made. This study has introduced a range of novel applications that combine digital manufacturing technology and conventional fabrication in the ceramic industry. The input image with the nuances of Islamic motifs with the format. JPG successfully raised to 2.5 D CAD models using ArtCAM software. To get the master pattern prints that precision, accurate and appropriate use of technology additive manufacturing with 3D Objet 30Pro. Conventional ceramics Manufacturing process technology at PT. Nuanza Porcelain Indonesia used to get ceramic products with Islamic nuances of Pattern puzzle flora. The integration of two technologies of CAD and RP this managed to get the design ceramic wall ornamental Islamic precision and detail with the dimensional error of less than 2.00 mm. Ceramic design results in accordance with Islamic ornamentation and was already installed in the mosque of Al Huda in 2018.

P. W. Anggoro (✉) · A. T. Yuniarto
Department Industrial Engineering, Faculty of Industrial Technology, University of Atma Jaya Yogyakarta, Jl. Babarsari 44, Yogyakarta, Indonesia
e-mail: pauluswisnuanggoro@ymail.com

M. Tauviqirrahman · J. Jamari · A. P. Bayuseno
Department Mechanical Engineering, University of Diponegoro, Jl Prof Sudarto, Tembalang, Central Java, Indonesia

K. B. Purwanto
Laboratory of Production System, Department Industrial Engineering, Faculty of Industrial Technology, University of Atma Jaya Yogyakarta, Jl. Babarsari 44, Yogyakarta, Indonesia

O. K. W. Widyanarka
PT Nuanza Porselen Indonesia, Dukuh Dedegan, Ngadirojo Village, Boyolali Ressiden, Ampel City, Central Java, Indonesia

U. Sabino et al. (eds.), *Proceedings of the 6th International Conference and Exhibition on Sustainable Energy and Advanced Materials*, Lecture Notes in Mechanical Engineering,
https://doi.org/10.1007/978-981-15-4481-1_51

Keywords Ceramic tiles · Islamic ornament · ArtCam · Additive manufacturing

1 Introduction

Ceramic art is one branch of art that processes ceramic materials. This art is used to create a work that is traditional to contemporary. Ceramics are coverage for all objects made of clay, which undergo a process of heat or combustion so that it hardens. In everyday life, ceramics are often used as household appliances, displays, electronic devices, building ornaments.

From the classical world that encircled the Mediterannean to the boundaries of Scandinavia and to the New World, tiles have been used as architectural fixtures and for the embellishment of interior walls, floors, and ceilings. Monochromatic tiles were the main means by which tin-glazing techniques were disseminated throughout much of the Medieval world [1]. Tile is an enduringly popular choice because it can famously low maintenance, durable, and easy to clean, tile is an enduringly popular choice. But lately, tile has shed its reputation as just a sturdy and practical design solution. A tiled surface can be a piece of artistry. "Designing with the tile" explaining the fundamental principles of color, pattern and texture and their impact in this research. Tiles can bring color, pattern, and texture to a room where it is lacking, or they can enhance attributes that already exist as reported by Renzi [2]. In the context of modern manufacturing product design according to [3–6] the use of computer-aided design (CAD)—based design technology is an absolute requirement to get quality, precise, accurate and precise ceramic products. CAD in the ceramic design process, engineers are used to generating 2D sketch images or photo images in .jpg format quickly, precisely and precisely into 3D model surfaces or solid models as customers want. The use of 3D CAD (Computer-Aided Design) tools is an important factor in shortening time to market and reducing product development costs. By adopting broad 3D CAD technology as reported by Refs. [7, 8], product development has moved from physical to digital mockups, and from 2D to 3D designs in recent decades. 3D CAD has become part of a fully digital development process that includes design, modeling, simulation, and tooling [9, 10]. For new designs, if digital forms of similar product models are already available in databases, search techniques 3D like what has been explained by Refs. [11–13] can be used to search for product models with similar design shapes and objectives. In this case, the new design can be accelerated by reuse, in whole or in part, from the previous design. 3D search and reuse techniques have been widely studied [14, 15]. Although there are already several commercially available 3D search engines [12], the search and development of new 3D models remains a very active research topic in the field of product design and retrieval areas, especially in the ceramic tile industry.

Computer-Aided Design (CAD) is a computer program used to draw a product or component in the form of 2D vectors and 2.5D models. CAD models usually go through several design modifications, each time requiring simulation of physical properties and functions before finally being produced into engineering products

[16]. The application of CAD allows the optimization of product concepts before the manufacturing process.

Artistic Computer Aided Manufacture (ArtCAM) makes it easy to produce highly complex artistic designs quickly and effectively [6]. ArtCAM is used in various manufacturing industries such as the souvenir industry, ceramics, jewelry, spare parts, etc. for designing. ArtCAM is suitable for forming 2D and 2.5D models with a high degree of detail. However, until now rarely found any paper or research that discusses this software application as a tool to solve the problem of a real ceramic wall design with Islamic patterns in the ceramic industry in Indonesia.

Ceramic tiles are arranged in various shapes and sizes used on the floor or wall. Ordinary ceramic tiles for ornaments in mosques, hotels, historic buildings, resorts, and others. Mosques as places of worship are buildings and spaces that are believed by believers to be sacred and holy places [17]. For centuries, geometric patterns of Islam/Islamic Geometric Pattern (IGP) have been used as decorative elements on walls, ceilings, doors, domes, and towers [18]. Geometry in Islamic art and architecture creates a basic pattern in design [19]. Ceramics are widely used as building ornaments in places of worship, so they can add high artistic value. Ornaments on mosque buildings usually contain a philosophy/history of the development of the design. Some previous studies [3, 17–21] have only talked about the ceramic design patterns of cultural aspects, history, design concepts, up to prototype. While this paper, really applies modern technology CARESystem to accelerate the design time, manufacturing and fabrication of wall ceramics precision, accurate and able to disassembling rapidly on the building area to accelerate the time Installation of ceramics in the field. This kind of research is very rare and has a novelty on the process design and manufacturing techniques of a ceramic puzzle with CARESystem Technology.

Most of the best ceramic industries in Indonesia produce large quantities of tiles for the global market. In recent decades the progress of material science and production engineering has led to significant advances in quality and productivity [22]. The application of digital computer technology in terms of design and manufacturing has contributed greatly to this development [6, 20, 23–25]. Some ceramic companies such as PT. Doulton Indonesia, PT. Nuanza Indonesia has also implemented this technology. However, at present the regulation of production systems for tableware and ceramic tile ceramics with high production volumes has undergone changes, some or all of them no longer use hand made technology in the design process of master pattern prints but have gone through digital technology integration such as digital printing with in-jet [21], digital CARESystem on orthotic shoes and ceramic tableware [6, 23, 24] and digital CAM-CNC ceramic. Through this sophisticated system, it allows an attractive aesthetic opportunity to effect the detailed contours of the surface and glaze on artistic ceramic tile walls that can be fulfilled according to customer orders.

This paper explains the stages of design and fabrication of ceramic walls with IGP ornaments from flora patterned batik patterns where the initial input from customer NPI is only in the form of photo images in the .jpg format. The output of detail and precision ornamental motifs were chosen to be worked on in this paper are Islamic nuances for the buildings and walls of the Al Huda Mosque in Jakarta. The selection and determination of Islamic tile ornaments by researchers are based on the results of tile designs by several previous researchers [1, 2, 17, 18].

2 Material and Method

Islamic tilework can be seen as a manifestation of cultural preoccupation with the covering of surfaces. The Islamic tile ornaments chosen for this paper are presented in Fig. 1 and have been installed on several walls of the Al Huda Mosque, Jakarta, Indonesia (see Fig. 7).

The generation of vector surface modeling from .jpg format photo files to 2.5D solid tiles component modeling in this paper using ArtCAMPro 2015 CAD software can be presented in Fig. 2.

This research originated from the problem of PT Nuanza Porcelain Indonesia in the manufacture of puzzle type wall ceramics which is still done manually (hand made) with the results not precise and incorrect when assembled. This technology also requires a design time of 10–12 months longer with the size of the wall ceramic image from the photo with the printed one not the same because it is not precise and accurate. The brainstorming results obtained the design of wall ceramics with the type of Islamic Geometrical Patterns (IGP) in the form of photos in the .jpg file format, as in Fig. 1. With the help of ArtCAM 2015R2, PowerSHAPE 2016 and Autodesk basic Netfaab, the vector format with.jpg format was 2.5D CAD model of IGP ornamental wall ceramics and designed based on puzzle patterns. The results

Fig. 1 Islamic geometric ornamen with floral patern with .jpg format

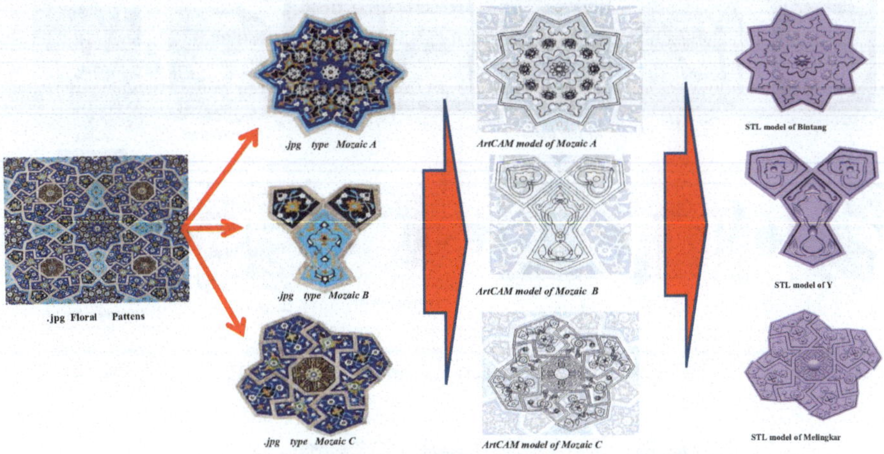

Fig. 2 Vector generation for geometric Islamic floral pattern using ArtCAM pro 2015

are presented in Fig. 3 the .STL format with 800–850 edge angles (see Fig. 8 at the bottom left). Giving edge angle is done so that when the pile ceramic is released from the core and the cavity can be released perfectly. The stages of vector generation of ceramic Tiles IGPs at ArtCAM 2015 are presented in Fig. 2. The 2.5D results of this ceramic model were then carried out by the verification process of geometry and solid surface modeling using basic Netfaab. The results are presented in Fig. 3 and are called master model prints of ceramic Tiles puzzle parts. This mold master finally fabricated the BOP (see Fig. 4) until the product was made into ceramic tiles as in Figs. 5, 6, and 7.

To get the results of part puzzle of ceramic tiles products with precision, accurate and precise IGP ornaments; Geometry measurements using a caliper dial with a tolerance of 0.01 mm until error deviations are obtained from the comparison of the dimensions between the dimensions of CAD drawings, prototypes of 3D printer results and pile prints before being burned with the final results (see Table 1 and Fig. 4).

In order for tile products designed with fabrication and fabrication to have perfect geometry and shape conformity [see Figs. 5 and 6] in terms of the deviation of errors that occur less than 2.00 mm then the procedure worked in this paper is to provide a magnification of a volume area of 15–20% on the 2.5D CAD ceramic model that

Fig. 3 The ceramic tile mold pattern master output from 3D print objet 30 Pro

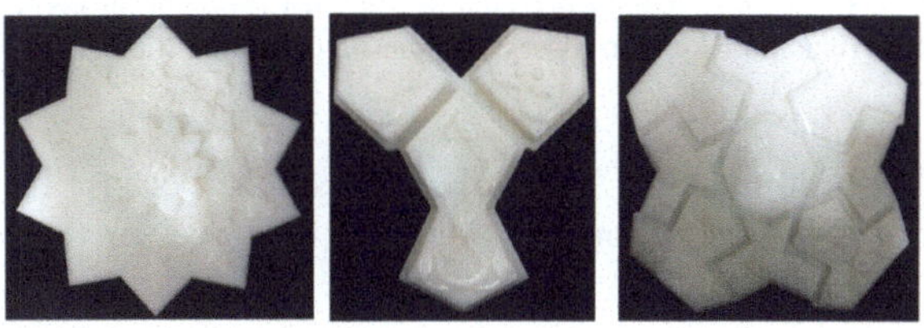

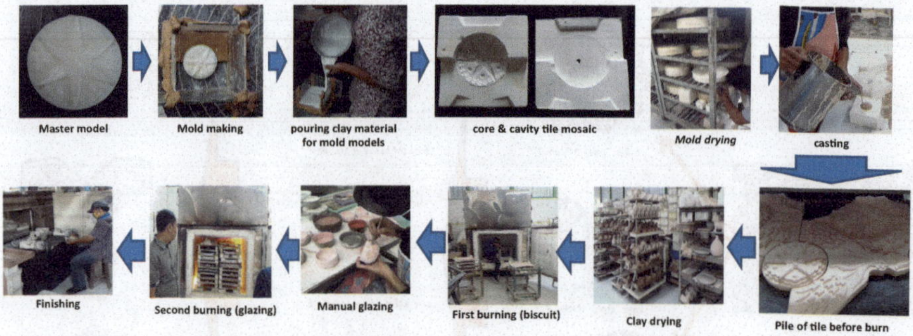

Fig. 4 Mold making and fabrication process ceramics tile at PT Nuanza Porcelain Indonesia (NPI)

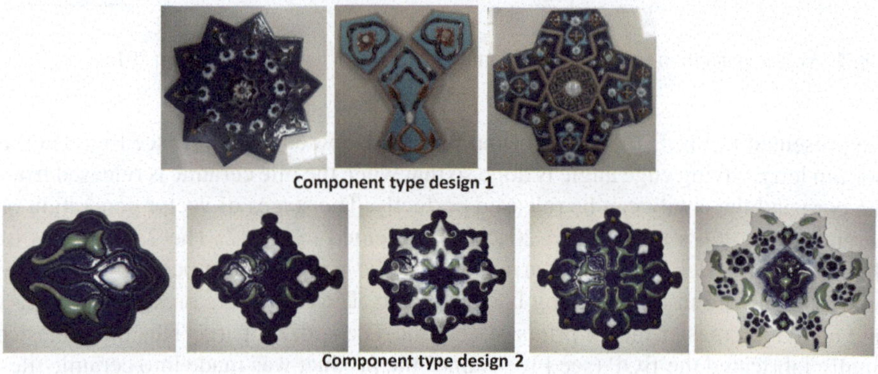

Fig. 5 Component tiles with floral patterns and geometric ornament

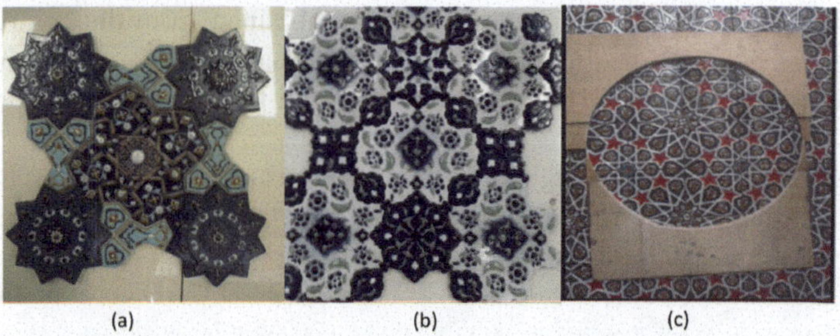

Fig. 6 Produk assembling Islamic tiles **a** Combination geometric ornament and floral patterns. **b** Syrian and Egyptian tiles. **c** Geometric ornament tiles

has been made at ArtCAM. This is done because based on the results of research conducted by researchers after the kiln process in the NPI oven generally the volume

Fig. 7 Some Islamic tiles from research have been installed on the walls of the Al Huda Mosque in Jakarta, Indonesia

of the final ceramic product will shrink by about 15% with the size present on CAD drawings. The final results of ceramics can be presented in Figs. 6 and 7.

Figure 5, if assembled into a single unit will form a wall decoration with Islamic ornamental ceramics as shown in Fig. 6 and after pairing it in the building of the Jakarta Al Huda mosque looks like in 7.

3 Result and Discussion

By the eighth century, the Islamic civilization had spread over a vast area that extended from India to North Arica and reaches up to Spain. Nourished by the cultural traditions of the conquered countries, it developed an astonishingly rich and varied aesthetic identify. The hot climate of the Islamic region and the extensive use of brick in the architecture made ceramic tiles a congenial and convenient form of embellishment for the inner and outer walls of both religious and secular building: glazed surfaces are cooling, and can be kept clean easily in a dusty environment [1, 2]. Geometric ornament, Floral patterns, calligraphic tiles, Persian monochrome tile, Seljuk or Abbasid tiles, Tile shape and styles, Il-Khnaid tile, Timurid Tiles, Safavid tiles, Syrian and Egyptian Tiles, and Iznik Tiles, is a variety of tiles that have been discussed in general by Refs. [1, 17, 18]. Base on Fig. 1, the images are included in the geometric ornament and Floral pattern categories and are an ornamental type of tiles so that they can be further processed to be vectored into a 2.5D solid ceramic wall model puzzle form using ArtCAM CAD. These two images will later be designed and fabricated into tiles that are arranged in the puzzle and mounted on the walls of the Al Huda Mosque, Jakarta Indonesia.

Table 1 Dimensional accuracy verification between 2.5D models, STL and RP of part safavid tiles

Point	2.5 Model on ArtCAM (mm)		2.5 D model with file STL (mm)		RP model (mm)		0,132Ceramic tiles (mm)		Error deviation of RP compare with ArtCAM (mm)		Error deviation RP to .STL (mm)	
	80°	85°	80°	85°	80°	85°	80°	85°	80°	85°	80°	85°
Part of cloudpart												
A	42.589	42.589	42.580	42.580	42.570	42.570	36.185	36.185	−0.020	−0.019	−0.010	−0.010
B	52.315	52.315	52.310	52.300	51.540	51.540	43.809	43.809	−0.770	−0.775	−0.770	−0.760
C	17.466	17.466	17.340	17.340	17.220	17.220	14.637	14.637	−0.250	−0.246	−0.120	−0.120
D	36.945	36.945	36.860	36.860	36.610	36.610	31.119	31.119	−0.340	−0.335	−0.250	−0.250
High	15.289	15.289	15.288	15.287	15.000	15.000	12.750	12.750	−0.290	−0.289	−0.290	−0.287
Part of sidepart												
A	110.815	110.815	110.610	110.700	110.550	110.570	93.968	93.985	−0.265	−0.245	−0.060	−0.130
B	95.801	95.801	95.750	95.790	95.690	95.710	81.337	81.354	−0.111	−0.091	−0.060	−0.080
C	78.534	78.534	78.500	78.510	78.320	78.330	66.572	66.581	−0.214	−0.204	−0.180	−0.180
D	92.243	92.243	92.210	92.240	92.190	92.230	78.362	78.396	−0.053	−0.013	−0.020	−0.010
E	64.333	64.333	64.310	64.320	64.240	64.260	54.604	54.621	−0.093	−0.073	−0.070	−0.060
F	72.100	72.100	72.000	72.000	71.870	71.890	61.090	61.107	−0.230	−0.210	−0.130	−0.110
High	17.246	17.246	17.245	17.245	17.170	17.170	14.595	14.595	−0.076	−0.076	−0.075	−0.075
Sidepart of corner												
A	16.240	16.240	16.190	16.220	16.030	16.180	13.626	13.753	−0.210	−0.060	−0.160	−0.040
B	135.753	135.753	135.270	135.490	135.040	135.220	114.784	114.937	−0.710	−0.533	−0.230	−0.270
C	17.252	17.252	17.110	17.100	16.730	16.720	14.221	14.212	−0.520	−0.532	−0.380	−0.380
D	117.115	117.115	117.050	117.000	116.750	116.680	99.238	99.178	−0.360	−0.435	−0.300	−0.320

(continued)

Table 1 (continued)

Point	2.5 Model on ArtCAM (mm)		2.5 D model with file STL (mm)		RP model (mm)		0,132Ceramic tiles (mm)		Error deviation of RP compare with ArtCAM (mm)		Error deviation RP to .STL (mm)	
Part of cloudpart												
	80°	85°	80°	85°	80°	85°	80°	85°	80°	85°	80°	85°
High	17.136	17.136	17.135	17.135	17.050	17.050	14.493	14.493	−0.090	−0.086	−0.090	−0.085
Part of center												
A	120.750	120.750	120.460	120.410	120.380	120.170	102.323	102.145	−0.370	−0.580	−0.080	−0.240
B	109.498	109.498	109.440	109.440	109.260	109.260	92.871	92.871	−0.240	−0.238	−0.180	−0.180
C	14.452	14.452	14.420	14.400	14.390	14.270	12.232	12.130	−0.060	−0.182	−0.030	−0.130
D	101.097	101.097	101.030	101.060	100.940	100.970	85.799	85.825	−0.160	−0.127	−0.090	−0.090
E	22.385	22.385	22.320	22.360	22.200	22.260	18.870	18.921	−0.190	−0.125	−0.120	−0.100
High	21.793	21.793	21.792	21.792	21.500	21.500	18.275	18.275	−0.290	−0.293	−0.290	−0.292
Part of white ornament												
A	56.713	56.713	56.700	56.690	56.680	56.690	48.178	48.187	−0.030	−0.023	−0.020	0.000
B	22.450	22.450	22.400	22.430	21.930	21.960	18.641	18.666	−0.520	−0.490	−0.470	−0.470
C	66.804	66.804	66.370	66.410	66.260	66.310	56.321	56.364	−0.540	−0.494	−0.110	−0.100
D	67.560	67.560	66.620	66.530	65.860	65.370	55.981	55.565	−1.700	−2.190	−0.760	−1.160
E	118.950	118.950	118.920	118.940	118.850	118.900	101.023	101.065	−0.100	−0.050	−0.070	−0.040
High	17.533	17.533	17.532	17.532	17.400	17.400	14.790	14.790	−0.130	−0.133	−0.130	−0.132

Figure 1 explains that the secret of Islamic art lies in the power of ornament. Geometry, floral patterns make up the decorative principles in market contrast to western art, which focus on nature and the human body. Ornament dissolves mass, forge space, engages, excites and surprises the human eye. Geometry is fundamental to Islamic ornament, just nature is organized on symmetrical principles and geometric patterning envolved to a degree of complexity and sophistication that had never before been seen (see Fig. 6). Two previous researchers [1, 18, 26] also noted that Islamic engineers or Islamic artists reproduced nature with a great deal of accuracy. Flowers and trees were often used as decorative motifs (see Fig. 1), but in the arabesque pattern, lines of vegetal ornamentation define space, playing a masterful game with colors and creating a three-dimensional effect (see Figs. 1 and 2). Figure 1 is the type of Iznik Tile. This is an exceptional example of the tiler's art in terms of quality and brilliance of color, demonstrating the sophistication of design attained during the third quarter of the sixteenth century [1, 2].

In decorative and architectural arts, ornaments are decorations that are used to beautify parts of a building or object. Ornaments can be carved from stone, wood or metal and can also be formed with clay. Ornaments found in mosque buildings usually form a certain pattern with Islamic designs. Islamic art uses many natural motifs such as flowers and trees with accurate designs. The strength of Islamic art lies in the ornaments that form geometry, flowers, patterns, and calligraphy. For centuries, the Islamic Geometric Patterns (IGP) have been used as decorative elements on walls, ceilings, doors, domes, and towers [18] and in this paper it has been shown that ornamental motifs in Fig. 1 can be developed become a form of ceramic wall with the characteristics of IGPs. Ornaments formed in general are in the form of Zillij motifs with recurring features and flipping through their basic shapes. Zillij is an Islamic ornamental art that repeats geometric shapes so that intertwined shapes and lines are intertwined (see Fig. 1). Plant motifs and flowers as shown in Fig. 1 have also long been the basis for decorating Islamic buildings [18, 26].

Figure 2, is the stage of the IGPs vector generation process with geometric ornament and Floral Pattern types from photo files in the .jpg format which are divided into several puzzles into 2.5D Artistic CAD with .STL format. This vector generation process is carried out using ArtCAMPro 2015 which has a very good ability to build an artistic 3D model according to the wishes of the engineer with a physical model or initial photo. This stage of vector development, in general, first divides photo files into puzzle parts that are ready to be raised into a 2.5D model (see Fig. 2).

This image was then continued by the researcher to verify the dimensions, surface and solid model with netffab software before the printing process was carried out with additive manufacturing technology on the 30Pro object 3D machine. This printout, then determined as the master model of ceramic Tiles made from verowhite which has been enlarged in volume by about 15% and can be presented in Fig. 3. Figure 4 explains the stages of fabrication of each master puzzle mold pattern of Islamic Ceramic tiles that is done by casting operators, glaze, burning at PT. Nuanza Porcelen Indonesia. After finishing the finishing process for each puzzle on the ceramic tile forming components, an Islamic ornamental puzzle wall ceramic product will be obtained (see Fig. 5) and after assembling it into one size 100 × 100 cm it will look like in Fig. 6. Installation results of all ceramic puzzles this ornament Islamic wall was successfully installed with great precision, good and accurate on the walls of the Al Huda Mosque in Jakarta, Indonesia (see Fig. 7). It can be seen that the results of the ceramics production and fabrication are following the results of the research that has been produced and developed by Refs. [3, 17, 18, 26–28].

Precision product results must have precise dimensional accuracy. The accuracy of dimensions in the design of wall ceramics is very important to compare from the initial dimensions to the results of the prints. This measurement aims to see and prove the success of the application of the reverse engineering method from the initial standard 2D to 2.5D (RP model) with the standards given by the PT. NPI of 2.00 mm. Dimension measurement is done by brainstorming with the creative team in the form of points, where each selected dimension is considered to represent several other dimensions. The measurement results with the caliper dial 0.01 mm produce several dimensions on the point of measurement as specified in Fig. 5 and are presented in Table 1.

The results showed that the deviation occurred below 1.00 mm for both design variations of 80° and 85°, while the standard given by the NPI was a maximum of 2.00 mm. Thus, the deviation between the 2.5D model and RP model meets the standard so that the model RP can be used as a master pattern template and can proceed to the ceramic tile fabrication stage (see Fig. 4) and the assembly process at PT. NPI until installation on all walls of Al Huda Mosque (see Fig. 7).

In summary, the stages of design—manufacturing—the fabrication of the distinctive Islamic floral wall tiles on this paper can be presented at Fig. 8. This stage proved to be able to bring the company to be able to compete in fighting overseas consumers will need artistic ceramic products with high selling value. Figure 8 next can be used as the basis for all local ceramic industry in Indonesia to be able to compete globally.

Ceramic tile produced at this research in an economical value is cheaper when compared with the manual technique that has been used by PT. NPI. According to information from the owner and Department of R&D PT NPI, it requires about 1–2 years for one type of precision and accurate ceramic mold pattern design. However, with CARESystem technology application in this paper for cases like in this paper, this can be produced in a month at least 10 variations of ceramic mold pattern design with Islamic pattern flora. This technology proved to accelerate the fabrication process of ceramic to the assembly stage on the production floor and installation in the mosque for less than 6 months.

4 Conclusion

Artistic 3D CAD-CAM such as ArtCAM2015 was applied to provide customized special ceramic tiles products that are specifically designed and manufactured for the Assembly of the Al Huda Mosque, Jakarta Indonesia. Ceramic Tiles products with ornament geometric and floral patterns exhibited precise surface contours and

Fig. 8 General stages of CARESystem application in the phase of design, manufacturing, and fabrication of ceramic wall in the local Indonesian ceramic industry

dimensional fit of the 2.5D CAD and 3D master pattern prints. All products provide good satisfaction for customers and PT NPI.

The Artistic CAD/CAM technology application with additive manufacturing on 3D printer machines to produce a master pattern of ceramic tile prints with IGP ornaments can provide precise, accurate ceramic end products with errors less than 1.00 mm and have been very well developed by researchers and PT NPI. Consumers also provide a very high appreciation for the results of ceramic tiles.

In the future, research on the usability of modern technology based on artistic CAD (Artcam, Vertec, brush), CAM (Powermill and Rhinoceros), and reliable CNC Milling machine as presented in this paper is able to complete practical solutions for local ceramic industry in Indonesia in order to compete with other competitors. The use of this modern technology will be able to produce ceramic products (tableware, tile wall, jewelry) containing the texture and ornaments with the characteristic of Indonesian native cultures such as batik, Islamic patterns, Java, Sumatra, animals, and plants. The result is in the form of good quality ceramic products, precision, accurate and truly presenting the characteristics of Indonesian culture and cultures.

The results of this research are also able to make PT NPI become the only local company in Central Java Indonesia that can compete at the Global level, especially in preparation to confront the industry 4.0.

References

1. Lang G (2004) Ten centuries of decorative ceramic 1000 tiles; 1st edition, Chronicle books LLC. ISBN: 0-8118-4235-5, USA
2. Renzi J (2009) The art of tile: designing with time-honored and new tiles; 1st edition, published in the United States by Clarkson Potter/Publisher, an imprint of the Crown Publishing Group, a division of Random House, Inc., New York, USA
3. Yao JG (2017) Application of computer-aided design in ceramic art design. 2017 International conference on manufacturing construction and energy engineering (IMCEE 2017). Hongkong, pp 252–256
4. Wang A-H, Sai S-T, Liu Y-M (2014) The high computer technology application study about the daily-use ceramic products design. IERI Procedia 10:184–189
5. Oancea G, Ivan NV, Pescaru R (2013) Computer-aided reverse engineering system used for customized products. Annals of MTeM for 2013 & proceedings of the 11th International MTeM Conference. Malaysia, pp 181–186
6. Anggoro PW, Bawono B, Sujatmiko I (2015) Reverse engineering technology in redesign process ceramics: application for CNN plate. Procedia Manufact 4:521–527
7. Ye XZ, Peng W, Chen ZY, Cai YY (2004) Today's students, tomorrow's engineers—an industrial perspective on cad education. Comput Aided Des 36:1451–1460
8. Ye X, Liu H, Chen L, Chen Z, Pan X, Zhang S (2008) Reverse innovative design—an integrated product design methodology. Comput Aided Des 40(7):812–827
9. Delchambre A (1996) CAD method for industrial assembly—concurrent design of products, equipment and control systems. Wiley, USA, Chichester
10. Hoffmann CM (1989) Geometric and solid modeling: an introduction. Morgan Kaufmann, USA, San Mateo
11. Iyer N, Kalyanaraman Y, Lou K, Jayanti S, Ramani K (2003) A reconfigurable 3D engineering shape search system part I: shape representation. In: DETC,'03
12. Tangelder J, Veltkamp R (2004) A survey of content-based 3D shape retrieval methods. In: International conference on shape modeling. pp 145–56
13. Funkhouser T, Min P, Kazhdan M (2003) A search engine for 3D models. ACM transactions on graphics. pp 83–105
14. Hong T, Lee K, Kim SC (2006) Similarity comparison of mechanical parts to reuse existing designs. Comput Aided Des 38(9):973–984
15. Pu JT, Jayanti S, Hou SY, Ramani K (2006) 3D CAD model retrieval based on multiple levels of detail. In: Pacific conference on computer graphics and applications
16. Zocca A, Colombo P, Gomes CM, Gunster J (2015) Additive manufacturing of ceramics: issues, potentialities, and opportunities. J Am Ceram Soc 98(7):1983–2001
17. Taib MZM, Rasdi MT (2012) Islamic architecture evolution: perception and behaviour. Procedia Soc Behav Sci 49:293–303
18. Abdullahi Y, Embi MR (2013) Evolution of Islamic geometric patterns. frontiers of architectural research. Front Archit Res 2:243–251
19. Othman HI (2012) Role of computer-aided design and computer-aided manufacturing technology in prosthetic implant restorations. Int J Dent Clin 4(4):22–34
20. Bechthold M (2016) Ceramic prototypes—design, computation, and digital fabrication. Informes de la Constr 68(544):1–11
21. Sanz V, Reig Y, Bautista Y, Ribes C, Edwards M (2012) Technical evolution of ceramic tile printing. J Imaging Sci Technol 56(5):50402-1-50402-7
22. Sanchez E, Garcia-Ten J, Sanz V, Moreno A (2010) Porcelain tile: almost 30 years of steady scientific-technological evolution. Ceram Int 36:831–845
23. Anggoro PW, Tauviqirrahman M, Jamari J, Bayuseno AP, Bawono B, Avellina MM (2018) Computer-aided reverse engineering system in the design and production of orthotic insole shoes for patients with diabetes. Cogent Eng 5(1):1–20

24. Anggoro PW, Tauviqirrahman M, Jamari J, Bayuseno AP, Wibowo J, Saputro YD (2019) Optimal design and fabrication of shoe last for ankle-foot orthotics for patients with diabetes. Int J Manuf Mater Mech Eng 9(2):92–99
25. Bechthold M, Kane AO, King N (2015) Ceramic material systems. Birkhäuser, Basel
26. Abdullahi Y, Embi MR (2015) Evolution of abstract vegetal ornaments in Islamic architecture. Int J Archit Res 9(2):31–49
27. Busheng L, Jing Fang H (2013) Development of ceramic three-dimensional design system based on sketch. Open Mechan Eng J 7:116–120
28. Koshnevis B (2001) Experimental investigation of contour crafting using ceramic materials. Rapid Prototyping J 7(1):32–41

An Optimization Study on Texture Depth for Bearing Sliders with Slip

M. Tauviqirrahman, M. L. Assaidiky, Paryanto, H. Indrawan, N. Cahyo, A. Simaremare, S. Aisyah, and Muchammad

Abstract Thrust bearing used to carry the external axial loads is a crucial part in the gas/steam turbine system of power generation plant. With demand for the high-speed thrust bearing and more efficiency savings, the texturing as well as the boundary slip applied on the bearing has been proven to improve the bearing tribological performance. In the present study, the modified Reynolds equation with slip based on the critical shear stress approach for solving the lubrication problem of textured bearing with double pockets is proposed. For obtaining effective computation, such equation is solved by finite volume method combined with tridiagonal matrix algorithm. An optimization model based on the exact optimization approach is employed to determine the best texture depth for the configuration of thrust pad for maximizing the load support. Results are also presented for various pocket shapes (i.e. rectangular, triangular and parabola). Numerical experiments show that in terms of load support, a significant improvement of 28% by optimization is achievable by optimizing the textured depth. It is also confirmed that the advantage of texture depth optimization is more obvious when the rectangular pocket shape is employed. The wide spans of findings here may provide a design reference for textured thrust bearing considering slip.

Keywords Lubrication · Numerical analysis · Optimization · Slip · Texturing

1 Introduction

Thrust bearing employed to carry the external axial loads from rotating journal is a crucial part in the turbine systems of power generation plant. The latest technological developments have emerged some new techniques such as coating, micro machining and laser surface texturing to improve the tribology performance (high load support but low friction at the same time) of the thrust bearing. With demand

M. Tauviqirrahman (✉) · M. L. Assaidiky · Paryanto · Muchammad
Mechanical Engineering Department, Diponegoro University, Semarang, Indonesia
e-mail: mtauviq99@yahoo.com

H. Indrawan · N. Cahyo · A. Simaremare · S. Aisyah
PT PLN (Persero) Research Institute, Jl. Durentiga 102, Jakarta Selatan 12760, Indonesia

U. Sabino et al. (eds.), *Proceedings of the 6th International Conference and Exhibition on Sustainable Energy and Advanced Materials*, Lecture Notes in Mechanical Engineering,
https://doi.org/10.1007/978-981-15-4481-1_52

for the high-speed thrust bearing and more efficiency savings, the surface texturing with or without the slip applied on the bearing has shown promising outcomes in improving the bearing performance. The potential of textured surface has attracted numerous workers around the word as noticed by many researchers. Buscaglia et al. [1] explored the effect of textures on the load-carrying capacity. Later, in terms of the friction coefficient Scaraggi et al. [2] investigated various surfaces with different micro hole depths focusing on how to minimize the friction. The authors noticed the optimal micro-hole depth values which are remarkable effective both in the hydrodynamic and mixed lubrication regime in minimizing the friction.

Recently, the exploration of the lubrication performance of textured thrust bearing has paid much attention by many researchers. Some new methods have been introduced to find the optimal texture shape in enhancing the performance. For example, the texture shape of star has been introduced by Uddin and Liu [3] for enhancing tribological performance. By optimization, they found that with respect to the minimization of the friction, the triangle effect is the most dominant feature. Later, based on genetic algorithm Zhang et al. [4] highlighted that the bullet texture as well as fish shape has better performance compared to conventional circular shape. In recent lubrication, the global-optimum textured geometry was studied in more detail in terms of the load support by Wang et al. [5]. The superiority of the developed hybrid method of GA-SQP was also revealed.

In addition to the textured surface, surface modification by modifying the wettability of the surface leading to the slip situation has attracted numerous researchers around the world. Employing the hydrophobic coating on the (textured) surface leading to the change of the wettability of the surface has proven to improve the lubrication performance. However, based on thorough literature survey, very few scholars studied the optimization of the slip-textured contact. Most of the previously published workers conducted the optimization of the texture based on pre-determined distributions and shapes which took time consumptive, for example [6–8]. In [6–8], there were uncertainty of global-optimum shapes was observed.

Extending the main frame of texture optimization, in the present work more attempts are required to provide the guideline of thrust bearing design. The main contribution of the present work is to numerically determine the optimal texture depth based on modified Reynolds approach combined with the exact optimization method. In particular, the slip-textured bearing with double pockets is studied in terms of the load support improvement.

2 Method

2.1 Governing Equation

A geometrical model of the slip-textured bearing with double pockets is reflected in Fig. 1. The artificial slip condition induced by the hydrophobic coating is engineered

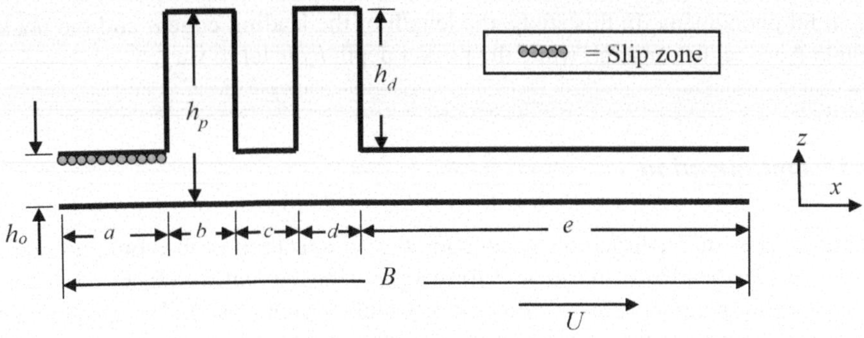

Fig. 1 Representation of double pockets of lubricated textured contact with slip applied at the contact inlet

to exist at the inlet of the contact based on the previously published works [8, 9]. The untextured lower wall moves with a sliding velocity, while for the upper wall with double pockets, the stationary condition is assumed. In this work, in order to solve the lubrication problem the modified Reynolds theory with slip assuming constant viscosity and incompressible and Newtonian flow is introduced and given as shown in Eq. (1) below.

$$\frac{\partial}{\partial x}\left(\frac{h^3}{12\eta}\frac{\partial p}{\partial x}\frac{h^2+4h\eta\alpha}{h(h+\eta\alpha)}\right)=\frac{U}{2}\frac{\partial}{\partial x}\left(\frac{h^2+2h\alpha\eta}{h+\eta\alpha}\right) - U\frac{\alpha\eta}{h+\alpha\eta}\frac{\partial h}{\partial x}+\frac{h}{2\eta}\frac{h\alpha\eta}{h+\alpha\eta}\left(\frac{\partial p}{\partial x}\frac{\partial h}{\partial x}\right) \quad (1)$$

where h refers to the fluid film thickness, α expresses the slip coefficient, p denotes the hydrodynamic pressure, U is sliding velocity and η refers to the lubricant viscosity. The interested reader for the derivation of Eq. (1) in more detail can refer to the work of Tauviqirrahman et al. [7, 8].

In this work, the load support of the bearing is of particular interest. The load support W is calculated by integrating the pressure over the computational domain.

2.2 Discretization

In the present study, a finite volume method combined with tridiagonal matrix algorithm is employed to solve the modified Reynolds equation (Eq. 1). The entire computational domain is assumed under the fully lubricated condition.

Based on the mesh independent study, the mesh number is approximately 5000–7000 nodes. Additionally, the refinement mesh is applied inside the texture to capture the fluid phenomena. In this study, the length of the leading-edge a and the pocket length b are set to constant, while the pocket depth h_d is let to vary.

2.3 Optimization

In the present study, during the calculation the optimization of the configuration of double pocket bearing with slip is performed by using the same reference parameters (i.e. working parameters as well as their boundary conditions). The physical and geometrical specification studied here are presented in Table 1

In this study, the objective function of the optimization calculation is to maximize the load support. Table 2 reflects the parameters for the optimization calculation in detail. It can be noticed that during the lubrication of the bearing system, two functional of requirements the load support should be met, as follows: (1) to support the load, and (2) to prevent the solid contacts. The optimization calculation aims to fulfill both functional objectives with various design parameters. For all following calculation, the computer numerical code including a numerical optimization code combined with the finite volume method is developed. In the present work, the exact optimization method is chosen.

Table 1 Parameters for calculation studied here

Parameter	Symbol	Value
Total length of bearing	B	0.02 m
Inlet length	a	0.003 m
First pocket length	b	0.0025 m
Valley length	c	0.003 m
Second pocket length	d	0.0025 m
Sliding velocity	U	1 m/s
Lubricant viscosity	η	0.01 Pa s
Atmospheric pressure	P_{atm}	100 kPa
Minimum film thickness	h_o	1 µm
Pocket depth	h_d	3 µm
Slip coefficient	α	0.02

Table 2 Detail parameter of optimization

Objective function	Maximize the load support
Design variable	Pocket depth Lower bond: 0.1 µm Upper bond: 10 µm
Constrain	Total bearing length

3 Results and Discussion

3.1 Validation

In order to validate the numerical code developed here, the comparison of hydrodynamic pressure between the present study and the published work of Muchammad et al. [9] is conducted and reflected in Fig. 2. In this section, the geometry as well as the input parameter studied here is the same with that in the reference [9]. It can be observed that the pressure profile matches well with the prediction of [9]. The prediction of the pressure lines coincides. It seems that two hydrodynamic pressure peaks are observed as expected. The peak pressure for the first pocket is higher than that for second pocket. The possible reason behind this is the fact that the presence of the slip encourages the larger pressure gradient due to the inlet section phenomena as discussed in more detail in [9].

In this work, in addition to the hydrodynamic pressure the validation of the code as well as the algorithm used here dealing with the result of the load support is also performed. The result shows that the predicted load support matches well with the result of Muchammad et al. [9]. The dimensionless load support predicted by the present study is 0.1466 or 0.2% higher compared with the result by Muchammad et al. [9]. Therefore, the numerical code developed here is acceptable and can be used to analyze other cases of the optimization of textured bearing.

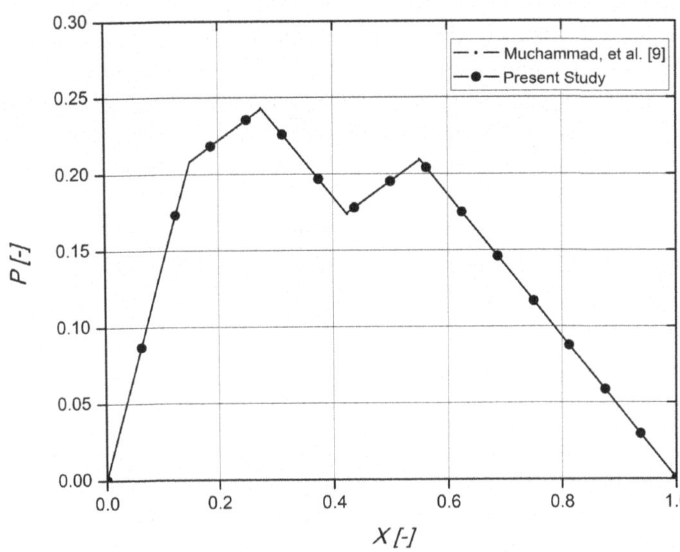

Fig. 2 Validation of the numerical algorithm

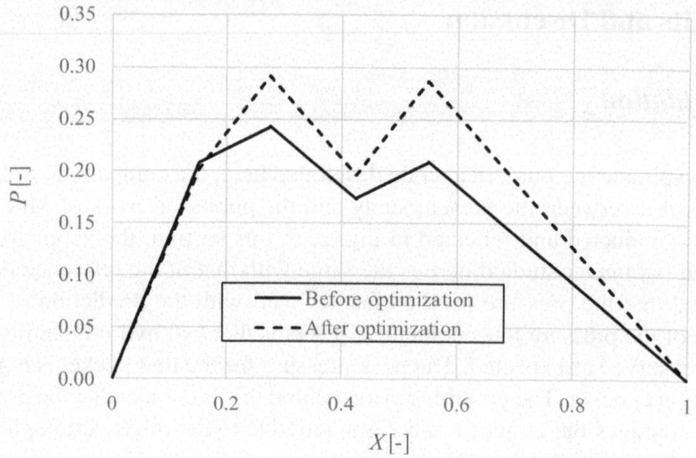

Fig. 3 Dimensionless hydrodynamic pressure distribution of textured bearing before and after optimization

3.2 Optimal Pocket Depth

The main objective of designing textured surface of the bearing is to obtain high load support at the operating conditions. The bearing load support can be determined by integrating the local pressure over the studied domain.

Figure 3 reflects the optimization result in comparison with the initial solution with respect to the hydrodynamic pressure. It can be found that the hydrodynamic pressure distribution enhances significantly along the textured area. The higher hydrodynamic pressure peaks are observed at the textured area in comparison to the initial bearing configuration. Based on the simulation result, it is found that the optimal pocket depth h_d leading to increased pressure profile is 0.7 μm or 300% lower than the initial pocket depth (see Fig. 4). With this textured configuration, the increased load support of up 21.40% can be achieved (see Fig. 5 for comparison).

3.3 At Varied Pocket Shape

In this section, the pocket shape of textured bearing with double pockets is varied dealing with the objective function of load support improvement. The pocket shape including the rectangular, triangular, and parabola are of particular interest. In the present work, the dimension of the pocket geometry for all pocket shapes is the same both in the pocket depth and the pocket length as shown in Table 1. Similar with the previous case, the slip situation is applied on textured pattern at the leading edge of the contact.

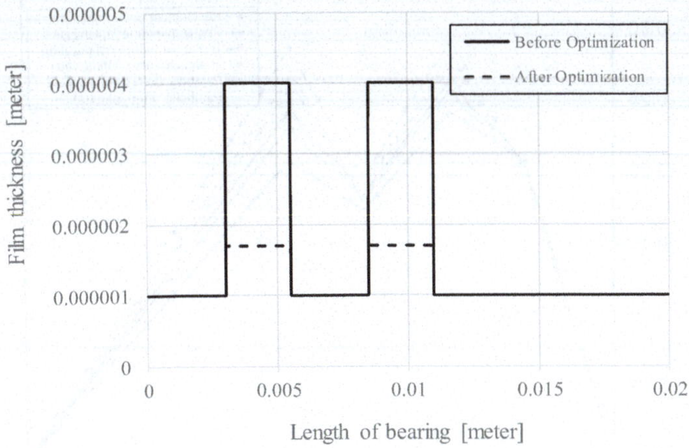

Fig. 4 The comparison of film thickness between the initial pattern and optimized pattern

Fig. 5 The predicted load support for initial pattern (before optimization) and optimized pattern (after optimization)

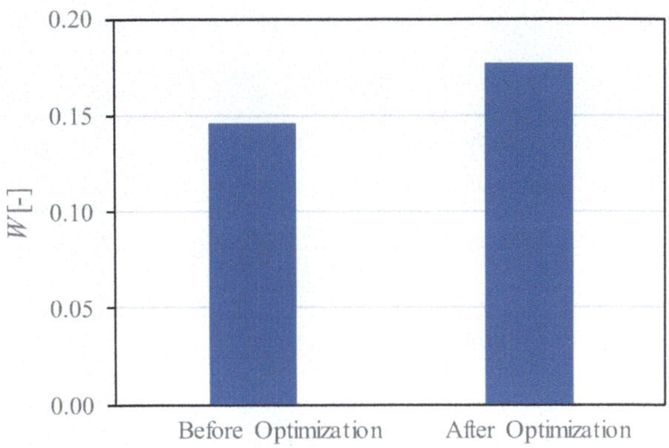

Figure 6 reflects the hydrodynamic pressure for various pocket shapes of textured bearing by comparing the result of "before" and "after" optimization. Based on Fig. 6, it can be found that with respect to the optimal shape of texture, the rectangular shape produces the highest performance after the performance process is conducted. The interesting result is that the optimization in the case of rectangular slip-textured bearing gives the significant increase in the load support. While for other shapes, i.e. triangular and the parabola, the increase in the pressure profile through optimization is quite lower.

Figure 7 depicts the comparison of the load support for several shapes of texture before and after optimization. The numerical results reveal that for all texture shapes, the optimization results in the load support improvement significantly. The highest enhancement of load support is achieved when the rectangular shape is used. It

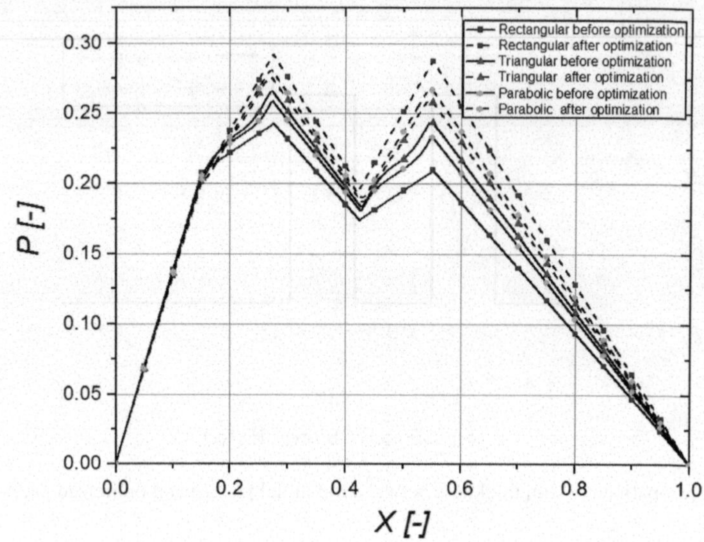

Fig. 6 The predicted hydrodynamic pressure for several textured bearing patterns

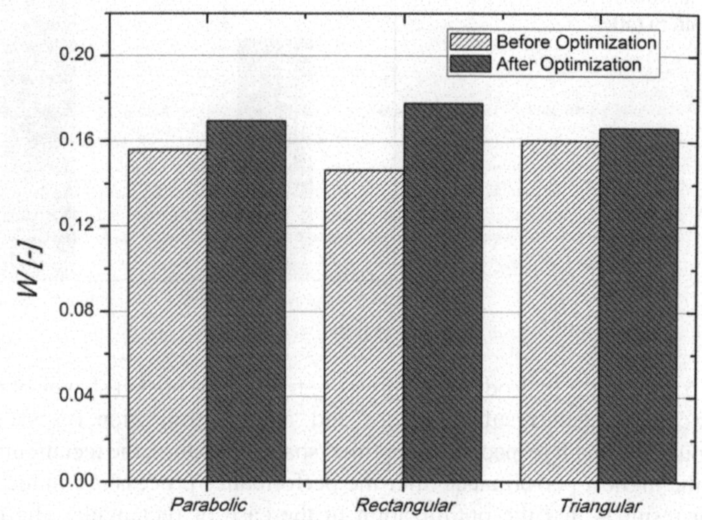

Fig. 7 The load support before and after optimization for case of textured bearings with double pockets for several shapes of texture

indicates that the rectangular shape should be considered as a promising way in enhancing the tribological performance.

4 Conclusion

This paper presents a study of lubricated bearing with double pockets combined with the slip boundary. The optimization process for the pocket depth of the bearing based on an exact optimization method was conducted focusing on how to enhance the load support. The modified Reynolds approach with slip was considered as a reference to solve the lubrication model. Several shapes of pocket including parabola, rectangular and triangular were compared using numerical modelling approach implemented via finite volume method.

Based on the discussion mentioned in the earlier section, some findings can be noticed. In the present work, it was highlighted that the numerical code developed including the optimization algorithm is successful to apply in solving the lubrication problem. For the case of slip-textured bearing with double pockets studied here, based on the optimization, the optimal pocket depth of the bearing maximizing the load support is 0.7 μm (or 300% lower than the initial depth).

The most interesting finding is that in comparison with other shapes (i.e. triangular and parabola), the rectangular pocket shape produces the highest performance. For the case of rectangular pocket, by optimizing the texture depth the improvement of the load support by around 28% can be achieved.

References

1. Buscaglia GC, Ciuperca I, Jai M (2007) On the optimization of surface textures for lubricated contacts. J Math Anal Appl 335(2):1309–1327
2. Scaraggi M, Mezzapesa FP, Carbone G, Ancona A, Sorgente D, Lugarà PM (2014) Minimize friction of lubricated laser-microtextured-surfaces by tuning microholes depth. Tribol Int 75:123–127
3. Uddin MS, Liu YW (2016) Design and optimization of a new geometric texture shape for the enhancement of hydrodynamic lubrication performance of parallel slider surfaces. Biosurface Biotribology 2:59–69
4. Zhang H, Hua M, Dong GZ, Zhang DY, Chen WJ, Dong GN (2017) Optimization of texture shape based on Genetic Algorithm under unidirectional sliding. Tribol Int 115:222–232
5. Wang W, He Y, Zhao J, Li Y, Luo J (2017) Numerical optimization of the groove texture bottom profile for thrust bearings. Tribol Int 109:69–77
6. Rao TVVLN, Rani AMA, Nagarajan T, Hashim FM (2012) Analysis of slider and journal bearing using partially textured slip surface. Tribol Int 56:121–128
7. Tauviqirrahman M, Muchammad, Jamari, Schipper DJ (2014) Numerical study of the load-carrying capacity of lubricated parallel sliding textured surfaces including wall slip. Tribol Trans 57(1):134–145
8. Tauviqirrahman M, Ismail R, Jamari, Schipper DJ (2013) Optimization of the complex slip surface and its effect on the hydrodynamic performance of two-dimensional lubricated contacts. Comput Fluids 79:27–43
9. Muchammad M, Tauviqirrahman M, Jamari J, Schipper DJ (2017) An analytical approach on the tribological behaviour of pocketed slider bearings with boundary slip including cavitation. Lubr Sci 29:133–152

Effect of the Surface Treatment on the Strength of Mixed Adhesive in Single Lap Joint Aluminum

Sri Hastuti, Neng Sri Suharty, and Triyono

Abstract Adhesive is a joining technology which is used in many transportation industries. The surface treatment and adhesive thickness are the important parameter which affect the adhesive joint strength. The mixed adhesive between silyl modified polymer (SMP) and epoxy (EP) adhesive is proposed to join aluminum. Aluminum alloy 5083 was used as adherent in this work. Its surface was cleaned using the sandblasting (SDB) method and then treated using chemical treatment of chromic-sulphuric acid (CSA) etch. The aluminum adherends were mixed adhesive joined with the standard of ASTMD 1002 in single lap joint (SLJ) type. The composition variation of mixed adhesive was 100%EP, 75%EP:25%SMP, 50%EP:50%SMP, 25%EP:75%SMP and 100%SMP. The adhesive thickness was varied of 0.2 mm, 0.4 mm and 0.6 mm. The strength of mixed adhesive was determined by the single lap shear test. The mixed adhesive joint without surface treatment was also performed for comparison. The surface with chemical treatment was finer than that without chemical treatment. It leads the adhesive strength of the joint with chemical treatment was lower than that of the joint without chemical treatment for all adhesive thickness levels. The adhesive thickness of 0.2 mm was the best thickness due to the achievement of the strongest adhesive strength. The small addition of SMP to EP adhesive will increase the adhesive joint strength.

Keywords Aluminum · Silyl modified polymer · Epoxy · Mixed adhesive · Surface treatment

1 Introduction

Joining technology like welding, rivet and adhesive has been used in many industries for years [1, 2]. It is commonly used in transportation industry such as automotive

S. Hastuti
Universitas Tidar, Jalan Kapten Suparman 39, Magelang, Indonesia
e-mail: hastutisrimesin@untidar.ac.id

N. S. Suharty · Triyono (✉)
Sebelas Maret University, Jalan Ir. Sutami No. 36A, Surakarta, Indonesia
e-mail: triyonomesin@uns.ac.id

U. Sabino et al. (eds.), *Proceedings of the 6th International Conference and Exhibition on Sustainable Energy and Advanced Materials*, Lecture Notes in Mechanical Engineering,
https://doi.org/10.1007/978-981-15-4481-1_53

[3], aerospace [4, 5] and naval [6] industry. Due to the need to reduce environmental impact of the transportation industry, the structural weight must be reduced in order to decrease the fuel consumption and increase the payload efficiency transportation [7]. To support this need, the appropriate joining system is adhesive joint. It is a joining process which uses adhesive placed between two material adherend surfaces and solidifies to produce an adhesive bond. It has many advantages such as light weight structural [8], lower fabrication cost [9], neatness, simple design, fatigue and corrosion resistance [10] and elimination of stress concentration [11]. However, it has some disadvantages such as long curing time, poor temperature and humidity resistance and requiring careful surface preparation before bonding [12]. There are many methods of adhesive joint which one of them is single lap joint [13]. It can be used for joining between metal to metal [14], composite to composite and metal to composite [15]. Accordingly, the enhancement of joint strength of single lap joint between metal to metal like aluminum to aluminum is important to be investigated.

The sandblasting process is considered as one of the most effective methods to control the level of surface roughness, increases the surface area of adherend, and increase joints resistance for metal [16, 17]. In general, blasted surfaces show higher surface energy values, which is supposed to enhance the adhesive joint strength [16–19]. Another method for treating the surface is chemical treatment. For aluminum, it is based on corrosion process from acid environments to help generating a new porous oxide layer. The high porous oxide layer could imply a high adhesive strength, increase the contact area, provide some degree of mechanical interlocking and tune adherend with adhesive [20]. The combination of laser pre-treatment and polyelectrolyte coating is also the treatment of the adherend surface. It can improve the adhesion strength [21].

Types of adhesive is in the range from very flexible and weak adhesive such as silyl modified polymer, to very stiff and strong adhesive such as epoxies [18, 19]. Adhesive joints will be weak but ductile if the weak adhesive is used. In the other hand, it will be strong but brittle if the strong adhesive is used. Weak adhesive is appropriate with ductile and flexible materials such as polymer while the strong adhesive is appropriate with stiff materials such as steel. For joining materials which have middle properties between flexible and stiff, weak and strong, such as aluminum, mixed adhesive between weak and strong adhesive have been proposed by previous studies [1, 3, 14, 22, 23]. They concluded that the best properties will be achieved in certain composition of the mixed adhesive. They have not evaluated the effects of the adherend surface treatment on the adhesive joint properties yet. So, in this work, both surface treatment and composition of mixed adhesive are proposed to be investigated. Mechanical properties of the mixed adhesive joint including shear strength, modulus elasticity and elongation were evaluated based on both mixed adhesive composition and surface treatment.

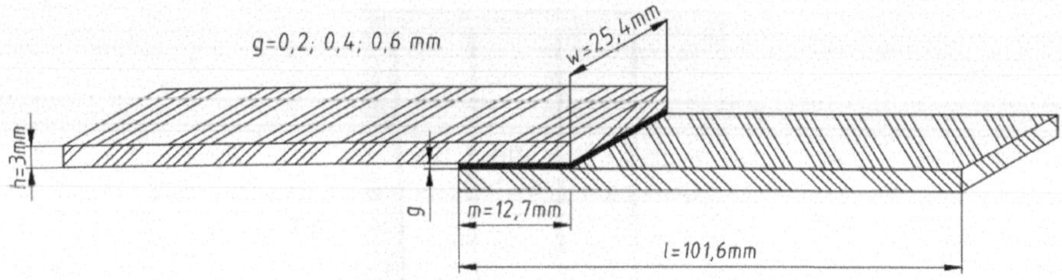

Fig. 1 Single lap joint specimen (ASTM D1002)

2 Experimental Method

The adherend material used in study was aluminum alloy 5083. Its surface was subjected the sandblasting and chemical treatment of chromic-sulphuric acid etch. The chemical treatment was done by a series of activities including dipping in the trichloroethylene solution for 30 min, washing using water, then dipping in the iso-propyl alcohol solution for 30 min, immersing in chromic-sulphuric acid etch (CSA) for 4 h and finally washing using distilled water. The CSA consisted of the H_2SO_4 solution (22.5 g), sodium dichromate (7.5 g) and distilled water (70 g). The adherends without chemical treatment were also prepared for comparison.

Two types adhesive, silyl modified polymer (SMP) and epoxy (EP) were used for representing the weak and strong adhesive. SMP adhesive was Simson ISR 70-05 from Bostik Inc., Netherlands. EP adhesive was bisphenol-A. The hardener was Polyaminoamide (PAA). They were then mixed with the composition variations of 100%EP, 75%SMP: 25%EP, 50%SMP:50%EP, 25%SMP:75%EP, and 100%SMP in weight fraction. The adhesive thickness was varied of 0.2 mm, 0.4 mm, and 0.6 mm. The mixed adhesive was prepared by using a stirrer with spatula at 60 rpm. Single Lap Joint (SLJ) specimens were manufactured at aluminum to aluminum with the mixed adhesive and given a pressure of 0.1 MPa. The post-curing process at 100°C for 100 min was performed. The procedure of adhesive joints was based on the ASTM D1002 standard. The configuration of joint specimens is shown in Fig. 1 where bonded dimension was 12.7 × 25.4 mm. The joints were characterized by shear test using universal testing machine of JTM UTS-210. Scanning Electron Microscope (SEM) of TESCAN Vega3 LMU was performed for investigating the fracture surface.

3 Results and Discussion

The chemical treatment of chromic-sulphuric acid (CSA) etch made the aluminum surface free from environmental contaminant. It caused a corrosion process which is aimed to remove the layer surface [20]. The macrograph of adherend surface with sandblasting process before and after CSA etch can be seen in Fig. 2. The surface treatment with sandblasting treatment made irregularities on the aluminum

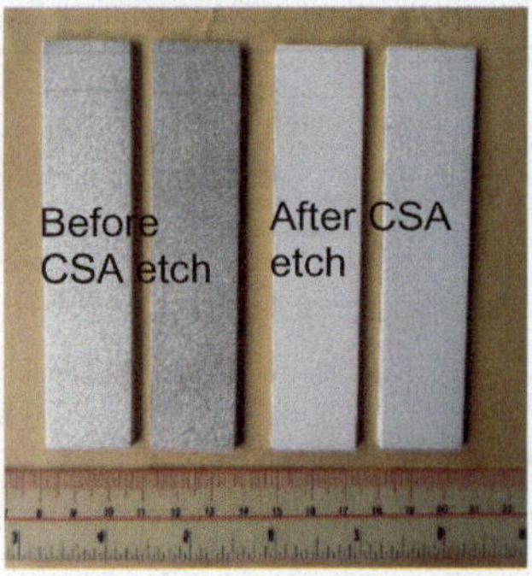

Fig. 2 The aluminum adherend

surface due to the shooting of silica sand particles [19] while the CSA etch led the aluminum surface be clean and also generated the fine porous. The surface peak of the sandblasting process was eroded by CSA etching so the surface has low roughness as seen in Fig. 3. The surface appearances with and without CSA etch were shown in Fig. 4.

Generally, the strength of the mixed adhesive joint on sandblasted surface without chemical treatment is higher than that with chemical treatment as seen in Fig. 5. The surface roughness is the main reason because the rough surface will lead the high adhesive force between adherend and adhesive. Mechanical interlocking of adhesive on the rough surface can be obtained easily [20]. The mixed adhesive of 75%EP:25%SMP is the best composition of the mixed adhesive where it has the

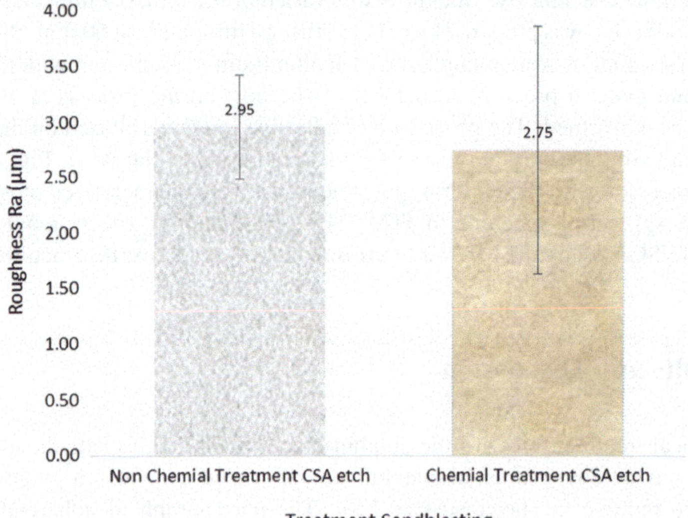

Fig. 3 Surface roughness

(a) Without chemical treatment.

(b) With chemical treatment

Fig. 4 The adherend surface

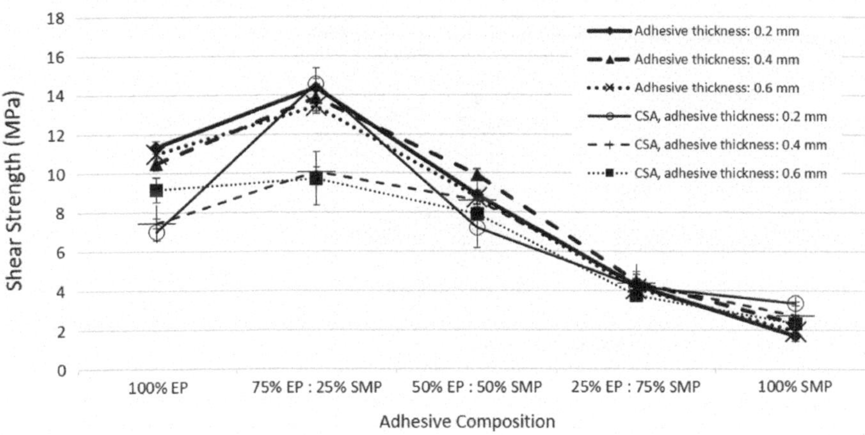

Fig. 5 A plot of adhesive joint strength for SLJ aluminum to aluminum

highest shear strength for all thickness levels. The mixed adhesive between little SMP and much EP gives better properties. Both adhesive and cohesive forces can be achieved simultaneously at this composition. Epoxy (EP) played a role in maintaining the cohesive force of adhesive while silyl modified polymer (SMP) played a role in maintaining the adhesive force between adherend and adhesive [14, 23]. Mixed failure mode between cohesive and adhesive failure mode will be occurred when adhesive and cohesive bonding was achieved on the adhesive joint. Furthermore, adhesive failure mode will occurre on the 100%EP adhesive joint specimens due to the strong of cohesive bonding of epoxy. Brittle fracture characteristic appears on this failure. The more addition of SMP adhesive to EP adhesive the less adhesive joint strength because of the low cohesive bonding of adhesive. In this condition, cohesive failure mode is usually occurred. The failure mode of mixed adhesive joints were shown in Fig. 6. The thickness of adhesive will affect the adhesive joint when the strong adhesive is used. It also can be seen in Fig. 5, where adhesive thickness of 0.2 mm achieved the best adhesive strength for 100%EP and 75%EP:25%SMP joints.

Figure 6 also shows that there are presences of porosities or air bubbles on fracture surfaces adhesive as well as presences ligaments. Their presences are caused by

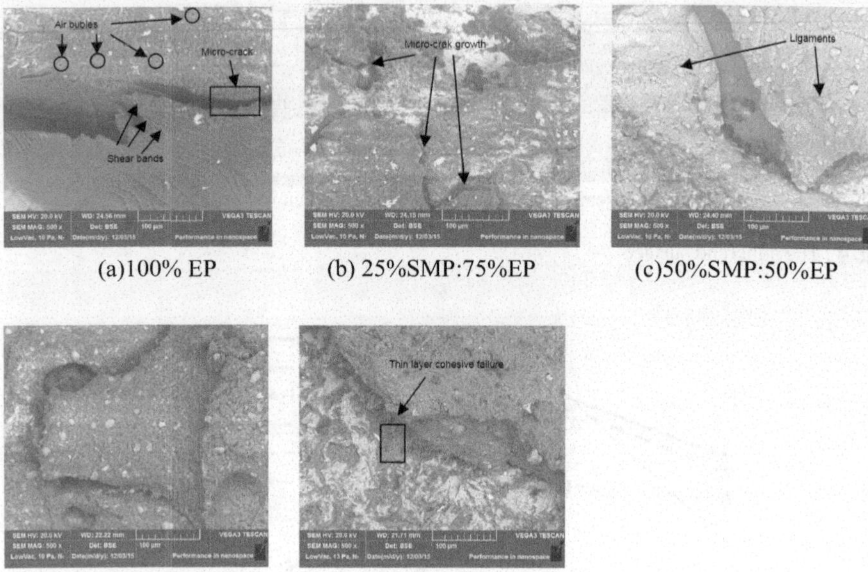

(a)100% EP (b) 25%SMP:75%EP (c)50%SMP:50%EP

(d) 75% SMP:25%EP (e) 100% SMP

Fig. 6 Failure surface of mixed adhesive joint

the micro cracking of the adhesive [24]. Porosity defect shape has less impact on reducing the strength of the adhesive joint strength [25]. The propogation of the crack is governed by the presence of a large porosities or air bubbles, shear bands, ligaments and micro-crack growth within adhesive.

Naturally, the elasticity modulus of materials is proportional to the strength while their elongation is inversely proportional to the strength. In this work, the elasticity modulus and elongation of the mixed adhesive joints followed the nature. The elasticity modulus of mixed adhesive joint with surface chemical treatment is lower than that of without surface chemical treatment. The more content of epoxy adhesive the higher elasticity modulus of the mixed adhesive joint as seen in Fig. 7a. It is because the epoxy adhesive has high stiffness, so the epoxy also has a high elasticity modulus [22]. Due to the nature of epoxy is brittle adhesive, the elongation of mixed adhesive joints with all levels of epoxy content is lower than that of SMP adhesive joint as seen in Fig. 7b. For 100%SMP joints, the thicker adhesive the higher elongation of adhesive joints. SMP adhesive is ductile adhesive, so its adhesive joints will have a high ductility.

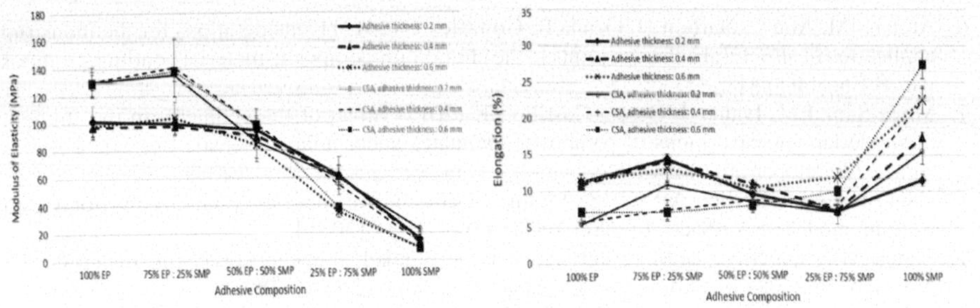

(a) (b)

Fig. 7 A plot of **a** Modulus elasticity. **b** Elongation

4 Conclusion

Effect of the surface treatment on the shear strength of mixed adhesive in single lap joint (SLJ) aluminum has been evaluated. The sand blasted surface roughness without chemical treatment was higher than that with chemical treatment. Sand blasting process led the irregularities due to the shooting of silica sand particles while chemical treatment eroded the irregularities formed by sand blasting process. Due to the fine surface, the adhesive strength of the joint with chemical treatment was lower than that of the joint without chemical treatment for all adhesive composition and thickness levels. The small addition of Silyl Modified Polymer (SMP) to epoxy (EP) adhesive up to 25% will increase the adhesive joint strength. For all adhesive composition and surface treatment, the adhesive thickness of 0.2 mm was the best thickness due to the achievement of the strongest adhesive strength.

Acknowledgements The authors gratefully acknowledge the financial support provided by Sebelas Maret University through MRG Research Grant 2016.

References

1. Martinsen K, Hu SJ, Carlson BE (2015) Joining of dissimilar materials. CIRP Ann Manuf Technol 64(2):679–699
2. Lambiase F, Paoletti A, Grossi V, Genna S (2017) Improving energy efficiency in friction assisted joining of metals and polymers. J Mater Process Technol 250(8):379–389
3. Akpinar IA, Gültekin K, Akpinar S, Akbulut H, Ozel A (2017) Experimental analysis on the single-lap joints bonded by a nanocomposite adhesives which obtained by adding nanostructures. Compos Part B Eng 110:420–428
4. Da Silva LFM, Ferreira NMAJ, Richter-Trummer V, Marques EAS (2010) Effect of grooves on the strength of adhesively bonded joints. Int J Adhes Adhes 30(8):735–743
5. Correia S, Anes V, Reis L (2018) Effect of surface treatment on adhesively bonded aluminium-aluminium joints regarding aeronautical structures. Eng Fail Anal 84(November 2017):34–45

6. Arenas JM, Alía C, Narbón JJ, Ocaña R, González C (2013) Considerations for the industrial application of structural adhesive joints in the aluminium-composite material bonding. Compos Part B Eng 44(1):417–423
7. Moya-Sanz EM, Ivañez I, Garcia-Castillo SK (2017) Effect of the geometry in the strength of single-lap adhesive joints of composite laminates under uniaxial tensile load. Int J Adhes Adhes 72(October 2016):23–29
8. Haghshenas M, Gerlich AP (2018) Joining of automotive sheet materials by friction-based welding methods: A review. Eng Sci Technol Int J 21(1):130–148
9. Banea MD, Da Silva LFM (2009) Adhesively bonded joints in composite materials: an overview. Proc Inst Mech Eng Part L J Mater Des Appl 223(1):1–18
10. Kahraman R, Al-Harthi M (2005) Moisture diffusion into aluminum powder-filled epoxy adhesive in sodium chloride solutions. Int J Adhes Adhes 25(4):337–341
11. Boutar Y, Naïmi S, Mezlini S, Ali MBS (2016) Effect of surface treatment on the shear strength of aluminium adhesive single-lap joints for automotive applications. Int J Adhes Adhes 67:38–43
12. Razavi SMJ, Ayatollahi MR, Nemati Giv A, Khoramishad H (2018) Single lap joints bonded with structural adhesives reinforced with a mixture of silica nanoparticles and multi walled carbon nanotubes. Int J Adhes Adhes 80(November 2017):76–86
13. Diharjo K, Firdaus Y, Raharjo WW, Tjahjana DDDP (2014) Improvement of strength on joint of GFRP/A1-5083 joint using epoxy/al-powder adhesive for electric vehicle car body. Proc 2014 Int Conf Electr Eng Comput Sci ICEECS 2014. (November):193–196
14. Triyono SH, Suharty NS (2017) Shear strength of the mixed adhesive joint silyl modified polymer-epoxy in single lap joint aluminum. J Mech Eng 4(1):235–248
15. Stuparu FA, Apostol DA, Constantinescu DM, Picu CR, Sandu M, Sorohan S (2017) Local evaluation of adhesive failure in similar and dissimilar single-lap joints. Eng Fract Mech 183:39–52
16. Zielecki W, Pawlus P, PerŁowski R, Dzierwa A (2013) Surface topography effect on strength of lap adhesive joints after mechanical pre-treatment. Arch Civ Mech Eng 13(2):175–185
17. Siryabe E, Renier M, Meziane A, Castaings M (2015) The transmission of lamb waves across adhesively bonded lap joints to evaluate interfacial adhesive properties. Phys Procedia 70:541–544
18. Dispersyn J, Hertelé S, De Waele W, Belis J (2017) Assessment of hyperelastic material models for the application of adhesive point-fixings between glass and metal. Int J Adhes Adhes 77(March):102–117
19. Hariss AF, Beevers A (1999) The effects of grit-blasting on surface properties for adhesion. Int J Adhes Adhes 19:445–452
20. Prolongo SG, Ureña A (2009) Effect of surface pre-treatment on the adhesive strength of epoxy-aluminium joints. Int J Adhes Adhes 29(1):23–31
21. Frenzel R et al (2015) Polyelectrolytes to promote adhesive bonds of laser-structured aluminium. Int J Adhes Adhes 61:35–45
22. Kim YJ, Labere J, Yoshitake I (2013) Hybrid epoxy-silyl modified polymer adhesives for CFRP sheets bonded to a steel substrate. Compos Part B Eng 51:233–245
23. Hastuti S, Suharty NS, Triyono (2017) Joint strength of mixed silyl modified polymer-epoxy adhesive on single lap joint etched aluminum. J Teknol 79(7–2):39–44
24. Boutar Y, Naïmi S, Mezlini S, Carbas RJC, da Silva LFM, Ben Sik Ali M (2018) Fatigue resistance of an aluminium one-component polyurethane adhesive joint for the automotive industry: effect of surface roughness and adhesive thickness. Int J Adhes Adhes 83(March):143–152
25. Heidarpour F, Farahani M, Ghabezi P (2018) Experimental investigation of the effects of adhesive defects on the single lap joint strength. Int J Adhes Adhes 80(August 2017):128–132

Thermal Stability of Bamboo Fiber with Virgin and Recycled High Density Polyethylene Matrix

Agung Prasetio, Indah Widiastuti, Budi Harjanto, Navira Alya Astadini, and Ryan Chandra Adiputra

Abstract Bamboo shows great potential as reinforcement material in polymer matrix. However, as a plant based fiber, its thermal stability should be carefully assessed for further application. It is expected that understanding the thermal stability of the composites will provide sufficient information in analyzing the feasibility for application under high temperature. This study aims to analyze the thermal behavior a bamboo-based composite using the Thermogravimetric Analysis (TGA) method with ASTM E-1131 standard. The composites were prepared by compression moulding using virgin and recycled high density polyethylene (HDPE) as the polymer matrix. For comparison purpose, the bamboo strips from inner and outer layer were investigated. Those strips were processed into particles with a size of 20 mesh and then treated with 5% of NaOH for 2 h priors to composite fabrication. The prepared composite consists of 10% bamboo loading and 90% of HDPE matrix. TGA analysis was performed using sample weight of 15–30 mg, a heating speed of 10 °C/min and a heating temperature of 30–600 °C. The results of this research show that the virgin HDPE composites had better thermal stability than r-HDPE composites. The highest thermal stability is on pure HDPE matrix composites with bamboo fiber from outer layer with a weight reduction of 5% at a temperature of 301.01 °C and a weight reduction of 10% at a temperature of 336.79 °C.

Keywords Thermal stability · HDPE · Thermogravimetric analysis

1 Introduction

The use of natural fiber composites is more desirable because besides the cost is relatively more affordable it is also environmentally friendly and easily renewable material. Fibers composites are used as reinforcement elements that greatly determine the mechanical properties of composites, because the mixture of fiber materials continues the load distributed by the matrix [1]. One of the many natural fibers available around us is bamboo. Until now, the use of bamboo as natural fiber has

A. Prasetio (✉) · I. Widiastuti · B. Harjanto · N. A. Astadini · R. C. Adiputra
Mechanical Engineering Education of Sebelas Maret University, Surakarta, Indonesia
e-mail: prasetioag@gmail.com

U. Sabino et al. (eds.), *Proceedings of the 6th International Conference and Exhibition on Sustainable Energy and Advanced Materials*, Lecture Notes in Mechanical Engineering,
https://doi.org/10.1007/978-981-15-4481-1_54

not been processed massively by bamboo craftsmen. Bamboo fiber is very potential to be used as a reinforcement of new materials in composites [2]. Even though the potential of bamboo in Indonesia is very high. In Indonesia there are 157 species of bamboo which constitute 10% of all bamboo species in the world. Throughout the world itself, bamboo species range from 1250–1350 species. Of all types of bamboo in Indonesia, 50% are endemic to native Indonesia and the other 50% are the result of cross-breeding with outside bamboo species [3].

In the manufacturing process of thermoplastic based composites with natural fibers as fillers or reinforcement, blending temperature becomes very important with the involvement of heat within the process. Analysis of the thermal properties of its components before blending is necessary to determine appropriate temperature mixing [4]. For composite products such as window frames and bathroom interiors, the thermal properties such as heat stability of composite products also become very important. For example the interior material of the bathroom must be stable in the presence of hot steam during shower use [5].

Various studies have been conducted to test the strength of polymer composites with natural fiber reinforcement. The test is carried out to determine the mechanical properties, dynamic properties, thermal conductivity and thermal properties. In a study of bamboo fiber reinforced polyester resin composites with an alumina composition of 0%, 5%, 10%, 15% and 20% using the hand lay-up method stated that the mechanical properties of composites such as tensile strength and flexural strength are strongly influenced by fiber composition. In the tensile and flexural test of the composite, there was a gradual increase in tensile and flexural strength of the composite with 0–15% alumina, but the tensile and flexural strength was reduced in composites with 20% alumina. Maximum tensile and flexural strength occurs in composites with 15% alumina [6]. Whereas in the research about the mechanical properties (flexural strength) of HDPE composites with bamboo fiber reinforcement with mass fractions of 5, 10, 15, 20, 30, 50, 60 and 70 wt% states that the maximum flexural strength occurs in a mass fraction of 30 wt% that is equal to 58.99 Mpa [7].

Several researches have been performed to analyze the thermal properties of natural fiber polymer composites. In a study conducted on the composite of polypropylene (PP) with rattan fiber nanoparticle filler, it was found that the thermal resistance of rattan skin fiber nanoparticle composite filler was comparable to glass fiber filler, because the dominant matrix had an effect on thermal resistance [8]. Whereas, in a study of four kinds of lignocellulosic fibers (LFs), namely, those from Chinese fir (*Cunninghamia lanceolata*), Taiwan red pine (*Pinus taiwanensis*), India-charcoal trema (*Trema orientalis*) and makino bamboo (*Phyllostachys makinoi*), were selected as reinforcements and incorporated into high-density polyethylene (HDPE) to manufacture wood-plastic composites (WPCs) by a flat platen pressing process and mixed with high density polyethylene (HDPE) to produce wood-plastic composites (WPC). The results showed that Taiwan's red pine showed a decrease in weight above 220 °C, while the decrease in weight of Chinese firs was around 60 °C higher, indicating the thermal properties of Chinese fir with a more stable HDPE matrix. These results reveal that the mechanism of thermal decomposition of lignocellulose materials is very different [9]. In another study on the thermal properties analysis of HDPE composites with argon nut skin reinforcement (ANS) as bio-filler, it was found that there was a shift on the degradation temperature of by inserting ANS particles into the HDPE matrix. Adding particles to the polymer matrix reduces the thermal stability of the entire material [10].

This study aims to analyze the thermal stability of recycled HDPE composites with bamboo fiber reinforcement. It is motivated by abundant available of waste HDPE as well as the potential of bamboo as reinforcement material in polymeric matrix.

2 Materials and Methods

2.1 *Thermoplastic Matrix*

In this study, the material for composite matrix is recycled HDPE obtained from a plastic waste collector. For comparison purpose, virgin HDPE in granule was also used. Both of the HDPE materials are grinded using a crusher machine to have finer size and then sieved with a 20 mesh size.

2.2 *Preparation and Treatment of Bamboo Fiber*

Firstly, bamboo strips (*Dendrocalamus asper*) were cut in 30 cm long with a saw, which taken from outer and inner layer of the culm. They were then beaten until the fibers separated with the lignin macro each other. The bamboo fibers were having a chemical treatment by soaking in 5% NaOH solution for 2 h to reduce the lignin content, so that the fibers can be easily separated from each other. The clean fibers were dried [11] in an oven for 12 h at 60 °C [12]. A crusher machine was utilized to have smaller size of bamboo fiber followed by a sieving process to have a 20 mesh size of fiber.

2.3 *Composition of Composite Materials*

The composites consist of a fixed mass fraction of 90% HDPE matrix and 10% bamboo fiber (BF) from inner part (BFI) and outer part (BFO) of bamboo culms. Table 1 shows the composition of BF-reinforced HDPE composites in each specimen.

Table 1 Composition of BF-reinforced HDPE composites

No	vHDPE (%wt)	rHDPE (%wt)	BFO (%wt)	BFI (%wt)
1	90	–	10	10
2	90	–	10	10
3	–	90	10	10
4	–	90	10	10

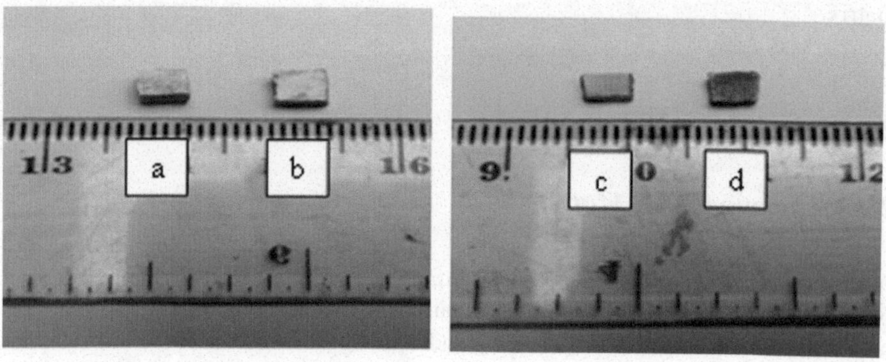

Fig. 1 **a** HDPE-Bamboo fiber from outer layer (HDPE-BFO), **b** HDPE-Bamboo fiber from inner layer (HDPE-BFI), **c** rHDPE-Bamboo fiber from outer layer (rHDPE-BFO), **d** rHDPE-Bamboo fiber from inner layer (rHDPE-BFI)

2.4 Composite Manufacturing

To have a thickness of 3 mm, the composites were prepared using a compression molding machine in total mass of 50 g. The specimens were manufactured at a pressure of 50 bar and a temperature of 150 °C with a holding time of 25 min [13]. The composite specimens after room temperature cooling and cut are shown Fig. 1.

2.5 Thermogravimetric Analysis (TGA)

Thermogravimetryc Analysis (TGA) aims to analyze the physical and chemical changes in the material by forcing the reaction with the use of heat. The test specimen will be heated gradually and measured how much the mass changes along with the increase in temperature.

Composite test specimens have a weight of 15–30 mg. The test is carried out at programmed temperature of 30–600 °C with a heating speed of 10 °C/min [9]. TGA testing is carried out based on the ASTM E1131 standard.

3 Result and Discussion

The results of TGA test in 4 (four) different specimens can be seen in Fig. 2. In Fig. 2 shows weight loss of composites in the temperature range of 30–600 °C.

In the area of temperatures below 250 °C, the composites weights are relatively constant since no significant weight loss was observed. Meanwhile, a clear drop of weight was seen in the temperature region of 250–475 °C. In this region, the HDPE-BFI and HDPE—BFO composites loss their weight of 12.26% and 18.67%, respectively. The recycled HDPE composites recorded different trend with the virgin HDPE as the weight loss of BFO based composites is smaller than the BFI composites.

In temperature regions above 475 °C, the mass of HDPE and rHDPE composites reinforced bamboo fibers continues to decline. Other study has reported that the bamboo fiber was more difficult to be decomposed than HDPE which inbihited mass reduction at temperatures 500–700 °C [14].

Mass reduction of polymeric materials occurs because of the release of hydrogen atoms from polymeric hydrocarbon bonds. The detachment of the hydrogen atom is due to the input energy that comes from heat. The release of hydrogen from hydrocarbon bonds will rise with increasing temperatures causing a continues mass reduction [15].

It can be known that different fiber composition resulted in different temperature resistance. The virgin HDPE composites were more resistant to temperature rather than the recycled HDPE. The HDPE—outer BF started to have thermal decomposition at 300 °C temperature. Using bamboo fiber derived from inner culm layer has no significant effect to the thermal decomposition temperature. In other hand, the composites based on recycled HDPE started to degrade in lower temperature. The weight has decreased from a temperature of 263 °C and 276 °C in HDPE-BFO and HDPE-BFI composites, respectively.

Table 2 shows the thermal stability of recycled HDPE and virgin HDPE composites with bamboo fiber reinforcement at a weight loss of 5% and 10%. The mass of HDPE-BFO specimens was drop by 5% at temperatures of 301.01 °C and 10% at

Fig. 2 Thermogravimetryc analysis of composite

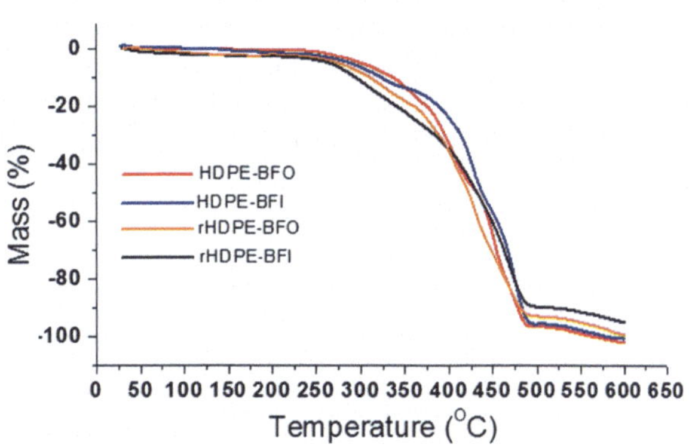

Table 2 Thermal Stability of 4 samples composite

Composite materials	Thermal stability	
	$T_{5\%}$ (°C)	$T_{10\%}$ (°C)
HDPE-BFO	301.01	336.79
HDPE-BFI	288.12	323.17
rHDPE-BFO	263.23	293.50
rHDPE-BFI	275.86	310.61

336.79 °C. Whereas, the HDPE-BFI specimens experienced a weight loss of 5% at temperatures of 288.43 °C and 10% at 323.17 °C. The 5% of weight reduction on rHDPE-BFO specimen is at 263.23 °C and the 10% is at 293.50 °C temperature.

A material is indicated possessing a high thermal stability if weight reduction occurs at higher temperatures [16]. Composites with pure HDPE matrix have higher thermal stability than recycled HDPE matrix, this is obtained based on Table 2 where the highest temperature at 5% weight reduction pure HDPE matrix is 301.01 °C while recycled HDPE matrix is 275.86 °C. Composites with Virgin HDPE-BFO having highest thermal stability as indicated by the highest temperature in which the materials lose 5% and 10% of their weight.

For comparison, a study on water hyacinth fiber reinforced-HDPE reported that 10% weight reduction occurred at a temperature of 278 °C [17]. Another report on epoxy resin matrix reinforced with glass fiber found that the material lose its 10% weight at 350 °C. It can be concluded that HDPE-bamboo fiber composites have moderate thermal stability compared to natural and synthetic fiber based composites.

4 Conclusion

According to the results of testing TGA (Thermogravimetric Analysis), composites with virgin HDPE matrix have better thermal stability than composites with recycled HDPE matrix. The highest thermal stability is on pure HDPE matrix composites reinforcing bamboo fiber from outer layer (HDPE-BFO) with a weight reduction of 5% at 298.16 °C and a weight reduction of 10% at temperatures of 334.97 °C. While the lowest thermal stability occurs in recycled HDPE matrix composite reinforcing bamboo fiber from outer layer (rHDPE-BFO) with a weight reduction of 5% at a temperature of 263.23 °C and a weight reduction of 10% at a temperature of 293.50 °C.

References

1. Sukoco BAB (2018) Komposit skin hybrid berpenguat serat bambu acak 50% dan serat rami anyaman 50% bermatriks polyester terhadap kekuatan tarik bending dan impact. Publikasi Ilmiah Universitas Muhammadiyah Surakarta, Surakarta
2. Efendi R (2017) Pengembangan komposit berbahan ebonit dengan kandungan sulfur 40 PHR yang diperkuat serat bambu untuk komponen otomotif. Publikasi Ilmiah Universitas Muhammadiyah Surakarta, Surakarta
3. Ami E, Hanum L, Dahlan Z (2017) Bamboo distribution in musi rawas district South Sumatera province
4. Sutiani (2003) Degradasi poliblend polistiren-pati menggunakan Bacteri pseudomonas fluorescens. Prosiding Seminar, bandung, departemen kimia FMIPA ITB
5. Yang HS et al (2005) Thermal properties of lignocellulosic filler-thermoplastic polymer biocomposites. J Therm Anal Calorim 82:157–160
6. Reddy SKB et al (2018) Mechanical properties of bamboo fabric with alumina as a filler material in polyester composite. Int J Pure Appl Math 119:911–916
7. Wang C et al (2017) Mechanical properties and prediction for nanocalcium carbonate-treated bamboo fiber/high-density polyethylene composites. J Mater Sci 52:11482–11495
8. Balfas A (2012) Analisis termal bionanokomposit filler serat kulit rotan. Skripsi Institut Pertanian Bogor, Bogor
9. Hung K et al (2017) Characterization of wood-plastic composites made with different lignocellulosic materials that vary in their morphology, chemical composition and thermal stability. J Polym 2017 9(726):1–22
10. Hl Essabir et al (2015) Morphological, structural, thermal and tensile properties of high density polyethylene composites reinforced with treated argan nut shell particles. J Bionic Eng 12:129–141
11. Haezer HE (2016) Analisa sifat akustik dan morfologi material komposit polypropylene berpenguat serat bambu dan rami. Tugas Akhir. Institut Teknologi Surabaya, Surabaya
12. Zhang K et al (2018) Thermal and mechanical properties of bamboo fiber reinforced epoxy composites. Polymers 10:608–616
13. Purnama KI (2016) Pengaruh temperatur hotpress terhdapap kekuatan mekanik komposit cantula-rHDPE. Skripsi Universitas Sebelas Maret, Surakarta
14. Zakikhani P et al (2016) Thermal degradation of four bamboo species. J Bioresour 11(1):414–425
15. Beyler CL, Hirschler MM (2002) Thermal decomposition of polymers
16. Zhang W, Li X, Yang R (2011) Pyrolysis and fire behavior of epoxy resin composites based on a phosphorus-ontaining polyhedral oligomeric silsesquioxane (DOPO-POSS). Polym Degrad Stab 96(10):1821–1832
17. Wirawan R et al (2012) Stabilitas termal komposit eceng gondok (Eichhornia crassipes) dengan matriks HDPE. Proceeding seminar nasional tahunan teknik mesin XI (SNTTM XI) & thermofluid IV Universitas Gadjah Mada (UGM), Yogyakarta

Effect of Slip Placement on the Performance of Textured Sliding Contact by CFD

M. Muchammad, M. Tauviqirrahman, J. Jamari, and M. M. Suryaman

Abstract In this paper, the effect of slip on the lubrication performance (i.e. load support, friction) is investigated by computational fluid dynamic (CFD) approach. Various slip placements applied on textured zone of sliding contact is of particular interest. Additionally, the comparison of the performance between the analysis with the cavitation model and that without cavitation model is conducted. It is found that the slip placement has a strong influence of the load support as well as the friction force. The slip is proven to enhance the performance (i.e. the high load support but low friction) of the bearing. It is also shown the inclusion of the cavitation model in the analysis leads to a significant difference of simulation results in comparison with the analysis without taking the cavitation model.

Keywords Cavitation · CFD (Computational fluid dynamic) · Slip · Texture

1 Introduction

Recently, the researches dealing with textured lubricated contacts was paid much attentions by wookers. This is due to the surface texturing is able to enhance lubrication characteristics. From the perspective of hydrodynamic lubrication theory, some important efforts have been directed toward exploring the mechanisms of textured features in more detail. The magnitude of the additional load support enabled by textured surface in hydrodynamic lubrication was highlighted by Yagi et al. [1]. The comparison of two boundary conditions, i.e. periodic boundary condition and atmospheric boundary condition. Later, Zhang et al. [2] employed a rectangular array of circle dimples to alter the film thickness profile. They found that the increased load support can be enhanced through appropriate arrangement of textures partially covering its sleeve. In recent lubrication, Shinde and Pawar [3] highlighted the optimal parameters of surface texturing in improving the performance of journal bearing.

M. Muchammad (✉) · M. Tauviqirrahman · J. Jamari · M. M. Suryaman
Mechanical Engineering Department, Engineering Faculty, Diponegoro University, Jl. Prof. H. Soedharto SH., Tembalang, Semarang 50275, Indonesia
e-mail: m_mad5373@yahoo.com

U. Sabino et al. (eds.), *Proceedings of the 6th International Conference and Exhibition on Sustainable Energy and Advanced Materials*, Lecture Notes in Mechanical Engineering,
https://doi.org/10.1007/978-981-15-4481-1_55

In addition to the use of textured surface, the slip surface became a popular approach to improve the lubrication performance. Interesting work dealing with the influence of large-area texture/slip surface on the performance of lubricated sliding contact was conducted by Lin et al. [4]. They found that texture/slip surface would not affect the pressure as well as load support when it locates at cavitation zone. Tauviqirrahman et al. [5] evaluated the location of the boundary slip on the textured surface based on modified Reynolds equation considering cavitation. Susilowati et al. [6] studied the influence of gap ratio of the parallel sliding contact on the performance of slip-textured bearing. The authors found that the gap ratio affect significantly the friction force and the volume flow rate. In general, combined method of texturing with the boundary slip gives more advantages compared to pure texturing with respect to the load support as well as the friction. Thus, the study of improvement of tribological performance via boundary slip and surface texturing and simultaneously have received an explosion of interest recently.

Starting from this main frame, the main of the present paper is to investigate the effect of slip arrangement of the textured bearing on the hydrodynamic pressure. The computational fluid dynamic (CFD) method is employed and coupled with the cavitation model. Two-phase cavitation model is adopted to obtain more realistic result.

2 Analysis

2.1 Governing Equation

In the present study, the lubrication problem is solved by the Navier-Stokes and continuity equations. The Navier–Stokes (N-S) equations are solved over the domain using a finite-volume method with the commercial CFD software package FLUENT®. The Navier–Stokes and the continuity equations can be expressed, respectively,

$$\rho \frac{Du_i}{Dt} = -\frac{\partial p}{\partial x_i} + \rho G_i + \frac{\partial}{\partial x_j}\left[2\eta e_{ij} - \frac{2}{3}\eta(\nabla . u_i)\delta_{ij}\right] \quad (1)$$

$$\nabla \cdot \mathrm{u} = 0 \quad (2)$$

Not like the previously published works in which the cavitation model is ignored, in the present study, the cavitation effect is taken into account. In FLUENT®, there are three available cavitation models: Schneer and Sauer model, Zwart-Gelber-Belamri model and Sighal et al. model [7]. In this study, the Zwart-Gelber-Belamri is employed due to their capability (less sensitive to mesh density, robust and converge quickly [7]).

In cavitation, the liquid-vapor mass transfer (evaporation and condensation) is governed by the vapor transport equation [7]:

$$\frac{\partial}{\partial t}(\alpha_{v}\rho_{v}) + \nabla.(\alpha_{v}\rho_{v}\mathrm{v}) = R_{g} - R_{c} \tag{3}$$

where α_v is vapour volume fraction and ρ_v is vapour density. R_g and R_c account for the mass transfer between the liquid and vapour phases in cavitation. For Zwart-Gelber-Belamri model, the final form of the cavitation is as follows:

$$\text{if}\quad p \leq p_{\upsilon},\quad R_{g} = F_{evap}\frac{3\alpha_{\text{nuc}}(1-\alpha_{\upsilon})\rho_{\upsilon}}{R_{\text{B}}}\sqrt{\frac{2}{3}\frac{\text{P}_{\upsilon}-\text{P}}{\rho_{\ell}}} \tag{4}$$

$$\text{if}\quad p \geq p_{\upsilon},\quad R_{c} = F_{cond}\frac{3\alpha_{\upsilon}\rho_{\upsilon}}{R_{\text{B}}}\sqrt{\frac{2}{3}\frac{\text{P}-\text{P}_{\upsilon}}{\rho_{\ell}}} \tag{5}$$

where F_{evap} = evaporation coefficient = 50, F_{cond} = condensation coefficient = 0.01, R_B = bubble radius = 10^{-6} m, α_{nuc} = nucleation site volume fraction = 5×10^{-4}, ρ_l = liquid density and p_v = vapour pressure.

2.2 *CFD Model*

In this section, textured bearing with single macro cavity is of particular interest as shown in Fig. 1. The model is inspired by Dobrica dan Fillon [8]. The textured bearing has a particular pattern as follows: $\lambda = 2$, $\rho_t = 0.5$ and $S = 1.0$. The slip arrangement is varied on top surface of bearing. The isothermal boundary condition is assumed. The corresponding Reynolds number Re is 57.3. The commercial software based on computational fluid dynamic ANSYS Workbench is used. For the computation, the hexahedral grid is used, as reflected in Fig. 2. The grid consists of multiples blocks to ensure the fine mesh in the computational domain. Table 1 shows the specification of the mesh used here.

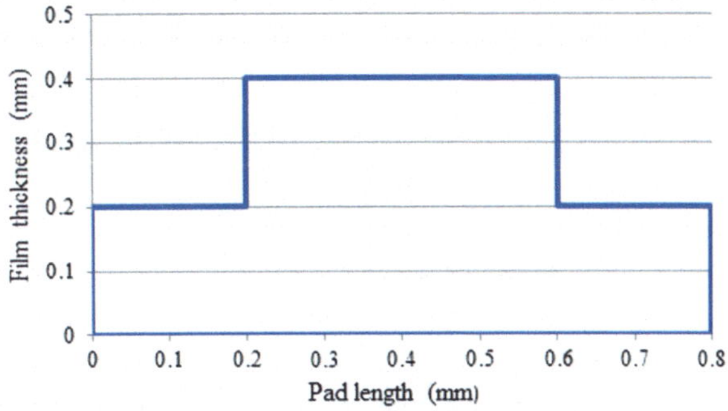

Fig. 1 Computational domain of textured bearing

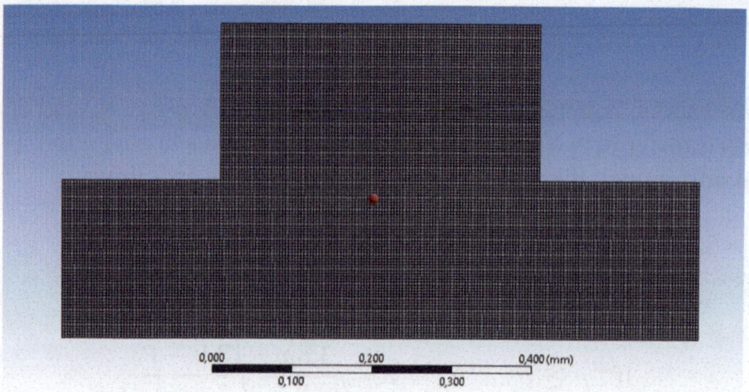

Fig. 2 Meshed model of textured bearing

Table 1 Mesh specification studied here

Method	Quadrilateral
Sizing	3.75×10^{-3} mm
Element number	169.96
Node number	17.28
Max skewness	2.99×10^{-3}
Min skewness	4.33×10^{-7}
Average skewness	1.02×10^{-3}

For following computations, the lubricant density ρ is 860 kg/m^3, the lubricant dynamic viscosity η is 0.03 Pa-s. For the vapour behavior, the parameter is as follows: the vapour density ρ_{vap} is 1.256×10 kg/m^3 and the vapour viscosity η_{vap} is 0.02556 Pa-s. Figure 3 shows the boundary condition adopted here. For two sides of the contact, the periodic boundary condition is adopted here, while for bottom surface, the sliding velocity is applied on that wall. It should be noted that in the present study, the proposed CFD model as well as the solution set up has been validated according to the reference. In detail, the validation is shown in Appendix. In the

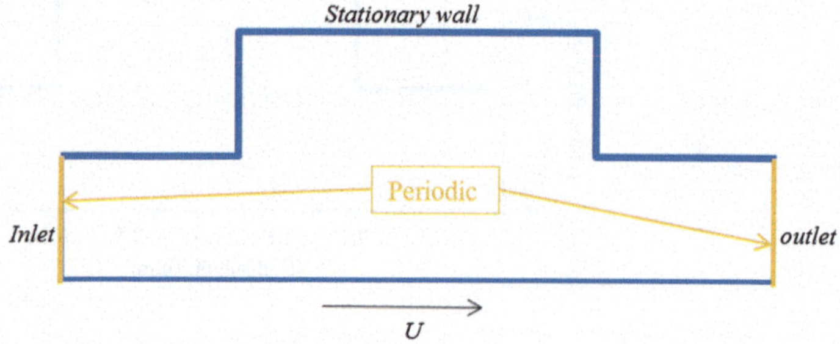

Fig. 3 Boundary condition adopted here

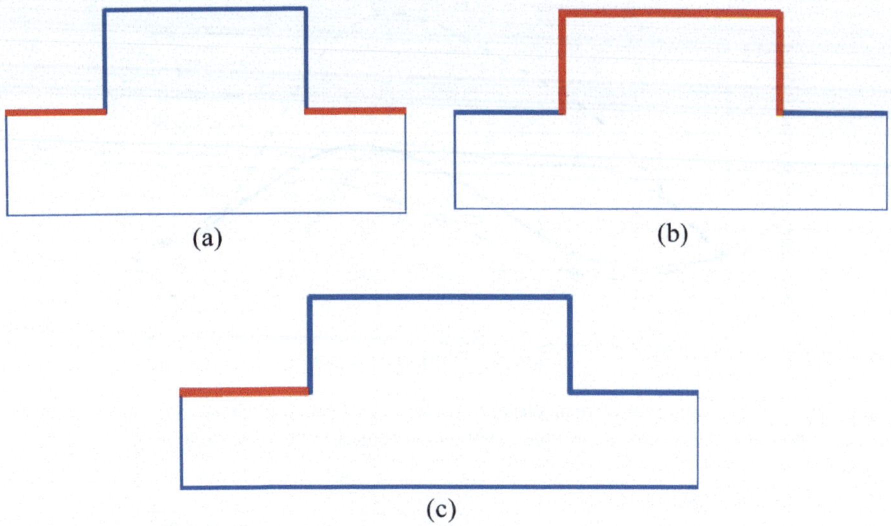

Fig. 4 Slip arrangement of textured bearing presented here **a** slip applied on texture valley (referred as Slip I), **b** slip applied on textured edges (referred as Slip II), **c** slip applied on the leading edge of the contact (referred as Slip III)

present work, the arrangement of slip boundary is of particular interest. It is hypothesized that the slip arrangements can alter the flow behavior during lubrication. Three slip arrangements are considered on textured lubricated contact, as reflected in Fig. 4.

3 Results and Discussion

Figures 5 and 6 show the hydrodynamic pressure for various slip surface conditions for the "no-cavitation" analysis and "cavitation" analysis. Based on these figures, dealing with the slip effect, it can be observed that applying the slip condition on a particular area has a different effect in comparison to the non-slip condition. The notable finding is that applying a slip to a certain area can eliminate the negative pressure generated under no-slip conditions. It can be seen that by applying the slip area at the leading edge of the contact can produce a pressure distribution that is entirely positive. Even though the application of slips in the texture edges (Slip I) is positively pressured. For Slip II (i.e. slip applied on texture valley), the slider bearing actually turns negative.

Dealing with the cavitation effect, the hydrodynamic pressure generated by several slip configurations is depicted in Fig. 6. It can be seen that the pressure distribution produced is slightly higher compared to conditions without cavitation modelling (see Fig. 5). It indicates that the cavitation can be considered as other mechanism of pressure generation which leads to additional lift generation. The simulation results

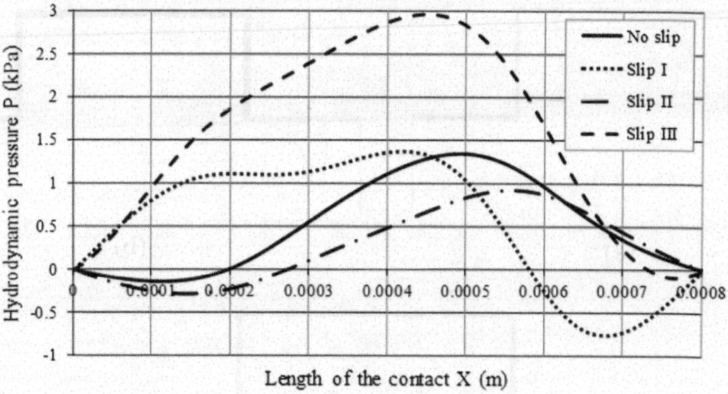

Fig. 5 The hydrodynamic pressure profiles of *textured slider bearing* at $Re = 57.3$ for various slip surface conditions. All results are evaluated without considering the cavitation model

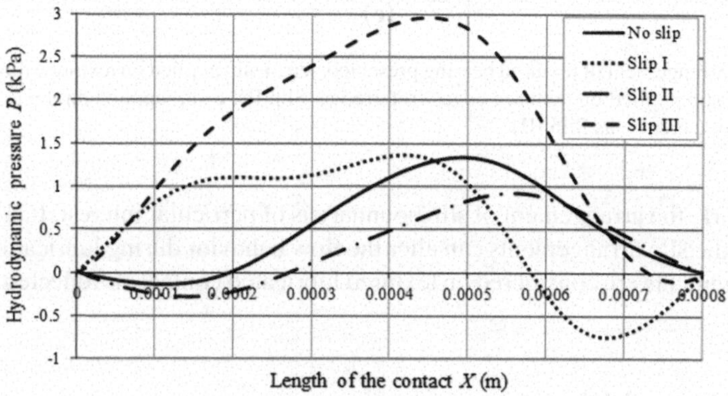

Fig. 6 The hydrodynamic pressure profiles of *textured slider bearing* at $Re = 57.3$ for various slip surface conditions. All results are evaluated considering the cavitation model

also reflect that the profiles have a similar trend quantitatively with the results without considering the cavitation model.

In order to represent the tribological performance, Figs. 7 and 8 reflects the load support and the friction force, respectively, by comparing two conditions, i.e. with cavitation model, and without cavitation model for various surface conditions. In terms of load supports as reflected in Fig. 7, it can be seen that applying slip at the leading edge of the contact gives the highest performance of load support. For the case of which the slip is applied on textured edges (Slip II), the lowest load support is achieved. As shown in Figs. 5 and 6, respectively, in the case of no-cavitation and cavitation, the Slip II pattern gives the lowest pressure profile when the lubricant enters the contact. Based on Fig. 7, it can be deduced that the cavitation produces the higher load support for all cases studied here. It also indicates that the chosen geometry of textured bearing encourages the existence of the cavitation phenomena.

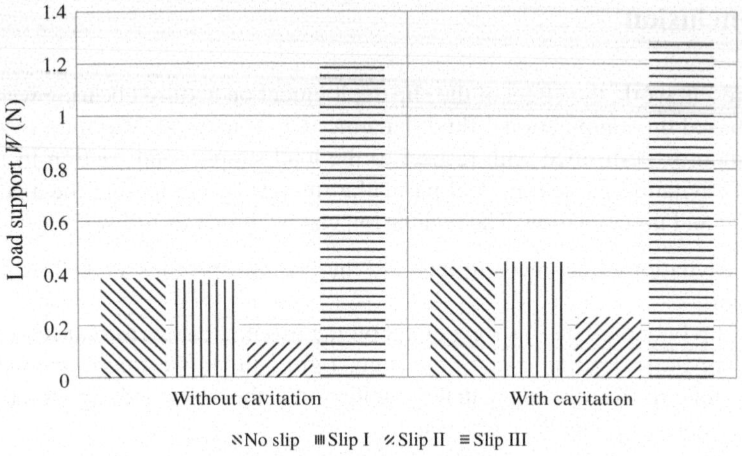

Fig. 7 The predicted load support for several surface conditions

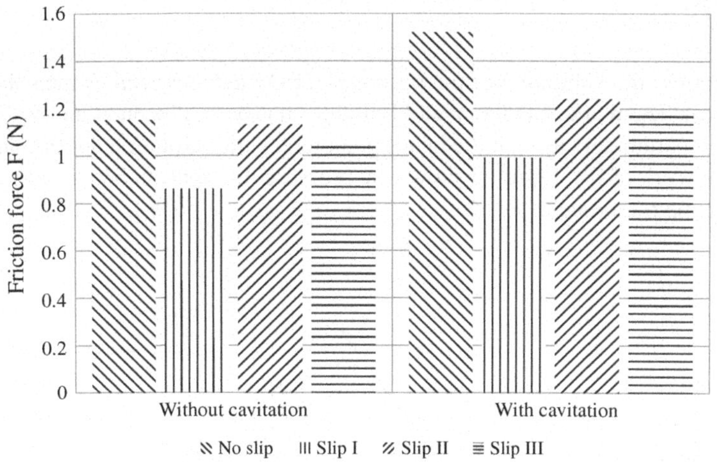

Fig. 8 The predicted friction force for several surface conditions

For complete analysis, in addition to the load support, the tribological performance of friction force predicted by the textured pattern with slip for different arrangements is reflected in Fig. 8. It can be found that the Slip I pattern (i.e. slip is applied on texture edges) gives the lowest friction. This is as expected because physically the slip boundary tends to reduce the shear stress. If the comparison is made with respect to the load support result, the best slip design dealing with the friction force reduction is different. Therefore, the optimization should be conducted to determine the best design of textured bearing with well-chosen slip arrangement so as the highest load support but low friction can be achieved at the same time.

4 Conclusion

In the present work, the effect of the slip deployment on textured bearing was investigated based on computational fluid dynamic (CFD) method. Variation of the slip arrangement is performed with respect to the load support and friction force. The effect of cavitation modelling is of particular interest by comparing the analysis of cavitation and no-cavitation. The conclusion can be drawn as follows:

1. The cavitation effect tends to increase the pressure profiles as well as the load support.
2. The slip boundary has a strong effect on the tribological performance.
3. The slip at the leading edge of the contact gives the highest performance of the load support. But, dealing with the friction force, the best slip design is the design when more slip boundary is applied.

Appendix

In this section, the validation of the proposed CFD model as well as the solution set up is shown. The work of Dobrica and Fillon [8] has been chosen as a reference. The simulation parameters as well the bearing geometry for validation are the same with that of the work of [8]. Figure 9 shows the pressure distribution between the present study and the reference [8].

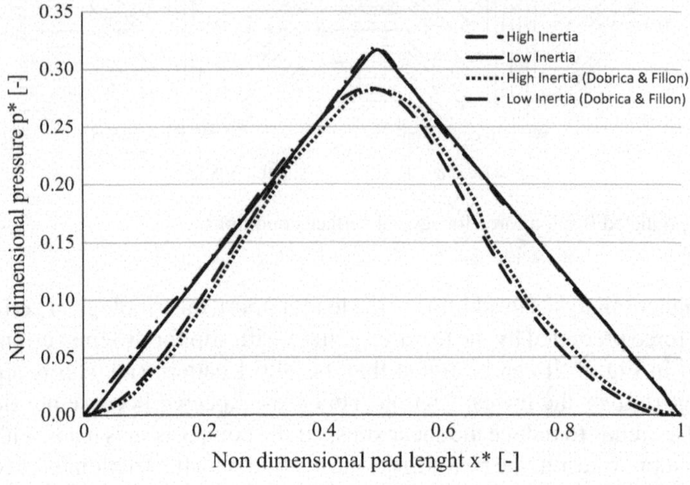

Fig. 9 The comparison between the pressure distribution between the present study and the reference [8]

References

1. Yagi K, Sato H, Sugimura J (2015) On the magnitude of load-carrying capacity of textured surfaces in hydrodynamic lubrication. Tribol Online 10(3):232–245
2. Zhang H, Dong G, Hua M, Guo F, Chin KS (2015) Parametric design of surface textures on journal bearing. Ind Lubr Tribol 67(4):359–369
3. Shinde AB, Pawar PM (2017) Multi-objective optimization of surface textured journal bearing by Taguchi based grey relational analysis. Tribol Int 114:349–357
4. Lin Q, Wei Z, Wang N, Chen W (2015) Effect of large-area texture/slip surface on journal bearing considering cavitation. Ind Lubr Tribol 67(3):216–226
5. Tauviqirrahman M, Ajie WK, Yohana E, Muchammad M, Jamari J (2016) A Study of slip position on improving the hydrodynamic lubrication performance of single-textured bearing using a mass conserving numerical approach. Int J Eng Technol 8(2):913–920
6. Susilowati S, Tauviqirrahman M, Jamari J, Bayuseno AP (2016) Numerical investigation of the combined effects of slip and texture on tribological performance of bearing. Tribol Mater Surf Interfaces 10(2):86–89
7. Muchammad M, Tauviqirrahman M, Jamari J, Schipper DJ (2017) An analytical approach on the tribological behaviour of pocketed slider bearings with boundary slip including cavitation. Lubr Sci 29(3):133–152
8. Dobrica MB, Fillon M (2009) About the validity of Reynolds equation and inertia effects in textured sliders of infinite width. Proc Ins Mech Eng Part J: J Eng Tribol 223(1):69–78

Effect of Reinforcement (Al_2O_3) Preheating on Hardness and Microstructure of Aluminum Matrix Composite

I. Setia, E. Surojo, and D. Ariawan

Abstract Aluminum matrix composites is aluminum matrix reinforced with metal, ceramic, or organic compound. The reinforcement is used to improve mechanical, thermal, or tribological properties of the matrix. Stir casting is an economical process to fabricate the aluminum matrix composites. Stir casting is conducted by mechanical stirring of the reinforcement particulate into metal liquid. The metal matrix composites resulted from the stir casting is affected by a homogeneous distribution of the reinforcement. The objective of this research is to investigate the effect of reinforcement preheating that has been on electroless coating on hardness and micro structure of aluminum metal matrix composite. This study used aluminum series 6061 as a matrix and Al_2O_3 as reinforcement. The temperatures of reinforcement preheating were varied at 200 °C, 300 °C, 400 °C, and 500 °C. Stirring temperature parameters of 700 °C ± 10 °C, melting temperature of 750 °C ± 10 °C, pouring temperature of 750 °C ± 10 °C, stirring times of five minutes, and stirring speed of 550 rpm. The results showed that the preheating Al_2O_3 reinforcement can reduce segregation. As the preheating temperature increases, reinforcement segregation decreases, resulting in an increase in the wettability of Al_2O_3 reinforcement in aluminum.

Keywords Aluminum · Metal matrix composite · Preheating

1 Introduction

Aluminum matrix composites (MMC) are made by dispersing reinforcement material such as metals, ceramics or organic compounds into aluminum matrix. The reinforcement is incorporated into aluminum to improve its properties such as strength, stiffness, and wear resistance. Aluminum matrix composites are widely used in several applications due to their high strength and hardness compared to other structural

I. Setia (✉) · E. Surojo · D. Ariawan
Department of Mechanical Engineering, Faculty of Engineering, Universitas Sebelas Maret, Surakarta, Indonesia
e-mail: irawansetia123@gmail.com

E. Surojo
e-mail: esurojo@ft.uns.ac.id

U. Sabino et al. (eds.), *Proceedings of the 6th International Conference and Exhibition on Sustainable Energy and Advanced Materials*, Lecture Notes in Mechanical Engineering,
https://doi.org/10.1007/978-981-15-4481-1_56

materials [1]. Aluminum matrix composites are applied in the automotive, aircraft, and electronics industries because of their light weight and high strength [2].

Stir casting is a metal processing method that is suitable for metal matrix composites due to its simplicity and low cost. The first step in stir casting is metal matrix melting. The main parameters of stir casting are stirring speed, stirring period, impeller blade angle, impeller size and impeller position [3]. Even though stir casting is simple and requires low cost in making metal matrix composites, it also has several problems.

One of the problems in the stir casting method is wettability of reinforcement. Wettability is the wetting capacity of the reinforcement by an aluminum matrix. Poor wettability results in non-uniform distribution of the reinforcement. One method to improve wettability and homogeneity of the reinforcement distribution is by reinforcement preheating [4].

2 Experimental Procedure

This study used aluminum 6061 series as a matrix and Al_2O_3 as reinforcement. The Al_2O_3 powder is the most widely used and cost-effective material having high strength and hardness, as well as excellent thermal conductivity, good size and shape stability, and good wear resistance [5]. The variations of the preheating temperature were 200 °C, 300 °C, 400 °C, and 500 °C. The Brinell hardness test was conducted using a load of 62.5 kgf and indenter diameter of 2.5 mm. The composition and mechanical properties of aluminum alloy 6061 are shown in Table 1 and Table 2, respectively.

The Al_2O_3 reinforcement was heated in oven with temperatures of 200, 300, 400, 500 °C $\pm$ 10 °C. Aluminum 6061 series was inserted into the furnace. After the furnace reached a temperature of 700 °C $\pm$ 10 °C, the preheated Al_2O_3 reinforcement was added into the molten metal, and then stirring process was began. The molten metal stirring process was carried out using equipment as shown in Fig. 1 with parameters: stirring temperature of 700 °C $\pm$ 10 °C, pouring temperature of 750 °C $\pm$ 10 °C, and stirring speed of 550 rpm. After stirring for five minutes, the molten metal was poured into the mold. After that, the casting was taken from the mold, and then cut into test specimen. There are five specimens for Brinell hardness test and microstructure examination as shown in Fig. 2. The microstructure of metal matrix composites (Al6061 + Al_2O_3) was examined using metallurgical microscope (Euromex Holland). The Brinell hardness test was carried out using Electronic Brinell Hardness Tester (Lauzhou Huayin) according to ASTM E10.

Table 1 Chemical composition of aluminum 6061 series [6]

Constituent	Si	Cu	Fe	Mn	Ni	Zn	Ti	Sn	Mg	Cr	Pb	Al
% Weight	0.43	0.24	0.43	0.139	0.05	0.006	0.022	0.001	0.802	0.184	0.204	Bal

Table 2 Mechanical Properties of aluminum 6061 [6]

Material	Ultimate tensile strength (MPa)	Yield strength (MPa)	Hardness (BHN)	Density (g/cc)
Al 6061	110–115	45–55	30	2.7

Fig. 1 Stir casting equipment

Fig. 2 Specimens of casting

3 Results and Discussion

3.1 Microstructure

Figure 3 shows the results of the microstructure observation. Figure 3(a, b), shows the presence of Al_2O_3 segregation in the specimen. This occured due to the agglomeration and inhomogeneous dispersion of the Al_2O_3 reinforcement. In this study,

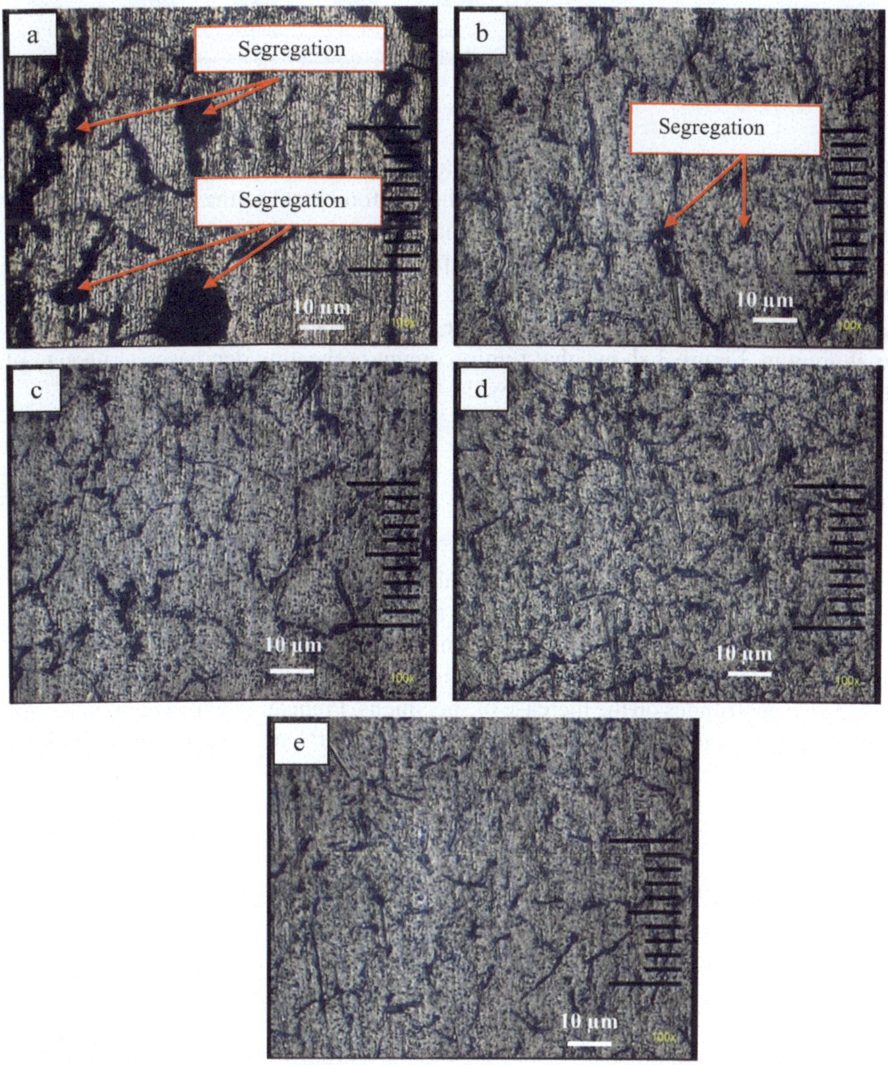

Fig. 3 Microstructure of the specimens: **a** without preheating; preheating at temperature of, **b** 200 °C, **c** 300 °C, **d** 400 °C, **e** 500 °C

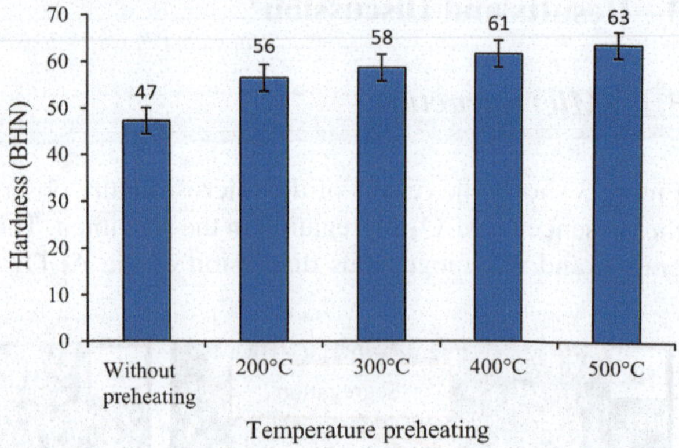

Fig. 4 Hardness of the specimens

the segregation that appeared was the agglomeration of Al_2O_3 reinforcement which occured due to the lack of wettability of the reinforcement to the matrix [6, 7]. One method to overcome segregation is by reinforcement preheating. In Fig. 3(c–e), it can be seen that the Al_2O_3 reinforcement preheating can reduce Al_2O_3 segregation. As the preheating temperature increased, there was a reduction in metal segregation because there was an increase in the wettability of the Al_2O_3 reinforcement in aluminum. Figure 3 shows that reinforcement preheating decreases grain size of aluminum and Al_2O_3 particles occupy the grain boundary.

3.2 *Hardness*

Figure 4 shows that reinforcement preheating increases the hardness of AMC specimen. The increase in the hardness value is due to the good reinforcement particle wettability, resulting in a strong bonding interface between the Al_2O_3 reinforcement and the aluminum matrix [8]. Mechanism of grain refinement hardening as shown in Fig. 3 also contributes on the increase of specimens. Preheating with a temperature of 500 °C resulted in the highest hardness value of 63 BHN, and reinforcement without preheating resulted in the lowest hardness value of 47 BHN.

4 Conclusions

From the conducted research, the following conclusions can be drawn: Al_2O_3 reinforcement preheating affect reinforcement segregation and matrix grain size. The reinforcement segregation and matrix grain size decrease with increasing temperature of reinforcement preheating. The increase in hardness value is because of matrix grain refinement and good wettability of the reinforcement in aluminum matrix. Reinforcement preheating with a temperature of 500 °C results in the highest hardness value of 63 BHN, and reinforcement without preheating has the lowest hardness value of 47 BHN.

References

1. James J, Ganesan M, Santhamoorthy P, Kuppan P (2018) Development of Hybrid Aluminum Metal Matrix Composite and Study of Property. Material Today Proceedings 5:13048–13054
2. Kevin KP, Sijo MT (2015) Effect of Stirrer Parameter of Stir Casting on Mechanical Properties of Aluminum Silicon Carbide Composite. International Journal of Modern Engineering Research (IJMER) 5:43–49
3. Mohit KS, Sahu RK (2018) Fabrication of Aluminum Matrix Composites by Stir Casting Technique and Stirring Process Parameters Optimization. Additional Information is Avaible at the end of the Chapter 7:112–126
4. Mahesh VP, Nair PS, Rajan TPD, Hubli RC (2011) Processing of Surface-Treated Boron Carbide-Reinforced Aluminum Matrix Composites by Liquid-Metal Stir-Casting Technique. J Compos Mater 45(23):2371–2378
5. Mithun BR, Madeva N, Auradi V, Bharath V (2017) Microstructure and Mechanical Properties of Cu-Coated A_2O_3 Particulate Reinforced 6061 Al Metal Matrix Composite. Material Today Proceeding 4:11015–11022
6. Juhua L, Zhengzhi Z, Di T, Ye N, Shufeng Y, Weining L (2017) Microstructural Homogeneity and Mechanical Property of Medium Manganese Steel with Mn Segregation Banding by Alternating Lath Matrix. Mater Sci Eng, A 711:175–181
7. Guoyun L, Hui X, Simeng L, Cunshan W, Qiang Z, Lijun S (2019) Quasi-Continuous-Wave Laser Surface Melting of Aluminium Alloy: Precipitate Morphology, Solute Segregation and Corrosion Resistance. Corros Sci 152:109–119
8. Phanibhushana MV, Chandrappa CN, Niranjan HB (2017) Study of Wear Characteristics of Hematite Reinforced Aluminum Metal Matrix Composites. Material Today Proceedings 5:3484–3493

The Properties of Nanofiber Membranes Made of Aloe Vera Gel Combined with Polyvinyl Alcohol

Harini Sosiati, Apriyanto, and Abdul Rahim Safarudin

Abstract Aloe Vera gel (AVG), and polyvinyl alcohol (PVA) are compatible materials with the human tissues. The current research studied the characteristic of nanofiber membranes made of AVG (aloe Vera filtrate/AVF) and aloe Vera extract (AVE) combined with PVA with varying AVF concentrations of 10, 20, and 30 wt.%), while AVE concentrations ranged from 0 to 5 wt.%. Changes in surface morphologies and tensile properties of the nanofiber membranes due to AVF and AVE concentrations and the correlation effect between them are the goals of this research. The fiber diameter and tensile strength of AVF/PVA membranes increased with AVF concentration. The AVF/PVA membrane with 30 wt.% AVF showed the highest tensile strength (3.58 MPa) and 233 nm in fiber diameter. The addition of 1% AVE concentration, significantly improved the tensile strength and modulus of AVE/AVF/PVA membrane: i.e., 8.78 MPa and 26.8 MPa, respectively, with a smaller fiber diameter around 124 nm. At AVE concentration higher than 1%, the tensile properties gradually decreased, and fiber diameter increased. As a result, the properties of the AVE/AVF/PVA membranes could be the candidate material for wound dressing due to their comparable properties to that of natural skin.

Keywords Aloe vera gel · PVA · Electrospinning, nanofiber membrane · Tensile properties

1 Introduction

Aloe Vera (AV) grows naturally in a tropical climate and low rainfall area, mainly in Asia, Africa, Europe, and America [1, 2]. Numerous AV-based industrial products, such as foods and cosmetics, have been well known. AV is also potentially used in medicines due to its significant antioxidant vitamins, which can improve wound healing [2]. Based on various beneficial substances contained in AV, and its properties

H. Sosiati (✉) · Apriyanto · A. R. Safarudin
Department of Mechanical Engineering, Faculty of Engineering, Universitas Muhammadiyah Yogyakarta, Jl. Brawijaya, 55183 Kasihan, Bantul, Yogyakarta, Indonesia
e-mail: hsosiati@ft.umy.ac.id

U. Sabino et al. (eds.), *Proceedings of the 6th International Conference and Exhibition on Sustainable Energy and Advanced Materials*, Lecture Notes in Mechanical Engineering,
https://doi.org/10.1007/978-981-15-4481-1_57

of antibacterial and anti-inflammatory, research and development of the AV-based materials have been extensively carried out.

Studies of AV-based nanofiber membranes have been reported [3–6]. Aghamohamadi et al. [3] have investigated the preparation and various characterizations of AV/polyvinylpyrrolidone (PVP) with different concentration of PVP, and AV acetate/PVP. AV powder and AV *acetate* are present in the crystalline phase. The addition of AV into the PVP increases the fiber diameter of AV/PVP membrane, but the addition of AV *acetate* reveals an adverse effect. However, by enhancing the concentration, both AV and AV acetate increase the fiber diameter of the membranes, and the membrane of AV acetate/PVP 6 wt% has shown the smallest fiber diameter (~95 nm). In this case, the fiber diameter does not linearly affect the tensile properties of the membranes, but the use of AV acetate increased the thermal stability of the membrane.

Studies on the properties of AV extract (AVE)/poly (vinyl) alcohol (PVA) nanofiber membranes have been previously investigated [4–6]. Inserting AVE into the PVA solution reduces the fiber diameter of the AVE/PVA membrane [4, 5] regardless of the molecular weight (Mw) of PVA. The addition of a similar AV concentration into the 10% PVA solution with a different Mw of PVA resulted in a different fiber diameter. The smallest one around 123 nm shown by using the highest Mw (85,000–124,000 g/mol) [4], whereas the use of Mw (22,000 g/mol) [5] and (66,000) [6] resulting in larger fiber diameter. The viscosity of the polymer solution as the spinning solution and other main parameters of the electrospinning process provide an impact on the morphology of membrane (fiber diameter and fiber structure) [5, 6], and also the mechanical properties of the membrane [5].

On the other hand, AV-based electrospun nanofiber membranes have also been investigated and developed for wound dressing and wound healing [7, 8], and tissue engineering [9, 10] applications. The membranes made of AV, chitosan (CS) and polyethylene oxide (PEO) reinforced polycaprolactone (PCL) [7], and AV filled in a mixture of PVA, PVP and polyethylene glycol (PEG) [8] have been investigated for wound dressing and wound healing. The AV/PCL membranes with some different AV concentrations studied for tissue engineering applications [9].

For wound dressing application, the material should be at least high in tensile strength, but not too high in tensile modulus as in the tensile properties of natural skin. Tensile and modulus strengths and elongation at break of the natural skin are ranging between 5–30 MPa, 4.6–20 MPa, and 35–115%, respectively. [7]. The membrane material having the tensile modulus much higher than that of natural skin tends to be easily peeled off from the skin surface. Inversely, the membrane would blend with skin when its tensile modulus is around the natural skin properties. However, a trend indicating that the higher the tensile strength, the higher the tensile modulus is generally shown in the previous results [7, 11].

Therefore, an effort to modify the membrane materials for achieving the tensile properties in the range of natural skin is still a challenge.

In the current research, both AV leaf gel and commercial AV extract (AVE) were used as reinforcing materials for PVA. Fabrication of the nanofiber membranes made of AV leaf gel in terms of AV filtrate (AVF) with PVA and AVE with AVF/PVA has been carried out to characterization and of the properties of membranes and compare the properties to those of the natural skin. The current discussion focuses on the effect of the AVF and AVE concentrations on the fiber structure and tensile properties of the produced membranes, and their correlation effect.

2 Experimental

2.1 *Materials and Preparation of the Spinning Solutions*

AV leaf gel, commercial AV extract powder, and PVA Gohsenol (PVOH/PVA, Mw: ~22,000 g/mol) were used as the primary component materials for fabricating the nanofiber membranes. The gel obtained from AV leaf was filtered to be the AV filtrate (AVF). In this case, filtration of the gel was repeated three times to make AVF in good quality. Ten % PVA solution as the matrix prepared by dissolving 10 g PVA powder into 100 g H_2O and was magnetically stirred at 80 ± 2 °C for one hour.

There are two spinning solutions: i.e., AVF/PVA and AVE/AVF/PVA. The first solution of AVF/PVA was prepared by blending AVF in the 10% PVA with varying AVF concentrations of 0, 10, 20, and 30% (w/w). Before preparing the second spinning solution of AVE/AVF/PVA, optimization of the tensile strength of AVF/PVA membranes is carried out. An optimum condition of the spinning solution resulted from the first process was then used as the matrix material reinforced with varying AVE concentrations (1, 3, and 5 wt.%). The spinning solutions of AVF/PVA and AVE/AVF/PVA were prepared similarly. Each of them was mixed by a magnetic stirrer while heated at 80 ± 2 °C for one hour and continued by stirring the solution at room temperature for two hours. The viscosity and electrical conductivity of those spinning solutions were measured.

2.2 *Preparation of the Nanofiber Membranes and Characterization*

All the spinning solutions of AVF/PVA and AVE/AVF/PVA were fabricated to be the nanofiber membranes at the optimum conditions: i.e., at applied voltage 15 kV, a fix distance from the needle tip to a collector plate (TCD) 12.5–16 cm, a syringe needle diameter of 8 mm and feed rate 0.5 ml/h. The process parameters were optimized based on the optical micrographs of the thin membranes. The membrane that shows defects-free fibers and straight-oriented fibers are known as reaching optimum parameters. The membranes resulted from the first spinning solutions were designated as Neat PVA, AVF-10/PVA, AVF-20/PVA, and AVF-30/PVA, whereas those from the other spinning solutions were known as AVF-30/PVA, AVE-1/AVF-30/PVA, AVE-3/AVF-30/PVA, and AVE-5/AVF-30/PVA.

The surface morphology of all membrane specimens, including the fiber diameter and fiber structure, were examined by scanning electron microscopy (SEM, Hitachi SU-3500). Fiber diameter in each membrane was measured from at least one hundred points by the ImageJ digital image analysis. The tensile test was conducted on all membrane specimens according to ASTM 882 using an ultimate tensile machine (UTM, Zwick Z0.5 Germany) at a crosshead speed of 10 mm/min and a gauge length of 20 mm. Five pieces of specimens are carried out for each tensile testing.

3 Results and Discussion

3.1 Morphological Characterization

SEM images of the nanofiber membranes (Fig. 1) exhibit that each membrane has a different fiber diameter and structure, leading to having a difference in properties. In the electrospinning process the characteristic of produced membranes is mainly affected by the properties of polymer solution as the spinning solution (viscosity which is related to the concentration, electrical conductivity), and processing parameters (applied voltage, TCD, the diameter of the spinneret, flow rate) [12, 13]. In general, increasing the spinning solution viscosity increases the fiber diameter [5, 11].

AV filtrate reinforced PVA membranes demonstrate comparatively small fiber diameter, and straight-oriented fibers in a neat PVA, AVF-10/PVA, AVF-20/PVA membranes, with average fiber diameter increases from 136 to 166 nm (Fig. 1a, b, c). The AVF-30/PVA (Fig. 1d), shows the most substantial fiber diameter (223 nm) among the AVF/PVA membranes, and straight-oriented fibers are not formed. The most significant difference shown in this membrane is strong bonding between the fibers. Their fiber diameters are present in the range from 0 to 300 nm (Fig. 2). Figure 3 shows the average fiber diameter for all membranes, which increases with both AVF and AVE concentrations. Those are generally corresponding to the enhancing viscosity and decreasing the electrical conductivity of the spinning solution. However, viscosity and electrical conductivity in the present study did not change linearly. The trend is the opposite of the general.

On the other hand, introducing a small amount of AVE with a concentration of 1% into the AVF-30/PVA changed the fiber structure and significantly decreased the fiber diameter (Fig. 1e). An increase of AVE concentration to 3 and 5%, however, remarkably improved the fiber diameter and changed the fiber structure. Among all nanofiber membranes, the combination of 30% AVF and 1% AVE in the PVA solution seemed to be the best. The distribution of its fiber diameter in the range from 0 to 100 nm is the highest, indicating that the membrane has the smallest average fiber size (124 nm). In contrast, the AVE-5/AVF-30/PVA membrane has the most sizable fiber size, so the distribution of the fiber size is mostly in the range from 200 to 300 nm and from 300 to 400 nm (Fig. 2).

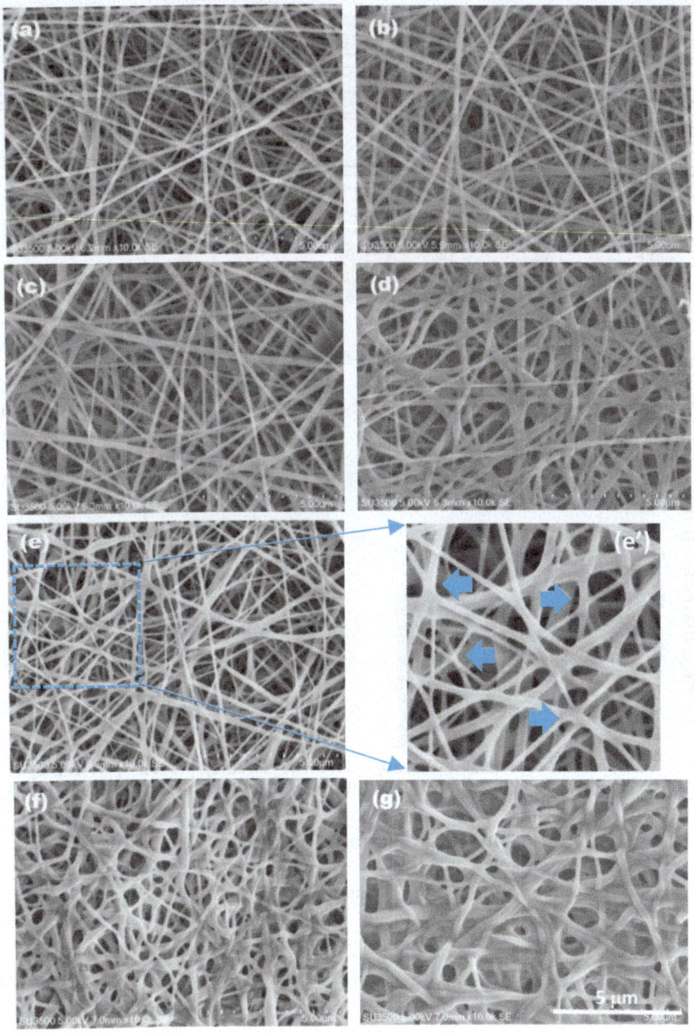

Fig. 1 SEM images of the electrospun nanofiber membranes of neat PVA (**a**), AVF-10/PVA (**b**), AVF-20/PVA (**c**), AVF-30/PVA (**d**), AVE-1/AVF-30/PVA (**e**), a magnified image of a square area marked in Fig. 1e (e'), AVE-3/AVF-30/PVA (**f**) and AVE-5/AVF-30/PVA (**g**)

The average fiber diameter of this membrane of 124 nm with the use of a low Mw of PVA (22,000 g/mol) is competitive with a fiber diameter of the produced membrane using high Mw of PVA (85,000–124,000 g/mol), namely 123 nm [4]. These results inform an essential point that is not a correlation effect between Mw of the polymer and fiber size of the produced nanofiber membrane. As mentioned above, the differences in electrospinning process parameters are the cause.

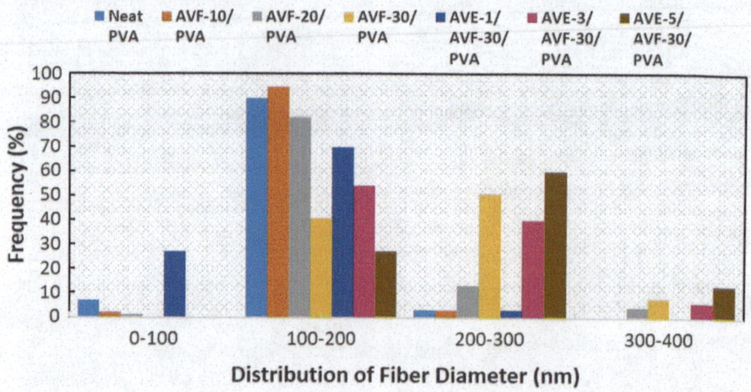

Fig. 2 Distribution of fiber diameter formed in the AVF/PVA and AVE/AVF/PVA membranes

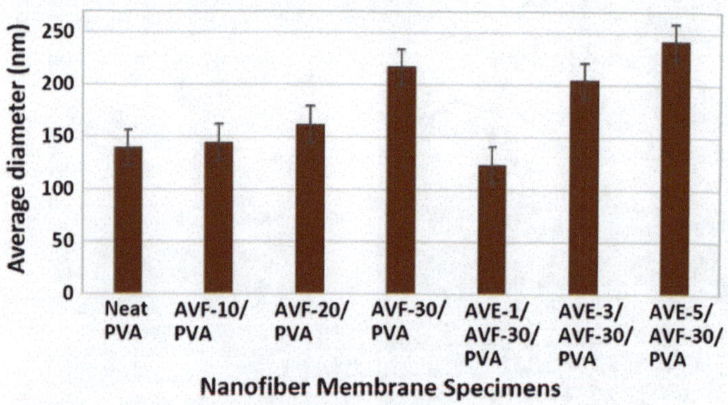

Fig. 3 Average fiber diameter of the AVF/PVA and AVE/AVF/PVA membranes

In this case, changes in the fiber diameters and structures are affected by the physical properties of the spinning solution. They also provide an impact on the mechanical property of the membranes discussed below.

3.2 *Tensile Properties of the AVF/PVA and AVE/AVF/PVA Membranes*

The tensile properties of all membrane specimens (Fig. 4) show that both tensile strength and tensile modulus of AVF/PVA are slightly higher than those of neat PVA. The maximum tensile strength of AVF-30/PVA membrane (3.58 $\pm$ 0.58 MPa) is lower than that of the natural skin (5–30 MPa), although the tensile modulus of 17.72 $\pm$ 1.58 MPa included in that indicated by the natural skin (4.6–20 MPa). Its

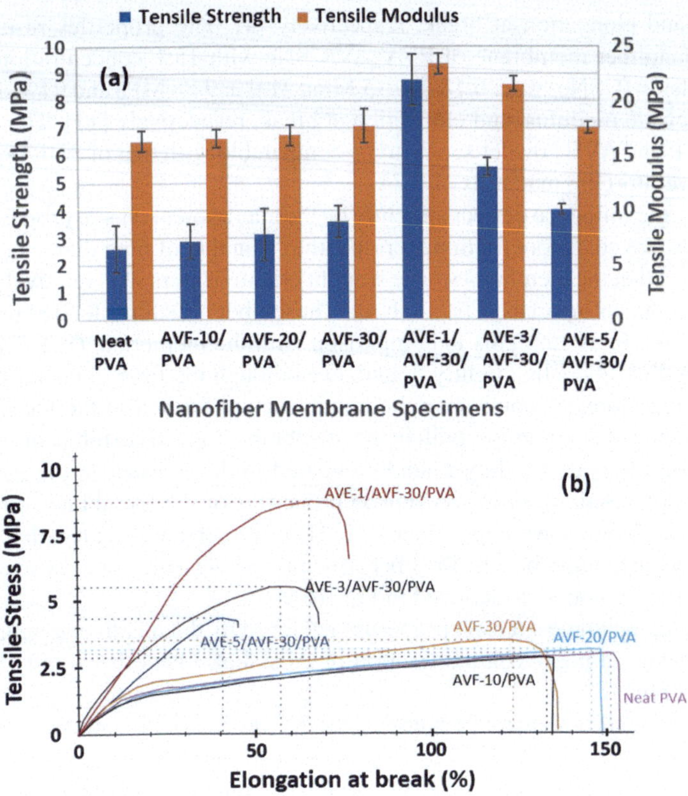

Fig. 4 Tensile strength and modulus (**a**), and tensile-stress versus elongation at break (b) of AVF/PVA and AVE/AVF/PVA membranes with different AVF and AVE concentrations.

elongation at break (135 ± 7.10%) is higher than that of the natural skin (35–115%) [7]. Therefore, the AVF/PVA membranes from this study are not recommended for wound dressing application owing to its comparatively low tensile strength.

Efforts to improve the tensile strength, which is balanced with the tensile modulus as the natural skin properties seemed to be complicated. Introducing a small amount (1%) of AVE in the AVF-30/PVA significantly improves the tensile strength from 3.58 ± 0.58 to 8.78 ± 0.94 MPa. The tensile modulus also increases from 17.72 to 23.39 MPa, but the elongation at break decreases from 135 ± 7.10 to 7575 ± 18.83% (Fig. 4a, b). Thus, comparing with the properties of the natural skin, tensile strength and elongation at break of AVE-1/AVF-30/PVA are included in the range of the natural skin properties, but the tensile modulus is slightly higher. An increase of AVE concentration noticeably reduced the tensile properties, but the elongation values are comparable, although they are limited.

The tensile properties of the AVE-1/AVF-30/PVA membrane, however, are higher in comparison with our previous works [4, 11]. The best values obtained from AV (4%)/PVA were 6.38 MPa, 34.75 MPa, and 125% for tensile strength, tensile modulus, and elongation at break, respectively [4]. The properties resulted from a hybrid nanofiber membrane of PVA_AV/CSNe with 15% concentration chitosan nano-emulsion (CSNe) were 6.18 ± 0.15 MPa, 21.9 ± 9.88 MPa and 97%, for tensile strength, tensile modulus, and elongation at break, respectively [11]. The combination of AVF and AVE powder seems to be comparable with that of AVE powder and CSNe for reinforcing materials of PVA.

Besides, the influence of fiber structure on the alteration of tensile properties of the membranes played a more prominent role rather than that of fiber diameter. It can be seen in Fig. 1 that the membranes have straight oriented fibers with relatively uniform fiber size in the entire areas (Fig. 1a, b, c). Those fiber structures tended to be lower tensile and modulus strengths in comparison with the membrane (Fig. 1d), having many cross-link fibers that tightly bound, even though the fiber diameter is slightly larger. Furthermore, by comparing the fiber structure formed in the membranes in Fig. 1d, f, g, the fiber structures built in the membrane (Fig. 1d) exhibits lower density of cross-link fibers that tightly bound compared to that present in the membranes in (Fig. 1f, g), resulting in higher tensile properties of the membranes in (Fig. 1f, g), even though they have larger fiber size. Therefore, the highest tensile properties achieved by the membrane with the fiber structure having a high density of cross-link fibers that tightly bound as depicted in Fig. 1e, e'.

Therefore, the highest tensile properties achieved by the membrane with the fiber structure having a high -volume fraction of cross-link fibers that tightly bound as depicted in Fig. 1e, e'.

A similar trend is shown in a study of the AV powder/PVP and AV acetate/PVP nanofiber membranes [3]. SEM micrographs of all membranes revealed the fiber structure with a straight-oriented fiber and having smaller fiber diameter compared to that of this work. However, the tensile strength resulted from that membranes is much lower than that of the present results.

4 Summary

The electrospinning technique has succeeded in fabricating the nanofiber membranes of PVA combined with AVF and AVE. The use of AVF shows a comparatively low tensile strength of the membrane, whereas adding small amounts of AVE improves the tensile properties significantly. Incorporation of 1% AVE powder with 30% AVF in PVA solution produces the nanofiber membrane with the tensile properties included to that of the native skin: i.e., 8.78 ± 0.94 MPa, 23.39 ± 0.94 MPa and 75 ± 18.83% for tensile strength, tensile modulus and elongation at break, respectively, which is recommended as a candidate for a wound dressing material. A significant finding from the current research is the fiber structure built in the nanofiber membrane has provided a prominent effect on the tensile properties rather than the fiber diameter, especially the fiber structure having a high density of cross-link fibers that tightly bound. This finding could be useful information related to nanofiber technology.

Acknowledgments The current study was partly supported by a research grant of "Multidisiplin" 2019, Universitas Muhammadiyah Yogyakarta, Indonesia with the contract no: 194/SK-LP3M/XII/2018. Our appreciative also goes to Dr Sinin bin Hamdan, Universiti Malaysia Sarawak, Malaysia, for his comments on the draft manuscript, and Mr Kunto Wandono for his assistance and consulting in the operation of the electrospinning machine.

References

1. Pegu AJ, Sharma MA (2019) Review on aloe vera. Int J Trend Sci Res Dev 3(4):35–40
2. Mangaiyarkarasi SP, Manigandan T, Elumalai M, Cholan P, Kaur R (2015) Benefits of aloe vera in dentistry. J Pharm Bioallied Sci 7:S255–S259
3. Aghamohamadi N, Sanjani NS, Majidi RF, Nasrollahi SA (2019) Preparation and characterization of aloe vera acetate and electrospinning fibers as promising antibacterial properties materials. Mater Sci Eng, C 94:445–452
4. Shukry NAA, Sekak KA, Ahmad MR, Effendi TJB (2014) Proceedings of the International Colloquium in Textile Engineering, Fashion, Apparel and Design 2014 (ICTEFAD 2014), vol 2014, pp 7–11
5. Sosiati H, Widodo AN, Nugroho AW (2018) The influence of aloe vera concentration on morphology and tensile properties of electrospun aloe Vera-PVA nanofiber. J Sains Materi Indonesia 19(4):157–162
6. Hikmawati D, Rohmadanik AR, Putra AP, Siswanto, Aminatun (2018) The effect of aloe vera extract variation in electrospun polyvinyl alcohol (PVA)-Aloe vera-based nanofiber membrane. J Phys Conf Ser 1120(1)
7. Miguel SP, Ribeiro MP, Coutinho P, Correia IJ (2017) Electrospun polycaprolactone/Aloe Vera_chitosan nanofibrous asymmetric membranes aimed for wound healing applications. Polymers 9(5):183-207.
8. Uslu İ, Keskin S, Gül A, Karabulut TC, Aksu ML (2010) Preparation and properties of electrospun poly (vinyl alcohol) blended hybrid polymer with aloe vera and HPMC as wound dressing. Hacettepe J Biol Chem 38(1):19–25
9. Salehi M, Farzamfar S, Bastami F, Tajerian R (2016) Fabrication and characterization of electrospun plla/collagen nanofibrous scaffold coated with chitosan to sustain release of aloe vera gel for skin tissue engineering. Biomed Eng Appl Basis Commun 28(5):1–8
10. Suganya S, Venugopal J, Agnes Mary S, Ramakrishna S, Lakshmi BS, Giri Dev VR (2014) Aloe vera incorporated biomimetic nanofibrous scaffold: a regenerative approach for skin tissue engineering. Iran Polym J 23(3):237–248 (English Edition)
11. Sosiati H, Fatihah WN, Yusmaniar, Nur Rahman MB (2018) Characterization of the properties of electrospun blended hybrid poly (vinyl alcohol)_Aloe vera/chitosan nano-emulsion nanofibrous membranes. Key Eng Mater 792: 74–79
12. Korycka P, Mirek A, Kramek-Romanowska K, Grzeczkowicz M, Lewinska D (2018) Effect of electrospinning process variables on the size of polymer fibers and bead-on-string structures established with a 23 factorial design. Beilstein J. Nanotechnol 9(1):2466–2478
13. Bera B (2016) Literature review on electrospinning process (a fascinating fiber fabrication technique). Imp J Interdiscip Res (IJIR) 2(8): 972–984

Numerical Investigation of the Sliding Contact of Tire Rubber Material Due to a Blade Sliding Indentation

B. Setiyana, J. Jamari, R. Ismail, S. Sugiyanto, and E. Saputra

Abstract The manufacturing technology of tire rubber material to improve its performance is still developing. Mechanical and tribological properties are considered in the rubber manufacturing. One of the important tribological properties is the braking capacity of the tire which is strongly associated with sliding contact. However, theoretical as well as experimental methods are still difficult to be applied to analyse these properties due to the unique behaviour of rubber material. This study proposes a numerical investigation of the sliding contact of rubber material due to a rigid blade sliding indentation using Finite Element Analysis (FEA). In this FE simulation, the rubber material is modelled as a hyper-elastic material with Money-Rivlin type for Strain Energy Function (SEF). There are three types of rubber material analysed, the first and second type are commonly used in vehicle tires i.e. vulcanized rubber 1 (R1) and vulcanized rubber 2 (R2), while the third type is a new product in the form of Solution Styrene Butadiene Rubber (S-SBR). Sliding indentation is carried out at a specified sliding speed with several depths and contact surface roughness. The simulation results shown are in the form of the rubber surface deformation, friction forces, and stress distribution. In general, the simulation results show that the S-SBR has a slightly higher coefficient of friction (COF) than the other types.

Keywords Blade · Rubber · Sliding · Tire

1 Introduction

The manufacturing technology of tire rubber material to improve its performance is still developing today. Parameters that need to be considered in improving rubber performance include mechanical and tribological properties [1, 2]. The use of rubber

B. Setiyana (✉) · J. Jamari · R. Ismail · S. Sugiyanto
Laboratory for Engineering Design and Tribology, Department of Mechanical Engineering, University of Diponegoro, Jl. Prof. Soedharto SH, Tembalang, Semarang 59275, Indonesia
e-mail: budisetiyana@yahoo.com

E. Saputra
Program of Mechanical Engineering (ME), University of Twente, Drienerloolaan 5, Postbus 217, 7500 AE Enschede, The Netherlands

U. Sabino et al. (eds.), *Proceedings of the 6th International Conference and Exhibition on Sustainable Energy and Advanced Materials*, Lecture Notes in Mechanical Engineering,
https://doi.org/10.1007/978-981-15-4481-1_58

for vehicle tires really requires good performance, including when the tires experience braking along driving. In general, good tires are those that have a high braking capacity. Braking capacity is closely related to friction contact between the tire and the road surface which is often expressed by the friction force and the coefficient of friction. A large coefficient of friction provides a large braking capacity too.

Experimentally, research on friction contacts in tire rubber has been widely carried out by some researchers. Many valuable results were obtained from the research, both on the prototype scale and laboratory scale [2–5]. The parameters produced are generally in the form of friction force, friction coefficient and abrasion wear. In general, the total coefficient of friction (COF) consist of adhesion and deformation that is obtained by dividing horizontal force induced by vertical force. The adhesion COF depends on the contact surface roughness, meanwhile, the deformation COF depends on the elastic force on deformed rubber surface [6]. However, the phenomenon of friction contact is very difficult to be discussed theoretically because of the unique rubber behaviour that are hyper-elastic and nonlinear stress-strain relationship, therefore numerical approaches are often used [7].

Sliding phenomena on the rubber surface using multi asperities as a counterface are still difficult to be analyzed theoretically. Thus, the analytic or numeric expression is usually started by using a single asperity. Experimentally, by using on a point contact with a sharp cone as a single indenter, abrasion on the rubber surface yields to a wear pattern, however, loss volume of the rubber material caused by abrasion does not practically occur [2]. Therefore, some researchers proposed a line-contact by using a blade indenter as counterface, thus, this method is much studied experimentally and analytically [8].

This paper investigates the phenomenon of friction contact between rubber specimen and a rigid blade indenter to obtain the frictional force, the coefficient of friction, deformation and the stresses that occur during friction. The investigation is carried out numerically based on the Finite Element Method (FEA). Simulations are carried out on the two types of rubber that are commonly used for tires i.e. vulcanized rubber type 1 (R1) and type 2 (R2), and a new product in the form of Solution SBR (Styrene Butadiene Rubber). The Solution SBR was reinforced with 80 phr (parts per hundred rubber) of highly dispersible silica. The FEA output is presented in the form of stress distribution, deformation contour and friction forces that occur during sliding. The final results obtained are the maximum stress and overall coefficient of friction that occur during sliding contact for each rubber type.

2 Method

This work is carried out numerically using a legal version of commercial finite element software package, ABAQUS 6.11 [9]. Rubber material is modeled as hyperelastic and considered in compressible. An experimental test for rubber material is required for obtaining Strain Energy Function (SEF) for FE simulation input data. The SEF data were constructed on Money-Rivlin version. It was obtained from stress-strain diagrams based on the results of tensile tests under elastic conditions. Therefore, the maximum stress generated in this study is presented in an elastic level so that crack initiation has not occurred yet.

Schematic illustration of a rigid blade that sliding on the rubber surface is depicted in Fig. 1a. Rigid blade indenter with 0.5 mm of tip radius slides on the rubber surface (elastomer). The boundary condition of the schematic drawing indenting system is depicted in this picture as well. Rubber specimens as high of 10 mm, width 20 mm, and thickness of 10 mm are modeled in the plane strain model. FE simulations are carried out in a constant sliding speed of 5 mm/s, maximum horizontal displacement of 4.0 mm with several data inputs namely rubber type, indentation depth and surface roughness. Rubber type is vulcanized rubber 1, vulcanized rubber 2 and solution SBR meanwhile the indentation depth selected is 0.5, 1.0 and 2.0 mm. Surface roughness is represented by the adhesion coefficient of friction (adhesion COF) and the given value is 0.0 and 0.5.

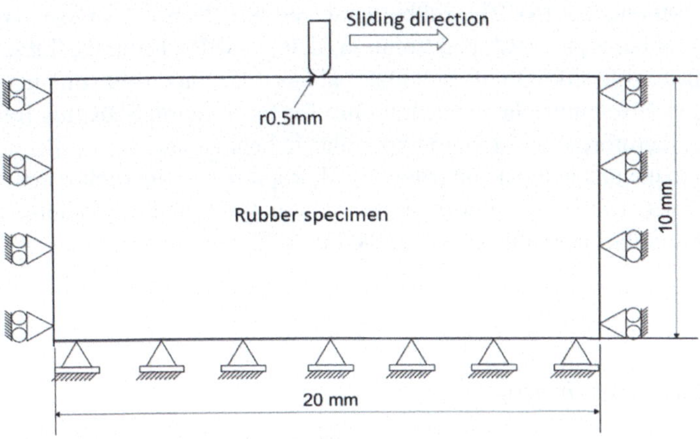

(a)

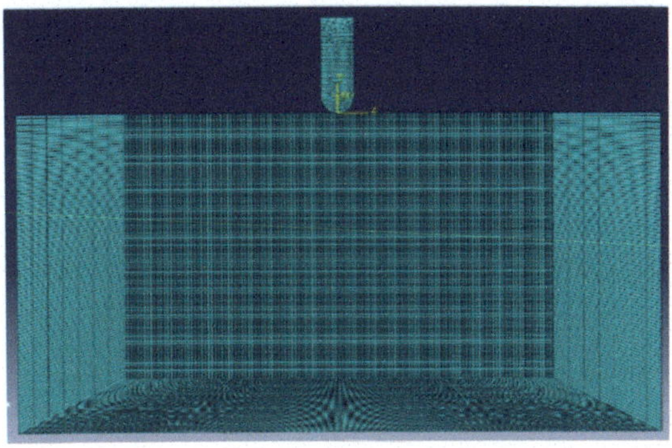

(b)

Fig. 1 Initial model of the rubber sliding. **a** Schematic model. **b** Generated mesh of FE

The FEA simulation is evaluated in three steps of indenter movement, namely starting in a stationary position (initial state), then continued in a moving state (sliding state) and finally in a stopped state (final state). The FEA output is presented in the form of stress field, contour deformation and the friction forces that occur during sliding. From this result, the maximum stress and coefficient of friction can be observed and identified. Finally, the maximum stress generated and friction coefficient can be described graphically with respect to various input data.

3 Results and Discussions

Regarding to the mechanical properties of rubber material, the tensile test results are given at Table 1 in the form of tensile strength and extension ratio when the test specimen is broken. The tensile test procedure was carried out based on ISO 37 Standard (1993 and 2015). The rubber analyzed is vulcanized rubber type 1 (R1), vulcanized rubber type 2 (R2) and Solution SBR (S-SBR). From the Table, it can be seen that the tensile strength of each material has difference value that may not very large. This is different to the extension ratio for the solution SBR that it has a very large value, therefore it has a more hyper-elastic behaviour.

Related to the mechanical and tribological properties of the rubber material along sliding contact, i.e. stress, deformation, friction force and coefficient of friction (COF), the simulation results based on the Finite Element Analysis (FEA) are given below.

3.1 Stress Distribution

Figure 2 shows the FEA output from the von Mises stress distribution and the contour of deformation surface of vulcanized R1 by the blade sliding. The FEA output with a sliding depth of 0.5 mm and adhesion COF of 0.5 is illustrated in these figures at the initial state (a), the sliding state (b) and the final state (c) respectively. In general, the highest contact stress regime is below the tip of the indenter. In the initial condition, the stress distribution is symmetrical because it is still in static condition. In addition, in the sliding state, there are two locations of high stress in the direction as well as in the opposite direction of the sliding. Meanwhile, in the final condition, the location

Table 1 Results of the tensile test for various rubber material

No.	Testing parameter	Test specimen		
		Vulcanized R1	Vulcanized R2	Solution SBR
1	Tensile strength (MPa)	15.90	18.10	19.69
2	Extension ratio at break	4.80	5.08	14.15

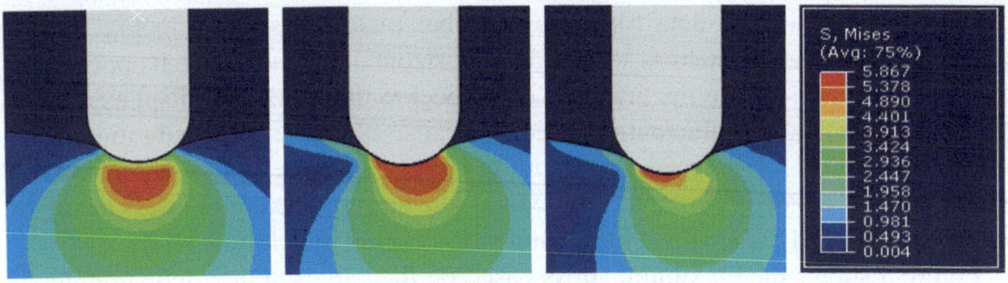

(a) (b) (c)

Fig. 2 The stress distribution of the rubber on the indentation depth of 0.5 mm and adhesion COF of 0.5 for vulcanizated R1. **a** Initial state. **b** Sliding state. **c** Final state

of the high stress close to the tip of the indenter that is shown in the opposite direction of the sliding. In the sliding and final state, the maximum stress is located at the back of the moving of the indenter tip.

Figure 3 shows a picture of the stress field for variations in rubber material under sliding state conditions. From the picture it can be seen that the largest stress occurs in vulcanized R1 that is equal to 5.613 Mpa and the lowest stress is in Solution SBR that is equal to 3.943 Mpa. The stress field indicates that the concentration of the high stress leads to the direction of the indenter movement.

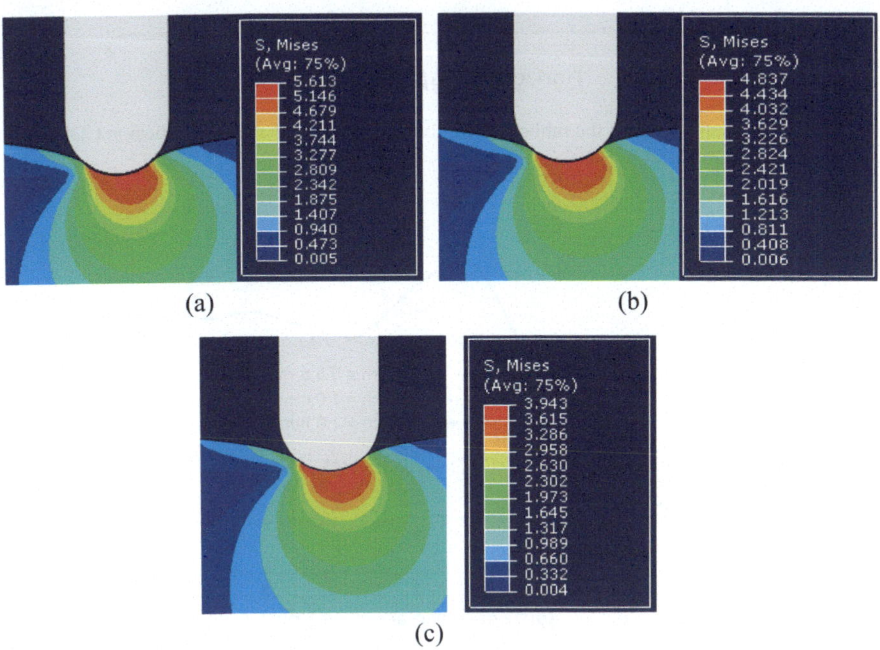

Fig. 3 The stress distribution of the rubber on the indentation depth of 0.5 mm and adhesion COF of 0.5 in the sliding state. **a** Vulcanizated R1. **b** Vulcanizated R2. **c** Solution SBR

The maximum stress data for various rubber material is given in Fig. 4. The maximum stress is presented as a function of horizontal displacement. It appears that the maximum stresses are fluctuating with respect to the horizontal displacement of the blade indenter. The fluctuation occurs because of self excited vibration along sliding contact. Force induced along sliding contact on the elastic material surface causes the self excited oscillation. In general, the largest maximum stress occurs in vulcanized R1 and the smallest maximum stress occurs in solution SBR.

Figure 5 shows the maximum stress based on the variation in indentation depth. Indentation with a depth of 2.0 mm gives a maximum stress value that is much higher than for depths of 0.5 and 1.0 mm. When the indentation depth is 2.0 mm, it appears that the stress occurs very fluctuative compared to lower depths. In general, this

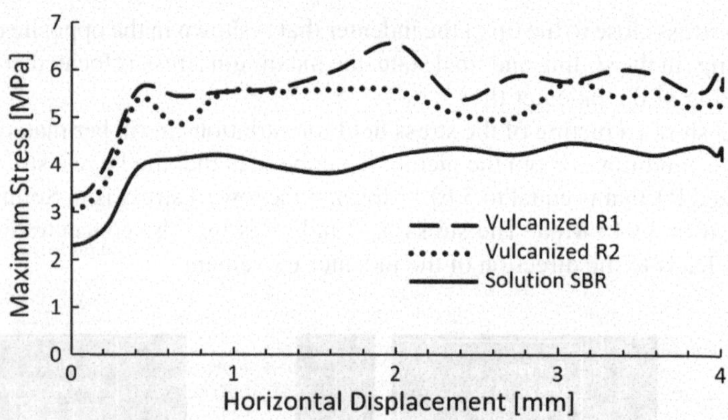

Fig. 4 The maximum stress of the rubber on the sliding depth of 0.5 mm and adhesion COF of 0.5 for various rubber material

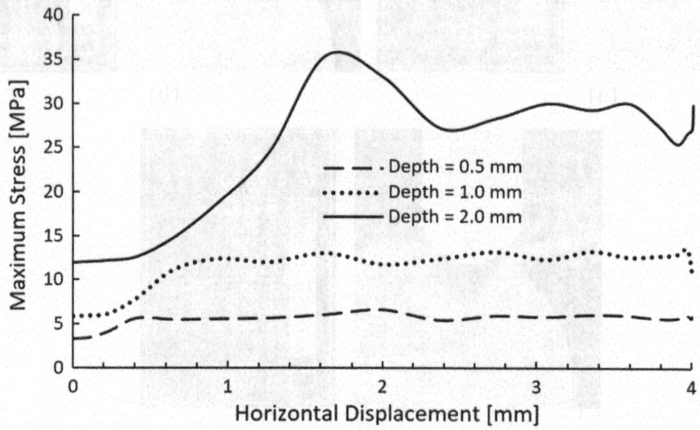

Fig. 5 The maximum stress of the rubber sliding for vulcanized R1 with the adhesion COF of 0.5 for various depth of sliding

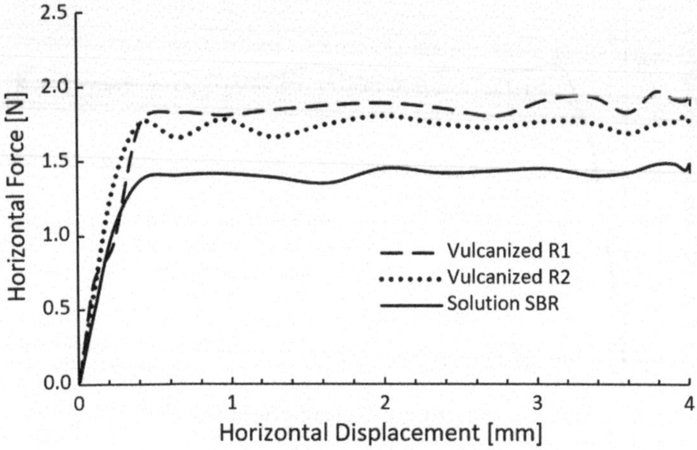

Fig. 6 The horizontal force of the rubber sliding on the indentation depth of 0.5 mm and adhesion COF of 0.5 for various rubber material

fluctuation phenomenon is related to the frictional contact nature of rubber, which is the occurrence of stick-slip [10, 11]. At first, the stick contact occurs between the rubber and the indenter until it reaches a high stress value, and then a contact slip occurs which causes the stress to drop. A peak stress of 35 MPa for a depth of 2.0 mm occurs when the indenter displaces around 1.3 mm from the initial position.

3.2 *Coefficient of Friction (COF)*

Figure 6 shows the horizontal force of the indenter during sliding on the rubber surface for variations in rubber material. It can be seen that vulcanized R1 has the largest horizontal force, meanwhile the solution SBR material has the smallest force. Figure 7 depicts the overall COF for the three types of material analyzed. In general, the overall COF values for the three materials have values that are not so different. The overall COF value for the S-SBR has the highest value around 0.60, while the vulcanized R1 material has the lowest value around 0.55. In general, it shows that the S-SBR material has a slightly higher overall COF than other rubber types with small horizontal or tangential force.

Overall COF with depth of indentation variations are given in Fig. 8. Overall COF for 0.5 mm depth have lower values than for 1.0 mm depth. But for indentation depth of 2.0 mm, COF values fluctuate, which is most likely due to the oscillation phenomenon of the rubber surface when under sliding contact. This phenomenon is often referred to as stick-slip occurrence which is common emerged in rubber sliding. The stick contact phenomenon occurs at the beginning of the sliding, where the tip of the indenter and the rubber surface are fused to move horizontally together before finally releasing the contact into a slip contact [11]. The figure shows that

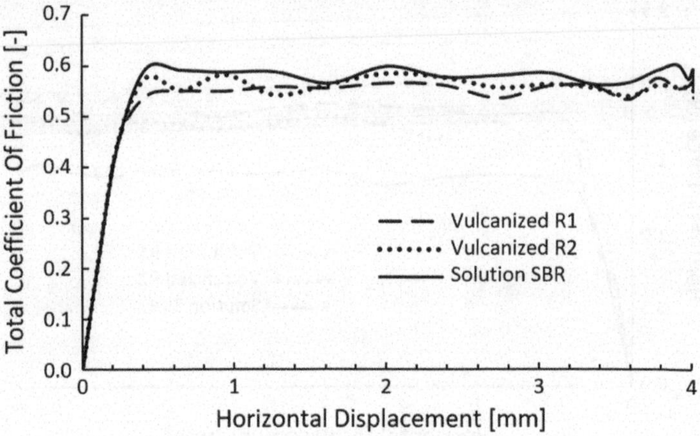

Fig. 7 The total Coefficient of Friction (COF) on the indentation depth of 0.5 mm and adhesion COF of 0.5 for various rubber material

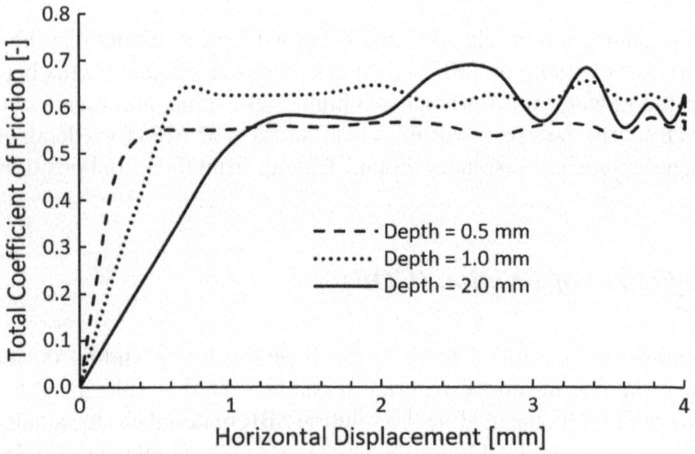

Fig. 8 The overall coefficient of friction of the vulcanizated R1 with the adhesion COF of 0.5 for various of depth of indentation

the larger depth of indentation also causes the larger sliding distance of the initial stick contact. This occurs because when a large indentation depth provides a large normal force and consequently require a large tangential force as well. This makes the sliding displacement during the initial stick contact becomes larger.

The overall COF values based on variations in surface roughness are given in Fig. 9. Smooth surface is expressed with adhesion COF of 0 and rough surfaces is expressed with adhesion COF of 0.5. For smooth surfaces, the overall value of COF is very small and does not fluctuate so much. Here the overall COF value is only influenced by the deformation COF. For rough surface, the overall COF highly

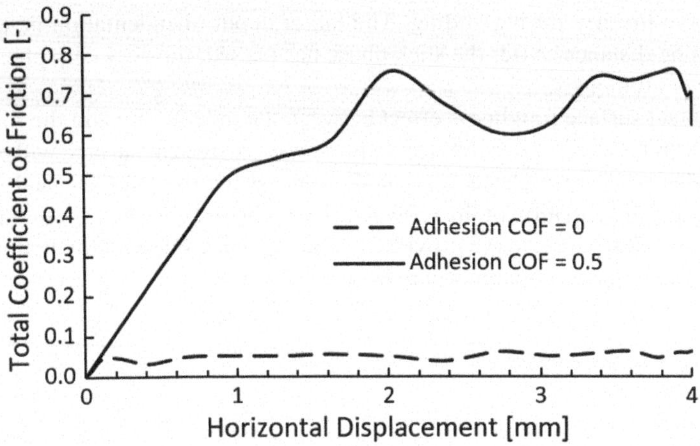

Fig. 9 The total coefficient of friction of the vulcanizated rubber type 2 with the depth of indentation 2.0 mm for various of adhesion COF

fluctuates which is started with stick contact. It is indicated by the overall value of the COF which increases linearly with respect to a sliding displacement of up to 1.0 mm. The rough surface provides a fluctuating value for the overall COF and the stick-slip phenomenon appears here, as is common emerged on the sliding contacts in rubber [11].

Based on the analysis results above, there are some results to note, namely that the solution SBR has more hyper-elastic behaviour than vulcanized material R1 as well as R2, but the value of the tensile strength is not so much different. Stress simulations show that for the same sliding indentation conditions, vulcanized R1 materials have the largest stress, while solution SBR have the lowest stresses. Likewise, the tangential force that occurs during sliding indentation, that the frictional force of vulcanized material R1 has the high friction force and SBR material solution has a low value. Stress simulations show that the deeper sliding indentation causes the larger stresses, however the stresses that occur during deep indentation have fluctuating values.

Regarding the overall coefficient of friction, the type of material does not have a significant difference in value. The coefficient of friction is closely related to the braking capacity of the tire, where a large coefficient of friction provides a large braking capacity as well. However, for SBR solution materials, despite having a slightly larger coefficient of friction than other materials, this material is very hyper-elastic so it has a very large deformation. This needs to be considered if this material will be applied as a vehicle tire.

Large overall coefficient of friction does not change much to the depth of indentation, but for deeper indentation, the overall coefficient of friction is fluctuating, as is the phenomenon of stick-slip which is common emerged in the rubber sliding. In addition, a large depth of indentation will provide a strong stick contact which is indicated by the large sliding distance during the stick contact. The stick condition at the beginning of the sliding indentation is indicated by the increased indenter friction force linearly during sliding. The larger depth of indentation is, the larger stick's sliding distance. After the stick phase has passed, the stick-slip phenomenon will occur continuously.

The contact surface roughness effect between the indenter tip and the rubber surface is very influential on the contact phenomenon. A smooth surface with adhesion COF equals to zero provides a very small value of the overall COF and it is only affected by the deformation of the rubber surface. On the other hand, a rough surface with adhesion COF equal to 0.5 provides a large overall COF with fluctuating values and the stick-slip phenomenon occurs here.

4 Conclusion

This paper investigates the phenomenon of friction contact between rubber specimen and a rigid blade indenter to obtain the frictional force, the coefficient of friction, deformation and the stresses that occur during friction. Simulations are carried out on the two types of rubber that are commonly used for tires i.e. vulcanized rubber type 1 (R1) and type 2 (R2), and a new product in the form of Solution SBR (Styrene Butadiene Rubber) that was reinforced with 80 phr (parts per hundred rubber) of highly dispersible silica. Sliding contact is carried out with specified indenter tip radius and sliding speed with variations in indentation depth and surface roughness.

Stress simulations show that the deeper sliding indentation causes the larger stresses, but the stresses that occur during deep indentation have fluctuating values. Regarding the overall coefficient of friction, in general the three types of material do not have significant differences in value. The value of the coefficient of friction is closely related to the braking capacity of the tire, where the larger coefficient of friction causes the larger braking capacity as well. However, for solution SBR material, even though it has a slightly larger coefficient of friction than other materials, this material is very hyper-elastic so it has a very large deformation. Such behaviour should be considered if this material will be applied as vehicle tires. It has been noted that stick-slip occurrence along rubber sliding is commonly emerged, especially for large depth of indentation and in a rough contact surface.

References

1. Gent AN (1992) Engineering with rubber, how to design rubber components, 3rd edn. Hanser Publication, Cincinati
2. Zhang SW (2004) Tribology of elastomer, tribology and interface engineering series. Elsevier, Amsterdam
3. Fukahori Y, Yamazaki H (1994) Mechanism of rubber abrasion—part 1: abrasion pattern formation in natural rubber vulcanizate. Wear 171:195–202
4. Khafidh M, Setiyana B, Jamari J, Masen M, Schipper DJ (2018) Understanding the occurrence of wavy track of elastomeric material. Wear 412–413:23–29

5. Setiyana B, Ismail R, Jamari J, Schipper DJ (2018) Analytical study of the wear pattern of an abraded rubber surface: interaction model. Tribol Mater Surf Interfaces, 1–7
6. Bowden FP, Tabor D (1954) The friction and lubrication of solids, vol I. Clarendon Press, Oxford
7. MSC Software Whitepaper (2010) Nonlinear finite element analysis of elastomer. MSC Software Corporation, Santa Ana
8. Uchiyama Y, Ishino Y (1992) Pattern abrasion mechanism of rubber. Wear 158:141–155
9. ABAQUS 6.11.: Standard user's manual. Dassault Systems Simulia Corp., USA (2011)
10. Fukahori Y, Yamazaki H (1994) Mechanism of rubber abrasion—part 2: General rule in abrasion pattern formation in rubber-like materials. Wear 178:109–116
11. Setiyana B, Ismail R, Jamari J, Schipper DJ (2016) Stick-slip behavior of a viscoelastic flat sliding along a rigid indenter. Tribol Online 11(3):512–518

Neuro-fuzzy Hysteresis Modeling of Magnetorheological Dampers

Julian Wisnu Wirawan, Seraf Steva Oryzanandi, Aji Masa'id, Fitrian Imaduddin, Ubaidillah, and Irfan Bahiuddin

Abstract This paper presents the neuro-fuzzy modeling approach to analyze the test results of MR damper. Every Electric current in provides to MR Fluid will have a different output. On the other hand, the meandering type valve has a different output and calculation. Therefore, the prototype of MR Damper that has been made was taken to a laboratory to test using Dynamic Testing Machine. The data test result will be analyzed using Neuro-Fuzzy. This paper aims to find a correlation between every variable is there in the testing of MR Damper. For the hysteresis modeling purpose, some parts of the data are taken as the training data source for the optimization parameters in the Neuro-Fuzzy model. The performance of the trained Neuro-Fuzzy model is assessed by validating the model output with the remaining measurement data and benchmarking. The assigned membership function results in a minimum error of 0.16 from 3000 epoch from 3 sets of data given as training data.

Keywords MR fluid · Damper · Neuro-fuzzy

1 Introduction

A magnetorheological (MR) fluid is a type of smart fluid in a carrier fluid, usually a type of oil. When subjected to a magnetic field, the fluid significantly increases its apparent viscosity, to the point of becoming a viscoelastic solid. Smart fluids or field responsive fluids are materials for which the viscosity can be reversibly controlled depending upon the value of the external field applied [1]. In the case of MR fluids, the existence of micron-sized magnetically soft particles suspended in the fluid medium causes the formation of particle aggregates and chains under the presence of a magnetic field [2, 3]. MR fluid is different from a ferrofluid, which

J. W. Wirawan · S. S. Oryzanandi · A. Masa'id · F. Imaduddin (✉) · Ubaidillah
Mechanical Engineering Program, Faculty of Engineering, Universitas Sebelas Maret, Surakarta, Indonesia
e-mail: fitrian@ft.uns.ac.id

I. Bahiuddin
Department of Mechanical Engineering, Vocational College, Universitas Gadjah Mada, Yogyakarta, Indonesia

U. Sabino et al. (eds.), *Proceedings of the 6th International Conference and Exhibition on Sustainable Energy and Advanced Materials*, Lecture Notes in Mechanical Engineering,
https://doi.org/10.1007/978-981-15-4481-1_59

has smaller particles. MR fluid particles are primarily on the micrometer-scale and are too dense for Brownian motion to keep them suspended (in the lower density carrier fluid). Ferrofluid particles are primarily nanoparticles that are suspended by Brownian motion and generally will not settle under normal conditions. As a result, these two fluids have very different applications.

An MR damper is a damper filled with magnetorheological fluid, which is controlled by a magnetic field, usually using an electromagnet. This effect allows the damping characteristics of the shock absorber to be continuously controlled by varying the power of the electromagnet. Fluid viscosity increases within the damper as electromagnet intensity increases. The induced structure of MR fluids can be deformed in three different ways, which are commonly known as flow mode, shear mode, and squeeze mode [4]. In all modes, the MR fluid is always situated between two magnetic pole plates, and the direction of the magnetic flux lines is always perpendicular to the plates in respect of the agreed fundamental chain structure [5]. In the literature, many variants of the term have been used by researchers regarding the flow mode, such as valve mode [4–6], Poiseuille flow mode [7], pressure-driven flow mode [7–9] or flow mode [10, 11]. In terms of application, the use of a valve as a fluid flow regulator in conventional devices is replaced by this concept of flow mode in MR fluid based-devices, such as servo valves, hydraulic controls, dampers, actuators and shock absorbers [12].

Aside from the MR fluid, the modeling of hysteresis phenomenon in an MR damper is also an interesting problem that has been discussed by many researchers. Hysteresis is the result of the memory effect in the force restoring mechanism, which depends not just on the instantaneous disturbances but also on the previous state of disturbances. In general, the hysteresis modeling in an MR damper can be divided into two different approaches, the parametric approach, and the non-parametric approach. The parametric approach is the modeling approach that characterizes the damper with combinations of several idealized physical elements such as springs and dampers. Since the assumption is set from the early stage of the modeling process, the parametric approach is typically represented in a more generic form of the equation with a fixed number of parameters, which can be adjusted to model many types of dampers by tuning the values of the parameters. The known method in the parametric approach is the Bouc-Wen model, the Dahl model, LuGre Model, Sigmoid Model. The non-parametric approach models the dynamics of the damper by employing analytical expressions rather than assuming a physical element like the one in the parametric approach. Although the approach will not produce a generic set of equations, many studies have revealed that the non-parametric models are generally more accurate and robust than the parametric models. Many types of models lie in this category, such as the polynomial model, Neural Network model, and the Neuro-Fuzzy model.

The purpose of this paper is to create non-parametric modeling of the damping force, which results in minimum error compared to the experimental data. The parametric approach is given to provide a better understanding of the difference between both models. The non-parametric model is performed using the Neuro-Fuzzy Designer toolbox from MATLAB software. The experiment data was used in this process, with the input parameter used was electric current, velocity, and displacement, while the output result used as validation data was damping force. The successfulness of the model was indicated by the slightest number of errors resulting from data other than the training data.

2 Governing Equations

A primary viscous damper working principle is to dissipate impact energy into thermal energy using the flow inside the valve. Thus it is stated that the MR valve is the crucial component of the MR Damper. The valve itself sets some parameters related to the pressure drop in both off-state or on-state conditions. Most of the parameters are the dimension-based parameters, which discussed further in the next section. Many types of research have been done to adjust these parameters, e.g., by combining multiple gap directions, such as annular or radial gaps. Therefore, a successful improvement of MR valve performance can significantly impact the development of other MR devices [12].

By the effect of the MR fluid, it is also possible to increase the pressure drop. This consideration directed towards the combinations between magnetic and non-magnetic components to increase the magnetic flux density in the system. Unfortunately, there are some setbacks by doing so because the MR fluid has a peak point, which limits its performance. The change in MR fluid rheological properties due to magnetic flux variation from the electromagnetic coil in the valve changes the fluid flow resistance and therefore changes the pressure drop so that the fluid flow can be slowed or even stopped [13].

The valve used in this research consists of two annular gaps and one annular gap. This structure was inspired from the work in [14] and was chosen with the aim of maximize the performance while minimizing the space needed. The materials of the valve vary, so it may result in the magnetic flux to be perpendicular to the MR flow to increase the efficiency of the magnetic flux density. Figure 1 shows the 3D model of the proposed MR valve.

The pressure drop prediction consists of two different equations based on the working conditions, which are the off-state condition and on-state condition. The off-state condition is the condition in which no magnetic field induced, which means pressure drop is only caused by the viscous effect. The on-state condition is a condition in which the magnetic field was induced by the effect of electric current flows through the coil and affecting the MR fluid, so pressure drop addition occurs by the effect of the yield stress of the MR fluid.

The calculation of the off-state condition consists of two formulas, which is the pressure drop for the annular and radial gap. These formulas are expressed by the following Eqs. (1 and 2).

$$\Delta P_{viscous,annular} = \frac{6\eta QL}{\pi d^3 R}. \quad (1)$$

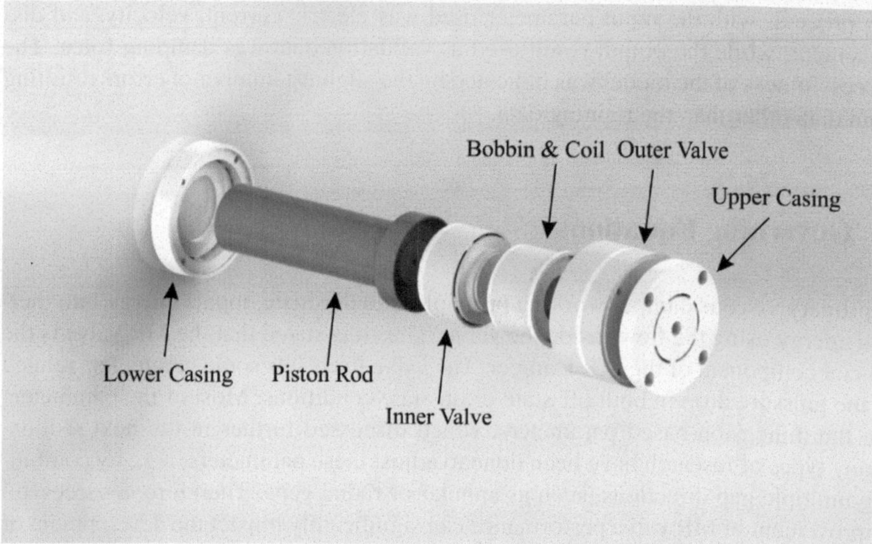

Fig. 1 3D model of the MR valve

$$\Delta P_{viscous,radial} = \frac{6\eta Q}{\pi d^3} ln\left(\frac{R_1}{R_0}\right) \tag{2}$$

where η is the viscosity of the fluid, Q is the fluid flow rate, L is the annular channel length, d is the flow channel gap size, R_1 is the outer channel radius, and R_0 is the inner channel radius. These are the variables that determine the viscous pressure drop. The flow rate of the fluid needs to be calculated first by using the following equation.

$$Q = \frac{\pi D^2 v}{4} \tag{3}$$

The calculation for the magnetic field effect on MR fluid consists of three steps. The first step was to determine the yield stress of the MR fluid in the effective area. In this case, MRF-132DG from Lord Corp. is used. This MR fluid was chosen due to its resistance to heat degradation, which is required for this type of use. The yield stress of the MR fluid is expressed by the following polynomial equation [4].

$$\tau_y(B) = \begin{cases} -58.92B^3 + 74.66B^2 + 35.74B - 3.387, \\ \quad for\ \tau_y(B) > 0 \\ 0,\ for\ \tau_y(B) \le 0 \end{cases} . \tag{4}$$

where τ_y is the yield stress of the MR fluid, and B is the flux density in Tesla. The magnetic simulation was conducted by FEMM to determine the flux density of each zone. After the simulations being conducted, the magnetic flux density on each zone was determined to be the function of this equation in each zone, respectively.

The next step was to determine the pressure drop caused by the yield stress of the MR fluid. There are two formulas used in this case, which is the pressure drop for the annular gap and the pressure drop for the radial gap. The following equation is used to determine the yield stress of the MR fluid [14].

$$\Delta P_{yield,annular} = \frac{c\tau_y(B)L}{d} \quad (5)$$

$$\Delta P_{yield,radial} = \frac{c\tau_y(B)}{d}(R_1 - R_0) \quad (6)$$

where c is the flow function coefficient, which is obtained by calculating the ratio between the field-dependent pressure drop and viscous pressure drop. The formula used to determine the flow function coefficient was expressed by the following equation [15].

$$c = \frac{2.07 + 12\,Qn}{(12Qn + 0.8\,\pi\,(Rd)^2\tau(B))} \quad (7)$$

The final step was to add the off-state condition with the yield stress of the MR fluid, so the final pressure drop was obtained. In order to be used in the analytical assessments, these equations were calculated for each zone and to be summed up for the final pressure drop. The parameters used in that equation will be illustrated in Fig. 2.

3 Simulation Work

The parameters of the MR valve are shown in Table 1. These parameters are to be used in the equations of the previous section, which divided into two subsections. The first section performs the magnetic flux simulation, while the second section explained the pressure drop prediction of the current design.

The result of the calculations will be determined with variations of the current input and piston velocity. The current input will vary between 0.5, 0.75, and 1 A at the peak, while the piston velocity will vary between 20, 50, 100, 300, and 600 mm/s. These variations are selected to observe the controllability of the damper itself.

3.1 Magnetic Flux Simulation

This subsection discusses the magnetic flux density simulation of the proposed MR valve using FEMM. The simulation was configured for the two-dimensional and

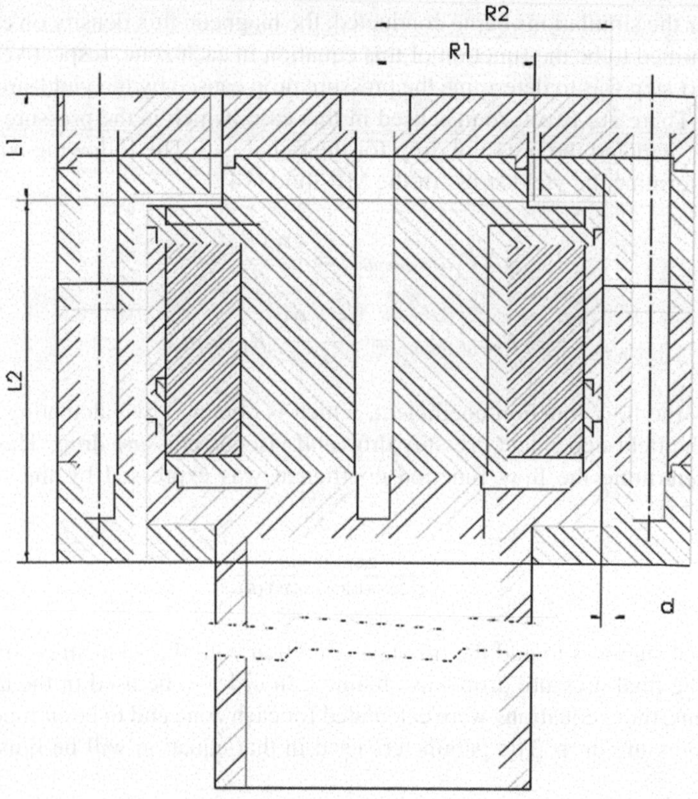

Fig. 2 Cross-section of the magnetorheological valve

Table 1 MR valve parameters

Parameters	Description	Units	Value
η (*MRF-132DG*)	Fluid viscosity	Pa s	0.112
d_a	Annular gap size	Mm	1
d_r	Radial gap size	Mm	1
L_1	Annular 1 gap length	Mm	8.5
L_2	Annular 2 gap length	Mm	29
R_1	Outer radius radial	Mm	12.5
R_0	Inner radius radial	Mm	18.5
D	Piston diameter	Mm	50

axis-symmetric condition to obtain the most accurate result. Figure 3 shows the meshed model and magnetic flux density of the MR valve when induced by 1 A electric current.

Figure 3, it can be observed that the inner of the valve has a dense flux magnetic while the outer of the valve has a sparse flux magnetic. This state is caused by the inner side of the valve has a smaller space than the outer side, and caused the flux density to be limited by the radius of the coil to the center of the valve. Another result of the simulation was the flux density versus flow path distance plot and shown in Fig. 4.

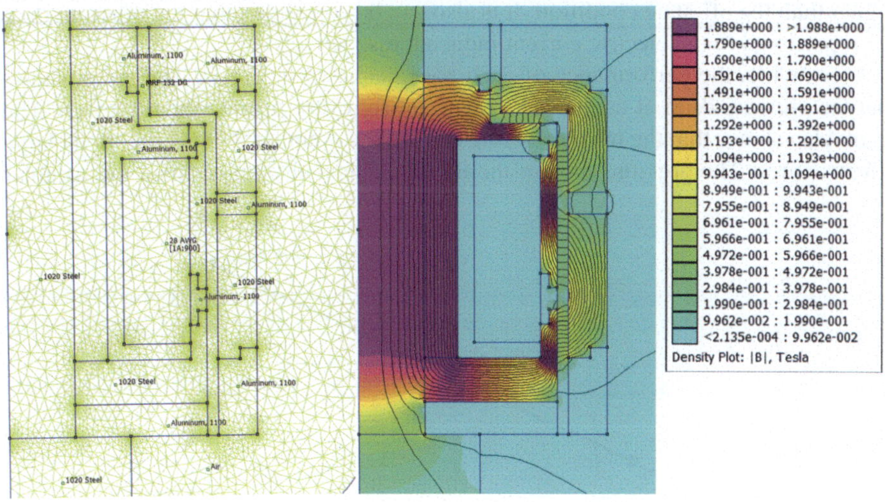

(a) (b)

Fig. 3 **a** Meshed model and **b** magnetic flux density of the MR valve on 1 A electric current induced using FEMM

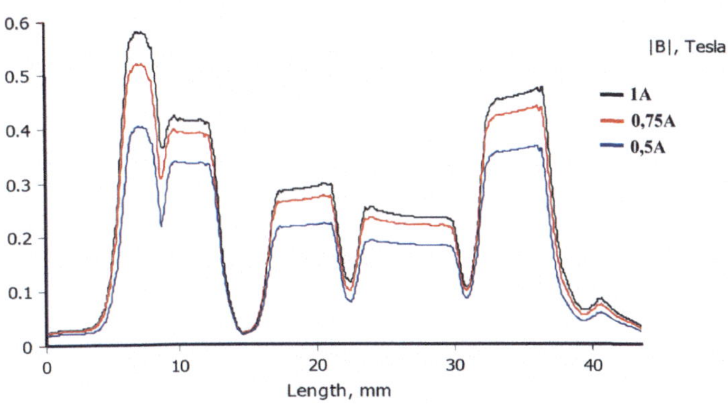

Fig. 4 Flux density versus gap length plot using FEMM

3.2 Pressure Drop Prediction

The calculation of these conditions used Eq. (1) for the annular gap and Eq. (2) for the radial gap. Piston velocity variation will be applied to these calculations. The result of the calculations is presented in Fig. 5. The figure shows that the pressure drop of the off-state condition proportional to the piston velocity.

For the on-state condition, the formulas used are Eq. (4) through Eq. (7). The results of the calculation are shown in Fig. 6 in the form of pressure drop versus piston velocity for each electric current induced. From the figure, it can be concluded that the pressure drop of the on-state condition, which has been calculated, is much higher than the off-state condition. It is also can be concluded that the pressure drop is proportional to the electric current induced, respectively.

Meanwhile, the performance of each zone is shown in Fig. 7. The result shows that the pressure drop of each zone is constant for every piston velocity given. The result also shows that the most affecting zone is the Annular 2 zone, due to its large effective area. The pressure drop of the Radial zone is also higher than the Annular

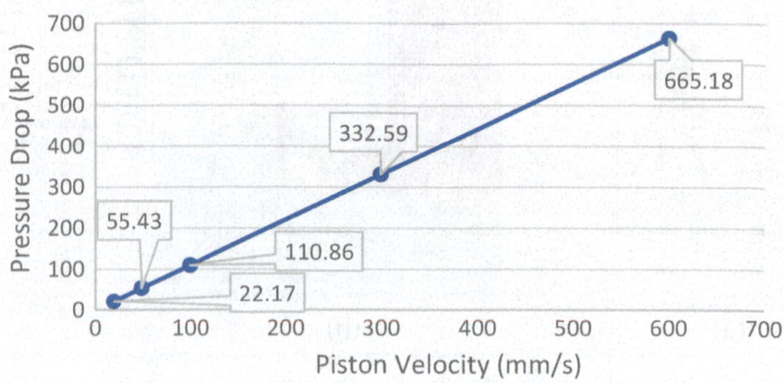

Fig. 5 Off-state pressure drop versus piston velocity of the MR valve

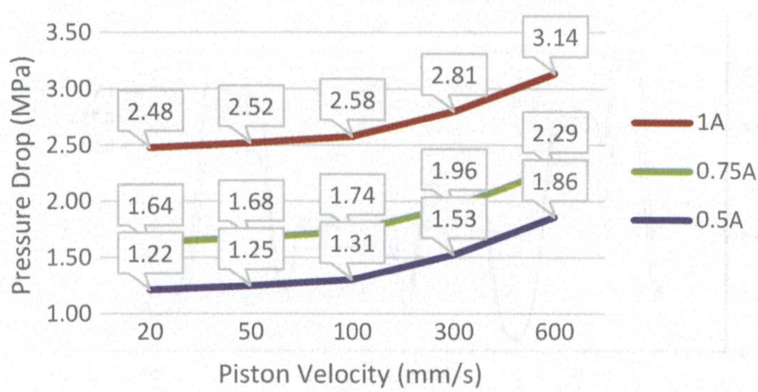

Fig. 6 On-state pressure drop versus piston for various electric current induced

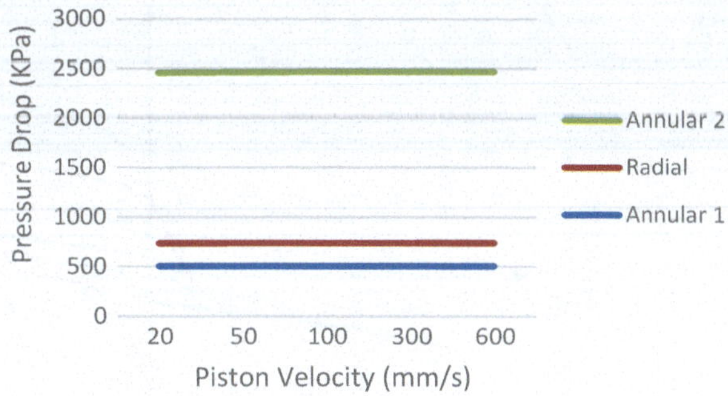

Fig. 7 Pressure drop versus piston velocity for each zone at 1 A

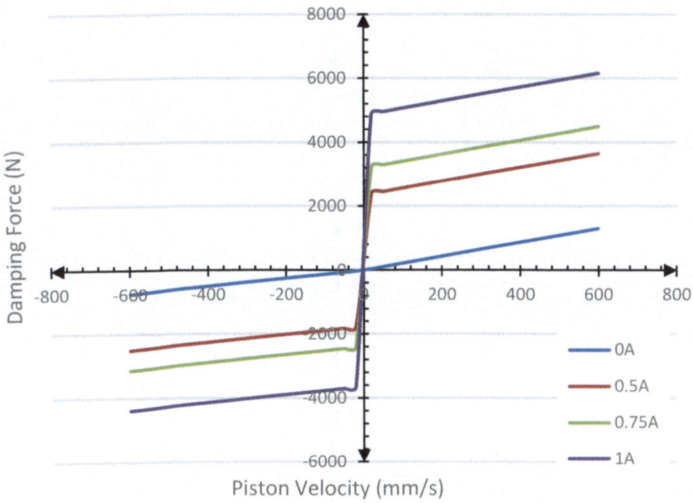

Fig. 8 Damping force versus Piston velocity predicted for each variation of current input

1, even though the Annular 1 has a larger effective area and magnetic flux density compared to the Radial zone. Thus, this result demonstrates that the radial gap indeed has better performance than the annular gap.

The damping force prediction of the MR damper on various current input is shown in Fig. 8. The increasing of current input directly proportional to the damping force, respectively.

4 Experimental Work

The experiment was conducted under five variations of frequency, 0.75, 1, 1.25, 1.5, and 1.75 Hz, with current input variation of 0, 0.4, and 0.8 A on each frequency. The experiment results are shown in Figs. 9, 10, 11, 12 and 13. The experiment result

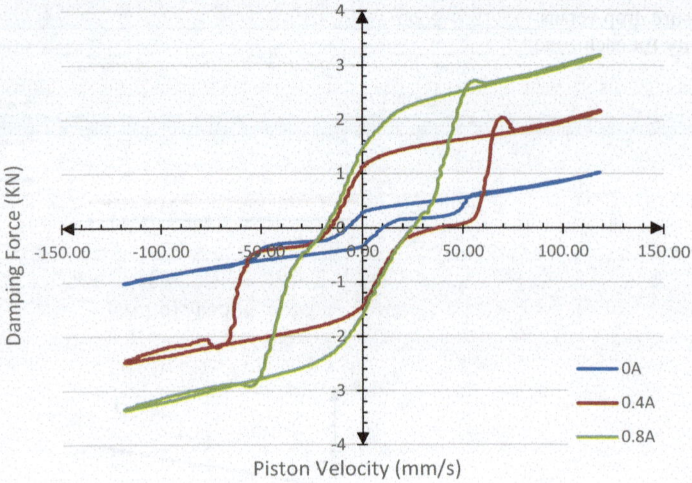

Fig. 9 Experimental data for 0.75 Hz

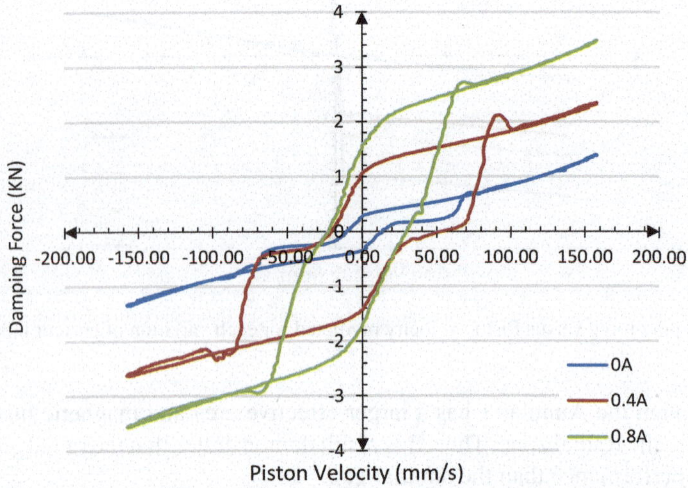

Fig. 10 Experimental data for 1 Hz

shown the hysteresis phenomenon on the MR damper, which cannot be predicted by using parametric modeling. Parametric modeling also cannot predict the damping characteristics over various working frequency.

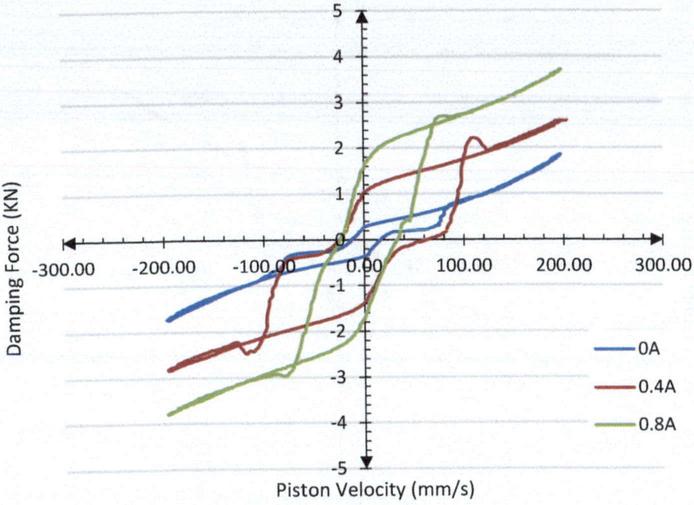

Fig. 11 Experimental data for 1.25 Hz

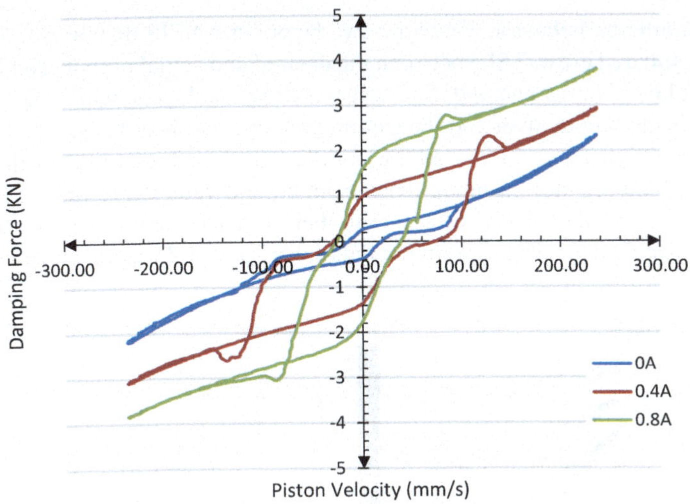

Fig. 12 Experimental data for 1.5 Hz

5 Neuro-fuzzy Model

To overcome the difficulties of modeling the damping characteristics of the MR damper, the non-parametric approach was conducted. One of the popular methods in the non-parametric approach is the Adaptive Neuro-Fuzzy Inference System (ANFIS). ANFIS combines the fuzzy system and artificial neural network algorithm

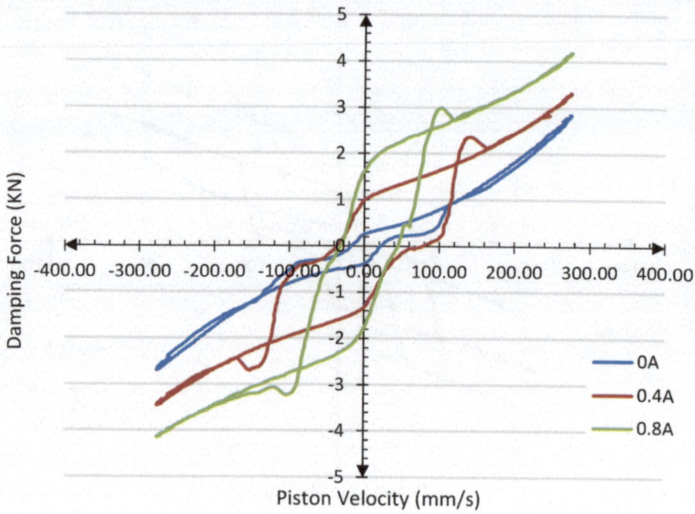

Fig. 13 Experimental data for 1.75 Hz

to mimic nonlinear behavior. There are five layers in ANFIS that aims to minimize the sum of squared error (SSE) between the desired and actual output. The first layer is the input layer, the second and third layers are the rule layers, while the fourth and fifth layer is the output layer and the summation layer, respectively.

The generic structure of the ANFIS is illustrated in Fig. 14, with three input variables, which are current input, displacement, and velocity. In order to simplify the inputs, each variable has a different number of membership functions according to the experiment data. The experiment data for 0.75, 1, 1.25, and 1.75 Hz was used

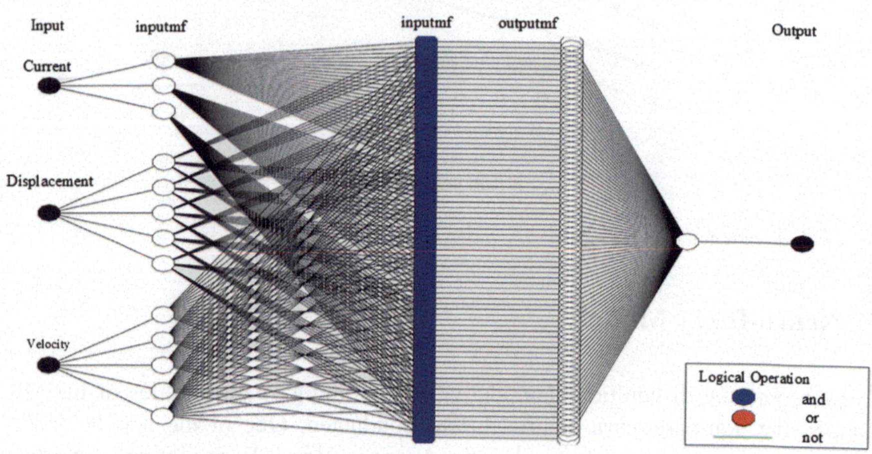

Fig. 14 Membership function structure

as training data, while the testing data used was 1.5 Hz. The comparison between experimental data and training results generated by ANFIS for 0.75 and 1.75 Hz are shown in Figs. 15 and 16. From the training process, the value of error progression within each epoch can be observed. After 3000 number of epochs, the final error value was observed 0.2. To check the accuracy of the trained data, new input data are tested against the trained data. The error value on each number of epochs is shown

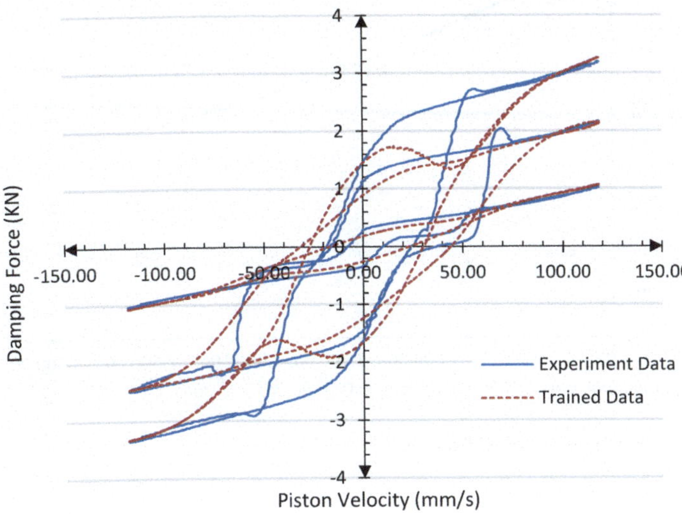

Fig. 15 ANFIS results for 0.75 Hz

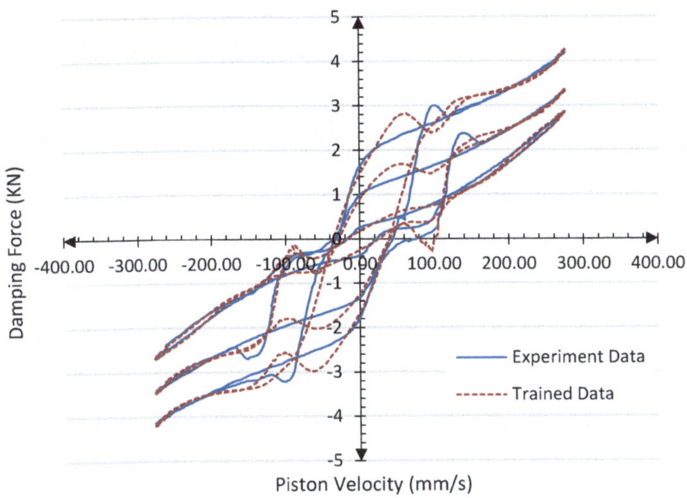

Fig. 16 ANFIS results for 1.75 Hz

in Fig. 17, and the estimation for the testing data, which not included in the training data in Fig. 18.

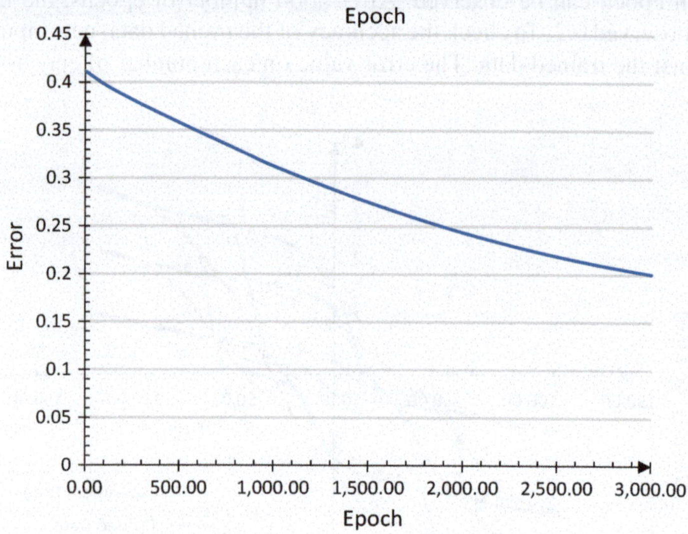

Fig. 17 Error versus number of epochs

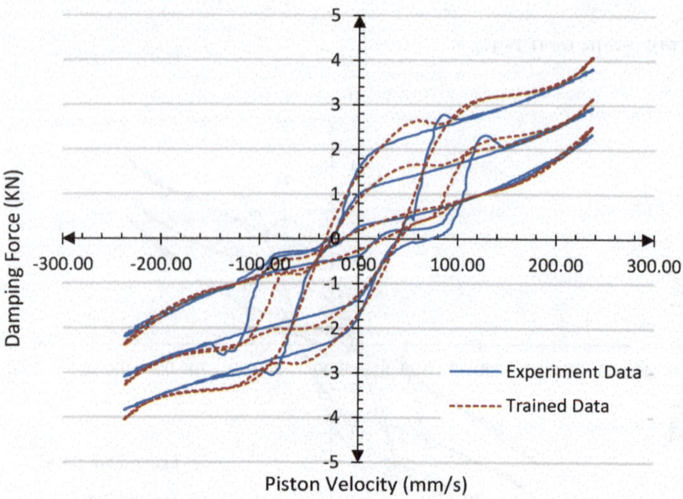

Fig. 18 ANFIS results for 1.5 Hz

6 Conclusion

The comparison between parametric modeling and non-parametric modeling has been discussed. The result of the parametric modeling can only predict the peak-to-peak damping force while lacking the ability to predict the hysteresis phenomenon on the MR damper. This limitation can be overcome by using non-parametric modeling by using several working parameters as input to define the output of damping force. The non-parametric estimation by neuro-fuzzy modeling results the error number of 0.2 for 3000 epoch, which still can be decreased by defining more complex membership functions and increasing the number of epoch. The comparison between experimental data used as training data with the trained data shows identical peak-to-peak damping force; dissimilar for the comparison for experiment data used for testing data with the trained data, the trained data result slightly higher peak-to-peak damping force. Overall, the non-parametric modeling results better characterization for the damping force and hysteresis phenomenon.

Acknowledgements The work presented in this study is funded by Universitas Sebelas Maret through International Collaboration Grant 2019 (ID:66502032019).

References

1. Sims ND, Stanway R, Johnson AR (1999) Vibration control using smart fluids: a state-of-the-art review. Shock Vib Dig 31:195–203. https://doi.org/10.1177/058310249903100302
2. Shiraishi T, Morishita S, Gavin HP (2004) Estimation of equivalent permeability in magnetorheological fluid considering cluster formation of particles. J Appl Mech 71:201–207. https://doi.org/10.1115/1.1667530
3. Ido Y, Yamaguchi T, Kiuchi Y (2011) Distribution of micrometer-size particles in magnetic fluids in the presence of steady uniform magnetic field. J Magn Magn Mater 323:1283–1287. https://doi.org/10.1016/j.jmmm.2010.11.022
4. Carlson JD, Jolly MR (2000) MR fluid, foam and elastomer devices. Mechatronics 10:555–569. https://doi.org/10.1016/S0957-4158(99)00064-1
5. Kulkarni P, Ciocanel C, Vieira SL, Naganathan N (2003) Study of the behavior of MR fluids in squeeze, torsional and valve modes. J Intell Mater Syst Struct 14:99–104. https://doi.org/10.1177/1045389X03014002005
6. Li Z-X, Xu L-H (2005) Performance tests and hysteresis model of MRF-04K damper. J Struct Eng 131:1303–1306. https://doi.org/10.1061/(ASCE)0733-9445(2005)131:8(1303)
7. Pappas Y, Klingenberg DJ (2005) Simulations of magnetorheological suspensions in Poiseuille flow. Rheol Acta 45:621–629. https://doi.org/10.1007/s00397-005-0016-8
8. Parlak Z, Engin T, Ari V, Sahin I, Calli I (2010) Geometrical optimisation of vehicle shock dampers with magnetorheological fluid. Int J Veh Des 54:371–392. https://doi.org/10.1504/IJVD.2010.036842
9. de Vicente J, Klingenberg DJ, Hidalgo-Alvarez R (2011) Magnetorheological fluids: a review. Soft Matter 7:3701–3710. https://doi.org/10.1039/c0sm01221a
10. Sung K-G, Choi S-B, Lee H-G, Min K-W, Lee S-H (2005) Performance comparison of MR dampers with three different working modes: shear, flow and mixed mode. Int J Mod Phys B 19:1556–1562. https://doi.org/10.1142/S021797920503058X

11. Imaduddin F, Mazlan SA, Ubaidillah, Zamzuri H, Fatah AYA (2016) Testing and parametric modeling of magnetorheological valve with meandering flow path, Nonlinear Dyn 85(1):287–302. http://dx.doi.org/10.1007/s11071-016-2684-6
12. Zhu X, Jing X, Cheng L (2012) Magnetorheological fluid dampers: a review on structure design and analysis. J Intell Mater Syst Struct 23:839–873. https://doi.org/10.1177/1045389X12436735
13. Wang DH, Ai HX, Liao WH (2009) A magnetorheological valve with both annular and radial fluid flow resistance gaps. Smart Mater Struct 18(11):115001
14. Fatah AYA, Mazlan SA, Koga T, Zamzuri H, Imaduddin F (2016) Design of magnetorheological valve using serpentine flux path method. Int J Appl Electromagnet Mech 50(1):29–44. https://doi.org/10.3233/JAE-150037
15. Imaduddin F, Mazlan SA, Idris MH, Bahiuddin I (2017) Characterization and modeling of a new magnetorheological damper with meandering type valve using neuro-fuzzy. J King Saud Univ-Sci 29(4):468–477, ISSN 1018–3647. https://doi.org/10.1016/j.jksus.2017.08.012

Thermal Spray Application for Improving the Mechanical Properties of ST 60 Carbon Steel Surfaces with Metcoloy 2 and Tafa 97 MXC Coatings

Z. Nurisna, S. Anggoro, and R. P. Wisnu

Abstract Thermal spray coating is a process of surface engineering technique in which the coating material is heated to melt then pushed by high-pressure flow of air as individual particle or droplets to a surface. In this study, thermal spray coating was used as an alternative to improve the material surface characteristic of worn industrial equipment, such as gears and bearing. Its quick and simple process could shorten the restoration process and thus could also minimize the lost. This research used ST 60 carbon steel as substrat. Coating material using Metcoloy 2 with the alloying elements of FeCrNiCSi, and the second material chosen Tafa 97 MXC with alloying elements of FeSiCrMnNiBWCTiC. Both coating materials were chosen because some of their alloying elements were characterized by a high property of hardness. The mechanical properties testing was done using pull of bonding test, micro-hardness test and ogoshi wear test. After the observation was conducted, the mechanical properties testing indicated that thermal spray coating process can improve the mechanical properties of material surface as resulted from hardness testing and wear resistance testing which showed an average improvement percentage of 61.22%. Whereas the highest improvement of the material surface having coating material with alloying elements of Tafa 97 MXC with the hardness increase percentage of 78.04%. Meanwhile, the wear resistance improvement showed a percentage of 86.10%. The increase on high pressure from 3 Bar to 5 Bar caused the hardness level to increase in as much as 16.99%. Whilst, the wear resistance increased in as much as 21.98%. The increase in air pressure could also lead to the increase of material coating adhesive property toward the surface substrate in as much as 5.86%.

Keywords Thermal spray coating · Carbon steel · Surface treatment

---

Z. Nurisna (✉) · S. Anggoro · R. P. Wisnu
Mechanical Engineering Department, Universitas Muhammadiyah Yogyakarta, Yogyakarta, Indonesia
e-mail: zuhrinurisna@umy.ac.id

U. Sabino et al. (eds.), *Proceedings of the 6th International Conference and Exhibition on Sustainable Energy and Advanced Materials*, Lecture Notes in Mechanical Engineering,
https://doi.org/10.1007/978-981-15-4481-1_60

1 Introduction

The development of the industrial world has been experienced impressive growth. This can be seen from the competition between industrial sectors of large, medium and small scale industries. Despite this growth is inevitable, many problems arise. One common problem that occurs is the damage that occurs in industrial equipment. Damage happened on equipment or facilities is something that cannot be eliminated in the industry. One factors affecting the damage is production. The continuous production process requires the industrial equipment keep continue to operate. Some cases on the equipment that continues to operate are equipment that work through the direct interaction with other materials, both slide and load. Moreover, the industry also continues to strive for routine care and maintenance of the equipment, but problems are still found, which is equipment experiencing wear and tear due to direct contact with other materials.

This wear problem is the main focus because it causes a decrease in production productivity which results in losses. Examples of equipment that often experiences wear and tear is driving systems (gears) and bearings. Presently happening in most industries to overcome the problem of wear and tear on their production equipment is to replace their equipment with new ones. Whereas replacing the equipment requires a long process and also has an impact on losses.

Thermal spray is a material coating technology, either for metal or non metal material, in which the materials (in the form of wire or powder) are melted by the heat sources. The melted coating material, in the form of molten or semi-molten, is sprayed on the base material/substrate using high pressure gas so that it is frozen and stuck to the substrate surface. The coating is formed when millions of particles attached overlap on the substrate [1–3]. Thermal spray coating is an alternative method that aims to improve the surface properties of materials experiencing wear problems. The fast and simple process can shorten the repair time therefore losses can be minimized.

The type of coating used for this wear application is a type of coating that has high hardness properties. In this research, coating material with the FeCrMnNiCSi (Metcoloy 2) alloy coating material was chosen because the chrome element of 13% is expected to increase the hardness and wear resistance of the material surface. The chrome element which is often used for hard chrome plating applications with thicknesses reaching 150 μm has hardness of more than 600 HVN suitable for applications in industrial equipment that is mobile and requires high abrasion resistance [4]. In addition then, the second chosen coating material was a coating material with FeSiCrMnNiBWCTiC alloy elements (Tafa 97 MXC). The addition of 1.57% boron element, 26% tungsten carbide and 6% titanium carbide were expected to further improve the surface wear resistance of the material. The elements of boron and tungsten carbide which are often used for applications in turbines and rocket nozzles as coating media due to their very high thermal resistance properties and hardness values reaching 1400–1500 HVN as well as tungsten carbide and titanium carbide elements which are often added to the manufacture hardmetal cutting tools such as the CNC machine chisel insert where the ability of the tool is resistant to wear, abrasion or scraping processes [5]. Moreover to increase the strength of the adhesive it coated with NiAl bond coat (Tafa 75B and Metco 405NS).

Furthermore, in the application of thermal spray coating, air pressure greatly affects the quality of the coating results. Air pressure is the only medium to deposit the coating material to a surface and this process will determine the surface characteristics of the substrate material [6]. Therefore in this study 3 and 5 Bar air pressure variables were used to determine the effect on the mechanical properties of the coating layer. This study also used spraying 3–4 times pass method to produce a thin layer in order to change the function of the coated material.

2 Experimental Method

Medium carbon steel ST 60 was used in this study. Then preparations were made for ST 60 steel by cutting 16 samples. This sample preparation was done to be used in testing. After sample preparation, the sandblasting process was carried out, before the sample was cleaned using alcohol to remove pollutants in oil form, paint and other organic compounds, then dried. After that, sandblasting was carried out on the surface of the substrate. There are two objectives for the sandblasting process. The first goal is to clean the surface of the substrate from rust and other impurities that cannot be cleaned using alcohol. The second goal is to provide a profile (roughness) on the surface of the substrate to be coated so due to better mechanical bonding of the coating material with the substrate. The material used for the sandblasting process was aluminum oxide (Al_2O_3) sand.

After the sandblasting process was done, it was checked again whether the sandblasting process had fulfilled the surface cleanliness standard and surface roughness standard, in which checking the level of dust that still attached to or left on the surface of the substrate and checking the level of surface roughness of the substrate. The sandblasting process has been carried out until it meets the standard of surface preparation of ISO 8502-3. Comparison of two feedstock materials was used in the coating process, the parameters of the nozzle distance, nozzle angle, electric current, and a constant voltage with variations in gas pressure were shown in Table 1.

The wire arc spray process used the Miller Delta Weld 602—MC95 machine which has 2 wires. One side of the wire electrified by negative current and one side of the wire was electrified by positive current, causing the two wires to spark and melt the feedstock material which was then given compressed air which caused the splash or melted feedstock material thrown from the spray gun and deposited on the substrate surface. This coating process used 4 types of feedstock material and produces 2 kinds of combinations.

The first feedstock material was a NiAl alloy element with the trademark Metco 405NS in form of wire which was then deposited on the surface of the substrate. This NiAl feedstock material became intermediate coating (the first layer) since the alloying element of nickel and aluminum has good adhesion for further coatings

Table 1 Process parameters of twin wire arc spray

Bond coat	NiAl (Metco 405MS)	NiAl (TAFA 75 B)
Top Coat	FeCrMnNiCSi (Metcoloy 2)	FeSiCrNiMnBWCTiC (TAFA 97 MXC)
Gas pressure (Bar)	3 and 5	3 and 5
Nozzle distance (mm)	200	200
Nozzle angle (°)	90	90
Electric current (A)	145	145
Voltage (V)	26.7	26.7

Table 2 The chemical composition of materials (wt%)

Element	Metcoloy 2	TAFA 97 MXC
Fe	Bal.	Bal.
Cr	13	14
Mn	0.5	0.55
Ni	0.5	4.5
C	0.35	–
Si	0.25	1.25
B	–	1.87
WC	–	26
TiC	–	6.0

which were above this first layer. After that the second feedstock was in form of FeCrMnNiCSi alloy element with the metcoloy 2 trademark in form of wire which was then deposited above the first layer and produces the first combination.

The second combination consisted of TAFA 75B trademark NiAl feedstock material in form of wire deposited on the surface of the new substrate. This first coating material also functioned as an intermediate coating since its solid elements have good adhesion ability for subsequent coatings, then feedstock material with the TAFA 97 MXC in form of wire was deposited over the NiAl TAFA 75B layer. The coating material composition specification are shown in Table 2.

3 Results and Discussion

The results of the Pull off layer test are shown in Fig. 1. The diagram below shows the left metco specimen which are specimens that have Ni80%–Al as bond coat and Metcoloy 2 as top coating layers. While right tafa specimens show specimens that have Ni95%–Al as bond coat and Tafa 97 MXC as top coating layers.

The adhesive strength of thermal spray coating caused by presence of porosity. Porosity was materialized on all coating thermal spray specimen. The void was

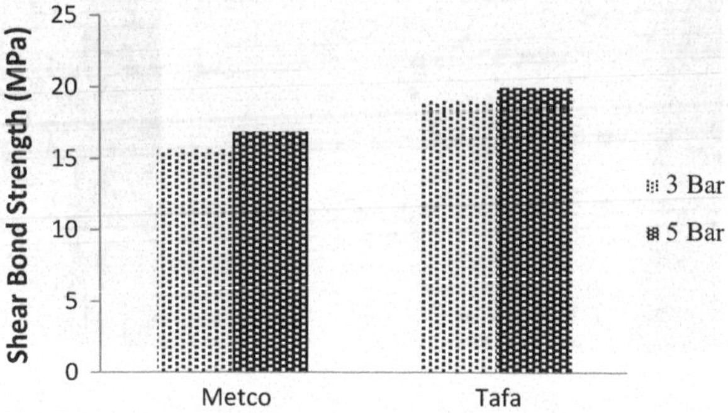

Fig. 1 Pull off layers test with variable air pressure

caused by the trapped air and gas when the liquid particles collided to the substrate during the arc spray process. The coating layer with small number of voids resulted in good substrate-coating binding strength and more homogenous coating layer. Another thing that also affects the adhesive strength of layers was due to the oxide compounds formed in the layer. Whereas the lower the oxide formed will increase the sticking strength of the coating material [7–9].

To support this analysis, visual observation using a microscope on the cross section of the specimen was needed to see more clearly the porosity formed at the interface boundary and coating results.

Figures 2 and 3 show the results of taking cross-section micro photos on each specimen, it can be seen the interface boundary between the surface of the substrate and the bond coating layer shown in Fig. 2. has greater porosity and oxide than the specimen shown in Fig. 3. This causes the lower adhesion strength of the specimen with a coating layer Metcoloy 2 where the value of each test is 15.55 MPa for variable air pressure of 3 Bar and 16.83 MPa for variable Bar air pressure of 5 Bar.

Figure 4 shows graphic of microhardness test results on the top coating layers. The graph shows a significantly difference. This can occur because of the alloy elements contained in the top coating layer differences. The addition of boron as much as 1.57%, WC 26.0% and TiC 6.0% are proven to further increase surface hardness.

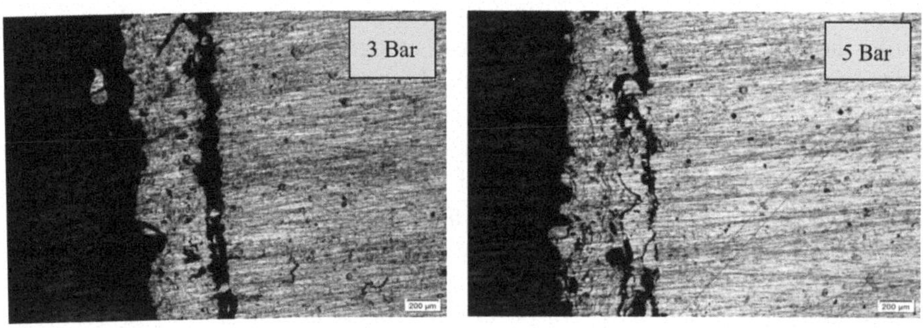

Fig. 2 The results of the cross section micro photos of metco specimens with variations in air pressure

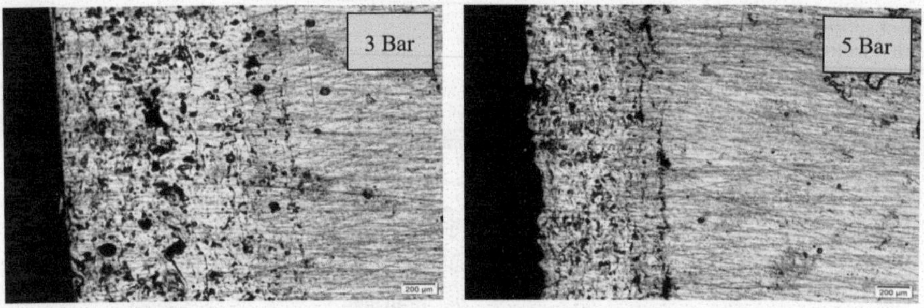

Fig. 3 The results of the cross section micro photos of tafa specimens with variations in air pressure

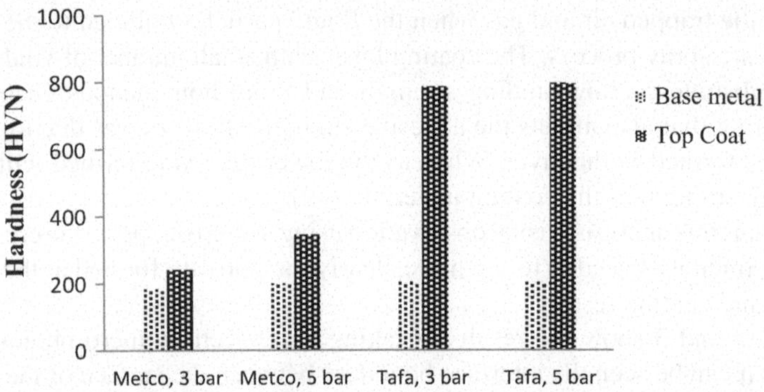

Fig. 4 The Bar diagram of the results of the microhardness test of the metco coating layer with the tafa coating layer, variable air pressure of 3 and 5 Bar

The use of boron and tungsten carbide material which often applied to turbines and rocket nozzles as coating media due to its very high thermal resistance properties also has very high hardness [10]. Whereas the elements of Titanium carbide (TiC) alloys have melting point of more than 3000 °C and have hardness of 9 of Mohs scale.

The difference of air pressure affects the hardness value shown in Fig. 5. The results show that the hardness value increases along with the increase of air pressure from 3 to 5 Bar. This happens because the higher the air pressure affects the air velocity level and splat impact strength. The liquid particles of the coating material produced will get smaller along with the increase of air pressure. Increasing air pressure also causes the liquid particles of the coating material to split because the collision gets narrower so as to produce finer structure and the density between the particles will increase. The value of coating surface hardness will increase along with the higher density [11].

The graph of wear test shown on Fig. 5. The wear values decrease consecutively starting from the original material which has wear values of 2.4764×10^{-9} mm^2/kg, then specimens Metcoloy 2 coatings with 3 Bar air pressure of $1.4304 \times$

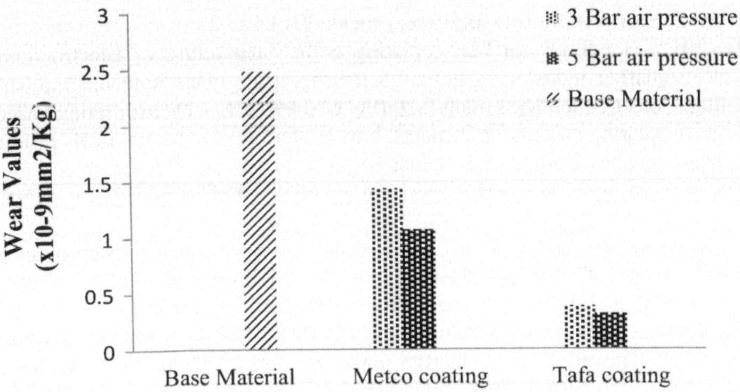

Fig. 5 Bar chart of wear test results on raw material, and specimens with variable air pressure of 3 and 5 Bar

10^{-9} mm²/kg, 5 Bar of air pressure of 1.0625×10^{-9} mm^2/kg and the smallest wear value is the specimen Tafa 97 MXC coating layer with a value of 0.3786×10^{-9} mm^2/kg at 3 Bar air pressure and a value of 0.3095×10^{-9} mm^2/kg at 5 Bar air pressure. Wear values decreasing indicate that the material has better wear resistance. Tafa 97 MXC coatings prove the higher value of material hardness increased the wear resistance. Besides that, the addition of 1.87% boron alloy elements, 26.0% tungsten carbide and 6.0% titanium carbide which have high hardness values are proven to increase the wear resistance of the material.

4 Conclusion

Carbon steel substrate were coated by thermal arc spray. The results from hardness test and wear test showed that coating layer Metcoloy 2 increased the hardness of 94% and the wear resistance of 49%, while Tafa 97 MXC increased the hardness of 298% and the wear resistance of 526%. Decreasing of air pressure, causing the obtainment of porous coatings, formed of thick lamellas or unmelted particles determining a low adherence of the coating.

References

1. Pawlowski L (2008) The science and engineering of thermal spray coatings, 2nd edn. Wiley Ltd, Chichester, UK
2. Berger L-M (2015) Application of hardmetals as thermal spray coatings. Int J Refract Metals Hard Mater 49:350–364. https://doi.org/10.1016/j.ijrmhm.2014.09.029
3. Malek MHA, Saad NH, Abas SK, Roselina NRN, Shah NM (2013) Performance and microstructure analysis of 99.5% aluminium coating by thermal arc spray technique. Procedia

Eng 68:558–565. https://doi.org/10.1016/j.proeng.2013.12.221
4. Panda H (2017) Handbook on Electroplating with Manufacture of Electrochemicals: how to start electroplating industry in india, electrochemicals manufacturing industry in india, most profitable electrochemicals manufacturing business ideas, electroplating based profitable projects, electroplating processing projects, how to start an electrochemicals production business, electroplating based small scale industries projects, new small scale ideas in electroplating processing industry, npcs, niir, process technology books. ASIA PACIFIC BUSINESS PRESS Inc.
5. Ko PL, Robertson MF (2002) Wear characteristics of electrolytic hard chrome and thermal sprayed WC–10 Co–4 Cr coatings sliding against Al–Ni–bronze in air at 21 °C and at −40 °C. Wear 252(11):880–893. https://doi.org/10.1016/S0043-1648(02)00052-2
6. Wang X, Heberlein J, Pfender E, Gerberich W (1999) Effect of nozzle configuration, gas pressure, and gas type on coating properties in wire arc spray. J Therm Spray Technol 8(4):565–575. https://doi.org/10.1361/105996399770350269
7. Nurisna Z, Muhayat TN, Wijayanta AT (2016) Effect of layer thickness on the properties of nickel thermal sprayed steel. AIP Conf Proc 1717(1):040012. https://doi.org/10.1063/1.4943455
8. Sampath S, Jiang XY, Matejicek J, Prchlik L, Kulkarni A, Vaidya A (2004) Role of thermal spray processing method on the microstructure, residual stress and properties of coatings: an integrated study for Ni–5 wt.%Al bond coats. Mater Sci Eng A 364(1):216–231. https://doi.org/10.1016/j.msea.2003.08.023
9. Sá Brito VRS, Bastos IN, Costa HRM (2012) Corrosion resistance and characterization of metallic coatings deposited by thermal spray on carbon steel. Mater Des 41:282–288. https://doi.org/10.1016/j.matdes.2012.05.008
10. Myalska H, Michalska JK, Moskal G, Szymański K (2017) Effect of nano-sized TiC powder on microstructure and the corrosion resistance of WC-Co thermal spray coatings. Surf Coat Technol 318:270–278. https://doi.org/10.1016/j.surfcoat.2017.01.078
11. Toma SL, Bejinariu C, Baciu R, Radu S (2013) The effect of frontal nozzle geometry and gas pressure on the steel coating properties obtained by wire arc spraying. Surf Coat Technol 220:266–270. https://doi.org/10.1016/j.surfcoat.2012.11.011

Analysis of Thermal Conductivity Properties of Recycled High Density Polyethylene Composite Materials Strengthened by Bamboo Fiber with Variations in Fiber Shapes

R. C. Adiputra, I. Widiastuti, D. S. Wijayanto, A. Prasetio, and N. A. Astadini

Abstract The problem of plastics waste in Indonesia is still a matter that needs to be considered, as evidenced by the high and continuous increase in the amount of plastics waste. One of solutions that can be applied is to reduce the amount of plastics waste in Indonesia by utilizing plastics waste into composite materials. Recycled high density polyethylene used as a matrix combined with *Gigantochloa apus* bamboo fiber as reinforcement. Composites were made with the fiber weight fraction of 20% with fiber shape variations that is particulate, inner bamboo woven, and mixed inner-outer woven. The hot press method was used in composite making. The research design was analytical with a cross-sectional approach. The material in this study were 20 mesh rHDPE as a matrix and *Gigantochloa apus* bamboo with a particle size of 20 mesh and woven fiber treated with soaking 5% NaOH for 2 h and dried using oven for 12 h at 60 °C. This composite can be applied as the roof of the house to decrease heat from the outside, it is necessary to test the thermal conductivity of composite materials. Conductivity test using ASTM E1225 standard. The particle fiber shape, inner woven, and mixed woven have 0.174, 0.185, and 0.202 W/m °K thermal conductivity value.

Keywords Composite · *Gigantochloa apus* · rHDPE · Thermal conductivity

1 Introduction

Every citizen in Indonesia contributes 0.52 kg of waste per day [1]. In 2016 the amount of landfill in Indonesia reached 65,200,000 tons per year with a population of 261,115,456 people [2]. HDPE material can be applied to composite material as matrix because it is considered to have properties that are stronger, harder, and more resistant to high temperatures [3]. In the industrial sector of structural components to automotive composite materials, there is a demand to constantly innovate in order

R. C. Adiputra (✉) · I. Widiastuti · D. S. Wijayanto · A. Prasetio · N. A. Astadini
Department of Mechanical Engineering Education, Faculty of Teacher Training and Education, Sebelas Maret University, Surakarta, Indonesia
e-mail: ryannchand@gmail.com

U. Sabino et al. (eds.), *Proceedings of the 6th International Conference and Exhibition on Sustainable Energy and Advanced Materials*, Lecture Notes in Mechanical Engineering,
https://doi.org/10.1007/978-981-15-4481-1_61

to find composite materials that have superior mechanical properties leading to rapid development of knowledge in the field of composite materials.

Natural fiber is an abundant resource that can be obtained directly in nature [4]. Bamboo fiber can be combined with a polymer matrix that is easy in the production process and relatively lower production costs compared to another matrix. The choice of bamboo fiber is considered good because it has good mechanical properties. One of bamboo that grows well and has good mechanical properties, namely apus bamboo (*Gigantochloa apus*), apus bamboo also known as tali bamboo. The tensile strength in apus bamboo reaches 151 MPa and the compressive strength is 335 MPa [4]. Widjaja et al. [5] stated that in Purwakarta, West Java per ha there were at least 150 apus bamboo (*Gigantochloa apus*).

According to the research of [6] the thermal conductivity carried out obtained the highest yield was 455.111 W/m °C at a temperature of 600 °C with a time of 6 h and the lowest was obtained 34.874 W/m °C in the sintering treatment 500 °C with a holding time of 1 h.

Research on the thermal conductivity of composite materials has not been widely carried out by researchers. The large amount of plastic waste and biological wealth in the form of bamboo in Indonesia made the researchers want to conduct research on the nature of thermal conductivity in composite materials reinforced bamboo fiber with HDPE plastic waste matrix.

Therefore, this study discusses the effect of particle fibers, inner woven, and mixed woven on the value of thermal conductivity of composite rHDPE reinforced bamboo fibers.

2 Materials and Methods

This study discusses the value of thermal conductivity of rHDPE composites with bamboo fillers. The composition of this composite material consists of 20 mesh rHDPE and particle and woven filler shapes. 20 mesh bamboo particles. The woven fabric consists of two types, woven bamboo inside and woven bamboo mixed with inside and outside. Bamboo fiber is treated alkaline with 5% NaOH solution for 2 h [7]. Bamboo fiber was dried using an oven for 12 h at a temperature of 60 °C [8].

Composites are made using a mass fraction of 80%:20%. Composite production using compression molding method with hot press machine with a temperature of 150 °C, pressing 50 bars for 25 min [9].

Testing materials using the ASTM E1225 standard (Standard Test Method for Thermal Conductivity of Solids) Using the Guarded-Comparative-Longitudinal Heat Flow Technique. Specimens for testing were made with dimensions of 40 mm × 4 mm and 40 mm × 2 mm using the OSK 4565-A Thermal Conductivity Testing Machine.

3 Results and Discussion

Variations in this study are three variations in the shape of fibers, namely particles, inner woven and mixed woven. Tests were carried out at three different temperatures namely 90, 100, and 110 °C.

Testing of thermal conductivity of rHDPE composites in fiber shape variations using three testing temperatures of 90 °C, 100 °C, and 110 °C, respectively. The average test for each temperature in Fig. 1 shows that the average test for composite rHDPE with particle fiber has the highest thermal conductivity value at the test temperature of 100 °C with a value of thermal conductivity of 0.199 W/m °K and has the lowest conductivity on the test 90 °C with 0.168 W/m °K, while for deep woven fibers has the highest conductivity value at the test temperatures of 100 and 110 °C with a thermal conductivity of 0.198 W/m °K and has the lowest thermal conductivity at 90 °C with 0.179 W/m °K, and mixed matting has 0.175, 0.208, and 0.187 W/m °K for each test (Fig. 2).

Test temperatures of 90–110 °C shows that there are differences but not significant. The use of temperature differences is intended to see the value of thermal conductivity in several conditions (Fig. 3).

The particle fiber has the lowest thermal conductivity value compared to other forms of fiber with 0.174 W/m °K, the shape of woven fiber has the highest thermal conductivity with 0.202 W/m °K, and variations in mixed woven have a value of thermal conductivity of 0.185 W/m °K.

The results above show that composite rHDPE with bamboo fiber has insulating properties with low thermal conductivity values, so it cannot deliver heat well. Surur [10] states that the value of wood thermal conductivity is 0.21 W/m °K. The wood conductivity with particle fiber rHDPE composites is quite close considering the conductivity of rHDPE with particle fibers of 0.174 W/m °K.

The isolative nature of the composite material rHDPE with bamboo fiber is suitable as a material to inhibit heat transfer. The use of hot press materials can be made for

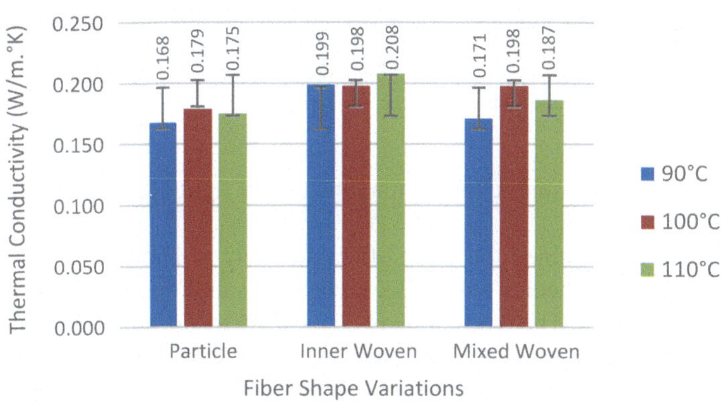

Fig. 1 Average thermal conductivity value with temperature test variations

Fig. 2 Variation of specimen

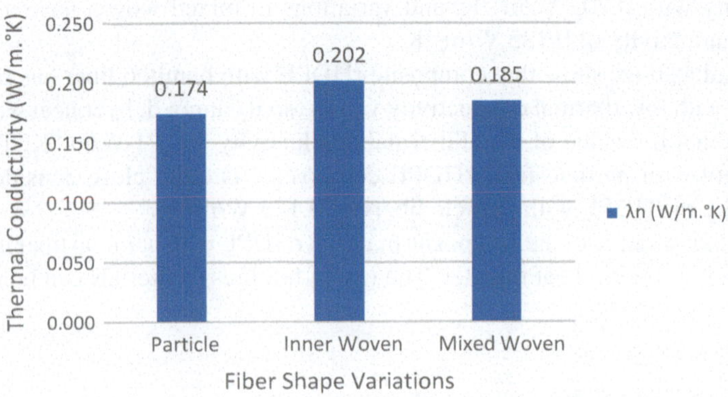

Fig. 3 Average thermal conductivity value

other applications, for example for the manufacture of lightweight steel replacement roofs.

Shah et al. [11] examined the value of conductivity of bamboo reinforced composites using bamboo species with different densities and by giving different treatments to bamboo fibers. The types of bamboo used are moso bamboo (*Phyllostachys pubescens*) and guadua bamboo (*Guadua angustifolia*) with different shapes, orientations, thicknesses, and fiber densities in each specimen. Measurements using Hot Disk from engineered bamboo products obtained data ranging between 0.20 and 0.35 W/m °K.

Furthermore [11] examined the compacted thermal conductivity of bamboo without air or polymer matrices showing values of 0.55–0.59 W/m °K. The results showed that the thermal conductivity of bamboo rHDPE composites was lower than that of Moso and Guadua bamboo species.

4 Conclusions

Research on the value of thermal conductivity of rHDPE composites with variations in the shape of bamboo fibers produces material with insulating properties that can inhibit the heat transfer rate. The use of the type of fiber affects the value of material thermal conductivity. Particle fibers have the lowest thermal conductivity with 0.174 W/m °K and the shape of the inner bamboo woven fiber has the highest thermal conductivity with 0.202 W/m °K.

References

1. Jambeck JR, Geyer R, Wilcox C, Siegler TR, Perryman M, Andrady A, Narayan R, Law KL (2018) Plastic waste inputs from land into the ocean. Science 347
2. Badan Pusat Statistik (2018) Statistik Lingkungan Hidup Indonesia 2018. Badan Pusat Statistik
3. Nurminah M (2002) Penelitian Sifat Berbagai Bahan Kemasan Plastik dan Kertas Serta Pengaruhnya terhadap Bahan Yang Dikemas
4. Eratodi IGLB (2017) Struktur dan Rekayasa Bambu. Universitas Pendidikan Nasional, Denpasar
5. Widjaja EA, Saefudin S, Hamzah (2005) Studi Populasi Bambu di Kebun Rakyat Kabupaten Purwakarta, Yogyakarta. Prosiding Seminar Nasional Perkembangan Perbambuan di Indonesia
6. Suarsana K, Astika IM, Suprapto L (2017) Karakterisasi Konduktivitas Termal dan Kekerasan Komposit Aluminium Matrik Penguat Hibrid SiCw/AL2O3. Jurnal Muara Sains, Teknologi, Kedokteran, dan Ilmu Kesehatan
7. Abral H (2010) Studi Kekuatan Tarik dan Sifat Fisik Serat Cyathea contaminans Sebelum dan Setelah Mengalami Perlakuan Alkali NaOH
8. Zhang K, Wang F, Liang W, Wang Z, Duan Z, Yang B (2018) Thermal and mechanical properties of bamboo fiber reinforced epoxy composites. Polymers (Basel) 10
9. Ratmanto A, Raharjo WW, Triyono T (2016) Pengaruh Tekanan Pengepresan terhadap Kekuatan Bending Komposit rHDPE Cantula. Prosiding SNST
10. Surur I (2007) Pemanfaatan Mikrokontroler Untuk Analisis Data Pada Alat OSK 4565-A thermal conductivity measuring apparatus
11. Shah DU, Bock MCD, Mulligan H, Ramage MH (2015) Thermal conductivity of engineered bamboo composites. J Mater Sci 51:2991–3002

Natural Weathering Effect on Mechanical and Physical Properties of Recycled High-Density Polyethylene Composite with Bamboo Reinforcement

N. A. Astadini, I. Widiastuti, B. Harjanto, R. C. Adiputra, and A. Prasetio

Abstract The utilization of plastic wastes to create polymer composite has become one of an effective solution to decrease the amount of untreated plastic wastes. The combination between polymer as a matrix with natural fiber as reinforcement in composite gives major advantages to its mechanical properties, but also few disadvantages that can reduce its potential application. An experiment was performed to investigate changes of composite surface quality and its mechanical properties before and after weather exposure. In this experiment, bamboo fiber reinforced recycled HDPE went through natural weathering for 1 and 2 months outside to analyze weathering effect on its physical and mechanical properties. The specimen consists of *r*HDPE as matrix and giant bamboo as reinforcement was made by compression molding. The bamboo fiber which previously treated with NaOH 5% was used as a composite's bottom layer while *r*HDPE as the top layer with a ratio of 20:80. The study showed that weather exposure gives no significant color changes nor quality degradation to the composite's surface. Mass and impact strength degraded with increasing of weathering duration. Moreover, composite tensile strength after experiencing 1-month weather exposure recorded higher than 0 month (without exposure).

Keywords Bio-composite · *r*HDPE · Bamboo · Natural weathering · Mechanical properties

1 Introduction

The use of polymer composite has escalated rapidly in the last few years due to its mechanical and physical properties. Polymer composite, especially with bio-based reinforcement, has been considered as advanced material replacing the common use of stainless steel and brass in the industrial and medical sectors. Lightweight, corrosion resistance, high strength, and durability are few among various advantages that

N. A. Astadini · I. Widiastuti (✉) · B. Harjanto · R. C. Adiputra · A. Prasetio
Universitas Sebelas Maret, Surakarta, Indonesia
e-mail: indahwied@uns.ac.id

N. A. Astadini
e-mail: naviraalya.a@gmail.com

U. Sabino et al. (eds.), *Proceedings of the 6th International Conference and Exhibition on Sustainable Energy and Advanced Materials*, Lecture Notes in Mechanical Engineering,
https://doi.org/10.1007/978-981-15-4481-1_62

give polymer composite an unlimited application even with geometric complexity. With the total of 396 million plastic has been produced in 2016 and 75% of them ended up wasted [1], plastic waste has the ability to cover the need for sustainability of renewable resources. Recycling plastic waste into polymer composite is also a partial environmentally friendly and reasonable cost solution that gives economic viability for various industrial sectors. The polymer as a matrix requires strong and low-density reinforcement to create an ideal composite with a combination of both properties [2]. Natural fiber, such as bamboo, provides the specified mechanical properties. Its utilization as reinforcement in polymer composite has drawn a lot of attention among other bio-based polymer composite.

As one of the fastest-growing plants in the world, bamboo can be harvested only in 3–5 years [3]. Bamboo has been used for various indoor and outdoor applications from housing materials to furniture and is attached to Southeast Asian culture. Its distributions are in tropical areas and native to warm temperatures. Bamboo has been proven to provide strength and ductility when combined with polymer composite, but not giving any specific superior mechanical properties that make them suitable for great tensile applications [4]. The presence of a wax layer on bamboo also lowers the interfacial bond between matrix and reinforcement, forcing an intervention by chemical substance [5].

Although better mechanical properties derive from the use of lignin natural fibers in composite [6], the hydrophilic characteristic from fibers and hydrophobia characteristics from polymer could jeopardize the composite itself. The contrast characteristics eventually lower down mechanical properties and potentially damage the composite faster [7]. With a variety of outdoor applications, temperature exposure, ultraviolet (UV) radiation, rain, and humidity are few important factors that determine composite longevity. Polymer obtains their mechanical properties from its long branching. The UV radiation from direct sunlight disconnects polymer branching and degrade lignin on bamboo. As a result, polymer would lose ductility and tensile strength, giving the composite high fracture possibility from brittleness. Aside from UV radiation, humidity increases the water absorption characteristic owned by bamboo and reduce load transfer capability of composite [5]. It is ironic how Southeast Asia and South America are biggest bamboo importer but the high humidity level due to its geographical location on Earth lowers the possible of wider outdoor applications.

This experiment analyzed the resistance of polymer composite with natural fiber reinforcement that survives under specific weathering. Using recycled High-Density Polyethylene (*r*HDPE) and bamboo, composite underwent natural weathering outdoor for 1 and 2 months. Changes in physical and mechanical properties were observed before and after the test, from surface gloss loss to tensile and impact strength.

2 Materials and Method

2.1 Materials

Grinded *r*HDPE was retrieved from local plastic collectors in Surakarta, Central Java, Indonesia. Only white and several transparent colors were used to create the specimen so it has homogenous color. *r*HDPE was washed with dish soap twice to reduce the smell and dried under the sun. The *r*HDPE has a crystallization temperature (Tc) of 113 °C, a glass transition temperature (Tg) of 120 °C, and melting temperature (Tm) of 180 °C [8]. HDPE as a hard plastic can provide maximum plastic properties and stiffness at the best level when processed between Tg and Tm [9].

Giant bamboo or dragon bamboo *(Dendrocalamus asper)* was retrieved from the local market in Surakarta, Central Java, Indonesia. The bamboo was on its harvest time (3–5 years) so it provided high strength and ductility from its outer parts [10]. The outer parts of bamboo were chopped into ±30 mm × 30 mm × 3 mm each and soaked in 5% NaOH aqueous solution for two hours on room temperature [7]. After the soaking process has completed, the bamboo fibers were taken out and washed on tap water to neutralized the pH level and dried with oven for 12 h on 60 °C [7]. Specimens were made using a hot press machine with a compression molding method. Both bamboo fibers and *r*HDPE were crushed until size of 5 mesh using crusher machine as shown in Fig. 1. The specimens used 80:20 *r*HDPE:bamboo ratio and layered on top of waxed astralon plastic. The mixture was heated until 175 °C and 50 bar pressure for 25 min. When done, specimens were cooled down on room temperature without removing from mold [8].

The specimen is a laminates composite which consists of two layers, namely the *r*HDPE which is brittle at the top layer and the bamboo which is ductile at the bottom layer. In the process of making specimens, an aluminum-based frame is used to maintain the thickness of the specimen when pressed using a heat press. To make specimens with a thickness of 3 mm, an aluminum frame of 120 mm × 100 mm

(a) (b)

Fig. 1 **a** Crushed bamboo particle, **b** crushed *r*HDPE

× 10 mm is used which is stacked three pieces and covered with a plastic top and bottom that has been given grease.

2.2 *Method*

Natural weathering was performed by exposing specimens to direct outdoor sunlight with angle 45° facing sun location against the earth. The specimens (see Fig. 2), were exposed for 0 month (without exposure), 1 and 2 months. Impact strength test used ASTM D4812 (unnotched) standard. Tensile strength test used ASTM D638 type V.

3 Results and Discussion

Natural weathering was performed to identify the effect of weather on mechanical properties and the changes in the physical condition of composite specimens both indoor and outdoor applications. Tables 1 and 2 show the meteorological data obtained from Agriculture Laboratory during the irradiation process.

Fig. 2 Composite specimen for natural weathering

Table 1 Monthly average temperature

Months	Temperature (°C)	Humidity (%)	Exposure (%)
	Mean		
April 2019	28.1	86	66.9
May 2019	28.4	83	65.6
June 2019	25.4	84	65.5

Table 2 Average evaporation

Months	Water height	Evaporation (mm)	Water temperature (°C)	Wind: 0.5 m
	Mean			
April 2019	58.5	3.8	28	6.3
May 2019	56.5	4.1	38	6.3
June 2019	56.0	4.1	28	6.3

3.1 *Impact Strength*

The pendulum was swung from an alpha height angle (α) of 135°. When the pendulum hits the specimen clamped at an angle of 0°, the specimen will break and the pendulum creates an angle of beta (β) which is the maximum point of the pendulum swing. Specimens used for impact strength testing are unnotched specimens which are generally used for sensitive material. The test was carried out five times for each type of specimen and obtained an average angle of beta (β) for each type of specimen. Table 3 below shows the data obtained during the composite specimen testing process.

In reliance on the fracture, the composite specimen experiences two types of fractures. Figure 3a is a composite net fracture that is dried for 2 months. The composite experienced bamboo loss which resulted in the composite specimen having a brittle

Table 3 Impact strength

Exposure duration	Average beta (β) angle	Impact energy (kJ)	Impact strength (kJ/m^2)
0 Month	118.17	0.20	3.39
1 Month	119	0.19	3.20
2 Months	120.17	0.17	2.95

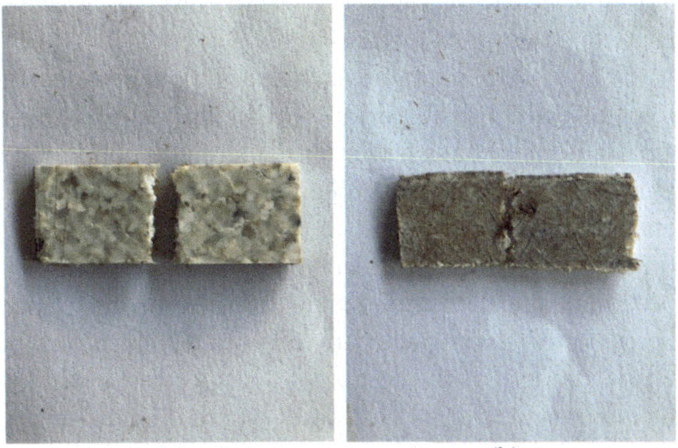

(a) (b)

Fig. 3 The fracture pattern of weathered composite for **a** 2 months, **b** 1 month

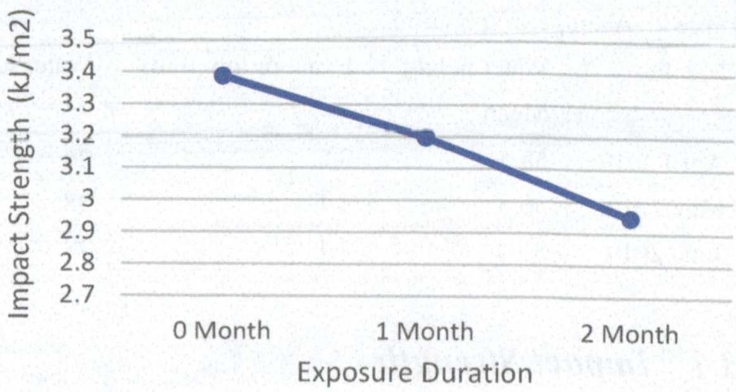

Fig. 4 Graph of the relation of impact strength to exposure duration

fracture due to the ductile nature that should be given from the bamboo reinforcement did not have a significant effect. In Fig. 2b, the composite which is dried for 1 month has bamboo still covering the bottom of the composite well. When the composite specimens are tested for their impact strength, the impact of the pendulum does not break the composite as a whole where the bamboo in the composite is not completely broken. This condition also appears in composite specimens that do not experience desiccation (0 months) (Fig. 4).

Referring to the graph, it can be concluded that the impact strength of the *r*HDPE bamboo-composite specimen decreases with the duration of exposure. The decrease that occurs has a small value because the exposure period is only 1 month each. The decrease in impact strength arises due to the effect of exposure to composite specimens at maximum conditions, where specimens experienced rain, strong winds, and heat during the drying process.

3.2 *Tensile Strength*

Composite tensile strength testing is carried out with a pull speed of 10 mm/min. The test is carried out 4 times in each variation of the specimen and 3 results are chosen to be the average. The figure below states the average of the tensile strength test data on composite specimens.

From Fig. 5a, it can be seen that the magnitude of the composite tensile strength is not affected by the period of the composite exposure. Composites without exposure (0 months) have lower tensile strength compared to other composite variations, which is 20,140 N/mm^2. Whereas composites that are dried for 1 month have better tensile strength compared to composites that are dried for 2 months. This is due to HDPE which obtains mechanical strength from its long polymer chain [2]. When the composite of 0 months (without exposure) undergoes tensile strength testing, the polymer chain owned by the composite is not perfect and resulted in a low tensile strength of the composite. In specimens that have been exposed for 1 month, the composite is at its best point where the tensile strength reaches 22,011 N/mm^2.

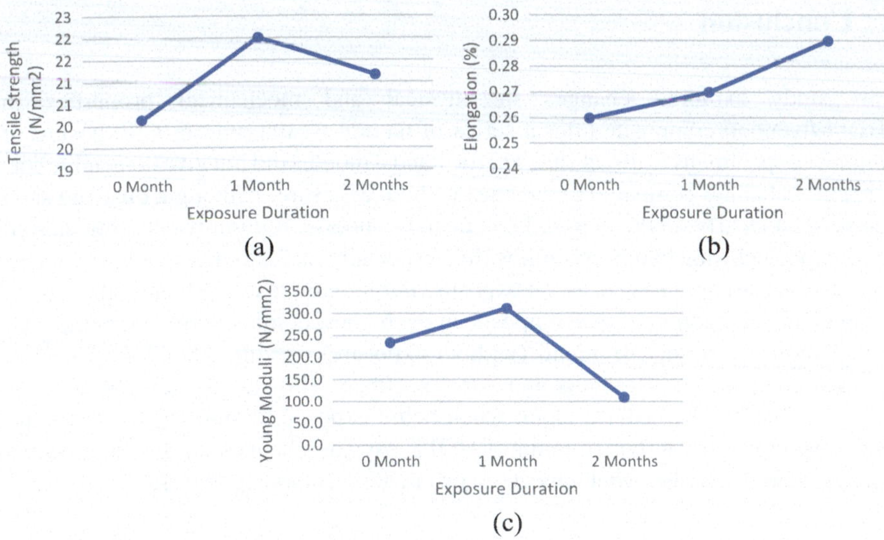

Fig. 5 Composite graph for **a** tensile strength, **b** elongation, **c** Young Moduli

However, when the composite was continued to cast for 2 months, the tensile strength decreased again due to cracks and changes in the microstructure of the composite.

After experiencing weather exposure, Young modulus' composite increased up to 1.5 times the value of Young modulus' composite that did not experience weather exposure. Young modulus composite values have decreased quite dramatically in specimens that have been exposed for 2 months. This shows that the presence of bamboo in the composite influences Young modulus value. A sharp decrease in composite specimens that have been exposed for 2 months has experienced a drastic reduction in bamboo due to loss during weathering process. Bamboo gives good stiffness to the composite and the stiffness decreases sharply in the composite specimen.

In addition to heat, exposure to UV radiation can also cause the breaking of polymer chains that cause micro-cracks in the composite [2]. However, the hydrophilic nature of bamboo causes the absorption of water and moisture in the composite to be higher and damages the structure of the composite. Referring to the weather data during testing, heavy rainfall allows water absorption and moisture to be an important factor in decreasing the tensile strength of composites. Fractures that occur on composites tend to be more rigid and brittle. Nevertheless, the three variations of the composite did not experience a fracture when reaching the ultimate tensile strength (UTS) point and experienced an extension and reduction in the thickness of the specimen that is generally experienced by inelastic material or often referred to as necking.

4 Conclusion

This study explored changes in physical and mechanical properties in *r*HDPE/bamboo composite after a series of natural weathering tests. Exposure of composite specimens to direct sunlight for 1 and 2 months did not give detectable surface color changes compared to the ones without exposure. Although, dust and sand can be seen clearly on the exposed composite's surface. Natural weathering caused the composite losing bamboo composition, especially on 2 months weathered specimen. Impact strength lowers as the exposure duration increases. Meanwhile, tensile strength and elongation of the composite at the highest point occurred on composite with 1-month exposure. 0-month (without exposure) specimen has higher tensile strength compared to 2 months exposure specimen. This due to sunlight and UV radiation that breaks *r*HDPE chain when being exposed to sunlight for too long. The breaking of the polymer chain of HDPE causing changes on the composite's microstructure and also what appears to be crack on composite's surface.

References

1. Dalberg Advisors (2019) Solving plastic pollution through accountability, vol 13. WWF—World Wide Fund For Nature, Gland, Switzerland
2. Brennan P, Fedor C, Pausch G (1988) Sunlight, UV and accelerated weathering. Paint Resin 58:17
3. Chandramohan D, Marimuthu K (2011) Drilling of natural fiber particle reinforced polymer composite material. Int J Adv Eng Res Stud 1(1):134–145
4. Dixit S, Goel R, Dubey A, Shivhare PR, Bhalavi T (2017) Natural fiber reinforced polymer composite materials—a review. Polym Renew Resour 8(2):71–78
5. Farrelly D (1984) The book of bamboo, vol 25. Sierra Club Books, California, USA
6. Hojo T, Xu Z, Yang Y, Hamada H (2014) Tensile properties of bamboo, jute and kenaf mat-reinforced composite. Energy Procedia 56(9):72–79
7. Wang B, Panigrahi S, Tabil L, Crerar W (2007) Pre-treatment of flax fibers for use in rotationally molded biocomposites. J Reinf Plast Compos 26(5):447–463
8. Wu TL, Chien YC, Chen TY, Wu JH (2013) The influence of hot-press temperature and cooling rate on thermal and physicomechanical properties of bamboo particle-polylactic acid composites. Holzforschung 67(3):325–331
9. Kai Z, Wang F, Liang W, Wang Z, Duan Z, Yang B (2018) Thermal and mechanical properties of bamboo fiber reinforced epoxy composites. Polymers 10(608):1–18
10. Malanit P (2009) The suitability of *Dendrocalamus asper* backer for oriented strand lumber. University of Hamburg

Effect of Fly Ash on the Mechanical Properties of Polyvinyl Chloride-Fly Ash Composite

A. W. Nugroho, M. K. P. Prasetyo, and C. Budiyantoro

Abstract Fly ash (FA), a fine powder obtained during the combustion of coal at a thermal power station, possesses several environmental issues, including atmosphere contamination and huge land area for dumping. One potential use of FA is as filler in polymers in order to improve their properties and to reduce production costs. Polyvinyl chloride (PVC) is a well-known thermoplastic and the most commonly used polymers in building industries due to its advantageous properties. This study investigated the influences of the FA as filler material on the mechanical properties of PVC/FA composites. PVC in powder form was mixed with addictive substances (Tribasic Lead Sulfate, Normal Lead Stearate, Calcium Stearate, and Stearate Acid) at a temperature of 100 °C. Following this, the FA was added to the mixed PVC in various compositions (0, 8, 10, 20, and 30 phr) and shook. The Hot Press Molding machine fabricated the PVC/FA composites under pressure of 13 MPa at a temperature of 200 °C for 300 s. Tensile and impact tests were carried out in accordance with ASTM D638-02A type IV and ASTM D6110-04, respectively. The results revealed that the incorporation of FA had significantly improved the tensile strength, Young's Modulus, and the elongation at break of the composites. The composite containing 8 phr FA showed the highest value of the respective properties. Those properties tended to decrease with the increase of FA content. The impact properties also showed a similar phenomenon. The value of impact strength and impact energy for the composite containing 10 phr were found the highest. Scanning electron microscopy images showed that FA particles were precipitated and mechanically interlocked in the PVC matrix. Particle agglomeration was found in the composites containing higher amounts of FA. The study indicates a remarkable potential for FA to produce useful PVC/VA composite.

Keywords Polyvinyl chloride · Fly ash · Filler · Mechanical properties · Composite

A. W. Nugroho (✉) · M. K. P. Prasetyo · C. Budiyantoro
Universitas Muhammadiyah Yogyakarta, Yogyakarta 55183, Indonesia
e-mail: ariswidyo.nugroho@umy.ac.id

U. Sabino et al. (eds.), *Proceedings of the 6th International Conference and Exhibition on Sustainable Energy and Advanced Materials*, Lecture Notes in Mechanical Engineering,
https://doi.org/10.1007/978-981-15-4481-1_63

1 Introduction

Pulverized coal is widely used to be burned as a heat source at a thermal power station to generate electricity accompanied by the solid residues. While the unburned residue drop to the bottom of the furnace, known as bottom ash (BA), the dust collection system extracts lighter fine ash particles from the flue gas called fly ash (FA). The composition of the remaining coal combustion is Silicon Dioxide (SiO_2), Aluminum Trioxide (AL_2O_3), Ferric Oxide (Fe_2O_3), Calcium Oxide (CaO), and others. Coal combustion produces ash, which has a percentage of 5–15% is BA, and 85–95% is FA. Due to the storage and handling difficulties in terms of environmental concern, the application of FA become challenges to be developed. Although FA is commonly dumped in lagoon and dams due to its many advantages, such as low cost, low density, smooth particle surface, good processability as filler material, and well-distributed internal stress [8]. It has been used in various engineering applications, for instance, soil stabilization, cement mixture, asphalt filler, concrete products, and tile making [20]. By doing so, it may reduce the usage of non-renewable natural resources and the replacement of materials that may be energy-intensive to fabricate. As filler in polymers and rubber is one potential use of FA to lower the fabrication cost and to improve certain properties. Those physical properties of FA make it an appropriate filler for rubber and polymer.

Many studies have also revealed the reinforcing effect of FA in polymer matrix composites. It was applied in various polymers such as epoxy [11], polyvinyl alcohol (PVA) [12], high density polyethylene (HDPE) [4], polyurea elastomer [16], polypropylene (PP) [13, 18, 19], polyetheretherketone (PEEK) [14], ethylene-octene copolymer (EOC) [1], acrylonitrile butadiene styrene(ABS) [10] and more recent polyvinyl alcohol (PVC) [8, 20]. In general, the addition of FA indicates an enhancement of the mechanical and thermal properties of the polymer composites.

Polyvinyl Chloride (PVC) is the second most common plastic material mainly used for fabricating pipe fittings, sheets, cables, etc. Pipes and fittings need good tensile strength and impact properties, short and long-term pressure sustainability. The use of PVC is growing fast in the research due to its properties such as low cost, low density, low thermal conductivity, improved acoustic damping properties, chemical stable, and excellent fire retardancy [5, 17, 21]. However, the low thermal stability and brittleness cause limitation of its application at high temperatures. Hence, enhancement in these properties of PVC will expand its application [2, 9]. Although many researchers have studied the impact and tensile properties of PVC-FA, most of them used injection molding or extruder machine to fabricate the composite. The present paper evaluated the effect of fly ash (FA) as a filler on the impact and tensile strength properties on PVC-FA composite fabricated using a hot press machine.

2 Experimental Method

FA was obtained from the thermal power station at a company in Cilegon Banten. The coal consumption was 300 tons, along with 30 tons FA by product. The company used coal as a fuel for power stations in steam generators unit to manufacture PVC plastic powders. The main constituent of FA consisted of silicon dioxide (SiO_2), aluminum trioxide (Al_2O_3), ferrite oxide (Fe_2O_3), and calcium oxide (CaO). The manufactured Polyvinyl chloride (PVC) required PVC K-65R powder and an additive compound, including tribasic lead sulfate as a heat stabilizer and normal lead stearate, calcium stearate, and stearate acid as a lubricant. Processing was conducted in two steps. In the first stage, those elements were formulated in the LAboplastomill mixer. The FA was added in the second stage of the composition, as shown by Table 1. The composition is expressed in part per hundred rubber (phr) unit.

The mixed powder was then poured into the flat mold and hot pressed using Hot Press Molding Collin P300E machine with two-stage for heating processes. The temperature was set up at 200 °C for 600 s in 60 Bar. Following this, the pressure was increased up to 130 bar for 300 s at a temperature of 200 °C. After that, the temperature was cooled to room temperature for 300 s with a pressure increase to 160 Bar. The PVC/FA plate composite being fabricated is shown in Fig. 1. Type specimens, as per ASTM D-638-02, were cut from the hot-pressed sheet. The specimens were tested

Table 1 Composition

No.	Code	Constituent as phr		
		PVC	Additives	FA
1	Phr0	200	11	0
2	Phr8	200	11	16
3	Phr10	200	11	20
4	Phr20	200	11	40
5	Phr30	200	11	60

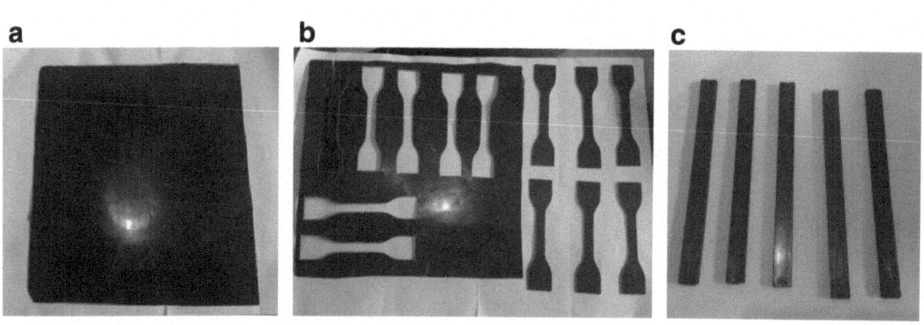

Fig. 1 PVC/FA composite plate (**a**), tensile test specimen (**b**), un-notched impact charpy specimen (**c**)

on a universal testing machine with a 10 kN load cell and a crosshead speed of 5 mm/min.

The notched Charpy Impact for a plastic material test as per ASTM D6110-04 was carried out using Zwick/Roell Impact Test Machine. The impact Charpy specimens were cut using profile cutting machines to avoid the micro crack. Five specimens were tested for each parameter of tensile or impact testing. Optical microscopy (OM) and Scanning Electron Microscopy (SEM) was used to examine the morphology of the composite.

3 Results and Discussion

3.1 Tensile Properties

Tensile properties were evaluated by tensile tests, including tensile strength, elastic modulus, and strain at break. Tensile strength and the elastic modulus of the PVC/FA composite as a function of the FA content are shown in Fig. 2.

Ultimate tensile stress initially found to increase on the addition of FA as expected to achieve a maximum of 64.57 MPa at 8 phr of FA but later decreases with the increase of FA content [3]. The addition of FA as filler encouraged inhomogeneity in the matrix, causing interference of filler in the mobility or deformability of the matrix. In addition, since the hardness of FA is higher than that of the PVC [8], the load is sufficiently transferred from the matrix into the filler. It may obstruct

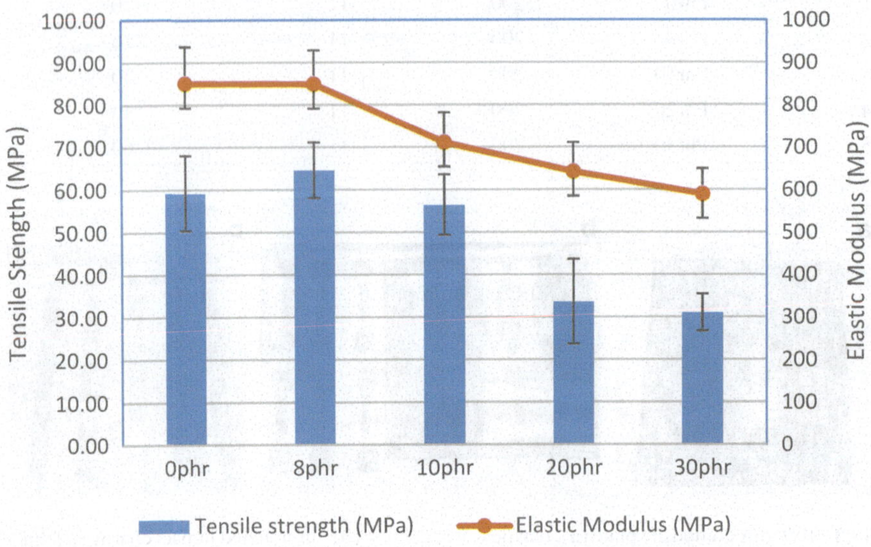

Fig. 2 Tensile strength and elastic modulus of the PVC/FA composites

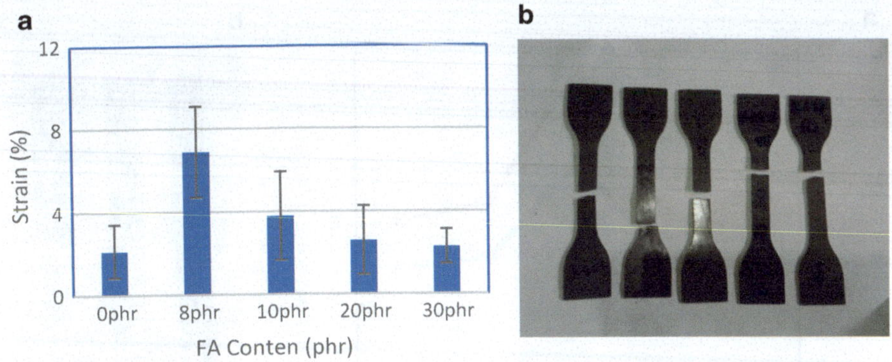

Fig. 3 The relation between FA content and strain at break (**a**), tensile specimen failure (**b**)

matrix deformation resulting in an increase of the tensile strength [15]. However, due to the absence of a coupling agent and higher probability of agglomeration of FA particle in the matrix at higher FA content, weak bonding between filler and matrix at higher FA content occurred, leading to a lack of stress transfer between them, resulting in a decrease in the tensile strength and reducing the elastic modulus as shown in Fig. 2. Those decrease may attribute to restriction of polymer chain movements leading to reduce in strain at break as shown in Fig. 3. Khoshnoud et al. [8] have demonstrated that tensile strength and elongation of composite decreases with increase in filler content. Correspondingly, incorporation of a small amount (8 phr) of filler improved the elastic modulus of the composites and later decreased at higher filler content. The tensile strength result is comparable to previous research [6], showing the tensile strength in the range of 23–38 MPa.

3.2 *Impact Strength*

The effect of FA content on the impact strength of the PVC/FA composites is shown in Fig. 4. In general, by adding FA filler, the impact strength and impact energy increased slightly.

It points out that at higher speed rate loading load could be transferred by the filler. It was because the FA particle occupied the interstitial between PVC particles. It tended to reduce void content, which led to an increase in the impact strength and impact energy. At 10 phr of FA filler, the impact strength and impact energy increased dramatically due to most void might be occupied by the filler. On the further addition of FA, there was inadequate wetting of FA by PVC. It tended to increase in void content, and thus impact energy and impact strength decreased and later remained nearly constants on further addition of FA. Previous research revealed that their impact strength and impact energy are in the range of 2.6–4.8 kJ/m^2 and 0.18–0.3 J, respectively [7], which is close to the current research. However, their values

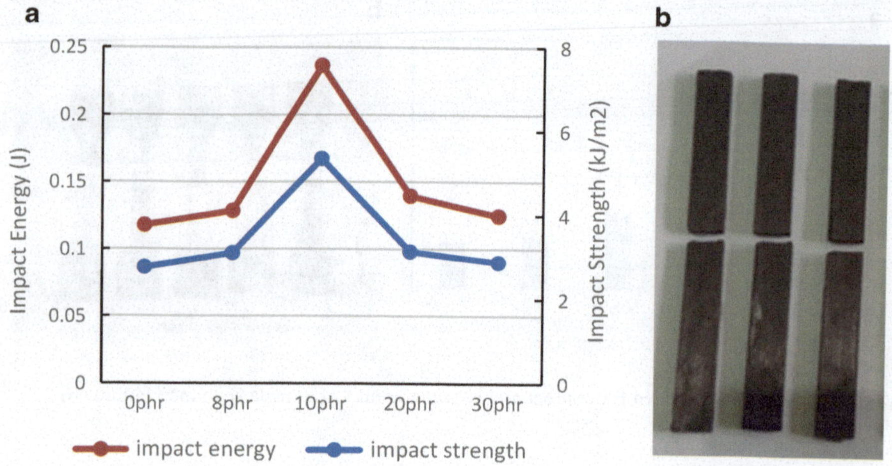

Fig. 4 Variation of impact strength of the PVC/FA composites with filler content (**a**), impact test specimen (**b**)

decreased along with the increase of the filler content. From an optical microscope image, it could be seen that the surface failure of the 8 phr content specimen showed gold-colored grains scattered, inhomogeneity, and agglomeration filler might exist. FA was spread evenly on the upper side of the specimen and was not completely mixed. Therefore, the impact test results were 8 phr variations lower than that of the 10 phr specimen. The surface failure on the v-notch side and impact hammer showed cleavage with smoother surface representing brittle failure (Fig. 5a). Optical micrograph of the surface 10 phr specimens of impact test showed rougher look like dimples (Fig. 5b), which was associated with less brittle failure. The gold-colored brown area representing FA was occasionally scattered on the sides of the specimen.

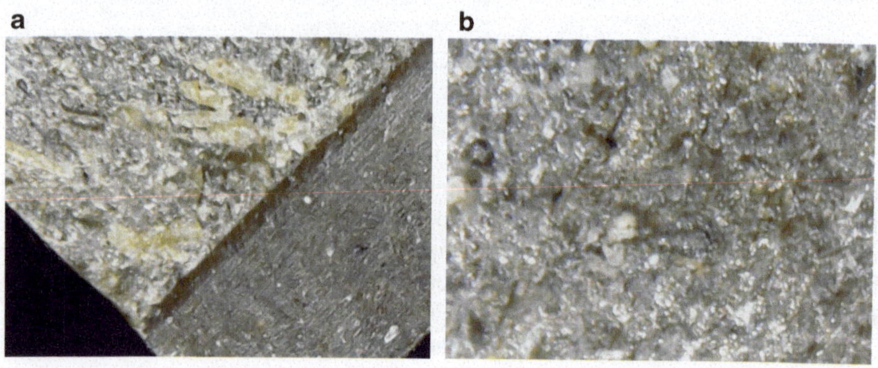

Fig. 5 Optical image of the failure impact testing specimen with FA filler 8 and 10 phr

Fig. 6 SEM micrograph of the PVC/FA with 10 phr of FA

SEM was applied to evaluate the morphology of the PVC/FA composites. Figure 6 presents the SEM micrographs of the PVC/FA composites with 10 phr of FA content. The morphological study revealed that FA possessed a smooth spherical surface providing more surface area for interaction. There was a good dispersion of filler particles in the polymer matrix. The excellent interaction between the matrix and the filler was also presented in the SEM image. In addition, the figure shows the size of the FA embedded in the PVC matrix. It could be seen that the FA particle does not diffuse into the matrix. The particle size of FA in the range of 2.58–7.11 µm in a spherical shape and rounded occupied the edges of the specimen. The FA particles were well dispersed in the polymer matrix. This image also supports those of the mechanical properties, as stated above.

4 Conclusions

The PVC/FA composite had already been fabricated using a hot press machine successfully. Inorganic filler, i.e., FA added to the PVC, improved the strength, rigidity, and impact strength at a certain amount and then decreased at a higher amount of the filler. The maximum strength and impact strength were achieved at different compositions, namely 8 phr and 10 phr, respectively. The tensile strength slightly increased due to the existence of the filler particle resulting in mechanical restraints. At higher speed loading, the presence of filler particles might adsorb the energy better. The morphological evaluation revealed that the filler particles in spherical shaped were well dispersed in the matrix and possessed an excellent interaction surface area.

References

1. Anandhan S, Sundar SM, Senthil T, Mahendran A, Shibulal G (2012) Extruded poly (ethylene-co-octene)/fly ash composites–value added products from an environmental pollutant. J Polym Res 19(3):9840
2. Awad WH, Beyer G, Benderly D, Ijdo WL, Songtipya P, del Mar Jimenez-Gasco M et al (2009) Material properties of nanoclay PVC composites. Polymer 50(8):1857–1867
3. Bose S, Mahanwar P (2004) Effect of flyash on the mechanical, thermal, dielectric, rheological and morphological properties of filled nylon 6. J Min Mater Charact Eng 3(2):65–89
4. Deepthi M, Sharma M, Sailaja R, Anantha P, Sampathkumaran P, Seetharamu S (2010) Mechanical and thermal characteristics of high density polyethylene–fly ash cenospheres composites. Mater Des 31(4):2051–2060
5. Eaves D (2004) Handbook of polymer foams. polimeri 25(6):1–2
6. Joshi PS, Marathe DS (2018) Experimental investigation of mechanical properties of impact modified polyvinyl chloride-fly ash composites. J Min Mater Charact Eng 7(1):34–47
7. Khoshnoud P, Abu-Zahra N (2015) Effect of cenosphere fly ash on the thermal, mechanical, and morphological properties of rigid PVC foam composites. J Res Updates Polym Sci 4(1):1
8. Khoshnoud P, Abu-Zahra N (2018) The effect of particle size of fly ash (FA) on the interfacial interaction and performance of PVC/FA composites. J Vinyl Add Tech 25(2):134–143
9. Khoshnoud P, Gunashekar S, Jamel MM, Abu-Zahra N (2014) Comparative analysis of rigid PVC foam reinforced with class C and class F fly ash. J Min Mater Charact Eng 2(06):554
10. Kulkarni M, Bambole V, Mahanwar P (2014) Effect of particle size of fly ash cenospheres on the properties of acrylonitrile butadiene styrene-filled composites. J Thermoplast Compos Mater 27(2):251–267
11. Kulkarni S (2002) Effects of surface treatments and size of fly ash particles on the compressive properties of epoxy based particulate composites. J Mater Sci 37(20):4321–4326
12. Nath DCD, Bandyopadhyay S, Boughton P, Yu A, Blackburn D, White C (2010) Chemically modified fly ash for fabricating super-strong biodegradable poly (vinyl alcohol) composite films. J Mater Sci 45(10):2625–2632
13. Nath DCD, Bandyopadhyay S, Yu A, Zeng Q, Das T, Blackburn D et al (2009) Structure–property interface correlation of fly ash–isotactic polypropylene composites. J Mater Sci 44(22):6078–6089
14. Parvaiz MR, Mohanty S, Nayak SK, Mahanwar P (2011) Effect of surface modification of fly ash on the mechanical, thermal, electrical and morphological properties of polyetheretherketone composites. Mater Sci Eng, A 528(13–14):4277–4286
15. Petchwattana N, Covavisaruch S (2013) Effects of rice hull particle size and content on the mechanical properties and visual appearance of wood plastic composites prepared from poly (vinyl chloride). J Bionic Eng 10(1):110–117
16. Qiao J, Amirkhizi AV, Schaaf K, Nemat-Nasser S (2011) Dynamic mechanical analysis of fly ash filled polyurea elastomer. J Eng Mater Technol 133(1):011016
17. Rabinovitch EB, Isner JD, Sidor JA, Wiedl DJ (1997) Effect of extrusion conditions on rigid PVC foam. J Vinyl Add Tech 3(3):210–215
18. Satapathy BK, Das A, Patnaik A (2011) Ductile-to-brittle transition in cenosphere-filled polypropylene composites. J Mater Sci 46(6):1963–1974
19. Sengupta S, Pal K, Ray D, Mukhopadhyay A (2011) Furfuryl palmitate coated fly ash used as filler in recycled polypropylene matrix composites. Compos B Eng 42(7):1834–1839
20. Sushma SP, Kumar AK (2014) Hardness and tensile testing of PVC and fly ash composite. In: Advance research and innovations in mechanical, material science, industrial engineering and management-ICARMMIEM, pp 273–277
21. Thomas N, Eastupy JR, Quirk P (2004) Rigid PVC foam. In: Formulating for sustainability, blowing agent and foaming process

Remaining Useful Life Estimation of the Motor Shaft Based on Feature Importance and State-Space Model

D. D. Susilo, A. Widodo, T. Prahasto, and M. Nizam

Abstract Induction motor is widely used in industry as a prime mover of the machine or mechanical equipment. The monitoring of this component is needed to assure that it works in its optimal performance and to prevent its sudden failure. The potential failures of the induction motor can be an electrical failure or mechanical failure. One of the mechanical failures is shaft failure. The failure of the motor shaft will cause the motor will not function properly. The objective of this paper is to estimate the RUL of the motor shaft based on Feature Importance (FI) and the state-space model. The FI of the feature is determined from the monotonicity and trendability criteria. Features with the high FI score then be used as the health indicator for RUL prediction. The RUL estimation was performed to the health indicator using the state-space model. The result shows that the state-space model can be used for the motor shaft RUL estimation satisfactorily before actual failure happened.

Keywords State-space model · Statistical feature · RUL estimation

1 Introduction

An induction motor plays an important rule in many industries. It is combined with the mechanical drive system as a gear train, belt-pulley or chain-sprocket to form an equipment drive system in the industry. This is due to its superiority as its robustness, simplicity in construction, reliability, and low cost [1]. Although the induction motor is undoubtedly reliable even though the occurrences of faults in it can not be avoided. The common failures of an induction motor are motor housing, motor stator, motor rotors, motor fans, motor bearings, motor insulation and windings, and motor shaft [2]. The failure of the induction motor can stop the system operation that leads to loss of production and sometimes causes safety problems. Therefore, the maintenance

D. D. Susilo (✉) · M. Nizam
Universitas Sebelas Maret, Surakarta 57126, Indonesia
e-mail: djoksus@uns.ac.id

A. Widodo · T. Prahasto
Diponegoro University, Semarang 50275, Indonesia

U. Sabino et al. (eds.), *Proceedings of the 6th International Conference and Exhibition on Sustainable Energy and Advanced Materials*, Lecture Notes in Mechanical Engineering,
https://doi.org/10.1007/978-981-15-4481-1_64

of an induction motor to ensure its performance becomes a priority in a modern maintenance policy.

Condition-based maintenance (CBM) that using machine run-time data to monitor the machine condition has been developed in modern maintenance strategy. There are three main processes in this strategy, namely system and signal processing, diagnostics and prognostics, and maintenance schedule. Signal processing covers data acquisition, data preprocessing, feature extraction and feature selection. Diagnostics refers to detecting, isolating, and identifying an incipient failure condition, while prognostics refers to the ability to predict the future condition of a failing component including their remaining useful life (RUL) [3].

Research in the field of prognostics of the induction motor and the mechanical drive system has been challenging in the last decades. Researchers use different signals and methods to obtain high accuracy of the RUL prediction. It is not an easy task to predict the future state of a system or component that experiences failure precisely. However, a hard effort has been doing by researchers to develop a prognostic method for the induction motor and mechanical drive system. Among them are presented here. Kraleti et al. [4] proposed a model-based prognostic for a three-phase induction motor that experiences insulation degradation and broken rotor. They used the neural network to estimate the motor's future state. Nguyen et al. [5] estimated the RUL of the induction machine under inter-turn fault using a particle filter model algorithm. Camci et al. [6] proposed a method to evaluate features goodness for bearing prognostics from the vibration signal. The rating was based on the monotonicity characteristic of the features. They found that RMS and standard deviation are good features for effective bearing prognostic. Sloukia et al. [7] estimated the bearing RUL based on RMS feature using the Support Vector Machine model and compared the results to the Mixture of Gaussian Hidden Markov Model. They found that the SVM model gave a better result than the HMM model. Zhang et al. [8, 9] deduced that energy ratio based on the AR model feature is the most effective feature for gearbox prognostic to accommodate the stationary and non-stationary conditions. Furthermore, they nominate new feature extraction methods to find features with a good degradation trend. They used the Sideband Index feature extracted from the vibration signal by the Narrowband Interference Cancellation technique. They found that this feature has better performance in gear prognosis than the traditional Sideband Index extracted from the time-synchronous average signal directly.

The prognostics process involves modeling of the degradation process using an appropriate model and then predict the future state of the machine/system using the model. One of the candidate models that can be used in prognostic is the state-space model. Bechhoefer et al. [10] presented a state-space model for predicting the RUL of a component based on a vibration signal. Feng et al. [11] applied the state-space model to predict the RUL of the lithium-ion batteries. Ni et al. [12] improved the state-space model more adaptive to predict the RUL of a planetary gearbox. Wang et al. [13] proposed an efficient prognostic method for battery RUL prediction based on state-space modeling with heterogeneity of noise variances. They used 26 lithium-ion batteries degradation data to illustrate how the proposed prognostic method works.

Most researches above focus on the prognostics of the failure of the components of the electrical and mechanical of an induction motor as stator winding, broken rotor, and rotor bearing. Rarely paper that examines the prognostics of the motor shaft, even though, this component has the potency to fail in the industrial practice. A motor shaft when working will be subjected to fluctuating loads of combined bending and torsion with various degrees of stress concentration. Moreover, a motor shaft usually has a geometry that causes stress concentrations such as the bearing support area and keyway. Continuous excessive torque loading and bending often lead to initial cracking in this area which can lead to motor failures such as cracks and keyway distortion [14, 15].

Therefore, maintenance practitioners must pay attention to the failure of the motor shaft to prevent failures that occur suddenly. This can be done by estimating the RUL of the motor shaft. This paper aims to develop a new method for estimating the RUL of motor shafts. The method covers the selection of the features as a prognostic indicator and the model for RUL estimation.

2 Methodology

This study tries to develop a motor shaft prognosis using individual features extracted from the vibration signal of an induction motor. The signals are acquired from the run to failure test. The run-to-failure test rig is shown in Fig. 1.

The test rig consists of a 3 phase 2 Hp induction motor with a maximum speed of 1500 rpm. The speed of the motor is controlled by an inverter. The loading system is gained from a hydraulic power pack through a motor loading system. The load type

Fig. 1 Motor run-to-failure test rig

is bending load. Two PCB 352C33 accelerometers are attached at the electric motor casing using a magnetic base to measure the vibration of the motor.

The process of the motor shaft RUL estimation follows the steps that figured in Fig. 2.

The raw vibration signal was preprocessed using Discrete Wavelet Transform (DWT). This method is considered as an effective tool to handle non-stationary signals, as it interprets the signal in time as well as frequency. The raw signals were processed with DWT 4-level decomposition.

In the Feature extraction step, 17 features including features in the time domain, frequency domain, and time-frequency domain are drawn out. They are mean, RMS, shape factor, skewness, kurtosis, crest factor, entropy estimation, entropy estimation error, histogram upper bound, histogram lower bound, frequency RMS, frequency center, Root variance frequency, and Second until Fifth Autoregressive coefficient.

The feature selection is done using the Feature Importance (FI) parameter. The FI parameter is obtained by combining the value of monotonicity and the trendability

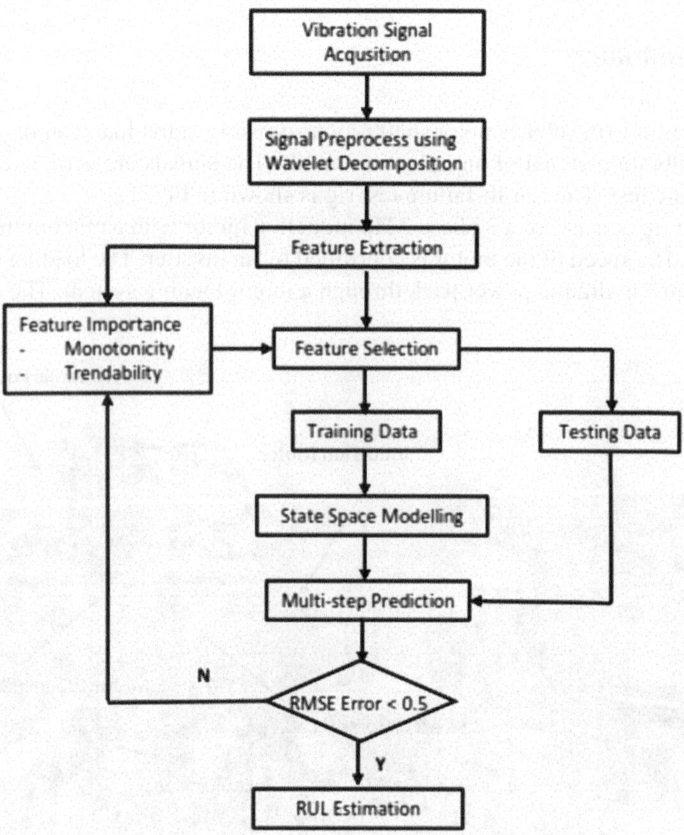

Fig. 2 Flow chart of the motor shaft RUL estimation

of each feature. Monotonicity characterizes the underlying positive or negative trend of the parameter while Trendability indicates the degree to which the parameters of a population of systems have the same underlying shape and can be described by the same functional form. The monotonicity and trendability are formulated in Eqs. (1) and (2) [14].

$$Monotonicity = \left| \frac{Num.of\, d/dx > 0}{n-1} - \frac{Num.of\ d/dx < 0}{n-1} \right| \tag{1}$$

$$Trendability = \left| \frac{Num.of d/dx > 0}{n-1} - \frac{Num.of\ d^2/dx^2 < 0}{n-2} \right| \tag{2}$$

The selected features are then divided into training data and testing data. The training data are used to build the state-space model, and the testing data are used to validate the RUL prediction. The number of training data is fifty percent of the feature data. The reason is at this time step, the feature data tends to increase its value.

The state-space models are models that use state variables to describe a system by a set of first-order differential or difference equations. State variables x(t) can be reconstructed from the measured input-output data. The state-space model structure can be used for quick estimation because it just requires to specify only one input and the model order n. The model order is an integer equal to the dimension of x(t) and relates to, but is not necessarily equal to, the number of delayed inputs and outputs used in the corresponding linear difference equation.

The most general state-space representation of a linear system with m inputs, p outputs, and n state variables is written in the following form:

$$\dot{x}(t) = A(t)x(t) + B(t)u(t) \tag{3}$$

$$\mathrm{y(t) = C(t)x(t) + D(t)u(t)} \tag{4}$$

where:

$t \in R$ denotes time
$x(t) \in R^n$ is the state (vector)
$u(t) \in R^m$ is the input or control
$y(t) \in R^p$ is the output
$A(t) \in R^{nxn}$ is the dynamic matrix
$B(t) \in R^{nxm}$ is the input matrix
$C(t) \in R^{pxn}$ is the output matrix
$D(t) \in R^{pxm}$ is the feedthrough matrix.

For the continuous invariant time, the state space model can be written as:

$$\dot{x}(t) = Ax(t) + Bu(t) \tag{5}$$

$$y(t) = Cx(t) + Du(t) \tag{6}$$

So that now, A, B, C, and D are constant and do not depend on t. The system represented by (5) and (6) is the first-order differential equation known as the state equation, and x(t) is the state vector and u(t) is the input vector. The Eq. (6) is referred to as the output vector.

The state-space model will be used to predict 100 and 200-time steps ahead of the motor shaft condition. This multi-step ahead prediction is intended to evaluate whether the model can be used for long time prediction or not. The prediction process is updated using a new measurement until the prediction value hits the predetermined failure threshold.

3 Results and Discussion

The motor shaft failure data was obtained from the run-to-failure experiment. The vibration signals were measured using two accelerometers attached to the motor casing. The test was stopped after the vibration amplitude reached more than four times from the initial amplitude as the failure indication. The shaft failure was found in the motor shaft after physical inspection. There is a distortion on the keyway which is on the shaft resulting in looseness at the shaft coupling. The keyway distortion is shown in Fig. 3.

Fig. 3 The keyway distortion on the motor shaft after run-to-failure testing

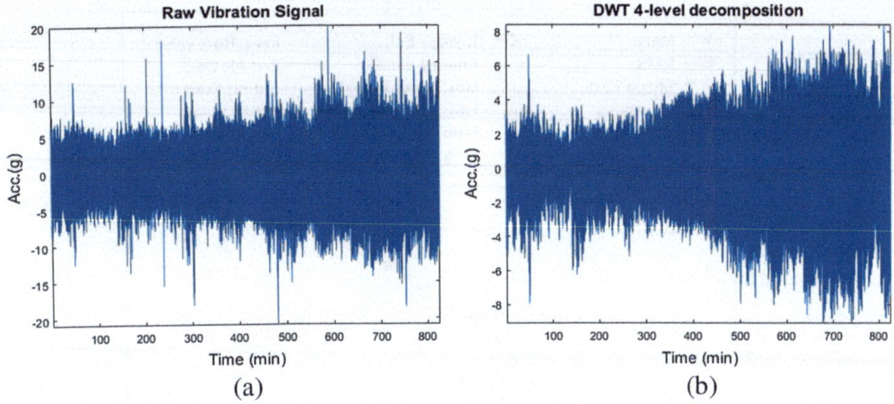

Fig. 4 **a** Raw vibration signal of the shaft failure, **b** DWT 4-level decomposition of the raw signal

The raw vibration signal of the motor and the preprocessed signal using Discrete Wavelet Transform decomposition is shown in Fig. 4a, b.

The next step is feature extraction. Features in the time domain, frequency domain, and time-frequency domain are extracted and evaluated to find their importance in the motor shaft RUL estimation. These features are then smoothed using a moving average filter. Each feature is calculated its monotonicity and trendability according to Formulas (3) and (4). The Feature Importance (FI) of the features is constructed from the superposition of the feature monotonicity and the trendability. The feature selection is done based on the monotonicity and trendability criteria. The FI score is evaluated on the amount of training data that vary, starting from 30% of training data to 90% of training data with an increase every 10%. The results are presented in Fig. 5.

From Fig. 5 it can be seen that by three features have a high FI score when compared to other features. They are RMS, Entropy estimation, and Histogram Lower bound features. It also shows that these three features tend to increase in value by increasing the amount of training data especially after the amount of training data reaches 50%. Therefore, these three features are selected as a prognostic indicator and will be used to estimate the RUL of the motor shaft. They are shown in Figs. 6, 7 and 8.

The state-space model is built based on the training data. It can be seen in Figs. 6, 7 and 8 that the features tend to increase their value after 50% of the motor life cycle or at 343 min. This point will be used as the prediction starting point (Ts). The state-space model obtained is used to predict 100 and 200-steps ahead. The prediction is updated with the new measurement data. The prediction is stopped when the predicted value hits the predetermined failure threshold that takes the maximum value of the features. The prediction is done using both original feature data and smoothed feature. The result for RMS feature is shown in Figs. 9, 10, 11 and 12.

Figures 9 and 10 present the motor shaft RUL estimation on the original RMS feature extracted from the vibration signal directly. Meanwhile, Figs. 11 and 12

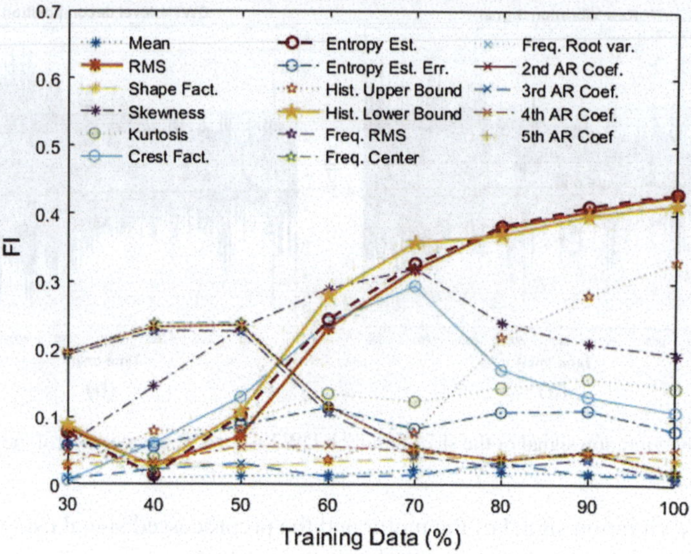

Fig. 5 The FI score of the features on different amount of training data

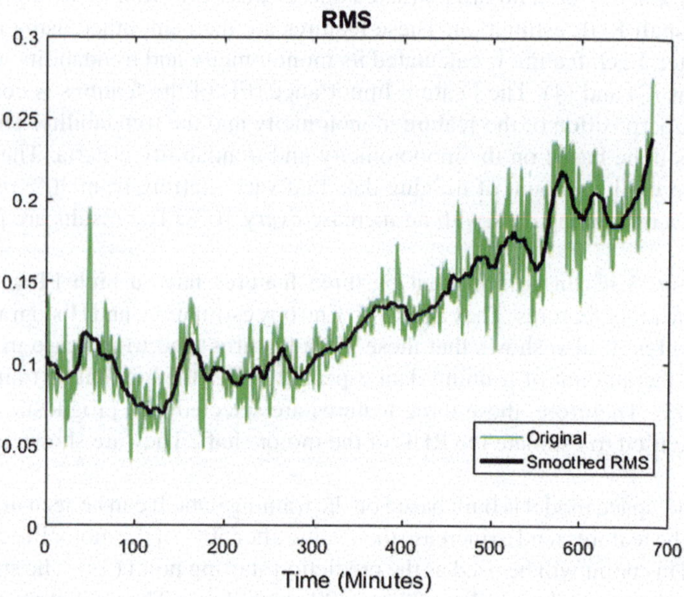

Fig. 6 The RMS feature

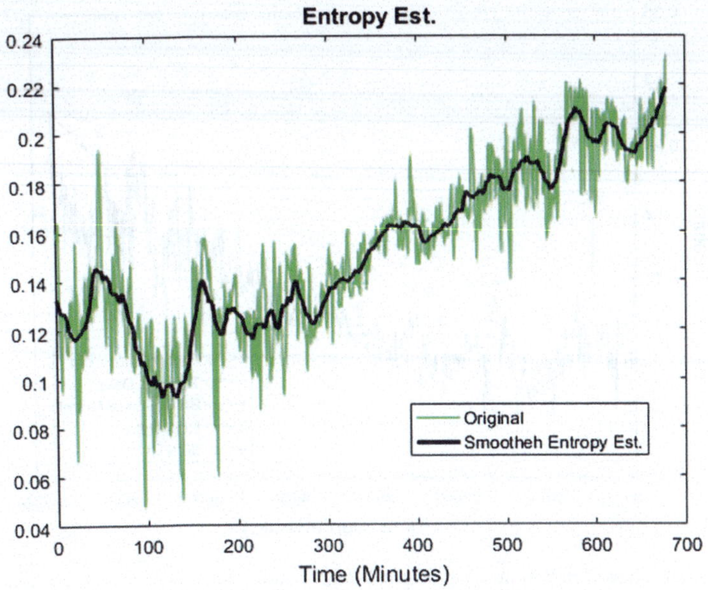

Fig. 7 The entropy estimation feature

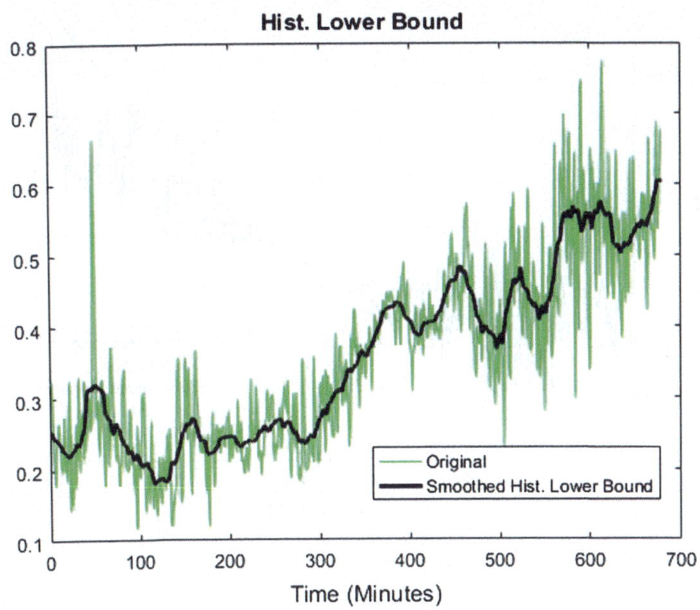

Fig. 8 The histogram lower feature

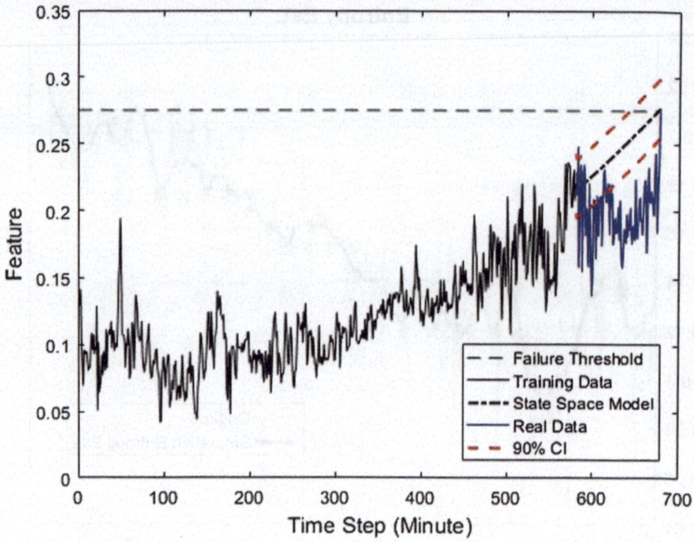

Fig. 9 100-steps ahead prediction using original RMS feature

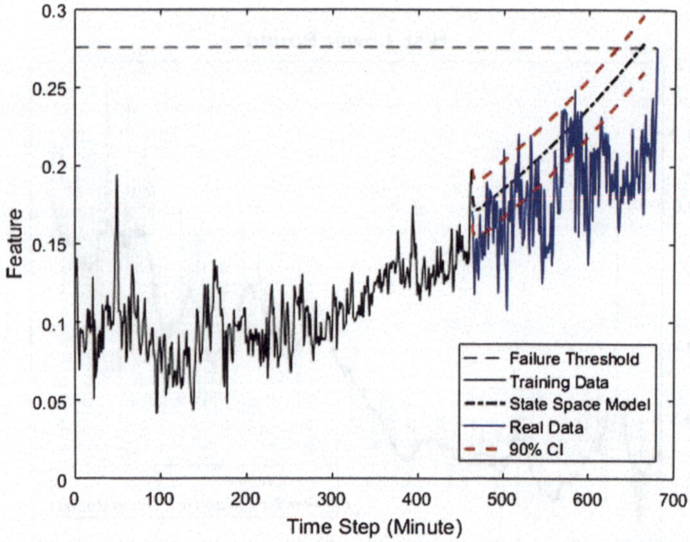

Fig. 10 200-steps ahead prediction using original RMS feature

presents the RUL estimation on the smoothed RMS feature. Although the RMS feature tends to increase after the 50% cycle of life to the end of life (EOL) the increase is not linear. So, it is not easy to predict appropriately the future condition of the motor shaft over time. However, the state-space model can be used to estimate

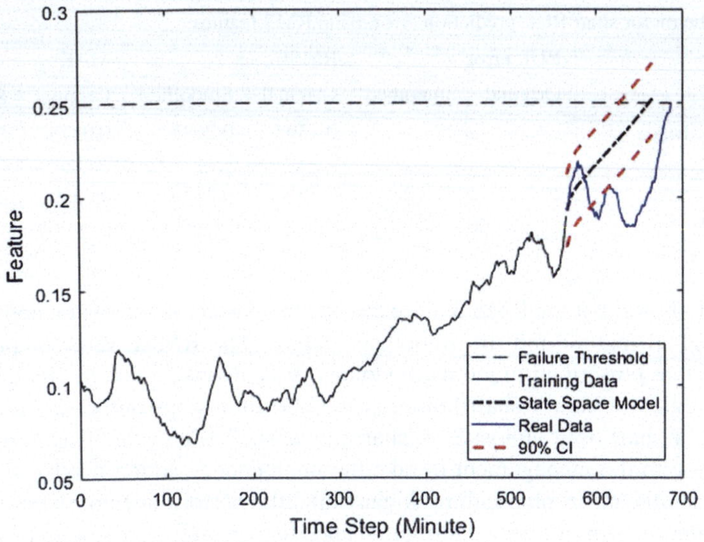

Fig. 11 100-steps ahead prediction using smoothed RMS feature

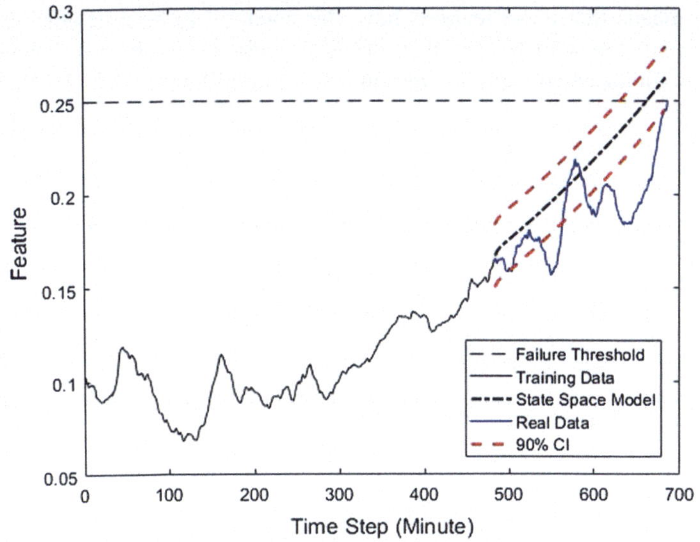

Fig. 12 200-steps ahead prediction using smoothed RMS feature

the motor shaft RUL quite satisfactorily. All of the RUL estimations predict the failure of the motor shaft just before the actual failure happened. The prediction performance of the motor shaft RUL estimation using the RMS feature is given in Table 1.

Table 1 The motor shaft RUL prediction error using RMS feature

	RUL error		RMSE		MAE	
	Original	Smoothed	Original	Smoothed	Original	Smoothed
100-steps ahead prediction	2	25	0.4595	0.2688	0.0475	0.0296
200-steps ahead prediction	18	25	0.3030	0.3323	0.0399	0.0243

Table 1 shows that the RMS feature and the proposed model can be applied both to the original feature and the smoothed feature. The RMSE error of prediction below 0.5 that prescribed in the methodology. It is also seen that the model can be used to predict the motor shaft failure at the 200-time steps before the end of life of the motor shaft or at almost 30% shaft motor shaft life cycle. This period gives enough time to the management to take the maintenance action needed to prevent the motor shaft catastrophic failure. Meanwhile, the performance of the motor shaft RUL estimation using Entropy Estimation and Histogram Lower Bound features are shown in Tables 2 and 3.

The RMS and Entropy Estimation feature combined with the state-space model can estimate the RUL of the motor shaft quite well. The RMSE and MAE close to zero. It means that these features have the potential to be a health indicator of the motor shaft that will be useful in the RUL estimation. However, the different condition is found when using histogram lower bound feature. The RMSE and the MAE are quite high. Therefore, the prediction model is far enough from the real data.

Table 2 The motor shaft RUL prediction error using entropy estimation feature

	RUL error		RMSE		MAE	
	Original	Smoothed	Original	Smoothed	Original	Smoothed
100-steps ahead prediction	73	32	0.1917	0.1224	0.0202	0.0143
200-steps ahead prediction	86	50	0.1698	0.1331	0.0165	0.0103

Table 3 The motor shaft RUL prediction error using histogram lower bound feature

	RUL error		RMSE		MAE	
	Original	Smoothed	Original	Smoothed	Original	Smoothed
100-steps ahead prediction	63	140	2.7850	1.3950	0.2857	0.1395
200-steps ahead prediction	32	54	2.3967	0.7186	0.1724	0.0578

4 Conclusion

This paper presents the estimation of the motor shaft RUL based on the feature importance and state-space model. The feature selection is based on the Feature Importance score constructed from the monotonicity and trendability criteria. The RMS, Entropy Estimation, and Histogram Lower Bound features are selected as a health indicator to bu used in the motor shaft RUL estimation based on their Feature Importance value. The state-space model is developed from the training data to estimate the RUL of the motor shaft. The results show that the RMS and Entropy Estimation feature combined with the state-space model performs quite well in motor shaft RUL estimation both using the original feature and smoothed feature. The RMSE and MAE of the prediction are below 0.5 that set in the methodology. Meanwhile, the Histogram Lower Bound feature is not recommended that shown by low RMSE and MAE. This RUL estimation that taken directly from the measure as a health indicator in RUL estimation of the motor shaft because its RMSE is quite large.

References

1. Ergin S, Uzuntas A, Gulmezoglu MB (2011) Detection of stator, bearing and rotor faults in induction motors. Procedia Eng 30:1103–1109
2. UE System Inc. Homepage. http://www.uesystems.com/news/the-50-failure-modes-of-electric-motors. Last accessed 2019/09/05
3. Vachtsevanos G, Lewis FL, Roemer M, Hess A, Wu B (2006) Intelligent fault diagnosis and prognosis for engineering systems. Wiley, New Jersey
4. Kraleti RS, Zawodniok M, Jagannathan S (2012) Model based diagnostics and prognostics of three-phase induction motor for vapor compressor applications. In: 2012 IEEE conference on prognostics and health management, pp 1–7
5. Nguyen V, Seshadrinath J, Wang D, Nadarajan S, Vaiyapuri V (2017) Model-based diagnosis and RUL estimation of induction machines under inter turn fault. IEEE Trans Ind Appl 53(3):2690–2701
6. Camci F, Medjaher K, Zerhouni, Nextoux P (2012) Feature evaluation for effective bearing prognostics. Qual Reliab Eng 29(4):477–486
7. Sloukia F, Bouarfa R, Medromi H, Wahbi M (2013) Bearing prognostics using mixture of gaussian hidden markov model and support vector machine. Int J Netw Secur Appl 5(3):85–97
8. Zhang X, Kang J, Zhao J, Chao D (2013) Features for fault diagnosis and prognosis of gearbox. Chem Eng Trans 33:1027–1032
9. Zhang X, Kang J, Bechhoefer E, Zhao J (2014) A new feature extraction method for gear fault diagnosis and prognosis. Maintenance Reliab 16(2):295–300
10. Bechhoefer E, Clark S, He D (2010) A state space model for vibration based prognostics. In: Annual conference of prognostics and health management society. Portland, Oregon, pp 1–8
11. Feng L, Wang H, Si X, Zou H (2013) A state-space-based prognostic model for hidden and age-dependent nonlinear degradation process. IEEE Trans Autom Sci Eng 10(4):1072–1086
12. Ni X, Zhang X, Sun F, Zhao J, Zhao J (2016) An adaptive state-space model for predicting remaining useful life of planetary gearbox. In: 2016 prognostics and system health management conference, Chengdu, pp 1–6
13. Wang D, Yang F, Zhao Y, Tsui KL (2017) Prognostics of lithium-ion batteries based on state space modeling with heterogeneous noise variances. Microelectron Reliab 75:1–8

14. Bonnet AH (2000) Root cause AC motor failure analysis with a focus on shaft failures. IEEE Trans Ind Appl 36(5):1435–1448
15. Raut SP, Raut LP (2014) A review of various techniques used for shaft failure analysis. Int J Eng Res Gen Sci 2(2):159–171

Preliminary Observation on Temperature Effect of Briquetting Cow Manure as a Solid Biofuel

N. M. M. Mitan and S. Badarulzaman

Abstract The depletion of fossil fuel turns the application of biofuel as an alternative fuel. Conversion of cow manure into briquette is the alternative to reutilize valuable chemical in cow manure. The aim of this study is to observe the suitable temperature for briquetting of cow manure as a solid biofuel. The process involved drying, pulverizing, carbonization and briquetting of cow manure. The carbonization of cow manure took place at three different temperatures; 300 °C (C300), 500 °C (C500) and 700 °C (C700). Some analysis on briquette are performed namely proximate analysis, compressive and caloric value tests. The calorific value for carbonization briquette of 300, 500 and 700 °C were 2434, 2402 and 2372 cal/g respectively. The carbonization of cow manure at 300 °C produced the highest calorific value of briquette. Therefore, preliminary observations indicated that carbonization at 300 °C able to densify the cow manure into briquette as a solid biofuel.

Keywords Biofuel · Temperature · Briquette · Cow manure · Carbonization · Calorific value

1 Introduction

Increasing of oil prices is the global problem in the world. Therefore, in order to overcome this problem, many green based technologies has been introduced. One of the green based technology is sustainable energy production. Sustainable energy is the supply of sustainable energy to meet the requirements of the current generation without reducing the ability of future generation. One of the related technologies is renewable energy. Renewable energy can be defined as energy derived from natural resources such as solar, biogas, hydro and biomass.

N. M. M. Mitan (✉)
Department of Chemistry, Faculty of Science and Computer, Universitas Pertamina, Jalan Teuku Nyak Arief, Simprug, Kebayoran Lama, Jakarta 12220, Indonesia
e-mail: nona.merry@universitaspertamina.ac.id

S. Badarulzaman
Department of Thermal-Fluids, Faculty of Mechanical Engineering, Universiti Teknikal Malaysia Melaka, Durian Tunggal, Melaka 76100, Malaysia

U. Sabino et al. (eds.), *Proceedings of the 6th International Conference and Exhibition on Sustainable Energy and Advanced Materials*, Lecture Notes in Mechanical Engineering,
https://doi.org/10.1007/978-981-15-4481-1_65

Cow manure is an undigested plant from bovine animal which is passed over the gut of animal. The manure is abundant with minerals. Developing countries apply the cow manure as a fertilizer, biogas production and thermal insulator [1].

As an animal-origin waste product, the daily intake, the body size and age may cause varies in amount and composition of cow manure. A 400 kg cow release 18–40 kg of manure (about 500–900 g biochemical oxygen demand). Besides that, cow manure has a high carbon content. The carbon to nitrogen ratio in cow manure is 20:1. However, cow manure can also pose health risks. Cow manure has the potential to carry primary pathogens such as bacteria and protozoa [2].

Potential and risk of cow manure as a cheaper fuel and higher health cost in rural area is observed by Pant [3]. Substituting cooking fuel by cow manure into biogas has several advantages, for instances decreased emission of black carbon and greenhouse gases [3].

Song et al. performed similar research on briquetting of cow-manure in the presence of various materials namely coal, potassium nitrate, citric acid, manganese dioxide, mixture of calmogastrin and molybdenum, sodium humate and red clay, and acidified calcium oxide. They revealed that cow manure briquette with the mass ratio of cow dung to coal is 1:3–4, the calorific value was 19.1 MJ/kg [4].

The current research is preliminary study about potential of cow manure to be modified as a solid biofuel in the briquette form. The study involved briquetting of cow manure, proximate analysis, calorific value and compressive test.

2 Methodology

2.1 *Materials and Preparation*

Cow manure was used as a raw material for briquette. It was dried under sunlight for seven days prior to process into a briquette. Furthermore, the dried cow manure was milled into 0.75 cm by centrifugal mill Retsch ZM 200. The carbonization of dried cow manure took place at 300, 500 and 700 °C after milling process (Fig. 1a, b). Briquetting of 10.0 g of carbonized cow manure was performed at 50 MPa by hydraulic press in a mold with inner diameter of 35 mm and outer diameter of 70 mm.

2.2 *Analysis of Cow Manure Briquette*

Cow manures were analyzed by proximate analysis, calorific value and compressive test. Proximate analysis was carried out according to American Society for Testing Materials (ASTM) D 3173-03, D 3174-02 and D 3175-02. These standards applied for moisture content, ash content and volatile matter respectively [5]. Calorific value

Fig. 1 **a** Dried cow manure, **b** carbonized cow manure

was examined by bomb calorimeter (IKA C 200) and compressive test by Universal Testing Machine Instron 5583.

3 Results and Discussion

As a candidate of solid biofuel, cow manure is required to fulfill the preliminary analysis such as proximate analysis and calorific value. The picture below shows the briquette of cow manure (Fig. 2). In proximate analysis, moisture content is important parameter to ensure the mechanical properties and heating value of the briquette. There are no significant differences in moisture content between three briquettes at various temperatures.

Ash content is residue remain after combustion completed. This content effect to the calorific value of the briquette. It was found that briquette C300 has the lowest ash content (49%). The graph shows that there has been slight increase the ash content in briquette C700.

To evaluate combustion characteristic of briquette, test of volatile matter is important aspect. Therefore, the volatile content test is performed on cow manure briquette as shown in Fig. 3. As temperature of carbonization of cow manure increased, the volatile matter decreased. The differences of volatile matter of the three briquettes

Fig. 2 Cow manure briquette

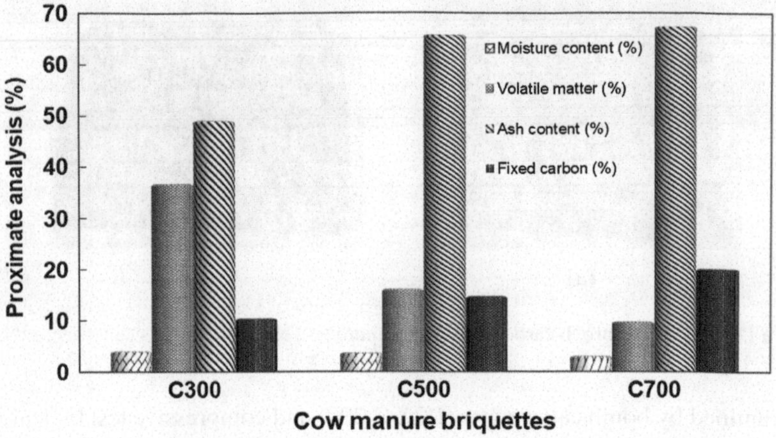

Fig. 3 Proximate analysis of cow manure briquettes at various temperature

are obviously. It means that the volatile gaseous also decrease as well. These gaseous indicate the ability of briquette to increase the duration of fire.

Carbonization of cow manures at three differences temperature produced briquette with the highest calorific value at 300 °C as depicted in Fig. 4. This study found that 2434 cal/g is produced from the C300 briquette.

In order to know the mechanical properties of briquettes, the compressive tests were performed for handling and logistic purposes. Compressive test is determined the maximum load that can be received by each briquettes produced. An important finding of compressive test at maximum load of C300 was 39.889 MPa.

Fig. 4 Calorific value of cow manure briquettes at various temperature

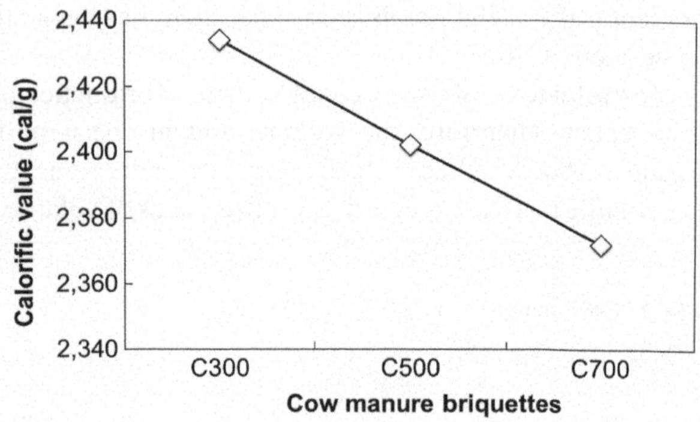

4 Conclusion

This study has shown that the briquetting of cow manure has potential as a solid biofuel. One of the more significant findings to emerge from this study produced the highest calorific value of 2434 cal/g was obtained from is that carbonization of cow manure at 300 °C. Further investigation of carbonization range temperature of cow manure is strongly recommended to optimize the calorific value and good mechanical properties of cow manure briquettes.

Acknowledgements The authors would like to express our gratitude to all staffs in Faculty of Mechanical Engineering, Universiti Teknikal Malaysia Melaka for their supports during this research. The partially of this result has been submitted to Universiti Teknikal Malaysia Melaka as a Final Year Project of Bachelor Degree.

References

1. Pahla G, Mamvura T, Ntuli F, Muzenda E (2017) Energy densification of animal waste lignocellulose biomass and raw biomass. S Afr J Chem Eng 24(24):168–175
2. Evans GM (2013) Biowaste and biological waste treatment. Earthscan, New York
3. Pant KP (2012) Cheaper fuel and higher health costs among the poor in rural Nepal. Ambio 41:271–283
4. Song A, Zha F, Tang XT, Chang Y (2019) Effect of the additives on combustion characteristics and desulfurization performance of cow dung briquette. Chem Eng Process Process Intensification 143
5. American Society for Testing and Materials (2002) Annual book of ASTM standards, designation D 3173, D 3174 and D 3175-02

Artificial Neural Network Modelling of Indoor CO_2 Reduction as Energy-Efficient Strategies

J. C. P. Putra and T. Susanto

Abstract Building consumptions has been the focus of numerous researchers since the introduction of building energy with zero carbon. However, many researches have indicated that excessive use of energy to produce comfort leads to problem associated with indoor air quality like high level of indoor CO_2. It is therefore essential to create a strategy to improve this condition. This paper presented both investigation and prediction regarding strategies to indoor CO_2 reduction as energy-efficient approaches. The research was consisted of two stages, in which the first stage was to monitor the existing of indoor CO_2 level and reduce it when exceed the standard. Whereas, the second stage was to predict the level of indoor CO_2. An approach of indoor CO_2 monitoring was carried out by smart sensor system through Arduino UNO Microcontroller and an Artificial Neural Network was involved to predict the level of indoor CO_2 during the application of smart sensor system. The results indicated that the application of smart sensor system was remarkable in diminishing indoor CO_2. Besides, the use of Artificial Neural Network involved network architecture of 4–30–30 with learning rate initial of 0.7 and activation function of relu. The model showed positive correlation which characterized by R^2 value that was 0.599. Ultimately, this research found the correlation between energy-efficient consumption in building and a strategy to reduce indoor CO_2 in order to create comfort and healthy environment. In addition, keep bear in mind that is important to find out the connection between the indoor environmental and its surroundings to create positive ambience that leads to sustainable development.

Keywords Energy-efficient building · Indoor CO_2 · Smart sensor system · Artificial neural network

J. C. P. Putra (✉)
Department of Civil Engineering, Universitas Bakrie, South Jakarta, Indonesia
e-mail: jouvan.chandra@bakrie.ac.id

T. Susanto
Department of Industrial Engineering, Universitas Bakrie, South Jakarta, Indonesia
e-mail: tri.susanto@bakrie.ac.id

U. Sabino et al. (eds.), *Proceedings of the 6th International Conference and Exhibition on Sustainable Energy and Advanced Materials*, Lecture Notes in Mechanical Engineering,
https://doi.org/10.1007/978-981-15-4481-1_66

1 Introduction

Indoor air pollutants have attracted a number of attentions during the years. It could bring a serious effect to health and lead to decreased work productivity [1, 2]. Various approaches that prominently used to improve indoor air quality are decreasing the pollution load, increasing the ventilation rate, and maintain indoor climate conditions [3]. By following those mechanism, it can enhance performance of employee which is characterized through performance of proof reading, addition work, and text typing [3]. However, those indicators are accomplished through experimental stage, and it is costly.

Subsequently, increasing ventilation rate is not solely effective in diluting indoor air pollutants and it is contradict with energy-efficient strategies. According to Bluyssen [4], a contaminant can be sourced from the utilization of HVAC system. It varies greatly based on to the type of construction, use and maintenance of the system [5]. In contrary, there are some valuables outcome of using HVAC system, especially its filter. The finding of research that was conducted by Zhao et al. [6] indicated that, application of filter in HVAC system both residential and commercial buildings were effectively in diluting existing ozone. It was approached through measuring the efficiency of ozone removal which is characterized by clean (unused) fiberglass, clean synthetic used both residential and commercial filters. The experiment was carried out in laboratory in order to control the environmental parameters. Even though this approach presented a valuable results but it is costly to do some experiments involving a numerous of parameters.

In 2008, Sofuoglu applied a method of Artificial Neural Network (ANN) to predict prevalence of building-related symptoms which subjected in office buildings [7]. In his research, a number of six indoor air pollutants and four comfort variables were considered as input to the network. As many as ten variables was optimized with one hidden layer and resulted to one target at output. A feed-forward network was involved to build a model of ANN which used back-propagation algorithm. It involves rate of momentum and learning rate variable. From that method, indicated that the application of ANN which adopted hyperbolic tangent function as the transfer function gives the significance of correlation as characterized by $R^2 > 0.5$ to predict prevalence of building-related symptoms in office buildings. Subsequently, Xie et al. [8] predicted indoor air quality using ANN based on previous studies. It involved six indoor air pollutants and three indoor comfort variables as input to the neural network. The algorithm of back-propagation was adopted in that research and also momentum rate and learning rate variables. In addition, as many as two-hidden-layer was used to optimize the process and is obtained high coefficient correlation with low mean root square error for the test set.

Moreover, Azid et al. [9] investigated regarding prediction of the level of air pollution by combining principal component analysis and artificial neural network. It engaged a number of eight parameters of air quality which is obtained from The Malaysian Department of Environment (DOE). Besides, the other approaches that was considered in Azid et al. [9] such as principal component analysis. The method was used to identify the source of pollutant in the area of study. The parameters were collected in seven monitoring stations during seven years (2005–2011). The combination both principal component analysis and artificial neural network were established to predicts the air pollutants index. It revealed that CH_4, NmHC, THC, O_3, and PM_{10} are the highest significant indicators. The combination of principal component analysis and artificial neural network showed valuable results in prediction air pollutants index by characterized by its value of R^2 and root mean square. Furthermore, Liu et al. [10] presented a study about exploring the potential relationship between indoor air quality and the concentration of airborne culturable fungi: a combined experimental and neural network modelling study [10]. The study aimed to provide an ultra-fast solution based on indoor air quality data which is comprised of indoor airborne concentrations, indoor/outdoor $PM_{2.5}$ and PM_{10} concentrations, indoor temperature, indoor relative humidity, and indoor CO_2 concentrations. It was only one layer involved regarding to the study of neural network. The results indicated that prediction from artificial neural network modeling has a good performance as characterized by its value of tolerance in the percentage of ±30%. Additionally, the significant value of the study is to present fast approach in determining the concentration of indoor airborne in order to be used in real-time assessment.

Putra et al. [11] carried out a prediction regarding indoor air quality which involved artificial neural network. It subjected at office buildings that is comprised of a number of experiments such as CO_2 measurements through Yes Plus LGA Meter. The data gathered from measurement was an input to neural network system. It classified into two categories, where the first was training data while the rest was the test data. A Levenberg-Method was applied to execute artificial neural network model. The results indicated that data preparation plays important role to obtain satisfying results. In addition, the selection of training method also contribute to the accuracy level of prediction model.

2 Experiments and Methods

2.1 Chamber

The utilization of chamber is aimed to model the existing of indoor environmental condition. It presented as in Fig. 1.

The Fig. 1 shows that the chamber used regarding to this study have the shape of house prototype in order to illustrate the distribution of indoor environmental condition in terms of CO_2, temperature, and relative humidity. Subsequently, to assess indoor air quality inside the chamber, it need to provide a sensor on it. Two sensors which are located both at outside and at inside of chamber are built on Arduino basis system. The smart sensor of indoor CO_2 built respecting to this study is shown as in Fig. 2.

Fig. 1 The chamber of house prototype

Fig. 2 Smart sensor installed on a chamber of house prototype

The serial of sensor is MHZ sensor which is can be connected to microcontroller of Arduino. It recorded a series of CO_2 measurement inside the chamber. Additionally, the sensors are located in these two location which are inside of chamber and outside of chamber because the results of measurement could also be influenced regarding to the outdoor air condition.

2.2 *Model Development of Artificial Neural Network*

According to Kalogirou [12], artificial neural network is excellent to solve the problem that subject to incomplete data sets, fuzzy or incomplete information, and for highly complex and ill-defined problems, where humans usually decide on an intuitional basis. In addition, there are some components of artificial neural network identified by Kalogirou [12] as follows:

1. Function approximation. Mapping of a multiple input to a single input is established. Unlike most statistical techniques, it can be done with adaptive model free estimation of parameters.
2. Pattern recognition and pattern association. It is a problem of pattern classification. ANNs can be effectively used to solve difficult problems in this field, for instance in sound, image, or video recognition. This task can even be made without an a priori definition of the pattern. In such cases, the network learns to identify totally new patterns.
3. Associative memories. It is the problem of recalling a pattern when given only a subset clue. In such applications, the network structures used are usually complicated, composed of many interacting dynamical neurons.

Furthermore, the mechanism of artificial neural network model is based on back propagation algorithm used in single hidden layer of artificial neural network model and comprising of a number of variables in hidden layers were developed respectively for training data purpose [13]. Subsequently, in order to conduct training stage, the activation function of sigmoid function was applied to the model [14].

According to this study, several aspects of indoor air quality that are considered as parameters are as follows: the indoor concentration of CO_2, temperature, and relative humidity, while the concentration level of CO_2 was appointed as output. As the data divided into two categories where the first is to training purpose, while the rest for data testing. Approximately, 80% of data were randomly selected and set as the training data, while the rest is automatically set as the testing data.

3 Results and Discussion

The simulation regarding Artificial Neural Network was built on python platform, once it done then a predictive model that can quickly and precisely evaluate the indoor CO_2, the model involved a number of approaches, such as:

1. Activation function regarding to this study is relu function.
2. The value of learning rate regarding to this study is 0.7.
3. The hidden layers involved in this study is two hidden layers which comprise of 30 nodes within each of it.

As the model built, the results from modelling are tabulated as in Table 1 and presented as in Fig. 3 respectively.

Table 1 The discrepancies between indoor CO_2 measured and indoor CO_2 predicted

No.	Indoor CO_2 measured (ppm)	Indoor CO_2 predicted (ppm)	Percentage of discrepancies (%)
1	1207	1450	20.13256
2	1883	1457	−22.62347318
3	1481	1686	13.84199865
4	2086	2264	8.533077661
5	1991	2268	13.91260673
6	1749	1679	−4.002287021
7	2704	2264	−16.27218935
8	1672	1429	−14.53349282
9	1211	1392	14.94632535
10	1801	1879	4.330927
11	2487	2244	−9.770808203
12	1551	1411	−9.026434558
13	1468	1741	18.59673025
14	2220	2250	1.351351351
15	2097	2283	8.86981402
16	1996	2283	14.37875752
17	1166	1608	37.90737564
18	400	499	24.75
19	1054	1447	37.28652751
20	1838	1407	−23.44940152

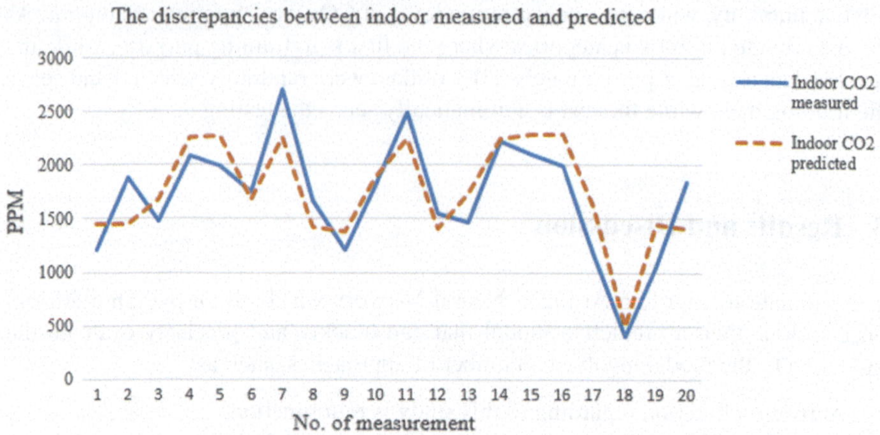

Fig. 3 The behavior of indoor CO_2 predicted in comparison with indoor CO_2 measured

The percentage of discrepancies between indoor CO_2 measured and indoor CO_2 predicted is in the range of −23.44 to 37.90% as tabulated in Table 1. Moreover, the illustration of those two discrepancies were presented in Fig. 3 and it could be found that the behavior from the prediction of indoor CO_2 tend to imitate the behavior of indoor CO_2 measurements. In addition, to evaluate the correlation between input parameters and output of artificial neural network model. It is essential to calculate R^2 and it's obtained that the value of R^2 is 0.59. This means that the input parameters of artificial neural network model has significant effect to predict indoor CO_2 as an output. Besides, it is expected that with an enormous of data will give more an accuracy in predicting either target or output of artificial neural network model.

Subsequently, the approach of this model is pretty applicable for real measurement due to all the input parameters used regarding to this study can be carried out in a short period. Even though the accuracy of prediction from artificial neural network is below than 90% due to the limitation of data scale, this approach is still be considered as fast calculation to obtain quick result. Besides, this is in line with the findings of Liu et al. [10] investigation regarding to the potential relationship between indoor air quality and the concentration of airborne culturable fungi: a combined experimental and neural network modelling study. However, the model is made by restricting to the certain condition of indoor environment and the consequence is that the model developed only applicable in the certain place. The solution emerged from Liu et al. [10] is by scaling up the variables of database based on various environment. Therefore, it is expected that quick evaluation method can shorten time of measurements [10].

4 Conclusion

The model of artificial neural network was built to this study in order to evaluate existing indoor CO_2 and predict it. It found that by using several indoor air quality parameters, a trained data of artificial neural network models, learning rate of 0.7, relu activation function, and using two hidden layers could immediately achieve the predicted of indoor CO_2. This means that, the approach of artificial neural network model can give insight to fast prediction regarding the real condition of indoor environmental quality. Finally, it is expected that the strategy to model the condition of indoor CO_2 could be applied to other assessment of indoor environmental quality parameters. Therefore, it should be come up with a model that supporting health environment and energy-efficient as well.

References

1. Billionnet C, Sherrill D, Maesano IA (2012) Estimating the health effects of exposure to multi-pollutant mixture. Ann Epidemiol 22(2)
2. Kosonen R, Tan F (2004) The effect of perceived indoor air quality on productivity loss. Energy Build 36:981–986
3. Wargocki P, Wyon DP, Fanger PO (2000) Productivity is affected by the air quality in offices. Healthy Build 1:635–640
4. Bluyssen PM (2009) Towards an integrative approach of improving indoor air quality. Build Environ 44(9):1980–1989
5. Bluyssen PM (2004) A clean and energy-efficient heating, ventilating and air-conditioning system: recommendations and advice. TNO, The Netherlands. ISBN 90-5986-009-8
6. Zhao P, Siegel JA, Corsi RL (2007) Ozone removal by HVAC filters. Atmos Environ 41:3151–3160
7. Sofuoglu SC (2008) Application of artificial neural networks to predict prevalence of building-related symptoms in office buildings. Build Environ 43:1121–1126
8. Xie H, Ma F, Bai Q (2009) Prediction of indoor air quality using artificial neural networks. In: 2009 fifth international conference on natural computation, pp 414–418
9. Azid A, Juahir H, Toriman ME, Kamarudin MKA, Saudi ASM, Hasnam CNC, Aziz NAA, Azaman F, Latif MT, Zainuddin SFM, Osman MR, Yamin M (2014) Prediction of the level of air pollution using principal component analysis and artificial neural network. A case study in Malaysia. Water Air Soil Pollut
10. Liu Z, Cheng K, Li H, Cao G, Wu D, Shi Y (2017) Exploring the potential relationship between indoor air quality and the concentration of airborne culturable fungi : a combined experimental and neural network modeling study. Environ Sci Pollut Res
11. Putra JCP, Safrilah, Ihsan M (2018) The prediction of indoor air quality in office room using artificial neural network. In: Human-dedicated sustainable product and process design: materials, resources, and energy—AIP conference proceedings, vol 20040
12. Kalogirou SA (2001) Artificial neural networks in renewable energy systems applications: a review. Renew Sustain Energy Rev 5(4):373–401
13. Wang JZ, Wang JJ, Zhang ZG, Guo SP (2011) Expert systems with applications forecasting stock indices with back propagation neural network. Expert Syst Appl 38(11):14346–14355
14. Liu Z, Liu K, Li H, Zhang X, Jin G, Cheng K (2015) Artificial neural networks-based software for measuring heat collection rate and heat loss coefficient of water-in-glass evacuated tube solar water heaters. PLoS One 1–16

Characterization of Microwave Absorber Material Based on Strontium Samarium Ferrite Produced by Hybrid Sol-Gel Method

M. Effendi, Untung, W. T. Cahyanto, and W. Widanarto

Abstract Several studies have been carried out to attain microwave absorber materials to reduce the level of exposure of electromagnetic waves that can damage various electronic devices and disrupt human health. This study aims to determine the characterization of microwave absorber material of strontium ferrite with samarium doping and made by using the hybrid sol-gel method, which is a combination of sol-gel and solid-state reaction. The initial stage was to process Fe_3O_4 into $Fe(NO_3)_3$ and then mix it with $Sr(NO_3)_2$ and Sm_2O_3. The mixture was made into a sol and then heated into a gel. After that, the sample was calcined to become powder. Then the sample powder is compacted and sintered at a temperature of 1100 °C. The XRD, VSM, and VNA characterization results show that the addition of samarium doping affects changes in microstructure, particle size, and magnetic properties. The final results show the formation of four phases, namely $SrFe_{12}O_{19}$ (hexagonal), Sr_2FeO_4 (tetragonal), $FeSmO_3$ (orthorhombic) and Sm_2Fe_{17} (rhombohedral) which have particle sizes of 21–23 nm and susceptibility values of $(3.5–7.3) \times 10^{-6}$ m^3/kg. Strontium ferrite samples with 0.15% mol of samarium doping concentration have the best microwave absorption ability of −19.523 dB at a frequency of 10.74 GHz.

Keywords Microwave absorber · Strontium ferrite · Samarium · Modified solid-state reaction method · Sol-gel

1 Introduction

The exposure level of electromagnetic waves from various frequencies increases significantly in line with technological developments. This phenomenon causes Electromagnetic Interference (EMI), which is a reflection of microwaves from multiple

M. Effendi (✉) · Untung · W. T. Cahyanto · W. Widanarto
Physics Department, Faculty of Mathematics and Natural Sciences, Jenderal Soedirman University, Jl. dr. Soeparno No. 61, Purwokerto 53123, Indonesia
e-mail: mukhtar.effendi@unsoed.ac.id

M. Effendi
Research and Development Center of New and Renewable Energy, Jenderal Soedirman University, Jl. dr. Soeparno, Purwokerto 53123, Indonesia

U. Sabino et al. (eds.), *Proceedings of the 6th International Conference and Exhibition on Sustainable Energy and Advanced Materials*, Lecture Notes in Mechanical Engineering,
https://doi.org/10.1007/978-981-15-4481-1_67

electronic devices. Therefore, Radar Absorbing Material (RAM) technology is currently being developed, which is a technology by creating new materials that can reduce reflections or absorb microwaves.

The metal material, which is often used as a microwave absorber, is the ferrite group. Further, iron sand can be extracted into natural ferrite material, which can be used as a microwave absorber material. However, the natural ferrite is only able to work within the MHz range. So it is necessary to utilize the type of microwave absorbing material with strong absorption and in a wide frequency range.

M-type ferrite material, which is a mixture of strontium (Sr) and ferrite ($SrFe_{12}O_9$), is a hard magnetic material that is widely used in the manufacture of permanent magnets, magnetic recording media, and microwave absorbers and it can be used in the GHz range [1]. The addition of rare earth to the microwave absorber material is proven to increase its performance up to the GHz range [2]. However, strontium ferrite still has a relatively small absorption frequency width. Fe^{3+} substitution with rare earth (RE) material can shift the resonant frequency and change the HA anisotropy field [3–5]. This phenomenon shows that the absorption properties of electromagnetic waves can be controlled by changes in the concentration of dopants used and the synthesis method [6–8]. This research uses Sm^{3+} rare earth material (RE) as dopant material and performing a hybrid sol-gel method to obtain the sample, which has a strong capability in microwave absorption and a wide frequency range.

2 Experimental

Synthesis of iron nitrate precursors done by mixing iron sand that has been through the process milling, nitric acid solution, and acid oxalate in a beaker-sized glass 500 mL. The mixture was then heated on a hot plate with a temperature of 90 °C and stirring at a magnetic stirrer with a speed of 650 rpm for 1 h.

$Sr(NO_3)_2$ powder and $Sm(NO_3)_3$ solution were put into a glass containing iron nitrate $Fe(NO_3)_3$ solution and heated with a temperature of 100 °C. The mixture is heated for 2 h and while stirring at 650 rotation per minute. The mole ratio used is $12Fe(NO_3)_3{:}(1 - x)Sr(NO_3)_2{:}(x)Sm(NO_3)_3$ with variations in the value of $x = 0$; 0.05; 0.1; 0.15 (mol).

During the heating process, the mixture was dropped by 48 mL ethylene glycol solution. The solution is then cooled to room temperature so that a sol will form. The formed sol was then added with ethanol while stirring using a magnetic stirrer for 1.5 h to form a gel. The gel is then heated at a temperature of 150 °C for 30 min to form a concentrated gel that is ready to be calcined. Calcination of concentrated gels was carried out at 350 °C for 2 h to produce strontium ferrite samarium powder.

Calcined Sm–Sr Ferrite powder then switches to the solid-state reaction method, namely the stage of compaction/compacting, pre-sintering, and sintering. In the compaction process, Sm–Sr Ferrite powder ±2 g was put into a mold placed on a field that is affected by an external magnet so that it experiences anisotropy during the compaction process. It also serves to rectify magnetic spins and as a process of pre-diffusion between atoms before the sintering process. Compaction process using a hydraulic press with a compressive load of up to 1 ton and held for 5 min.

The next process is the heating process. The heating process consists of two stages, namely the process of pre-sintering and sintering—a pre-sintering process with a temperature of 800 °C for 1 h. Pre-sintering aims to strengthen the material and so that the sample does not experience significant shrinkage. When pre-sintering, the crystal structure has formed, but in this condition, the bond between the powder particles was not strong/still fragile. Then the sintering process was carried out at a temperature of 1100 °C for 3 h so that the atoms in the Sm–Sr Ferrite material underwent a diffusion process. The next process is the characterization of XRD, VSM, SEM, and VNA.

3 Results and Discussion

X-ray Diffraction (XRD) characterization was used to identify the phase, crystal structure, and particle size of the Sm–Sr Ferrite sample. XRD data is in the form of the peak intensity of the diffraction pattern and the position of the diffraction pattern (angle 2θ). The peak of diffraction is adjusted to the International Center for Diffraction Data (ICDD) so that processing diffraction patterns can be shown as in Fig. 1.

Based on Fig. 1 which shows the diffraction pattern of Sm–Sr Ferrite samples starting at an angle position of 2θ from 5° to 90°, four phases are formed: Strontium Iron Oxide ($SrFe_{12}O_{19}$), Strontium Iron Oxide (Sr_2FeO_4), Iron Samarium Oxide

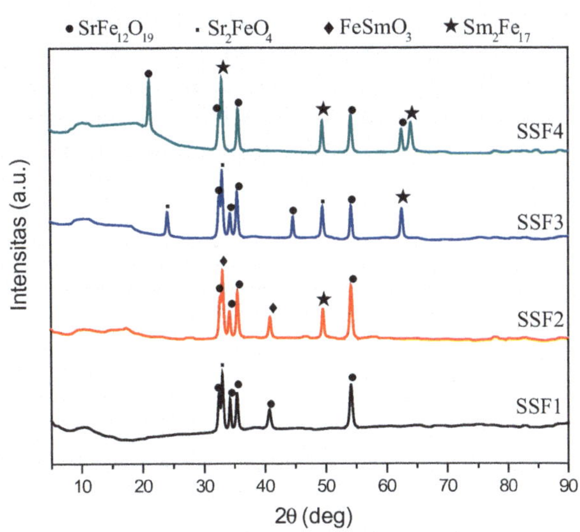

Fig. 1 XRD pattern of Sm–Sr-ferrite sample

($FeSmO_3$), and Iron Samarium (Sm_2Fe_{17}). The diffraction pattern of the SSF1 sample, which is a sample without doping (x = 0) is shown in black. In the diffraction pattern, SSF1 formed six peaks and two phases, namely Strontium Iron Oxide ($SrFe_{12}O_{19}$) and Strontium Iron Oxide (Sr_2FeO_4). Strontium Iron Oxide $SrFe_{12}O_{19}$) is found at four peaks in position 2θ, namely 32.3564°; 34.2082°; 35.3612°; 40.6702°; and 54.0650° with hexagonal crystal structure (α = β = 90° and γ = 120°). The $SrFe_{12}O_{19}$ phase has cell parameters a = b = 5.8868 Å, and c = 23.0370 Å, and the cell volume is 691.38 Å. Strontium Iron Oxide with the chemical formula Sr_2FeO_4 formed at a peak at position 2θ is 32.9399° and has a tetragonal crystal structure (α = β = γ = 90°). This phase has cell parameters a = b = 3.8642 Å and c = 12.3968 Å with a cell volume of 185.11 Å. In SSF1, $SrFe_{12}O_{19}$ formed due to strontium nitrate mixed with ferrite nitrate at doping 0 mol, the crystalline phase formed in the form of the $SrFe_{12}O_{19}$ phase and has the highest peak number and high intensity. While the Sr_2FeO_4 phase is a transition phase in the formation of the $SrFe_{12}O_{19}$ phase, the phases in the SSF1 sample are doping samples shown in Table 1. Identification of phases and crystalline systems refers to the International Center Diffraction Data (ICCD).

In the same way, the analysis was also carried out for other samples. In this study, Sm^{3+} doping can be mixed well with ferrite, as can be seen in SSF2 samples, which have Iron Samarium Oxide ($FeSmO_3$) and Iron Samarium (Sm_2Fe_{17}) phases, and in SSF3 and SSF4 samples have Iron Samarium (Sm_2Fe_{17}) phase, where the Iron Samarium (Sm_2Fe_{17}) phase has the highest intensity value and the highest number of peaks in the SSF4 sample. Strontium material can also be mixed well with ferrite can be seen in every sample made SSF1, SSF2, SSF3, and SSF4 has the Strontium Iron Oxide ($SrFe_{12}O_{19}$) with the highest intensity value in the SSF1 sample.

In Fig. 2, the black line is the hysterical curve of the SSF1 sample (without doping). The SSF1 sample has a magnetization value (Ms) of 12.02 emu/g, remanent magnetization value (Mr) of 3.97 emu/g, and its coercivity field value (Hc) of 639.29 Oe and Hmaks of 20.227 kOe. Then, the red line is the hysteresis curve of the SSF2 sample (sample with a doping concentration of 0.05 mol). The SSF2 sample had a value of Ms of 8.30 emu/g, an amount of Mr of 2.86 emu/g, and the value of Hc of 734.06 Oe and Hmaks of 20.280 kOe. The blue hysteresis curve is a curve for SSF3 samples (samples with a doping concentration of 0.1 mol). SSF3 sample has a value

Table 1 Crystal phase of SSF1 sample

Crystal phase	Crystal structure	2θ Position (°)
Strontium Iron Oxide ($SrFe_{12}O_{19}$)	Hexagonal	32.3564
		34.2082
		35.3612
		40.6702
		54.0650
Strontium Iron Oxide (Sr_2FeO_4)	Tetragonal	32.9399

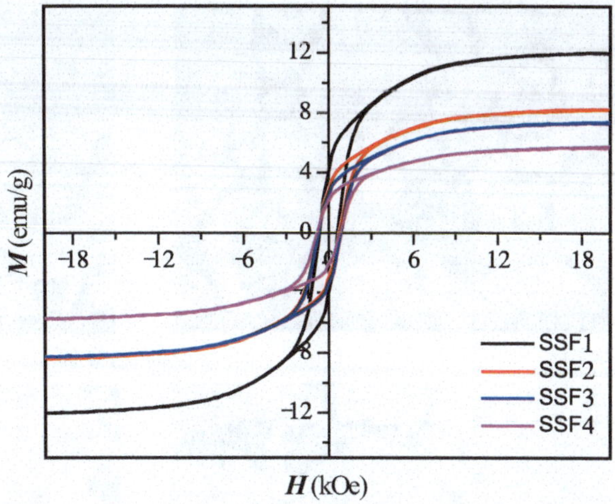

Fig. 2 Magnetic hysteresis curve of Sm–Sr-ferrite sample

of Ms of 7.42 emu/g, a value of Mr of 2.70 emu/g, and the value of Hc of 787.98 Oe and Hmaks of 20.262 kOe. And the green line is the hysteresis curve of the SSF4 sample (samples with a doping concentration of 0.15 mol). SSF4 sample has a value of Ms of 5.79 emu/g, a value of Mr of 2.11 emu/g, and the value of Hc of 804.95 Oe and Hmaks of 0.263 kOe. A compilation of the analysis the results of is shown in Table 2.

The next step is to analyze the morphology of the samples by acquiring the SEM image capture. The SEM image of the samples that have been obtained is shown in Fig. 3.

The SEM image shows that the addition of Sm^{3+} causes a change in particle size. The particle size of Sm–Sr-ferrite magnetic material appears larger with the addition of Sm^{3+}. Nevertheless, the existing porosity is maintained. These two things seem to influence the material characteristics in the capability of microwaves absorption. The next examination, that is, the microwave absorption capability of the sample was performed by analyzing the VNA spectrum (Fig. 4).

The reflection loss represents the microwave absorption capability, which shows the value of absorbed microwave by the sample. The reflection loss calculated by the equation [9].

Table 2 Magnetic properties of Sm–Sr-ferrite

Sample	M_S (emu/g)	M_r (emu/g)	H_C (Oe)	H_{maks} (kOe)	χ_m (m^3/kg)	μ_r (m^3/kg)
SSF1	12.02	3.97	639.29	20.227	7.4658×10^{-6}	1.00000746
SSF2	8.30	2.86	734.06	20.280	5.1425×10^{-6}	1.00000514
SSF3	7.42	2.70	787.98	20.262	4.6030×10^{-6}	1.00000460
SSF4	5.79	2.11	804.95	20.263	3.5918×10^{-6}	1.00000359

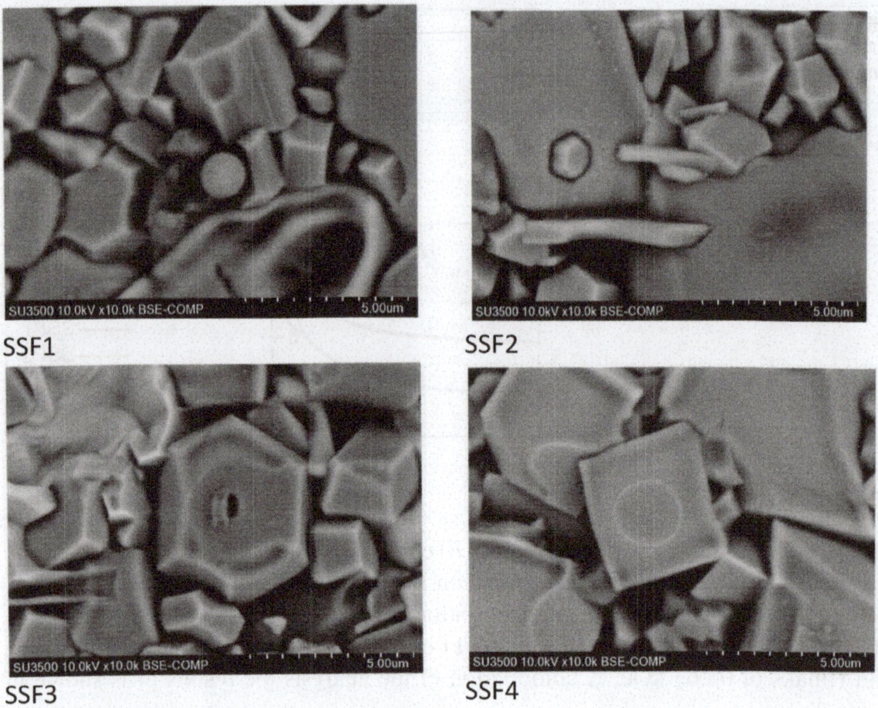

Fig. 3 The SEM image of the Sm–Sr-ferrite sample

Fig. 4 The VNA spectrum of the Sm–Sr-ferrite sample

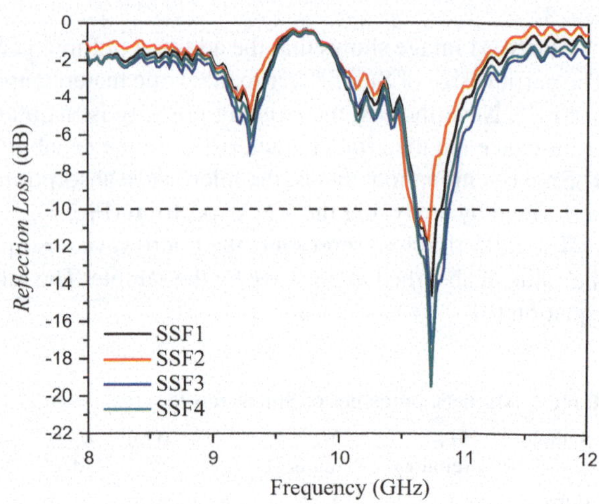

Table 3 Reflection loss and absorption value of Sm–Sr-ferrite samples

Sample	Frequency (GHz)	Effective absorption bandwidth (GHz)	Reflection loss (dB)	Absorption (%)
SSF1	10.74	0.18	−14.631	96.56
SSF2	10.74	0.12	−11.541	92.98
SSF3	10.74	0.24	−17.104	98.05
SSF4	10.74	0.22	−19.523	98.88

$$RL = 20\log\left|\frac{Z_{in} - Z_0}{Z_{in} + Z_0}\right| \quad (1)$$

Reflection loss relates to impedance in (Z_{in}) and impedance 0 (Z_0), which corresponds to permittivity (electrical property, ε), permeability (magnetic property, μ), and also material thickness. These factors influence the microwave absorption capability [5]. For the RL value less than −10 dB, it means that only 10% of the microwave energy is reflected, and 90% of it is absorbed. The corresponding frequency range over which the RL is smaller than −10 dB is defined as the effective absorption bandwidth [10]. Base on the reflection loss data spectrum (Fig. 4), the analysis results of the effective absorption bandwidth and the percentage of the microwave absorption capability exposure in Table 3.

Based on this result (Table 3), the sample, which has the highest doping concentration value of 0.15 mol, has the highest percentage of wave absorption, which is 98.88% of the microwave can be absorbed. The table also shows that as the concentration of doping material increases in the Sm–Sr Ferrite material, it has a smaller value of reflection loss. However, the SSF2 sample with a doping concentration of 0.05 mol had a lower RL value than the sample with a doping sample (SSF1). The widest effective absorption bandwidth achieved by SSF3 samples with a value of 0.24 GHz. The absorption of microwaves in the coating and powder samples has different values, where the Sm–Sr Ferrite material samples with compacted powder form have better RL values. This is influenced by the thickness of the sample. The compacted powder sample has a thickness of up to 1 mm thickness. So that the thicker a sample, the better the absorption value [11].

The increasing concentration of doping in Sm–Sr Ferrite material affects the crystalline phase formation and confirmed by the results of XRD pattern data. Samples of SSF1–SSF4 show the results of the increasing Sm_2Fe_{17} phase diffraction peak, and the intensity value increases. Increasing doping concentration also affects the value of Ms, which has decreased along with the increasing Sm^{3+} concentration. The smaller Ms value indicates better absorption [6]. The permeability value of the sample also decreased or approached the value of one, according to the increase in doping concentration. So that the permeability value is close to one, has a better microwave absorption value.

4 Conclusion

Sm–Sr-Ferrite microwave absorber material has been made using the modified solid-state reaction method with variations in the concentration of doped Sm^{3+}; that is 0; 0.05; 0.1; 0.15 mol. The addition of Sm^{3+} doping affects changes in microstructure, particle size, and magnetic properties. A total of four compound crystalline phases were formed, with $SrFe_{12}O_{19}$ formed in each sample, and Sm_2Fe_{17} increased with increasing doping concentrations. Furthermore, there are four different crystal structures, namely hexagonal, tetragonal, orthorhombic, and rhombohedral, with a particle size in a range of 21–23 nm. This Sm–Sr-Ferrite material has a magnitude of χ_m in order 10^{-6}, which means it is a super paramagnetic material and has a slender magnetic property of the hysteresis curve, which indicates that the Sm–Sr-Ferrite material is a soft magnetic material. The SSF4 sample with an Sm^{3+} doping concentration of 0.15 mol has the best microwave absorption characteristic with RL value reaching −19.523 dB at a frequency of 10.74 GHz with an effective absorption bandwidth of 0.22 GHz.

Acknowledgements Acknowledgements were conveyed to the Directorate General of Higher Education of the Indonesian Ministry of Research, Technology, and Higher Education for financial support.

References

1. Chen N, Yang K, Gu M (2010) Microwave absorption properties of La-substituted M-type strontium ferrites. J Alloy Compd 490:609–612
2. Widanarto W, Khaeriyah S, Ghoshal SK, Kurniawan C, Effendi M, Cahyanto WT (2019) Selective microwave absorption in Nd^{3+} substituted barium ferrite composites. J Rare Earths
3. Li C-J, Wang B, Wang J-N (2012) Magnetic and Microwave Absorbing Properties of Electrospun Ba(1 − x)LaxFe$_{12}$O$_{19}$ Nanofibers. J Magn Magn Mater 324(7):1305–1311
4. Pawar RA, Desai SS, Tamboli QY, Shirsath SE, Patange SM (2014) Ce 3 þ incorporated structural and magnetic properties of M type barium hexaferrites. J Magn Magn Mater 1–5
5. Kurniawan B, Laksmi W, Sahara NA (2018) Microwave absorption properties of La 0.8 Ca 0.2-xAgxMnO$_3$ (x = 0.05; x = 0.15) synthesized by sol-gel method. J Phys Conf Ser 1011:0–5
6. Widanarto W, Ardenti E, Ghoshal SK, Kurniawan C, Effendi M, Cahyanto WT (2018) Significant reduction of saturation magnetization and microwave-reflection loss in barium-natural ferrite via Nd^{3+} substitution. J Magn Magn Mater 456:288–291
7. Effendi M, Pratama DP, Cahyanto WT, Widanarto W (2019) The effects of sintering temperature on structure, magnetic properties and microwave absorption capability of manganese-natural ferrite. J Phys Conf Ser, IOP J Phys Conf Ser 1153(1):012060
8. Effendi M, Rahmah FS, Susanti L, Rochman NT (2019) Preparation of monodisperse polystyrene spheres by physical method. IOP Conf Ser Mater Sci Eng 509(1)
9. Cheng YL, Dai JM, Zhu XB, Wu DJ, Yang ZR, Sun YP (2009) Enhanced microwave absorption properties of intrinsically core/shell structured La 0.6Sr 0.4 nanoparticles. Nanoscale Res Lett 4(10):1153–1158

10. Chen J, Liu M, Yang T, Zhai F, Hou X, Chou KC (2017) Improved microwave absorption performance of modified SiC in the 2–18 GHz frequency range. Cryst Eng Comm 19(3):519–527
11. Luo J, Xu Y, Gao D (2014) Synthesis, characterization and microwave absorption properties of polyaniline/Sm-doped strontium ferrite nanocomposite. Solid State Sci 37:40–46

Combustion Performance and Exhaust Emission Analysis of Spent Bleaching Earth (SBE) Oil as Burner's Fuel

M. Afzan, A. M. Ithnin, and W. Jazair

Abstract Spent bleaching earth (SBE) is a type of hazardous solid waste generated during the bleaching process of crude palm oil (CPO). Despite years of studies being done on how best to manage the waste, the issue is still largely unsolved and has created massive economic and environmental problems. SBE is generally disposed by dumping onto landfills because it is by far the cheapest method. However, due to the high quantities of water insoluble substance; mostly free fatty acids (FFA) and peroxide as well as heavy metals, the waste decomposes very slowly and thus becoming a serious hazard to the environment that can cause water and soil pollution as well as fire risk. Recent breakthrough in studies have allowed significant amount of residual oil contained in SBE to be extracted and this has opened up new avenues to tackle the SBE dumping issue. By extracting the oil, it can be used as an alternative fuel in lieu of fossil fuels in power generation. This will inadvertently create a value added product that can mitigate the environmental hazards of SBE dumping but also reduce the cost of handling and disposing the waste. However, at present the research on SBE is limited to the advancement of residual oil extraction technique. There is no research that focus on the evaluation of SBE oil as a substitute to fossil fuels. As such this paper will evaluate and determine the combustion performance and exhaust emission of SBE oil as a source of fuel for burner. The combustion performance in terms of CO_2, CO, NO_x and flame temperature will be compared with neat diesel under the same conditions. SBE oil shows some promising combustion performance, since it produces no SO_x due to the absence of Sulphur, emits lower CO_2 than diesel, while releases higher CO than diesel. The higher amount of CO produced by SBE oil can be largely attributed by the high viscosity of the oil. The high viscosity and density of SBE oil greatly affects the fuel spray which in turns causing a poor atomization and combustion hence the high amount of CO emission.

Keywords Spent bleaching earth · Residual oil · Burner fuel · Emission

---

M. Afzan (✉) · A. M. Ithnin · W. Jazair
Advanced Vehicle System (VSE) Laboratory, Malaysia Japan International Institute of Technology (MJIIT), Universiti Teknologi Malaysia (UTM), Kuala Lumpur, Malaysia
e-mail: m.afzan@graduate.utm.my

U. Sabino et al. (eds.), *Proceedings of the 6th International Conference and Exhibition on Sustainable Energy and Advanced Materials*, Lecture Notes in Mechanical Engineering,
https://doi.org/10.1007/978-981-15-4481-1_68

1 Introduction

Palm oil is the most produced and consumed vegetable oil in the world. This is because the oil possesses some unmatched and desirable qualities over other vegetable based oil like soybean and rapeseed oil. The palm tree of the species *Elaeis guineensis* from the *tenera* variety, a hybrid between the dura and *pisifera* which is considered as the most efficient crop bearing oil requires only 0.26 ha of land to produce 1 ton of oil [1–3]. As a comparison, soybean and rapeseed require 2.22 and 2 ha of land to produce equivalent yields respectively [4].

Countries like Malaysia and Indonesia have slowly increased the percentage of palm oil in a biodiesel mixtures to reduce the dependency of fossil based fuels [5]. However, it is not without drawbacks. Even though palm oil requires the least amount of land to produce a given volume of oil, the environmental disasters that have been plaguing the industry is still a main concern. Rapid deforestation and waste generation have been steadily increasing and this has attracted lobbyists, environmentalists and the general public to protest towards the industry.

A study has claimed that to produce 1 ton of oil, the industry will also produce about 9 tons of solid and liquid waste across the downstream and upstream of the industry [6]. More studies are needed in order to reduce the waste margin to reduce its impacts towards the environment. However, the focus has always been to produce higher yields of palm oil species and the importance of reducing waste is somewhat overlooked. In the plantation, most of the waste is organic and such can be easily reused as fertilizer after composting. The same can't be said with waste generated in the milling and refinery plant.

One of the major waste produced at the refinery is spent bleaching earth (SBE) which is a type of waste generated during the bleaching process of CPO refinery. The process which utilizes bentonite clay will strip the oil of any impurities, pigments as well as moisture. Due to the adsorptive nature of the minerals, significant amount of residual oil (RO), as much as 30% by volume [7–9], will also be inadvertently discarded along the SBE as well. With proper technique, this oil can be extracted and put to good use instead of dumping it on the environment.

The oil extracted from SBE have several defining characteristics such as having very high FFA content as well high content of impurities. Residual oil can have as high as 40% of FFA and this is promising as it can potentially be used as a source of fuel for burners [8, 10–12].

The advantage of using the SBE oil is *double whammy* as not only it can reduce the environmental hazards with SBE dumping, it can also be used directly as source of burner's fuel without the need for further treatment or conversion into biodiesel that can be expensive and time consuming. Theoretically, there should be absolutely no problem in utilizing SBE oil as a burner's fuel since it has been established that the calorific value of SBE oil is about 38 MJ/kg which proves that it possess a comparable amount of hydrocarbons to neat diesel [8]. However, as of writing, there is no study being conducted to explore the usage of the oil directly to the burner especially in terms of its combustion performance. This paper will evaluate the combustion and exhaust emission analysis of direct use of SBE oil in term of its CO, CO_2, and NO_x as well flame temperature and to compare it with neat diesel.

2 Materials and Methodology

2.1 SBE Oil

The SBE oil was supplied by a local company called Eco Oils Sdn Bhd which specializes in extracting the residual oil from the SBE. The oil is dark in color and appear to be very viscous in room temperature. Several laboratory tests have been conducted to ascertain the primary physical and chemical properties namely; viscosity, density, calorific value as well as the amount of free fatty acids (FFA) presence in the oil.

Calorific value was determined by utilizing a bomb calorimeter model ECO made by Cal2K while the method and formula used to determine the percentage of FFA is in accordance to American Oil Chemical Society (ACS) Ca 5a-40 [13]. Solution of NaOH was titrated into the mixture of SBE oil sample with Isopropanol and three drops of phenolphthalein indicator. The titration process will be stopped once the SBE oil sample turned its color from orange into pink for at least 10 s. Then, by using the formula below, FFA percentage of the SBE oil is calculated. Since SBE oil is derived from palm oil which mainly consists of palmitic acid, the value 25.6 corresponds to the palmitic acid molar mass.

$$(25.6 \times \mathrm{M} \times \mathrm{V})/\mathrm{M_s} \tag{1}$$

where;

M molarity of NaOH
V volume of titrated NaOH (mL)
M_s mass of SBE oil sample (g).

2.2 Burner and Measurement Devices

The type of burner used is a small, low power industrial burner model KV-03 which specializes on burning waste oil that have at least 33 MJ/kg of calorific value. The burner comes equipped with fuel pump, manually adjustable air damper as well as a preheating element. The detail specification of the burner is listed in the Table 1.

The emission analyzer used is from Ecom model J2KN pro and is able to detect and calculate typical combustion gases like CO, CO_2 and NO_x. The emission analyzer probe will be inserted into the sampling port at the chimney and the data is recorded

Table 1 Burner specifications

Burner model KV-03	
Oil consumption	2–4 L/h
Power output	14–50 kW
Maximum output power	15,000–40,000 kcal/h
Nozzle	(1 × 1.0 mm) Brass syphon
Tube diameter	89 mm
Heating pipe	1 kW
Motor power	180 W
Power supply	220 V/50 Hz

Table 2 Emission analyzer specifications

Ecom J2KN pro		
Measurement	Range	Accuracy
Carbon Monoxide (CO)	0–10,000 ppm	±2% (measured)
Carbon Dioxide (CO_2)	0—fuel maximum CO_2	Calculated
Nitric Oxide (NO_x)	0–5000 ppm	±5% (measured)

and analyzed. The specifications and accuracy range of the emission analyzer is listed in Table 2.

2.3 *Experiment Set-Up*

The burner is retrofitted into a tubular combustion chamber that comes equipped with a chimney at the other end as shown in Fig. 1. There is a sampling port located at the chimney that allows insertion of probes and sensors that will record and monitor the combustion gases.

Prior to the experiment, a pre-testing has been conducted to determine the pre-heating temperature. This is important to ensure smooth burner operation with stable combustion as well as to avoid a waste of energy. Then, the diesel and SBE oil is burned as per the stoichiometric combustion. The velocity of air that corresponds to the opening segments of the damper have been measured by using an anemometer. The damper which is responsible to supply primary air for combustion have been adjusted accordingly to reflect the amount of air needed for stoichiometric combustion. The manipulation of the Air Fuel Ratio (AFR) is done by only adjusting the air volume while the fuel feed rate remains constant.

Temperature probes of the K-type thermocouple have been placed into the combustion chamber at five different equidistant locations in relative to each other. To ensure accuracy, the thermocouples have been calibrated accordingly. During the

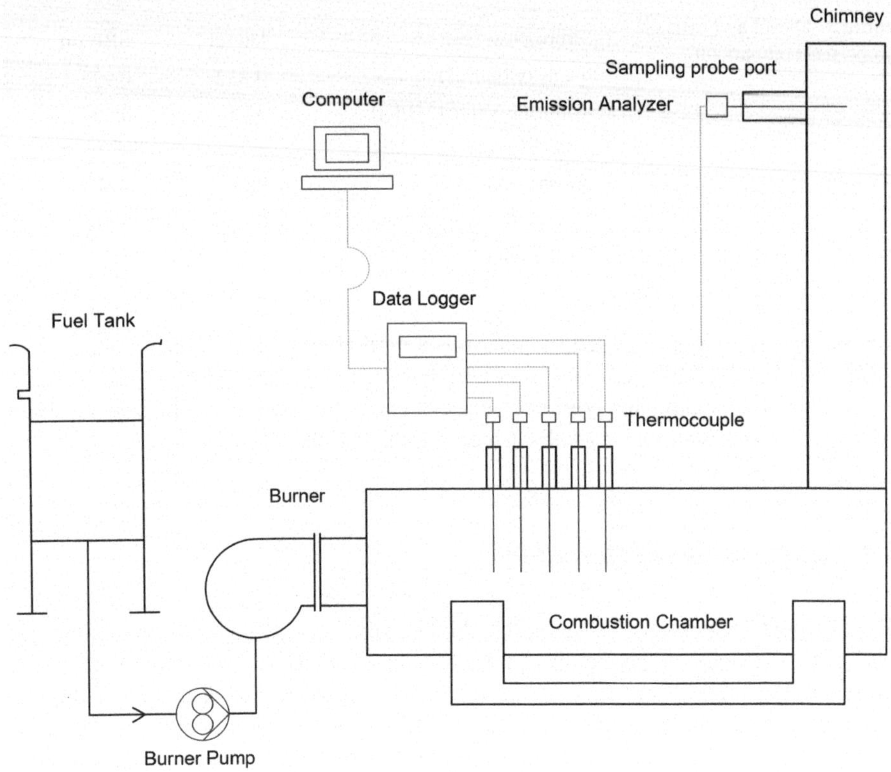

Fig. 1 Burner and combustion chamber setup

combustion, the probes will be ensured to be located right in the middle of the flame so that it can measure the flame temperature accurately.

Emission analyzer which measures emission products has been used by placing the probe sensor into the chimney sampling port.

3 Results and Discussions

The properties, combustion performance as well as the flame temperature of SBE oil is compared to neat diesel as shown in Table 3. The density of SBE oil is 0.923 g/m^3 which is comparable to other published studies [8, 12] which is significantly higher than neat diesel at 0.832 g/m^3. There is also a huge disparity of viscosity between SBE oil and neat diesel. At 40 °C, SBE oil is very viscous, about 25 times higher, at 93.8 Mpa s compared to the diesel at about 3.7 Mpa s

The calorific value of SBE oil was determined to be at about 38.8 MJ/kg while the neat diesel is 45.5 MJ/kg. This is impressive as SBE oil contains about 85% of

Table 3 Diesel and SBE oil properties comparison

Properties	Diesel	SBE oil
Density (kg/m^3)	0.832	0.923
Calorific Value (MJ/kg)	45.5	38.81
Viscosity at 40 °C (Mpa s)	3.7	93.8
Flash point (°C)	93.9	200
Ash content (% of mass)	0.1	0.617
FFA content (%)	–	31

the potential energy possessed by neat diesel while being considered as waste. The flash point for the SBE oil is also much higher than diesel at 200 °C.

On the other hand, the presence of ash content in SBE is significant at about 0.617% by mass while the diesel's ash content is merely 0.1%.

3.1 Combustion Performance

The CO and CO_2 emission of SBE oil is shown in the Figs. 2 and 3. The data indicates that SBE oil combusts poorly as opposed to diesel. SBE oil emits about 2.7 times more CO than diesel and this can be largely attributed to the viscosity of the oil. Even after the preheating, the viscosity of SBE oil is still significantly higher than diesel. A very high viscosity of oil or fuel will cause a problem in a way that the droplets size becomes bigger and thus affecting combustion. Smaller droplets in the range 25–35 μm is desirable as the total surface area of the droplets for combustion is optimum, decreasing the formation of carbon monoxide due to the incomplete

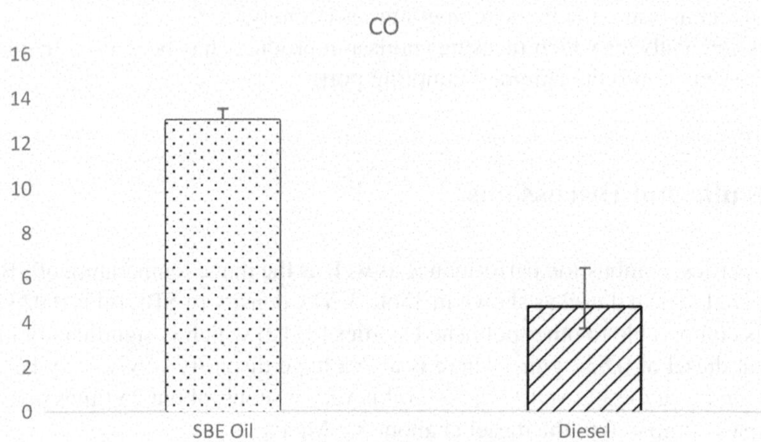

Fig. 2 CO emission

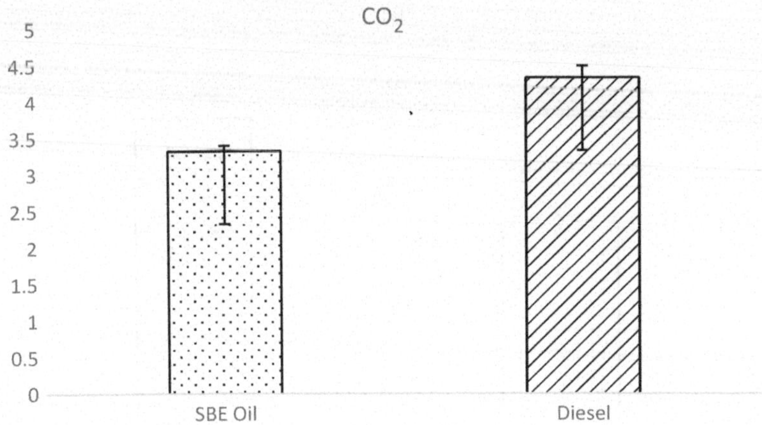

Fig. 3 CO_2 emission

combustion as well as reducing the burning time. This is proven by several studies that looked into the effect of burning high viscosity oil against the droplets size [14–18]. A general conclusion that can be drawn from the studies is that combustion performance will be greatly affected by the fuel viscosity regardless of the type of atomizers used.

On the other hand, the CO_2 emission of SBE is lesser than diesel in a factor of 0.77–1. While at first glance it looks promising in term of the reduction of CO_2 emission as opposed to diesel, the data also indicates that SBE oil burns rather poorly. In this instance, it is caused by the combination of the high viscosity and the lack of retention times. In essence, due to the geometry of the combustion chamber and the draft induced by the burner fan blower, the carbon content of the oil lack the time needed in order to be fully oxidized and thus failed to be fully converted into CO_2. As a result, more CO is produced which then escapes to the chimney.

In regards to NO_x emission, SBE oil produces 37.4% less NO_x compared to diesel as shown in Fig. 4. This is largely attributed by the lower flame peak temperature of SBE oil as opposed to diesel which is shown in Fig. 5 which is due to the lower energy content of SBE oil. Since the formation of NO_x is largely dependent of the combustion temperature, the lower temperature of the combustion resulting to lower NO_x emission.

Figure 5 reveals that the flame profile of SBE oil is less uniform than diesel. The thermocouples which have been fixed at an equal distance of each other is used to measure the flame temperature by ensuring a direct contact at the center axis of the flame. The temperature of the flame keeps dropping significantly the further the flame propagates. At point 5, the SBE oil flame temperature drops to about 490 °C which is about 49.3% lower than at the point 1. Meanwhile the diesel flame temperature is more consistent and uniform throughout as at point 5 it only drops to 747 °C which account to about 20.5% of temperature drops. This might be caused by the bigger droplets size produced by the pressure jet atomizer which is not built and designed

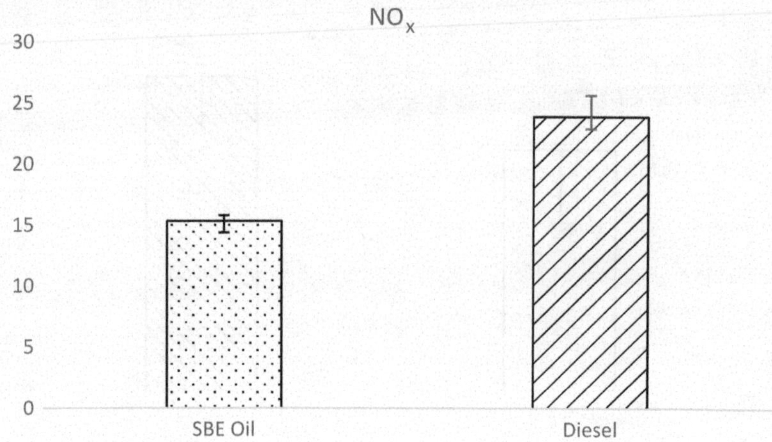

Fig. 4 NO_x emission

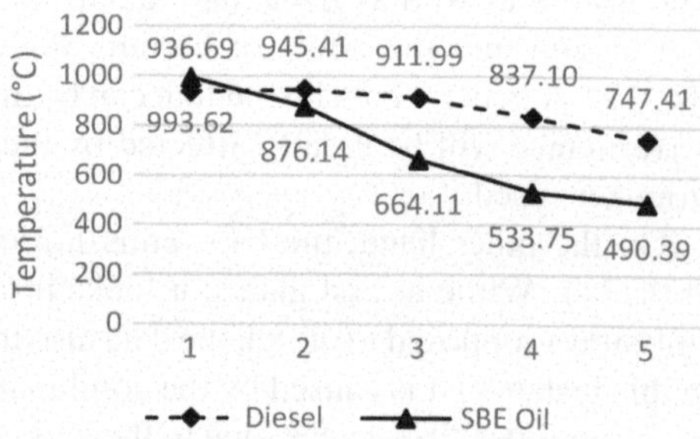

Fig. 5 Flame temperature at 5 equidistant points along the combustion chamber parallel to the flame length. Point 1 is the nearest to the burner injector nozzle while 5 is the furthest

to run a heavy viscosity oil like SBE. High viscosity oils will always produce big droplets which tend to escape the combustion zone as reported by several studies [19–21].

4 Conclusions

SBE oil produces lesser amounts of CO_2 and NO_x while generates more CO than Diesel. It doesn't produces any SO_x because the absence of Sulphur. These results show that the oil have a potential as an alternative fuel for burner. However, the viscosity of the oil remains a challenge that needs to be taken into account to reduce the formation of CO_2 and CO, as the high viscosity of the oil is the chief factor of the slightly worse combustion performance than diesel.

References

1. Singh RP, Ibrahim MH, Esa N, Iliyana MS (2010) Composting of waste from palm oil mill: a sustainable waste management practice. Rev Environ Sci Biotechnol 9(4):331–344
2. Loh SK, Cheong KY, Choo YM, Salimon J (2015) Formulation and optimisation of spent bleaching earth-based bio organic fertiliser. J Oil Palm Res 27(1):57–66
3. Japir AA, Salimon J, Derawi D, Bahadi M, Al-Shuja'a S, Yusop MR (2017) Physicochemical characteristics of high free fatty acid crude palm oil. OCL 24(5):D506
4. Malaysian Palm Oil Board (2018) The Malaysian palm oil facts [Online]. Available: www.mpob.gov.my
5. Szulczyk KR, Atiqur Rahman Khan M (2018) The potential and environmental ramifications of palm biodiesel: evidence from Malaysia. J Clean Prod 203:260–272
6. Abdullah N, Sulaim F (2013) The oil palm wastes in Malaysia. Biomass Now Sustain Growth Use
7. Mu B, Wang A (2019) Regeneration and recycling of spent bleaching earth. In: Handbook of ecomaterials, vol 5, pp 3147–3167
8. Kheang LS, May CY (2006) A study of residual oils recovered from spent bleaching earth: their characteristics and applications. Am J Appl Sci 3(10):2063–2067
9. Sapawe N, Hanafi MF (2018) Analysis of the pyrolysis products from spent bleaching clay. Mater Today Proc 5(10):21940–21947
10. Chung C, Eidman V (1997) Economic use of spent bleaching earth. Inf Int News Fats Oils Relat Mater 8(7):739–742
11. Fahmil ASQRM, Gumbira-Sa'id E, Suryani A (2014) Biodiesel production from residual palm oil contained in spent bleaching earth by in situ trans-esterification. Environ Asia 7(2):30–35
12. Kheang LS, May CY (2007) Residual oil from spent bleaching earth (sbe) for biodiesel and biolubricant applications
13. Mahesar SSA, Sherazi STH, Khaskheli AR, Kandhro AA (2016) Analytical approaches for free fatty acids assessment in oils and fats
14. Al Omari SAB, Hamdan MO, Selim MY, Elnajjar E (2019) Combustion of jojoba-oil/diesel blends in a small scale furnace. Renew Energy 131:678–688
15. Che Mat S, Idroas MY, Teoh YH, Hamid MF (2019) Optimisation of viscosity and density of refined palm oil-melaleuca cajuputi oil binary blends using mixture design method. Renew Energy 133:393–400
16. Du E, Cai L, Huang K, Tang H, Xu X, Tao R (2018) Reducing viscosity to promote biodiesel for energy security and improve combustion efficiency. Fuel 211:194–196
17. Park SH, Youn IM, Lim Y, Lee CS (2013) Influence of the mixture of gasoline and diesel fuels on droplet atomization, combustion, and exhaust emission characteristics in a compression ignition engine. Fuel Process Technol 106:392–401
18. Wang X, Huang Z, Kuti OA, Zhang W, Nishida K (2011) An experimental investigation on spray, ignition and combustion characteristics of biodiesels. Proc Combust Inst 33(2):2071–2077
19. Tzanetakis T, Moloodi S, Farra N, Nguyen B, McGrath A, Thomson MJ (2011) Comparison of the spray combustion characteristics and emissions of a wood-derived fast pyrolysis liquid-ethanol blend with number 2 and number 4 fuel oils in a pilot-stabilized swirl burner. Energy Fuels 25(10):4305–4321
20. Mahfouz A, Gad MS, El Fatih A, Emara A (2018) Comparative study of combustion characteristics and exhaust emissions of waste cooking-diesel oil blends. Ain Shams Eng J
21. Ghorbani A, Bazooyar B, Shariati A, Jokar SM, Ajami H, Naderi A (2011) A comparative study of combustion performance and emission of biodiesel blends and diesel in an experimental boiler. Appl Energy 88(12):4725–4732

Ceramic Jewelry with Texture and Ornament Islamic Pattern and Batik Indonesia—Design, Manufacturing, and Fabrication

P. K. Fergiawan, P. W. Anggoro, A. T. Yuniarto, K. B. Purwanto, and O. D. W. Widyanarka

Abstract The development of science and technology has become important to grow the manufacturing industry in Indonesia and this is characterized by high consumer demand for customized product designs that are precise, accurate, and detailed on complex ornaments or textures and have high selling value in the industrial market. This paper discusses and demonstrates the application of modern computer-aided engineering system (CARESystem) technology in the process of designing—manufacturing ceramics jewelry at the ceramic company PT. Indonesian Nuanza porcelain. Batik and Islamic patterns are chosen as the basic texture for pendant products. The design process uses Zbrush to get 3D CAD models for ceramics jewelry. The process of manufacturing optimization on CNC machines using computer-aided manufacturing (CAM) software PowerMill 2016 and Rhinoceros 4.0. The results of the research show that the optimal processing time and quality of machining in CNC machines is the use of a toolpath strategy horizontal roughing, and parallel finishing in Rhinoceros 4.0 for roughening and semi-finish processes. Toolpath strategy model area clearance, optimize constant z, and step and shallow finishing at PowerMill 2016 for the finishing process. Production time can be increased to twice compared to manual technology.

Keywords Jewelry ceramics · CARESystem · CAD · CAM · CNC

P. K. Fergiawan · P. W. Anggoro (✉) · A. T. Yuniarto
Department Industrial Engineering, Faculty of Industrial Technology, University of Atma Jaya Yogyakarta, Jl. Babarsari 44, Yogyakarta, Indonesia
e-mail: pauluswisnuanggoro@ymail.com

K. B. Purwanto
Laboratory of Production System, Department Industrial Engineering, Faculty of Industrial Technology, University of Atma Jaya Yogyakarta, Jl. Babarsari 44, Yogyakarta, Indonesia

O. D. W. Widyanarka
PT Nuanza Porselen Indonesia (PT. NPI), Dukuh Dedegan, Ngadirojo Village, Boyolali Residence, Ampel, Central Java, Indonesia

U. Sabino et al. (eds.), *Proceedings of the 6th International Conference and Exhibition on Sustainable Energy and Advanced Materials*, Lecture Notes in Mechanical Engineering,
https://doi.org/10.1007/978-981-15-4481-1_69

1 Introduction

The Ministry of Industry of the Republic of Indonesia in 2018 stated that the development of science and technology became important in order to grow the manufacturing industry and this was marked by high consumer demand for customized, accurate, and detailed design of customized products against ornaments or complex textures and had high selling points in the industrial market [1]. Manufacturing industries include machine tools, aircraft, automotive, ceramics and so on. The ceramics industry in Indonesia, in general, is an industry that is growing quite rapidly and promising, especially industries that make ceramic products that give different touches with the addition of textures or ornaments with varied motifs [1]. Ceramics took from Greek, namely "KERAMIKOS" which has the meaning of burning material. In general, ceramics have refractory properties that can maintain their properties which are useful in very severe conditions due to high temperatures and contact with corrosive materials compared to superalloy materials which tend to have brittle properties. This property arises due to the mixing process of kaolin, quartz, feldspar in the heat treatment process that occurs at high temperatures (firing) [2]. The use of ceramics in everyday life is increasingly developing and varied and influences the lifestyle of the users. This happens because ceramic materials can produce products that are unique, competitive, precise, accurate and can be combined with various colors and other aspects of life according to the development of current technological advancements [3].

Manufacturing of engineering ceramics extends beyond the pure processing science which is specific for different materials. Manufacturing must equally respect economic and technical boundary conditions to select a processing cycle capable of producing ceramic components based on a set of requirements predefined by the application in terms of reproducibility, performance, and cost. The science of ceramic manufacturing in a competitive global environment must thus combine elements of basic material science, engineering and business economics [5].

Previous researchers [4] successfully combined artistic ceramic process technology to create jewelry manually which was not based on artistic computer-aided design (ArtCAD). This merger connects the unique technological variations (narikomi technique) from artistic ceramics with metal material in the development of the bird ring jewelry product. But the design process from it is not based on the use of artistic CAD technology that is capable of producing designs jewelry in many variations and precision and accuracy. The product formed is indeed unique and artistic, but if done in bulk, the geometric quality that the customer wants is impossible to achieve.

This research wants to demonstrate the ability of modern artistic technology based on a computer-aided reverse engineering system (Art CARESystem) in the process of designing—manufacturing—fabricating ceramics jewelry in Indonesian ceramic companies. The products produced in this research are jewelry necklaces with texture motifs typical of Islam and Indonesian batik but are produced in large quantities with texture geometry that is precise, accurate and complex in detail. This paper also explains the real problem conditions faced by PT Nuanza Porcelain Indonesia, an Indonesian national ceramics company that seeks to develop rapidly as a creative industry to produce unique ceramic products, has high selling value, is able to compete in the global market and always prioritizes the aspects of the womb local Indonesian by giving unique characteristics of textures and ornaments to each product (ceramic tableware, ceramic tile, and ceramic jewelry).

2 Material and Method

3D CAD artistic models of ceramic jewelry done in this paper are shown in Fig. 1. Three materials used in the processing of jewelry ceramic products, namely yellow gypsum, white gypsum, and china ash bone (see Fig. 2).

Gypsum material with a size of 350 mm × 250 mm × 15 mm if you experience the combustion process with a temperature of 1250–1400 °C, the characteristics change with a smooth, white, and hard texture [2]. Gypsum which is burned with low temperature is called soft gypsum (white) while the one that is burned with high temperature is called hard gypsum (yellow). Gypsum is a material that is often used in the ceramic industry because this material has a good strength structure [2]. Yellow gypsum is used to create jewelry ceramic mold master patterns, while white gypsum is used to make jewelry ceramic mold patterns. The china ash bone [2] material is white, with a smooth profile, and is strong with transparency. This material has fewer plastic properties which result when the material is processed by casting techniques,

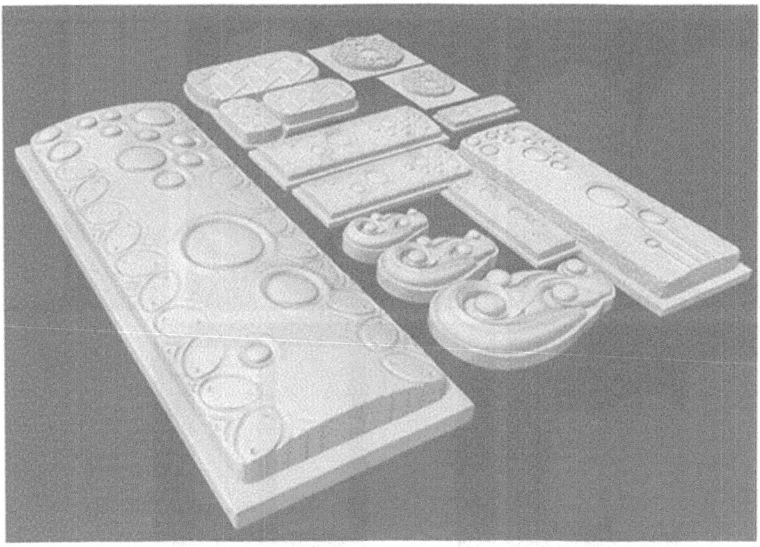

Fig. 1 3D CAD master pattern of jewelry ceramic molds

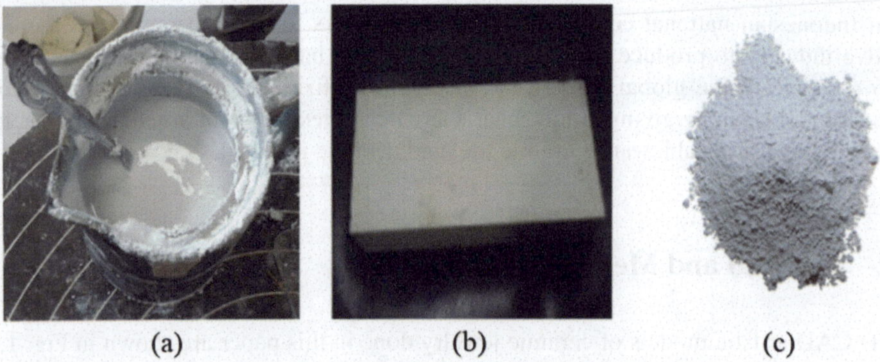

Fig. 2 Material of jewelry ceramic: **a** white gypsum, **b** yellow gypsum, **c** China ash bone

this material has difficulty. This material is used to make the pendant through the combustion process at a temperature of 1040–1080 °C.

Fabrication of jewelry ceramic products in this paper starts from the stage of making 3D designs using CAD z-brush software. The output generated is a 3D CAD model in the .stl file format. Providing work axes in 3D CAD models is necessary for research so that designs in CAD can be manufactured using CAM software (Rhinoceros 4.0 and PowerMILL 2016). Work axis granting verification is done using Netfab 2017. In this paper, the master machining process of the Jewelry Ceramic mold pattern (see Fig. 3) uses a CNC machine (see Fig. 4a) and two type cutter milling,

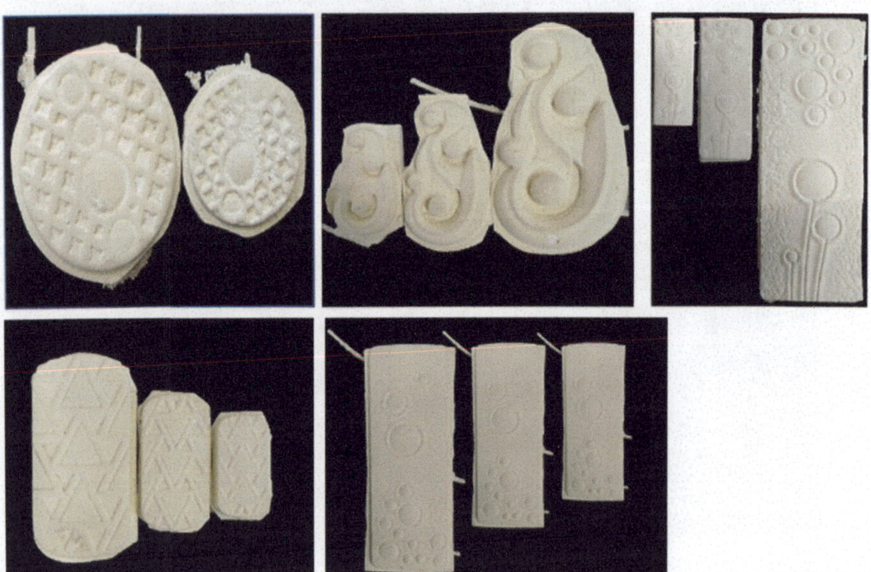

Fig. 3 Master pattern of jewelry ceramic molds

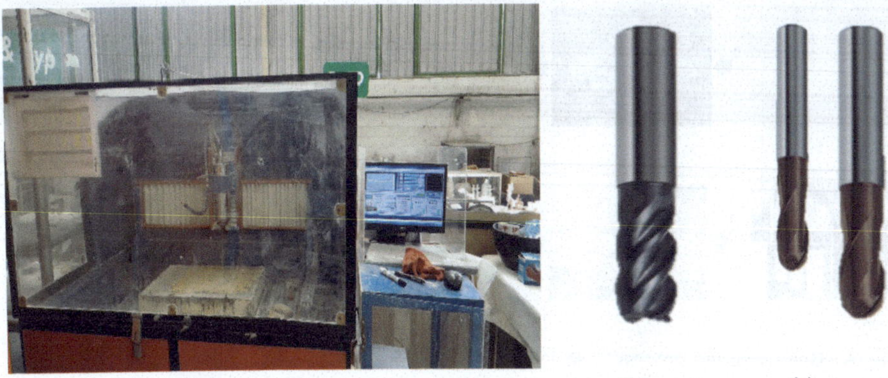

(a) (b) (c)

Fig. 4 Machine and tooling research: **a** CNC machine; **b** End Mill; **c** Ballnose

namely Endmill (see Fig. 4b) and Ballnose (see Fig. 4c). This mold pattern will later be poured white gypsum material into the pattern of jewelry ceramic prints (see Fig. 5). The pattern of the mold is then in the oven to be hard and can be used repeatedly.

The process of pouring china ash bone (liquid) into the pattern of jewelry ceramic prints takes 15 min and is heated for 4 h until it becomes a pendant. The final stage of the fabrication process is to assemble the pendant with Korean rope, and other additional accessories. The stages of the design, manufacturing, and fabrication of ceramic jewelry in this paper are presented in Figs. 6 and 7.

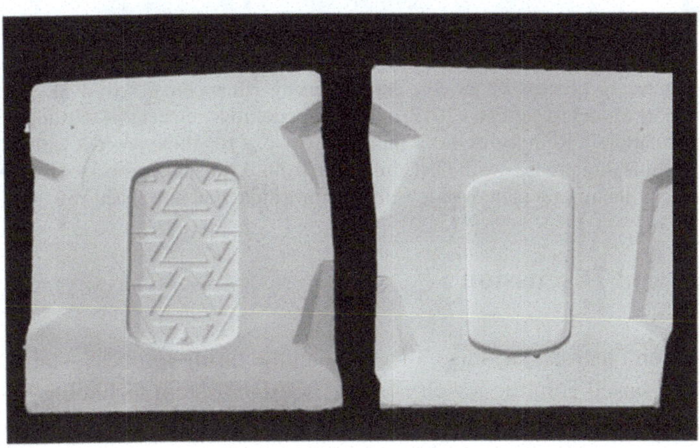

Fig. 5 Patterns of jewelry ceramic molds

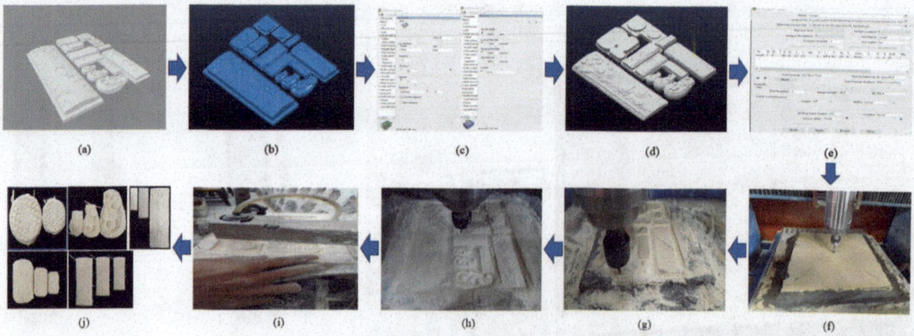

Fig. 6 Manufacturing process of jewelry ceramic using PowerMILL 2016: **a** 3D CAD Jewelry Ceramic, **b** import Model to PowerMill 2016, **c** input parameter condition, **d** simulation in PowerMill 2016, **e** process NC-Code, **f** roughing process with CNC milling retrofit, **g** semi-finishing process with CNC milling retrofit, **h** finishing process with CNC milling retrofit, **i** cutting process master patterns of jewelry ceramic molds with sawing machine, **j** results of the master pattern of jewelry ceramic mold that have been cut

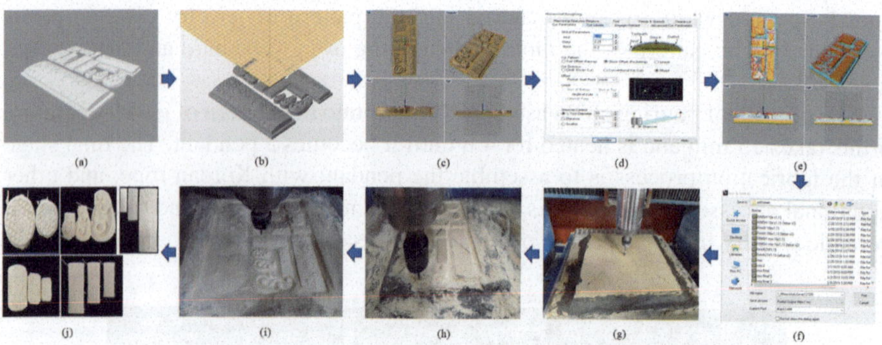

Fig. 7 Manufacturing process of jewelry ceramic using Rhinoceross 4.0: **a** 3D CAD Jewelry Ceramic, **b** import Model to NetFabb 2017, **c** import model to Rhinoceros 4.0, **d** input parameter, **e** condition simulation in Rhinoceros 4.0, **f** input NC-Code, **g** roughing process with CNC milling retrofit, **h** semi-finishing process with CNC milling retrofit, **i** finishing process with CNC milling retrofit, **j** results of the master pattern of jewelry ceramic mold that have been cut

3 Result and Discussion

This paper demonstrates a stage of the manufacturing process—fabrication of CARESystem-based ceramic jewelry [3, 11]. CARESystem technology infrastructure consists of two units, namely the mold pattern master formation unit (see Fig. 3) including Artistic CAD (Zbrush); Manufacturing optimization uses two CAM software (PowerMILL 2016, Rhinoceros 4.0) and retrofits CNC machines. The second unit is ceramic fabrication technology with kiln kitchens (see Fig. 8). The textures of the Islamic pattern [5] and Indonesian Batik have more floral patterns displayed by researchers on both surface jewelry ceramics (see Figs. 3 and 5). Figure 2.5D CAD

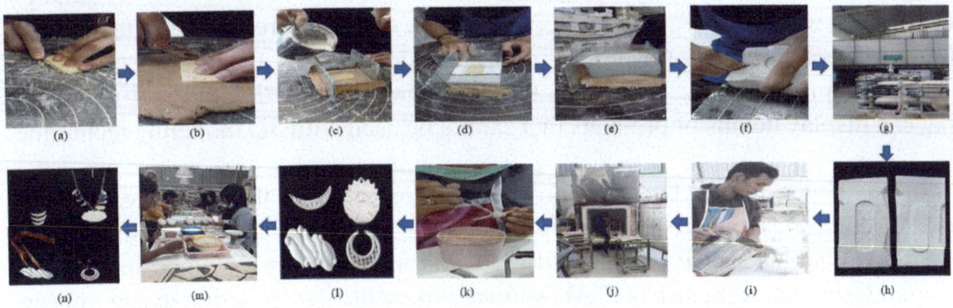

Fig. 8 Fabrication process of jewelry ceramic: **a** cutting process master patterns of jewelry ceramic molds with the contours, **b** cribbing process master patterns of jewelry ceramic molds on clay, **c** White Gypsum pouring process on the master pattern of jewelry ceramics molds (Core), **d** cribbing process the bottom of master patterns of jewelry ceramic molds, **e** White Gypsum pouring process on the master pattern print (Cavity), **f** process of making a Champers and locks in a molds, **g** Combustion Process 1, **h** Core and Cavity Jewelry Ceramic products, **i** China ash bone pouring process on the jewelry ceramic molds, **j** Combustion process 2, **k** glazing process, **l** Leontine of jewelry ceramic, **m** assembly process, **n** jewelry ceramics product

texture models can be displayed with perfect results thanks to the use of Zbrush CAD artistic software (Fig. 7).

The stages of jewelry ceramic design take patterns with Islamic pattern textures as they were successfully introduced by [6] and presented in Fig. 9, and Indonesian batik is done using CAD Zbrush 4R7 software. This software has advantages in the scaling process quickly, precisely and artistically compared to other CAD software: artCAM, Autodesk fusion 360, solid work, master cam. ZBrush is an artistic CAD software that can support users can draw illustrations in the form of 2.5 dimensions and 3

Fig. 9 Examples 3D CAD model with Z-brush

dimensions where the techniques applied in this software are sculpting techniques. Sculpting techniques are the process of product design such as sculpting sculptures, artistic ornamental objects (see Fig. 9) and having the function of creating objects that can display details of products that cannot be used with 3D modeling techniques [7].

In this paper, the CAD design process is different from that done by Kutsenko and Arventyeva [4] where here the direct physical model was designed using Zbrush by a CAD engineer at PT. NPI until 3D CAD models were obtained in the .STL format. Using the right artistic CAD software like Zbrush can optimize and improve the product before the machining process. The form of products with ornaments or textures as desired by the customer can be quickly done by the CAD engineer until the manufacturing stage in CNC or 3D printer machines. High-tech computer technology can increase the speed of 50–70% in the development of ceramic products [8]. The care system application in this paper can improve the design time of ceramic jewelry products less than 50% better than with manual methods.

This paper also demonstrates the stages of the CAM-based manufacturing optimization process using PowerMill 2016 and Rhinoceros 4.0 (see Figs. 6 and 7). The use of PowerMill was conducted by researchers to obtain a surface quality of jewelry better than Rhinoceros 4.0 as had been done by [11, 12] in the optimization of manufacturing insole made from EVA Foam rubber. Modern CAM systems (PowerMILL; Solid edge; MasterCAM, SolidCAM) can produce tool paths based on the conditions of the most constant geometric bonds that may occur in addition to conventional predetermined trajectory divisions [9, 10]. The selection of PowerMill in this paper is due to the needs of the NPI which wants the surface quality to be smoother and more assertive, in addition to the many choices of tool path strategies displayed by PowerMILL compared to other CAM software.

In this paper, the path strategy tool used is divided into three stages, namely roughing, semi-finishing, and finishing [11, 12]. Model area, Raster Finishing Clearance, and Step and shallow were chosen as the optimal path strategy tool in the execution of the master pattern master ceramics jewelry. These three strategies have also been used by [12] with success in carrying out shoe last made from Ebalta wood with the results of 95–98% geometry following the 3D CAD model shown. As a comparison to show the quality of machining results, we also use the path strategy tool from Rhinoceros 4.0, i.e.: Horizontal Roughing, and parallel finishing [13, 14]. Deepa and Jayesh [13] also uses these two Rhinoceros strategies in their research in ring jewelry machining processes based on CAD 3D jewelry. Toolpath of the two software simulations is able to display the best workmanship quality in CNC machines, but after testing on retrofit CNC machines in PT. NPI, proven Roughing Horizontal strategy, and parallel finishing of Rhinoceros used for roughing and semi-finish work to catch up time the manufacturing process and dept of cut are high, while the pursuit of the target surface is smoother and the contours of the texture built can be in accordance with the 2.5 CAD image used Step and shallow at Power mill 2016 as a toolpath strategy for Finishing.

CNC Retrofit machines (see Fig. 4a) are used in this paper to process yellow gypsum as raw material into master mold patterns (see Fig. 2b) ceramics jewelry with a variety of textures that can be displayed. For the milling chisel movement to be controlled properly by CAM and able to get the textural quality as desired, a communication bridge is needed in the form of a post-processor on a CNC machine. This bridge is in charge of reading the NC Code [15–20] which has been copied by CAM. The language displayed in this NC Code can move the cutter milling to process the grinding of workpieces according to the toolpath designed by CAM engineer PT. NPI.

Ceramics jewelry is a type of product with complex and complex contour detail. To obtain the surface geometry quality that is following the results of the drawing, machine tools are needed with good manufacturing subtractive technology. The use of CNC high-speed machining machines as has been done [8, 15] in this paper is used to maximize engine speed to obtain products with a fast time and the best product quality. The CNC HSM engine is usually in the range of RPM 10,000 and above, while the CNC retrofit in this paper is capable of reaching 23,000 rpm. Antunes uses HSM to obtain ceramic tableware and sanitary products, while research is being carried out at this time to obtain master pattern prints for ceramics. This research also demonstrates the application of machining process techniques using subtractive manufacturing technology which aims to reduce machining time and the quality of machining results on CNC product surfaces (texture and ornaments produced in more detail, precision, and precisely by following 3D artCAD images).

Mater mold pattern (see Fig. 5) then carried out the fabrication process with modern casting technology at PT. NPI as shown in Fig. 8. The final result of the fabrication process is a pendant (see Fig. 10) which is then assembled with several supporting components such as Korean rope and accessories become pendant necklaces (see Fig. 8n). The Glassing technique is also used in this paper to beautify the shape of the pendant so as to increase the selling value of the product (see Fig. 8k). The work on ceramics jewelry using manual technology at PT. NPI currently takes 7–10 days per one design model (design-manufacturing-fabrication). After the use of modern

Fig. 10 Examples of jewellery ceramic pendants

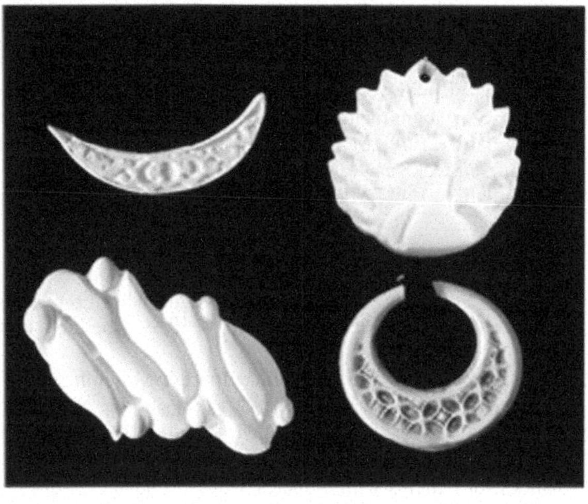

CARESystem technology, the process of jewelry ceramic takes only 3 days, resulting in a reduction in the production time of around 50% and an increase in production to double than before.

4 Conclusion

The modern CARESystem technology in this paper can increase product design—manufacturing—ceramics time around 50% compared to the manual technique that has been used by PT. NPI and other traditional ceramics industries in Indonesia.

Products made with CARESystem technology can produce products with detailed and precise contours when mass-produced, when compared to using manual technology. The results of this research can also be used as a reference in the construction of other ceramic-based products such as tableware, sanitary, wall ornaments.

Acknowledgements We would like to gratefully thank you for PT. Nuanza Porcelain Indonesia, Boyolali, Central Java and Naruna Porcelain Studio in Salatiga, Central Java Indonesia and SIBAD Undip Group Research, Semarang, Central Java Indonesia that already provide full support in the form of infrastructure support CAD, CAM, RE and CNC during the design-fabricate ceramic and developed process as well as the writing of this paper.

References

1. Kementrian Perindustrian Republik Indonesia (2014) Keramik dan pusaran teknologi. https://kemenperin.go.id/artikel/6656/Keramik-dalam-Pusaran-Teknologi. Access on July 2019
2. Budiyanto S, Rohmat S, Fajar P, Taufiq EY (2008) Kriya Keramik Jilid 3. Direktorat Pembinaan Sekolah Menengah Kejuruan, Direktorat Jenderal Manajemen Pendidikan Dasar dan Menengah, Department Pendidikan Nasional, Jakarta
3. Oancea G, Ivan NV, Pescaru R (2013) Computer-aided reverse engineering system used for customized products. In: Annals of MTeM for 2013 and proceedings of the 11th international MTeM conference, Malaysia, pp 181–186
4. Kutsenko LE, Arventyeva NA (2016) Mixed technologies of artistic ceramics processing for the jewelry manufacturer. Tomsk Polytechnic University, Russia
5. Abdullahi Y, Embi MR (2015) Evolution of abstract vegetal ornaments in Islamic architecture. Int J Architect Res 9(2):31–49
6. Abdullahi Y, Embi MR (2013) Evolution of Islamic geometric patterns. Front Architect Res 2, 243–251
7. Spencer S (2011) ZBrush character creation: advanced digital sculpting, 2nd edn. Sybex, USA
8. Wang A, Sai S, Liu Y (2014) The high computer technology application study about the daily-use ceramic products design. Jingdezhen Ceramic Institute, China
9. iMachining—Die Revolution in der CNC-Bearbeitung. https://www.solidcam.com. Access on 28 July 2019
10. Planen und programmieren mit virtueller Maschine. Tebis Technische Informations systeme AG. Maschine + werkzeug, 10/2018. Access on 28 July 2019
11. Anggoro PW, Tauviqirrahman M, Jamari J, Bayuseno AP, Bawono B, Avellina MM (2018) Computer-aided reverse engineering system in the design and production of orthotic insole shoes for patients with diabetes. Cogent Eng 5(1):1–20

12. Anggoro PW, Bawono B, Wijayanto A, Jamari J, Bayuseno AP (2016) Parameter optimization of strategies at CNC milling machines Rolland Modela MDX 40R CAM against surface roughness made insole shoe orthotic eva rubber foam. Int J Mech Mech Eng 6(4):96–102
13. Deepa SG, Jayesh JD (2013) Adaptability of CAD/CAM for jewellery making industry using method comparison technique. Int J Latest Trends Eng Technol (IJLTET) 3(1):44–58
14. Gatti R (2017) Introduction to Rhinoceros 4.0 The ARC. Rhinoceros NURBS. Edizioni FAG Milano
15. Antunes Simoes JFCP, Cole TJ, Cheshire DG, Anthonio RP (2003) Analysis of multi-axis milling in an anthropomorphic robot, using the design of experiments methodology. J Mater Process Technol 135:235–241
16. Stenberg V (2015) Student CNC guard. KTH Royal Institute of Technology, Stockholm, Sweden
17. Wise D, Anderson M (2011) Secrets of Zbrush experts: tips, techniques, and insights for users of all abilities, 1st edn. Course Technology PTR
18. Baban M, Baban CF, Buidos T, Stanasel I (2015) A reverse engineering approach for product development. Romanian Association of Nonconventional Technologies, Romania, pp 12–17
19. Bagci E (2009) Reverse engineering applications for recovery of broken or worn parts and re-manufacturing: three case studies. Adv Eng Softw 40:407–418
20. Lopez CI, Pinillos JC, Moreno JC (2013) Comparison between two design methods implants, based on reverse engineering, design, and engineering technologies, BIO CAN/CAD/CAE. Universidad Industrial de Santander, Escuela de Diseno Industrial, Bucaramanga, Colombia

Improvement of Space Tube Frame for Formula Student Vehicle

H. Hazimi, U. Ubaidillah, R. Alnursyah, H. Nursya'bani, B. W. Lenggana, and Wibowo

Abstract Formula Student is a prestigious competition for engineering student which enables students doing design, analysis, and fabricating a full-scale formula-style vehicle. One of the most important parameters of the formula vehicle is power to weight ratio. Based on this parameter, the vehicle must have the highest power with lowest vehicle weight. Chassis have a big role in the vehicle weight; thus, it must have as low weight as possible but maintaining good strength. This article focuses on the development of third-generation chassis based on the previous generation. The chassis strength is firstly evaluated using finite element analysis. During simulation, the strength of 3 types of chassis (1st, 2nd, and 3rd generations) are studied their torsional strength. The obtained results are then compared to each chassis weights. Based on this approach, the ratio between torsional stiffness and weight of all chassis can be compared. All chassis are designed using same material and joint method, whilst the differences are amount of frame member, dimension, and structure layout. The simulation results show that the torsional stiffness to weight ratio is 15.39, 24.03, and 25.12 Nm/deg kg respective to 1st, 2nd, and 3rd generation. It can be concluded that the 3rd generation has the highest stiffness to weight ratio so far.

Keywords Chassis · Torsional · Stiffness · Formula

1 Introduction

1.1 Formula Student

Formula Student competes student teams to design, fabricate, and develop formula-style vehicles. Vehicles should have high performance and durability to finish all competition events [1]. All vehicle systems are supported by structural system called

---

H. Hazimi · U. Ubaidillah (✉) · R. Alnursyah · H. Nursya'bani · B. W. Lenggana · Wibowo
Mechanical Engineering Department, Faculty of Engineering, Universitas Sebelas Maret, Jl. Ir. Sutami 36A, Kentingan, Jebres, Surakarta 57126, Indonesia
e-mail: ubaidillah_ft@staff.uns.ac.id

U. Sabino et al. (eds.), *Proceedings of the 6th International Conference and Exhibition on Sustainable Energy and Advanced Materials*, Lecture Notes in Mechanical Engineering,
https://doi.org/10.1007/978-981-15-4481-1_70

chassis [2]. Chassis have to connect front and rear suspension rigidly; the attachment point must minimally deform as the vehicle maneuvering [3]. Two main types of chassis are space frame and monocoque. Space frame uses welded structure to withstand loads and monocoque is the type where loads are supported by external panels [1, 4].

1.2 Chassis

The space frame is very popular in Formula Student because this type has a lower cost than another, simple design and easily constructed. The frame design has several rules which are ruled by international Formula SAE Rules [1] such as the dimension, material, and joint method [4]. And then monocoque frame is used by joining composite panel with basic space frame structure such as roll hoops and bulkhead.

1.3 Torsional Stiffness

Longitudinal Torsion is one main load that generates chassis deformation [2]. Chassis ability to withstand the twisting load called torsional rigidity. The largest loads magnitude that transmitted through chassis is torsional load. The resistance to the torsional load is torsional stiffness [5]. The force is established when maneuvering in cornering and crossing undulating road surface [6]. Torsional stiffness should be high enough to withstand the forces to ease the handling for a race car, including Formula Student car [7]. The effect of torsional deformation is on lateral load transfer distribution between the rear and front axle [8]. During a turn a rigid chassis will cause the rear and front roll angle identical. Allowing torsional deformation in the chassis will redistribute some amount of weight transfer between the rear and front tires [9].

Torsional stiffness can be obtained by calculating the moment given to chassis equivalent to the deformation in the front arm's chassis. The moment given can be obtained using Eq. 1 [10] which multiple the force given by the distance from the center of chassis. The deformation of chassis in distance unit is calculated to find the deformation in angle unit using Eq. 2 [10]. Then the torsional stiffness can be found using Eq. 3 [10] by dividing the moment given by the deformation. The equation symbol is explained in Fig. 1.

$$T = m \times g \times L = (F_1 + F_2)w \quad (1)$$

$$\theta = \tan^{-1} \frac{a - b}{L} \quad (2)$$

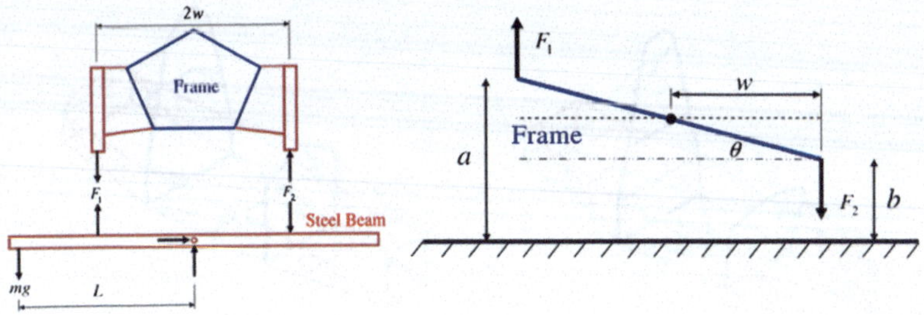

Fig. 1 Equation model [10]

$$K = \frac{T}{\theta} \tag{3}$$

2 Chassis Design

The basic purpose of chassis design is to ensure the loads will be transferred via nodes, so there will no bending forces in chassis [11]. To achieve good chassis design, the design has to combine effective and efficient chassis member triangulation [12, 13]. This article discusses about 3 chassis generation of Formula Student car to analyze the torsional stiffness. The design of the chassis from first generation to third generation if shown in Fig. 2. As shown in Fig. 2, those chassis use space frame with different amount of member and dimension but with the same weld joint method and material. The differences between the first (a) and second (b) generation are little higher from 1265 to 1085 mm, reduced frame member from 62 to 55 pieces, the thickness of some members are also different as shown in Fig. 2, and the structural form of the rear section is simpler. Then, the differences between the second (b) to the third (c) generation are the third generation higher than the second from 1085 to 1200 mm, its reason is to set the driver in upright position to reduce the number of frame member. The members of the frame reduce in the all section from 55 to 50 pieces, and overall simpler form and it has shorter dimension for reaching a lower moment [14, 15]. The difference between three models are the width to minimize moment load, overall member to minimize weight, and the structure to optimize strength as shown in Table 1.

Solidworks is used to design the chassis, this software gives support for student access using some features design such as 3D sketch, weldment, and structural member. Then, ANSYS is used to simulate the chassis because of the customization of mesh, results, and display.

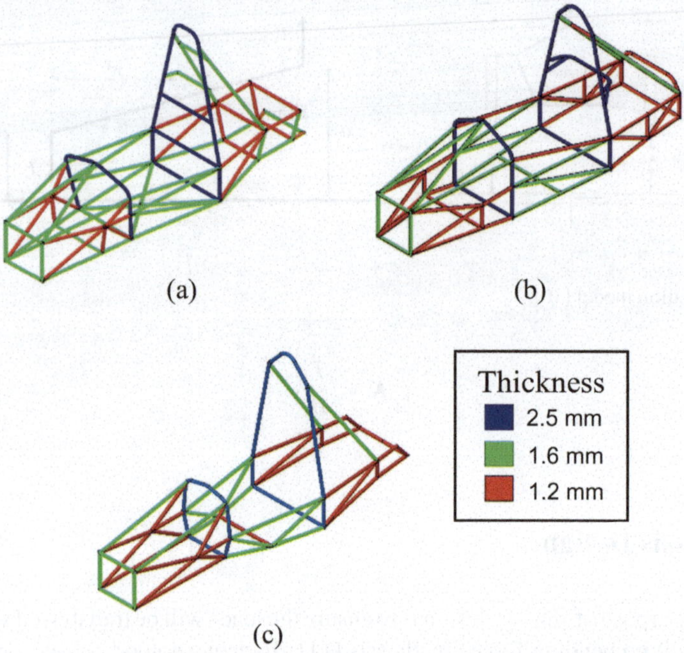

Fig. 2 Chassis design; **a** 1st generation; **b** 2nd generation; **c** 3rd generation

Table 1 Three generation chassis difference

Chassis	Width (mm)	Weight (kg)	Number of overall members
First generation	490	30.8	62
Second generation	485	29.8	55
Third generation	320	23.2	50

3 Material

There were many popular materials in Formula Student such as Chromium-Molybdenum steel (Chromoly), SAE AISI 1018, and STKM 11A [3]. Plain carbon steel and AS 4130, commonly known as chrome-moly because of its made from chromium and molybdenum content [12]. STKM 11A was used for the chassis material because of some reason; the properties regulation and the density. Firstly, STKM 11A comply to the regulation for tube strength. And then secondly, STKM11 A is lighter than chromoly as we can see at Table 2.

Table 2 Material physical properties [16]

Parameter	Chromoly 4130 Steel	STKM 11A
ρ, Density (kg m^{-3})	78,000	77,000
E, Young's modulus (Gpa)	210	190
$\sigma_{y,}$ Yield strength (Mpa)	480	290

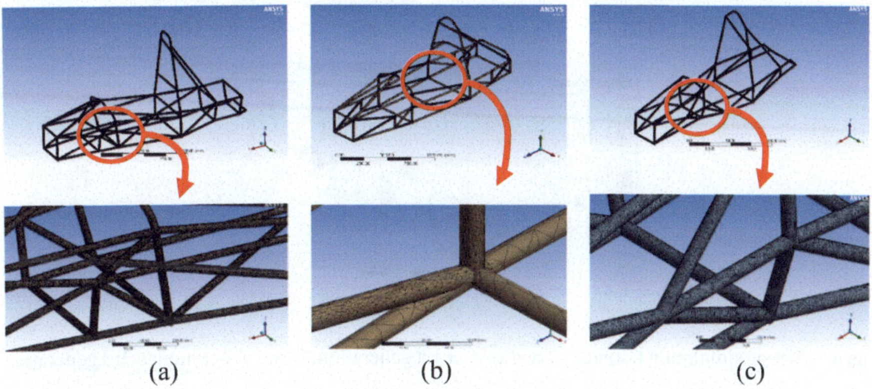

Fig. 3 Chassis simulation mesh; **a** 1st generation; **b** 2nd generation; **c** 3rd generation

4 Simulation Setup

4.1 Meshing

All three models were simulated using the same meshing parameter which are standard mechanical triangle shape with 0.272 transition ratio, 5 maximum layer, and 1.2 growth rate. As shown in Fig. 3, the meshing size and transition is nearly the same size. The first generation (a) has 1,698,650 elements, the second generation (b) has 103,698 elements and the third generation (c) has 1,157,162 elements.

4.2 Boundary Condition

The boundary condition to analyze the chassis in twisted moment are placing fixed support at the rear arms mount, and then the forces in the front arms mount were placed in opposite direction at the right and left arms resulting in twisted moment [17]. The boundary conditions are based on the real condition when the vehicle was maneuvering inn the corner and when the vehicle crossing uneven road [17]. the forces given was 750N at each side as the maximum forces from the vehicle's weight

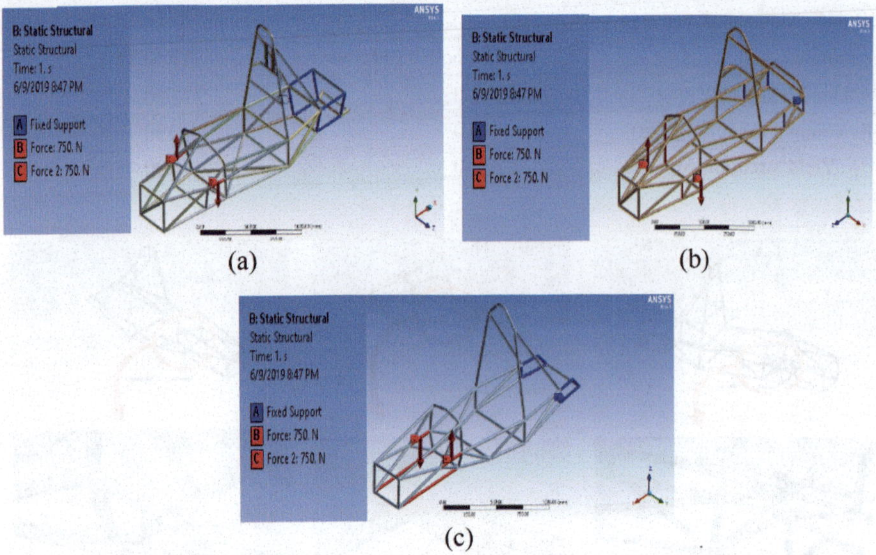

Fig. 4 Chassis simulation boundary condition; **a** 1st generation; **b** 2nd generation; **c** 3rd generation

(300 kg) divided by four wheels. The fixed supports were placed in the rear arms as a twisted point [17].

As shown in Fig. 4, the boundary condition to do the simulation was to make the chassis loaded with torsional load, this load can be made by using two forces at each front arm with opposite direction to make the moment about its longitudinal axis and then the rear arms were fixed to make the chassis act as a cantilever with one end fixed and the other end free [18]. This boundary condition simulates the condition when the chassis take the highest possible torsional loads. The forces magnitude was the maximum vehicle total mass divided by four wheels. The maximum vehicle total mass was found 300 kg, so the forces magnitude was 750N at each.

The first generation (a), second generation (b), and third generation (c) has the same front arms position, so the load point will be at the same location. The differences are the number of chassis member that support the arms. The fixed point also same because the rear arms are placed at the same location between 3 generation.

5 Results and Discussion

The torsional load magnitude depends on the dimension of the chassis. As shown in Eq. 1, the torsional load magnitude is different at each chassis when the length of the load from the centerline (L) is different with the same forces (F) applied. The calculation of torsional load magnitude is shown in Table 3.

Table 3 Chassis deformation at arms calculation

Chassis	Chassis width (mm)	Torsional load (Nm)
First generation	490	735
Second generation	485	727.5
Third generation	320	480

$$T = m \times g \times L = (F_1 + F_2)w \tag{1}$$

The simulation shows 2 different parameters to define the chassis torsional stiffness which are; total deformation and safety factor as shown in Fig. 5, the results then calculated using Eq. 3 so that the chassis torsional stiffness value will be found.

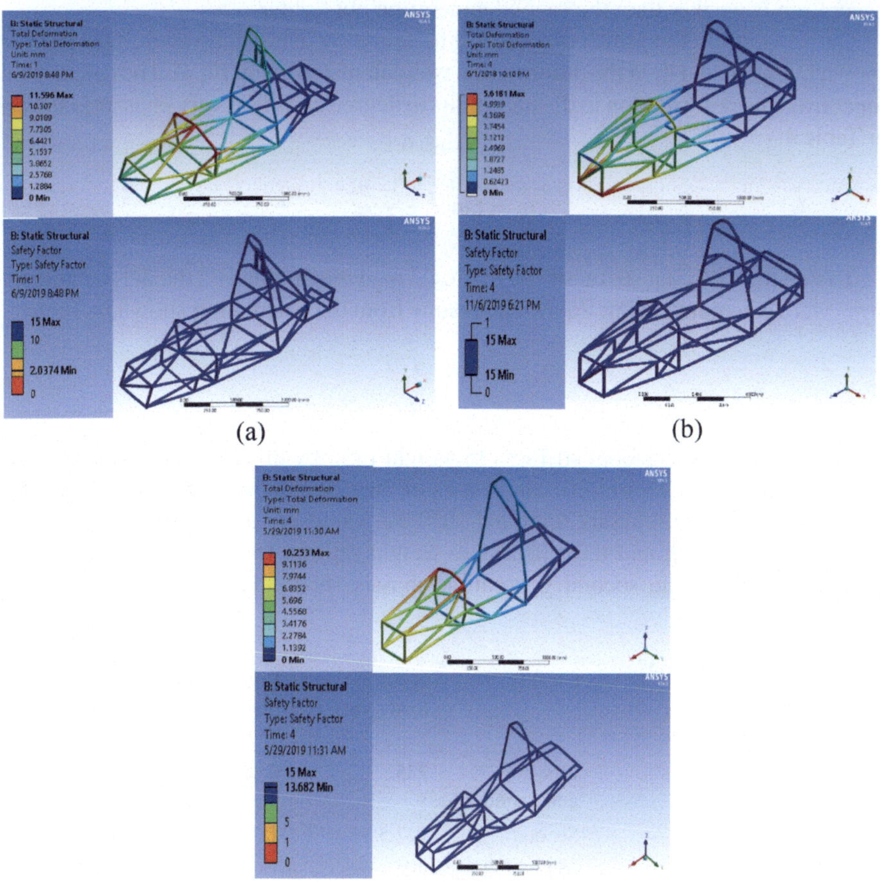

Fig. 5 Chassis simulation results; **a** 1st generation; **b** 2nd generation; **c** 3rd generation

Table 4 Chassis deformation at arms calculation

Chassis	Deformation (mm)	Deformation (deg)
First generation	6.6	0.82
Second generation	4.3	1.01
Third generation	2.3	1.55

The first generation (a) shows total deformation 10 mm and safety factor 2 with 32.8 kg of weight. The results show that the chassis was rigid enough but the weight was too heavy. And then the second generation (b) with 29.8 kg of weight shows the results 5.6 mm total deformation and 12.9 safety factor. The second-generation chassis is stronger and lighter than the first one but the safety factor is too big indicating lack of effectiveness in weight. Lastly, the third-generation chassis shows total deformation 10.2 mm and safety factor 13.7 with 23.2 kg of weight. This generation is weaker than the second generation but lighter.

Deformation in the arms position will be found by using deformation legend bar shown in Fig. 5, and then the deformation was calculated using Eq. 2. The calculation for converting deformation in distance units to degree units for every chassis is shown in Table 4.

$$\theta = \tan^{-1}\frac{a - b}{L} \tag{2}$$

To find the torsional stiffness, the torsional load magnitude and the deformation at arms were calculated using Eq. 3. The results from the calculation shown in Table 5.

$$K = \frac{T}{\theta} \tag{3}$$

Finding chassis torsional stiffness to weight ratio by dividing torsional stiffness divided to the chassis's weight. The results of the calculation shown in Table 6.

After calculation, chassis torsional stiffness value for the first generation was found 473.9 Nm/deg and the stiffness to weight ratio is 15.4 Nm/deg kg and then he calculation for the second generation chassis shows the torsional stiffness is

Table 5 Chassis torsional stiffness calculation

Chassis	Torsional load (Nm)	Deformation (deg)	Torsional stiffness (Nm/deg)
First generation	735	0.82	473.9
Second generation	727.5	1.01	716
Third generation	480	1.55	582.7

Table 6 Chassis torsional stiffness to weight ratio

Chassis	Torsional stiffness (Nm/deg)	Chassis weight (kg)	Torsional stiffness to weight ratio (Nm/deg kg)
First generation	473.9	30.8	15.4
Second generation	716	29.8	24
Third generation	582.7	23.2	25.1

716 Nm/deg, and the stiffness to weight ratio is 24 Nm/deg kg. The torsional stiffness was increased by almost 2 times from the first generation. This is because in the second generation chassis, the chassis member that create moment has been shortened causes the moment to be smaller with the same forces applied. Also, the frame member in the second generation was positioned upright unlike the first generation. The chassis inclination will affect the force distribution to the other chassis member [5, 7].

Lastly, the third generation chassis torsional stiffness is 582.8 Nm/deg and the stiffness to weight ratio is 25.1 Nm/deg kg. Torsional stiffness was reduced from the second generation but the safety factor shows that the chassis is safe. The chassis weight is reduced by 6 kg resulting higher torsional stiffness to weight ratio. Comparing to Oshinibosi [19], his 14th generation chassis has 330 Nm/deg chassis stiffness with 11.11 Nm/deg kg torsional stiffness to weight ratio. Also, compared to Singh [18], his old chassis have 250 Nm/deg torsional stiffness to weight ratio, with 6.94 Nm/deg kg torsional stiffness to weight ratio, also his new chassis have 615.98 Nm/deg with 19.2 Nm/deg kg torsional stiffness to weight ratio.

6 Conclusion

A space tube frame of FSAE car has been investigated its torsional stiffness through finite element simulation. From the results of simulation and calculation, it can be concluded that the chassis have been improved at every generation. The increasing between torsional stiffness to weight ratio from first generation to second generation is 9.4 Nm/deg kg and then from second generation to third generation, the torsional to weight ratio is increased by 1.1 Nm/deg kg. Improving torsional stiffness to weight ratio of the chassis can be obtained by removing unnecessary frame member and placing frame member in the critical area. The space tube frame can further be considered as stiff enough. Based on the simulation, the structure can be recommended for further prototyping in full scale. After fabrication, the future opportunity is verification of stiffness through experimental work using torsional stiffness test apparatus.

Acknowledgements Authors greatest appreciation goes to KEMENRISTEKDIKTI for providing *Program Kreativitas Mahasiswa* and Universitas Sebelas Maret as funding resources.

References

1. SAE-International (2018) Formula SAE Rules 2019, 2.1. Society of Automotive Engineers International, USA
2. Mahesh GG, Lavanya D, Ajay V, Yuvaraj C (2014) Design and analysis of a single seater race car chassis frame 2(8):12–23
3. Das A (2013) Design of student formula race car chassis. Int J Sci Res 14(4):2319–7064. ISSN (Online Index Copernicus Value Impact Factor)
4. Salter JI (2011) Design, analysis and manufacture of 2011 REV Formula SAE vehicle chassis. School of Mechanical and Chemical Engineering The University of Western Australia
5. Leptos M (2005) Design and manufacture of formula SAE chassis and body
6. Gaffney III E, Salinas A (1997) Introduction to Formula SAE suspension and frame design
7. Riley WB, George AR (2002) Design, analysis and testing of a Formula SAE car chassis. SAE Tech Pap (724)
8. Siegler BP, Butler L, Deakin AJ, Barton DC (1999) The application of finite element analysis to composite racing car chassis design. Sports Eng 245–252
9. Danielsson O, Cocana AG (2015) Influence of body stiffness on vehicle dynamics characteristics in passenger cars. CRC Press
10. Limwathanagura T, Sithananun C, Limchamroon T, Singhanart T (2012) The frame analysis and testing for student formula. Int J Mech Mechatronics Eng 6(5):998–1002
11. Milliken DL, Milliken WF (1994) Race car vehicle dynamics. Warrendale, PA
12. Charubhun S, Rodkwan W (2004) Design of the space frame racing car front clip and rear clip for torsional rigidity. The 18th Conference of the Mechanical Engineering Network of Thailand
13. Soo AM (2008) Manufacturing, and verification of a steel tube spaceframe chassis for Formula SAE
14. Bin Pokaad AZ, Hudha K, Nasir MZBM, Ubaidillah (2011) Simulation and experimental studies on the behaviour of a magneto rheological damper under impact loading. Int J Struct Eng 2(2):164–187
15. Ahmad Khairi MH, Mazlan SA, Ubaidillah, Ku Ahmad K, Choi S-B, Abdul Aziz SA (2017) The field-dependent complex modulus of magneto rheological elastomers consisting of sucrose acetate isobutyrate ester. J Intell Mater Syst Struct 28(14):1993–2004
16. ASM Handbook Committee (1990) ASM handbook: properties and selection: irons, steels, and high-performance alloys, 1st ed. ASM International, Ohio
17. Velie HD (2015) Chassis torsional rigidity analysis for a Formula SAE racecar
18. Singh RP (2010) Structural performance analysis of formula sae car. J Mek 31:46–61
19. Oshinibosi A (2012) Chassis and impact attenuator design for formula student race car

Mapping of Circulating Rate to Determine Non-mechanic Valve Operation in Dual Fluidized Bed Gasifier Cold Flow Model

N. Aklis, T. A. Rohmat, and H. Saptoadi

Abstract The number of vehicles increases every year. It causes the energy supply in the transportation sector to rises. Currently, oil is the main energy of the transportation sector. On the contrary with demands, the production of petroleum decreases. To prevent the energy security problem renewable energy should be promoted. One of the alternative resources that can substitute for oil is biomass. Biomass can be converted into synthetic liquid fuels by the Fischer-Tropsch process by first being converted into high-quality syngas. Dual Fluidized Bed Gasifier (DFBG) is one of promoting technology to convert biomass into a high-quality gas. To ensure the DFBG work properly, a description of characteristics of circulating material is needed. One important factor that affects the circulating material is parameter operation of the non-mechanic valve. The aim of this paper is to study characteristics of the circulating rate of the particle with variations of gas supply in the non-mechanic valve. In the present study DFBG with the configuration of the riser as a combustor and bubbling fluidized bed as a gasifier has been constructed. To connect the reactor the L-valve and the loop-seal were used as the lower connection and the upper connection respectively. The influence of gas supply velocity in L-valve and loop-seal on the solid circulating rate was inspected. The result shows that inventory or high of bed and L-valve has a significant effect on the circulating rate of particle. The rate of particle changed when passing through the riser, cyclone, and loop-seal. The rate decreased when passing the cyclone and loop-seal so if the rate of the particle in the inlet of loop-seal is higher than in the outlet the particle will be stuck in loop-seal and the installation cannot be work properly.

Keywords DFBG · Circulating rate · L-valve · Loop-seal

N. Aklis (✉)
Student of Doctoral Program of Mechanical and Industrial Engineering Department, Universitas Gadjah Mada, Jl. Grafika no. 2, Yogyakarta, Indonesia
e-mail: nur.aklis@ums.ac.id

Mechanical Engineering Department, Universitas Muhammadiyah Surakarta, Jl. a. Yani Tromol Pos 1 Pabelan, Surakarta, Indonesia

T. A. Rohmat · H. Saptoadi
Mechanical and Industrial Engineering Department, Universitas Gadjah Mada, Jl. Grafika no. 2, Yogyakarta, Indonesia

U. Sabino et al. (eds.), *Proceedings of the 6th International Conference and Exhibition on Sustainable Energy and Advanced Materials*, Lecture Notes in Mechanical Engineering,
https://doi.org/10.1007/978-981-15-4481-1_71

1 Introduction

According to the Central Bureau of Statistics of Indonesia (BPS), from 2012 to 2017, the number of vehicles in Indonesia has increased annually by an average of 6.37% [1]. In contrast, the production of oil fuel as the primary energy for the vehicle in the same period tended to drop. The imbalance between domestic demand and production of fuel oil causes greater dependence on imports, which in turn will affect energy security. Regarding the environmental issue of using fossil energy, renewable energy can be promoted to replace fossil fuel. As an agricultural country, Indonesia has abundant biomass resources. Merriam-Webster defines biomass as waste from plants and animals used as a source of energy [2]. Several methods are converting solid biomass into liquid fuels, namely fast pyrolysis, direct liquefaction, transesterification, bio-oil from Algae and Fischer Tropsch (F-T). Compared to others, F-T is the most suitable for processing automotive synthetic liquid fuels. It is caused by the F-T can produces low emissions fuel [3]. The F-T is a series of chemical reaction converts carbon monoxide and hydrogen into liquid hydrocarbon fuels. Figure 1 shows the process of converting biomass into liquid fuel through F-T. It can be seen that the gasification stage is an essential step in BTL. The F-T uses gasification product as a raw material. Gasification is a thermo-chemical conversion of solid fuels such as; coal, petroleum, coke, plastic, biomass, and waste to valuable gases through partial oxidation at high temperatures using gasification agents. The F-T process requires high-quality syngas. Gasification is the oldest biomass thermochemical conversion technology. It changes the chemical structure of biomass at a temperature of 500–900 °C using gasification agents (e.g. water, oxygen, CO_2, steam or a mixture of these components) [4].

Currently, Dual Fluidized Bed Gasifier (DFBG) has been developed to produce high-quality gas. Figure 2 explains the working principle of dual fluidized bed gasifier. It can be seen that there are two regions in DFBG: the gasification and the combustion area. Char and bed material is circulated to the combustion reactor and burned to generate heat and flue gas. The heated bed materials flow to the gasification reactor and supply the heat required in the endotherm process. DFBG enables producing hydrogen rich-gas with high heating value because using the steam as an agent and is an external heat supply obtained from the residual charcoal gasification process. The reactor dual fluidized bed gasifier can be arranged in the different

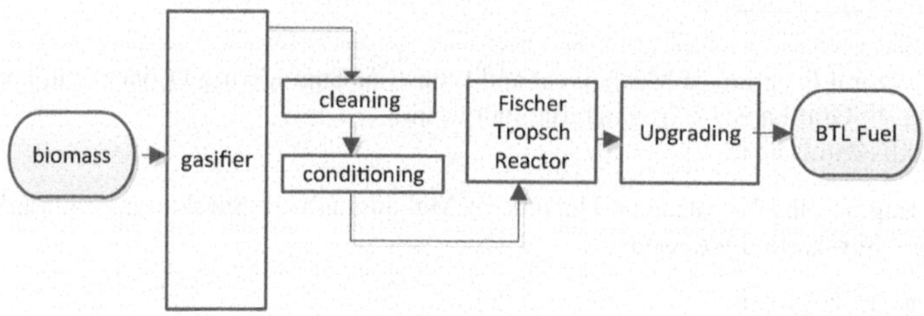

Fig. 1 Stages of biomass conversion into biomass to liquid fuels [5]

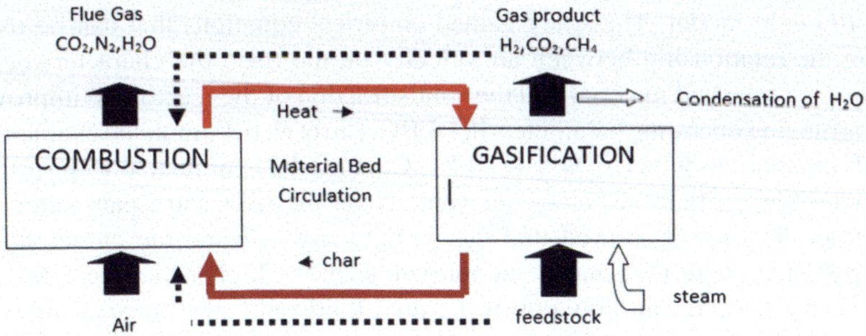

Fig. 2 Working principle of dual fluidized bed gasifier [7]

regime as follow; (a) Both reactors work on the fluidized bed bubbling regime, (b) Both reactors are risers that work on circulating fluidized bed regimes, (c) Internal circulating fluidized bed, (d) circulating fluidized bed (combustion) with a fluidized bed bubbling (gasification) and (e) Riser (gasification) with fluidized bed bubbling (combustion). The combination circulating fluidized bed as a combustion reactor and gasification with fluidized bed bubbling gasification is widely used due to longer residence time solid fuel in gasifier reactor and higher conversion of solid fuel [6].

To ensure the material circulates properly, DFBG uses a connecting system which usually consists of a combination of a standpipe and non-mechanical valves such as loop seal and L-valve. The non-mechanical valve must be able to control the circulation of material bed and provide a gas seal. Circulation of bed material and char is an essential parameter in DFBG operations. If the amount of char leading to the combustor is too less the temperature of the system will be low, and heat is insufficient to supply the endothermic process, conversely if too many, char will produce excessive heat and will reduce gasification efficiency [8].

Kamarkar and Datta [9] investigated the hydrodynamic characteristics of DFBG cold models to find the effect of primary and secondary velocity on riser and particle size on the solids circulation rate. The study was conducted on the experiment using the L-valve as the connecting reactor. The experimental results are compared with the results of mathematical modelling. The results showed that the aeration velocity and superficial velocity affects the solids circulation rate. The solids circulation also is affected by particle size where the smaller particle size causes the circulation rate to rise. Shrestha et al. [10] conducted a study of solids circulation rate in a dual fluidized bed cold model using a loop-seal at the top and L-valve at the bottom. The study focused on the effect of air velocity on the riser and the aeration air velocity on the L-valve on the material circulation rate. The results show that the air velocity of the riser affects the velocity of the material circulation where the riser air velocity increases will increase the rate of circulation of solids, but at a certain point, the rate of circulation of solids will decrease after reaching the maximum point. The same result was shown in the test of the effect of aeration velocity on the L-valve. To determine the design and operating parameters of DFBG, Lim et al. [11] investigated the correlation between operating parameters and hydrodynamic characteristics on the DFBG cold model. The study gained empirical equations that can be used to estimate the relationship between air velocity on the riser, the characteristics and particle size of the bed material and the solids fraction of the reactor. To improve the way in arranging operating parameters in DFBG, Liu et al. [6] simulated characteristic of solids circulations with CFDs. The DFBG Cold Model is simulated with variations in particle size, particle inventory, air velocity on the riser and steam velocity on the gasifier. The result showed that the particle size affected the circulation rate of the particle, where the smaller the particle size, the higher the circulation rate. However, the particle can flow out of the dual fluidized bed if its sizes too small. Particle circulation rate is also influenced by inventory where the more significant the inventory will increase the rate. The simulation on the effect of air supply in the riser showed that sand would be accumulated in the loop seal and cyclone area when the air velocity is too high.

Non-mechanic valve in DFBG is one of the crucial components to ensure solid circulate properly. Several studies have been carried out by researchers, both experiments and simulations. However, the study discusses the effect of parameters on non-mechanic valves in pairs is rare. In Practice, the result of the study can be used as an alternative DFBG operating option. This paper discusses the mapping of solids circulation rates at several points in the DFBG cold-flow reactor model with variations operation of L-valve and loop-seal velocity.

2 Experiment

2.1 Description of DFBG and Test Material

The study was conducted in the DFBG cold flow model, as shown in Fig. 3. It consists of bubbling fluidized bed gasifier and riser as a combustor connected with cyclone, loop-seal in top and L-valve in the bottom. Table 1 shows the dimension of each DFBG component. In the top of the gasifier, there is the sand feeder to supply sand to the gasifier. All of the components of DFBG are made of steel.

2.2 Material and Method

Silica sand was used as bed particles and their properties are presented in Table 2.

The DFBG uses air as the fluidization fluid. Riser gets from the blower and compressors used for gasifiers, L-valves and loop seals. Gasifier and riser velocities were constant while the velocity of the L-valve and loop-seal were varying. Table 3 shows the operation parameter of the air velocity.

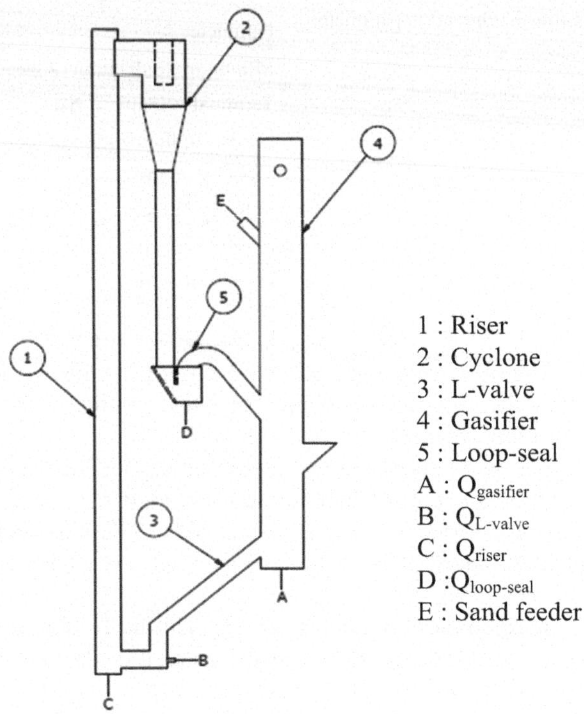

Fig. 3 Installation of DFBG cold model

Table 1 Dimension of DFBG component

Parameter	Value (mm)
Height of the riser	3000
Diameter of the riser	114
Height of gasifier	2000
Diameter of gasifier	200
Height of stand pipe	91.5
Diameter of stand pipe	76
Length of loop-seal	170
Width of upper Loop-seal	182
Width of lower loop-seal	110
Diameter of upper connection	76
Diameter of lower connection	76

Table 2 Particles properties

Parameter	Value	Unit
Size range	210–580	μm
Density	2200	kg/m^3
Bulk density	1390	kg/m^3

Table 3 Operation parameter

Parameter	Value	Unit
Minimum fluidization of gasifier	0.206	m/s
Terminal velocity of riser	4.54	m/s
Minimum fluidization of loop-seal	0.063	m/s
Velocity of gasifier	0.216	m/s
Velocity of riser	5.67	m/s
L-valve velocity	0.068–0.159	m/s
Loop-seal velocity	0.148–0.198	m/s

The minimum fluidization velocity of the gasifier and loop-seal are determined from the experiment, while the riser terminal velocity is determined from the empirical equation.

Below the riser, there is a removable grid. Measurement of sand circulation rate initially conducted when the riser and loop-seal are in iddle. The velocity at the L-valve are varied at 0.068, 0.1134, and 0.1587. The sand from L-valve is collected in the bag placed at the bottom of the riser and measured with a digital scale per ten seconds.

To keep the inventory constant, the sand feeder valve is opened. The amount of sand added is determined from the average sand rate that exits the l-valve when tested with no addition sand. The data obtained is the rate of circulation in the loop-seal. The second data retrieval is carried out for data collection of the rate of sand flowing out of the standpipe or entering the loop-seal. In this condition, the riser is in constant velocity. The last data collection is carried out the rate of flow of sand flowing out of the loop-seal or returning to the gasifier. In this step, a loop seal is connected to the gasifier and riser. The loop seal velocity used are 0.1482, 0.1647, 0.1811, 0.1893 and 0.1975 m/s.

3 Results and Discussion

3.1 Circulation Rate of Particle in L-Valve

Figures 4 and 5 show the effect L-valve velocity on the particle circulation. Figure 4 shows the particle accumulation per minutes, while Fig. 5 shows its circulation rate. It can be seen that increasing L-valve velocity caused total mass of particle increase. The results are in accordance with results from several previous studies. The data shows that the average circulation rate of particle in L-valve is 76.97 g/s, 86.95 g/s, 97.53 g/s for 0.068 m/s, 0,1134 m/s, 0.1587 m/s respectively.

The highest of rate of particle is resulted in first time (0–10). The rate will decrease and then tends to flat.

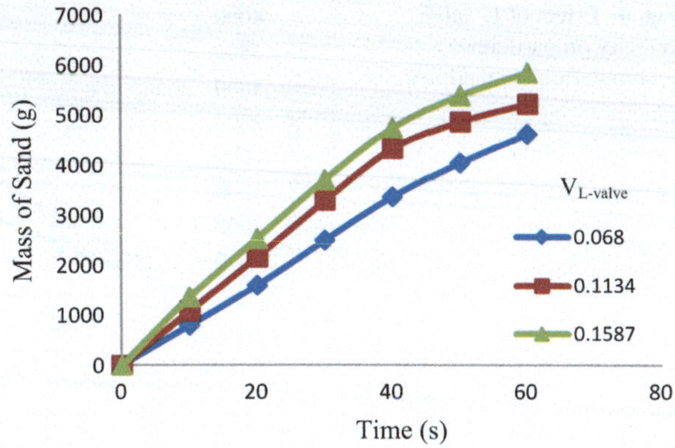

Fig. 4 Effect of L-valve velocity on particle accumulation

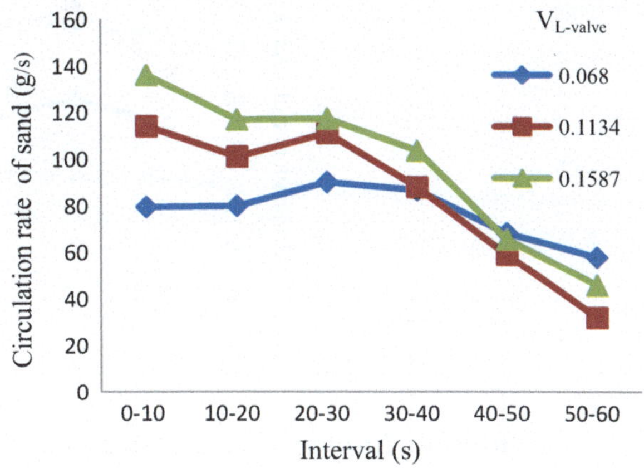

Fig. 5 Effect of L-valve velocity on particle rate

3.2 *Circulation Rate of Particle on Standpipe*

Figures 6 and 7 show the effect L-valve velocity on the particle circulation in standpipe. Figure 6 shows the particle accumulation per minutes, while Fig. 7 shows its circulation rate. The rate of the particle flow in standpipe increased when the L-valve velocity increase. The average mass rate of particle of 0.068 m/s, 0.1134 m/s, 0.1587 m/s were 35.93 g/s, 48.83 g/s, 55.70 g/s respectively. If it is compared with the particle rate flow out from the L-valve, the rate of particle in cyclone decrease 36–54%.

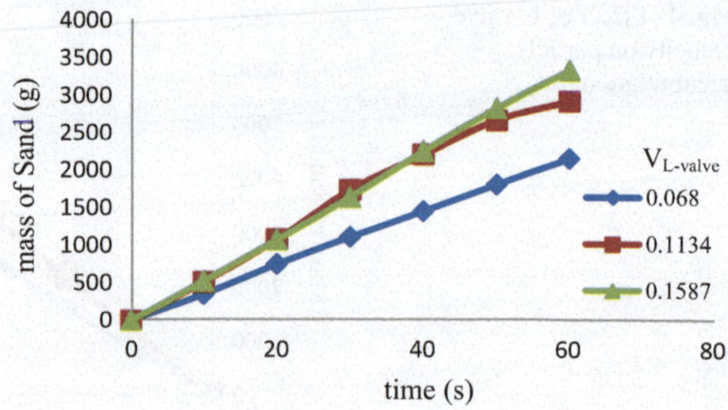

Fig. 6 Effect of L-valve velocity on particle accumulation in standpipe

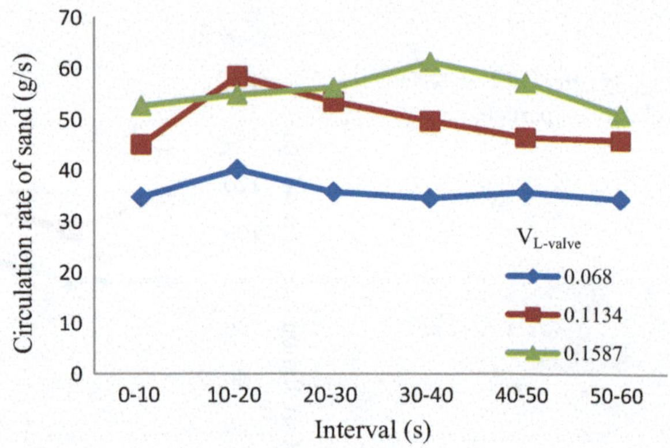

Fig. 7 Effect of L-valve velocity on particle rate in standpipe

3.3 *Circulation Rate of Particle on Loop-Seal*

Figures 8 and 9 show the effect of loop-seal velocity on the particle circulation in output of loop-seal. Figure 8 shows the particle accumulation, while Fig. 9 shows

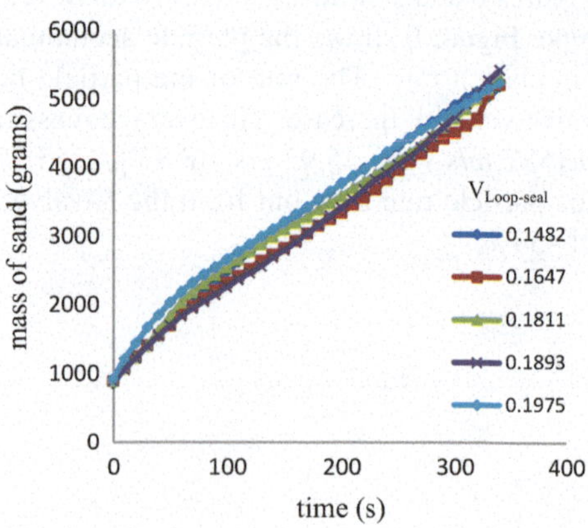

Fig. 8 Effect of loop-seal velocity on particle accumulation

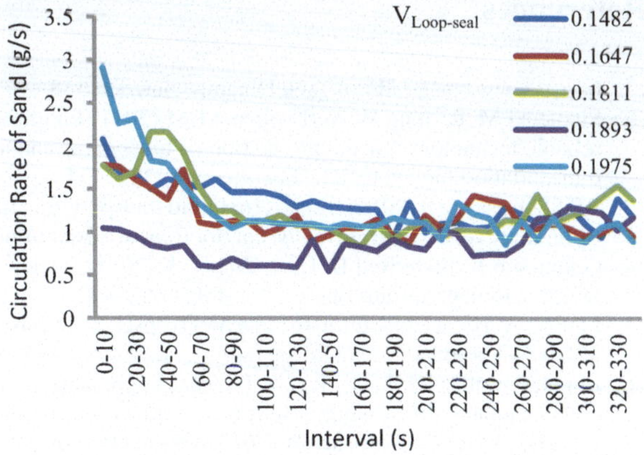

Fig. 9 Effect of loop-seal velocity on circulation rate particle

its circulation rate. The data just explain the particle rate with L-valve 0.068 m/s. In the L-valve velocity of 0.1134 and 0.1587 m/s particles accumulate in the inlate pipe before 60 s and caused stuck. This phenomenon is in line with result that explained by Liu et al. [6]. Figure 9 shows that there is no significant effect caused be the varies of loop-seal velocity. It can be take the average rate are 1.275 g/s, 1.273 g/s, 1.297 g/s, 1.334 g/s, 1.271 g/s for 0.1482 m/s, 0.1647 m/s, 0.1811 m/s, 0.1893 m/s and 0.1975 m/s respectively. Compared with the rate of particle inlet of loop-seal, the rate outlet is smaller than the rate in inlet of loop-seal.

4 Conclusion

The study of the effect of operating parameters on the recirculation rate on the dual fluidized bed gasifier cold flow model has been carried out by an experimental method. It was found that the L-valve velocity affected the rate of solids circulation where increasing velocity causing the circulation rate to rise. The rate of particles in L-valve dropped in the standpipe and dropped again at the loop-seal outlet. If the circulation rate in the standpipe is too large, the particle will be accumulated in the loop-seal, and the flow will be stuck. The increase in air velocity in the loop seal does not affect the rate of circulation at the loop seal if the particle rate at the loop-seal inlet is low.

Acknowledgements The authors would like to acknowledge the support from Lembaga Pengelola Dana Pendidikan (Educational fund management institution) LPDP Indonesia Government which has provided scholarships through Beasiswa Unggulan Dosen Indonesia Dalam (BUDI DN) 2016.

References

1. https://www.bps.go.id/linkTableDinamis/view/id/1133
2. Siedlecki M, de Jong W, Verkooijen AHM (2011) Fluidized bed gasification as a mature and reliable technology for the production of bio-syngas and applied in the production of liquid transportation fuels-a review. Energies 4(3):389–434
3. Ail SS, Dasappa S (2016) Biomass to liquid transportation fuel via Fischer Tropsch synthesis—technology review and current scenario. Renew Sustain Energy Rev 58:267–286
4. Göransson K, Söderlind U, He J, Zhang W (2011) Review of syngas production via biomass DFBGs. Renew Sustain Energy Rev 15(1):482–492
5. Lappas A, Heracleous E (2016) Fischer-Tropsch synthesis: biomass-to-liquids. In: Handbook of bio fuel production, vol 2e, pp 449
6. Liu H, Cattolica RJ, Seiser R (2017) Operating parameter effect on the solid circulation rate in CFD simulation of a dual fluidized-bed gasification system. Chem Eng Sci 169:235–245
7. Kern S, Pfeifer C, Hofbauer H (2013) Gasification of wood in a dual fluidized bed gasifier: influence of fuel feeding on process performance. Chem Eng Sci 90:284–298
8. Murakami T, Xu G, Suda T, Matsuzawa Y, Tani H, Fujimori T (2007) Some process fundamentals of biomass gasification in dual fluidized bed. Fuel 86:244–255
9. Karmakar MK, Datta AB (2010) Hydrodynamics of a dual fluidized bed gasifier. Adv Powder Technol 21(5):521–528
10. Shrestha S. Ali BS, Jan BM, Diana M, Hamid B, Sheikh KEl (2015) Hydrodynamic characteristics in cold model of dual fluidized bed gasifiers. Powder Technol 286:246–256
11. Tzeng M, Saw W, Pang S (2015) Effect of fluidizing velocity on gas bypass and solid fraction in a dual fluidized bed gasifier and a cold model. Particuology 18:58-65

Studies on Kinetics and Optimum Agitation of Phenolic Compound Extraction from Intact Red Sorghum

D. Y. Susanti, W. B. Sediawan, M. Fahrurrozi, and M. Hidayat

Abstract Quantitative analysis on the phenolic extraction from red sorghum grain for production antioxidant material in agitated vessel has been studied. The process is imagined to be controlled by the mass transfer of the functional compounds from surface of material to the bulk of solvent. Concentrations of the compound in extract are affected by agitation speed and extraction time. The purposes of this research are to evaluate the kinetics and to determine the optimum values of agitation speed and extraction time to raise the maximum performance of phenolic extraction from whole sorghum grains. The kinetic studies evaluated the accuracies of the second order kinetics model and the mechanistic model in predicting the process. The optimization was examined using two level factorial design of Response Surface Methodology based on data observed in the phenolic extraction using distilled water as solvent at 60 °C of temperature; 10:1 of sorghum-solvent mass ratio; 300, 400 and 500 rpm of agitation speed and 90, 120, 150 min of extractions time. In this case, the mechanistic model shows higher accuracy than the second order kinetics. The concentration of phenolic compound in the extract was predicted to be maximum at 407.1 rotations per minute (rpm) and 120.9 min of extraction time. The values of determination coefficient (R^2) was 94.79%. For the purpose of generalization, the kinetic model is quantitatively describe the proposed extraction process.

Keywords Kinetic · Optimum · Agitation · Phenolic · Extraction · Sorghum

1 Introduction

Antioxidant is believed to be a reactive compound which has ability to inhibit the initiation and propagation process in the chain of oxidation process caused by reactive

D. Y. Susanti · W. B. Sediawan (✉) · M. Fahrurrozi · M. Hidayat
Department of Chemical Engineering, Faculty of Engineering, Universitas Gadjah Mada, Yogyakarta 55281, Indonesia
e-mail: wbsediawan@ugm.ac.id

D. Y. Susanti
Department of Agricultural and Biosystems Engineering, Faculty of Agricultural Technology, Universitas Gadjah Mada, Yogyakarta 55281, Indonesia

U. Sabino et al. (eds.), *Proceedings of the 6th International Conference and Exhibition on Sustainable Energy and Advanced Materials*, Lecture Notes in Mechanical Engineering,
https://doi.org/10.1007/978-981-15-4481-1_72

oxygen species (ROS) and free radical species. Recently, natural antioxidants gain high preference because of their safety, affordability than synthetic antioxidants [butylated hydroxyl toluene (BHT) and butylated hydroxyl anisole (BHA)] which are considered to pose risks to health [1].

Extraction of phenolic compounds from natural resources is useful to meet the growing need of food antioxidants and herbal medicines. The ability of phenolic compounds to capture free radicals in living material systems causes their capability to prevent oxidative damage for extending the food shelf life and reducing the risk of oxidative diseases [2]. The phenolic compounds derived from secondary metabolites are potential to be processed as natural antioxidant or herbal medicine which is recognized more effective, safe and eco-friendly than the synthetic product [2–4].

As potential source of phenolic compounds, the red sorghum plant is easy to cultivate and the grain has phenolic compound 4.1–8.9 mg gallic acid equivalents/g of grain, embedded in its pericarp [5]. Sorghum grain has been proven to contain some advantageous phytochemical compounds as antioxidant include polyphenols, tannins, sterols and anthocyanins [6]. The phenolic compound is recognized for its antioxidant activity [7]. Therefore, the process development in collecting phenolic compounds from red sorghum grain became an challenging object [8].

Several studies have developed methods in extraction and separation of phenolic compounds from natural sources [9, 10]. The methods which have been developed including maceration, UAE, MAE etc. [7, 10]. The most efficient and effective process is expected can be applied in designing antioxidant production systems. Water extraction has been widely developed to replace the use of organic solvents in the extraction of antioxidant bioactive compounds in antioxidant extraction for food additives and medicinal compounds [11]. As a polar solvent with high number of hydrogen bond in its structure [12], water is nontoxic, cheap, and highly available [13]. Furthermore, water has better affinity to the solute [14] and is more efficient than nonpolar solvent [8]. Phenolic extraction using water as solvent can be an alternative method applied in every level of industry as a green technology without less negative effects on the biomaterial and environment. Temperature treatment, agitation and extraction cycles can improve water capacity in extracting the embedded materials [11].

This paper is focused on examining the effect of stirring and time on the achievement of phenolic extraction from intact red sorghum grains using water as solvent in agitated vessel. The proposed extraction which was applied for intact sorghum was expected to be a simple alternative method which is affordable, safe, efficient and effective to replace the extraction from sorghum bran and the use of organic solvent. Sorghum grains remaining after extraction is more suitable for food because in less phenolic compound, the sorghum taste would be favorable. The performance of phenolic extraction and the correlation between extraction rate, rotational agitation speed and extraction time were the main attention in this discussion [15, 16].

As commonly observed in previous researches, quantitative approaches of process has been analyzed to describe the process. The quantitative approaches consist of kinetic studies and process optimization [14, 17]. The kinetic studies compare two models, set-up by different application of mathematical techniques [9], which are second order kinetic and mechanistic model. This study is expected to suggest kinetics model which is more accurate in predicting process and in reviewing various parameters. The parameters are considered to affect the process and can be applied in industrial scale process design. Furthermore, optimization of the parameter is also expected to be useful for designing the extraction process systems. The optimization is needed to prevent excessive use of energy and maintain the integrity of grains. So, the quantitative approach is very important to get better accuracy in determining the most efficient and effective extraction conditions in designing extraction system.

The aim of this research are to evaluate the kinetics and to determine the optimum values of agitation speed and extraction time in achieving the maximum performance of phenolic extraction from whole sorghum grains.

2 Methodology

2.1 Material and Chemical

Phenolic compounds were extracted from intact red sorghum grains (*Sorghum bicolor* L. Moench) derived from Wonogiri using aquadest as solvent. The chemicals in quantitative analysis of total phenolic concentration (TPC) consisted of Folin Ciocalteu reagent, gallic acid, sodium bicarbonate were obtained from Merck Co. (Darmstadt, Germany).

2.2 Preparation and Extraction Procedure

Material preparation involved cleaning, grain drying and grain characterization were conducted to red sorghum grain before the extraction. In grain characterization, the radius of grain was measured using caliper and the initial TPC value in solid matrix were determined using the method described in Sect. 2.3. The extraction process was set at 60 °C of temperature during 150 min of extraction time. The mass ratio of solvent and grain (w/w) was 10:1. Extraction was conducted at 300, 400, and 500 rpm of agitation speed using laboratory scale vessel. The TPC value in aqueous extract was measured every 30 min of interval during the extraction by taking samples.

2.3 Determination of Total Phenolic Content (TPC) in Aqueous Extract

Total phenolic concentration (TPC) was determined by the Folin Ciocalteu method with little modification [18]. Briefly, 0.5 mL sample extract was mixed with 2.5 mL

Folin Ciocalteu reagent (0.2 N). After that, 2 mL Na_2CO_3 (7.5%) was added to each sample. After 5 min incubation at 50 °C, the absorbance were read at 760 nm using a spectrophotometer UV-VIS (Genesys 10S). The results were expressed as milligram gallic acid equivalent per mL sample (volume mg GAE/mL).

3 Kinetic Formulation

3.1 Second Order Kinetic Model

The empirical model applied in evaluation of the profile of total phenolic concentration (TPC) in aqueous extract during extraction was second order extraction, shown in Eq. 1 [11, 13, 19].

$$\frac{dC_P}{dt} = k_P \cdot (C_{Pe} - C_P)^2 \tag{1}$$

where C_P (g/cm^3) = the TPC in solvent; C_{Pe} (g/cm^3) = the equilibrium TPC in extract; and k_P (cm^3 solvent/g/min) = the constant parameter of extraction rate.

Then, the Eq. 2 was obtained from the integration of Eq. 1 under the initial condition: $C_{P0} = 0$ at $t = 0$. The determination of k_p was conducted by Eq. 3.

$$C_P = \frac{C_{Pe}^2 \cdot k_P \cdot t}{1 + C_{Pe} \cdot k_P \cdot t} \tag{2}$$

$$\frac{t}{C_P} = \frac{1}{k_P \cdot C_{Pe}^2} + \frac{t}{C_{Pe}} \tag{3}$$

From the slope and intercept of Eq. 3, the value of C_{pe} and k_p can be determined.

3.2 Mechanistic Model

The mechanistic model was built based on the following assumptions. The mass of solid matrix was assumed to be constant during the extraction. The grains was assumed to be spherical and their radius is uniform. The overall mass transfer was assumed to be controlled by mass transfer of solute from the pericarp to solvent. It was not controlled by the intra particle diffusion, since the pericarp layer was relatively thin. The material balance of solute produces Eq. 4.

$$\frac{dVC_P}{dt} = \frac{dm_s \cdot X}{dt} = N_b \cdot S \cdot k_c \cdot \left(C_P^* - C_P\right) \tag{4}$$

The equilibrium concentration of total phenolic was approximated by Eq. 5.

$$C_P^* = H \cdot X. \tag{5}$$

Since the total solute is concern, so the solute material balance of the entire system results in Eq. 6.

$$X = X_0 + \frac{V}{m_s} \cdot C_{P0} \tag{6}$$

The combination of Eqs. 4, 5 and 6 gave Eq. 7.

$$\frac{dC_P}{dt} = \frac{N_b \cdot S \cdot k_c}{V}\left[H \cdot \left(X_0 + \frac{V}{m_s} \cdot C_{P0}\right)\right] - \frac{N_b \cdot S \cdot k_c}{V}\left(\frac{H \cdot V}{m_s} + 1\right) \cdot C_P \tag{7}$$

In the above equations, N_b = the number of grains extracted in solvent; t (min) = time; k_c (cm/min) = the mass transfer coefficient; V (mL) = solvent volume; m_s (g) = mass of grains; H (g pericarp/cm^3 solvent) = coefficient of distribution. S (cm^2) = the area of pericarp surface; X (g/g solid sample) = the total phenolic content in solid matrix; X_0 (g/g solid sample) = the initial total phenolic content in solid matrix

The proposed mechanistic model in Eq. 7 was solved analytically to produce phenolic concentration (C_P) in extract as the function of time. The values of parameters involved (k_c and H) were evaluated by curve fitting. The sum of squares of error was minimized by varying the value of k_c and H.

3.3 *Optimization Using Response Surface Methodology (RSM)*

The experimental optimization of the variable involved (agitation speed and time) was also done in phenolic using a set of the Design of Experiment (DOE) in Minitab® 18 Statistical Software, which is recognized to be better, faster and more accurate in decision making. The experimental design was based on two factors. The extraction was conducted at 10 of sorghum-solvent ratio and 60 °C of temperature. The parameter involved the rotation speed and extraction time in 3 level design. The simulation was set at 400, 500 and 600 of rpm and 90, 120 and 150 min or extraction time to get the maximum value of TPC.

4 Result and Discussion

4.1 The Total Phenolic Content in Aqueous Extract During Extraction

The profile of total phenolic compound (TPC) in aqueous solvent as the accumulation of released compound from the pericarp was shown in Fig. 1.

Figure 1 showed that the TPC value in aqueous extract increases faster at the early stage of extraction and incline to be slower at end stage of extraction. This phenomena is conceivable since at the early stage, the difference of solute concentration in the solvent and the equilibrium concentration is large (the driving force of extraction rate depends on the difference). The TPC value reaches maximum at 3.818 mg/mL for 300 rpm; 3.972 mg/mL for 400 rpm; and 3.781 mg/mL for 500 rpm. Figure 1 also showed that after around 100 min the TPC value is almost constant, and then slightly decreases after 120 min. The constant values maybe due to the equilibrium condition and the slight decrease of TPC value is caused by degradation of TPC. It is recommended that the extraction can be stopped after 120 min. At that time, the equilibrium was assumed to be attained. The equilibrium value indicated the saturation capacity of the solvent.

The effects of agitation speed can be evaluated from the TPC profile and the height of maximum TPC value in each rpm. It turned out that the higher extraction rate and the higher amount of solute extracted after 120 min were achieved at higher agitation speed. The maximum TPC value was 3.972 mg/mL raised at 400 rpm of agitation speed. This phenomenon is conceivable since the higher agitation speed promotes higher turbulence, so the mass transfer rate from surface of the solid to solution would be higher. However, it can be expected that at too high agitation speed, the effect of agitation speed becomes less since at very high agitation speed, the movement of particles will be almost equal to the movement of liquid as result the slip velocity between particle and liquid becomes relatively low causing the low mass transfer rate.

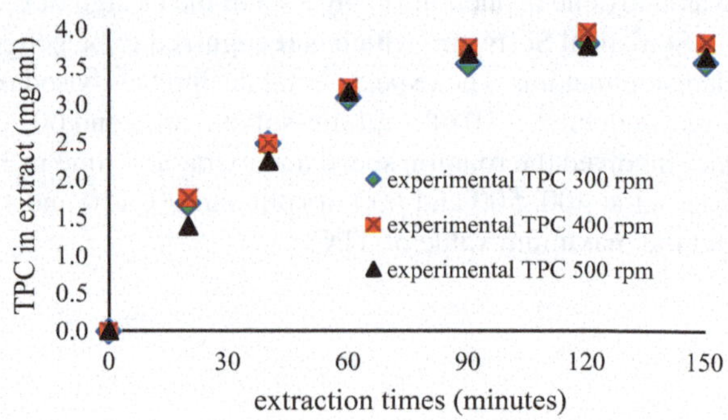

Fig. 1 The total phenolic concentration (TPC) in solvent during extraction

4.2 *The Kinetic Studies*

4.2.1 The Second Order Kinetic

It can be seen on the Fig. 1, the TPC in solvent (C_P) increases faster at the beginning and then slower, leading to an equilibrium concentration (C_{Pe}). From the slope and intercept of Eq. 7, the parameter values of second order kinetic model in phenolic extraction process obtained from the observed data were shown in Table 1.

The trend at Fig. 2 suggests that the second order kinetics is qualitatively correct. The fitness of the model will be further evaluated in more quantitative way. Figure 2 also shows that even though the R^2 of the correlation between the calculated

Table 1 The parameters value of phenolic extraction in second order kinetic model

Rpm	k_P $\left(cm^3_{solvent}\ g_P^{-1}\ min^{-1}\right)$	C_{Pe} $(g\ cm^{-3})$	C_{Pe} $(mg\ mL^{-1})$	SSE
300	15.867	4.032×10^{-3}	4.032	4.69×10^{-5}
400	13.146	4.333×10^{-3}	4.333	4.91×10^{-5}
500	10.670	4.264×10^{-3}	4.264	4.45×10^{-5}

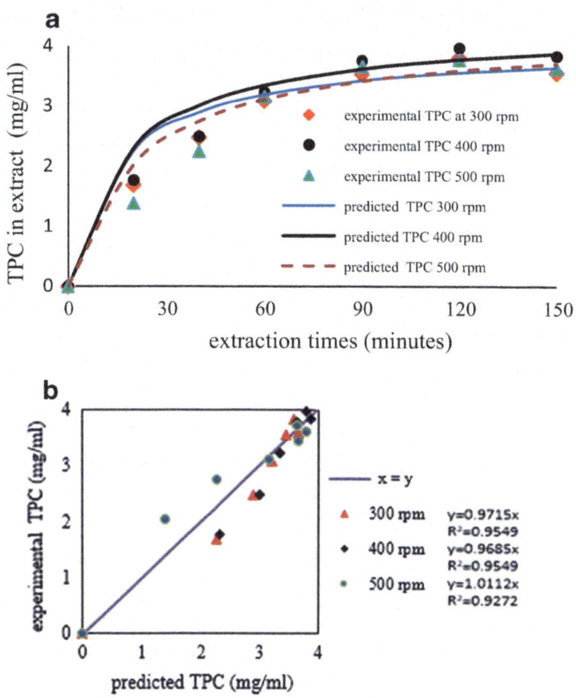

Fig. 2 **a** The predicted and experimental TPC from the second order kinetic model, **b** the correlation of the predicted (x) and the experimental (y) TPC from the second order kinetic model

Table 2 The parameters value of phenolic extraction in mechanistic model

Rpm	k_c (cm^{-1} min^{-1})	H (g pericarp cm^{-3} solvent)	SSE
300	13.049×10^{-3}	22.302×10^{-3}	3.263×10^{-08}
400	13.374×10^{-3}	22.899×10^{-3}	6.884×10^{-08}
500	11.907×10^{-3}	22.328×10^{-3}	1.416×10^{-07}

results and the experimental data is relatively good (>0.9), the visual comparison concludes that the model cannot well quantitatively describe the extraction phenomena. Significant deviations were observed especially at the early stage of extraction.

4.2.2 Mechanistic Model

The mechanistic model in this paper is built entirely from the concept of mass transfer based on the transfer mechanism of the phenolic compounds from the solid surface of pericarp to the solvent. This model is theoretically more comprehensive than second order kinetics, since this model takes into account the grain dimension, the mass of grains and the volume of solvent, so the model written will be easier to be applied for designing larger scale of extraction systems.

Equation 7 was solved at the extraction condition exactly the same as the experiment, in which 1252 red sorghum grains (35 g) were extracted in 350 mL of water. On the average, the grain has the radius of 0.19 cm; the pericarp thickness of 0.01 cm, the initial TPC of 216.7 mg/g grain. The curve fitting result the values of parameter as shown in Table 2.

The accuracy of the mechanistic model in predicting the TPC value during extraction was illustrated in Fig. 3. It can be visually observed that the mechanistic model proposed showed good agreement with the experimental data. The values of gradient and R^2 are calculated to be very close to 1, so the mechanistic model is statistically accurate. Furthermore the H values were seem to be constant at various agitation speed. This results are conceivable since the H value should not be affected by agitation speed.

The value of k_p at 400 rpm is higher than the one at 300 rpm, while the value of k_p at 500 rpm is lower than the one at 400 rpm. This phenomenon shows consistency with the related theory in which the agitation speed will increase the value of k_p at relatively slow speed, but at relatively high speed, the agitation speed decreases the value of k_p, as mentioned in the previous discussion on second order model.

Based on the mechanistic model approached, it can be concluded that the maximum value of the extraction rate expressed by the mass transfer coefficient (k_p) had been reached at around of 400 rpm. The value became the limit of the agitation speed before causing the particle movement follow the fluid motion and reducing the relative particle velocity which triggered the interfacial mass transfer.

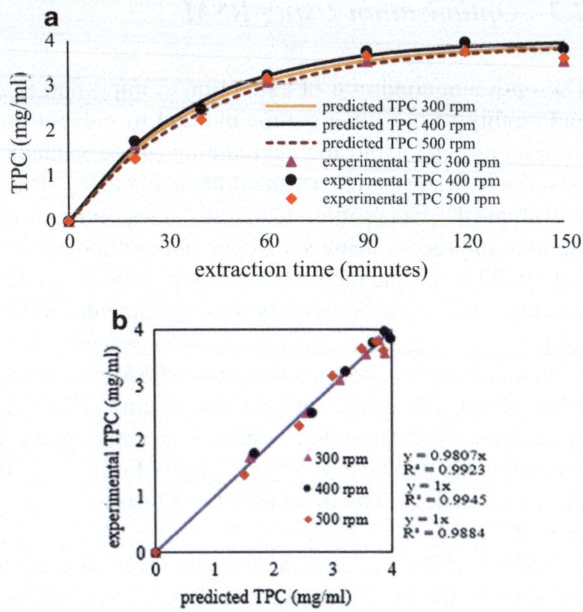

Fig. 3 **a** The predicted and experimental TPC from the mechanistic model, **b** the correlation of the predicted (x) and the experimental (y) TPC from the mechanistic model

Furthermore, from the visual comparison of the predicted value of the mechanistic model results and the second order kinetic results in Fig. 4, the mechanistic model has the better accuracy in predicting TPC in aqueous extract.

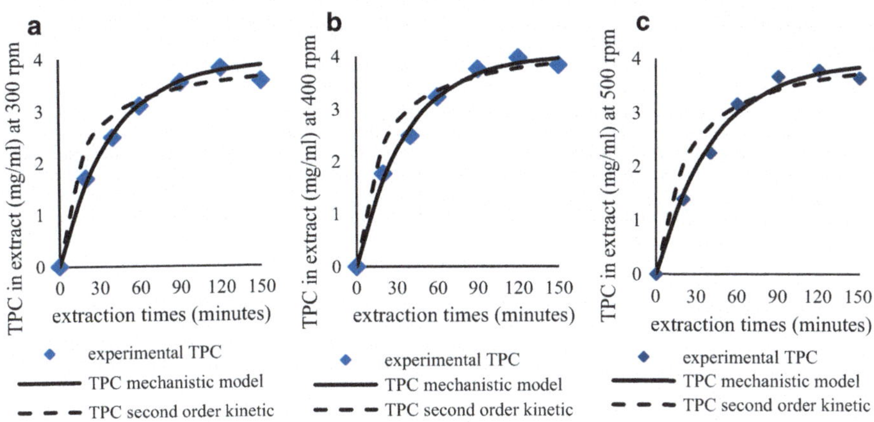

Fig. 4 The comparison between the second order kinetic model and the mechanistic model in this case

4.3 Optimization Using RSM

The optimum condition of extraction is important to reach the maximum quantity and quality of bioactive compound and to prevent the oversupply of energy, time of agitation and avoid the degradation of the valuable compound. Optimization is commonly performed experimentally using RSM. This method is commonly applied to evaluate the data profile and to find out the optimum condition of some independent variable in process using statistical and mathematical method [1]. Several research used RSM as simple method to develop, improve, determine the optimum condition and the correlation of some process parameters in limited of resources and times [20].

In this study, the optimization using RSM was applied to determine the optimum value of agitation speed and extraction time. This value is important as parameter in designing the extraction system to get maximum TPC yield. The optimization was run using two level factorial design of minitab in the limit range for optimizing shown in Table 3. The experimental data fitted in the optimization were shown in Table 4.

RSM describe the empirical model for predicting the TPC correlated with rpm and time by the Eq. 8 with the determination coefficient (R^2) of 0.9479.

Table 3 Experimental range and level of independent variable

Independent variables	Unit	Range and level		
		−1	0	−1
Agitation speed (X_1)	rpm	300	400	500
Extraction time (X_2)	min	90	120	150

Table 4 Experimental data for optimization using RSM

Experiment	Independent variables		Dependent variables (R = Response)
No.	X_1 agitation speed (rpm)	X_2 extraction time (min)	TPC (mg/mL)
1	300	90	3.54216
2	300	120	3.81825
3	300	150	3.56714
4	400	90	3.76340
5	400	120	3.97240
6	400	150	3.83770
7	500	90	3.66536
8	500	120	3.78142
9	500	150	3.63763

$$TPC = -2.422 + 0.01592X_1 + 0.0524X_2 - 0.000019X_1X_1 - 0.000209X_2X_2 - 0.000004X_1X_2 \quad (8)$$

In Eq. 8, X_1 (rpm) is the agitation speed, X_2 (min) is the extraction time and TPC (mg/mL) is the total phenolic concentration in extract.

The correlation of TPC and each variable was also analyzed using pareto chart in Fig. 5. By the large proportion described, individually the agitation speed and the extraction time has significant effect to TPC in extract, however their interaction effects are not significant.

The contour plot and the three dimensional plot of TPC (surface plot) were shown in Fig. 6.

The optimizer of RSM resulted that the best condition for gaining the maximum TPC was at 407.01 rpm and 120.9 min of extraction (Fig. 7). The condition is recognized as the best condition which is effective for extraction. This is the limit

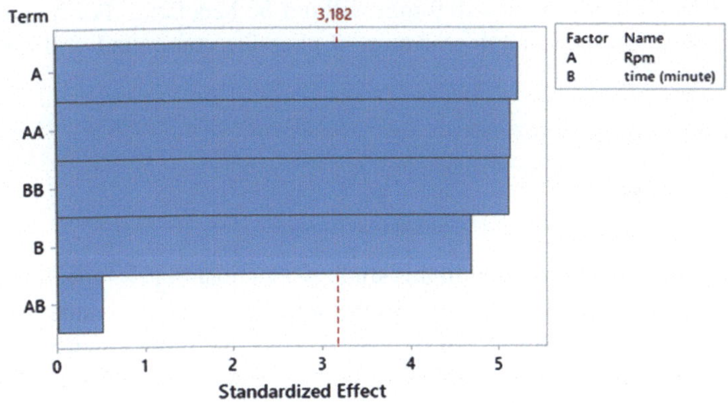

Fig. 5 Pareto chart of standardized effect of rpm and time to TPC

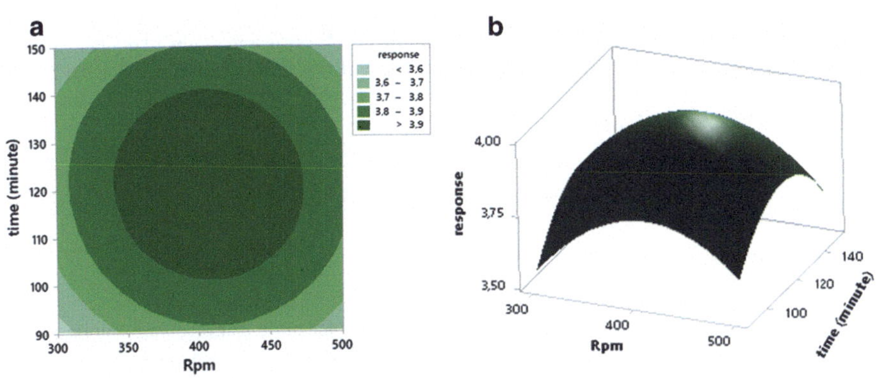

Fig. 6 The contour plot and the surface plot of TPC value

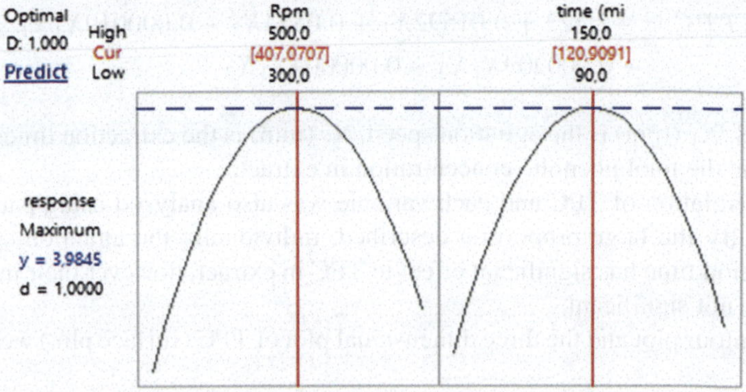

Fig. 7 The optimum condition (rpm and time) and the maximum TPC

value on the increasing of agitation speed and extraction time. It suggested also that extraction time of higher than 120.9 min will not be beneficial. For longer time, it can be logically foreseen that degradation of phenolic compound will possibly be encountered.

5 Conclusions

The mechanistic model proposed in this study works quantitatively well to describe the phenomena in phenolic extraction from intact red sorghum grains using water as the solvent in agitated vessel and is better than second order kinetics. This result showed that the mass transfer of solute from the surface of grains to the liquid solvent has significant effects on the extraction phenomena. Furthermore experimental optimization supported by RSM suggested that the optimum condition for phenolic extraction was at 407.01 rpm and 120.9 min.

Acknowledgements The authors are very grateful to "Direktorat Riset dan Pengabdian Masyarakat, Direktorat Jenderal Penguatan Riset dan Pengembangan, Kementerian Riset, Teknologi dan Pendidikan Tinggi" for research funding through "Hibah Penelitian Disertasi Doktor" with agreement number 2900/UN1.DITLIT/DIT-LIT/LT/2019.

References

1. Gani SSA, Azahar NF, Mokhtar NF (2018) Response surface optimization of high antioxidative extraction from *Curcuma zedoaria* leaves. J Adv Res Fluid Mech Therm Sci 43(1):90–103
2. Rahim NA, Zakaria N, Dzulkarnain SMH, Mohd N, Azahar ZM, Abdulla MA (2018) Antioxidant assay of alstonia angustifolia ethanolic leaf extract. J Adv Res Fluid Mech Therm Sci 1(1):80–86

3. Rahim NFAR, Norhayati M, Abdullah N, Talib BA, Jihan N, Dusuki S (2018) Optimization of the antioxidant properties of the polyherbal formulations. J Adv Res Fluid Mech Therm Sci 1(1):16–25
4. Mokhtar N, Nordin MFM, Morad NA (2018) Total phenolic content, total flavonoid content and radical scavenging activity from *Zingiber zerumbet* rhizome using subcritical water extraction. Int J Eng 31(8):1421–1429
5. Dykes L, Rooney LW, Waniska RD, Rooney WL (2005) Phenolic compounds and antioxidant activity of sorghum grains of varying genotypes. J Agric Food Chem 53(17):6813–6818
6. Liu L, Chen L, Abbasi AM, Wang Z, Li D, Shen Y (2018) Optimization of extraction of polyphenols from *Sorghum moench* using response surface methodology, and determination of their antioxidant activities. Trop J Pharm Res 17(4):619–626
7. Luo X, Cui J, Zhang H, Duan Y, Zhang D, Chen G, Cai M (2018) Ultrasound assisted extraction of polyphenolic compounds from red sorghum (*Sorghum bicolor* L.) bran and their biological activities and polyphenolic compositions. Ind Crops Prod 112(301):296–304
8. Pascal Agbangnan CD, Tachon C, Dangou J, Chrostowska A, Fouquet E, Sohounhloue DCK (2012) Optimization of the extraction of sorghum's polyphenols for industrial production by membrane processes. Res J Recent Sci 1(4):1–8
9. Yasemi M, Rahimi M, Heydarinasab A, Ardjmand M (2016) Optimization of microfluidic gallotannic acid extraction using artificial neural network and genetic algorithm. Chem Prod Process Model 12(1)
10. Rezaei S, Najafpour GD, Mohammadi M, Moghadamnia AA, Kazemi S (2016) Formic acid and microwave assisted extraction of curcumin from turmeric (*Curcuma longa* L.). Int J Eng 29(2):145–151
11. Ruslan MSH, Ganeson T, Hasan M, Idham Z, Siti HMS, Zaini MAA, Morad NA, Yunus MAC (2014) Kinetic study of catechin extracted from Areca catechu seeds using green extraction method. Asia-Pacific J Chem Eng 9(5):743–750
12. Amiri ZN, Najafpour GD, Mohammadi M, Moghadamnia AA (2019) Subcritical water extraction of bioactive compounds from ginger (*Zingiber officinale Roscoe*). Int J Eng 31(12):1991–2000
13. Kusuma HS, Mahfud M (2015) Preliminary study: kinetics of oil extraction from sandalwood by microwave-assisted hydrodistillation. ASEAN J Chem Eng 15(2):62–69
14. Sayar S, Abidin ZZ, Yunus R (2012) Optimisation of solid liquid extraction of jatropha oil using petroleum ether. Asia-Pacific J Chem Eng
15. Oediyani S, Ariyanto U, Febriana E (2019) Effect of concentration, agitation, and temperature of Pomalaa limonitic nickel ore leaching using hydrochloric acid. IOP Conf Ser Mater Sci Eng 478(1)
16. Zhang P, Cheng Q, Tang K, Qiu Y, Xu W, Jiang P, Dai G (2017) Study on kinetics of reactive extraction of propranolol enantiomers by multiple linear regression method. Asia-Pac J Chem Eng 12(4):551–560
17. Sereewatthanawut I, Lisawadi S, Prasittisopin L (2018) Mathematical modelling of molecular separation processes in aggressive solvent systems. Chem Prod Process Model 1–15
18. Torun M, Dincer C, Topuz A, Sahin–Nadeem H, Ozdemir F (2015) Aqueous extraction kinetics of soluble solids, phenolics and flavonoids from sage (Salvia fruticosa Miller) leaves. J Food Sci Technol 52(5):2797–2805
19. Harouna-Oumarou HA, Fauduet H, Porte C, Ho YS (2007) Comparison of kinetic models for the aqueous solid-liquid extraction of Tilia sapwood a continuous stirred tank reactor. Chem Eng Commun 194(4):537–552
20. Sönmez O, Gözmen B, Çevik T, Giray ES (2016) Optimization of solvent extraction process of some Turkish coals using response surface methodology and production of ash-free coal. Asia-Pacific J Chem Eng

An Overview of Interface/Interphase Modification in Functional Composites

D. F. Smaradhana, E. Surojo, and R. Alnursyah

Abstract Interphase (or termed interface) is an important part of composites placed between the fibre and the matrix. This review highlights the role of the interface, the methods to modify interface and the purpose of modifying interface. This review also shows that modifying interface of composites does not only improve the mechanical properties but also enhance the functions of the composites. Composites with abilities such as shape memory capability and in situ monitoring can be produced through interface modification. The potential applications in the future are also discussed.

Keywords Interface · Interphase · Functional composites

1 Introduction

Fibre composites do not only consist of fibre and matrix but also interphase which exists between the fibre-matrix. The interphase is a volume between the fibre and the matrix where coating and matrix are diffused into each other's domain in which the physical, mechanical and chemical properties of the interphase are different from either the bulk matrix or the fibre [1–5] while the interface in fibre reinforced polymer (FRP) composites is a boundary separating the fibre, the interphase and the matrix. According to Kim and Mai [4], the definition of the interphase can be used for the interface as well since both cannot be separable. Figure 1 represents a model of the interphase region adapted from Jones [3].

The interface has a major role in affecting the composite properties since the load will be transferred from the matrix to fibres through the interface. The simultaneous interaction of chemical and physical force at the interface combined with the interphase region produces the level of adhesion, called interfacial bond, between the fibre and the matrix [3, 4, 6]. Understanding the level of adhesion between those two constituents is important since the interfacial bond can influence various aspects of composites behaviour such as toughness, strength, interlaminar shear strength and failure modes [7, 8].

D. F. Smaradhana · E. Surojo (✉) · R. Alnursyah
Mechanical Engineering Department, Universitas Sebelas Maret, Surakarta, Indonesia
e-mail: esurojo@ft.uns.ac.id

U. Sabino et al. (eds.), *Proceedings of the 6th International Conference and Exhibition on Sustainable Energy and Advanced Materials*, Lecture Notes in Mechanical Engineering,
https://doi.org/10.1007/978-981-15-4481-1_73

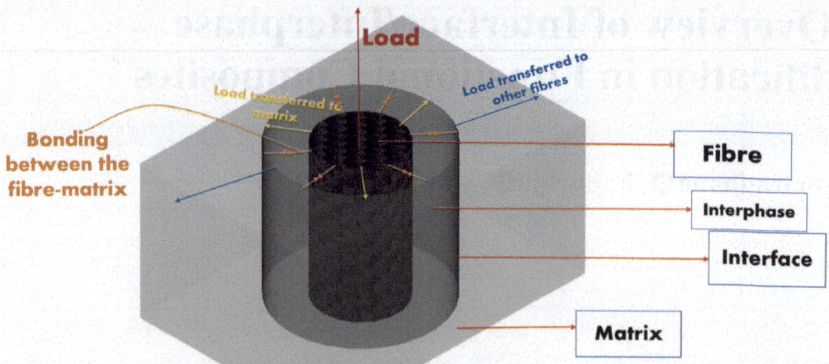

Fig. 1 The schematic of the fibre, the interface, the interphase and the matrix in FRP composite

Modifying interface of composites can be done, not only by giving surface treatments to the fibre before impregnated to the resin [7–9] (see Fig. 2) but also by inserting the interleaf film in between the adjacent plies in laminated composites [10–12], as shown in Fig. 3. Initially, modification of interface by giving polymer sizings is conducted to protect critical surface flaws on fibres as depicted in Fig. 4. Furthermore, between the 1980s and 1990s, the most common works were conducted regarding modified interface to improve the toughness of composites [13–15] as composites are very brittle. However, the more recent works have shown that controlling interface of composites is not only to enhance its mechanical properties, but also to create more functional materials, for example, the application in stiffness control in morphing skin and structural health monitoring (SHM) by multiscale composites.

Therefore, the objective of this review is to investigate the range of roles and the state of the art of the controlled interface in composites that have been published to examine the potential applications in producing functional composites including shape memory composites and structural health monitoring.

2 Modified Interface in Shape Memory Composites

Morphing skin is used to cover morphing structure which needs high flexibility materials that can change its shape at least in of two ways on demand [17] which is usually called shape memory materials. Furthermore, the demand for this shape memory materials in the aerospace application is the reason to develop shape memory composites having stiffness control [18]. Shape memory composites can be produced through modification of composites' interface which can be done by two methods: giving thermoplastic coating on fibre surface and inserting thermoplastic interleaf film in between the adjacent plies.

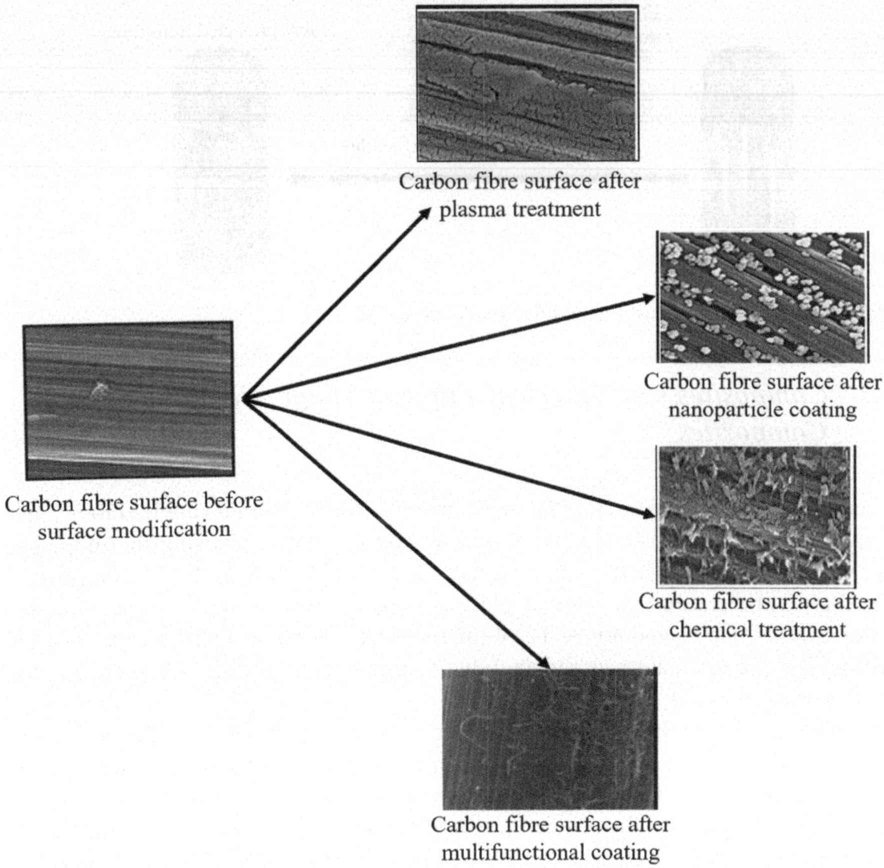

Fig. 2 Surface modification techniques obtained from Bijwe et al. [16] reproduced with the permission of Wiley

Fig. 3 Composites with interleaf film

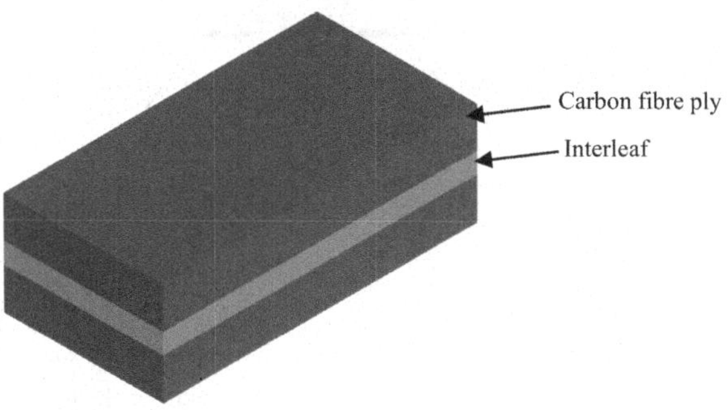

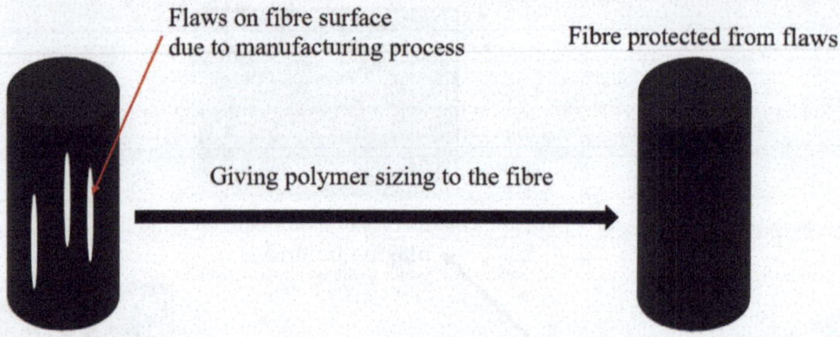

Fig. 4 Initial purpose of surface modification on fibre

2.1 Composites with Interleaf Films in Shape Memory Composites

According to Robinson et al. [19], thermoplastic interleaf can be applied to achieve composites with controllable stiffness. The concept is by choosing thermoplastic polymer which has glass transition temperature (Tg) lower than that of composites. The thermoplastic interleaf was then placed in between the adjacent plies in laminated composites. Once heated above Tg of the interleaf but below Tg of composites, the stiffness of the laminate was reduced due to the laminae sliding relatively to each other as shown in Fig. 5.

Maples et al. [20] inserted thermoplastic polymer—polystyrene—between adjacent plies of carbon fibre reinforced polymer (CFRP) which resulted a high reduction of flexural stiffness (98% reduction) once heated at 120 °C. This was reversible as the laminate could return to the original position without damage when heated at 120 °C without applied load. Raither et al. [21] used elastomer interleaf layers attached in

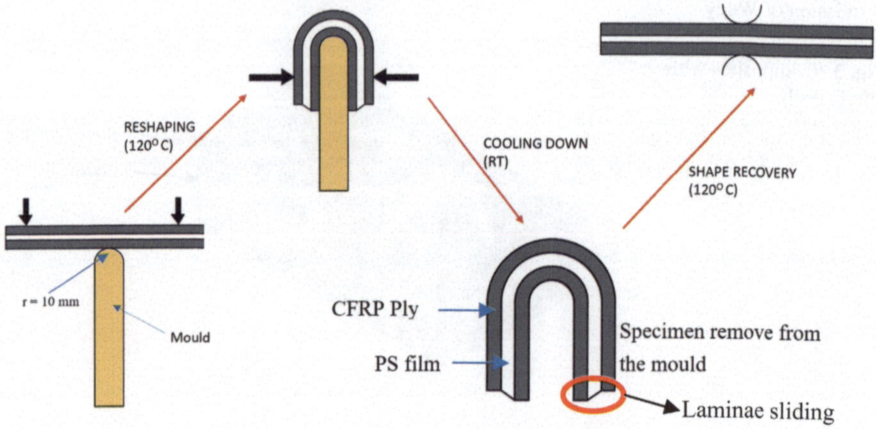

Fig. 5 Shape memory composites concept using interleaf film

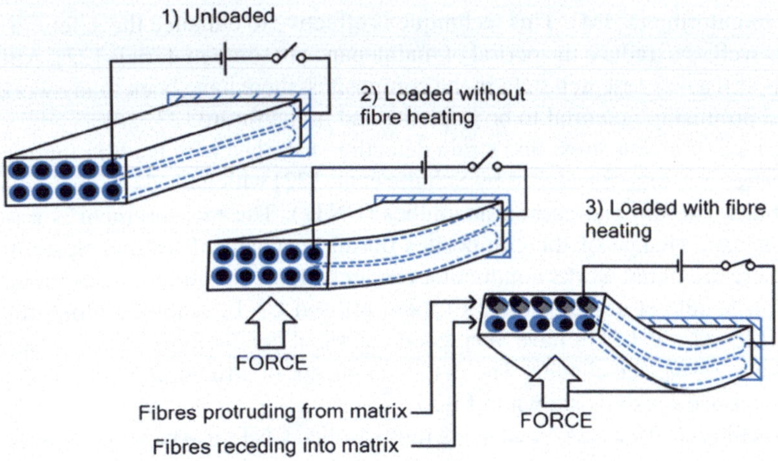

Fig. 6 Shape memory concept using fibre coating method [18] reproduced with the permission of ACS

multidirectional CRFP laminate which indicated a reduction of stiffness by a factor 10 once heated above elastomer Tg. From the results, it can be concluded that the use of interleaf layer to control stiffness of the composites is successful.

2.2 *Fibre Coating in Shape Memory Composites*

According to Tridech et al. [18], a thin polyacrylamide (PAAm) was electrocoated on the surface of carbon fibres before impregnating to epoxy resin. The shape memory concept in this method is almost similar to interleaved composites by heating the composites above Tg of PAAm (but below Tg of epoxy) to allow the fibres to slide within the matrix. Current was used in this work to heat the specimen. 88% reduction in flexural stiffness was obtained in this research, and the CFRP was fully recovered once cooled without any damage and loss in properties. Though the stiffness reduction was not as high as that of the composites with interleaf film, the future application of this composite is promising. Figure 6 shows the concept of fibre coating to produce shape memory composites.

3 Modified Interface in Structural Health Monitoring

One of the developments of multifunctional composites is the possibility of doing in situ monitoring of its structural health. This technique is usually termed structural health monitoring (SHM). This technique is effective to enhance the safety of structures as well as to reduce the period of maintenance in some materials [22]. Although SHM is still a new research field in the composites, there have been many works that showed promising potential to be implemented in the future [23, 24].

The idea of in situ stress and strain detection, together with health monitoring in composites was first introduced by Schulte et al. [22] which used conductive matrix filler, black carbon, and carbon nanotubes (CNTs). The measurement is based on the resistance change of the composites during mechanical testing. Several years later, there are many works conducting research about structural health monitoring based on Schule et al. [22]. Most of them utilised CNTs scattered along the bulk matrix [25–27], as CNTs have very good electrical conductivity which is sensitive to measure resistance change. The detailed concept of structural health monitoring using interface sensor is shown in Fig. 7.

Glass fibres (GF) were coated using multi-walled CNTs in a wet coating subjected to tensile testing by Rausch et al. [26]. The matrix used was polypropylene matrix (thermoplastic). The results showed that the initial resistance of CNTs-coated GF yarns depends on the weight fraction of the CNT and the yarn length. If the coatings were below 10%, more discontinuous content occurred on the glass fibres' surface which resulted in higher resistances.

Gao et al. [27] deposited multi-walled carbon nanotube (MWNT) onto GF surface to create composite with in situ multifunctional sensors in which epoxy was used as the matrix. MWNTs deposited influenced the formation of electrically semiconductive MWNT-glass fibres and fibre/polymer interphase. It also showed that unidirectional composites depicted ultrahigh semiconducting electrical properties and an ultralow electrical percolation threshold. It also revealed that the increase of the temperature would decrease the resistance monotonically.

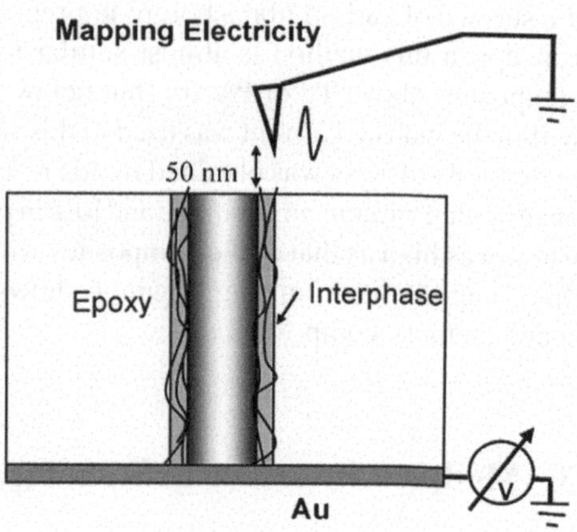

Fig. 7 Concept of interphase sensors [27] reproduced with the permission of Wiley

4 Discussion and Concluding Remarks

The function, the ways to modify the interface and the state of the art of the controlled interphase have been reviewed. The main function of interface is to transfer loads between the fibre and the matrix. Moreover, the interface can be modified in several ways to create more advanced composite materials. The modified interface is able to not only increase the bonding of the fibre-matrix and mechanical properties of composites but also to invent more functional composites such as shape memory composites (stiffness control) and in situ measurements.

In terms of shape memory composites, it is clear that all literatures reviewed have a similar concept, utilising different thermoplastic polymer which has lower Tg than the matrix to coat the fibres or to be inserted in between adjacent plies in the laminated composites. In order to control the stiffness, heating is applied throughout the composites. The most significant effect was obtained with the use of polystyrene as the interleaf which was almost 100% reduction of flexural stiffness. As the suggestion for the potential of stiffness control, applied transverse load shall be investigated later in stiffness control application because the major problem of the composites is anisotropy. The use of woven fibres with the combination of coating or interleaf in stiffness control application can be considered in which it has managed to improving the toughness.

Regarding in situ monitoring ability, the use of interface for the region in monitoring structural health resulted in more appropriate results than the use of the bulk matrix though there are still several weaknesses such as discontinuous coating and the temperature dependency once using thermoset matrix. The use of other thermoplastic matrices can be considered combined with MWNT. The potential of other fibres such as natural fibres shall be taken into account for the next research.

As suggestion, it is likely to be investigated composites having more than one functionality, for example, composites having ability to perform in situ monitoring together with shape memory capability which will produce more effective materials.

Acknowledgements The research was supported by Ministry of Research, Technology, and Higher Education of the Republic of Indonesia through the Fundamental Research Grant with Contract No. 719/UNS27.21/PN/2019.

References

1. Drzal LT (1990) The role of the fiber-matrix interphase on composite properties. Vacuum 41:1615–1618
2. Drzal LT (1990) The effect of polymeric matrix mechanical properties on the fiber-matrix interfacial shear strength. Mater Sci Eng, A 126:289–293
3. Jones FR (1996) Interphase formation and control in fibre composite materials. Key Eng Mater 116–117:41–60

4. Kim JK, Mai YW (eds) (1999) Engineered interfaces in fiber reinforced composites. Elsevier
5. Pitchumani R (2012) Long-term durability of polymeric matrix composites. In: Long-term durability of polymeric matrix composites, p 77
6. Hull D, Clyne TW (1996) An introduction to composites. Cambridge Solid State Science Series
7. Tiwari S, Sharma M, Panier S, Mutel B, Mitschang P, Bijwe J (2011) Influence of cold remote nitrogen oxygen plasma treatment on carbon fabric and its composites with specialty polymers. J Mater Sci 46:964–974
8. Sharma M, Bijwe J (2011) Influence of fiber-matrix adhesion and operating parameters on sliding wear performance of carbon fabric polyethersulphone composites. Wear 271:2919–2927
9. Sharma M, Gao S, Mäder E, Sharma H, Wei LY, Bijwe J (2014) Carbon fiber surfaces and composite interphases. Compos Sci Technol 102:35–50
10. Grail G, Pimenta S, Pinho ST, Robinson P (2015) Exploring the potential of interleaving to delay catastrophic failure in unidirectional composites under tensile loading. Compos Sci Technol 106:100–109
11. Altus E, Ishai O (1990) The effect of soft interleaved layers on the combined transverse cracking/delamination mechanisms in composite laminates. Compos Sci Technol 39:13–27
12. Ishai O, Rosenthal H, Sela N, Drukker E (1988) Effect of selective adhesive interleaving on interlaminar fracture toughness of graphite/epoxy composite laminates. Composites 19:49–54
13. Hancox NL, Wells H (1977) The effects of fibre surface coatings on
14. Mai YW (1983) Controlled interfacial bonding on the strength and fracture toughness of Kevlar and carbon fibre composites. J Mater Sci Lett 2:723–725
15. Drzal LT, Rich MJ, Koenig MF, Lloyd PF (1983) Adhesion of graphite fibers to epoxy matrices: II. The effect of fiber finish. J Adhes 16:133–152
16. Bijwe J, Sharma M (2013) Carbon fabric-reinforced polymer composites and parameters controlling tribological performance. Wear Adv Mater 1–60
17. Thill C, Etches J, Bond I, Potter K, Weaver P (2008) Morphing skins. Aeronaut J 112:117–139
18. Tridech C, Maples HA, Robinson P, Bismarck A (2013) High performance composites with active stiffness control. ACS Appl Mater Interfaces 5:9111–9119
19. Robinson P, Maples H, Gaite O, Smith S, Bismarck A (2013) Carbon fibre reinforced epoxy composites with variable stiffness for use in morphing aerostructures. ICCM19
20. Maples HA, Robinson P, Bismarck A (2012) High performance carbon fibre reinforced polymer composites with controllable stiffness: can it be done ? 24–28
21. Raither W, Bergamini A, Gandhi F, Ermanni P (2012) Adaptive bending-twist coupling in laminated composite plates by controllable shear stress transfer. Compos Part A Appl Sci Manuf 43:1709–1716
22. Schulte K, Baron C (1989) Load and failure analyses of CFRP laminates by means of electrical resistivity measurements. Compos Sci Technol 36:63–76
23. Black S (2019) Structural health monitoring: composites get smart. http://www.compositesworld.com/articles/structural-health-monitoring-composites-get-smart. Last accessed 2019/08/03
24. Alexopoulos ND, Bartholome C, Poulin P, Marioli-Riga Z (2010) Structural health monitoring of glass fiber reinforced composites using embedded carbon nanotube (CNT) fibers. Compos Sci Technol 70:260–271
25. Rausch J, Mäder E (2010) Health monitoring in continuous glass fibre reinforced thermoplastics: tailored sensitivity and cyclic loading of CNT-based interphase sensors. Compos Sci Technol 70:2023–2030
26. Rausch J, Mäder E (2010) Health monitoring in continuous glass fibre reinforced thermoplastics: manufacturing and application of interphase sensors based on carbon nanotubes. Compos Sci Technol 70:1589–1596
27. Gao SL, Zhuang RC, Zhang J, Liu JW, Mäder E (2010) Class fibers with carbon nanotube networks as multifunctional sensors. Adv Funct Mater 20:1885–1893

The Investigation of Nozzle Arch Variations Against the Water Inflow to the Runner of Banki Turbine Based on CFD

Sirojuddin, L. K. Wardhana, O. Rizky, R. Ibnawati, and Junior R. Syahri

Abstract Banki turbine is a cross flow type water turbine, the construction is simple and suitable for micro hydro power plants. Increasing power efficiency is a very important point. One of the method to increase power is by controlling the water inflow to the runner. This research aimed to investigate variations of the nozzle arch against the water inflow of the banki turbine's runner in order to obtain good flow trajectory of water from the first to the second stage of runner blades without striking the shaft. The design of bottom runner case was round and the top was rectangular. The nozzle arch 2D geometry were formed in AutoCAD software, while the 3D geometry using Solidworks. Flow analysis and simulation using CFD software. The nozzle arch were made into five variants, N-1 variant radius was 200 mm, N-2 = 225 mm, N-3 = 250 mm, N-4 = 275 mm, and N-5 = 300 mm. The water discharge was 2 m^3/min with head 5.5 m, the angle of water inlet was 16° from vertical direction. It was found that the best variant were N-3, where the flow trajectory and the velocity from the first stage to the second stage of the blade did not strike the shaft.

Keywords Investigation · Variation · Nozzle arch · Runner · Flow trajectory

1 Introduction

Banki turbine is known as Mitchell-Banki turbine or cross-flow type turbine, usually for micro-hydro scales generally up to 100 kW applications. This turbine is suitable for remote areas. The advantages are relatively low cost, easy transportation, simple assembly, and maintenance.

Power efficiency was a point for the researcher to be studied. One of the methods to increase power was controlling the water inflow to the runner.

Choi et al. [1] investigated the effect of turbine structural configuration on the performance and internal flow characteristics of the cross-flow turbine. The result was the shape of the nozzle gives the greatest effect on turbine performance and internal flow, but the flow simulation still struck the shaft. De Andrade et al. [2] analyzed the

Sirojuddin (✉) · L. K. Wardhana · O. Rizky · R. Ibnawati · J. R. Syahri
Universitas Negeri Jakarta, Jakarta, Indonesia
e-mail: Sirojuddin@unj.ac.id

U. Sabino et al. (eds.), *Proceedings of the 6th International Conference and Exhibition on Sustainable Energy and Advanced Materials*, Lecture Notes in Mechanical Engineering, https://doi.org/10.1007/978-981-15-4481-1_74

velocity and pressure fields of the cross-flow within the runner and to characterize for the different runner speeds. Absolute flow velocity angles are obtained at runner entrance for simulations with and without the runner. Flow recirculation in the runner inter blade passages and shocks of the internal cross-flow cause considerable hydraulic losses by which the efficiency of the turbine decreases significantly. The CFD simulations results were compared with experimental data and were consistent with global performance parameters. The flow at each different runner speed still struck the shaft. Sammartano et al. [3] optimized design of the Banki-Michell turbine with a novel two-step procedure. In the first step, the initial and final blade angles, the outer impeller diameter and the shape of the nozzle are selected using a simple hydrodynamic analysis, based on a very strong simplification of reality. In the second step, the inner diameter, as well as the number of blades and their shape, are selected by testing single options using computational fluid dynamics (CFD) simulations, starting from the suggested literature values. Good efficiency was attained not only for the design discharge but also for a large range of variability around the design value. The result was that the presence of the shaft affects the reduction of inefficiency. The flow of water still struck the shaft that could decrease efficiency. Kaunda Chiyembekezo et al. [4] investigated, using a numerical method, the flow profile in the turbine at the best efficiency point and at operating conditions that are away from the best efficiency point. Actual pictures of the flow were taken to compare the actual flow with the numerically determined flow profile. Turbine efficiency results from the previous performance evaluation experiment conducted on the model crossflow turbine were compared with the numerically obtained efficiency results. It had been observed that the numerically obtained flow profile compares favorably with the actual flow pictures. Pockets of negative pressures and flow circulation had been observed in the flow profile. At the constant head and valve opening, the velocity profile was found to vary more with runner speed than the pressure profile. The numerically obtained flow pattern showed positions where the flow gives maximum efficiency. Therefore, the study had shown that the numerical method was a superior design tool for Crossflow investigated the flow profile in the Crossflow turbine. This simulation only made a variation of rotation of runner but not variated the arc of nozzle radius. Saini and Singal [5] optimized design approach of cross-flow turbine based on CFD for 5 kW capacity under the different operating conditions with head 10 m and discharge 0.072 m^3/s. This design did not variate the arc of the nozzle radius. Efficiency reached 66.67% when calculated from 150 N m torque. Patel and Nirav Oza [6] analyzed the CFD flow simulation on a Cross-flow turbine with head 16 m and discharge 0.315 m^3/s. The flow trajectory not so good from the first stage to the second stage of the blade. Nozzle arch not variated. Chichkhede et al. [7] investigated the effect of design parameters on crossflow turbine flow velocity using CFD simulation with different nozzle opening. Simulation results showed that the flow struck the shaft at all types of the nozzle opening. Popescu et al. [8] studied numerically and experimentally to identify why the banki turbines' efficiency was poor at low rotational speeds. Simulation results showed that at low rotational speeds the flow struck the shaft caused lower efficiency compared to at high rotational speeds with a flow that did not strike the shaft. This research did not variate the arc of nozzle radius. Adhikari and Wood [9] reviewed a study of high-efficiency crossflow turbine design. It was found that many recent studies had been carried out but there are misconceptions of design principles to achieve high efficiency.

Mostly researcher before did not make research on the arch of the nozzle combined with guide vane with ratio arch radius and diameter of the runner. Therefore this research would focus on the variation of the nozzle arch against the inflow of water to the runner using the CFD Simulation to obtain a good flow trajectory and do not strike the shaft so that the efficiency would be maximized.

2 Method

This research was conducted at the Design Laboratory of Mechanical Engineering, Jakarta State University. Flow chart can be seen in Fig. 1.

2D geometry design created in Autocad, while the 3D in Solidworks. Flow simulation run in CFD Solidworks flow simulations. The nozzle were made in 5 variants with different arch (Figs. 2 and 3).

Design parameters and data of turbine can be seen in Table 1 and boundary condition can be seen in Table 2.

2.1 Governing Equation

Flow Simulation solves the Navier-Stokes equations, which are formulations of mass, momentum and energy conservation laws for fluid flows. The equations are supplemented by fluid state equations defining the nature of the fluid, and by empirical dependencies of fluid density, viscosity and thermal conductivity on temperature. Inelastic non-Newtonian fluids are considered by introducing a dependency of their dynamic viscosity on flow shear rate and temperature, and compressible liquids are considered by introducing a dependency of their density on pressure. A particular problem is finally specified by the definition of its geometry, boundary and initial conditions.

Flow Simulation is capable of predicting both laminar and turbulent flows. Laminar flows occur at low values of the Reynolds number, which is defined as the product of representative scales of velocity and length divided by the kinematic viscosity. When the Reynolds number exceeds a certain critical value, the flow becomes turbulent, i.e. flow parameters start to fluctuate randomly.

Most of the fluid flows encountered in engineering practice are turbulent, so Flow Simulation was mainly developed to simulate and study turbulent flows. To predict turbulent flows, the Favre-averaged Navier-Stokes equations are used, where time-averaged effects of the flow turbulence as the Reynolds stresses appear in the equations for which additional information on the flow parameters are considered, whereas the other, i.e. large-scale, time-dependent phenomena are taken into account

Fig. 1 Flow chart

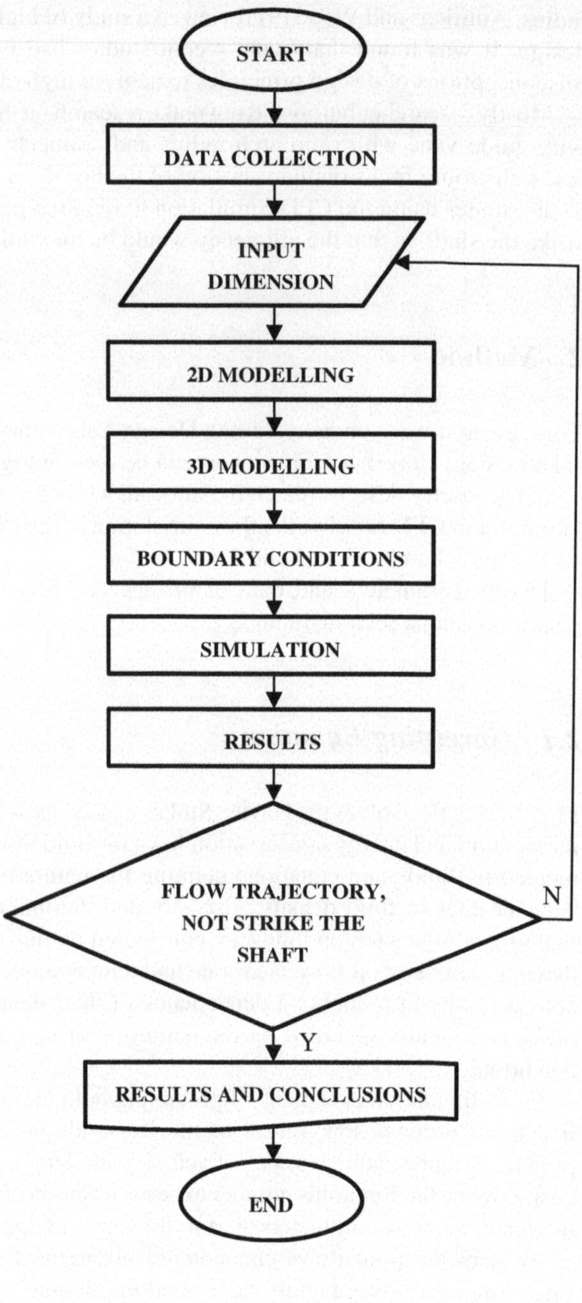

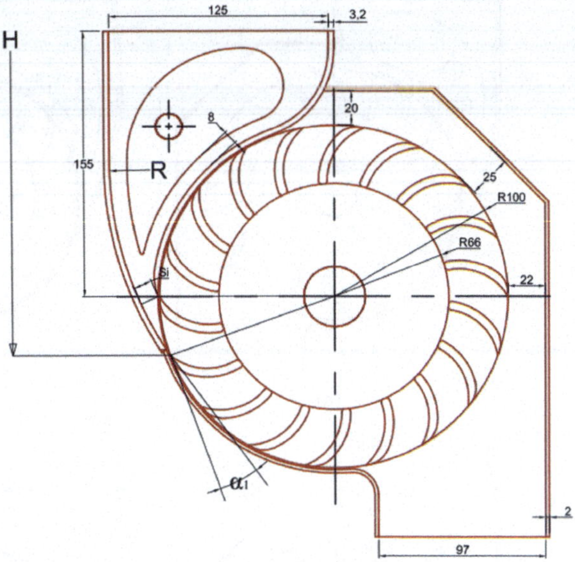

Fig. 2 Dimension of turbine

directly. Through this procedure, extra terms known must be provided. To close this system of equations, Flow Simulation employs transport equations for the turbulent kinetic energy and its dissipation rate, the so-called k-ε model [10].

The k-ε model can be derived from the incompressible Navier-Stokes equations [11].

$$\rho\left(\frac{\partial u_i}{\partial t}+\sum_j uj\frac{\partial u_i}{\partial x_j}\right)=-\frac{\partial p}{\partial x_i}+\eta\nabla^2 u_i \tag{1}$$

where u(x, t) represents the velocity vector field, p(x, t) is the pressure field, p(x, t) is the pressure field, ρ is the density constant, η is the dynamic viscosity, and $\upsilon=\frac{\eta}{\rho}$ is the kinematic viscosity.

Noting Navier-Stokes are derived from the equations for conservation of mass, momentum, and energy, we have that

$$\frac{\partial\rho}{\partial \mathrm{t}}+\sum_{\mathrm{j}} u_j\frac{\partial_\rho}{\partial \mathrm{x_j}}=\rho\sum_j\frac{\partial u_j}{dx_i}=0 \tag{2}$$

Applying statistical averaging to (N S) produces the Reynolds equations:

$$\rho\frac{\partial\overline{u_i}}{\partial t}+\sum_j\left(\rho\overline{u_j}\frac{\partial\overline{u_i}}{\partial x_j}+\frac{\partial\overline{u_i^{'}}}{\partial x_j}u_j^{'}\right)=-\frac{\partial\overline{p}}{\partial x_i}+\sum_j\frac{\partial\overline{\tau_{ij}}}{\partial x_j} \tag{3}$$

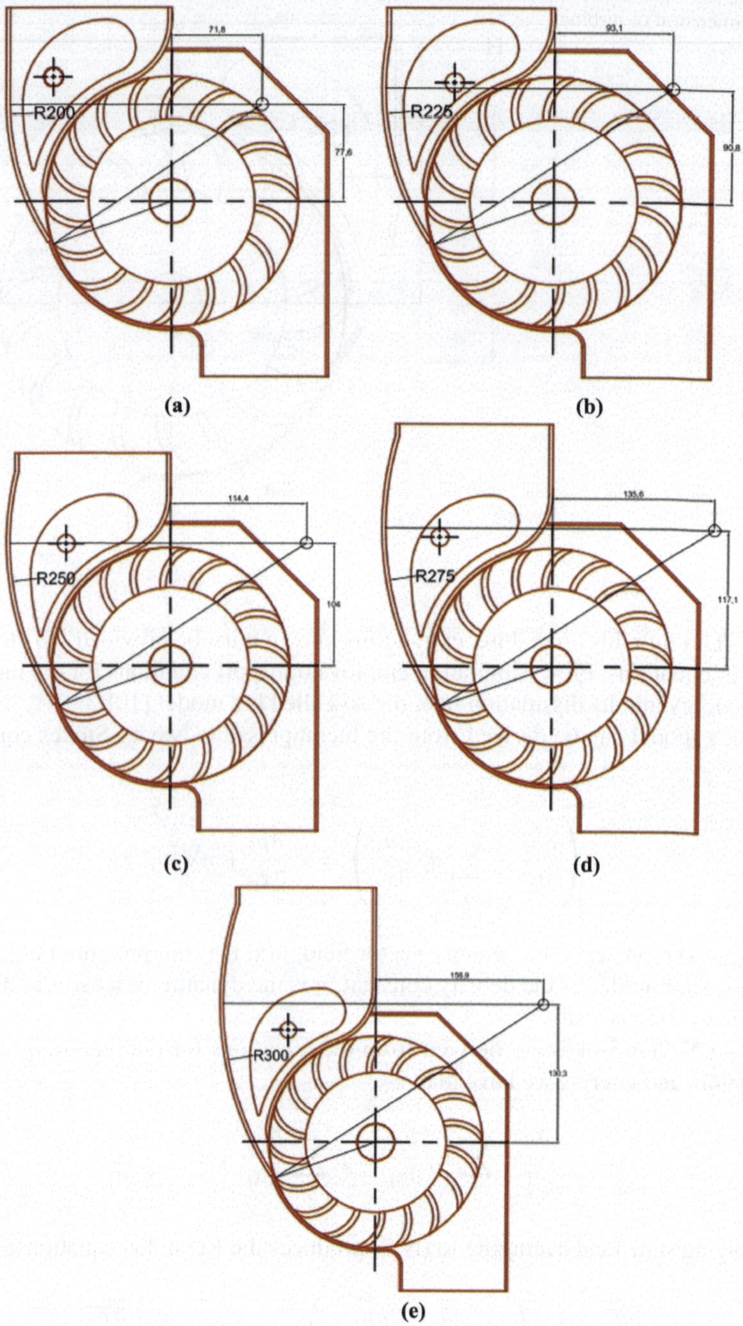

Fig. 3 Variant **a** N-1, $R_1 = 1\ D_1$, **b** N-2, $R_2 = 1.125\ D_1$, **c** N-3, $R_3 = 1.25\ D_1$, **d** N-4, $R_4 = 1.275\ D_1$, **e** N-5, $R_5 = 1.5\ D_1$

Table 1 Design parameters and data of turbine

Design parameter	Data calculation	Design dimension	Description
α_1	16°	16°	Angle of entrance velocity
β_1	30°	30°	Angle of blade
D_1	200 mm	200 mm	Outer diameter of runner
D_2	131.97 mm	132 mm	Inner diameter runner
n	18 pieces	18 pieces	Number of blades
V_1	10.18 m/s	10.18 m/s	Absolute velocity of water
L	200 mm	200 mm	Width of wheel/runner
H	5.5 m	5.5 m	Head of water
N	468.84 rpm	470 rpm	Runner speed
Q	2 m^3/min	2 m^3/min	Water discharge
Ds	30 mm	30 mm	Shaft diameter
Df	52 mm	52 mm	Hub diameter
S_1	16 mm	16 mm	Thickness of jet

where $u = \overline{u} + u'$ written in the mean plus fluctuation decomposition,

$$\overline{\tau_{ij}} = \eta\left(\frac{\partial \overline{u_i}}{\partial x_j} + \frac{\partial \overline{u_j}}{\partial x_i}\right), \eta = \nabla^2 u_i = \sum_j \frac{\partial \tau_{ij}}{\partial x_j} \quad (4)$$

Table 2 Boundary conditions

Item	Value
Turbulence model	K-epsilon
Rotation	Non-rotation
Material	Water
Density	998.2 kg/m^3
Discharge	0.0333 $m^{3/}s$
Reference axis	Y
Gravitation	9.81 m/s^2
Calculation type	Steady state
Environment pressure	101,325 Pa
Temperature	293.20 K
Turbulence parameters	Boundary layer parameters

3 Results and Discussions

3.1 Results

Flow trajectory based on the results of the solidworks flow simulation for different variants were shown in the Fig. 4.

3.2 Discussions

From Fig. 4 velocity contour, it could be seen that the flow trajectory of water for variants N-1 and N-4 still struck the shaft. While variants N-2, N-3, and N-5 the flow trajectory quite good and water did not strike the shaft.

4 Conclusions

From simulation above could be concluded as follows:

1. The arch nozzle radius affected the flow trajectory of the water from the first and second stage of the blades runner and whether the flow struck the shaft or not.
2. Variants N-2, N-3, and N-5 the flow trajectory quite good and water did not strike the shaft. The best variant was N-3 with a 250 mm radius compared to others because flow trajectory was very good and the flow did not strike the shaft.

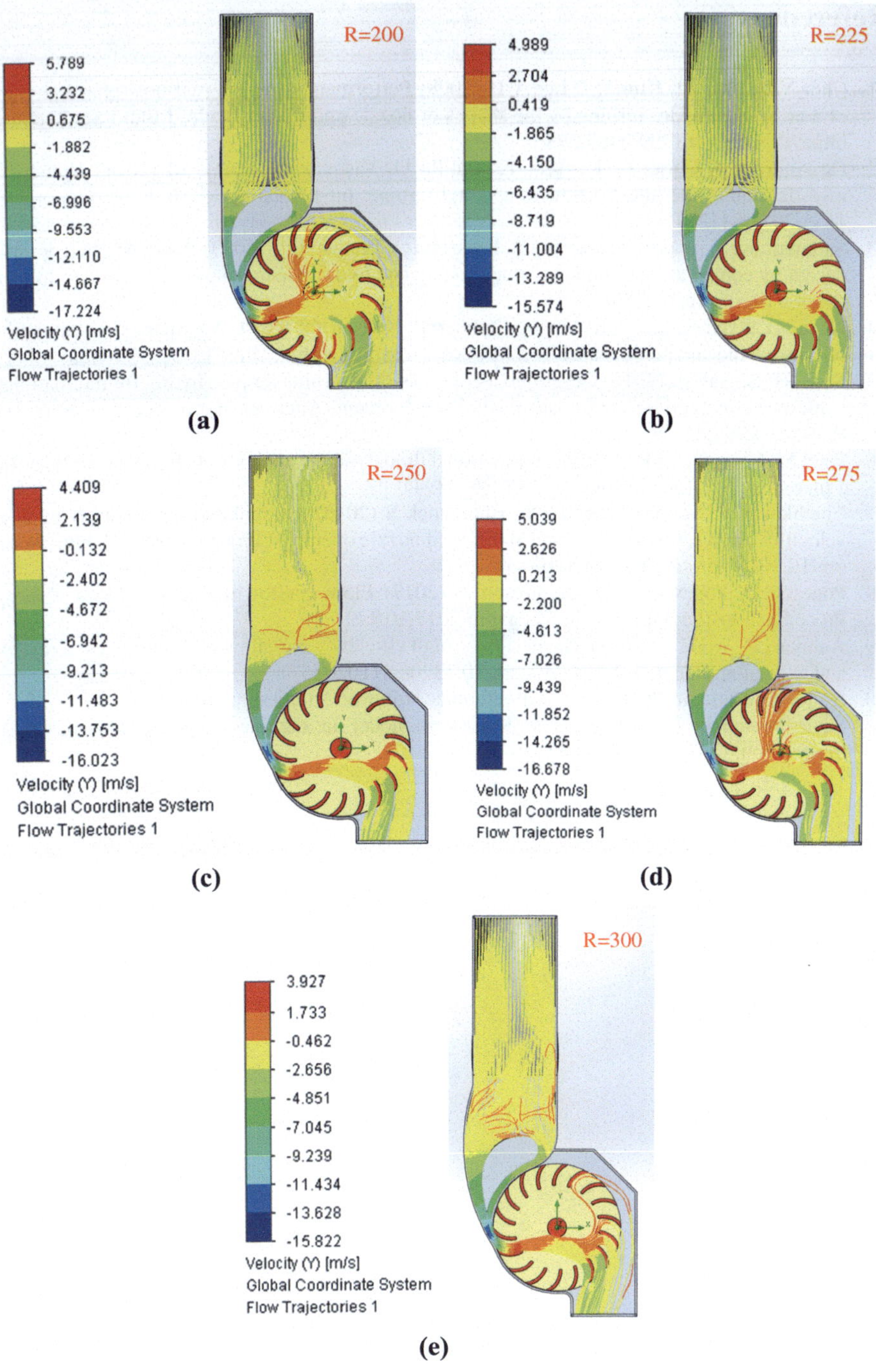

Fig. 4 Velocity contour variant **a** N-1, **b** N-2, **c** N-3, **d** N-4, **e** N-5

References

1. Choi Y-D, Lim J-I, Kim Y-T, Lee Y-H (2008) Performance and internal flow characteristics of a cross-flow hydro turbine by the shapes of nozzle and runner blade. J Fluid Sci Technol. https://doi.org/10.1299/jfst.3.398
2. De Andrade J, Curiel C, Kenyery F, Aguilln O, Vásquez A, Asuaje M (2011) Numerical investigation of the internal flow in a Banki turbine. Int J Rotating Mach. https://doi.org/10.1155/2011/841214
3. Sammartano V, Aricò C, Carravetta A, Fecarotta O, Tucciarelli T (2013) Banki-Michell optimal design by computational fluid dynamics testing and hydrodynamic analysis. Energies. https://doi.org/10.3390/en6052362
4. Kaunda Chiyembekezo S, Kimambo Cuthbert Z, Nielsen TK (2014) A numerical investigation of flow profile and performance of a low cost crossflow turbine. Int J Energy Environ
5. Saini RP, Singal SK (2015) Development of cross-flow turbine for pico hydro. In: International conference on hydropower for sustainable development. Alternate Hydro Energy Centre, IIT Roorkee, Uttarakhand
6. Patel M, Nirav Oza KP (2016) Computational fluid dynamic analysis of cross flow turbine. Int J Innov Res Sci, Eng Technol 5(9):18936–18944
7. Chichkhede S, Verma V, Gaba VK, Bhowmick S (2016) A simulation based study of flow velocities across cross flow turbine at different nozzle openings. Procedia Technol. https://doi.org/10.1016/j.protcy.2016.08.190
8. Popescu D, Popescu C, Dragomirescu A (2017) Flow control in Banki turbines. Energy Procedia. https://doi.org/10.1016/j.egypro.2017.10.272
9. Adhikari R, Wood D (2018) The design of high efficiency crossflow hydro turbines: a review and extension. Energies. https://doi.org/10.3390/en11020267
10. Systemes Dassault (2016) Technical reference solidworks flow simulation
11. Scott-Pomerantz CD (2004) The K-epsilon model in the theory of turbulence. Ph.D. dissertation. University of Pittsburgh

Preparation of Anode Active Material by Utilizing of Silica from Geothermal Sludge for Li-Ion Battery Application

H. Widiyandari, A. S. Wijareni, R. Ardiansyah, B. Purnama, and A. Purwanto

Abstract Geothermal sludge (GS) from Dieng geothermal power plant has high Silica (SiO_2) content. Silica is a functional material in industrial application. This work reported preparation of anode active material from GS silica material for li-ion battery application. The anode active material was prepared by a series of processes as follows: acidification with 3% HCL, silica extraction with 3 M NaOH and 1 M HCL, silica purification with 37% HCL, metallothermal, and combustion. Chemical composition of the samples after metallothermal process (M1) contains the pure Si of 2.26% and increased after combustion process (M1P) to 10.60%. The electrochemical measurement was conducted using a cylindrical full cell battery 18650 with NCA as cathode. The specific discharge capacity for samples of M1P, M1 and graphite are 761.16, 605.56 and 314.45 $mAhg^{-1}$, respectively. These results show that GS can be used as a raw material for anode active materials in li-ion battery applications.

Keywords Geothermal sludge · Anode · SiO_2 · Metallothermal · Li-ion batteries

1 Introduction

Geothermal exploration carried out at this time gives the by-product of silica. Silica is a by-product of geothermal exploration which is very abundant. This abundant silica can cause environmental degradation due to the deposition of silica. The presence of silica deposition occurs when geothermal heat increases in the pipe so that chemical barriers are formed from chemical compounds in the pipe due to physical changes when the geothermal fluid is exploited. The equilibrium of chemical compounds is disturbed due to physical changes that affect the solubility. When there is a drastic decrease in temperature resulting in a decrease in the quality of the solubility of silica

H. Widiyandari (✉) · A. S. Wijareni · R. Ardiansyah · B. Purnama
Department of Physics, Faculty of Mathematics and Natural Science, Universitas Sebelas Maret, Surakarta 57126, Indonesia
e-mail: hendriwidiyandari@staff.uns.ac.id

A. Purwanto
Department of Chemical Engineering, Faculty of Engineering, Universitas Sebelas Maret, Surakarta 57126, Indonesia

U. Sabino et al. (eds.), *Proceedings of the 6th International Conference and Exhibition on Sustainable Energy and Advanced Materials*, Lecture Notes in Mechanical Engineering,
https://doi.org/10.1007/978-981-15-4481-1_75

and carbonate. When changes in pressure in the reservoir surface cause a decrease in water content in the geothermal fluid which makes compounds such as silica and carbonate in a state of saturation which continues at the deposition stage. Specifically, the silica compound after the deposition step continues the polymerization stage into a silica polymer then crusts. This crusty silica becomes sludge in geothermal energy processing [1].

Currently, the use of silica in the geothermal mining area is in making silica brick. In addition, the use of silica in the industry is usually in the form of silica sand but it also exists on a finer scale, silica flour or converted to cristobalite. By chemical nature, solid silica has a high melting point which is related to bonds between atoms. A large amount of silica sludge from ex-mining is currently used as a new product useful. Silica can be used as a mixture of concrete, composite in marble and production of glass. In addition, other uses are the production of silicates, sand filtration, refractory, paint and plastic fillers, silicon carbide, and materials for electronics. The silica contained in sodium silicate is converted into synthetic zeolite used in washing powder or silica flour for paint formulations [2]. The purification of silica can produce pure silicon (Si). Silicon are known as promising anode materials for li-ion batteries because they have a high theoretical capacity. In these electrodes, reactions such as silicon can provide as much as 5–10 times higher discharge capacity than graphite. Si as an electrode material can intrinsically increase the rate of ionic velocity. Another advantage of this electrode is that lightweight reaches 10% of the total weight of the battery. The electrode is positioned in a certain position it will increase corrosion resistance and increase flexibility [3, 4]. Thus, the objective of this study is to prepare the anode active material containing silicon using raw material from geothermal sludge silica.

2 Methodology

2.1 Materials and Characterization

The materials used to prepare the anode active material was geothermal sludge, Chloride Acid (Merck), Natrium Hidroksida (Merck), aquades, Magnesium (Merck). The materials used to assembly the battery is $AgNO_3$, NCA, Monolayer Polypropylene, Acetylene black, Polyvinylidene Fluoride and *N*-Methylpyrolydinone.

The morphologies of sample were characterized by scanning electron microscopy (SEM, JEOL JSM-636oLA), X-ray diffractometer (XRD, Bruker D2 Phaset 2nd Gen) and X-Ray fluorescence (XRF, Bruker S2 Ranger).

2.2 Synthesis of Anode Active Material from Geothermal Sludge

Figure 1 shows the schematic diagram for the preparation of anode active material by utilizing GS as a raw material. 200 g of GS soaked in hot water for 2 h of heating. After that XRF was analyzed. 25 g of washed sample was even added with 500 mL 3% HCL in a three neck flask. The mixture was refluxed for 2 h, then filtered and the residue was washed with hot distilled water to neutral pH. The results were dried in oven for 24 h and at 100 °C. 150 mL of 3 M NaOH was added to the sample from the acidification process. Samples were mixed with a magnetic stirrer at a speed of 150 rpm and a temperature of 65–70 °C with a time of ±3 h. Then filtered and obtained filtrate. The result filtrate with silica content was separated and a concentration of 1 M HCL was added to pH at level 7. The result filtrate was left for 18 h and washed the precipitate with hot distilled water. The precipitate is dried in an oven at 100 °C for 4 h. 10 g of silica samples resulting from the silica extraction

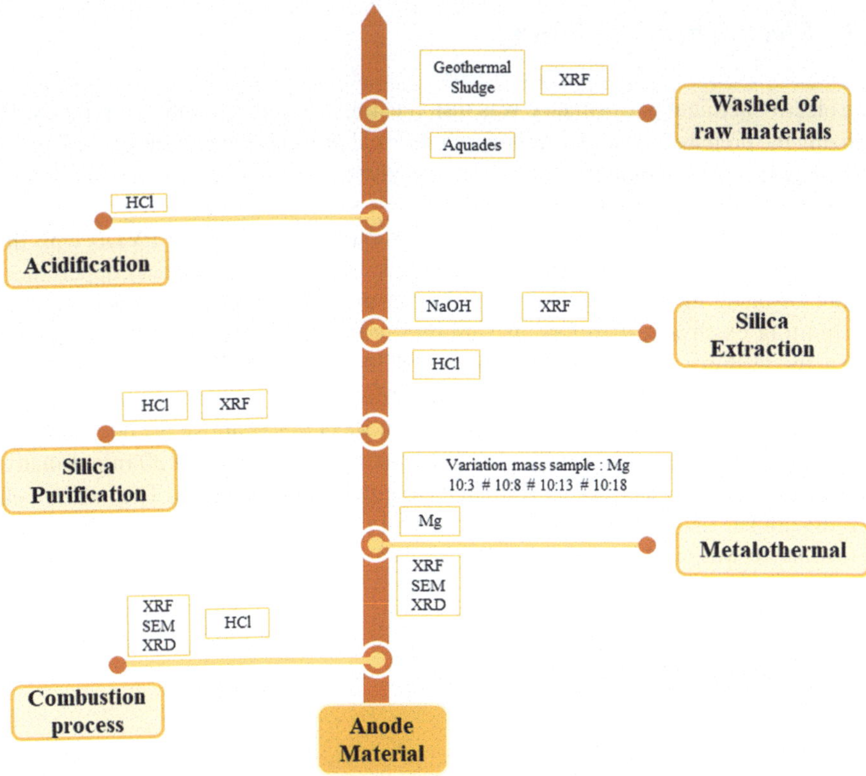

Fig. 1 Schematic diagram of anode active material production

process were placed in an erlenmeyer and 20 mL 37% HCL solution was added and heated to 90 °C hotplate for 3 h. The result was dried in an oven at 110 °C for 3 h.

The sample results were analyzed by XRF. The results of the purification sample were placed in a crucible ceramic. The metallothermal step was to add magnesium to the powder phase in a ratio of 10:8 silica/Mg. Mixed with the solid-state method to be homogeneous then heated for 3 h at 650 °C. The process was repeated at a variation of ratio of 10:3; 10:13 and 10:18 silica/Mg. The sample results that have been through the reduction process were analyzed by XRF. The best results from all four variations will be analyzed by XRD and SEM. After going through the reduction process, the best sample from the reduction process was put into a beaker and added 150 mL of HCL 2 M. Samples were heated at 80 °C while stirring with a magnetic stirrer for 3 h after which the samples were cooled and filtered with filter paper obtained by solids. The solid is washed with distilled water to a neutral pH then dried and weighed. This final combustion process was carried out to remove MgO compounds and analyzed by XRF, XRD and SEM.

2.3 *Electrochemical Testing*

The electrochemical performance was tested using a cylindrical full cell type 18650 utilizing the prepared samples after metallothermal process, after combustion process and graphite as the counter anode. The acetylene black (AB) and polyvinylidene fluoride (PVDF) were used as a conductive agent and binder, respectively. The anode slurry was prepared by mixing the prepared samples with AB and PVDF with the ratio of 7:1:2 in deionized water. The anode slurry was coated on a copper sheet by doctor blade method. To assure full Li-ion transfer, the cathode was chosen as the limiting electrode; therefore, it was selected as the basis mass of calculation. The cathode slurry was prepared by mixing NCA samples with AB and PVDF with the ratio of LNMC:AB:PVDF = 92:3:5 in *N*-Methyl-2 pyrrolidone (NMP) solvent. The cathode slurry then placed on an aluminum sheet that is above the doctor blade. Then, the sample put into furnace with a temperature of 180 °C for 30 min. Finally, filling the cells with electrolyte LiPF6 in the argon flowed glove box. The battery cell was assembled to be tested for electrochemical performance. The charge-discharge performance was tested using an NEWARE Battery analyzer and BTS software.

3 Result and Discussion

3.1 *Silica Extraction*

At the initial stage, sample preparation from geothermal sludge obtained XRF analysis results with a percentage above 1%, namely Silica 89.73%, Cl 3.90%, P_2O_5 1.90%

and SO_3 1.41%. The next step is acid leaching and in this process reflux treatment occurs to reduce metal oxides, namely ferroxide (Fe_2O_3) from 0.40 to 0.16%. The reaction in the acid leaching process is seen in Eq. 1.

$$Fe_2O_{3(s)} + 6HCl_{(aq)} \rightarrow 2FeCl_{3(aq)} + 3H_2O_{(l)} \quad (1)$$

The precipitation method occurs at the stage of silica extraction with 3 M NaOH solution. Geothermal sludge samples from acid leaching results with 3% HCL solution will be dissolved into NaOH solution. The reaction between silica and NaOH can be seen in Eq. 2.

$$SiO_{2(s)} + 2NaOH_{(aq)} \rightarrow Na_2SiO_{3(aq)} + H_2O_{(aq)} \quad (2)$$

From Eq. 2 the reaction product is sodium silicate (Na_2SiO_3) in liquid and colorless form with basic properties. The next step is the reaction between Na_2SiO_3 and HCL where an exchange between the Na^+ ion and H^+ ion with the reaction in Eq. 3. In this equation, the hydrolysis and condensation of the silicate acid reaction from silica extraction from the dieng geothermal sludge occurs

$$Na_2SiO_{3(aq)} + 2HCl_{(aq)} \rightarrow H_2SiO_{3(s)} + 2NaCl_{(aq)} \quad (3)$$

$$H_2SiO_{3(s)} \rightarrow SiO_{2(s)} + H_2O_{(g)} \quad (4)$$

The next step is silica purification where this process is carried out to increase silica content and reduce metal impurity. In this case, the percentage of compounds obtained from the XRF process are aluminum oxide (Al_2O_3) and ferroxide (Fe_2O_3) compounds with a percentage of 3.5 and 0.16% to 0.17 and 0.07%. The percentage of impurities can be reduced due to the reaction with strong acids which is 37% HCL with pH 1 which serves to reduce compounds containing oxides. The silica yield from the silica purification process was 94.84%. This happens because the process of diffusion of Na^+ ions and other anions causes silica levels to increase.

3.2 *Isolation of the Anode Material from Geothermal Sludge*

The metallothermal method was carried out to isolate the anode material from geothermal sludge in this study using magnesium to reduce silica with the following variations: 10:3; 10:8; 10:13 and 10:18. This variation is in accordance with research conducted [5], with a variation of 1:0.8; 1:1.5; 1:2 and 1:2.5. Equation 5 shows the reaction between silica and magnesium.

$$SiO_{2(s)} + 2Mg_{(s)} \rightarrow 2MgO_{(s)} + Si_{(s)} \quad (5)$$

Table 1 Quantification of the products of the reduction reaction using XRF

Sample name	Mass ratio	Elemental concentration (%)		
	Silica:Magnesium	Mg	O	Si
M1	10:3	44.43	38.31	2.26
M2	10:8	44.3	37.86	1.32
M3	10:13	41.52	36.62	0.59
M4	10:18	42.8	36.97	0.63

The best variation obtained is 10:3 where the results get the most silicon out of the 4 other variations. The variation results can be seen in Table 1 which shows the product quantification of the reduction reaction using XRF.

Magnesium (Mg) was chosen because it can reduce active metals and this is based on the Elliangham diagram. According the Elliangham diagram found that Mg had the lowest melting point at 922 K and the highest at 1378 K [6]. The reaction between magnesium and oxygen produced magnesium oxide (MgO) with enthalpy of standard molar formation (ΔHf^{o}) and exothermic reaction occurred and the entropy value (ΔS) is negative so the reaction is exothermic. Then it will produce a spontaneous reaction at a temperature of 922 K which corresponds to the gibbs (ΔG) energy value obtained.

The XRF characterization in Table 1 shows the percentage results on the elements Magnesium (Mg), Oxygen (O) and Silicon (Si) formed in all four variations where the XRF analysis results have a single element standard. From Table 1 the biggest element is magnesium rather than silicon. Referring to Eq. 5, magnesium must bind the oxygen contained in silica. But the results obtained are not in accordance with Eq. 5 because magnesium as an active metal that easily reacts with oxygen. So that in this study, magnesium does not optimally bind the oxygen contained in silica but binds oxygen in the environment. This is same with other research where this research [7] was carried out purification of silicon using magnesium with nitrogen gas flowing in confined spaces and without special care. The results obtained show that with closed space and nitrogen gas flowing, the percentage of purity is higher at 64% compared to no specific treatment at 59%.

3.3 *The Combustion Process Using HCL*

Equation 6 shows the reaction between HCL and Magnesium which describes the reaction that occurs at the combustion stage.

$$2HCl_{(aq)} + \mathrm{Mg}_{(s)} \rightarrow \mathrm{MgCl}_{2(aq)} + \mathrm{H}_{2(g)} \tag{6}$$

M1 sample is taken to enter this stage with the best XRF yield reference for the highest amount of silicon from the previous process. After that, the M1 sample is

Table 2 Quantification of samples before and after the combustion process

Sample name	Elemental concentration (%)		
	Mg	O	Si
M1	44.43	38.31	2.26
M1P	29.02	40.03	10.60

processed into the M1P sample. So by using the XRF analysis the results of the concentration of the M1 and M1P samples can be compared in Table 2.

Table 2 shows an increase in pure silicon yield from 2.26 to 10.60%. At this stage, the aim is to reduce the MgO compounds found in the sample. Based on the XRF results the MgO compound in the M1 sample was 73.68% and the M1P sample was 48.13%. Figure 1 is the results of the XRD test with geothermal sludge samples that have not yet been treated. Figure 1 shows that there is a sloping highest peak so that it shows that this material is amorphic at $2\theta = 21.56°$. This is consistent with research before that peaks in the range of $2\theta = 22°$ [8] and research titled characterization of silica nanoparticles synthesized by thermo-mechanical route with toppings in the range of $2\theta = 21.54°$ [9].

Figure 2 shows the XRD pattern of prepared material. The sample after metallothermal process (M1) shows that the diffraction peak is at an angle of $2\theta = 42.92°$ and $2\theta = 62.3°$. This is the peak of MgO so that a crystal size of 52.06 nm is obtained. The XRD results were obtained at a peak of $2\theta = 42.84°$ and $2\theta = 62.24°$ [10]. In metallothermal samples with a ratio of silica to magnesium is 10:3 (M1), the percentage of MgO is 73.68% in the XRF test. Figure 1 in the sample after the combustion stage (M1P) obtained 2 highest peaks namely at $2\theta = 18.58°$ and $2\theta = 38.02°$ with a crystal size of 41.78 nm. From Fig. 2, the results of the XRD analysis were matched with JCPDS (Joint Committee on Powder Diffraction Standards) data. In the XRD M1 analysis results obtained diffraction peaks which have similarities with JCPDS No. 87-0653 which shows MgO crystals. MgO has a cubic crystal system with a =

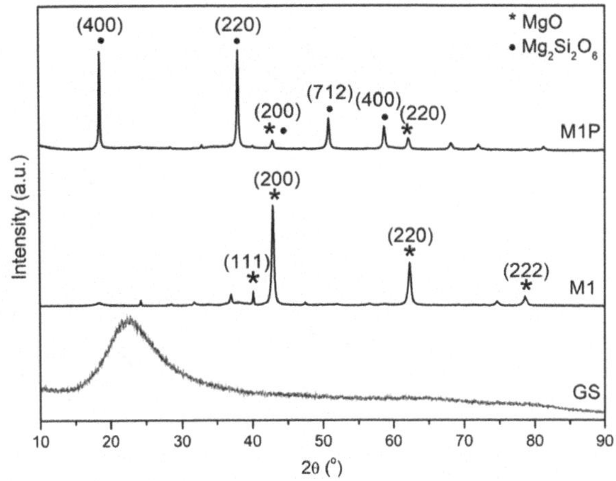

Fig. 2 XRD pattern of the raw materials and products

b = c and α = β = γ = 90 °C. In the XRD M1P analysis results obtained diffraction peaks which have similarities with JCPDS No. 80-0275, Magnesium Silicate ($Mg_2Si_2O_6$). $Mg_2Si_2O_6$ has a crystalline system which is orthorhombic with a ≠ b ≠ c and α = β = γ = 90 °C. It can be seen from Fig. 1 that is on the M1P graph when compared to M1 that in the MgO compound there is a significant decrease seen from the peak in the M1 sample which decreases at the peak of the M1P sample. At the combustion stage, it is known that the yield of other compounds is $Mg_2Si_2O_6$ which is listed on the XRD results. So the reaction of the formation of $Mg_2Si_2O_6$ can be seen from Eq. 7 and there is the formation of other compounds seen in Eqs. 8, 9 and 10.

$$3SiO_{2(s)} + 2Mg_{(g)} \leftrightarrow 2MgSiO_{3(s)} + Si_{(s)} \tag{7}$$

$$4MgSiO_{3(s)} + 2Mg_{(g)} \leftrightarrow 3Mg_2SiO_{4(s)} + Si_{(s)} \tag{8}$$

$$Mg_2SiO_{4(s)} + 2Mg_{(g)} \leftrightarrow 4MgO_{(s)} + Si_{(s)} \tag{9}$$

$$Si_{(s)} + 2Mg_{(s)} \leftrightarrow Mg_2Si \tag{10}$$

SEM characterization was performed on M1 and M1P samples. This test was carried out to determine the morphology of the metallothermal samples after going through the heat treatment process at a temperature of 650 °C for 3 h and the morphology in the 1P sample occurred under the magnesium burning process with HCL solution.

Figure 3 shows the SEM images of M1 and M1P samples. Figure 3a shows the surface morphology of M1 with the highest percentage of compounds is MgO. It can be seen that the surface of M1 sample looks more homogeneous compared to Fig. 3b. Figure 3b which is at the first highest peak is $Mg_2Si_2O_6$ which shows that the surface has agglomeration compared to Fig. 3a. In Fig. 3b visible pores in the morphological

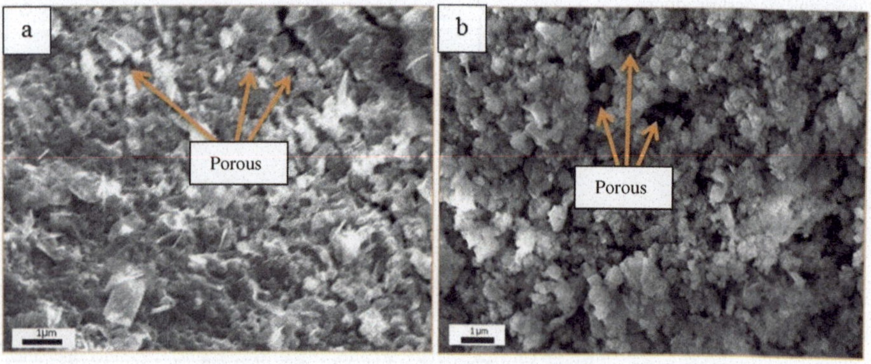

Fig. 3 SEM images of the samples, **a** M1 and **b** M1P

Table 3 Value of specific charge-discharge capacity on variation battery performance anode

Anode variation	Specific charge capacity (mAh/g)	Specific discharge capacity (mAh/g)	Initial efficiency (%)
M1P	1489.71	761.16	51.09
M1	906.79	605.56	66.78
Graphite	416.41	314.45	76.11

structure that look more clearly when compared to Fig. 3a. The existence of this pore is due to the void of space so that the release and stretching of porogen particles were the main factors due to heating. Figure 3a, b have pores but Fig. 3b has more visible pores that have a large diameter than Fig. 3a. In Fig. 3a the predicted porogen particle is MgO because of its polar properties, according to research before [11]. In Fig. 3b it is predicted that the porogen particle is $Mg_2Si_2O_6$ which is seen from XRD data also has polar properties. It is also known that the difference in the ratio at the metallothermal stage affects the surface area, diameter and diameter of the pore formed. This pore greatly affects the distribution of lithium ions and electrons during the battery charge-discharge process. In electrodes, pores can facilitate the distribution of lithium ions and electrons [12].

3.4 *Electrochemical Property of the Battery Cell*

Tests carried out with a 18650 cylindrical battery cell and the value of the charge and discharge capacity can be seen in Table 3.

Figure 4 shows the charge and discharge graphs for the 3 anode variations that have been made. On the graphite graph, it is seen that the capacity lost between charge and discharge is the smallest compared to M1 and M1P. The M1P sample has the highest specific capacity as evidenced by the longest M1P graph. In Table 3 when compared with the research before [13] with silicon anode batteries and NCA cathodes the capacitance value was 1039 mAh/g and research before [14] with a capacity value of 1211.15 mAh/g. From the XRD results, it is known that the highest peak known from the JCPDS results is the $Mg_2Si_2O_6$ peak. XRD data is needed to determine the compounds contained in the sample. The theological value of pure silicon capacity is 4200 mAh/g so when compared to the results obtained that is 761.16 mAh/g is still not suitable. However, this can be considered good because silicon at the purification anode is not 100% pure. The reaction that occurs in the battery charging and discharging process can be seen in Eq. 11.

$$LiNi_{0,8}Co_{0,15}Al_{0.15}O_2 + 4Si \leftrightarrow Li_{1-x}Ni_{0,8}Co_{0,15}Al_{0,05}O_2 + Li_xSi_4 \quad (11)$$

The intercalation mechanism that occurs in lithium and silicon in the first release process, where the voltage drop occurs. The deeper lithium intercalation, the lithium

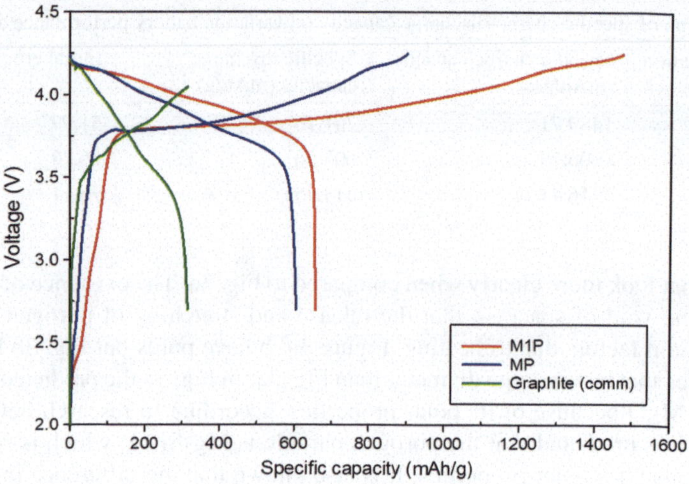

Fig. 4 Graph of charge-discharge li-ion batteries at different anode active materials

ion which reacts with silicon will increase and form a mixture between lithium-silicon which the end result is LixSi4. In this process silicon as an electron receiver is elastic enough to receive many electrons. But unfortunate as a result the volume at the battery anode will increase and cause battery swelling. The effect of this enlarged volume will damage the silicon electrode itself and the active ingredient will lose the contact reaction with electricity which in this case is an electron so that the lithium-ion de-escalates and a large irreversible capacity. When the lithium-ion de-escalation process occurs, the solid electrolyte from the surface of the silicon will break and an increase in volume

will make the electrolyte layer thicker. This solid electrolyte formation will consume lithium ions which cause great irreversibility. So that there is a loss of electrical contact which is marked by the discharge value which has decreased from the value of the charger. The unstable solid electrolyte layer causes silicon and lithium-silicon alloy to come into direct contact with the electrolyte resulting in loss of capacity value. When compared with graphite anode batteries, the capacitance value of silicon is greater because the theoretical capacity of graphite itself is 372 mAh/g. From Table 3 the graphite anode, when compared with the theoretical value, includes results that are close to the theoretical value. Anode from graphite material can already be found in the market and needs quite a lot, but the value of capacitance which is considered relatively small, a lot of research has been done to find other anode substitute graphite. Graphite is known to have good electrical conductor properties, has mechanical properties such as metal and is lightweight. Charge and discharge reactions can be seen in Eq. 12 [15].

$$LiNi_{0,8}Co_{0,15}Al_{0,15}O_2 + 6C \leftrightarrow Li_{1-x}Ni_{0,8}Co_{0,15}Al_{0,05} + Li_xC_6 \qquad (12)$$

From Eqs. 12 it can be seen that graphite material has a structure composed of layers of graphene so that lithium ions allow intercalation reactions to occur. So if the weight between the anode and the cathode is the same, the advantage of the graphite anode is that it can accommodate more lithium ions than the lithium ions released from the cathode during the charging process. The graphite structure forms a hexagonal ring in which every 3 valence electrons in the C atom will bind to 3 covalent electrons from the other C atom. Related to the electrical conductivity of graphite is influenced by electrons which are not covalent-bound. Orbital overlap occurs because the electrons are spread evenly in each of the C atoms. So when given a potential difference, the movement of electrons is delocalized towards the anode so that these electrons produce an electric current. The next variation is the anode with M1 sample. Where the most composition is Mg in accordance with the XRF results obtained, namely 44.43% and 2.26% silicon. Besides these two elements, there are also impurities. The theoretical capacity value of magnesium is unknown, so in its calculation using the cathode as a reference, namely NCA. When compared with garfit, the M1 sample can also be used as a better anode material because it is seen from the higher capacitance value of graphite which can store more electrons. This fairly good result is likely the result of suitable impurities that can be an anode material so that it helps in accommodating electrons which are intercalated from the cathode. The charge and discharge reaction can be seen in Eq. 13.

$$\mathrm{LiNi_{0,8}Co_{0,15}Al_{0,15}O_2} + a(\mathrm{MgO}) \leftrightarrow \mathrm{Li_{1-x}Ni_{0,8}Co_{0,15}Al_{0,05}O_2} + \mathrm{Li}_x(\mathrm{MgO})_a \quad (13)$$

From Eq. 13 in the reaction, the amount of MgO compound which is excited in the reaction is unknown because the substances contained in the metallothermal sample are very complex. But it is known from the XRF results that the percentage composition of the MgO compound is 73.68%. It is known that magnesium has an atomic number 12 with a 2–8–2 configuration so that there are 2 valence electrons. So when 2 electrons intercalate from the anode to the cathode there is 1 valence electron from lithium that enters the anode. It is known that lithium has an atomic number 3 on the periodic table with configuration 2-1. It can be said that magnesium can hold more than the electrons it receives. Currently, research on magnesium batteries is underway to find out the potential of batteries other than lithium batteries.

Figure 5 shows the performance value of specific battery capacity with anode variation. It is known that M1P, M1 and graphite samples have a retention ratio of 27.57, 28.08 and 78.82%. It appears that the best retention ratio is owned by the graphite. This retention ratio shows the level of material stability at the electrodes. So it can be concluded that the anode material from geothermal sludge can be calculated as sustainable anode material. But additional ingredients are needed to improve.

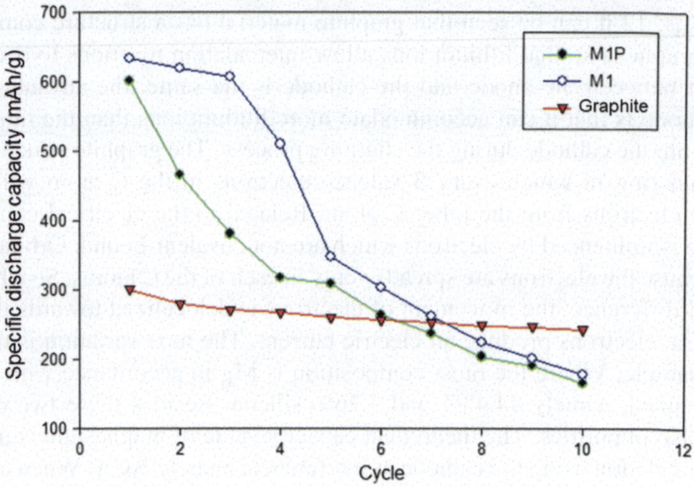

Fig. 5 Graphs of battery specific capacity performance on anode variations

4 Conclusion

The anode material from geothermal sludge silica material with 10.60% silicon content has been successfully prepared using metallothermal method. Based on XRD characteristics after the metallothermal (M1), crystalline MgO crystals are formed which dominate, then after the combustion process (M1P) MgO crystals are reduced and $Mg_2Si_2O_6$ crystals are formed. The specific capacity of li-ion battery with M1P anodes from geothermal sludge material is 761.16 mAh/g and a retention ratio of 27.57%. In li-ion battery with anode M1, a specific capacity of 605.56 mAh/g and a retention ratio of 28.08% was obtained. The li-ion battery with graphite anodes that have been mass-produced obtained a specific capacity of 314.45 mAh/g and a retention ratio of 78.82% after 10 cycles.

Acknowledgements The authors thank to Ministry of Research, Technology and Higher Education of the Republic of Indonesia for the financial support through the research grant of "Penelitian Dasar", Contract number: 719/UN27.21/PN/2019, Cornellius Satria Yudha for the assistance in the measurement of electrochemical property and discussion.

References

1. Brown K, Dunstall K (2000) Silica scaling under controlled hydrodynamic conditions. In: Proceedings world geothermal congress 2000, Japan
2. Europe IMA (2013) Recycling industrial minerals. AISBL, Belgium
3. David L, Bhandavat R, Barrera U, Singh G (2016) Silicon oxycarbide glass-graphene composite paper electrode for long cycle lithium-ion batteries. J Nat Commun 7:10998

4. Qin S, Mori D, Takeda Y, Yamamoto O, Imanishi N (2017) Silicon-carbon composite electrode materials prepared by pyrolysis of a mixture of manila hemp, silicon powder and flake artificial graphite for lithium batteries. J Energy 10(1803)
5. Andarini N, Haryati T, Yulianti R (2018) Pemurnian Silikon (SI) Hasil Reduksi Silika dari Fly Ash Batubara. Jurnal Berkala Sainstek 6:49–54
6. Rao YK (1985) Stoichiometry and thermodynamics of metallurgical processes. Cambridge University Press, Cambridge
7. Purwati AI (2015) Pemurnian silikon asal arang sekam padi menggunakan magnesium dan gas nitrogen. Skripsi, Program Studi Kimia, Institut Pertanian Bogor, Bogor
8. Sapei L, Padmawijaya KS, Sutejo A, Theresia L (2015) Karakterisasi silika sekam padi dengan variasi temperature leaching menggunakan asam asetat. Jurnal Teknik Kimia 9:2–8
9. Singh YP, Abhishek, Pawar SJ (2017) Characterization of silica nano-particles synthesized by thermo-mechanical route. Int J Adv Res Sci Eng 6
10. Alvionita N, Astuti (2017) Sintesis nanopartikel magnesium oksida (MgO) dengan metode presipitasi. Jurnal fisika Unand 6:1–7
11. Lazarov VK, Plass R, Poon H-C, Saldin DK, Weinert M, Chambers SA, Gajdardziska-Josifovska M (2005) Structure of the hydrogen-stabilized MgO(111)-(1X1) polar surface: integrated experimental and theoretical studies. J Phys Rev 9:11–18
12. Suwandi GRF, Dwi AI, Haekal M, Prama EM, Iskandar F (2013) Optimasi resistansi ohmik baterai li-ion melalui kombinasi bahan elektroda berpori dan elektrolit. In: Conference: seminar national material, Bandung
13. Sternad M, Forster M, Wilkening M (2016) The microstructure matters: breaking down the barriers with single crystalline silicon as negative electrode in Li-ion batteries. J Sci Rep 6:6
14. Liao D, Kuang X, Xiang J, Wang X (2018) A silicon anode material with layered structure for the lithium-ion battery. J Phys 986
15. Lundgren CA, Xu K, Jow R, Allen J, Zhang SS (2017) Springer handbook of electrochemical energy. Springer, Berlin

Microstructure, Optical, and Electrical Properties of Barium Titanate ($BaTiO_3$) and $Ba_{1-x}Nd_xTiO_3$ Thin Films Deposited by Chemical Solution Deposition (CSD) Method

R. P. Rini, A. U. L. S. Setyadi, F. Nurosyid, and Y. Iriani

Abstract Barium titanate ($BaTiO_3$) and $Ba_{1-x}Nd_xTiO_3$ thin films used by chemical solution deposition method and prepared with spin-coating technique. The thin films of $BaTiO_3$ and $Ba_{1-x}Nd_xTiO_3$ were grown on a quartz and silicon (Si) substrate. The research aim is to get the best of mole variation of Nd doping to ferroelectric material properties. Doping Nd was varied at 5, 10% in $BaTiO_3$. The thin films were annealed at 850 °C with one hour holding time. The crystal structure, optical and electrical properties of $BaTiO_3$ and $Ba_{1-x}Nd_xTiO_3$ have been investigated. The microstructure of $BaTiO_3$ and $Ba_{1-x}Nd_xTiO_3$ were analyzed using XRD equipment, which showed that $BaTiO_3$ and $Ba_{1-x}Nd_xTiO_3$ crystal had a perovskite shape. The addition mole dopant (x) neodymium (Nd^{3+}), the diffraction angle of the XRD results were seen shifting to the right. As the mole dopant (x) Nd increased, so crystallinity from the $Ba_{1-x}Nd_xTiO_3$ thin films decreased. The crystallite size of $BaTiO_3$, $Ba_{0.95}Nd_{0.05}TiO_3$, and $Ba_{0.9}Nd_{0.1}TiO_3$ were 70; 21; 28 nm. The addition mole dopant (x) Nd^{3+} resulted in smaller grain sizes but it did not significantly affect thickness. The greater the addition mole dopant (x) Nd, the smaller the band-gap produced. The results of the IV characterization showed that the $Ba_{1-x}Nd_xTiO_3$ thin films had a response to light and photovoltaic effects.

Keywords $BaTiO_3$ · Neodymium · Microstructure · Optical properties · CSD

1 Introduction

The development of science and technology is increasingly advanced, particularly in Physics in the area of electronics. The growth of thin films is mostly done and applied in the electronics field such as solar cells. Solar cells are the use of solar energy as

R. P. Rini · A. U. L. S. Setyadi · F. Nurosyid · Y. Iriani (✉)
Physics Department, Graduate Program, Universitas Sebelas Maret, Jl. Ir. Sutami 36A Kentingan, Surakarta 57126, Indonesia
e-mail: yofent_iriani@staff.uns.ac.id

F. Nurosyid · Y. Iriani
Physics Department, Faculty of Mathematics and Natural Sciences, Universitas Sebelas Maret, JL. Ir. Sutami 36A Kentingan, Surakarta 57126, Indonesia

U. Sabino et al. (eds.), *Proceedings of the 6th International Conference and Exhibition on Sustainable Energy and Advanced Materials*, Lecture Notes in Mechanical Engineering,
https://doi.org/10.1007/978-981-15-4481-1_76

an alternative energy that is made by converting solar energy into electrical energy. Ferroelectric is a dielectric material that is able to change the electric polarization spontaneously due to the application of an external electric field [1]. The development of ferroelectric thin film research is increasingly becoming a great potential issue since it can be applied to solar cells [2]. The electric field in the ferroelectric solar cells can control and regulate the response of solar cells [3]. When ferroelectric material is applied to solar cells under ultraviolet lighting, this ferroelectric material will have a high output voltage i.e., above 1 kV [2].

$BaTiO_3$ is one of the materials that attract the attention of many researchers because its ferroelectric nature has a high dielectric constant and a large electro-optical coefficient [4]. The $BaTiO_3$ thin film has an ABO_3 perovskite structure, where A is located at the corner of the cube, B is located on the diagonal of the cube space, and O is located at the diagonal plane of the cube [5]. Perovskite $BaTiO_3$ has five types of crystal structure including hexagonal, cubic, tetragonal, orthorhombic, and rhombohedral. Each of which depends on the temperature of the phase [6]. $BaTiO_3$ can be made in the form of thin films using several deposition methods such as vacuum and non-vacuum methods. The vacuum method consists of Laser Ablation, Chemical Vapor Deposition (CVD), PVD, Ion Plating [7]. The non-vacuum method consists of Screen Printing, Dip Coating, Electrophoresis, Electrodeposited, Spray Pyrolysis, and Spin Coating [5], sol-gel or Chemical Solution Deposition (CSD) [6]. CSD is a thin-film manufacturing method that combines physical and chemical methods. The steps of making thin films by using CSD method include making solutions, conducting spin coating process with a certain rotational speed, then conducting annealing process [8].

$BaTiO_3$ has a wide gap energy ($Eg > 3$ eV) [3]. To use the $BaTiO_3$ as solar cell material, it is necessary to reduce the energy gap by giving doping. Materials that can be used for doping administration such as Ni^{2+}, Er^{3+}, Mn^{4+}, Zn^{2+}, Hf^{4+}, Fe^{2+}, Ce^{4+}, W^{6+}, Gd^{3+}, Nd^{3+}, Sr^{2+} and Co^{2+} [3]. Atoms in perovskite materials can be doped using other atoms provided in which the radius of the atom is almost the same with the radius of the atom replaced [8]. The addition of neodymium (Nd^{3+}) can replace Ba^{2+} in the position of *perovskite* A. This happens since the atomic radius Nd^{3+} (0.995 Å) is smaller than Ba^{2+} (1.35 Å) and the radius of Ti^{4+} (0.68 Å) [8]. The addition mol dopan neodymium (Nd^{3+}) in the making process of $BaTiO_3$ is ion donor (donor doping) that donates excess valence electrons to ion Ba^{2+} [8]. The doping material given can affect the crystal structure, microstructure, and optical properties of a material [3]. In this paper presented $BaTiO_3$ substituted ionic Nd^{3+} [8].

In this research, $BaTiO_3$ thin film is doped with (Nd^{3+}) use the chemical solution deposition (CSD) method with the spin coating technique. The solution $BaTiO_3$ and $Ba_{1-x}Nd_xTiO_3$ was grown on quartz and silicon (Si) substrates deposited using spin coating with a rotational speed of 5000 rpm. The samples annealing process used 850°C with a holding time of 1 h. Samples were characterized using X-Ray Diffraction (XRD) equipment used to determine the microstructure, Scanning Electron Microscopy (SEM) used to determine the grain size and the thickness of the film, UV-Vis used to determine the amount of absorbance and transmittance of thin films, and a Keithley I–V Meter used to determine the amount of current and voltage of thin films.

2 Experiment

The thin films of $BaTiO_3$ and $Ba_{1-x}Nd_xTiO_3$ used by chemical solution deposition method (CSD). Doping Nd was varied at 5, 10 in $BaTiO_3$. The material used in this research was *Barium Acetate* [$Ba(CH_3COO)_2$] 99.00%, *Neodymium Acetate* [$Nd(CH_3COO)_3$] 99.9%, *Acetic Acid* [(CH_3COOH)] 100%, *Titanium Isopropoxide* [$Ti(OC_3H_7)_4$] 97%, *Ethylene Glycol* [($HOCH_2CH_2OH$)] 100%. There were several steps of making $BaTiO_3$ and $Ba_{1-x}Nd_xTiO_3$ solutions including preparing tools and materials and weighing ingredients to mixing materials. The solution was deposited on the Quartz and silicon (Si) substrate, prepared with spin-coating technique at a rotational speed of 5000 rpm. The samples were annealed at 850 °C with a holding time of 1 h. The material properties of the samples were identified using XRD equipment for microstructure testing. The XRD equipment had a Lamda Cu (1.5406 Å). Samples were characterized using SEM to determine the morphology and the thickness of the samples. In addition, samples were characterized using UV-Vis spectrophotometer to determine the amount of absorbance and transmittance of thin films. Ultimately, samples were characterized using a Keithley I–V Meter to determine the amount of current and voltage of thin films.

3 Result and Discussion

3.1 Characterization of Microstructure

The XRD characterization results was show on Fig. 1. Based on Fig. 1, it can be seen that there are almost no other phases that affect the crystal structure of $BaTiO_3$ thin film and $Ba_{1-x}Nd_xTiO_3$. In this experiment, increasing mole dopant Nd succeeded in replacing Ba, therefore the diffraction pattern produced by $Ba_{1-x}Nd_xTiO_3$ was similar to $BaTiO_3$ pattern.

Based on the XRD results, with the addition mole dopant Nd, the diffraction angle is seen shifting to the right. These results are consistent with what has been done by Sandi et al. [9]. The ionic Nd^{3+} can replace Ba^{2+} in the perovskite structure. Along with the addition mole dopant Nd, it was also seen that the intensity value produced from XRD was getting higher. In accordance with the research of Iriani et al. the additional mole dopant Nd^{3+} on the $BaTiO_3$ thin films can increase the intensity of value formed [8].

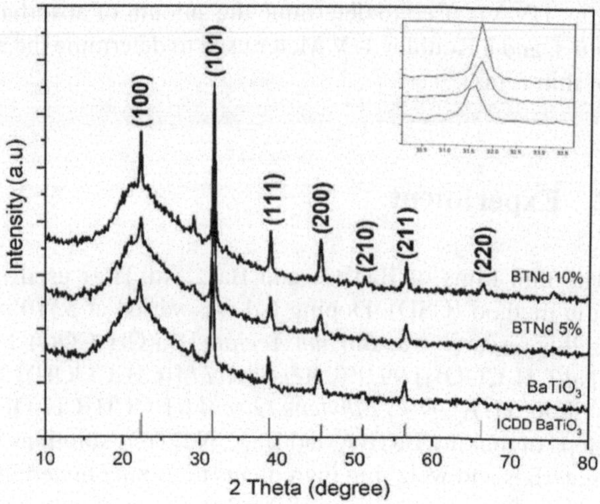

Fig. 1 Diffraction pattern of $BaTiO_3$ and $Ba_{1-x}Nd_xTiO_3$

Table 1 Crystallinity and crystallite size of $BaTiO_3$ and $Ba_{1-x}Nd_xTiO_3$ thin films

Sample	Crystallinity (%)	Crystallite size (nm)
$BaTiO_3$	83.33	70
$Ba_{0.95}Nd_{0.05}TiO_3$	80.31	21
$Ba_{0.90}Nd_{0.10}TiO_3$	80.27	28

The calculation of the crystallinity samples can be seen in Table 1. The addition mole dopant Nd causes the crystallinity of the thin film decrease. The crystallinity of the samples is calculated from the magnitude of the intensity of the diffraction peaks formed. The crystallinity of the samples is a picture of the level of regularity of the crystals that are formed in each film.

The crystallite size of $BaTiO_3$ and $Ba_{1-x}Nd_xTiO_3$ thin films can be calculated using the following Scherrer equation [3]:

$$D = \frac{k\lambda}{\beta \cos\theta} \quad (1)$$

Table 2 Lattice parameters of $BaTiO_3$ and $Ba_{1-x}Nd_xTiO_3$ thin films from the calculation result

Sample	Lattice parameters (Å)		Tetragonality
	a = b	c	
$BaTiO_3$	4.003	4.020	1.004
$Ba_{0.95}Nd_{0.05}TiO_3$	4.003	4.020	1.004
$Ba_{0.90}Nd_{0.10}TiO_3$	4.016	4.030	1.003

where the value of β is the value of FWHM, θ is the diffraction angle, k is the constant of Scherrer which is at 0.9, and λ is the wavelength of X-ray. The calculation of the crystallite size samples can be seen in Table 1.

Based on the additional mole dopant Nd, the result of crystallite size dwindled. In accordance with studies that have been carried out by Gaur et al. that the crystallite size decreases with the additional mole dopant Nd [8]. The result of the crystallite size ranges from 70 to 28 nm.

The refined XRD results using the GSAS software can be seen in Fig. 2. This software processes with a refinement process so that the crystal structure and lattice parameters a, b, and c of the thin film can be identified.

The lattice parameter values from the GSAS software refinement results can be seen in Table 3. The result shows that thin film $BaTiO_3$ and $Ba_{1-x}Nd_xTiO_3$ has a tetragonal structure since it has a lattice parameter value $a = b \neq c$ and the angle $\alpha = \beta = \gamma = 90°$. The results of refinement $Ba_{1-x}Nd_xTiO_3$ thin film diffraction pattern with

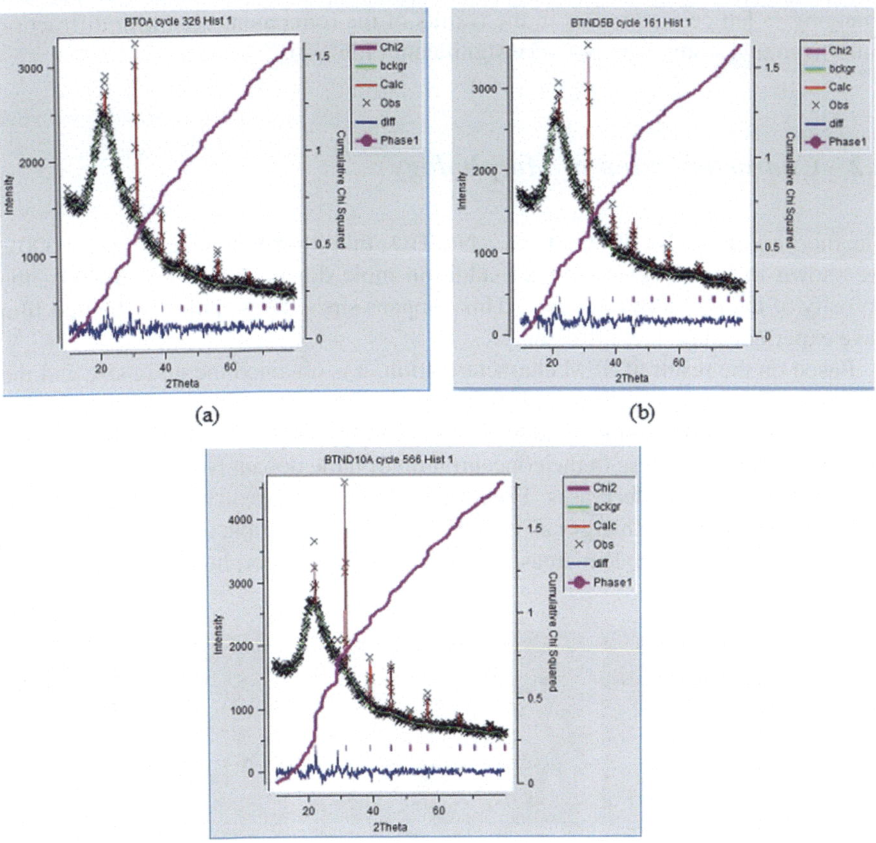

Fig. 2 The refinement of $BaTiO_3$ and $Ba_{1-x}Nd_xTiO_3$ diffraction patterns

Table 3 Lattice parameters of $BaTiO_3$ and $Ba_{1-x}Nd_xTiO_3$ thin films from the result of the refinement software GSAS

Sample	Lattice parameters (Å)		Tetragonality
	a = b	c	
$BaTiO_3$	3.992	4.018	1.006
$Ba_{0.95}Nd_{0.05}TiO_3$	3.983	4.008	1.006
$Ba_{0.90}Nd_{0.10}TiO_3$	3.984	4.002	1.004

a variation mole dopant Nd, show the alteration of lattice parameter because of dopant Nd^{3+} into $BaTiO_3$. Decrease in the $Ba_{1-x}Nd_xTiO_3$ lattice parameter with respect to the initial lattice parameter $BaTiO_3$ caused the diffraction angle of the $Ba_{1-x}Nd_xTiO_3$ thin film with a variation mole dopant Nd has shifted from the $BaTiO_3$ diffraction angle on ICDD which has a certain field orientation (*hkl*). It can affect the distance between atoms in the crystal and cause changes in parameters including grid a, b, and c. These results are in accordance with research conducted by Iriani et al. [8]. The changing in lattice parameters in the results of the refinement, thin film diffraction patterns in this study were not very significant (Table 2).

3.2 *Characterization of Morphology*

The morphology of $BaTiO_3$ and $Ba_{1-x}Nd_xTiO_3$ thin films with magnitude 80,000× are shown in Fig. 3. Based on the addition mole dopant Nd in the $BaTiO_3$, the porosity of the thin film decreases. This happens since the granules in the thin film have experienced perfect diffusion.

Based on the result of SEM characterization, it is obtained the grain size and the thickness of $BaTiO_3$ and $Ba_{1-x}Nd_xTiO_3$ that can be seen in Table 4. The grain size of $BaTiO_3$ thin films cannot be calculated since the granules are completely diffused. When there is an increase in the concentration of mole dopant Nd in the thin film of $BaTiO_3$, the grain size decreases. The results of this study are in accordance with the research conducted by Zhang et al. [10]. However, in the sample of $Ba_{0.90}Nd_{0.10}TiO_3$, the grain size of the sample increased. The reason for the morphological change was

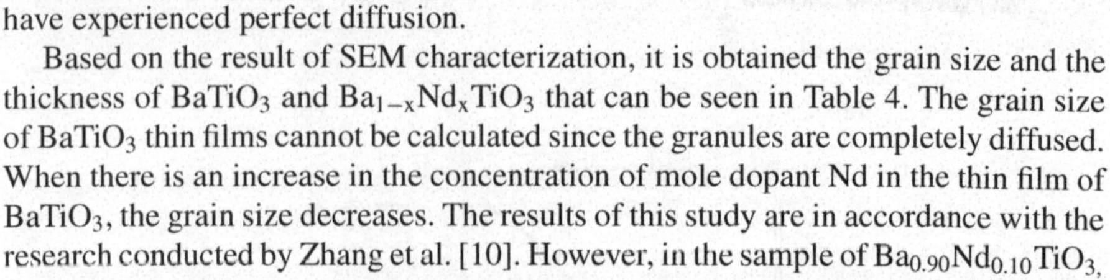
(a) (b) (c)

Fig. 3 SEM image of **a** $BaTiO_3$, **b** $Ba_{0.95}Nd_{0.05}TiO_3$, **c** $Ba_{0.90}Nd_{0.10}TiO_3$

Table 4 Grain size and thickness of $BaTiO_3$ and $Ba_{1-x}Nd_xTiO_3$ thin films

Sample	Grain size (nm)	Thickness (nm)
$BaTiO_3$	–	633
$Ba_{0.95}Nd_{0.05}TiO_3$	103	300
$Ba_{0.90}Nd_{0.10}TiO_3$	126	335

not determined. It can only be assumed that the grain boundaries in the SEM results are clearly visible so that the number of grain size that can be calculated and the average is increasing.

The thickness was got from the results of cross-section measurements on the $BaTiO_3$ and $Ba_{1-x}Nd_xTiO_3$ thin film, as presented in Table 4. The thickness of $BaTiO_3$ and $Ba_{1-x}Nd_xTiO_3$ films affects the addition mole dopant Nd. When $BaTiO_3$ was given an additional mole dopant Nd, the thickness of the film got thinner. However, the more mole concentrations dopant Nd was added, the thickness of the film got thicker. As the results of research from Iriani et al. [8]. The thickness of $Ba_{0.95}Nd_{0.05}TiO_3$ is 300 nm, $Ba_{0.90}Nd_{0.10}TiO_3$ is 335 nm, and the thickness of $BaTiO_3$ pure is 633 nm.

3.3 *Characterization of Optical Properties*

The calculation of band-gap is using the method of *tauch-plot* with the following equation:

$$(\alpha h\upsilon)^2 = A(h\upsilon - E_g) \tag{2}$$

where $h\upsilon$ is the photon energy, A is a constant, α is the coefficient of absorption, and E_g is the optical band-gap [3].

The transmittance spectrum of quartz substrate grown by $BaTiO_3$ and $Ba_{1-x}Nd_xTiO_3$ is presented in Fig. 4. It can be seen that the higher the concentration mole dopant Nd is given, the lower the peak transmittance value. Based on Fig. 4, it is found that the thin film of $BaTiO_3$ reaches the highest transmittance value. The amount of this transmittance value is inversely proportional to the value of the absorbance value.

The transmittance value of $BaTiO_3$ thin film in the visible light area is greater if mole dopant Nd is increased. The addition mole dopant Nd can increase the absorption of visible light in the thin film of $BaTiO_3$. The transmittance of the $Ba_{1-x}Nd_xTiO_3$ film decreases with the decreasing wavelengths to a certain extent. Waves of absorption from thin films $Ba_{1-x}Nd_xTiO_3$ are bigger if it is compared to $BaTiO_3$. This result is in accordance with the research by Fu et al. [11].

Based on UV-Vis and cross-section test results, the band-gap value is obtained. The band-gap values of the $BaTiO_3$ and $Ba_{1-x}Nd_xTiO_3$ thin films are presented in Table 5. The results show that the band-gap of $BaTiO_3$ thin films is greater than

Fig. 4 Transmittance of $BaTiO_3$ and $Ba_{1-x}Nd_xTiO_3$ thin films

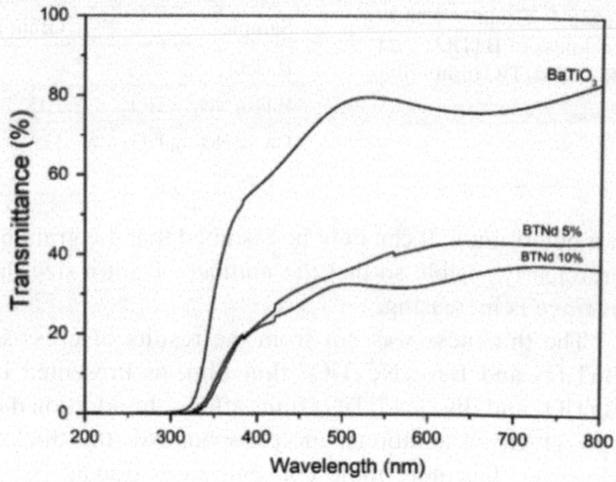

Table 5 Energy gap of $BaTiO_3$ and $Ba_{1-x}Nd_xTiO_3$ thin films

Sample	Energy gap (eV)
$BaTiO_3$	3.64
$Ba_{0.95}Nd_{0.05}TiO_3$	3.00
$Ba_{0.90}Nd_{0.10}TiO_3$	3.00

$Ba_{1-x}Nd_xTiO_3$. This happens since the addition mole dopant Nd can reduce the band-gap of the thin film. In this study, the Nd^{3+} ionic replaces the Ba^{2+} ion on the A-side so that it can cause the narrowing of the conduction band-gap. This is in accordance with the research of Jiang et al. [3]. In this study, the value of the band-gap of the thin film of $BaTiO_3$ and $Ba_{1-x}Nd_xTiO_3$ is around 3.00–3.64 eV for the overall concentration of mole.

3.4 *Characterization of Electrical Properties*

The characterization of electrical properties (current–voltage) from $BaTiO_3$ and ($Ba_{1-x}Nd_xTiO_3$) thin films was got by varying mole dopant Nd^{3+} using Keithley I–V meters. Figure 5 shows the current and voltage relationship curves of $BaTiO_3$ and $Ba_{1-x}Nd_xTiO_3$ films in dark and light conditions which indicate that the thin film has diode characteristics. The curve shows that in the $Ba_{1-x}Nd_xTiO_3$ film with Si substrate (p-type semiconductor), a connection has occurred. This curve shows that there is an induced current flow (photocurrent) so that the $Ba_{1-x}Nd_xTiO_3$ film shows the response to the light. The $Ba_{1-x}Nd_xTiO_3$ film when it is irradiated (in light conditions) has more conductive properties if it is compared to that when it is in dark conditions due to the absorption of photon energy. The result of the photon energy charge can be separated and collected by the cathode and anode. The results

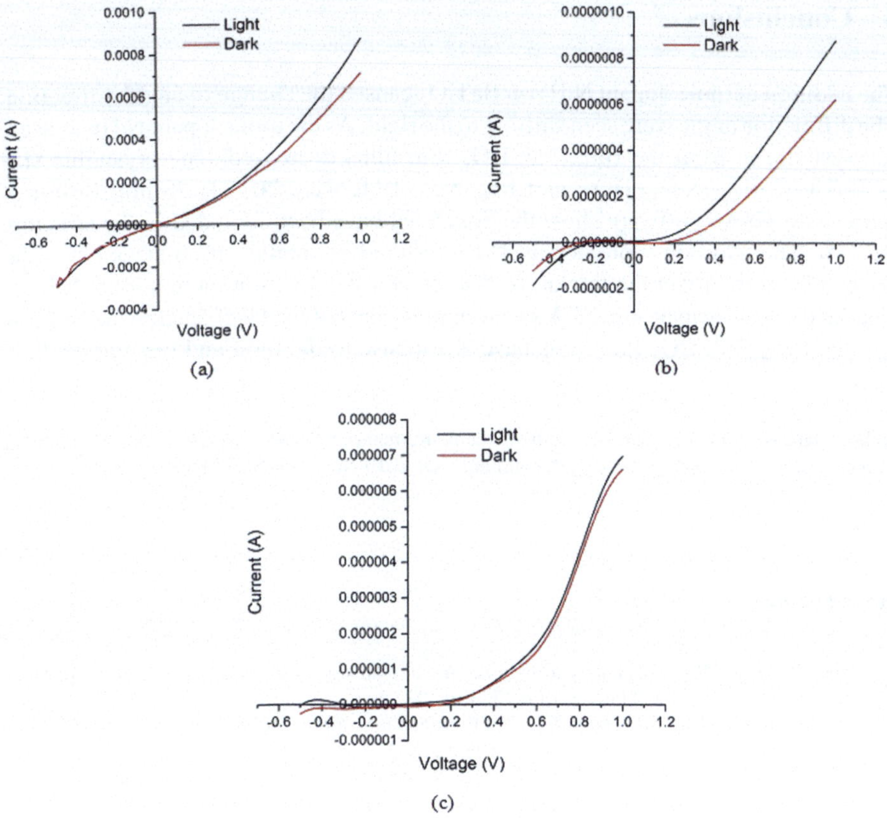

Fig. 5 Characterization curve I–V of $BaTiO_3$ and $Ba_{1-x}Nd_xTiO_3$ thin films

of this study are in accordance with the research conducted by Wang et al. [12]. The amount of I_{sc} and V_{oc} value from $BaTiO_3$ and $Ba_{1-x}Nd_xTiO_3$ film are shown in Table 6. Based on Table 6, it can be seen that the additional mole dopant Nd^{3+} can increase the value of I_{sc} and V_{oc}. The results are in accordance with the research conducted by Jiang et al. [3].

Table 6 I_{sc} and V_{oc} value of $BaTiO_3$ and $Ba_{1-x}Nd_xTiO_3$ thin films

Sample	I_{sc} (mA)	V_{oc} (V)
$BaTiO_3$	2.14×10^{-9}	1.78×10^{-5}
$Ba_{0.95}Nd_{0.05}TiO_3$	6.88×10^{-9}	2.12×10^{-1}
$Ba_{0.90}Nd_{0.10}TiO_3$	1.09×10^{-8}	1.52×10^{-2}

4 Conclusions

The addition of mole dopant Nd^{3+} on $BaTiO_3$ caused the change of angel diffraction. The diffraction angle were seen shifting to the right. As the mole dopant Nd increased, so crystallinity from the $Ba_{1-x}Nd_xTiO_3$ thin films decreased. The crystallite size of $BaTiO_3$, $Ba_{0.95}Nd_{0.05}TiO_3$, and $Ba_{0.9}Nd_{0.1}TiO_3$ were 70; 21; 28 nm. Result of refine with GSAS software show the dopan Nd have been included in the structure $BaTiO_3$. The addition mole dopant Nd^{3+} resulted in smaller grain sizes but it did not significantly affect the thickness. The greater the addition mole dopant Nd, the smaller the band-gap produced. Moreover, the results of the IV characterization show that the $Ba_{1-x}Nd_xTiO_3$ thin films have a response to the light and the photovoltaic effects.

Acknowledgements The authors wish to thank Kementrian Riset, Teknologi and Pendidikan Tinggi for financial support through Penelitian Dasar 2019 No 719/UN27.21/PP/2019.

References

1. Iriani Y, Yasin SR (2018) Fabrication of barium titanate doped strontium using coprecipitation method. IOP Conf Ser: Mater Sci Eng 333:012047
2. Cai W, Fu CL, Lin ZB, Deng XL (2011) Vanadium doping effects on microstructure and dielectric properties of barium titanate. Ceram Int 37:3643–3650
3. Jiang W, Cai W, Lin Z, Fu C (2013) Effects of Nd-doping on optical and photovoltaic properties of barium titanate thin films prepared by sol–gel method. J Mater Res Bull 48:3092–3097
4. Kumbhar SS, Mahadik MA, Chougule PK, Mohite VS, Hunge YM, Rajpure KY, Moholkar AV, Bhosale CH (2015) Structural and electrical properties of barium titanate ($BaTiO_3$) thin films obtained by spray pyrolysis method. Mater Sci Poland 33(4):852–861
5. Vijatovic MM, Bobic JD, Stojanovic BD (2008) History and challenges of barium titanate: Part II. Sci Sinter 40:155–165
6. Li W, Xu Z, Chu R, Fu P, Hao J (2009) Structure and electrical properties of $BaTiO_3$ prepared by sol–gel process. J Alloy Compd 482:137–140
7. Irzaman PY, Apipah ER, Noor I, Alkadri A (2015) Characterization of optical and structural of lanthanum doped $LiTaO_3$ thin films. Integr Ferroelectr 167:137–145
8. Iriani Y, Jamaludin A, Nurhadi N (2016) Deposition barium titanate ($BaTiO_3$) doped lanthanum with chemical solution deposition. J Phys: Conf Ser 776:012064
9. Sandi DK, Supriyanto A, Jamaluddin A, Iriani Y (2016) The influences of mole composition of strontium (x) on properties of barium strontium titanate ($Ba_{1-x}Sr_xTiO_3$) prepared by solid state reaction method. Am Inst Phys Conf Proc 1710:030006.1–030006.4
10. Zhang W, Cao L, Wang W, Su G, Liu W (2013) Effects of neodymium doping on dielectric and optical properties of $Ba_{(1-x)}Nd_xTi_{1.005}O_3$ ceramics. Ceram Silik 2:146–150
11. Fu C, Cai W, Lin Z, Jiang W (2014) Photovoltaic effects of bismuth ferrite and Nd-doped barium titanate thin films prepared by sol–gel method. Mater Sci Forum 787:347–351
12. Wang W, Liu F, Lau CM, Wang L, Yang G, Zheng D (2014) Field-effect $BaTiO_3$-Si solar cells. Appl Phys Lett 104:123901

Investigating the Effect of Layer Thickness on the Product Quality of PLA Manufactured by 3D Printing Technique

H. Sukanto, D. F. Smaradhana, J. Triyono, and P. Wicaksono

Abstract Three dimensional printing (3DP) has become a popular technique to create prototype with a complex shape. The demands for the quality of the product including dimensional and shape accuracy become the reason that makes 3DP operated with slow speed. However, the products of 3DP still need a further finishing process to obtain good quality grades, so there is opportunity to execute 3DP with higher speed. 3DP technique with high speed process varying the layer thickness was investigated in this work. Fused Deposition Modeling (FDM) method was utilized with filament of polylactic acid (PLA) as the printing material. The speed, extruder temperature and bed temperature were determined at 80 mm/s, 200 °C and 60 °C respectively. Infill type was selected using line of 100% and nozzle diameter of 5 mm. The ratio of layer thickness and nozzle diameter were calculated to vary layer thickness which is 20, 40, 60 and 80%. This work aims to understand the effect of layer thickness to the 3DP product quality as well as its mechanical properties. The results show that the increase of layer thickness enlarged the roughness of vertical surface in the range of 6–11 μm. The increase of layer thickness also affected on the dimension deviation in horizontal and vertical direction at 0.12% and 0.03% respectively.

Keywords 3D printing · Layer thickness · Polylatic acid · Infill pattern

1 Introduction

Additive manufacturing (AM) or three dimensional printing (3DP) is one of the most techniques for producing a functional object with complex internal shapes and structures. 3DP also able to significantly shortening the complication of planning route and shop floor scheduling of production process [1, 2]. It constructs a part through layer by layer method as piling papers up to become a rim. *3DP* explores coming up challenges and chances in additive manufacturing, ranging from new establishments of processes and materials to new simulation and design tools [3]. 3DP is predicted to have a dominant role in the future industrial revolution. It will

H. Sukanto (✉) · D. F. Smaradhana · J. Triyono · P. Wicaksono
Mechanical Engineering Department, Universitas Sebelas Maret, Surakarta, Indonesia
e-mail: herusukanto@staff.uns.ac.id

U. Sabino et al. (eds.), *Proceedings of the 6th International Conference and Exhibition on Sustainable Energy and Advanced Materials*, Lecture Notes in Mechanical Engineering,
https://doi.org/10.1007/978-981-15-4481-1_77

make technology cheaper and more accessible to anyone. It may seem like science fiction being able to print out an artificial limbs, prosthetic noses and even human organs, but this is fast becoming a scene of people everyday reality by utilizing 3DP for applications within the medical industry [4].

There are several models or methods that were able to create a product by AM or 3DP. Derived from the standard process, AM's have been classified into seven categories; they are material extrusion, photo polymerization, powder bed fusion, binder jetting, sheet lamination, material jetting and directed energy deposition [5]. Extrusion based AM called fused deposition modeling (FDM) is already available in desktop and known for being low-cost and capable of high printing speeds as compared to other 3D printing models [6]. FDM process technology enables the design to be changed with a simple file modification, and therefore leaves more freedom in terms of parts design and shape. The FDM uses the motion of a three-axis machine to lay a melted material filament down on the piece during the production. The filament will solidify immediately when it comes into contact with the previous layer. The materials used for the filament normally are wax, polylatic acid (PLA), polyamide, polypropylene and acrylonitile butadiene styrene (ABS) [7].

Quality of products resulted by FDM process have been investigated on mechanical properties considering to printing parameters. In general, the optimum parameters were applied by engineers to design the best parts with minimum weight and maximum strength, minimum processing time and energy consumption in order to fulfill the customers requirement [8]. Mechanical properties including tension, flexural and impact strengths, of 3DP specimens came from PLA material have been revealed with the level of 48.7 MPa, 157.3 MPa, and 13.7 kJ/m^2 respectively [9]. Another investigation found that specimens have just 10–73% of the strength of the samples produced by injection molding depending on infill (sometime called as hatch pattern or raster) orientation [10]. Furthermore, the internal geometrical structures of samples tested include full, honeycomb, drills and stripes. It was found that the specimen with the honeycomb pattern showed the highest tensile strength at fracture, as well as specific tensile strength at break [11].

2 Materials and Method

The PLA filament from e-Sun with 1.75 mm diameter was selected as material that would be melted and extruded down to the bed for constructing specimens via nozzle with hole diameter of 0.5 mm. The specimens were built by 3DP machine from Creality Ender 3.0 and the G-code program corresponding to size and shape of specimen was generated by Cura 3.6. The layer thicknesses were measured as ratio to the nozzle hole diameter which were presented as percentage. They were set at the value of 20, 40, 60 and 80% to the nozzle hole diameter of 0.5 mm and treated as variables to be observed. Thickness is expressed in terms of the ratio of the nozzle hole diameter to create a perception that it is not justified to set the thickness of the layer beyond the nozzle hole size. A layer thickness of more than 100% of nozzle hole will produce an overhang layer and this is forbidden in the concept of the 3DP process.

The main printing process parameters were generated by Cura 3.6 including printing speed, bed temperature, extruder temperature and infill density which are pointed at 80 mm/s, 60 °C, 200 °C and 100% respectively. During G-code execution on the machine, the environment temperature was maintained at 27 °C for ensuring homogeneous cooling rate overall specimens.

The physical quality of the specimens are specified in quantity from measurements of density, surface roughness and dimensional deviation. The density measurement process refers to the ASTM D792. Meanwhile the roughness and the deviation of dimension involve two directions of measurement namely the vertical direction to measure them in the direction of the thickness of the layer (wall) and the horizontal to measure the surface roughness (infill roughness).

For the purposes of mechanical testing, the shape and size of the specimen are made according to the tensile and bending test standards of ASTM D638 and ASTM D790. Figure 1 shows the shape of the tensile and bending specimen after it has been printed on the 3DP machine. The specimen hardness property is investigated by durometer at the level unit of Shore D which each sample is taken at five points.

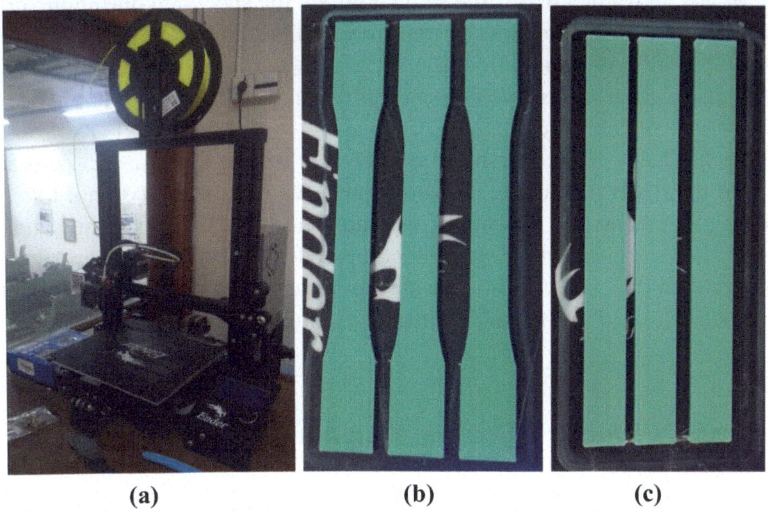

Fig. 1 **a** Ender3 Creality 3DP machine, **b** tensile test specimen, **c** flexural test specimen. *Note* The images scale is neglected

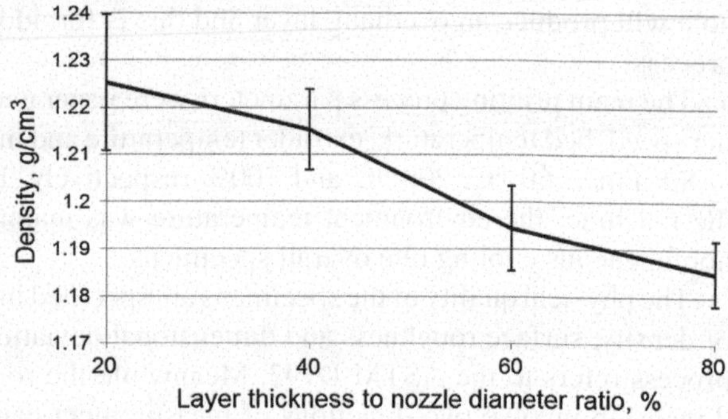

Fig. 2 The graphically trend of layer thickness to the specimens density

3 Results and Discussion

3.1 Layer Thickness Effects on Samples Quality

The mass intensity or density of the specimens which they are resulted by 20% thickness layer depict as the highest value. The phenomenon has relation to the amount of nozzle traffic crosses the same point on the specimen during the printing process. The short distance between the surface of the nozzle and the last layer and its relatively soft conditions allow the surface of the nozzle to apply pressure to the layer, making the specimens denser. Figure 2 shows the trend line of specimens densities versus nozzle hole diameter.

Air gap has been defined as the space between the tracks or raster. Zero air gap means that the roads just touch; a positive gap means the tracks do not touch; and a negative gap means that two tracks overlap. The −0.003 air gap indicates more dense and shows higher tensile strength compared with the zero air gap, also an air gap smaller than this value is not advisable as there will be excess material build up at the nozzle of the 3d printer, and on the part being printed [12].

3.2 Layer Thickness Effect on Roughness

3D printing technology creates several non-ideal effects on the designed structure which are not well studied so far to the best of our knowledge. The 3D printing strategies create a surface roughness over the printed device. Although high resolution tips for 3D printing are available, usually they lower the 3D printing speed significantly and increase the cost of production [13].

The effect of layer thickness as the ratio of the nozzle hole diameter in the vertical direction (specimen wall) is relatively no difference at each level of the ratio. The average roughness rate (Ra) fluctuates in the range of 0.5–3.8 μm, as shown in Fig. 3. Fifteen specimens have been tested which they display a uniform roughness pattern.

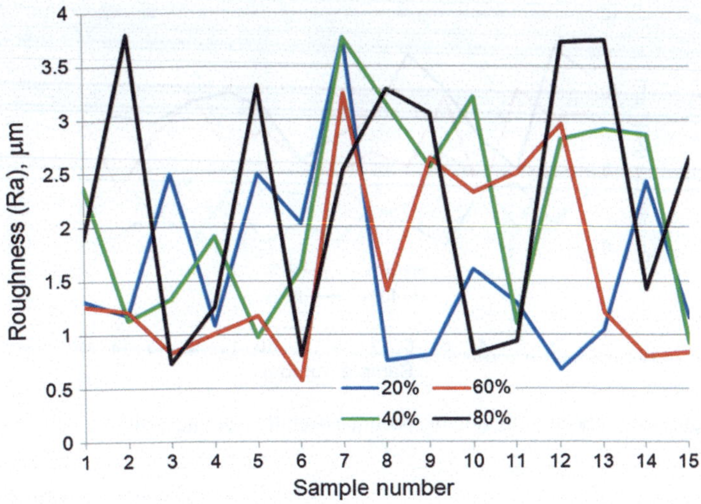

Fig. 3 The roughness profile of wall specimens

The wall surface profile formed of all specimens shows closely same on topography as well, as shown in Fig. 4.

Surface roughness in the horizontal direction (flat plane) is dominantly determined by the infill setting when building the G-code program. However, the layer thickness also has the potential to affect the surface quality of flat plane. Heavier thickness will create a concave curve between the paths of the molten filament to be deeper so that the roughness profile will also increase, as shown in Fig. 5. Layer thickness of 20% of the nozzle hole size produces the smoothest surface (average roughness rate 6–7.5 μm) compared to the others thicknesses. The ideal schematic formation

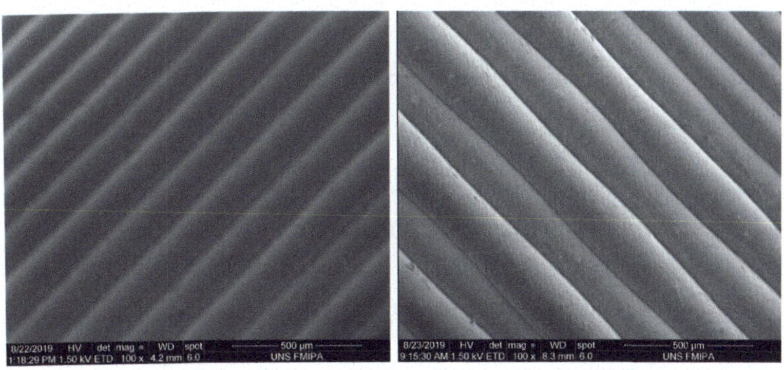

(a) (b)

Fig. 4 Wall surface profile of specimens with thickness layer ratio of **a** 20%, **b** 40% to hole nozzle diameter obtained by SEM

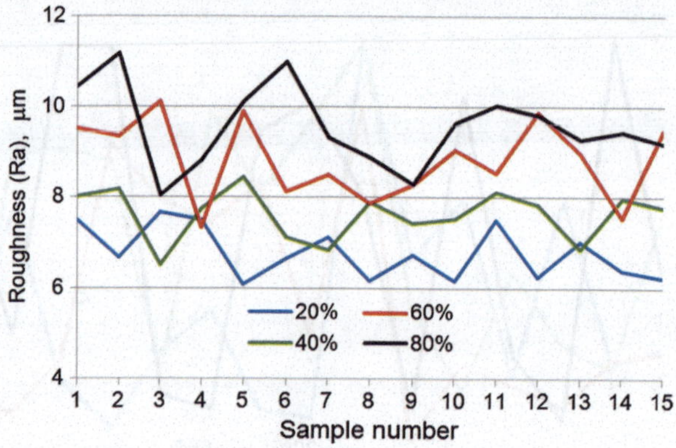

Fig. 5 Roughness number distribution of specimen overall layer thickness

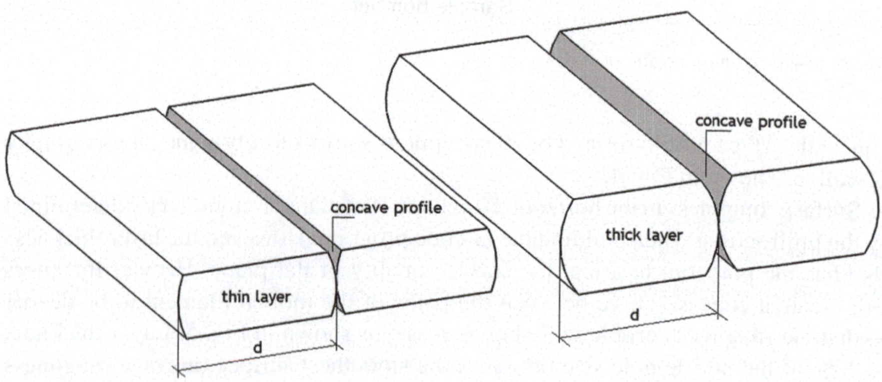

Fig. 6 Schematic concave curve creation between adjacent track of molten filament. Deep of concave profile indicates the roughness flat plane

of a flat surface profile in a specimen with different layer thickness is illustrated in Fig. 6.

3.3 Layer Thickness Effect on Dimension Deviation

One possible reason for the improved dimensional precision and reliability of frameworks fabricated by the 3DP technology is the achievement of 100% of the product's density at the final printing procedure [14]. The accuracy of the dimension in the vertical direction or shape of the wall shows a better precision than the horizontal direction. The dimensions of the specimens wall are distorted from the designed

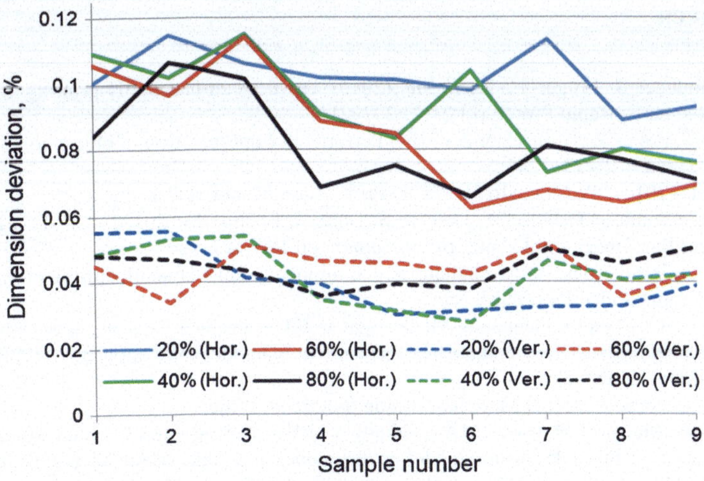

Fig. 7 Distortion of specimen volumetric dimension

dimensions to become bigger or less than planned. During printing process the outermost layer of the specimen wall is most likely pushed out by the molten filament of the inside while it is still soft. This phenomenon results in the addition of dimensions to the specimen wall. While in the direction of thickness, it is possible the lower layer to experience perfect solidification before being crushed by the upper layer. The condition of the already solid layer is able to hold perfectly the next layer without deformation of shape and size occur. Figure 7 shows the value of specimen volume distortion that it lies in the range of 0.03–0.12%.

4 Conclusion

This research reveals that the layer thickness takes account on product density. The thinner layer tend to create a denser product than the thicker one. In consideration to the shape and size, the thickness of the layer does not have much effect on dimensions especially in the vertical direction of the product.

Acknowledgements This research was founded by PNBP UNS 2019.

References

1. Glasschroeder J, Prager E, Zaeh MF (2015) Powder-bed-based 3D-printing of function integrated parts. Rapid Prototyp J 21(2):207–215
2. De Tre G, Hallez A, Bronselaer A (2014) Performance optimization of object comparison. Int J Intell Syst 29(2):495–524
3. Lipson H (2014) "Welcome to 3DP", 3D Print. Addit Manuf 1(1):1
4. Atala A, Cronin L, Dawood A, Defilice S, Derisi J, Ebeling M (2015) How 3d printing will save your live. Image Publishing, Bounermouth, pp 123–127
5. Gao W et al (2015) The status, challenges, and future of additive manufacturing in engineering. CAD Comput Aided Des 69:65–89
6. De Leon AC, Chen Q, Palaganas NB, Palaganas JO, Manapat J, Advincula RC (2016) High performance polymer nanocomposites for additive manufacturing applications. React Funct Polym 103:141–155
7. Radi B, El Hami A (2016) Material forming processes. Willey
8. Majeed A, Muzamil M, Awan MZ, Siddiqui A (2019) Developing of a manufacturing cycle architecture for fused deposition modeling technique. Int J Light Mater Manuf 10–14
9. Raj SA, Muthukumaran E, Jayakrishna K (2018) A case study of 3D printed PLA and its mechanical properties. Mater Today Proc 5(5):11219–11226
10. Song Y, Li Y, Song W, Yee K, Lee KY, Tagarielli VL (2017) Measurements of the mechanical response of unidirectional 3D-printed PLA. Mater Des 123:154–164
11. Galeta T, Raos P, Stojšić J, Pakši I (2016) Influence of structure on mechanical properties of 3D printed objects. Procedia Eng 149(June):100–104
12. Ahn SH, Montero M, Odell D, Roundy S, Wright PK (2002) Anisotropic material properties of fused deposition modeling ABS. Rapid Prototyp J 8(4):248–257
13. Honari MM, Mirzavand R, Saghlatoon H, Mousavi P (2018) Investigation of the 3d printing roughness effect on the performance of a dielectric rod antenna. IEEE Antennas Wirel Propag Lett 17(11):2075–2079
14. Noort RV (2012) The future of dental devices is digital. Dental Materials 28(2):3–12

A Review on Aluminum Arc Welding and It's Problems

I. Habibi, Triyono, and N. Muhayat

Abstract The latest trends in the automotive world have shifted from conventional materials to lightweight materials such as aluminium. Aluminum was chosen because it is lighter than steel, this is because the specific gravity of aluminum is smaller than the weight of steel which is 2.70 g/cm^3 and iron 7.86 g/cm^3. The same volume and sufficient strength make aluminum used in the land and air transportation industry The big challenge for this industry is to make vehicles that are fuel efficient and environmentally friendly, one way to save fuel is to reduce the weight of the material. Aluminum welding is the most important thing in the joining process. Aluminum weldability has many disadvantages, because aluminum is very reactive. Common defects in aluminum welding are porosity, cracked shrinkage and large heat affected zone (HAZ). Something that affects weldability in aluminum is high thermal conductivity, high shrinkage in solidification, formation of oxides on the surface, high solubility of hydrogen under melting conditions. Some of these problems become problems in conventional processes (GTAW and GMAW) to new processes such as laser beam welding and cold metal transfer. Until now the defects in the aluminum welding process have been studied by researchers. to facilitate the understanding of readers in aluminum problems and their resolution, this review is needed.

Keywords Aluminum · Welding · CMT · Hydrogen · Porosity

1 Introduction

In the modern industry, especially the automotive industry, the use of lightweight and strong materials greatly supports vehicle efficiency [1]. The efficiency of a vehicle can be increased due to the reduced weight of the vehicle [2]. The same vehicle with a lighter weight will naturally consume less fuel. Aluminum has a third density

I. Habibi (✉) · Triyono (✉) · N. Muhayat
Department of Mechanical Engineering, Faculty of Engineering, Universitas Sebelas Maret, Surakarta, Indonesia
e-mail: habibigoh@gmail.com

Triyono
e-mail: triyonomesin@uns.ac.id

U. Sabino et al. (eds.), *Proceedings of the 6th International Conference and Exhibition on Sustainable Energy and Advanced Materials*, Lecture Notes in Mechanical Engineering,
https://doi.org/10.1007/978-981-15-4481-1_78

lighter (2.70 g/cm^3) than carbon steel (7.86 g/cm^3). Therefore the use of aluminum in the industry is more popular.

Welding is an important joining process and leads to chemical, physical and mechanical changes in aluminum alloys. Chemical changes can occur due to different types of filler and base metal, physical changes can occur due to the influence of heat generated from the welding process. This thermal cycle also causes changes in the base metal, weld metal, and heat affected zone (HAZ).

The welding process for aluminum that is commonly used is the gas and arc welding process. Arc welding is more often used because the area affected by heat is smaller so that the heat that is applied is more focused and the welding speed can increase. This is very influential because the conductivity of aluminum is high (205.0 W/m K), four times higher than steel (50.2 W/m K) [3].

Deficiencies in aluminum welding besides high thermal conductivity there are several other factors namely: high hydrogen solubility in the welding process [4], increase in high solubility when solidification starts at temperatures close to 660 °C [5], low solubility near nothing in the solidification process [6]. There have been many attempts by researchers to solve the problem of aluminum arc welding. To facilitate the reader in the aluminum arc welding problem, a review article is needed.

2 Welding Process for Aluminum

2.1 GTAW (Gas Tingsten Arc Welding)

The welding process of gas tungsten arc welding (GTAW) and gas metal arc welding (GMAW) are widely used to join aluminum in various applications such as the automotive industry. Corrosion resistance of aluminum alloy welds is affected by aluminum alloys, filler alloys and the welding techniques used. The cause of galvanic cells that cause corrosion is the potential difference between base metal, weld metal, and HAZ. Potential differences can also occur due to uneven changes in microstructure [7].

The GTAW welding process uses a wire filler that is outside the welding arc. Figure 1 shows the GTAW welding scheme. This welding process requires both hands to weld so it is more difficult to form good welding.

In aluminum welding, hydrogen porosity is a common problem. The high stability of aluminum oxide creates an instant oxide film on the surface of the wire and base metal. The surrounding humidity reacts with surface oxides which form water-containing chemical compounds. Because of this natural process, welding wires that are not stored properly or welding grooves without wires before welding can contribute to moisture in the welding arc. Humidity causes the hydrogen content to enter the welding process and causes porosity in the freezing [5].

TIG and MIG welding processes are arc welding using shielding argon gas. For special needs it is possible to mix argon and helium in certain ratios. It is important

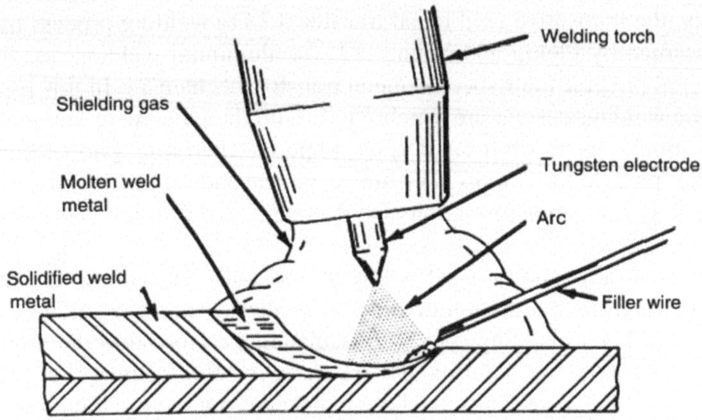

Fig. 1 GTAW process scheme

to keep shielding gas from containing hydrogen (from water/steam) or oxygen. Both of these gases can cause porosity in welding because it dissolves in weld pool and is trapped during the compaction process. The oxide membrane contains water vapor which turns into hydrogen and oxygen when heated. The oxide membrane must be removed during the welding process. Whereas the fusion point is much higher than aluminum fusion point so it cannot be liquefied [8].

Shielding gas in aluminum welding uses argon and helium. Shielding gas can use pure argon, pure helium and a mixture of both. The use of pure argon is more popular in both GMAW and GTAW welding for aluminum. This gas is not good for health if inhaled continuously. Disease caused can be severe and lead to death, for that it is strongly recommended during the welder welding process using standard equipment [9–11].

2.2 *GMAW (Gas Metal Arc Welding)*

Metal inert gas (MIG) or gas metal arc welding (GMAW) is an arc welding technique that is widely used in industry. High productivity, flexibility and low cost make this process suitable for a variety of industrial applications. This technique uses consumable wire electrodes that melt in an electric arc between the cut and the tip. Arc and weld pool are protected by protective gas produced from nozzles. Continuous innovation has improved the quality and productivity of welds. The main evolution is MIG pulses, which use pulse welding currents. This feature controls the formation of droplets and their transfer via arcs. Over the past decade, new electronic power supplies have allowed precise control of the welding current. Thus, the deposition of molten metal can be modulated according to the welding current signal parameters.

Recently, the innovative cold metal transfer (CMT) welding process by Fronius has been studied by Pickin and Young [12] for aluminum welding and by Robert Talalaeva et al [13] for their specific metal transfer mechanism. In this process, the pulses in the welding current are synchronous with the pulse wire feeding. Regular wire oscillation helps to drop the dorplet properly. Therefore, part of the welding process takes place at low energy. The impact of heat and fumes is reduced as shown by Furukawa [14], spatter formation also decreases and dilution can be controlled properly.

Another mode that can be used is to combine pulse currents with CMT, Puls-Mix CMT6 which increases productivity. With their specific properties, the MIG process is suitable for cladding applications, cladding requires low delution and low penetration. In contrast to the usual TIG process, in MIG welding the main part is used to melt the filler wire. Thus, the base metal does not require too much heat and lower penetration than the TIG process [15].

The main problem in welding aluminum is the decrease in mechanical properties in the HAZ. Microstructure changes cause loss of hardening and other mechanical strength in the HAZ region of heat treatable aluminum alloys [16]. This can be avoided with the correct welding design. Alternatively, the dual-torch opposite GTAW process can be used to make a balanced weld [16–18].

2.3 CMT (Cold Metal Transfer)

Cmt is defined as 'cold metal transfer', this process refers to GMAW with different heat input ratio and is commonly called the short arc process [12]. The CMT process illustrates the revolution of welding techniques in equipment and usability. This process is a new process and many unexplored applications [15]. CMT has a long history, in 2002 this process began to be recognized and researched to obtain applications and optimal solutions to touch to the real industry. CMT technology has revolutionized the welding process and produced aesthetic welds, besides the deposition of contaminated metals and low heat input is also an advantage of this process [19].

The difference between the CMT and MIG/MAG processes is the cutting method and the mechanism of the mechanism that did not yet exist. During the welding process, temperature variations in welding and the parent metal have an important role in the characteristics of the material, the residual stresses in the dimensions and the accuracy of the welds results [13]. This method provides a deposition method for material deposition and low heat input by combining an innovative feeder wire system with high-speed digital control. The wire feed rate and the arcing phase cycle are controlled to melt the globular of filler wire and base metal [19].

There are two main features of the CMT process: one is at the point of short circuit with low current corresponding to a low heat input, another is the short circuit occurrence in a stable controlled manner [19]. This peculiarity of the CMT process translates into the elimination of the costly activities associated with cleaning the piece of molten metal spatter. Another significant advantage of the CMT process is perfect arc length management; in the traditional process, the arc length is managed by means of measurement of the weld voltage.

The third significant advantage of CMT is the increased ability to compensate for high edge coupling tolerances. The problem, encountered when welding thin thicknesses with large gaps between the edges to be joined, is represented by the relatively high heat input, so high indeed as to melt the edges before the weld bead has been formed [15].

In the CMT process, when the electrode wire tip makes contact with the molten pool, the servomotor of the 'robacter drive' welding torch is reversed by digital process control. This causes the wire to retract promoting droplet transfer which is depicted in Fig. 2. During metal transfer, the current drops to near-zero and thereby any spatter generation is avoided. As soon as the metal transfer is completed, the arc is re-ignited and the wire is fed for- ward once more with set welding current reflowing [14]. A typical CMT welding electrical signal cycle can be defined as the period required to deposit a droplet of molten electrode into the weld pool. The analysis of current and voltage waveform is essential to study the energy distribution of different phases in droplet transfer process [20, 21].

The retraction of the wire during the short circuiting phase plays an important role, as it leads to prevention of spatter generation and also produces better weld bead aesthetics. The Cold Metal Transfer Welding has a wide variety of applications such as crack repair welding, additive manufacturing, composite joint pin fabrication, and cladding.

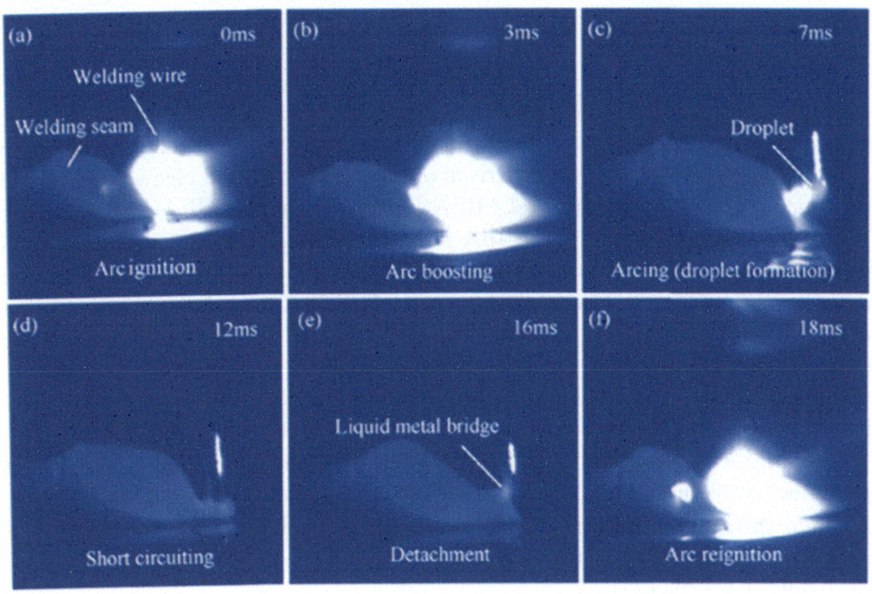

Fig. 2 Dorplet transfer image

3 Defects in Aluminum Welding

The porosity identified in laser welding of aluminum alloys consists mainly of four types: hydrogen pores, porosity caused by the collapse of unstable keyholes, porosity due to entrapment of gases by surface turbulence and shrinkage porosity [22, 23].

Hydrogen is the main cause of pores during the welding of aluminium. The hydrogen solubility of iron is compared to that of aluminium. The significant surge in solubility when solidification takes place (at about 660 °C) and the low rate of solubility in the solid state show that hydrogen in aluminium (if the rate of cooling is too rapid) can easily be frozen in, thereby creating pores. The solidification range can be considered as a yardstick for determining the formation of pores. If there is no solidification range or only a slight one, the aluminium welding deposit solidifies so rapidly that degassing cannot fully take place, thereby causing porosity [5]. Heat affected zone liquation cracking, Solidification cracking, Oxide inclusions, Loss of alloying elements. There are five main reasons for the occurrence of pores and lack of fusion in the welding of aluminium materials:

- High thermal conductivity
- Good hydrogen solubility in the molten state
- A noticeable surge in solubility when solidification occurs at about 660 °C
- Low solubility in the solid state
- No or virtually no solidification range.

4 Solution for Minimizing Defect

This study proposes a modification to the MIG-IEA joint so that sound and aesthetic welds may be obtained in a single welding pass when welding plates of 6061-T6 with 12.7 mm in thickness. A comparison of the macrostructural characteristics and mechanical behaviour of the novel MIEA joint with the MIG-IEA and the single V groove joints showed that the MIEA joint produces uniform weld beads with a reduced HAZ as a result of the low heat input. This reduction in thermal affection minimizes the overaging effect of the base metal leading to a mechanical strength that fulfills the acceptation criteria for 6061-T6 welded joints and speeds up the welding process [16].

When Pavan kumar et al. welded thin aluminium alloy sheets using filler, which is of same composition as of base metal, the weld exhibited a quasi-binary composition. This composition is potentially less susceptible to solidification cracking, controlled fusion line, narrower heat affected zone (HAZ) and reduced intermetallic phase area. The microstructures for different weld parameters revealed fine recrystallization at the joints. A uniform distribution of grains and its size in weld HAZ and base metal was distinctly visible [24].

Song et al. conducted joining of 5A06 aluminum alloy and AISI stainless steel using various filler metals (i.e., pure Al 1100, ER4043 and ER4047); the results showed that Si additions had great effect on preventing the build-up of IMC layer and minimizing its thickness.

Results indicate that CMT provides better joint tensile strength and ductility compared to metal inert gas welding. The softness in the heat affected zone was very low, but a big hardness gap was recognised in the welded metal compared to base metal. The joint had mechanical property coefficients of 77, 60 and 69% for yield strength, ultimate tensile strength and elongation respectively. [25] Additionally, it could be concluded that CMT process achieved mechanical characteristics comparable to FSW and LBW since these processes use low heat input as well. We should mention here the advantages of CMT in comparison with other welding processes with respect to its higher welding speeds, simplicity and easy automation, low skill required for operation, availability, cost and economic issue. The Laser-CMT hybrid welding produces welds with better mechanical properties and aesthetics than the Laser welding and Laser-MIG hybrid welding.

References

1. Gould BYJE (n.d.) Joining aluminum sheet in the automotive industry–a 30 year history
2. Sun J, Yan Q, Gao W, Huang J (2015) Materials and design investigation of laser welding on butt joints of Al/steel dissimilar materials. Mater Des 83:120–128. https://doi.org/10.1016/j.matdes.2015.05.069
3. Xiong J, Zou S (2019) Active vision sensing and feedback control of back penetration for thin sheet aluminum alloy in pulsed MIG suspension welding. J Process Control 77:89–96. https://doi.org/10.1016/j.jprocont.2019.03.013
4. Praveen P, Yarlagadda PKDV (2005) Meeting challenges in welding of aluminum alloys through pulse gas metal arc welding. J Mater Process Technol 164–165:1106–1112. https://doi.org/10.1016/j.jmatprotec.2005.02.224
5. Prakash J, Tewari SP, Srivastava BK (n.d.) Shielding gas for welding of aluminium alloys by TIG/MIG welding-a review 1:690–699
6. Mezrag B, Benachour M (2015) Control of mass and heat transfer for steel/aluminium joining using cold metal transfer process 20:189–199. 10.1179/1362171814Y.0000000271
7. Rao KS, Rao KP (2004) Pitting corrosion of heat-treatable aluminium alloys and welds : a review 57:593–610
8. Zheng W, He Y, Yang J, Gao Z (2018) Hydrogen diffusion mechanism of the single-pass welded joint in welding considering the phase transformation effects. J Manuf Process 36:126–137. https://doi.org/10.1016/j.jmapro.2018.09.026
9. Antonini JM (2003) Health. Health Eff Weld 33:4–10. https://doi.org/10.1080/01449290310001624338
10. Hariri A, Leman AM, Yusof MZM (2013) Experimental study on welding fumes exposure from aluminum metal inert gas (MIG) process 701:382–386. www.scientific.net/AMR.701.382
11. Antonini JM (n.d.) Health effects associated with welding, Elsevier. https://doi.org/10.1016/b978-0-08-096532-1.00807-4
12. Pickin CG, Young K (2006) Evaluation of cold metal transfer (CMT) process for welding aluminium alloy 11:583–586. https://doi.org/10.1179/174329306x120886
13. Talalaev R, Veinthal R, Laansoo A, Sarkans M (2012) Cold metal transfer (CMT) welding of thin sheet metal products 243–250. https://doi.org/10.3176/eng.2012.3.09

14. Taylor P (2010) New CMT arc welding process—welding of steel to aluminium dissimilar metals and welding of super-thin aluminium sheets new CMT arc welding process—welding of steel to aluminium dissimilar metals and welding of super-thin aluminium sheets 37–41. https://doi.org/10.1533/wint.2006.3598
15. Taylor P, Lorenzin G, Rutili G (2009) The innovative use of low heat input in welding: experiences on 'cladding' and brazing using the CMT process 37–41. https://doi.org/10.1080/09507110802543252
16. Ambriz RR, Barrera G, García R, López VH (2009) A comparative study of the mechanical properties of 6061-T6 GMA welds obtained by the indirect electric arc (IEA) and the modified indirect electric arc (MIEA). Mater Des 30:2446–2453. https://doi.org/10.1016/j.matdes.2008.10.025
17. Zhang YM, Pan C, Male AT (2000) Improved microstructure and properties of 6061 aluminum alloy weldments using a double-sided arc welding process 31:2537–2543
18. Zhang BYMZB (1999) Welding aluminum alloy 6061 with the opposing dual-torch GTAW process 202–206
19. Pickin CG, Williams SW, Lunt M (2011) Characterisation of the cold metal transfer (CMT) process and its application for low dilution cladding. J Mater Process Tech 211:496–502. https://doi.org/10.1016/j.jmatprotec.2010.11.005
20. Selvi S, Vishvaksenan A, Rajasekar E (2018) Cold metal transfer (CMT) technology—an overview. Def Technol 14:28–44. https://doi.org/10.1016/j.dt.2017.08.002
21. Sun Z, Lv Y, Xu B, Liu Y, Lin J, Wang K (2015) Investigation of droplet transfer behaviours in cold metal transfer (CMT) process on welding Ti-6Al-4V alloy. https://doi.org/10.1007/s00170-015-7197-9
22. Cao X, Jahazi M, Immarigeon JP, Wallace W (2006) A review of laser welding techniques for magnesium alloys. J Mater Process Technol 171:188–204. https://doi.org/10.1016/j.jmatprotec.2005.06.068
23. Taylor P, Cao X, Wallace W, Immarigeon J, Poon C (2007) Materials and manufacturing processes research and progress in laser welding of wrought aluminum alloys. In: II metallurgical microstructures, research and progress in laser welding of wrought, pp 37–41. https://doi.org/10.1081/amp-120017587
24. Pavan N, Vendan SA, Shanmugam NS (2016) Investigations on the parametric effects of cold metal transfer process on the microstructural aspects in AA6061. J Alloys Compd 658:255–264. https://doi.org/10.1016/j.jallcom.2015.10.166
25. Elrefaey A (2015) Effectiveness of cold metal transfer process for welding 7075 aluminium alloys 20:280–286. https://doi.org/10.1179/1362171815y.0000000017

Analytical Calculation, Numerical and Hydrostatic Test as a Validation of Material Strength of the New RX-450 Rocket Motor Tube

Setiadi, B. Wicaksono, A. Riyadl, Bagus H. Jihad, and A. Apriyanto

Abstract The RX-450 rocket motor tube Hydrostatic Test was carried out at the end of 2016. The test was carried out at the place where the tube was made, with the aim is validating the tube's strength against workload and there was no leakage as the result. The test results show that at a pressure of 120 bar the tube can withstand the pressure and does not leak. To further strengthen the validation of the test results, analytical and numerical calculations are carried out using structural analysis software. The analytical method is done by considering the tube structure as a thin cylinder closed at the end, so a mathematical approach is carried out to the membrane theory and bending stress on the shell structure. Analytical and numerical calculation results show that the stress caused by a workload of 120 bar, the tube structure is safe against yield strength with safety factors close to 1.5. Another important validation is the tensile test of materials at B2TKS Puspiptek Serpong, as a comparison of its mechanical properties. From the certificates received with the type of cylindrical material, mention that KH 406 has a Yield Strength value 1320 MPa and Tensile Strength value is 1620 MPa. The results of tensile tests at B2TKS Puspiptek Serpong were 92.4% of datasheet value.

Keywords Hydrostatic test · Work load · Numerical calculation

1 Introduction

1.1 RX-450 Rocket Motor Tube

The rocket motor tube is one component of the rocket that contains rocket fuel known as solid propellant which is also a place for burning rocket fuel which causes the rocket to have a forward thrust. Therefore, the material of a motor tube must have

Setiadi · B. Wicaksono (✉) · A. Riyadl · B. H. Jihad · A. Apriyanto
Rocket Technology Center—LAPAN, Jakarta, Indonesia
e-mail: bagus_wicaksono2002@yahoo.com

Setiadi
e-mail: seti1159@gmail.com

U. Sabino et al. (eds.), *Proceedings of the 6th International Conference and Exhibition on Sustainable Energy and Advanced Materials*, Lecture Notes in Mechanical Engineering,
https://doi.org/10.1007/978-981-15-4481-1_79

certain specifications, especially in terms of the strength of the tube structure because during propellant combustion takes place, the rocket motor tube will experience high pressure and temperature. And so the calculation of analysis and numerical strength, hydrostatic test of the tube structure becomes important to do so that it can be determined from the beginning to define whether a rocket tube material is feasible or not to used as a rocket motor tube from its structural strength.

Another important thing to validating the strength of the rocket motor tube material structure is by doing a tensile test of the material as a comparison to the mechanical properties of the material from the certificate we received. From the factory certificate we received, the type of material is KH-406 [1]. Looking at its chemical composition, this material is closely identic with 30CrMnSiA Steel [2]. Furthermore, things related to the strength of the material structure are its tensile strength and its yield strength values.

1.2 Background and Purpose of the Hydrostatic Test

It is mentioned in another paper or in previous paper that hydrostatic test is commonly used as a necessary step to validate the pipeline to distribute the high pressure gas or liquid such as Benefits and Limitation Testing by John F. Kiefner and Willand A. Maxey 2013, Collective effects of Leakage, Temperature Changes and Entrapped air During Hydrostatic Testing by Dr. Lawrence Matta 2017, Studies on Maximum Hydrostatic Maximum Pressure for New Pipeline by Pengchao Chen, Hong Zhan, Hong Ju Wang, 2011. This paper presents the study on maximum hydrostatic testing pressure and gives the theoretic and experimental results to demonstrate the fitness of the tube material which used as the rocket tube. At the end of 2016, a new RX-450 motor tube was obtained to replace the old one. When it was delivered to the LAPAN rocket technology center hydrostatic testing was carried out at the place of making the tube.

The purpose of hydrostatic testing is to find out the a mechanical structure changes as the effect of a material statical load and also to verify the rocket tube structure design [3]. Basically, hydrostatic test is a pressure test to evaluate the quality of a pipe or other component, where the test object is filled in with water and its pressure is raised to reach a certain pressure value, then held for a certain duration to determine the strength of the material from the result of applying a large pressure approaching the limit the strength of the material. Water is used as a vehicle for pressure transfer because water is a type of fluid that is not compressible so that it will provide an optimal impact on material that contains such high pressure. If the material being tested for its strength is unable to withstand the pressure of the water fluid, the material will crack or break and experience fluid leakage, which is an indicator that the material tested has exceeded its strength capability. This test is carried out at pressures above the normal working conditions of the test object. In the context of a rocket motor tube, this test aims to determine the strength of the motor tube material when resisting pressure from the inside. Hydrostatic testing needs to be done to test the rocket motor tubes made by manufacturers that are different from previous manufacturers. The rocket tube structure is required to have a strong enough character to withstand the internal pressure generated by burning solid propellants. Therefore, in its design it takes careful calculation and needs to be validated by testing its strength. In this hydrostatic test only pressure parameters are measured which affect the structure of the tube material, while the measurement of other parameters such as strain will be carried out on other tests and the other chapter or discussion.

2 Methodology

2.1 Test Implementation

The step of hydrostatic implementation as follow [4]:

- A 4 m long rocket motor tube with flanges on both ends, then the flanges are connected with a round plate with bolt connections so the tubes can be tightly closed.
- The tube is slightly removed then filled with water. After the two ends of the tube are closed tightly then connected with the hose to the pressure pump.
- The tube containing water is pumped in stages until it reaches a pressure of 12 Mpa.
- After the pressure on the tube reaches 12 Mpa, the pressure is held for 10 min.
- The result is the tube does not crack or leak.

And the fotograph documentation of the Hydrostatic Test is as shown in Fig. 1 [5].

In order to validate and make sure the strength of tube structure, the analytical calculation and strength testing preparation or measurement is necessary to define

a. **b.**

Fig. 1 **a** Hydrostatic test preparation on a 4 m length tube; **b** both ends of the tube are given a flange and closed with a round plate wether this material is proper to used as rocket tube or not as shown in Fig. 1. In this case, the hydrostatic test, and the statical test is done to filtering the strength quality and also to make sure that this material is strong enough to have a treatment of high pressure and high temperature during the burning time is running.

Hydrostatic testing is a pressure test to evaluate the quality of a pipe or other component, where the test object is filled with water and the water pressure is raised to reach a certain pressure value, held for a certain duration to determine the strength of the material from the impact of treatment giving a large pressure approaching the limit the strength of the material with water media as a vehicle for pressure transfer because water is a type of fluid that is not compressible so that it will provide an optimal impact on the material that holds the water fluid that contains a large pressure and if the material tested strength is not able to withstand the pressure of the water fluid then the material it will crack or break and experience fluid leakage, which is an indicator that the material being tested has exceeded the value of its ability to withstand the pressure. During the hydrostatic test, the pressure is supplied to the steel tube. This test is carried out at pressures above the normal working conditions of the specimens [6]. In the context of a rocket motor tube, this test aims to determine the strength of the motor tube material when resisting pressure from the motor tube by liquid fluid. Hydrostatic testing needs to be done to test the rocket motor tubes made by manufacturers that are different from previous manufacturers. The type of material used in this rocket motor tube is steel metal with a thickness of 4 mm. The structure of the rocket tube is required to have a character strong enough to withstand the internal pressure generated by burning solid propellants as well as URL [7]. Therefore, in its design it takes careful calculation and needs to be validated by testing its strength.

Hydrostatic testing is basically done to represent the real phenomenon when used for burning Propellant burning time but this hydrostatic test does not involve the burning of solid propellant because it takes into account the safety factor of the personnel involved in the hydrostatic test process.

The Hydrostatic Test is also a validation of the theoretical strength calculation of the material. Theoretically or calculations on paper as Fig. 2, by involving the calculation of the material strength of the radial direction and the longitudinal direction, the results obtained calculations that state that this material is safe to be used to withstand pressures up to 120 bar with a material thickness of 4 mm.

The mechanism of the hydrostatic testing process is by connecting the steel tube to a pressure conditioner through a rubber hose that is able to withstand pressures of up to ±450 bar, which first ensures that the entire volume of the tube is fully filled with water without leaving room volume for air. After this condition is reached, the cylinder pressure can be gradually increased through the Pressure Conditioner by leveraging the pressure lever on the pressure conditioner. With each increase in the pressure value of 10 bar, the test is holded for 5 minutes for each 10 bar pressure value increment to be visually checked for leaks and phenomena of tube cracking along the test material and then to continue adding the pressure value and so on until the test is stopped at 120 bar as shown at figure above. And during the Hydrostatic test, the value of the pressure applied to the test tube is numerically recorded on the graph in the data acquisition recorder.

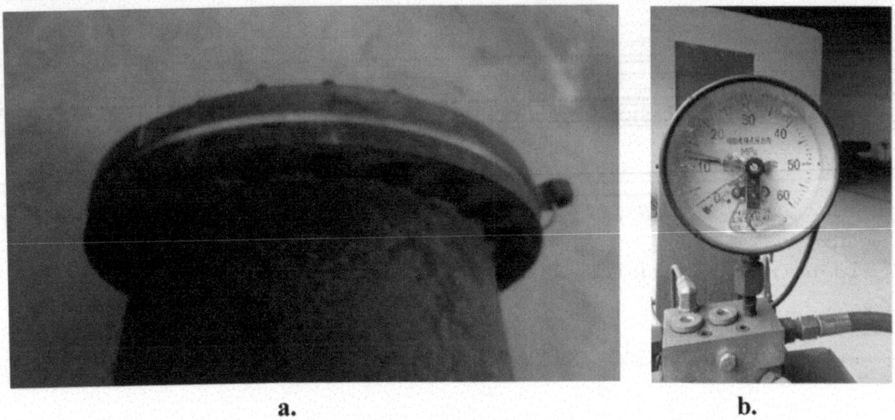

Fig. 2 **a** The tube after filled with water and closed tightly and then pressured gradually; **b** pressure gauge shows the number 12 Mpa

2.2 *Basic Theory with Test Models*

Stress analysis on the motor tube is done by taking into account the discontinuous area on the motor tube and the cap or lid.

At a sufficient distance from the junction region, membrane theory is quite accurate. For cylinders the radius of a produces a longitudinal direction of the cylinder and a hoop. If you only take into account the membrane stress, you will get a disconnect. Thus at that connection a Q_0 shear force and M_0 bending moment are evenly distributed along the circumference [8] as shown in Fig. 3.

Thus the Bending moment M_x and Shear Force Q_x equation become :

$$(M_x)_{x=0} = -D\left(\frac{d^2w}{dx^2}\right)_{x=0} = M_0 \tag{1}$$

Fig. 3 The bending force direction illustration during the hydrostatic test

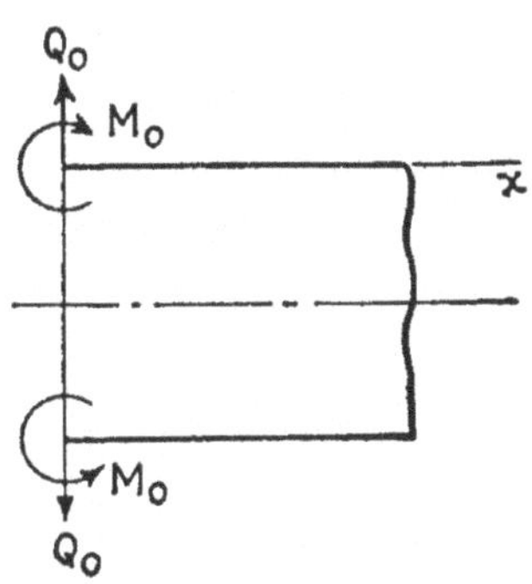

$$(Q_x)_{x=0} = \left(\frac{dM_x}{dx}\right)_{x=0} = -D\left(\frac{d^3w}{dx^3}\right)_{x=0} = Q_0 \tag{2}$$

Thus the final expression for deflection w is

$$w = \frac{e^{-\beta x}}{2\beta^3 D}[\beta M_0(\sin\beta x - \cos\beta x) - Q_0\cos\beta x] \tag{3}$$

The maximum deflection is obtained at the load end, where

$$(w)_{x=0} = -\frac{1}{2\beta^3 D}(\beta M_0 + Q_0) \tag{4}$$

2.3 Application of Analytical Hydrostatic Testing Formulas

If the ends of the shell are built in, they cannot move out, and local bending occurs at the edges as Fig. 4. If of the length L of the shell is sufficiently large, we can investigate this bending, the moment M_0 and the shear force Q_0 being determined from the condition that the deflection and the slope along the built-in edge $x = 0$. According to these condition, the equation become [9]:

$$-\frac{1}{2\beta^3 D}(\beta M_0 + Q_0) = \delta \tag{5}$$

$$\frac{1}{2\beta^2 D}(2\beta M_0 + Q_0) = 0 \tag{6}$$

$$D = \frac{E.h^3}{12(1 - v^2)} \tag{7}$$

$$\beta^4 = \frac{E.h}{4a^2.D} = \frac{3(1 - v^2)}{a^2.h^2} \tag{8}$$

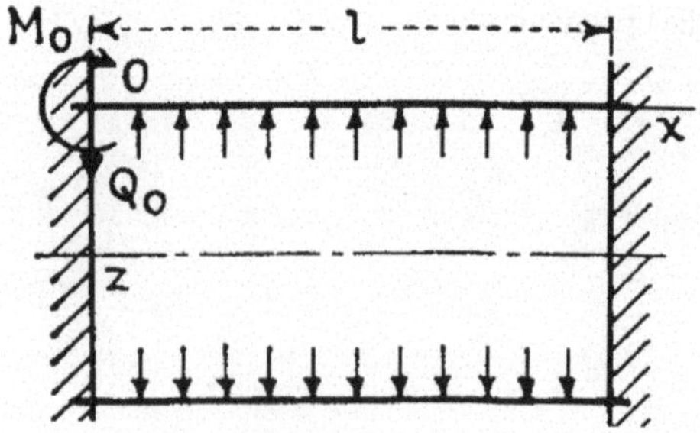

Fig. 4 The force direction inside the tube of hydrostatic test

where

β is constant value at calculation,
D is flexural rigidity equation for a thin elastic plate.
a is tube radius
h is tube thickness
E is Elastic Modulus of material,
δ is increases value of cylinder radius,
ν is poisson ratio.

$$\beta^4 = \frac{3(1-\upsilon^2)}{a^2h^2} \tag{9}$$

Combining the maximum bending stress produced by M_x with the membrane stress, gives:

$$(\sigma_x)_{max} = \frac{a.p}{2h} + \frac{3}{4}\frac{a.p}{h\sqrt{3(1-\upsilon^2)}}\zeta\left(\frac{\pi}{4}\right) = 1.293\frac{a.p}{2h} \tag{10}$$

This stress which acts at the outer surface of the cylindrical shell is about 30% larger than the membrane stress acting in the axial direction. In calculating stress in the circumferential direction in addition to the membrane stress *pa/h,* the hoop stress caused by the deflection w as well as the bending stress produced by the moment $M_\varphi = v.M_x$ must be considered. In this way stress obtained at the outer surface of the cylindrical shell become

$$\sigma_t = \frac{a.p}{h} - \frac{E.\omega}{a} - \frac{6\upsilon}{h^2}M_x = \frac{a.p}{h}\left[1 - \frac{1}{4}\theta(\beta x) + \frac{3\upsilon}{4\sqrt{3(1-\upsilon^2)}}\zeta(\beta x)\right] \tag{11}$$

Taking $\upsilon = 0.3$ and using numerical value of the function $\theta(\beta x)$ and $\delta(\beta x)$ [8], can be written :

$$(\sigma_t)_{max} = 1.032\frac{a.p}{h} \text{ at } \beta x = 1.85 \tag{12}$$

2.4 *Numerical Calculation*

Nastran Software
This section will present an analysis of the RX-450 motor tube strength using the finite element method, with the help of MSC software, Patran Nastran 2018. The material used as a material for making tubes is KH 406. The type of analysis is

Fig. 5 A Ansys software analysis method illustration

static linear with an internal load of 12 MPa (average operational pressure). The tube geometry specifications used in the analysis are: Tube length 4 meters, outer diameter 457 mm and tube thickness 4 mm.

Then its material properties, Yield Strength 1320 Mpa, Ultimate tensile strength 1620 Mpa and Modulus of elasticity 205,939 Mpa.

Ansys Software

ANSYS is an engineering software based on finite element analysis capable to design products and create simulations that predict the product's performance. This software allows engineer to analyze detail visualization of where structures bend and twist, and simulate the distribution of stresses.

In this paper, ANSYS will be used to simulate and analyze the stress in the vessel walls due to the internal pressure resulted during hydrostatic test as Fig. 5. The result of the simulation analysis using ANSYS will be compared to the analytical method.

3 Results and Discussion

3.1 Analytical Calculation Results

In calculating stress in circumferential direction, the hoop stress caused by deflection w as well as bending stress produced by the moment M_{Θ} must be considered and the equation of stress become

$$\sigma_t = \frac{a.p}{h}\left[1 - \frac{1}{4}\theta(\beta x) + \frac{3v}{4\sqrt{3(1-v^2)}}\zeta(\beta x)\right] \tag{13}$$

$$\sigma_x = \frac{a.p}{2h} + \frac{3}{4}\frac{a.p}{h\sqrt{3(1-v^2)}}\zeta(\beta x) \tag{14}$$

With value $\beta x = 1.85$, the value σ_t and σ_x become :

$\sigma_t = 1.032\ a.p/h = 7012\ \text{kg/cm}^2$
$\sigma_x = 0.56865\ a.p/h = 3864\ \text{kg/cm}^2$
The combined stress $(\sigma t^2 + \sigma x^2)^{1/2}$ produces the value:
$\sigma_{tot} = 1.1783\ \text{ap/h} = 7990\ \text{kg/cm}^2$
$\sigma\, vonmisses = (\sigma_t^2 + \sigma_x^2 - \sigma_t . \sigma_x)^{1/2} = 6930\ \text{kg/cm}^2$
$\sigma_{yield} = 13{,}200\ \text{kg/cm}^2$
$\sigma\, vonmisses < \sigma_{yield} \rightarrow$ Safe.

3.2 *Numerical Calculation Results*

Numerical analysis is done using MSC software and Patran Nastran with Hexahedral 8 mode and size is 5 mm. The result of software calculation form is table converted into graphical form as shown below for the value of Hoop stress, longitudinal stress and the von Misses.

Nastran Software

The calculation was covering along the surface of the tube and based on Fig. 6, we get the highest value of Hoop Stress, Von Misses, and Longitudinal Stress at the edge of the tube as follow:

Hoop Stress = 707.9 Mpa
Von Misses = 616.2 Mpa
Longitudinal Stress = 371.9 Mpa.

Ansys Software

The structure model is made by Solidwork software and the modelling for analysis was done using Ansys software. The hexa element was chosen for the mesh. Hexa

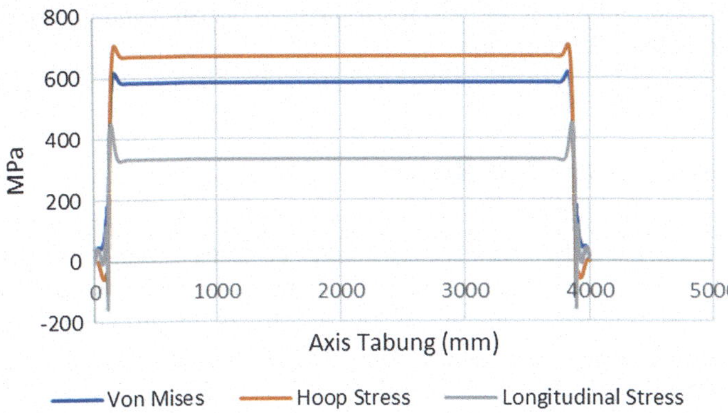

Fig. 6 A Von Mises stress distribution on the surface of the RX450 tube using Nastran software

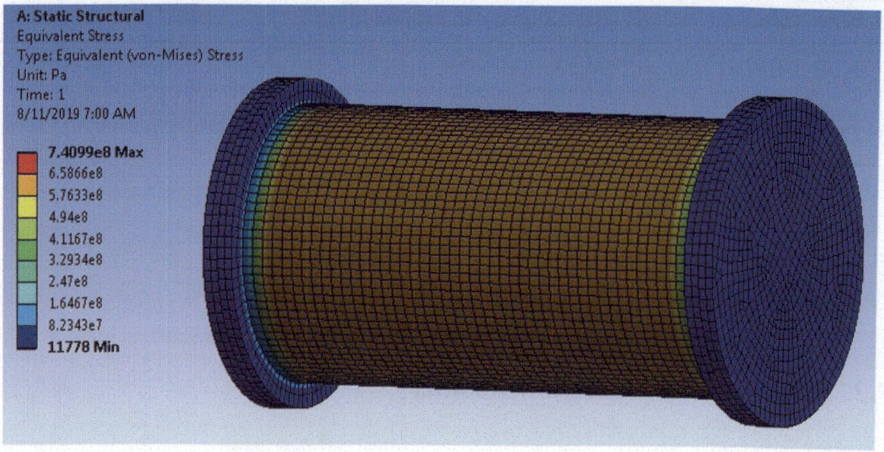

Fig. 7 Ansys software analysis result

element gives more accurate result in finite element analysis. In the modelling for analysis, assumed that both end of the tube are fixed and the pressure is applied inside the tube, perpendicular to the surface of the tube. The internal pressure inside the tube is 120 bar. The result of the simulation is depicted in Fig. 7. From the simulation, the maximum stress occurred on the tube is about 740 Mpa.

3.3 Hydrostatic Test Result

The tube is given a pressure of 120 Bar or equivalent to 12 MPa with water media, the result is there were no cracking or leakage in the test specimen by visual inspection. In this case, those phenomenon indicating that this material is fine and proper to use at motor rocket tube of RX450 to resist the high pressure and high temperature during the propellant burning time.

3.4 Material Tensile Test Result

Tests were conducted at B2TKS Puspiptek Serpong at the end of 2006. There were 3 specimens of KH-406 material tested for static tensile strength as shown in Fig. 8, the average yield of tensile strength was 1497.5 Mpa or 15,260 kg/cm^2 [10]. From the factory certificate we received a tensile strength value of 1620 MPa.

Thus the results of the tensile test at Serpong Puspiptek gave a yield of 92.4% of the value of tensile strength and yield strength of the factory certificate, or the yield strength value of 1220 Mpa or 12,434 kg/cm^2.

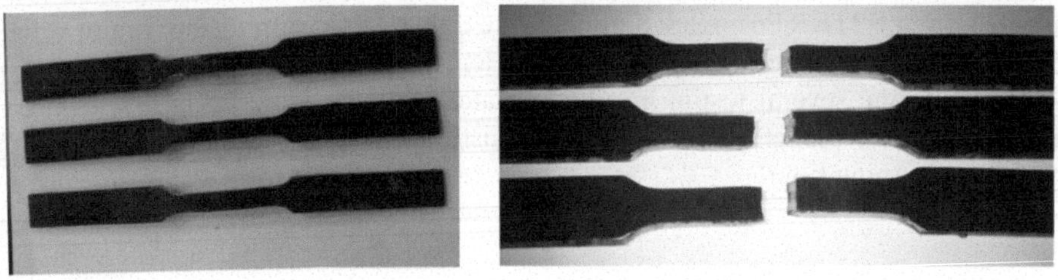

Fig. 8 A tube material specimen of tensile test result

3.5 *Discussion of Results*

From the several chapter above we get some result as follow:

- Based on hydrostatic test result shown that if the result is fine, indicated from the visual if there was no leak or crack during the test so the material is able to used as Rocket tube RX450.
- From analytical calculation result show that result is fine where $\sigma_{\text{von misses}}$ value is less than σ_{yield} value. So based on this result, this tube material is safe to use as rocket tube RX450.
- Based on Numerical calculation, we get the maximum value of each parameter as follow:

  Hoop Stress maximum value is 707.9 Mpa
  Von Misses maximum value is 616.2 Mpa
  Longitudinal Stress maximum value is 371.9 Mpa.

- Based on Tensile test result we conclude that this material able to used as Rocket Tube RX450 since the Tensile test value result is still under limit value which the result of tensile test strength value is only give a 92.4% of maximum yield mentioned in material certificates sheet.

4 Conclusion

Equation that used to calculate the strength value of tube material have been developed and also used to evaluate the material strength. Analytical and numerical calculation results show that the stress caused by a workload of 120 bar, the tube structure is safe against yield strength with safety factors close to 1.5.

The hydrostatic test has been done and the results show that at a pressure of 120 bar the tube can withstand the pressure without any leak.

From the certificates received by the type of cylindrical material are KH 406 with a Yield Strength of 1320 MPa and Tensile Strength of 1620 MPa. The results of tensile tests at B2TKS Puspiptek Serpong were only 92.4% from its maximum value of material datasheet certificate.

Based on the several testing result, indicates that this material is fine to used as RX450 motor rocket tube to hold and resist the high pressure during the solid propellant burning time.

References

1. Dalian Chlorate Co. Ltd. (2016) KH 406 China Xi-An aerospace chemical propulsion plant: certificate of origin mechanical performance, size tube
2. Surface hardening of 30CrMnSiA Steel using Continuos Electron Beam, Aug 2017
3. Report of Hydrostatic test for Rocket Tube Assy of R-Han 122B, PT PINDAD, 20 Sept 2018
4. Hydrostatic Test of Aluminium Tube 6061 Report, 2018, at strength of structure technology workshop, Tangerang
5. Factory Acceptance Test (FAT) Material Tube KH406 Report, November 2016, Workshop Xi'an Hongqing Industrial Area, China
6. Papavinasam S (2014) Corrosion control in the oil and gas industry. Elsevier, Oxford. https://doi.org/10.1016/C2011-0-04629-X
7. Neal CM (1964) United State Patent No. US3377841A. Accessed from https://patents.google.com/patent/US3377841A/en?q=~patent%2fUS4149404A [18 March 2019]
8. Timoshenko S, Woinowsky-Krieger S (1987) Stheory of plate and shells, 2nd edn. Classic Textbook Reissue. Mc Graw Hill Book Company, New York
9. Bednar HH (1986) Pressure vessel design handbook, 2nd edn. Van Nostrand Reinhold, New York
10. Balai Besar Teknologi Kekuatan Struktur Puspiptek, December 2016. *Laporan Uji Tarik Material Tube RX-450*

The manufacturer's authorised representative in the EU is Springer Nature Customer Service Centre GmbH, Europaplatz 3, 69115 Heidelberg, Germany. If you have any concerns regarding our products, please contact ProductSafety@springernature.com

Printed and bound by CPI Group (UK) Ltd, Croydon, CR0 4YY
30/09/2026
02225557-0019